P9-AEZ-636

Gauge Interactions

Theory and Experiment

THE SUBNUCLEAR SERIES

Series Editor: ANTONINO ZICHICHI, *European Physical Society, Geneva, Switzerland*

1. 1963 STRONG, ELECTROMAGNETIC, AND WEAK INTERACTIONS
2. 1964 SYMMETRIES IN ELEMENTARY PARTICLE PHYSICS
3. 1965 RECENT DEVELOPMENTS IN PARTICLE SYMMETRIES
4. 1966 STRONG AND WEAK INTERACTIONS
5. 1967 HADRONS AND THEIR INTERACTIONS
6. 1968 THEORY AND PHENOMENOLOGY IN PARTICLE PHYSICS
7. 1969 SUBNUCLEAR PHENOMENA
8. 1970 ELEMENTARY PROCESSES AT HIGH ENERGY
9. 1971 PROPERTIES OF THE FUNDAMENTAL INTERACTIONS
10. 1972 HIGHLIGHTS IN PARTICLE PHYSICS
11. 1973 LAWS OF HADRONIC MATTER
12. 1974 LEPTON AND HADRON STRUCTURE
13. 1975 NEW PHENOMENA IN SUBNUCLEAR PHYSICS
14. 1976 UNDERSTANDING THE FUNDAMENTAL CONSTITUENTS OF MATTER
15. 1977 THE WHYS OF SUBNUCLEAR PHYSICS
16. 1978 THE NEW ASPECTS OF SUBNUCLEAR PHYSICS
17. 1979 POINTLIKE STRUCTURES INSIDE AND OUTSIDE HADRONS
18. 1980 THE HIGH-ENERGY LIMIT
19. 1981 THE UNITY OF THE FUNDAMENTAL INTERACTIONS
20. 1982 GAUGE INTERACTIONS: Theory and Experiment

Volume 1 was published by W. A. Benjamin, Inc., New York; 2–8 and 11–12 by Academic Press, New York and London; 9–10 by Editrice Compositori, Bologna; 13–20 by Plenum Press, New York and London.

[SUBNUCLEAR SERIES]

Gauge Interactions

Theory and Experiment

Edited by

Antonino Zichichi

European Physical Society
Geneva, Switzerland

PLENUM PRESS • NEW YORK AND LONDON

Library of Congress Cataloging in Publication Data

International School of Subnuclear Physics (20th: 1982: Erice, Sicily)
Gauge interactions.

(The Subnuclear series; v. 20)
"Proceedings of the 20th course of the International School of Subnuclear Physics . . . held August 3-14, 1982, in Erice, Sicily, Italy"—T.p. verso.
Bibliography: p.
Incudes index.
1. Gauge fields (Physics)—Congresses. 2. Nuclear reactions—Congresses. 3. Hadrons—Scattering—Congresses. 4.Yang, Chen Ning, 1922- . I. Zichichi, Antonino. II. Title. III. Series.
QC793.3.F5I58 1982 530.1'43 84-13290
ISBN 0-306-41738-3

Proceedings of the 20th Course of the International School of Subnuclear Physics on Gauge Interactions: Theory and Experiment, held August 3-14, 1982, in Erice, Sicily, Italy

A Division of Plenum Publishing Corporation
233 Spring Street, New York, N.Y. 10013

Printed in the United States of America

PREFACE

In August 1982 a group of 104 physicists from 70 laboratories of 31 countries met in Erice to attend the 20th Course of the International School of Subnuclear Physics.

The countries represented at the School were: Argentina, Australia, Austria, Belgium, Brazil, Bulgaria, Canada, Chile, China, Czechoslovakia, France, the Federal Republic of Germany, Greece, India, Iran, Israel, Italy, Japan, Morocco, the Netherlands, Norway, Poland, South Africa, Spain, Sweden, Switzerland, Turkey, the United Kingdom, the United States of America, Yugoslavia, and Zimbabwe.

The School was sponsored by the Italian Ministry of Public Education (MPI), the Italian Ministry of Scientific and Technological Research (MRSI), the Sicilian Regional Government, and the Weizmann Institute of Science.

This year, on the occasion of the 60th birthday of Chen Ning Yang, the "Ettore Majorana" Centre decided to pay tribute to the outstanding scientific achievements of one of the most prominent scientists of our time, by dedicating the 20th Course of the International School of Subnuclear Physics to a review of the present status of one of the fields of physics where Chen Ning Yang has contributed most profoundly: gauge interactions. The theoretical foundations and the most recent developments were presented by Chen Ning Yang. The most general consequences of a gauge interaction -- supersymmetry -- with its theoretical aspects and the experimental implications were discussed by Sergio Ferrara and Demetrios Nanopoulos. It is at present believed that, in the description of nature, all gauge forces must suffer spontaneous symmetry breaking. The basic particles in the game are the so-called Higgs particles. An attempt to derive Higgs particles from gauge fields was given by Hagen Kleinert.

A special case of a gauge theory is Quantum Chromodynamics (QCD). One of the most direct predictions of QCD is the existence of glueballs. Sam Lindenbaum has reported two examples of such "new" hadronic states recently discovered at Brookhaven. The successes and difficulties of QCD in describing the basic interactions among quarks and gluons in the field of purely hadronic phenomena were discussed by Peter Landshoff. Roberto Odorico dealt with an application of QCD to the case of "heavy flavour" production in strong interactions.

A review of the experimental results on heavy flavour production in hadronic processes was presented by Francis Muller. The comparison of QCD with hadron production in (e^+e^-) interactions was given by James Branson, whilst Franz Eisele reviewed how the experimental results on deep inelastic phenomena compare with this gauge theory of the strongest forces of nature.

The most successful example of the unification of gauge forces is the electroweak case, whose predictions can be tested in many reactions and, in particular, in the (e^+e^-) annihilations. The "standard" version of the electroweak model was confronted with experimental data by James Branson and Peter Duinker, in the study of (e^+e^-) final states. Theoretical alternatives to the "standard" electroweak model were discussed by John Sakurai.

One of the most spectacular results of subnuclear physics has been the discovery of the heavy lepton by Martin Perl at SLAC. The heavy leptons remain in the forefront for understanding the basic symmetry structure of quarks and leptons. The review of the experimental status in this fascinating field of physics was given by Peter Duinker.

Photon scattering at very high energies and photon-photon interactions were described in lectures by Clemens Heusch and Christoph Berger. Berger discussed the "structure" of the photon in terms of QCD predictions.

I myself presented a study of the multiparticle hadronic systems produced in high-energy soft (pp) interactions, based on a new method (i.e. the subtraction of leading particle effects) which allows us to discover universality features in (pp) interactions, in the multiparticle hadronic systems produced in (e^+e^-) annihilations, and in deep-inelastic phenomena.

The new results from the CERN ($p\bar{p}$) ISR and collider data were confronted by André Martin with the exact results of his newly discovered theorems.

In the field of future experimental facilities for European physics, Günther Wolf discussed the LEP experimental programme and proposals.

The Closing Lecture on "The requirements of a basic physical theory" was given by Paul Adrien Maurice Dirac.

I hope the reader will enjoy the book as much as the students enjoyed the School -- and the most attractive part of it, the Discussion Sessions. Thanks to the Scientific Secretaries, these Discussions have been reproduced as closely as possible to the real event.

During the various stages of my work I have enjoyed the collaboration of many friends in Erice and Geneva; their contributions are highly appreciated.

Antonino Zichichi
Geneva, 1982

CHEN NING YANG

CONTENTS

OPENING LECTURE

Gauge Fields. 1
C.N. Yang

THEORETICAL LECTURES

Gauge Fields (Lecture 2). 19
C.N. Yang

Gauge Fields (Lecture 3). 39
C.N. Yang

Supersymmetry and Unification of Particle Interactions. . . . 65
S. Ferrara

Physical Consequences of Global and Local Supersymmetry. 109
D.V. Nanopoulos

Testing QCD in Hadronic Processes 157
P.V. Landshoff

QCD Predictions for Heavy Flavour Production. 197
R. Odorico

Exact Results for pp and $p\bar{p}$ Diffraction Scattering at High Energies 245
A. Martin

SEMINARS ON SPECIALIZED TOPICS

Photon Scattering at Very High Energies -- OR: How Does the Photon Evolve? 265
C.A. Heusch

Higgs Particles from Pure Gauge Fields. 301
H. Kleinert

Detectors Proposed for LEP. 327
G. Wolf

REVIEW LECTURES

Review of High Energy e^+e^- Physics 369
J.G. Branson

Status of and Search for New Leptons at PETRA 455
P. Duinker

The Photon Structure Function 523
Ch. Berger

Status of Deep Inelastic Phenomena. 555
F. Eisele

Evidence for Explicit Glueballs from the Reaction $\pi^- p \rightarrow \phi\phi n$. 615
S.J. Lindenbaum

Hadroproduction of Heavy Flavours 659
F. Muller

High-Energy Soft (pp) Interactions Compared with (e^+e^-) and Deep-Inelastic Scattering 701
A. Zichichi et al.

Special Session on Symmetries and Gauge Invariance. 725
A. Zichichi et al.

The Glorious Days of Physics: Professor Dirac's Birthday 741
A. Zichichi et al.

CELEBRATION OF C.N. YANG'S BIRTHDAY

Greetings to Frank Yang in Mid-Course (August 12, 1982) . . . 747
E. Teller

CLOSING LECTURE

The Requirements of a Basic Physical Theory 757
P.A.M. Dirac

CLOSING CEREMONY

Prizes and Scholarships, etc. 771

PARTICIPANTS. 775

INDEX 787

GAUGE FIELDS

Chen Ning Yang

Institute for Theoretical Physics
State University of New York
Stony Brook, New York 11794

In this lecture I'll describe the origin of the concept of gauge invariance and gauge fields.

Early History

Einstein's discovery of the relationship between gravitation and the geometry of space-time stimulated work by many great geometers: Levi-Civita, Cartan, Weyl, and others. In his book, Raum, Zeit und Materie (space, time and matter), Weyl[1] attempted to unify gravity and electromagnetism through the use of the geometric concept of a space-time-dependent scale change. The basic idea is summarized below.

	dx_μ	
	• ⟶	•
scale	1	$1 + S_\mu dx^\mu$
f	f	$f + (\partial f/\partial x^\mu)dx^\mu$
scale change	f	$f + (\partial/\partial x^\mu + S_\mu)f dx^\mu$

In the summary above, the first line indicates how the scale changes in going from a point x^μ to a neighboring point $x^\mu + dx^\mu$ of space-time. The second line shows how a function of space-time changes as a result of the change in argument from x^μ to $x^\mu + dx^\mu$. Finally, if the scale change is applied to the function f, one obtains at $x^\mu + dx^\mu$ the product

$$(f + \partial f/\partial x^\mu)dx^\mu(1 + S_\mu dx^\mu).$$

Expanding to first order in the small displacement gives the last line in the summary. The increment in f is, then

$$(\partial_\mu + S_\mu)f\, dx^\mu. \tag{1}$$

Weyl tried to incorporate electromagnetism into a geometric theory by identifying the vector potential A_μ with a space-time-dependent S_μ, generating scale changes as described. This attempt proved, however, unsuccessful.

One of the critics of Weyl's space-time-dependent scale change was Einstein, who argued that if one brings a clock around in a loop in four dimensional space-time, Weyl's scale change would produce a change of the speed of the clock; "the speed of a common clock would depend on its history." "it seems to me that the basic hypothesis of the theory is unfortunately not acceptable. Its depth and boldness, however, will certainly fill every reader with admiration."

In 1925, the concepts of quantum mechanics emerged. A key concept in quantum mechanics is the replacement of the momentum p_μ in the classical Hamiltonian by an operator:

$$p_\mu \rightarrow -i\hbar(\partial/\partial x^\mu).$$

For a charged particle, the replacement is

$$p_\mu - (e/c)A_\mu \rightarrow -i\hbar[\partial/\partial x^\mu - i(e/\hbar c)A_\mu]. \tag{2}$$

Weyl's identification would be correct if one makes the replacement

$$S_\mu \rightarrow -i(e/\hbar c)A_\mu.$$

In other words, instead of a scale change

$$(1 + S_\mu dx^\mu),$$

one considers a phase change

$$[1 - i(e/\hbar c)A_\mu dx^\mu]=\exp[-i(e/\hbar c)A_\mu dx^\mu], \quad (3)$$

which can be thought of as an imaginary scale change. This was discussed in Fock[2] and London[3]. Weyl put all of these expressions together[4] in a remarkable paper (which also first discussed the two-component theory of a spin-1/2 particle) in which the transformation of the electromagnetic potential

$$A_\mu \rightarrow A'_\mu = A_\mu + \partial_\mu \alpha \qquad \text{(second-type transformation).} \quad (4)$$

and the associated phase transformation

$$\psi \rightarrow \psi' = \psi \exp(ie\alpha/\hbar c) \qquad \text{(first-type transformation),} \quad (5)$$

of the wave function of a charged particle were explicitly discussed[5].

Although the phase change factor (Equation 3) is no longer a scale factor, Weyl kept the earlier terminology that he used in 1918-1920 and called both the transformation (Equation 4) and the associated phase change of wave functions "gauge"transformations.

Once it became clear that it was not a"gauge transformation" that is important, but a phase transformation, the objection of Einstein mentioned above could also be explained away: the speed of a clock is not dependent on its history. Its quantum phase is, but that does not affect its speed.

Non-Abelian Gauge Fields

With the discovery of many new particles after World War II, physicists explored various couplings between the "elementary particles." Many possible couplings can be written down, and the desire to find a principle to choose among the many possibilities was one of the motivations for an attempt to generalize the gauge principle for electromagnetism. The point here is that for electromagnetism, the gauge principle determines, all at once, the way in which any particle of charge qe, a conserved quantity, serves as a source of the electromagnetic field. Because the isotopic spin $\vec{I}$ is also conserved, a natural question was, "Does there exist a generalized gauge principle that determines the way in which $\vec{I}$ serves as the source of a new field?"

Another motivation for an attempt at generalization is the observation that the conservation of $\vec{I}$ implies that the proton and the neutron are similar. Which to call a proton or, indeed, which superposition of the two to call a proton, is a convention that one can select arbitrarily (if the electromagnetic interaction is switched off). If one requires this freedom of choice to be independent for observers at different space-time points, that is, if one requires localized freedom of choice, one is led to a generalization of the gauge principle.

These two motivations were, of course, intertwined and led[6,7] quite naturally to the formulation of non-Abelian gauge fields. The main point is that one rewrites (5) as

$$\psi \to \psi' = G\psi \tag{5'}$$

and writes $\psi = \begin{pmatrix} \psi_p \\ \psi_n \end{pmatrix}$,

and G is 2 x 2 unimodular unitary SU_2 isotopic spin rotation. One also replaces

$$\partial_\mu - (ie/\hbar c)A_\mu \to \partial_\mu - i\varepsilon B_\mu$$

where B_μ is a 2 x 2 hermitian matrix representing the SU_2 gauge field. One wants to have a simple transformation for $(\partial_\mu - i\varepsilon B_\mu)\psi$. I.e.

$$(\partial_\mu - i\varepsilon B'_\mu)\psi' = G(\partial_\mu - i\varepsilon B_\mu)\psi. \tag{6}$$

It follows that

$$B'_\mu = GB_\mu G^{-1} + i\varepsilon^{-1}G(G^{-1}),\mu \tag{7}$$

where $,\mu$ means derivative with respect to x_μ. Since G is an element of SU_2, $\mathrm{Tr}G(G^{-1})_{,\mu} = 0$. Thus we can consistently take

$$\text{Trace } B_\mu = 0. \tag{8}$$

The next step is to define the field strengths $F_{\mu\nu}$. It is easy to show that $B_{\mu,\nu} - B_{\nu,\mu}$ does not have simple transformation properties under (7). But

$$F_{\mu\nu} \equiv B_{\mu,\nu} - B_{\nu,\mu} + i\varepsilon(B_\mu B_\nu - B_\nu B_\mu) \tag{9}$$

does have simple transformation properties under (7):

$$F_{\mu\nu}' = GF_{\mu\nu}G^{-1} \tag{10}$$

We remark that (5') and (7) → (10) all reduce to the usual electromagnetic equations if B_μ is 1×1 real and G is 1×1 unitary.

The gauge field part of the Lagrangian is thus simply

$$\mathcal{L} = -\frac{1}{16\pi}\int tr(F_{\mu\nu}F^{\mu\nu})d^3x.$$

For interaction with a spin or field or other fields, generalizations[7] are easy, in exactly the same way as in Maxwell's theory.

Integral Formalism

In electrodynamics one can define a phase factor along any path AB:

$$\Phi_{AB} = \exp\left[-i\frac{e}{\hbar c}\int_A^B A_\mu dx^\mu\right]. \tag{11}$$

B

A

If $f_{\mu\nu} \neq 0$, Φ_{AB} is not just dependent on the end points A and B, but also the path. Thus the name path-dependent phase factor, or non-integrable phase factor[8]. The basic properties of Φ_{AB} are

(i) Group property: $\Phi_{ABC} = \Phi_{AB}\Phi_{BC}$

(ii) Infinitesimal path: $\Phi_{A,A+dx} = 1 - \frac{ie}{\hbar c} A_\mu dx^\mu.$

These can be easily generalized if the phase factor, instead of being an element of U(1), is an element of a Lie group $\mathcal{G}$. The two properties are then

X

(1)' Group property $\Phi_{ABC} = \Phi_{AB}\Phi_{BC}$ (12)

(ii)' Infinitesimal path: $\Phi_{A,A+dx} = 1 - b_\mu^i dx^\mu X_i.$ (13)

The last term in (13) has to be proportional to dx^μ and to the generators X_i of the group. $b_\mu^i(x)$ is the coefficient.

Starting from (12) and (13) one can develop[8,9] an integral formalism of gauge fields which is more intrinsic and elegant than the differential formalism outlined earlier. It is more intrinsic because, for example, formula (9) results from the formalism in a natural way. It is, mathematically speaking, equivalent to the concept[10] of connections on fiber bundles.

<u>General Relativity and Gauge Field</u>

The integral formalism permits an identification of Levi-Civita's parallel displacement as a phase factor. In this identification the Christoffel symbols $\{^{\lambda}_{\mu\nu}\}$ become the gauge potentials, and the Riemann curvature tensor $R^{\alpha}_{\beta\mu\nu}$ become the field strengths. With this identification what corresponds to Maxwell's equations would contain first order derivatives of $R^{\alpha}_{\beta\mu\nu}$, i.e. third order derivatives of $g_{\mu\nu}$. In fact for matter-free gravity the equation[8] would become

$$R_{\alpha\beta;\gamma} - R_{\alpha\gamma;\beta} = 0. \tag{14}$$

where ; is the usual Riemannian covariant derivative. (14) is not identical to Einstein's equation for matter-free gravity.

This line of development is not complete, because the coupling between gravity and matter has not been made into an integral part of the geometrical theory. (Nor has that been done in Einstein's theory.) Deep general questions remain as well as technical ones. Should there be torsion? What is the role of spin? What is the role of the metric? Is supersymmetry or supergravity the key? I believe this is an area of research at once challenging and full of promise.

<u>Symmetry Dictates Interactions</u>

With the successes of the Glashow-Weinberg-Salam theory of electroweak interactions, the renormalizability proof of gauge theories by t'Hooft, and the developments of QCD, it is now universally accepted that all interactions are based on the gauge principle. I had coined the phrase <u>Symmetry Dictates Interactions</u> to describe this view about the role that symmetry plays in the conceptual structure of nature's fundamental forces, a view that is a dominant theme of contemporary physics.

One reason that symmetry plays such an important role is that in the language of the physics of today, <u>symmetry is good for renormalizability</u>. However, future physics will understand the concepts of renormalizability and of symmetry, it is hard to believe that some aspects of this statement will not survive.

It is remarkable that in this century, symmetry has gradually come to be a central theme of microscopic physics. It attained this position through a series of subtle evolutions in the concept of symmetry itself and in its phenomenological manifestations. It is my firm belief that this evolutionary process has not come to an end, and further meaning of the concept of symmetry, with perhaps new mathematical structures, will develop in the coming years.

References

1. H. Weyl, Raum, Zeit und Materie, 3rd edit. Springer Verlag, Berlin-Heidelberg, New York (1920).
2. V. Fock, Z. Phys. 39:226 (1927).
3. F. London, Z. Phys. 42:375 (1927).
4. H. Weyl, Z. Phys. 56:330 (1929).
5. W. Pauli, Handbuch der Physik, 2nd edit. Vol. 24(1): 83 Geiger and Scheel (1933); W. Pauli, Rev. Mod. Phys. 13:203 (1941). The terminology of the first and second type gauge transformations seem to be first defined in this 1941 review paper.
6. C. N. Yang and R. Mills, Phys. Rev. 95:631 (1954).
7. C. N. Yang and R. Mills, Phys. Rev. 96:191 (1954).
8. C. N. Yang, Phys. Rev. Lett. 33:445 (1974).
9. C. N. Yang, Proc. Sixth Hawaii Topical Conf. Particle Phys. (1975).
10. T. T. Wu and C. N. Yang, Phys. Rev. D12:3845 (1975).

DISCUSSION

CHAIRMAN: C.N. YANG

Scientific Secretaries: L. Castellani and J.E. Nelson

DISCUSSION

- *CASTELLANI:*

Are there any restrictions on the choice of gauge group?

- *YANG:*

This is an open question. Clearly one would like to use Lie groups. The energy is only positive-definite for compact Lie groups. This may or may not be a handicap.

- *CASTELLANI:*

Can one construct a gauge theory where the transformations on the fields are realized nonlinearly?

- *YANG:*

The gauge theory itself depends only on the group structure and not on the particular representation. Now your question perhaps relates to whether one can have spacetime-dependent fields which are complicated representations of Lie groups. That is a very interesting consideration, and is of course deeply connected with other possibilities, like, for example, complex manifolds. The gauge idea has generated so much enthusiasm that there is a tendency to think any additional mathematics should come from the same considerations. I think this perhaps is a blind spot. Maybe different mathematical structures should come into the theory.

- *CASTELLANI:*

I would like to recall another earlier attempt at unification of gravity and electromagnetism, namely that of Kaluza and Klein (1921) where pure gravity in 5 dimensions reduces to 4-dimensional gravity and electromagnetism. This approach was later generalized to arbitrary gauge groups (De Witt, 1963). Dimensional reduction is now extensively used and seems to be an essential tool in theories where spacetime and internal symmetries are mixed.

- *YANG:*

I am certainly enthusiastic about broadening the general approach to symmetry, and dimensional reduction is one way.

- *ZICHICHI:*

The gauge groups "chosen" by Nature are U(1), SU(2), and SU(3). Is there any theoretical reason for these being the gauge groups?

- *YANG:*

There is no generally accepted view about this and I do not believe that an cogent conclusion has yet come out in any discussion.

- *MUKHI:*

My question concerns the possibility that existing gauge theories, i.e. SU(3), SU(2)xU(1), are effective theories for composites. Do you think local gauge invariance suggests that this is, or is not, the case?

- *YANG:*

I have no good answer to this question. In principle it is difficult to exclude the possibility that the "elementary" particles are composite after all.

- *MUKHI:*

Do you think it likely that an underlying theory with global or only local Abelian invariance could lead to the "usual" non-Abelian theories as effective theories?

- *YANG:*

In general the structure of a non-Abelian theory is tighter, and in some ways prettier. I would doubt that a fundamental theory can be Abelian.

- *VAN DER SPUY:*

In your lecture you said that "symmetry dictates interactions". What about broken symmetries?

- *YANG:*

Broken symmetry is of course very important. One would like a theory in which broken symmetry comes also from symmetry. There have been a number of attempts in this direction but I don't think any has been generally accepted. The prevailing way of introducing a broken symmetry through the Higgs field is thought by many people, including myself, to be rather "ad hoc". My view is that the Higgs field is an important development, perhaps not unlike the Fermi theory (of beta decay) which played an important role in physics, but is unlikely to be the final story.

- *DOBREV:*

What are the main problems for free gravity as a gauge theory?

- *YANG:*

It is obvious that gravity is deeply related to gauge theories, but exactly how is however still unclear. The higher order differential equations that I discussed this morning arise from one specific point of view.

The quantisation of the gravitational field is a very important problem, as yet unresolved. The beauty of the left-hand side of Einstein's equations impresses everyone, as it did Einstein himself. He considered the right-hand side of his equations as haphazard. He wanted, as we do, to move the right-hand side to the left-hand side. That is, we all would like to see the coupling of gravity to other fields an integral part of a unifying principle, which will hopefully have enough symmetry to solve the problems inherent in the quantisation of the gravitational field.

- *DOBREV:*

Can you describe some classical solutions of gravity (presumably without cosmological constant) in terms of analytic vector bundles like, for example, the self-dual solutions of Yang-Mills theory?

- *YANG:*

In the case of self-dual gauge fields in recent years it has indeed been shown that the connections are related to more general analytic bundles: Similar work has been done in the case of gravity, but I am not an expert on it.

- *VAN BAAL:*

Can you explain what you had in mind when you said that new mathematics might come into physics, in particular complex manifolds?

- *YANG:*

How complex manifolds may come into physics I do not know precisely. In the past the dependent variable has been a number or a field. We have learned enough, especially from ideas like harmonic mapping, that, as in the nonlinear σ - model, it is much better to think of the dependent variable as a point (with co-ordinates) on a manifold. From there we enter very easily into the concept of complex manifolds. Whether or not that is the correct route I cannot say, but I am so impressed with its beauty that I feel it may have a role to play in physics.

When Dirac wrote his famous paper of 1931 on the magnetic monopole, he wrote about three disjointed discussions which were at first glance quite unrelated. In the first, Dirac said he believed that progress of modern physics is dependent on the continual introduction of new mathematics. Secondly he gave the history of the hole theory and charge conjugation, including the discussions of Tamm and Oppenheimer regarding the necessity of the antiproton and the antielectron. Finally he discussed the magnetic monopole. I hazard the guess that he put these three things together because in each case he was thinking of new mathematics.

Of course it is extremely difficult to evaluate, but I made an order of magnitude estimate and found that only a few percent of mathematics really comes into physics. So, if you take any new mathematics, you are not likely to be successful.

- *LYKKEN:*

Do you have an opinion about the programme of Mandelstam, in which he tried to completely reformulate Yang-Mills theory and gravity in terms of the path-ordered phase factors introduced in your talk? Could this be a useful alternative approach?

- *YANG:*

I am not familiar with the papers of Mandelstam, and cannot really give an opinion on his work. Path-dependent phase factors are an alternative to the differential description of gauge theories, and are in many senses more intrinsic, in the same way that Lie groups are more intrinsic than Lie algebras. However, if the topological aspects are given, one can equivalently use Lie algebras. The same holds for gauge fields: the differential formalism is equally good. There is no need of the path-dependent formalism when the global structure is known. But in second-quantised theories, the path-dependent formalism is more natural, especially if you use Feynmann path-integrals.

- *VAN DE VEN:*

Regarding the question about the choice of symmetry group, if one requires maximal internal symmetry based on a Lie group, one arrives at the non-compact group $GL(n,\mathbb{C})$. You have considered non-compact $SL(2,\mathbb{C})$ theory and found problems with positivity of the energy. The $GL(n,\mathbb{C})$ theories of Cahill are however unitary and positive-definite. Have you changed your opinion of non-compact gauge theories since your work on $SL(2,\mathbb{C})$?

- *YANG:*

I still do not know how to make sensible use of a non-compact group. But I am now, a priori, less resistant to the idea of non-compactness. I am not familiar with the $GL(n,\mathbb{C})$ work you mentioned.

- *TUROK:*

Could you comment further on symmetry breaking independent of Higgs fields, in particular the approaches to technicolour and your own formulation in terms of sections?

- *YANG:*

The section approach is not an alternative to symmetry breaking. I have not studied in depth the technicolor discussions in the literature. But I am prejudiced: although the Higgs field mechanism is remarkably successful, I cannot believe it is the final story.

- *DUBNIČKOVÁ:*

Although local gauge theories are good for renormalisability, there are problems with quantisation of non-Abelian gauge theories. These problems are not present in non-local field theories. Why are they so unpopular?

- *YANG:*

Non-Local field theories have been considered by many people. To my knowledge no one has made any real sense out of it, but of course this does not mean it is wrong.

- *HOFSÄSS:*

I believe there exists a paper by Klein (around 1940) which already contains some aspects of non-Abelian gauge invariance. Could you comment on how much he knew about this subject?

- *YANG:*

I believe the paper was presented by Klein at a League of Nations conference in Warsaw. The paper was then completely forgotten. I know that Pauli never mentioned it, while he did talk about the paper of 1932 by Schrodinger. Someone discovered Klein's paper three or four years ago, and subsequently I received a copy. Klein did have some elements of the non-linear term in the gauge interaction. It is interesting to ask how he wrote it down, because he had no isospin gauge transformations. (He used the expression "gauge transformation" in his article, but I think he referred to the electromagnetic transformations.) It seems that he wrote down a non-linear coupling like the (B,B) commutator term from the connection commutator in five dimensional general relativity as Schrodinger had done in a paper of 1932, (that he (Klein) referred to,) which studied space-time dependent representations of the γ matrices.

Klein's paper is interesting to look at, but is difficult to understand. It produced very little impact at the time and Klein did not seem to have pursued it himself. Maybe I should ask Prof. Wigner if he remembers such a paper.

- *WIGNER:*

I certainly do not remember it.

- *HOFSÄSS:*

Did Klein write down the correct action without realizing its invariance?

- *YANG:*

I did not understand it enough to answer this question. Some equations, like (B,B) terms, look familiar.

In this respect I should make the following remark; in M.E. Mayer's book (with W. Drechsler) it is stated that a reading of the Yang-Mills paper clearly indicates that we knew about the concept of connection. This is totally wrong. We knew nothing about connections and in the 1960s when I finally understood that the nonlinear term is exactly the same as that in the curvature tensor, I was greatly thrilled. The truth is that we were trying very hard to generalize Maxwell theory, and did not know at that time the geometrical meaning of the Maxwell equations.

- *KREMER:*

Is local gauge symmetry a property of nature or only of its description? I always thought that one introduces redundant degrees of freedom in order to have a convenient description of nature.

- *YANG:*

I am not sure I understand the philosophy of your question. The use of symmetry in physics is quite recent, first in the last century. Pierre Curie wrote interesting and important articles about the concept of symmetry in physics, because he was studying the structure of crystals. Around 1890 the discovery of the 230 space groups came as a great revelation. That is the first use of symmetry.

That development, however interesting, did not exert a profound influence on twentieth century physics. What did exert an influence is the relationship between conservation laws and symmetry, which was considered at the beginning of this century, and generated a number of papers, culminating in the famous paper of Emmy Noether around 1918. In quantum mechanics this relationship is even more important because quantum mechanics is a linear theory, so that e.g. elliptical orbits as well as circular ones can be treated by symmetry arguments. The importance of symmetry in quantum mechanics, in the existence of quantum numbers, selection and intensity rules, became generally recognized in the 1930s, 1940s and 1950s, pioneered by the works of Wigner and of Weyl. One may call that development the second use of symmetry in physics.

Then in the mid 1950s it was found that the discrete symmetries P, C, and T are related in local field theory through the CPT theorem. It was further found, to everybody's surprise, that these discrete symmetries, though related to space-time, are not strictly observed. That discovery generated great excitement and deeper penetration into the symmetry properties of fundamental physics. That was the third phase in our evolving understanding of symmetry.

The next phase began with what I described this morning. The first person to notice that charge conservation is related to a symmetry was Hermann Weyl. But his first concept was a global one. At the same time he developed the gauge idea which was born of geometrical considerations. By 1929, local phase invariance, as related to local conservation laws, was part of the general scheme. That is the theory which, when generalized to non-Abelian gauge fields, exerts now a tremendous influence. This development leads to the principle that symmetry dictates interactions. The gradual recognition of this principle is the fourth phase of our understanding of symmetry.

But to interpret symmetry as only gauge symmetry is perhaps too narrowminded. One should broaden its possibilities, and one of my favourite possibilities is the use of complex manifolds. In any case I believe it unlikely that there will not be a fifth phase of our understanding of symmetry.

- *WIGNER:*

Through a gauge transformation we only get a different description of the same state. Therefore it is quite natural that if the time development of two states differs by a gauge transformation, then the state descriptions themselves are gauge transformable into each other. This is quite different from the other symmetries which connect the time development (and other properties) of physically different states, and not different descriptions of the same state.

- *YANG:*

That is a very important observation which is; can we produce an experimental verification of the fact that we should make the phase variable local, rather than deducing it from complicated calculations? In this respect gauge symmetry is different from the usual global symmetries. The (p - eA) coupling gives a dynamical theory whose predictions agree with experiment. [The only experimental discussion that is related to it is the Bohm - Aharonov experiment, which really goes to the heart of the physical meaning of electromagnetism.)] But that justification, you say, is different from the usual experimentally direct justification for symmetry.

- *WIGNER:*

In my opinion one of the greatest achievements of Newton was the separation of initial conditions and laws of nature. Does this separation still survive (in gauge theories)?

- *YANG:*

There are two possible realms in which to discuss this question, one in the domain of elementary particle physics, another in the domain of cosmology. My opinion is that for a long time to come, as far as elementary particle physics is concerned, the separation of initial conditions and equations of motion will continue as in the past. I am not an expert in cosmology.

- *MARTIN:*

Gauge invariance may lead to observable symmetries like the flavour symmetries SU(2), SU(3) which we now understand as a consequence of colour symmetry.

- *KAPLUNOVSKY:*

The non-integrable phase approach is used to define lattice gauge theories. Which came first?

- *YANG:*

The concept of non-integrable phase factors was first introduced in Dirac's paper of 1931. After that there were a number of papers, including those of Mandelstam, Bialynicki - Birula and me. Wilson's development of lattice gauge theories is also independent.

- *LOEWE:*

You stressed the importance of gauge fields as phase fields in connection with non-relativistic quantum machanics. In that theory there are Abelian realisations of this phase. Do you know an example of a non-Abelian phase in non-relativistic quantum mechanics or solid state physics?

- *KLEINERT:*

Concerning this question, the best examples in solid state theory are the theory of spin glasses and the theory of defects in solids, studied in great detail some fifteen years ago. This is maybe also a good example for Mukhi's question, when he asked whether there is something sacrosanct about gauge theories. From these examples we learn that whenever there are massless long-range excitations they can be re-written in terms of gauge fields.

- *FERRARA:*

I think that one motivation for gauge symmetry in local field theory is the fact that gauge theories are the only consistent way of describing interactions of particles with spin higher than $\frac{1}{2}$. This is true for all known cases; Yang-Mills theory for spin 1, supergravity (i.e. local supersymmetry) for spin $^3/_2$, gravity for spin 2. In fact all these theories can be derived from the requirement of consistency of the interactions.

GAUGE FIELDS (LECTURE 2)

Chen Ning Yang

Institute for Theoretical Physics
State University of New York
Stony Brook, New York 11794

The magnetic monopole in quantum physics was first discussed in a brilliant paper[1] by Dirac in 1931. Conceptually it is related to[2] a topological property of the U(1) gauge field, (i.e. electromagnetism). Many experimental searches for the monopole have been made, and up to this year, the result[3] has been negative. Recently, however, Cabrera[4] reported an event which he interpreted as possibly due to the passage of a magnetic monopole of the Dirac magnetic charge

$$g = \hbar c(2e)^{-1}. \tag{1}$$

through his apparatus. Such a monopole has been conjectured in the GUT, with a mass $\sim 10^{16}$ GeV and moving slowly. Cabrera's experiment thus generated intense new interest in the magnetic monopole. In this and the next lecture I shall concentrate on the monopole. Since I had already talked about the subject in[2] the 1976 Erice School, I'll avoid all matters covered in 1976.

Cabrera's Experiment

Cabrera's experiment[4] utilizes two important but simple ideas:

(a) In usual electrodynamics, because of the equation $\nabla \cdot H = 0$, the flux through any surface

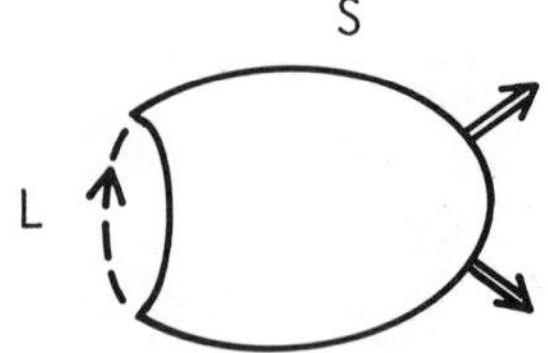

Figure 1 Surface S Bounded by Loop L

S bounded by the loop L (Fig.1) is dependent only on L and is independent of S. However, if there are monopoles, this statement becomes incorrect. The flux is still independent of S as long as S is moved without passing through a monopole. When S is moved through a monopole of strength g, however, the flux suddenly changes by $\pm 4\pi g$. Thus in Fig. 2, the flux Φ_1 through surface S_1 is ~ 0 and that through S_2 is

$$\Phi_2 = \Phi_1 - 4\pi g.$$

Hence with the presence of a magnetic monopole, the flux through a loop L has two values, one value for all surfaces (S_2) behind the monopole which we shall call Φ_{back}, and one value for all surfaces (S_1) in front of the monopole, which we shall call Φ_{front}.

(b) Now replace the loop L in Figure 2 by a superconducting wire of finite thickness. Since $\dot{H} = -\nabla \times E$ at any space-time point not on the world line of the monopole, the flux in front of the monopole has the time rate of change

$$\dot{\Phi}_{front} = \iint \dot{H} \cdot d\sigma = - \oint E \cdot d\ell \quad \text{(inside wire)},$$

which is zero inside a superconductor. Similarly $\dot{\Phi}_{back} = 0$ Thus both Φ_{front} and Φ_{back} are independent of time:

$$\Phi_{front} \quad \Phi_{back} = \text{independent of time}$$

$$\Phi_{front} - \Phi_{back} = 4\pi g. \tag{2}$$

In Cabrera's experiment, before the arrival of the monopole

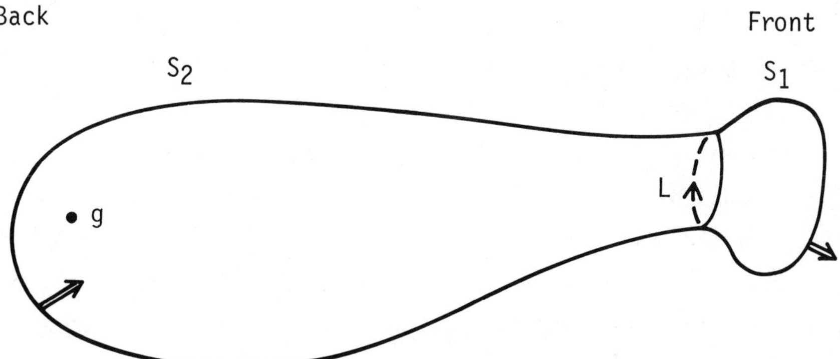

Figure 2 Surfaces S_1 and S_2 both bounded by loop L. The fluxes Φ_1 and Φ_2 are defined as positive in the direction of the arrows.

the background magnetic field strength has been much diluted through a special technique, so that

$$\Phi_{front} \sim 200\Phi_o, \tag{3}$$

where

Φ_o = unit magnetic flux = $hc/(2e) = 2\times10^{-7}$ cgs units.
With the passage of a monopole of strength (1) through the loop, after the monopole is gone, the flux in the loop is Φ_{back}, and the increase of flux is

$$\Phi_{back} - \Phi_{front} = -4\pi\hbar c(2e)^{-1} = -2\Phi_o. \tag{4}$$

(Actually, since Cabrera has a four-loop coil, $\Phi_{front} \sim 800\ \Phi_o$ and $\Phi_{back} - \Phi_{front} = -8\Phi_o$.) Although this increase is only ~1% of (3), Cabrera's system is stable enough[4] to be sensitive to a charge of flux of this order of magnitude.

The magnetic field distribution is sketched in Figures 3 and 4 for which the initial background field is assumed to be zero, so that $\Phi_{front} = 0$. Figure 3 gives the magnetic lines of force when the monopole is at the center of the loop, and figure 4 the magnetic lines of force later when the monopole is already quite far from the loop. Notice that in both cases there is a circle Q on which $\vec{H} = 0$.

Cabrera reported[4] that during a run of 151 days one event was recorded, on February 14, 1982, in which a flux jump of $8\Phi_o$ was observed. He considered this a candidate event for a Dirac monopole. He is planning now further experiments with three mutually perpendicular coils each about four times as large as in the original experiment.

Cabrera's experiment has excited great interest and many other laboratories are planning low temperature or counter experiments to look for slow heavy magnetic monopoles. A number of counter experiments would entangle with the question of how such a monopole interacts with matter, how it interacts

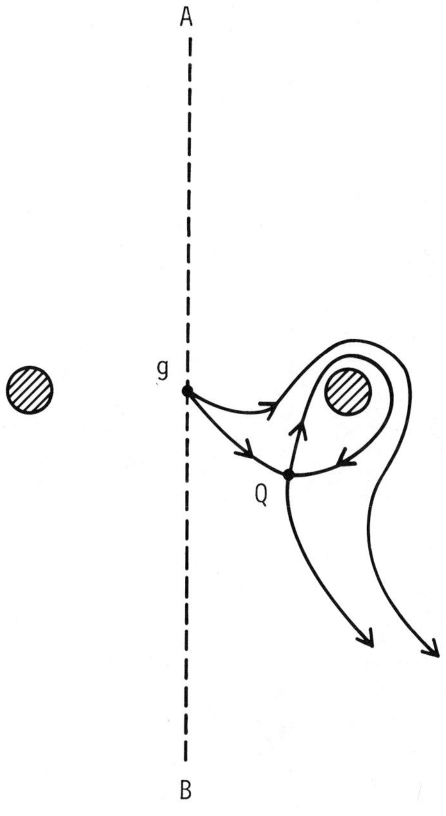

Figure 3 Magnetic lines of force. Monopole g is at center of superconducting loop which is represented by shaded circles. The axis of the loop is the dotted line AB, along which g moves downwards. At Q the magnetic field is zero.

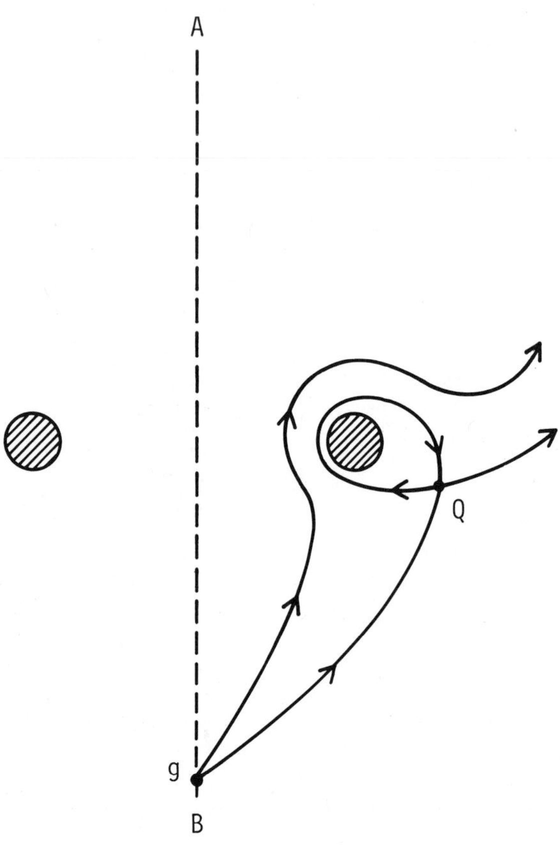

Figure 4 Magnetic line of force. Monopole g has moved past the center of superconductivity loop which is represented by shaded circles. The axis of the loop is the dotted line AB, along which g moves downwards. At Q the magnetic field is zero.

with electrons and with nucleons.

To study such questions one would have to solve, for example, the Schrodinger equation

$$\frac{1}{2m}(p-eA)^2\psi = E\psi$$

for an electron in the field of a heavy magnetic monopole. Here A is the vector potential due to the magnetic monopole. In my lecture here in Erice in 1976 I showed that the vector potential would have to have a string of singularities. The reason for this is topological. The string of singularities has caused great confusion in the theory of the magnetic monopole for many years. In 1976 Wu and I found that by introducing the concept of "sections", the string of singularities problem can be entirely avoided[2]. (In fact we found that the string of singularities does not really exist.) The concept of sections is a part of the mathematical theory of fibre bundles which is a profound branch of mathematics vigorously being developed in the last forty years. Using these ideas there resulted very naturally "monopole harmonics" which are beautiful generalizations of ordinary spherical harmonics.

The monopole harmonic $Y_{q,\ell,m}$ (where $q = eg$) reduces to the usual spherical harmonic when $q = 0$:

$$Y_{o,\ell,m} = Y_{\ell,m}.$$

For any given q, integral or half integral, the collection of all $Y_{q,\ell,m}$ form a complete orthonormal set of wave sections, again a generalization of an important theorem about the completeness and orthogonality of the usual spherical harmonics. In fact, almost all known theorems about spherical harmonics can be generalized to monopole harmonics. These are discussed in reference 5, where the useful theorem about

$$\int Y_{\ell,m} Y_{\ell',m'} Y_{\ell'',m''} d\Omega$$

and the theorem about spherical harmonics addition are generalized.

Also discussed in reference 5 is a symmetry between the indices q and m:

$$Y_{q,\ell,m} = Y_{m,\ell,q} \qquad \text{in region a}$$

$$Y_{q,\ell,m} = Y_{m,\ell,q} \exp\left[2i\phi(m-q)\right] \qquad \text{in region b.}$$

This is very interesting since physically q(=eg)and the angular momentum component m seem to have little to do with each other.

The Dirac quantization rule (1) is also understood in a more natural way using[2] the section formalism. In particular the confusion about factors of 2 never arises with this formalism.

Using monopole harmonics it was shown[2] that for the case of a charge e moving around a monopole of strength g, if

$$2eg \equiv 2eg/\hbar c = \text{odd integer}, \tag{5}$$

the monopole harmonics have total angular momenta that are equal to (integer + $\frac{1}{2}$). This result appears at first sight to be quite strange, since the motion is entirely orbital. To resolve this puzzle we remark that a rotation around the z axis of the monopole field is <u>not</u> simply the transformation

$$\phi \rightarrow \phi' = \phi + \Delta. \tag{6}$$

In fact, the infinitesimal rotation is <u>not</u> simply

$$-i\frac{\partial}{\partial\phi}. \tag{7}$$

This fact follows from the expressions for the angular momentum:

$$\vec{L} = \vec{r} \times (\vec{p} - \frac{e}{c}\vec{A}) - eg\,\frac{\vec{r}}{r}\,. \tag{8}$$

In particular its z component is[2]

$$L_z = -\,i\frac{\partial}{\partial\phi} - eg \text{ in region a,}$$

$$L_z = -\,i\frac{\partial}{\partial\phi} + eg \text{ in region b,} \tag{9}$$

In (9), the $-i(\partial/\partial\phi)$ terms came from the term $(\vec{r} \times \vec{p})$ of (8). That this cannot be the z component of angular momentum is clear: $\vec{r} \times \vec{p}$ is not even an operator in the Hilbert space of wave sections, as discussed in Ref. [2], where it was proved that (8) is the correct expression for the angular momentum.

Since (7) is not L_z, (6) cannot be the operator for a rotation around the z axis by an angle Δ. The correct operator for such a rotation is[5]

$$e^{i\Delta L_z} = e^{-ieg\Delta} e^{\Delta\partial_\phi} \qquad \text{(region a)},$$

$$= e^{ieg\Delta} e^{\Delta\partial_\phi} \qquad \text{(region b)}. \qquad (10)$$

This _is_ an operator on the Hilbert space, since it is a function of L_z, which is an operator.

We can now also answer the question of whether the rotational invariance of the system is for the group SU_2 or SO_3. It is SU_2, as we can see from (9), since the eigenvalue of L_z is (integer $+eg$) which is $=$(integer $+\frac{1}{2}$)) because of (5). Alternatively we arrive at the same conclusion by putting $\Delta = 2\pi$ in (10), obtaining

$$e^{i2\pi L_z} = (-1).$$

REFERENCES

1. P.A.M. Dirac, Proc. Roy. Soc. A133:60 (1931).
2. Tai Tsun Wu and Chen Ning Yang, Nuclear Physics B107:365 (1976); Chen Ning Yang, in _Understanding The Fundamental Constituents of Matter_, ed. A. Zichichi, p.53 (Plenum Press, 1978)
3. See e.g., the review by G. Giacomelli, in _Monopoles in Quantum Field Theory_, eds. N.S. Craigie, P. Goddard and W. Nohm, p. 377 (World Scientific, 1982).
4. B. Cabrera, Phys. Rev. Letters 48:1378 (1982).
5. Tai Tsun Wu and Chen Ning Yang, Phys. Rev. D16:1018 (1977).

D I S C U S S I O N

CHAIRMAN: C.N. YANG

Scientific Secretaries: C. Devchand, V. Dobrev and S. Mukhi

- *DOBREV:*

You said that in front of the moving monopole there are no flux lines. Doesn't that contradict the usual picture of magnetic poles? For instance, imagine the movement of two ordinary magnets with opposite poles towards each other.

- *YANG:*

This question probably derives from a misunderstanding. What I said is, if we have a loop in the usual case without magnetic monopoles, you can say what the flux through the loop is. You need not specify the surface over which you measure the flux. But if there are magnetic monopoles, then you can have the surface in front of the monopoles or behind the monopole. In fact you can wrap the surface around the pole a number of times. There is thus a discrete infinite set of possible surfaces and you have to specify which one of this set you are talking about, in order to have a well-defined flux.

I said that the flux in front of the monopole is zero because, first, this flux is well-defined, second, this well-defined flux through the superconducting loop, in front of the pole does not change in time. So the flux inside is stuck. Now when the monopole is at $-\infty$ this flux is zero, so it is zero always.

- *DOBREV:*

- What is the strength of the magnetic fields which are safely excluded in the Cabrera experiment compared with the magnetic monopole strength?

- *YANG:*

When the monopole goes through, it generates 8 units of flux. But the background is of the order of 800 units of flux. That is what is worrisome. Now you can ask why he does not lower the background. The answer is that it is very difficult. In fact he already has about the lowest field anybody has achieved on earth. (Without his method of expelling magnetic lines, these would have been millions of units of flux.)

- *DOBREV:*

You showed us that there is a symmetry in m and q in the Y - function in the "a" region. What is the corresponding symmetry in the "b" region?

- *YANG:*

In the "b" region there is a similar formula with a trivial phase factor in front of it.

- *MUKHI:*

You commented that introducing monopoles is an elegant way of quantizing charge. However if we are really interested in 'tHooft-Polyakov solutions in unified theories then charge is already quantized by the unifying group. Does this mean we have lost one of our aesthetic motivations for introducing monopoles?

- *YANG:*

You are asking whether we still have the motivation like Dirac had in 1931. The motivation indeed has changed.

- *MUKHI:*

On the question of a "fermionic" wave function for a spinless e - g system, I believe there was some work by Goldhaber on this. Could you comment?

- *YANG:*

It is related but it is not the same. I did not discuss that aspect because it has subtleties in it which are not completely clear to me. The point I was making this morning has nothing to do with the boson or fermion nature of a complex system, which entangles with the question of the tails of the wave functions. And it is precisely because of that that I think those discussions, for example of Goldhaber, are very interesting, but I do not feel I have completely understood all the subtleties about it.

- *DEVCHAND:*

Can the theorems you talked about this morning, in particular the one stating that the monopole harmonics form a complete orthonormal set, be generalized to the case of particles with isospin in the field of a non-Abelian monopole?

- *YANG:*

I have looked into this question. In 5-dimensional space they can, but not in the usual 3-dimensional space. The 5-dimensional generalization was published in the Journal of Mathematical Physics in about 1978. It is very elegant mathematics but I do not know how to use it in physics.

- *KREMER:*

How does one get the estimate of $10^{-3}c$ for the monopole velocity?

- *YANG:*

This estimate is based on astrophysical arguments. The moving monopoles are accelerated in the intergalactic magnetic fields. If a certain monopole density is assumed, large velocity monopoles would absorb too much energy and the magnetic field would have been quenched. This is the kind of reasoning by which they arrived at the velocity $10^{-3}c$.

- *KAPLUNOVSKY:*

Monopoles are gravitationally bound to the galaxy and their non-gravitational interactions are very weak (monopoles are very heavy). Thus monopoles should have a velocity (relative to the galaxy) of a typical star, which is $10^{-3}c$. They cannot be slower unless there is some additional mechanism of monopole acceleration (none is known). Since the speed of the earth relative to the galaxy is also of order $10^{-3}c$, monopoles would move with typical velocity $10^{-3}c$ also relative to the earth.

- *KREMER:*

It is known that monopoles occur as solutions of classical Yang-Mills-Higgs systems. But how do we know that these monopoles appear as particles in a quantum field theory?

- *YANG:*

I think you are as able, or unable, to answer that question as I am. When you write down a field theory which has classical solutions, these solutions are stationary points of the action integral. It is not proven that they have to be particles in field theory, but it is highly suggestive. I do not think anybody has a better answer to that.

- *CASTELLANI:*

One of the problems of the Cabrera experiment is the possibility of a change in the self-inductance of the coil. Could one circumvent this problem by placing two (or more) independent coils in the cylinder, each with its own "squid"? The coils should not be too far apart, so that a monopole passing through the first would also pass through the second.

- *YANG:*

In the first place, the zero of the current is difficult to find. What he observed is a jump of 8 units of flux and it is produced by a corresponding jump of the current. With the three coils which are orthogonal to each other some of your effect would be there because of mutual inductance. Your suggestion is also a good one.

- *PERNICI:*

The Dirac monopole is pointlike, while the 'tHooft-Polyakov monopole has a spatial extension. Does the Lipkin-Weisberg-Peshkin problem exist also for the 'tHooft-Polyakov monopole?

- *YANG:*

No, of course not. This particular problem exists only for the Dirac monopole. If you have a very small region in which you have a complex structure which is the origin of this outside magnetic force then that particular problem disappears. Of course what happens in this small region is complicated. To my knowledge electron motion in that region has not been studied.

- *GOCKSCH:*

During the lecture, two ways of describing a monopole in terms of A_μ were described. One required a string, the other didn't. Do you prefer the latter (section) picture over the string? Was the string ever thought to be a physical thing, i.e. could one experimentally distinguish between a monopole with a string and one without?

- *YANG:*

About whether I prefer the section description of the magnetic monopole over the one with one map but with a singularity, the answer is clear - I prefer the section picture. There were very complicated discussions first introduced by Wentzel, who pointed out that if you have the string then there is a return flux. He raised the suggestion that it has to be subtracted out and this led to endless difficulties. I personally definitely think that the only reason that we were able to do some rather complicated calculations, as you will see next time, is because we have adopted the section idea.

The string was never thought to be physical.

- *MUKHI:*

I just want to point out that if you have an infinitely thin string carrying the flux from a magnetic monopole then you cannot measure the "Bohm-Aharonov" effect. The quantization condition precisely implies that there is no phase difference in this case.

- *YANG:*

Yes, but that does not tell you that there are no other effects. Also, as I said this morning, the string complication was what caused the confusion about factors of 2 and 4 which persisted for a long time. With the string eliminated there are no such difficulties.

- *VAN DE VEN:*

The idea that particles with fractional electric charge would exist in nature in the form of quarks is by now very familiar. What does this imply for Dirac's formula $eg = 1/2$? Suppose you shoot a monopole into a nucleon where it would see a fractionally charged quark, what would this imply for the quantization condition?

- *YANG:*

As we know, the current thinking about the fractional charge is that there is confinement. If Cabrera did discover a magnetic monopole of the usual Dirac strength, and if the same laboratory discovers a fractional electric charge, they will have some reconciling to do, and they are worried about it. These things are very complicated and one would also have to analyse each experiment and see whether it can stand scrutiny. I would find it extremely difficult to believe, if the monopole $g = (2e)^{-1}$ exists, that there are free fractional charges in nature - because of the Dirac quantization condition. As to the quarks inside a hadron, if Cabrera's discovery is confirmed, the precise meaning of localized fractional charges inside a hadron will have to be re-examined.

- *VAN DER BIJ:*

Shrinking the overlap region and shifting it to the pole, we seem to get back a string. Does this mean the string picture and the section picture are equivalent?

- *YANG:*

The first observation is correct. If you just expand region "a" and shrink region "b" to zero then of course the region "a" potential would have a string singularity. I would say that the two pictures are not equivalent because all the confusion in the past was due to the fact that the section picture was not developed.

- *VAN DER BIJ:*

Have you looked into taking the limit?

- *YANG:*

You do not need to take a limit. You can in fact shrink it. But it is a bit like when there is a "ghost" and you are afraid of it. You try to sweep it under the rug by making the thing very small. But if you look in fact under the rug everything is fine.

- *HOFSÄSS:*

I want to make a comment on the question whether the section-formulated monopole and the monopole with string are physically equivalent. I think there is a simple argument which shows they are not: if one chooses two half-spheres as the sections on which one specifies the gauge fields, such that they just touch at the equator, one has a description which uses just one gauge field for the whole space. This field then changes by a step function across the equator. It should them be connected to the gauge with the string singularity via a singular gauge transformation, so it should be physically different.

- *VAN BAAL:*

In Abelian gauge theories we need matter fields to enforce a fiber bundle structure (single-valuedness of the matter field). This is not so in non-Abelian gauge theories (apart from the role of the centre of the group for a simple group, also giving rich topological structure). So pure U(1) gauge theories without matter fields are not really properly defined, but, for example, pure SU(N) (actually $SU(N)/Z_N$) are.

- *YANG:*

I don't understand the question.

- *SCHÄFER:*

There is the claim that the magnetic monopole is very heavy. Can you explain why one thinks so?

- *YANG:*

The scale of energy of Grand Unified theories is of the order of 10^{14} GeV. The mass of the 'tHooft-Polyakov monopole is about equal to that divided by α which is 1/100, so you get about 10^{16} GeV.

- *LAVIE:*

Couldn't the equivalence between Y_{q1m} and Y_{m1q} be traced to some simple argument like that the interaction between them is proportional to the product of "m" - the magnetic quantum number - and q = eg which is proportional to the strength of the monopole?

- *YANG:*

I am not sure I understand your explanation, but there have been a lot of papers discussing this particular subject already. If you have a spherical top, there is a laboratory z-axis angular momentum which defines "m" and an internal z-axis angular momentum which defines q = eg and there is a symmetry between them. Many people have discussed this possible relationship and indeed mathematics-wise it looks very similar. But I think the understanding of it does not have the clarity which would be required if you can not only understand it but can take off from it and do more things. So I think it's worth further looking into. I do not know whether this direction is more or less along your thinking. 'eg' of course does create, by the crossed electric and magnetic fields, an angular momentum, so this symmetry does have some relationship with internal and external angular momentum.

- *KAPLUNOVSKY:*

If Cabrera found a monopole, the flux is about 1 per square inch per year or 10^{-8} cm^{-2} sec^{-1}. For a velocity around $10^{-3}c$ one has n around 10^{-15} cm^{-3} so their mass density is bigger than that of nucleons. Problems remain even if they are bound to the galaxy and are much greater if they are uniform in the universe (it would be closed by more than is observed).

- *YANG:*

This is one of those hotly debated questions between physicists and astronomers. Cabrera is aware of the problem of the implications of the density of such heavy magnetic monopoles in relation to non-observed masses in the universe. He claims that there is no contradiction but I know that there are people who dispute this. As we all know, the astronomers deal with order-of-magnitude statements with some plus or minus signs in the exponent. While the matter of the Cabrera experiment should be discussed in terms of these astronomy discussions, the matter can only be settled by repeating the experiment or by doing similar experiments.

- *KAPLUNOVSKY:*

Electrons have a magnetic dipole moment. Do you expect monopoles to have an electric dipole moment?

- *YANG:*

Yes, if the magnetic monopole has spin 1/2. Presumably one will want it to have spin 1/2 because that seems to be a more basic particle. This is a very complicated problem and I will come to it when I discuss the problem of a monopole interacting in field theory with an infinite sea of electrons. Both of these problems, the monopole and the sea, are due, of course, to Professor Dirac.

- *TUROK:*

In a grand unified monopole there is a Higgs field as well as the gauge fields. As far as your treatment of an electron in the field of a monopole, the Higgs field if long-range would alter the results drastically. However at short range the singularities at r = 0 would probably be smoothed out.

- *YANG:*

The question of the 'tHooft-Polyakov monopole with a Higgs field and the electron motion in the field of that has not been tackled yet, to my knowledge. What I was telling you today and I'll finish telling you tomorrow is the Dirac magnetic monopole interacting with the Dirac electron.

- *KLEINERT:*

I have a naive question. Isn't the question of the existence of a string an experimental question rather than a theoretical one? In other words, I got the impression that your section formulation has so much beauty that if Prof. Zichichi does an experiment and finds a monopole with a string attached, you have to resign? Or what happens? Since Prof. Dirac is here, and we all would like to understand the history of the confusion of the string, could he be so kind and tell us what he thought about the physical reality of the string?

- *DIRAC:*

I am inclined to think that it is not real. It is just a mathematical fiction.

Could I make a comment about an earlier question: what was the motivation for the monopole? The motivation was to try to find an explanation for the number $\hbar c/e^2$, which has the value 137, and I completely failed in finding any reason for this number. I would like to say now that I believe it is necessary to find an explanation for this number before we will make any real progress in physics. I think this is a most important problem, and perhaps the most important problem, and physicists are neglecting it altogether, and I'm very disappointed by that. The motivation was to get an explanation for this number, and that motivation failed. It gave the monopole as a sort of by-product and people have taken it up and worked it out extensively. But I think that perhaps they are working it out too extensively without having an explanation of the number 137. Thank you.

- *YANG:*

Could I ask Prof. Dirac a question? Your really very original

1931 paper on the magnetic monopole had three different parts in it. The last part was on the magnetic monopole. But you started with a one-page discussion on the necessity to continually introduce new mathematical concepts into fundamental physics, and you took the position that that is inevitable and will go on. In fact a very interesting additional statement you made, which is very perceptive, is that as time goes on the experiments will become more and more difficult, and therefore one will have to rely more and more on finding this new mathematics not through the experimental route but through the route of judgement of mathematical beauty. Then in the next couple of pages you went into a discussion of the history of the infinite sea and what we now call charge conjugation, and you referred to the work of Tamm, Weyl and Oppenheimer, saying that there must be an anti-electron and an anti-proton. Then, at last, you came to the magnetic monopole. My question is, why did you put those three things into one paper? They have nothing to do with each other.

- *DIRAC:*

I quite agree. I think it was just that I didn't write very many papers and wanted to crowd together several things into one paper. But with regard to your first question, I believe that some new mathematics is needed to explain the number 137. We will not get the explanation just by developing existing ideas based on existing mathematics. We have to break out into some new mathematics and I was taking as an example the breaking out which introduced spins to the electron wave equation. Previously people were working just with the Klein-Gordon equation. They might have gone on working with it, but they would never have made any fundamental progress. You had to have the new idea of bringing in the spins, and some similar quite new idea is needed, maybe some new dynamical variables of a kind which are not yet considered by physicists. They will have to be brought in; they perhaps play a fundamental role in the basic interactions, and one will not get them by further experiments, just as one would not have got spins into the electron wave function from further experimental knowledge. It would have to come from some mathematical development and the guiding principle is mathematical beauty.

- *YANG:*

Could I ask a second question, which is related to what you have just said. What is your view about the possibility that quaternions play a central role?

- *DIRAC:*

I have considered it quite a bit, but have not had any success with it. That is the sort of new mathematics that one ought to study, but I don't think that it is the right new mathematics, just because I have tried it and failed. But I don't want to discourage other people from trying it, because other people may succeed where I failed.

- *YANG:*

Could I press this point a little further? Which large domain of mathematics do you perceive this new idea to be in, analysis, geometry, algebra or topology?

- *DIRAC:*

I think analysis, probably complex variable theory. Complex variable theory is so beautiful that I feel that nature must have made good use of it, and, very likely, we need to make stronger use of it than we've done up to the present.

GAUGE FIELDS (LECTURE 3)

Chen Ning Yang

Institute for Theoretical Physics
State University of New York
Stony Brook, New York 11794

In this lecture I wish to discuss several problems concerning the Dirac monopole.

Monopole-electron Interaction

Using monopole harmonics, the scattering of a Dirac electron by a fixed Dirac monopole has been[2] explicitly evaluated. The Hamiltonian of the system is

$$H = \alpha\cdot(-i\nabla - eA) + \beta M. \tag{1}$$

$$H\psi = E\psi.$$

We choose the standard representation

$$\alpha = \begin{pmatrix} 0 & \sigma \\ \sigma & 0 \end{pmatrix}, \qquad \beta = \begin{pmatrix} 1 & 0 \\ 0 & -1 \end{pmatrix} \tag{2}$$

and try to find a wave function

$$\psi = \begin{bmatrix} f(r)\ \xi_{jm} \\ g(r)\ \xi'_{jm} \end{bmatrix}$$

where f and g are radial wave functions respectively for the "big" and "small" components. This procedure is entirely similar to the usual one for the Dirac equation in a central potential, except of the vector operator p-eA satisfy in the present case

$$[\,[A,B],C\,] + \text{cyclic permutation} = -4\pi e g\,\delta^3(r) \qquad (5)$$

if A, B, C are taken to be these three components. The fact that the right hand side of (5) is not zero shows that the vector operator p-eA is not properly defined. Lipkin, Weisberger and Peshkin further pointed out that for the Schrodinger equation the wave function vanishes at the origin where

$$\delta^3(r)\,\psi^*\psi = 0,$$

so that for the Schrodinger equation the operator p-eA is in effect properly defined. For the Dirac equation, because the wave function does not vanish for type (3) angular distribution, the Hamiltonian is not defined.

This difficulty was resolved in reference 1 by adding a very small extra magnetic moment κ so that the Hamiltonian becomes

$$H_{new} = H - \kappa e g \beta \vec{\sigma}\cdot\vec{r}(2Mr^3)^{-1}. \qquad (6)$$

The small extra magnetic moment prevents the electron from reaching the monopole. It does not change anything for types (1) and (2) wave functions, but damps the wave function of type (3) near r = 0. Other methods of avoiding the difficulty were discussed in references 3 and 4.

The Lipkin-Weisberger-Peskin difficulty is related to the problem of the classical orbit of an electron moving toward a magnetic monopole at an impact parameter = 0. (In other words, for a head-on collision.) For such an orbit, although the electron suffers no acceleration since $\vec{v} \times \vec{H} = 0$ along the orbit, it <u>does not pass through</u> the monopole, but is bounced back.

To appreciate this we first demonstrate that the orbit of an electron in the magnetic field of a fixed monopole always lies on a circular cone whose apex is at the position of the monopole, a beautiful result due to Poincare. First, v^2 does not change with time since the force $\vec{v} \times \vec{H}$ is always perpendicular to $\vec{v}$. Next the total angular momentum for two facts: (i) for a given total j, there are in the usual case always two possible values of ℓ for the spin-angle wave function ξ and ξ', $\ell = j \pm \frac{1}{2}$, (see Fig. 1). In the present case, $q \neq 0$, and Fig. 2 shows that for $j \geqq |q| + \frac{1}{2}$, there are two values of ℓ, but for $j = |q| - \frac{1}{2}$, there is only one value of $\ell = |q|$. (ii) Parity conservation holds for the normal case, but not for (1).

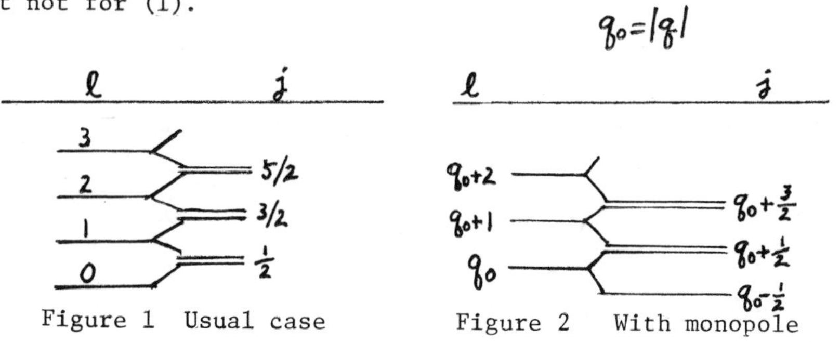

Figure 1 Usual case Figure 2 With monopole

Taking these into consideration, it was found that[1] there are now three types of wave functions. Type (1) and type (2) are for $j \geqq |q| + \frac{1}{2}$ and type (3) is for $j = |q| - \frac{1}{2}$. For all these types the radial wave functions f and g satisfy differential equations which are quite similar. For type (3) these equations are

$$(M-E)f - iq|q|^{-1}(\partial_r + r^{-1})g = 0,$$

$$-iq|q|^{-1}(\partial_r + r^{-1})f-(M+E)g = 0. \tag{3}$$

For types (1) and (2) the radial wave functions can satisfy the boundary condition

$$f = g = 0 \text{ at } r = 0. \tag{4}$$

But for type (3) wave functions, there are no nonvanishing solutions of (3) that satisfy (4). In other words, the Hamiltonian (1) is not really defined for type (3) angular dependence.

This difficulty had already been discussed by Lipkin, Weisberger and Peshkin[2] who pointed out that the x,y,z components

$$\vec{J} = \frac{r \times m_o \vec{v}}{\sqrt{1-v^2}} - eg\vec{r}r^{-1} \tag{7}$$

is independent of time. Thus in Fig. 3 the side $\vec{J}$ is fixed. The other two sides are perpendicular to each other and one of them has a fixed length $|eg|$. Thus the point P can

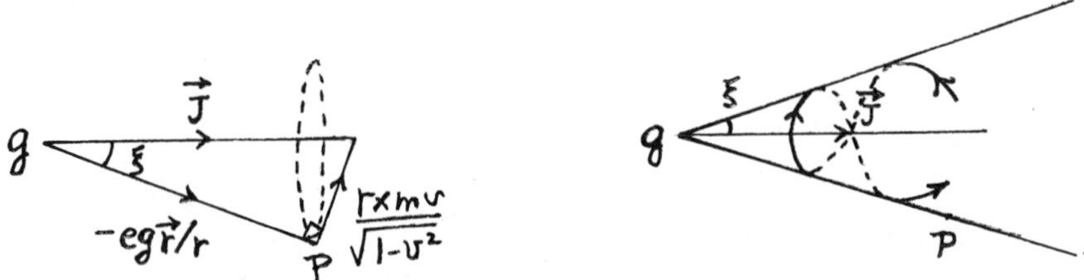

Figure 3   Diagram for total angular momentum $\vec{J}$

only rotate around $\vec{J}$, along the dotted circle. We have thus proved that the electron, which is on the line gP, must move on a circular cone with half aperture angle ξ given by

$$\tan \xi = \left| \frac{1}{eg} \frac{r \times mv}{\sqrt{1-v^2}} \right| . \tag{8}$$

For a head-on collision, r x mv = 0. Thus ξ = 0. Now for very small ξ, the cone is very pointed, and the orbit spirals in and spirals out. (Fig. 3.) We have thus proved that for a head-on collision, the electrons move along a half-line and bounce back at the monopole.

Using the new Hamiltonian (6) the scattering amplitude of an electron by a heavy monopole was computed in reference 1. The result is in agreement with the expectation that for small scattering angles θ, the cross-section is[1] given by

$$\frac{d\sigma}{d\Omega} = \frac{g^2 v^2}{Z^2 e^2} \left(\frac{d\sigma}{d\Omega}\right)_R$$

where

$$\left(\frac{d\sigma}{d\Omega}\right)_R = \frac{Z^2 e^4}{4k^2 v^2} \left(\sin \frac{\theta}{2}\right)^{-4}$$

is the Rutherford scattering cross-section. For larger scattering angles θ the cross-section is different from the Rutherford scattering by a complicated factor. These large angle scatterings are important for the determination of δ-ray counts along the path of a heavy monopole in traversing through matter.

Electron-monopole Bound States

The bound state problem for the electron in the field of a fixed magnetic monopole was solved in references 5 and 6. For type (3) wave function the problem is similar to the Sturm-Liouville theory, but for types (1) and (2) the problem is an eigenvalue problem of four coupled linear first order differential equations and requires new techniques. We will outline this new technique as follows.

We consider the new Hamiltonian (6) and consider types (1) and (2), i.e. type A, wave function:

$$\psi_{jm}=r^{-1}\begin{pmatrix} h_1\xi^{(1)}_{jm} + h_2\xi^{(2)}_{jm} \\ -i[h_3\xi^{(1)}_{jm} + h_4\xi^{(2)}_{jm}]\kappa q/|\kappa q| \end{pmatrix}. \tag{9}$$

where all notations follow references 1 and 5. As was shown in reference 5, the radial wave functions $h_i(r)$ satisfy the following equation

$$\Omega^{(o)}\, h^{(o)} = 0, \tag{10}$$

where $$h^{(o)} = \begin{bmatrix} h_1 \\ h_2 \\ h_3 \\ h_4 \end{bmatrix}$$

$$\Omega^{(o)} = \partial_\rho - \mu a_3\rho^{-1} + b_1^{-2} + A_o a_1 b_1 + iB_o a_1 b_2, \tag{11}$$

$$r = |\kappa q|\rho(2M)^{-1},$$

$$\mu = [(j+\tfrac{1}{2})^2 - q^2]^{1/2} > 0,$$

$$A_o = \kappa q/2 \neq 0, \; B_o = \kappa q E(2M)^{-1}$$

and a_1, a_2, a_3 and b_1, b_2, b_3 are two sets of Pauli matrices. The boundary conditions are

$$\lim_{\rho\to o} h^{(o)}(\rho) = \lim_{\rho\to\infty} h^{(o)}(\rho) = 0. \tag{12}$$

Eq. (10) is an eigenvalue equation for four ordinary differential equations of the first order in four unknowns. It is thus a generalization of the Strum-Liouville problem. For the Sturm-Liouville problem the fundamental trick which allows for an elegant analysis of the eigenvalues is the definition of a phase angle (related to the logarithmic derivative of the wave function) which is monotonic with respect to the energy. It turns out that the eigenvalue problem (10) allows for a similar analysis through the definition of two phase angles.

Multiplying (10) on the left by ia_1b_2 we obtain

$$[(ia_1b_2)\partial_\rho - \frac{\mu}{\rho}a_2b_2 + \frac{a_1b_3}{\rho^2} + A_ob_3 - B_o]\, h^{(o)} = 0. \tag{13}$$

Choosing a representation where a_1, a_3, b_1, b_3 are real symmetrical and a_2, b_2 are equal to i times real antisymmetrical matrices, we find (13) to be of the form

$$[\omega_o\partial_\rho + V_o]h^{(o)} = B_oh^{(o)} \tag{14}$$

where

$$\omega_o = \left(\begin{array}{cc|cc} & & & 1 \\ & & 1 & \\ \hline & -1 & & \\ -1 & & & \end{array}\right)$$

and V_o is real symmetrical.

It is convenient to make a further similarity transformation with

$$T = \begin{pmatrix} 1 & 0 & & \\ 0 & 1 & & \\ & & 0 & 1 \\ & & 1 & 0 \end{pmatrix}:$$

$h^{(1)} = Th^{(o)}$, $\omega_1 = T\omega_o T$, $V_1 = TV_o T$.

Then $[\omega_1 \partial_\rho + V_1]h^{(1)} = B_o h^{(1)}$ (15)

where $$\omega_1 = \left(\begin{array}{cc|cc} & & 1 & \\ & & & 1 \\ \hline -1 & & & \\ & -1 & & \end{array}\right)$$

and V_1 is real symmetrical.

Eq. (15) is now ready for an analysis which is similar to the Sturm-Liouville theory. We consider a finite internal $\rho = a$ to $\rho = b$ and first take the boundary conditions to be

$$\lambda = K_a \mu \text{ at } \rho = a. \tag{16a}$$

$$\lambda = K_b \mu \text{ at } \rho = b, \tag{16b}$$

where λ and μ are the upper and lower 2 components of $h^{(1)}$:

$$H^{(1)} = \begin{pmatrix} \lambda \\ \mu \end{pmatrix}. \tag{17}$$

We further assume K_a and K_b to be real symmetrical 2 x 2 matrices. That such conditions are relevant for the boundary condition (12) can be demonstrated after a simple analysis.

The key concepts in the generalized Sturm-Liouville analysis of (15) is contained in the following:

If Ψ is a real 4 x 2 matrix consisting of two columns each of which satisfied (15) and (16a), and are linearly independent, then

$$\Psi \omega_1 \Psi = 0 \text{ for all } \rho.$$

Further writing

$$\Psi = \begin{pmatrix} \xi \\ \eta \end{pmatrix},$$

where ξ and η are 2 x 2 matrices, then for any ρ, $\xi\eta^{-1}$ is real symmetrical and $\frac{\partial}{\partial B_o} \xi\eta^{-1}$ is negative definite or semi-definite, if the determinant of η is not zero. For the case when the determinant of η is zero at a point $\rho = \rho_o$, the analysis can be easily extended. We can now define two phase angles from the eigenvalues of $\xi\eta^{-1}$. Both phase angles would then be monotonic in E.

The rest of the anlaysis closely resembles the usual Sturm-Liouville theory. Through such an analysis we obtain the bound state diagram for angular momenta. $j > |q| - \frac{1}{2}$ as illustrated in Fig. 4.

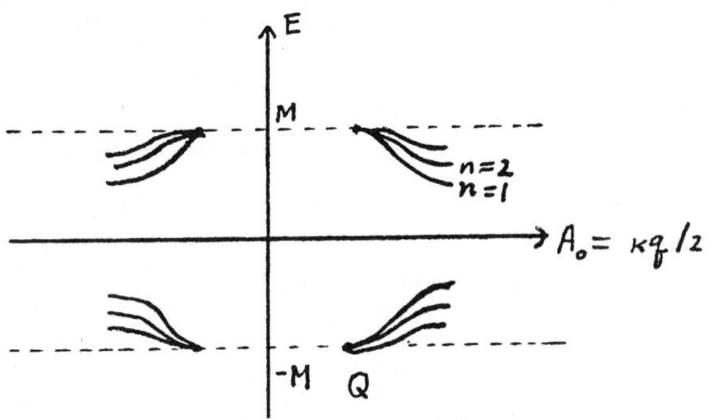

Figure 4 Bound-state energy for $j > |q| - \frac{1}{2}$. Solid lines show bound-state energies E_n as functions of A_o. The dashed lines are thresholds outside of which there are only scattering states. $E = 0$ is a bound state for all $A_o \neq 0$. The curves are symmetrical with respect to a reflection in either the A_o or the E axis. Each integral value of $n > o$ gives rise to four curves. Only those for $n = 1$, 2 and 3 are shown. At the end point Q, the value of A_o is $\frac{1}{2}(\mu^2 - \frac{1}{4})$ where μ is given in the text.

Plasma of e^+e^- Pairs Around Monopole

It was pointed out in reference 5 that with infinity many $E = 0$ bound states there will be a plasma of e^+e^- pairs around the monopole. The argument is as follows. The creation of a e^+e^- pair and putting them into $E = 0$ states is energetically favorable, since the Coulomb energy between the e^+ and e^- would be negative. Thus there would be infinitely many pairs of e^+e^- around a monopole, so that each $E = 0$ state is,on the average, half occupied by an electron, and therefore half occupied by a positron. The Coulomb energy of this plasma is, however, very difficult to estimate.

Classical Lagrangian for e-g-γ System

In classical quantum electrodynamics without magnetic monopoles, the Maxwell equations and the Lorentz force equations for the motion of charge particles of charge e are both derivable from an action principle. One starts with an action integral

$$\mathcal{A} = -(16\pi)^{-1}\int d^4\xi f_{\mu\nu} f^{\mu\nu} - m\int ds + e\int A_\mu(x)dx^\mu \tag{18}$$

where the last two integrals are along the world-lines of the charged particles. One then varies $A_\mu(\xi)$ and writes down the condition for $\delta\mathcal{A}/\delta A_\mu = 0$. That results in Maxwell's equations. One then varies the world lines and writes down $\delta\mathcal{A}/\delta x_\mu = 0$. That results in Lorentz' force equasions.

With a e-g-electromagnetism system, can one similarly formulate an action principle classically?

The answer is yes, according to reference 7. To formulate the action integral, one first will have to define not one A_μ but two A_μ's, in regions a and b in 4-space, related to each other by a gauge transformation. The action integral is now

$$\mathcal{A} = -(16\pi)^{-1}\int d^4\xi f_{\mu\nu} f^{\mu\nu} - m\int ds - M\int dS + e\oint A_\mu(x)dx^\mu \tag{19}$$

where the dS term is along the monopole world lines and the last integral has to be carefully defined since A_μ has two values $(A_\mu)_a$ and $(A_\mu)_b$ in the overlap between regions a and b. With the proper definition, variation of $\mathcal{A}$ again leads to the Maxwell equations and Lorentz force equations for both the charged particles and for the monopoles.

Second Quantized e-g-γ System

The action integral (19) has the very remarkable property that it is only definable modulo $4\pi eg/c$:

$$\mathcal{A} = \mathcal{A} \text{ (modulo } 4\pi eg/c). \tag{20}$$

This remarkable fact was already stated[8] by Dirac in 1948, but previously I never had understood his reasoning. With the formulation of (19), the ambiguity (20) is now clear: It resides in the ambiguity in the definition of the last integral, which in turn is related to the nontrivial nature of the fibre bundle that is necessary for the description of monopoles in interaction with charged particles.

Because of (20), if one tries to second quantize the e-g-γ system by Feynman's path integrals:

$$\exp[i\,\mathcal{A}/\hbar] \ d(\text{path}) \tag{21}$$

the ambiguity (20) implies[7] that (21) is ill-defined unless

$$4\pi eg/c\hbar = 2\pi(\text{integer}),$$

which is exactly Dirac's quantization rule.

The evaluation of (21) is, however, a nontrivial matter. Instead of pursuing this route, Tu, Wu and I had[9] second quantized the e-g-γ system through separating out the longitudinal field, after Fermi's theory of the second quantized e-γ system.

The theory of the second quantized e-g-γ system is thus in the same stage of development as QED in, say, the nineteen thirties. How about renormalization? We have looked into this problem but have met with some essential difficulties. These reside in the fact that we are unable to find the generalization to the e-g-γ case of the Feynman diagram calculation. I believe this is a deep difficulty that will require new ideas for solution.

REFERENCES

1. Y. Kazama, Chen Ning Yang and Alfred S. Goldhaber, Phys. Rev. D15:2287 (1977)
2. H. J. Lipkin, W. I. Weisberger and M. Peshkin, Ann.Phys. (N.Y.) 53:203 (1969).
3. Alfred S. Goldhaber, Phys. Rev. D16:1815 (1977).
4. Dai Xian-Xi and Ni Guang-Jiong, Physica Energiae Fortis et Physica Nuclearis 2:225 (1978).
5. Y. Kazama and Chen Ning Yang, Phys. Rev. D15:2300 (1977).
6. Chen Ning Yang, in Monopoles in Quantum Field Theory, p.237, Edited by N. S. Craigie, P. Goddard and W. Nahm, (World Scientific, 1982).
7. Tai Tsun Wu and Chen Ning Yang, Phys. Rev. D14:437 (1976).
8. P.A.M. Dirac, Phys. Rev. 74:817 (1948). See also J.Schwinger, Particles, Sources and Fields (Addison-Wesley), vols. 1 and 2.
9. Tu Tung-sheng, Wu Tai-Tsun and Yang Chen-ning, Scientia Sinica 21:317 (1978).

DISCUSSION

CHAIRMAN: C.N. YANG

Scientific Secretaries: C. Devchand, V. Dobrev and S. Mukhi

- *HOFSÄSS:*

The classical action for the monopole-electron system contains a term which couples the electric charge to the gauge field. Then for this expression to be independent of Q one needs that the two gauge fields A_μ^a and A_μ^b have to be gauge copies of each other in the overlap region. But then one may rewrite this piece of the action without the additional B(Q) term, if one instead extends the integration over the world line to cover the overlap of the two regions twice, once in region a and once in region b. This generates the B(Q) term. It seems to me that the total action $\mathcal{a}$ can be written as the sum of two independent actions, each being defined in one of the sections only.

- *YANG:*

This section business is necessary when it is not possible to think of one gauge, when you have to think of two gauges at the same time. The situation is similar to a case in which you go into a building with two floors, neither of them covering the whole ground.

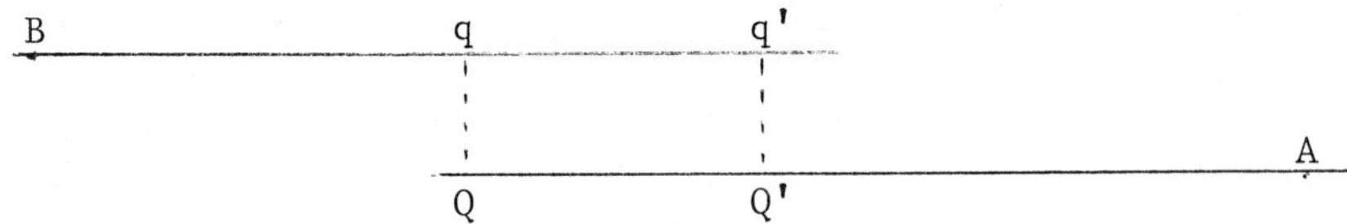

Imagine a person goes from point A (not covered by the upper floor) to point B (with no lower floor under it), and he can go up via one of many staircases at Q, or Q' or.. . The classical action has a term

$$\int_A^B A_\mu dx^\mu ,$$

which we would replace by

$$\int_A^Q (A_\mu)^a dx^\mu + \int_q^B (A_\mu)^b dx^\mu ,$$

since neither $A_\mu{}^a$ or $A_\mu{}^b$ covers the whole path A to B. But this expression is not correct since if we choose Q'q' as the staircase to go upstairs, rather than Qq, we would obtain a different expression. Instead

$$\int_A^q (A_\mu)^a dx^\mu + \beta(Q) + \int_q^B (A_\mu)^b dx^\mu$$

is independent of the choice of staircase.

The simultaneous dealing with many, many layers is precisely what the mathematicians have mastered in dealing with manifolds. If you have a situation where one floor covers it, then you do not have to do this; but I was talking about precisely the case where this is needed.

- *HOFSÄSS*:

You have presented the second quantized version of a system of electrons and monopoles with fixed particle numbers. If one wants to give up this restriction and go to a quantum field theory, do you think one can still use the concept of sections?

- *YANG*:

Of course one has to generalize ordinary field theory so that it includes the section idea. The sections would have infinitely many layers, but I do not think this is a difficulty. The generalization from a finite number of particles to the case of the infinite sea or equivalently to a field theory is similar to Fermi's (Rev. Mod. Phys. 1932) generalization to ordinary field theory. For the younger people here, I may mention that this Fermi paper was the paper which really made people understand what QED is. Before that QED was thought to exist, but it was only after Fermi's paper that people really understood what it is, because there the extra degrees of freedom were eliminated. For the present case we follow the same procedure.

- *VAN DE VEN*:

You told us that you couldn't define sensible Feynman rules. Are all the three fundamental fields second quantized or just the electromagnetic field? Secondly, I would like to know if it is clear that one should quantize the monopole field - is there any fundamental reason for doing so?

- *YANG*:

I was referring to the Feynman rules of a second quantized theory with all three fields quantized. As you know, Dirac was the first person to quantize the elctromagnetic field. He pointed out that if you only quantize the finite number of degrees of freedom of the hydrogen atom, but not the electromagnetic field, then you get into contradictions. Once you quantize a piece of an interacting system, every degree of freedom must be similarly quantized. If one does not quantize part of the system, then after this part couples to the quantized part, one would produce nonquantized action in the quantized part.

- *DOBREV*:

I also want to ask about the second quantization. You said that your propagator has a longitudinal and a transverse part, but have you tried an antisymmetric part? For instance in a conformally invariant field theory with electric and magnetic charges there is such a propagator.

- *YANG:*

I do not know how to compute the propagator. We have Heisenberg equations of motion. They are covariant, relativistic and gauge covariant. So we have a dynamical system which has been second quantized but we do not know how to compute quantities from it in any simple way.

- *DOBREV:*

I could not understand the asymmetry in the picture of the bound states for the type-3 solutions. Is there a physical reason for that?

- *YANG:*

There is no a priori reason why there must be a symmetry for $K \to -K$. This cannot be explained very simply. It comes from looking at the differential equations. The sign of K does play a role because it gives the sign of the extra magnetic moment, the interaction of which, at extremely small distances, is very strong, according to Dirac's equation. So that sign does play an essential role.

- *MUKHI:*

In the original formulation, Dirac's quantization condition is a quantum mechanical principle. However, last year Coleman gave a series of lectures (at Erice and elsewhere) on monopoles, in which he stressed the fact that it's a purely classical condition. Could you comment on that?

- *YANG:*

Well I don't understand how a quantization condition which explicitly contains $\hbar$ is not quantum mechanical. But I've explained that in a classical e-g-Maxwell system the action cannot be defined except modulo $4\pi eg$. That is a pre-quantum mechanics statement discovered by Dirac.

- *MUKHI:*

Yes, that is what suggested this question. But in the case of a non-Abelian gauge theory, does one get a condition from the classical formulation, for example for the 'tHooft-Polyakov solution? I assume this is what Coleman must have meant.

- *YANG*:

In the case of non-Abelian gauge fields, the magnetic monopole (the generalization of Dirac's monopole) is not the 'tHooft-Polyakov monopole. That is something I have examined a little bit. It is a very complicated subject and I'm confused about it.

- *MUKHI*:

Although, as you said, not much is known about renormalization of monopole theories, I think several years ago there began a controversy over whether eg = 1/2 referred to the bare or the renormalized e and g. What is your opinion?

- *YANG*:

There were discussions which were very confusing to me and, I think, to everybody. I can only guess, just because the monopole-electron dynamical system has been quantized in such a nice form, the whole thing might make sense. So perhaps it is renormalizable and if that is the case I can only understand the statement that it is the renormalized e times the renormalized g which is quantized.

- *MUKHI*:

There's a formulation, due to Zwanziger, whose renormalization has been studied ...

- *YANG*:

I do not understand Zwanziger. That does not mean he is wrong.

- *GOCKSCH*:

Could you repeat the arguments that lead to the e^+e^- plasma surrounding the monopole? I want to understand if it is physical, and if so, whether it could be used to detect a magnetic monopole. I'm thinking of γ emission in a steady state.

- *YANG*:

If there is a monopole somewhere, it can capture any number

of electrons and positrons. Since every angular momentum state has $E = 0$ bound states for e^+ and e^-, which is energetically favourable, the monopole would be surrounded by a plasma. If the monopole is discovered the plasma is bound to be there because of these bound states. The effect of this plasma is complicated. However, the plasma does not change the magnetic flux. No motion of the electrons and positrons will change the total magnetic flux ($4\pi g$). So you will still have the original g, but with additional electromagnetic effects beyond the bare g. This is a tantalizing and important question if the monopole exists.

- *ITOYAMA:*

Could you explain the difficulty in the path integral quantization of the e-g-γ system?

- *YANG:*

The path integral with our action in the exponent is a very complicated and confusing thing with all the sections in it.

- *LAVIE:*

I'd like to suggest that the difficulties encountered in getting the Feynman rules for the e-g-γ system arise because of the absence of a rigorous relativistic quantum mechanics with a Hilbert space and a positive-definite norm.

- *YANG:*

When I said that we could not find the Feynman rules, I meant that we made guesses and none of them worked. If one considers the usual Feynman propagator, it contains a longitudinal and a transverse part. The transverse part presumably twists E and H when it goes from the electron to the monopole, and this could be correctly handled. However the longitudinal part cannot be separated and handled in a covariant manner. This may have very deep roots, but at the calculational level one cannot even guess the correct expression.

- *DOBREV:*

Have you tried to add an antisymmetric piece to the usual propagator?

- *YANG:*

We tried all kinds of possibilities, and they did not work. After trying for some time, we now feel that it can't be done; no such rules exist. We do not have a rigorous proof of this. However, our guesses were very natural.

- *KAPLUNOVSKY:*

Did you try ghosts? Perhaps the theory is not ghost-free like ordinary QED.

- *YANG:*

Maybe.

- *LAVIE:*

May I mention that in relativistic quantum mechanics where we do get something like Feynman rules, they come from a Born series in the potential, and only in special cases can one distinguish the propagator from the vertex functions. This might be one of the cases where you cannot do this.

- *YANG:*

I cannot comment on that.

- *KREMER:*

Would the fact that monopoles form bound states with ordinary matter affect the way monopole search experiments should be done?

- *YANG:*

The very complicated state of affairs even in the vacuum around the monopole was what I was discussing. When the monopole complex enters matter there are interactions of the plasma with the electrons, and, more importantly, the monopole itself, with such a hugh magnetic field, would wreck the atom and even the nuclei it goes through in a very complicated way. The problem of the interaction of a heavy Dirac monopole with matter, specially when the monopole is slow and gets stopped in matter, is a fascinating but difficult question.

- *KREMER:*

In GUT models monopoles occur quite naturally. So in order to have a quantum field theory with monopoles, should one not quantize the gauge and Higgs fields and consider the monopole to be a bound state of these fields?

- *YANG:*

Your question concerns the fact that the GUT monopole is a monopole with structure. To deal with questions like renormalization one has to consider this complex structure. The structure, however, is presumably very small, so outside it, the monopole would behave just like a Dirac monopole. If you want to apply what we've tried to the GUT monopole, you certainly also have to study the internal structure in addition. However, my point is that you still need to study the pure Dirac monopole case, since most of the GUT monopole is the Dirac monopole; most of the space around it is affected by it in exactly the same way as it would be affected by the Dirac monopole.

- *BRANSON:*

At the Paris conference, three experiments reported on searches for slow monopoles. Two of them, for example, used stacks of scintillators. They find a much smaller flux than Cabrera, though there's an open question concerning the calculation of the ionization in the scintillators due to slowly moving monopoles.

- *YANG:*

Which are the three groups?

- *BRANSON:*

One was from Michigan (Larry Sulak et al.)? Then there was Dave Ritson, and the SUDAN I group also have some results from their proton decay detector. This is brand new data instigated by the Cabrera experiment.

- *YANG:*

Thank you. This is the first time I have heard about this.

- *BRANSON:*

Would there also be a quark plasma around monopoles due to electromagnetic interactions?

- *YANG:*

Presumably there would be complications similar to the case of the e^+e^- plasma, but we have not worried about this yet.

- *S. ERREDE:*

In the context of grand unified theories, there have been discussions in the literature on the plasma or "condensate" of fermion-antifermion pairs surrounding the monopole. In general, these $f\bar{f}$ pairs were of all types of charged fermions – e,μ,τ,u,d,s,c,t,b, – with a characteristic radius $R \sim 1/m_f$. For grand unified monopoles, the interactions of this condensate with normal hadronic matter will in general violate baryon and lepton number – i.e., catalyze baryon decays with cross-sections of order of a few millibarns (i.e. hadronic interaction cross-sections). Would you please comment on this?

- *YANG:*

Very recently I have heard about work by Rubakov and also by Callan on this. They seem to be getting very large cross-sections, but I have not come to any conclusions about it.

- *S. ERREDE:*

Some time ago, a paper in Physical Review (I do not remember the author) discussed the relation of monopoles and the CPT theorem. In particular, the CPT theorem was replaced by a CPTM theorem, where M is a magnetic charge conjugation. Would you comment on how this would affect the interactions of a magnetic monopole in the context of either conventional electromagnetic interactions with matter, or of GUTS (baryon/lepton-number violating interactions).

- *YANG:*

It is obvious that you will have to have the monopole conjugation too. However, I do not immediately see any experimental implications of this. This would of course also depend on the way in which the electromagnetic interaction is assigned to the monopole.

- *S. ERREDE:*

As an experimentalist, one of my long time worries has been that if, for some reason, magnetic charge is not conserved, a grand unified monopole could decay to a real X or Y boson. Thus we may be 10^{10} years too late to see them.

- *DEVCHAND:*

Could you explain why the extra magnetic moment term in the Hamiltonian kills the Lipkin-Weisberger-Peshkin difficulty?

- *YANG:*

The extra term introduces a very divergent term of the form $1/r^2$ into what is equivalent to the potential in the Dirac equation, with a very small coefficient. If you consider the solution of the radial equation, you find that the only way to make sense of these equations is to quench them at the origin. This is what the extra term does. You could alternatively add some δ - function type of term to bounce it back. This has been tried with the same result. These are examples of what is called the completion of a non-self-adjoint operator in the von Neuman sense, and one can do it in various ways with the same results.

- *LOEWE:*

Can one apply Dirac's method of quantizing constrained systems to the e-g-γ system?

- *YANG:*

We have not tried it. I do not know what you would get if you use the Dirac method of eliminating superfluous degrees of freedom. The 1948 paper of Dirac was somewhat along that direction, and it was there that the statement that the action integral is defined only modulo $4\pi eg$ appeared. When I read the paper as a graduate student, I thought there was something very deep in it, but I still do not understand the arguments Dirac used in that paper. It is also interesting that it took 17 years from 1931 when Dirac proposed the monopole to 1948 when he wrote the quantization paper. During these 17 years nobody else seems to have touched the second quantization problem. I do not know why. Perhaps Professor Dirac knows.

- *DIRAC:*

I do not remember the details of that paper and I would not like to answer the question without looking them up. But I have the belief that this second paper did not contain any deep results which were not contained in the first.

- *YANG:*

In other words, you do not think the fact that the classical action is only defined modulo $4\pi eg$, which you found in the second paper, is a deep result?

- *DIRAC:*

No, I wouldn't say it's a deep result.

- *LOEWE:*

How can you speak about renormalization if the most elementary Feynman graphs in the theory are not known?

- *YANG:*

That's a good question. It is obvious that there has to be renormalization because even the part without g has the e-γ interaction, and this is renormalized. So, that there exists renormalization is obvious, I think. But since we do not have a computational rule, its study is stalled.

- *TUROK:*

Could you give your opinion of attempts to formulate the quantum field theory of monopoles as a theory of solitons - based on the Sine-Gordon-Thirring model equivalence? In particular, while the pointlike approximation is valid if the Higgs field is massive, if it is massless the monopoles really are spread out and the approximation cannot be valid.

- *YANG:*

I cannot comment on what you've just said about monopoles. On the other question, I've always subscribed to the view that there is a relationship between excitations in a system with many degrees of freedom and elementary particles. The elementary particles are really excitations of the vacuum. We, of course, inherited this idea from Dirac. After his infinite sea picture, the vacuum is no longer simple, it is an infinitely complicated system with excitations which are elementary particles. This, of course, is deeply related to the soliton idea. But I do not know whether one gains any insight by using such a picture. In other words I subscribe to such a picture but don't know how to use it.

VAN DER SPUY:

Does parity nonconservation play a role in the problem of constructing Feynman diagrams?

- *YANG:*

Parity is not conserved only in the following sense: if you have just an isolated monopole and electron, then parity is not conserved. But for the full interacting field theory, it is conserved. Now whether that is the source of our inability to write down Feynman diagrams I never thought about. I would doubt it, but it is an interesting point.

- *CASTELLANI:*

I am puzzled about why you have to guess your Hamiltonian. What prevents one from deriving it from the Lagrangian in the usual way? Also, why do you have to guess the propagators and the vertices in the second quantized e-g-γ system? Once the Lagrangian is written down in terms of the fields describing e, g and A, one should in principle be able to read off the Feynman rules.

- *YANG:*

Because that Lagrangian is a bit different from the usual one. Usually we have a Lagrangian and we have a rule for calculating the Hamiltonian. In this case the Lagrangian already contains the section idea. So it is not straightforward. However, I am certain that is can be computed. In other words if you apply some subtlety I am sure that the usual procedure, properly generalized, would allow you to handle a Lagrangian with a section idea in it and obtain a Hamiltonian. But that is not the route we followed. The one we followed is more satisfactory because as in e-γ interaction, which only became crystal clear in Fermi's paper, we've arrived at this crystal clear case all at once by guessing. But you are entirely right, one should pursue the Lagrangian formalism via some generalization of the canonical quantization method, and I'm sure it would give the same result. As to the difficulty of deriving the Feynman rule, it originates with the fact that there is no interaction representation.

- *DIRAC:*

Could I make a comment about the Lagrangian. The Lagrangian is essentially a classical theory. There is a general theory of the Lagrangian only when the dynamical variables all commute. If you bring in non-commuting quantities like spins and more complicated things, there is no general Lagrangian theory. The important thing in quantum mechanics is the Hamiltonian. You have to get the right Hamiltonian and then you can go ahead and you can forget all about the Lagrangian. The Lagrangian is not much help except in special cases, and I would not put too much importance on it.

- *YANG:*

Do you have some comments about the relative fundamental nature of the usual second quantization procedure versus the Feynman path integral second quantization procedure, which uses the Lagrangian?

- *DIRAC:*

I think this Feynman path method would not work in general cases where you have special spins, special elements of a group coming in. It works only in special cases.

- *ETIM:*

I have a comment about that. In curved spaces for instance if one writes down the path integral, there are ambiguities about certain terms that would come in, due to non-commuting operators. You have a classical Lagrangian and when you write the Feynman path integral, you have induced potentials coming in. You have to rearrange the operators in a certain way. So in curved space, there are problems.

- *CASTELLANI:*

You would have these problems also with canonical quantization, not only in Feynman path integrals. There, there are ordering problems between P's and Q's.

- *YANG:*

I would like to make two comments. These are extremely speculative but I will be bold and make them. I think that the canonical second quantization methods are not as good as the Feynman path integral method. It is true that the Feynman path integral is not fully understood. The very definition of the path integral method is not understood. But there is something about it which catches one's imagination. I feel that it encapsulates the fundamental essence of second quantization. A related speculation is the following. In the Feynman path integral, what you write is an action $\mathcal{A}$ divided by $\hbar$. I would guess that if quaternions do come into Physics in a basic way, it will be $(i\alpha/\hbar)$ which will in fact be important. What is the difference? exp $(i\alpha/\hbar)$ is a phase factor. If quaternions are important, this could be replaced by a pure imaginary quaternionic phase, exp $(i\alpha + j\beta + k\gamma/\hbar)$, where i, j and k are quaternion units. And the imaginary nature of the exponent is the thing which makes wave interference work. In the usual case, the exponent divided by i is α, the real action. In the quaternion case, the exponent is imaginary, but one would not be able to define a real action α out of it. I have tried this but so far I have no results. I feel that if quaternions do come into Physics in a major way, and I would make a small bet that they will, then I would think that this is the way in which they will come into the picture. But as I said this is extremely speculative.

- *ETIM:*

But the path integrals are not well-defined. They are just symbols.

- *YANG:*

Of course. But if you wait for perfectly defined things you will never make progress.

- *WIGNER:*

My question is somewhat similar to that of Dr. Kremer. What puzzles me is that if we confine a pair of positive and negative magnetic monopoles in such a way that their distance from each other is restricted to be between 30 and about 60 $\hbar/Mc$ (where M is the mass of the monopole) the total energy of the pair can become negative. This is a contradiction to the principle of finiteness of energy supply if the Lorentz invariance is valid. There must be an error in this argument if the monopoles with the large assumed charge exist.

- *YANG:*

That is indeed a serious and difficult problem which must be faced if the monopole is experimentally found.

- *DIRAC:*

May I ask why it's a catastrophe if the total energy is negative? I don't see any objections to it; when you have gravitational systems it could be that the total energy is negative.

- *WIGNER:*

If the total energy is negative, it can be increased arbitrarily by putting the system into motion. I do not believe that there are systems with total negative energy. It would mean that an infinite amount of energy can be exhausted from it, and this follows from Lorentz invariance.

- *DIRAC:*

Is it so that the total energy of the universe has to be positive? I don't know whether that is settled.

- *WIGNER:*

It certainly is positive. We are all heavy and our total mc^2 is pretty big.

SUPERSYMMETRY AND UNIFICATION OF PARTICLE INTERACTIONS

Sergio Ferrara

CERN
Geneva, Switzerland

INTRODUCTION

I will describe supersymmetric Yang-Mills theories for particle interactions coupled to N = 1 supergravity. I will give the most general Lagrangian which fulfils the requirement of local supersymmetry and the Higgs and superHiggs effects will be studied, the latter being responsible for the gravitino mass generation.

I will show that supergravity effects are important when the primordial supersymmetry breaking energy scale M_S is intermediate between the weak interaction scale $M_W = 0\ (100\ \text{GeV})$ and the Planck scale $M_P = 0\ (10^{19}\ \text{GeV})$

$$M_S = O(\sqrt{M_W M_P}) = O(10^{10} - 10^{11}\ \text{GeV})$$

Models for low-energy particle physics based on supergravity-induced supersymmetry breaking will be considered.

LOW-ENERGY EFFECTS OF N = 1 SPONTANEOUSLY BROKEN SUPERGRAVITY

Since its discovery[1] it has been realized that supersymmetry is not only an extremely elegant symmetry of Lagrangian field theories but it also provides, at the quantum level, examples of model field theories with exceptional ultraviolet properties.

These properties are related to the so-called non-renormalization theorems[2] of supersymmetric field theories and are connected to the fact that naive power counting no longer holds in these theories due to additional cancellations of ultraviolet divergences between boson and fermion loops.

The final goal of supersymmetric field theories is to provide a consistent theory of quantum gravity and at the same time to achieve the ultimate unification of all fundamental interactions.

Whilst we may have to wait quite some time for having solution (if any) to these formidable problems, supersymmetry may be relevant at present (or next) available accelerator energies for less ambitious purposes which however have nowadays a great theoretical importance.

One of these problems is the so-called hierarchy problem[3] of grand unified theories and of the standard model of electroweak and strong interactions. In these theories quadratic divergences associated to elementary scalar Higgs fields or, in other words, the absence of symmetries protecting scalar fields from acquiring a mass, spoil the naturalness of having light Higgs particles.

Another theoretical problem, which is also a naturalness problem, is to explain the smallness of the CP violation in strong interactions, related to the actual extremely small value of the so-called θ-angle[4].

Supersymmetry promises to give a solution to these theoretical puzzles and in doing so it predicts a dramatic change in low energy physics at energies well beyond the TeV region by the introduction of a pletora of new particles which should be possible to produce in the forthcoming accelerators.

In this scenario phenomenologically acceptable models of electroweak and strong interactions based on spontaneously broken supersymmetry seem to require[5] a primordial supersymmetry breaking scale M_S, defined as the vacuum energy of a matter system, as large as

$$M_S \simeq 10^{10} - 10^{11}\ \text{GeV}$$

In this regime the effects of supergravity couplings may play a crucial role for particle phenomenology. In fact in local supersymmetry the effective supersymmetry breaking scale, which controls the splitting of low-energy particle multiplets, is the gravitino mass, given by the universal relation[6]

$$m_{3/2} = \frac{1}{\sqrt{3}} \kappa M_S^2 \qquad (k = \text{gravitational constant} = M_P^{-1})$$

This mass, for an intermediate breaking scale $M_S \simeq 10^{10}$ GeV is comparable to the weak vector boson masses $m_{3/2} \simeq O(M_W) \simeq O(100\ \text{GeV})$.

Under these circumstances, it is natural to ask whether gravitational effects become important at low energies $M_W \ll M_S \ll M_P$. The coupling of the N = 1 supergravity Lagrangian to an arbitrary Yang-Mills system[7,8] gives a definite answer to this problem. Remarkably enough, inspection of the matter supergravity coupled system shows that supersymmetry spontaneously broken at a scale $M_S = O(\sqrt{M_W M_P})$ behaves at low-energy as a globally supersymmetric theory, explicitly but softly broken[7,9] at a scale $m_{3/2} = O(M_W)$. More interestingly, in a large class of models[10] the gravitino mass $m_{3/2}$ induces the breaking of SU(2) × U(1) and therefore one is able to determine M_W in terms of $m_{3/2}$.

To be concrete, we consider the N = 1 pure supergravity Lagrangian[11]

$$\mathcal{L}_{SG} = -\frac{1}{2K^2} e R - \frac{1}{2} \varepsilon^{\mu\nu\rho\sigma} \bar{\Psi}_\mu \gamma_5 \gamma_\nu D_\rho \Psi_\sigma \tag{1}$$

(D_ρ denotes the spinor covariant derivative with spin $^3\!/_2$ contorsion contribution to the spin $^1\!/_2$ connection) and couple it to an arbitrary (renormalizable) Yang-Mills system, which in flat space is described by the following Lagrangian[12]

$$\begin{aligned}\mathcal{L} = &-\frac{1}{4} F_{\mu\nu}^{\alpha\,2} - \frac{1}{2} \bar{\lambda}^\alpha \not{D} \lambda^\alpha - \frac{1}{2} \bar{\chi}_{Li} \overleftrightarrow{\not{D}} \chi_R^i - \frac{1}{2} |D_\mu z_i|^2 \\ &- \frac{1}{2} |g^i|^2 - \frac{1}{8} \tilde{g}^2 (z^{*i} T_i^{\alpha\, j} z_j)^2 \\ &+ \left(g_{/2}^{\prime\prime ij} \chi_{Li} \chi_{Lj} - i\tilde{g}\, T_j^{\alpha\, k} z_k \bar{\lambda}_R^\alpha \chi_R^j + \text{h.c.} \right)\end{aligned} \tag{2}$$

where (V_μ^A, λ^A) are the vector and spinor components respectively of a generic Yang-Mills multiplet, and (z_i, X_{iL}) are the scalar and spinor components of chiral multiplets transforming according to some arbitrary representations of the gauge group G with representation matrices T_i^{Aj} (i,j = 1, ..., dim Rep.G, A = 1, ..., dim G).

The arbitrariness of the supergravity couplings consists, in general, of a non-canonical modification of the chiral kinetic part[7,8]

$$e\, \mathcal{G}_i^{\prime\prime\, j}(z, z^*)\, D_\mu z_j\, D_\nu z^{*i} g^{\mu\nu} \tag{3}$$

as well as of a non-canonical modification of the quadratic Yang-Mills part[7]

$$-\frac{1}{4} e\, \mathrm{Re} f_{\alpha\beta}(z)\, F_{\mu\nu}^\alpha F_{\rho\sigma}^\beta g^{\mu\rho} g^{\nu\sigma} \tag{4}$$

The function G(z,z*) is real and G invariant

$$\mathcal{G}^{,j} T^{\alpha i}_{j} z_i = z^{*j} T^{\alpha i}_{j} \mathcal{G}'_i \qquad \left(\mathcal{G}^{,i} = \frac{\partial \mathcal{G}}{\partial z_i} \right) \tag{5}$$

and the function $f_{\alpha\beta}(z)$ is analytic and transforms as the symmetric product of the adjoint representation of G.

The functions $\mathcal{G}(z,z^*)$ and $f_{\alpha\beta}(z)$, because of supersymmetry, proliferate in all interaction terms and in particular in the complete scalar potential which determines the tree-level vacuum state of the theory[7]

$$V(z,z^*) = -e^{-\mathcal{G}}\left(3 + \mathcal{G}^{,i} \mathcal{G}^{-1}{}^{,j}_{i} \mathcal{G}'_j\right) + \frac{1}{2} \tilde{g}^2 \operatorname{Re} f^{-1}_{\alpha\beta} \left(\mathcal{G}^{,i} T^{\alpha j}_{i} z_j\right)\left(\mathcal{G}^{,k} T^{\beta \ell}_{k} z_\ell\right) \tag{6}$$

$\tilde{g}$ is the gauge coupling constant and we have set the gravitational coupling k = 1.

We remark that the arbitrariness of the supergravity coupling of the Yang-Mills system is related to the fact that we cannot demand renormalizability in order to fix the matter interaction when gravity is present. In particular, some interaction terms, suppressed by inverse powers of the Planck mass, can give a sizeable contribution, even at low energies, if some scalar fields develop v.e.v's of order $O(M_P)$. This in fact is what generally happens if the cosmological constant is fine-tuned to zero. As a consequence the low-energy coupled supergravity Yang-Mills system contains interaction terms which remain finite[7,9,13,14] in the global limit $k \to 0$ if the gravitino mass is kept finite in the same limit.

In order to make contact with particle physics, we later restrict the analysis to a particular class of couplings which we will call the "minimal" coupling of matter to supergravity[7]. These couplings are defined by demanding that the scalar and Yang-Mills kinetic terms have a canonical structure

$$\mathcal{G}''^{\,i}_{j} = -\frac{1}{2}\delta^{i}_{j} \quad , \quad f_{\alpha\beta}(z) = \delta_{\alpha\beta} \tag{7}$$

for all values of the scalar fields, z_i, z^{*i}. Equations (7) imply that the complete Lagrangian of the matter-supergravity system has the same arbitrariness, in curved space, as a globally supersymmetric Yang-Mills theory, i.e. a superpotential function g(z) for the chiral multiplets. The superpotential g(z) is related to the function $\mathcal{G}(z,z^*)$ by the relation

$$\mathcal{G}(z,z^*) = -\frac{1}{2}|z_i|^2 - \log|g(z)|^2/4 \tag{8}$$

The G invariance of $\mathcal{G}$ implies that

$$g'^{i}(z)\,T_i^{\alpha j} z_j = 0 \tag{9}$$

There is an exception to Eq. (9) which corresponds to the so-called Fayet-Iliopoulos term[15] which is possible for U(1) factors of G. In this case, the superpotential g(z) satisfies the weak R-symmetry condition[13,8]

$$g'^{i}(z)\,z_i q_i = \xi g(z) \tag{10}$$

where ξ is the Fayet-Iliopoulos constant and q_i are the U(1) charges of the chiral multiplets with scalar components z_i.

For arbitrary functions $\mathcal{G}(z,z^*)$, the scalar part of the matter supergravity system defines a σ model on a Kahler manifold[16,8]. The Kahler metric $\mathcal{G}''^{\,j}_{i}$ defines the Kahler potential J up to a Kahler transformation

$$\mathcal{G}(z,z^*) = J(z,z^*) - \log \frac{g(z)}{2} - \log \frac{g^*(z^*)}{2} \tag{11}$$

Equation (11) means that the superpotential function g(z) can always be reabsorbed in the Kahler potential J by means of an appropriate Kahler gauge choice provided $\langle g(z)\rangle \neq 0$. In fact, under Kahler transformations one has

$$\begin{aligned} J(z,z^*) &\to J(z,z^*) + f(z) + f^*(z^*) \\ g(z) &\to g(z)\, e^{f(z)} \end{aligned} \tag{12}$$

The minimal coupling defined by Eq. (7) corresponds to a flat Kahler manifold. The complete Lagrangian for the coupled supergravity-Yang-Mills system is

$$e^{-1}\mathcal{L} = e^{-1}\mathcal{L}_B + e^{-1}\mathcal{L}_F \tag{13}$$

with

$$\mathcal{L}_B = \mathcal{L}_{BK} + \mathcal{L}_P \tag{14}$$

$$\mathcal{L}_F = \mathcal{L}_{FK} + \mathcal{L}_M \tag{15}$$

Where $\mathcal{L}_B$, the pure bosonic part of the Lagrangian, splits into a part $\mathcal{L}_{BK}$ which contain derivatives of boson fields and a part $\mathcal{L}_P$ which defines the scalar potential

$$\begin{aligned} e^{-1}\mathcal{L}_{BK} = -\frac{1}{2}R + \mathcal{G}''^{\,i}_{\;j}\, D_\mu z_i\, D^\mu z^{*j} - \frac{1}{4}\,\mathrm{Re} f_{\alpha\beta}\, F^\alpha_{\mu\nu} F^{\mu\nu\beta} \\ + \frac{i}{4}\,\mathrm{Im} f_{\alpha\beta}\, F^\alpha_{\mu\nu} \tilde{F}^{\mu\nu\beta} \end{aligned} \tag{16}$$

$$e^{-1}\mathcal{L}_P = -V(z,z^*)$$

Analogously $\mathcal{L}_{FK}$ are those fermionic terms which contain at least one (covariant) derivative on the fields and $\mathcal{L}_{F,M}$ is the fermionic part without derivative terms.

$$
\begin{aligned}
e^{-1}\mathcal{L}_{FK} = {} & \frac{1}{2}\,\mathrm{Re}\, f_{\alpha\beta}\Big(-\frac{1}{2}\bar\lambda^{\alpha}\not{D}\lambda^{\beta}+\frac{1}{2}\bar\lambda^{\alpha}\gamma^{\mu}\sigma^{\rho\sigma}\psi_{\mu}F^{\beta}_{\rho\sigma} \\
& -\frac{1}{2}\bar\lambda^{\alpha}_{L}\gamma_{\mu}\lambda^{\beta}_{R}\,G'^{i}D_{\mu}z_i\Big)-\frac{i}{8}\,\mathrm{Im}\, f_{\alpha\beta}\,e^{-1}D_{\mu}\big(e\,\bar\lambda^{\alpha}\gamma_5\gamma_{\mu}\lambda^{\beta}\big) \\
& -\frac{1}{2}f^{i}_{\alpha\beta}\bar\chi_{Li}\,\sigma\cdot F^{\alpha}\lambda^{\beta}_{L}-\frac{1}{4}e^{-1}\bar\psi_{\mu}\gamma_5\gamma_{\nu}D_{\rho}\psi_{\sigma}\,\varepsilon^{\mu\nu\rho\sigma} \\
& +\frac{1}{8}e^{-1}\varepsilon^{\mu\nu\rho\sigma}\bar\psi_{\mu}\gamma_{\nu}\psi_{\rho}\,G'^{i}D_{\sigma}z_i-G''_{i}{}^{j}\,\bar\psi_{\mu L}\not{D}z^{*i}\gamma^{\mu}\chi_{Lj} \\
& -\bar\chi_{Li}\not{D}z_j\,\chi^{k}_{R}\Big(G'''_{k}{}^{ij}+\frac{1}{2}G''^{i}_{k}G'^{j}\Big)+G''^{\;i}_{j}\,\bar\chi_{Li}\not{D}\chi^{j}_{R}+\text{h.c.}
\end{aligned}
\qquad (17)
$$

$$
\begin{aligned}
e^{-1}\mathcal{L}_{FM} = {} & e^{-G/2}\,\bar\psi_{\mu R}\sigma^{\mu\nu}\psi_{\nu R}+\frac{1}{4}e^{-G/2}G'^{l}\,G''^{-1}{}_{l}{}^{k}\,f^{*}_{\alpha\beta k}\bar\lambda^{\alpha}_{R}\lambda^{\beta}_{R} \\
& +e^{-G/2}\big[G''^{ij}-G'^{i}G'^{j}-G'^{l}G''^{-1}{}_{l}{}^{k}\,G'''_{k}{}^{ij}\big]\bar\chi_{Li}\chi_{Lj} \\
& -\frac{i}{2}\tilde g\,G'^{i}\,T^{\alpha}_{i}{}^{j}z_j\,\bar\psi_{L}\cdot\gamma\lambda^{\alpha}_{R}-e^{-G/2}G'^{i}\,\bar\psi_{R}\cdot\gamma\chi_{Li}- \\
& -g\frac{i}{2}\,\mathrm{Re}\, f^{-1}_{\alpha\beta}\,f^{k}_{\beta\gamma}\,G'^{i}T^{\alpha}_{i}{}^{j}z_j\,\bar\chi_{Lk}\lambda^{\gamma}_{L}+2i\tilde g\,G''^{\;j}_{i}T^{\alpha}_{j}{}^{k}z_k\,\bar\lambda^{\alpha}_{R}\chi^{i}_{R} \\
& +\frac{1}{32}G''^{-1}{}_{l}{}^{k}\,f^{l}_{\alpha\beta}\,f^{*}_{\gamma\delta k}\,\bar\lambda^{\alpha}_{L}\lambda^{\beta}_{L}\bar\lambda^{\gamma}_{R}\lambda^{\delta}_{R}+\frac{3}{32}\big(\mathrm{Re}\, f_{\alpha\beta}\,\bar\lambda^{\alpha}_{L}\gamma_m\lambda^{\beta}_{R}\big)^2 \\
& +\frac{1}{8}\,\mathrm{Re}\, f_{\alpha\beta}\,\bar\lambda^{\alpha}\gamma^{\mu}\sigma^{\rho\sigma}\psi_{\mu}\,\bar\psi_{\rho}\gamma_{\sigma}\lambda^{\beta}+\frac{1}{2}f^{i}_{\alpha\beta}\big(\bar\chi_{Li}\sigma^{\mu\nu}\lambda^{\alpha}_{L}\cdot \\
& \cdot\bar\psi_{\nu L}\gamma_{\mu}\lambda^{\beta}_{R}+\frac{1}{4}\bar\psi_{R}\cdot\gamma\chi_{Li}\,\bar\lambda^{\alpha}_{L}\lambda^{\beta}_{L}\big)+\frac{1}{8}G''_{i}{}^{j}\,\bar\chi^{i}_{R}\gamma_{d}\chi_{Lj}\cdot \\
& \big(\varepsilon^{abcd}\,\bar\psi_{a}\gamma_{b}\psi_{c}-\bar\psi_{a}\gamma_5\gamma^{d}\psi_{a}\big)+\frac{1}{16}\bar\chi_{Li}\gamma^{\mu}\chi^{j}_{R}\,\bar\lambda^{\delta}_{R}\gamma_{\mu}\lambda^{\gamma}_{L}\cdot \\
& \big(-2G''^{\;i}_{j}\,\mathrm{Re}\, f_{\gamma\delta}+\mathrm{Re}\, f^{-1}_{\alpha\beta}\,f^{i}_{\alpha\gamma}\,f^{*}_{\beta\delta j}\big)+
\end{aligned}
\qquad (18)
$$

(continued)

$$+\frac{1}{16}\bar{\chi}_{Li}\sigma_{\mu\nu}\chi_{Lj}\bar{\lambda}^{\gamma}_{L}\sigma^{\mu\nu}\lambda^{\delta}_{L}\,\mathrm{Re}f^{-1}_{\alpha\beta}f^{\prime i}_{\alpha\gamma}f^{\prime j}_{\beta\delta}+\frac{1}{16}\bar{\chi}_{Li}\chi_{Lj}\bar{\lambda}^{\gamma}_{L}\lambda^{\delta}_{L}\cdot$$

$$\cdot\left(-4G^{\prime\prime\prime ij}_{k}G^{\prime\prime -1\,k}_{\ell}f^{\prime\ell}_{\gamma\delta}+4f^{\prime\prime ij}_{\gamma\delta}-\mathrm{Re}f^{-1}_{\alpha\beta}f^{\prime i}_{\alpha\gamma}f^{\prime j}_{\beta\delta}\right)+\left(-\frac{1}{2}G^{\prime\prime\prime\prime ij}_{k\ell}+\right. \tag{18}$$

$$\left.+\frac{1}{2}G^{\prime\prime\prime ij}_{m}G^{\prime\prime -1\,m}_{n}G^{\prime\prime\prime n}_{k\ell}-\frac{1}{4}G^{\prime\prime i}_{k}G^{\prime\prime j}_{\ell}\right)\bar{\chi}_{Li}\chi_{Lj}\bar{\chi}^{k}_{R}\chi^{\ell}_{R}+h.c.$$

The Einstein scalar curvature R in Eq. (16) as well as the fermion covariant derivatives in Eq. (17) contain spin 3/2 torsion terms through the Lorentz connection $\psi_{\mu mn}(e,\psi)$ which appears in the definition of the spinor covariant derivatives D_μ of Eq. (17).

The Lagrangian defined by Eq. (13) is a local supersymmetrc density under the following local supersymmetry transformations of the fields

$$\delta e_\mu^{\ m}=\frac{1}{2}\bar{\epsilon}_L\gamma^m\psi_{\mu R}+h.c.$$

$$\delta\psi_{\mu L}=\left(\partial_\mu+\frac{1}{2}\omega_{\mu mn}(e,\psi)\sigma^{mn}\right)\epsilon_L+\frac{1}{2}\sigma_{\mu\nu}\epsilon_L G^{\prime\prime\ j}_{i}\bar{\chi}^{i}_{R}\gamma^{\nu}\chi_{Lj}$$
$$+\frac{1}{2}\gamma_\mu\epsilon_R e^{-G/2}+\frac{1}{4}\psi_{\mu L}\left(G^{\prime i}\bar{\epsilon}_L\chi_{Li}-G^{\prime}_{i}\bar{\epsilon}_R\chi^{i}_{R}\right)$$
$$-\frac{1}{4}\epsilon_L\left(G^{\prime i}D_\mu z_i-G^{\prime}_{i}D_\mu z^{*i}\right)+\frac{1}{4}\left(g_{\mu\nu}-\sigma_{\mu\nu}\right)\epsilon_L\bar{\lambda}^{\alpha}_{L}\gamma^{\nu}\lambda^{\beta}_{R}\,\mathrm{Re}f_{\alpha\beta}$$

$$\delta B^{\alpha}_{\mu}=-\frac{1}{2}\epsilon_L\gamma_\mu\lambda^{\alpha}_{R}+h.c. \tag{19}$$

$$\delta\lambda^{\alpha}_{R}=\frac{1}{2}\sigma^{\mu\nu}\hat{F}^{\alpha}_{\mu\nu}\epsilon_R+\frac{1}{2}i\epsilon_R\,\mathrm{Re}f^{-1}_{\alpha\beta}\left(-\tilde{g}G^{\prime i}T_{i}^{\beta j}z_j+\right.$$
$$\left.+\frac{i}{2}f^{\prime i}_{\beta\gamma}\bar{\chi}_{Li}\lambda^{\gamma}_{L}-\frac{i}{2}f^{*\prime}_{\beta\gamma i}\bar{\chi}^{i}_{R}\lambda^{\gamma}_{R}\right)-\frac{1}{4}\lambda^{\alpha}_{R}\left(G^{\prime i}\bar{\epsilon}_L\chi_{Li}-G^{\prime}_{i}\bar{\epsilon}_R\chi^{i}_{R}\right)$$

$$\delta z_i=\bar{\epsilon}_L\chi_{Li}$$

$$\delta\chi_{Li}=\frac{1}{2}\hat{\not{D}}z_i\epsilon_R-\frac{1}{2}\epsilon_L e^{-G/2}G^{\prime\prime -1\ j}_{i}G^{\prime}_{j}-\frac{1}{8}\bar{\lambda}^{\alpha}_{R}\lambda^{\beta}_{R}G^{\prime\prime -1\ k}_{i}f^{*\prime}_{\alpha\beta k}$$
$$+\frac{1}{2}\epsilon_L G^{\prime\prime -1\ k}_{i}G^{\prime\prime\prime j\ell}_{k}\bar{\chi}_{Lj}\chi_{L\ell}+\frac{1}{4}\chi_{Li}\left(G^{\prime}_{j}\epsilon_R\chi^{j}_{R}-G^{\prime j}\bar{\epsilon}_L\chi_{Lj}\right)$$

In order to describe the Higgs and superHiggs effects we confine our attention to the scalar potential term given by Eq. (16) and the quadratic part in the fermion fields $\mathcal{L}_F^{(2)}$ which can be read from Eq. (18)

$$\begin{aligned} \mathcal{L}_F^{(2)} = {} & e^{-G/2}\bar{\psi}_{\mu R}\sigma^{\mu\nu}\psi_{\nu L} - \bar{\psi}_R\cdot\gamma\left(e^{-G/2}G^{,i}\chi_{Li} - \tfrac{i}{2}\tilde{g}G^{,i}T_i^{\alpha j}z_j\lambda_L^{\alpha}\right) \\ & + e^{-G/2}\left[G^{,ij} - G^{,i}G^{,j} - G^{,\ell}G^{,-1k}_{\ell}G_k^{,ij}\right]\bar{\chi}_{Li}\chi_{Lj} \\ & - 2i\tilde{g}G_i^{,k}z^{*j}T_j^{\alpha i}\bar{\lambda}_L^{\alpha}\chi_{Lk} \\ & + \tfrac{1}{2}f_{\alpha\beta}^{,k}\left(\tfrac{1}{2}e^{-G/2}G^{,\ell}G^{,-1\ell}_{k}\bar{\lambda}_L^{\alpha} + i\tilde{g}\,\mathrm{Re}\,f^{-1}_{\alpha\gamma}G^{,i}T_i^{\gamma j}z_j\bar{\chi}_{Lk}\right)\lambda_L^{\beta} + \mathrm{h.c.} \end{aligned} \tag{20}$$

The Goldstone fermion is uniquely defined by the combination of the fermion fields which couple to the gravitino field ψ_μ. It is given by

$$\begin{aligned} & -\bar{\psi}_R\cdot\gamma\,\eta_L \\ & \eta_L = e^{-G/2}G^{,i}\chi_{Li} - \tfrac{i}{2}\tilde{g}G^{,i}T_i^{\alpha j}z_j\lambda_L^{\alpha} \end{aligned} \tag{21}$$

where the scalar dependent factors which defined η_L are meant to be computed at the extremum of the scalar potential

$$V^{,i} = \frac{\partial V}{\partial z_i} = 0 \tag{22}$$

To discuss models for particle physics, it is of special interest to discuss the Higgs and superHiggs effects in a Minkowski background, so we also demand the cancellation of the cosmological constant at the extremum defined by Eq. (22)

$$V(z,z^*) = 0 \quad , \quad V^{,i}(z,z^*) = 0 \tag{23}$$

As already anticipated, we now confine our analysis to the case of a "minimally" coupled Yang-Mills system so that Eqs. (7) hold. This analysis can also be generalized to arbitrary Kahler potential functions J(z,z*).

In the "minimal" case the scalar potential (6) becomes

$$V(z,z^*) = e^{-\mathcal{G}}[2\mathcal{G}^{,i}\mathcal{G}_{,i} - 3] + \frac{1}{2} D^\alpha D^\alpha$$
$$D^\alpha = \tilde{g}\, \mathcal{G}^{,i} T^{\alpha}{}_{i}{}^{j} z_j = -\frac{1}{2}\tilde{g}\, z^{*i} T^{\alpha}{}_{i}{}^{j} z_j \tag{24}$$

Note that the definition of D^α through $\mathcal{G}^i$ automatically includes possible Fayet-Iliopoulos terms for U(1) factors as it can be easily checked by the use of Eq. (10) [17] and the fermionic bilinear terms given by Eq. (20) become

$$\mathcal{L}_F^{(2)} = e^{-\mathcal{G}/2}[\bar{\psi}_{\mu R}\sigma^{\mu\nu}\psi_{\nu R} - \bar{\psi}_R\cdot\gamma\tilde{\eta}_L - \frac{2}{3}\bar{\tilde{\eta}}_L\eta_L] +$$
$$+ \bar{\chi}_{Li}\mathcal{M}^{ij}\chi_{Lj} + 2\bar{\chi}_{Li}\mathcal{M}^{i\alpha}\lambda^\alpha_L + \bar{\lambda}^\alpha_L\mathcal{M}^{\alpha\beta}\lambda^\beta_L + h.c. \tag{25}$$

and the fermionic spin ½ mass matrix, after subtraction of the Goldstino mode, becomes

$$\mathcal{M}^{ij} = e^{-\mathcal{G}/2}[\mathcal{G}^{,ij} - \frac{1}{3}\mathcal{G}^{,i}\mathcal{G}^{,j}]$$
$$\mathcal{M}^{i\alpha} = -\frac{i}{3}\mathcal{G}^{,i}D^\alpha - i\, D^{,\alpha i} \tag{26}$$
$$\mathcal{M}^{\alpha\beta} = -\frac{1}{6}e^{\mathcal{G}/2}D^\alpha D^\beta$$

The gravitino mass term is given by

$$\mathcal{M}^2_{3/2} = e^{-\mathcal{G}/2} \tag{27}$$

From Eqs. (25), (26) and (27) and the vector boson masses we get the model independent mass sum rule[7]

$$\sum_{J=0}^{3/2}(-)^{2J}(2J+1)\, m_J^2 = (N-1)(2m^2_{3/2} - k^2 D^\alpha D^\alpha) - 2\tilde{g}_\alpha D^\alpha \mathrm{Tr}\, T^\alpha \tag{28}$$

Inspection of Eq. (28) already shows that if $m_{3/2} \simeq O(M_W)$ one gets a shift of the scalar-fermion mass differences, which invalidates the sum rules of global supersymmetry[18] obtained by setting $m_{3/2} = 0$ and $k = 0$, and which only hold if supersymmetry breaking occurs so that $m_{3/2}/M_W << 1$.

In order to exhibit models in which the Higgs and superHiggs effects co-exist we introduce a chiral multiplet, singlet under the gauge group G, whose scalar component we denote by z. We call y_i the scalar components of all chiral multiplets other than z.

We construct a class of models by defining the following overall superpotential[19]

$$g(z,y_i) = g(z)\left[1 + \frac{g(z)^{q-2}}{\langle g(z)\rangle^{q-1}} h(y)\right] \tag{29}$$

Then we insert the superpotential $h(z,y_i)$ into the supergravity scalar potential and use the singlet z to cancel the cosmological constant and to give the substantial contribution to the gravitino mass

$$V_{,z} = 0 \quad , \quad V_{,y_i} = 0 \quad , \quad V = 0 \tag{30}$$

with $\langle z\rangle = O(M_P)$ and $\langle y_i\rangle/M_P \ll 1$. Under these circumstances, which also assume that the function h(y) does not contain dimensional parpmeters $O(M_P)$, after shift of the scalar goldstino by its v.e.v. $z \to \hat{z} \to z - \langle z\rangle$, the scalar field $\hat{z}$ decouples from the low-energy theory described by the multiplets with scalar components y_i, and for $M_P \to \infty$ and $m_{3/2}$ fixed one gets the following effective potential[9,10]

$$V(z,y_i) = V(z) + V(y_i) + \frac{1}{2}(D^\alpha)^2 + O\left(\frac{1}{M_P}\right) + \dots \tag{31}$$

For the class of functions given by Eq. (29) one gets

$$V(y_i) = \left|\frac{\partial h}{\partial y_i} + m_{3/2}\, y^{*i}\right|^2 + (A-3)\, m_{3/2}\, (h + h^*) + \frac{1}{2}(D^\alpha)^2 \tag{32}$$

where $A = 3 + \sqrt{3}\,(q-2)$

$$V(z) = m_A^2 A^2 + m_B^2 B^2 \qquad m_A^2 + m_B^2 = 4 m_{3/2}^2 \tag{33}$$

in terms of the gravitino mass given by

$$m_{3/2} = \kappa^2 \left|g(\langle z\rangle)\right| e^{\kappa^2 \langle z\rangle^2/2} + O\left(\frac{1}{M_P}\right) \cdot \quad (G^{i}{}_{j} = -\delta^{i}_{j}) \tag{34}$$

The remaining part of the coupled Yang-Mills-supergravity system contains the normal interactions of a globally supersymmetric system and additional non-renormalizable interactions which are suppressed by inverse powers of M_P. Note that in order to derive Eq. (32) we have assumed that $\langle g(\langle z\rangle)\rangle$ is $O(\mu M_P^2)$ where μ is a "low-energy" scale, so that $m_{3/2} \to O(\mu)$ for $M_P \to \infty$. This corresponds to an intermediate supersymmetry breaking scale

$$M_S = \left(\sqrt{3}\, m_{3/2}/\kappa\right)^{1/2} = O\left(10^{10} - 10^{11}\ \text{GeV}\right) \tag{35}$$

From Eq. (32) we see that after the superHiggs effect, in the limit $M_P \to \infty$ with $m_{3/2}$ fixed, the scalar potential acquires new terms, beyond the ones dictated by global supersymmetry

$$V_{SUSY}(y_i) = \left|\frac{\partial h}{\partial y_i}\right|^2 + \frac{1}{2}(D^\alpha)^2 \tag{36}$$

which are "soft" in a supersymmetric sense[20]

$$\Delta V = m_{3/2}^2 |y_i|^2 + m_{3/2}\left(\tilde{h}(y) + \tilde{h}^*(y^*)\right) \tag{37}$$

with

$$\tilde{h}(y) = \frac{\partial h}{\partial y_i} y_i + (A-3)h(y)$$

for scale invariant potentials $\hat{h}(y) = Ah(y)$ and we recover the parametrization of Ref. 10. For A = 3 (q = 2) the scalar potential is always semi-positive definitive

$$V(y_i) = \left| \frac{\partial h}{\partial y_i} + m_{3/2} y^{*i} \right|^2 + \frac{1}{2} (D^\alpha)^2 \tag{38}$$

and we recover the parametrization of Ref. 21. Because of the particular form of ΔV in the limit $M_P \to \infty$, one gets field-independent mass sum rules for the y fields

$$\sum_{J=0}^{1} (-)^{2J} (2J+1) m_J^2 = 2 N m_{3/2}^2 - 2 \tilde{g}_\alpha D^\alpha \, Tr T^\alpha \tag{39}$$

N denotes the number of chiral multiplets y_i and the last term can only occur for U(1) factors of G. The general form of the scalar square mass matrix becomes

$$\mathcal{M}_S^2 = \mathcal{M}_{S,\,SUSY}^2 + \Delta \mathcal{M}_S^2 \tag{40}$$

with

$$\Delta \mathcal{M}_S^2 = \begin{pmatrix} m_{3/2}^2 \delta_i^{\ k} & m_{3/2} \tilde{h}''_{ik} \\ m_{3/2} \tilde{h}^{*'',ik} & m_{3/2}^2 \delta^i_{\ k} \end{pmatrix} \tag{41}$$

In the supergravity-induced supersymmetry breaking the gravitino mass $m_{3/2}$ and the Goldstino scalar field z trigger an explicit but soft supersymmetry breaking in the flat limit $k \to 0$. The main physical information from the breaking is contained in Eq. (41). The low-energy broken theory is defined by the y_i superpotential

$h(y_i)$ of the globally supersymmetric unbroken theory and by the parameters of the pure gravitational sector. For the class of models defined by Eq. (29) they are the dimensionless parameter A and the gravitino mass $m_{3/2}$. The former appears in the definition of $\tilde{h}(y_i)$ and the latter in the definition of V.

Remarkably enough, the terms produced by the supergravity corrections can be rewritten most elegantly in superspace language, in terms of a chiral spurion superfield[22] whose only non-vanishing F-component is $F = m_{3/2}$

$$\rho = (0, 0, m_{3/2}) \tag{42}$$

In terms of the η and the matter chiral superfields y_i the previous soft-breaking terms become

$$Y_i(\rho+\rho^*)Y^{*i}\Big|_D \; ; \; (A-3)\rho\, h(y_i)\Big|_F \tag{43}$$

Note that the parametrization of Ref. 21 corresponds to the vanishing of the second elementary matter-spurion vertex. It gives rise to universal mass splittings for fields not having v.e.v's, like the quark and lepton superfields, of the form[21]

$$m_s^{\pm} = |m_f \pm m_{3/2}| \tag{44}$$

where m_s, m_f are the scalar and fermion particles, members of the same chiral multiplet.

Most interestingly it has recently been shown that the scalar potential defined for A = 3 is completely (one-loop) stable, including radiative gravitational corrections, as far as quadratic divergences are concerned[23].

A particularly appealing property of the corrected scalar potential given by Eq. (32) is that the gravitino mass scale $m_{3/2}$, responsible for the supersymmetry breaking, can also trigger the

Higgs effects, i.e. it can set up the scale of the weak interactions, by spontaneous breakdown of $SU(2) \times U(1) \to U(1)_{e.m.}$. If the low-energy superpotential $h(y_i)$ does not contain an intrinsic scale, a necessary condition for the $SU(2) \times U(1)$ breaking to occur has been shown[10] to be $A \geq 3$. If the low-energy superpotential comes from a GUT theory, this does not need to be the case and $A - 3$ can be negative as well, as in the models of Ref. 24.

To be more explicit let us consider, for instance, the following low-energy Higgs superpotential[22]

$$h(y_i) = \alpha Y H H^c + \beta/_3 Y^3 \tag{45}$$

where Y is an $SU(2) \times U(1)$ singlet and H, $\bar{H}$ are two Higgs doublets with opposite hypercharge. Quark and lepton multiplets also have normal super-Yuakwa interactions with Higgses through ordinary trilinear superpotential terms.

If we consider the simplest parametrization A = 3, one immediately realizes that there are $SU(2) \times U(1)$ breaking minima with

$$\langle H \rangle, \langle H^c \rangle, \langle Y \rangle \propto m_{3/2} \tag{46}$$

The low-energy spectrum of the resulting spontaneously broken electroweak theory gives rise to the following particle spectrum. A first set of states is given by the charged and neutral vector supermultiplets associated with the $W^{\pm}$ and Z^0 vector bosons with mass splitting $O(m_{3/2})$. These scalar components are the opposite parity partners of the would-be Goldstone bosons with masses

$$M_{\pm}^2 = M_W^2 + 4m_{3/2}^2 \quad , \quad M_0^2 = M_Z^2 + 4m_{3/2}^2 \tag{47}$$

while their fermionic components, together with the gauginos associated with the $W^{\pm}$, Z^0 bosons, give rise to two charged (Dirac) fermions and two neutral Majorana fermions with masses

$$m^2_\pm(1,2) = M^2_W + \frac{1}{2} m^2_{3/2} \left(1 \pm \sqrt{1+4M^2_W/m^2_{3/2}}\right)$$

$$m^2_0(1,2) = M^2_Z + \frac{1}{2} m^2_{3/2} \left(1 \pm \sqrt{1+4M^2_Z/m^2_{3/2}}\right) \tag{48}$$

The $W^\pm$ and gravitino masses are related as follows

$$M^2_W = \frac{\tilde{g}^2}{\alpha^2}(1-x)m^2_{3/2} \quad , \quad x = \beta/\alpha \tag{49}$$

There are four more neutral scalars with masses

$$M^2_0(2,3) = \frac{1}{2}(1-x)m^2_{3/2}\left[5 - x \pm \sqrt{9-10x+x^2}\right]$$

$$M^2_0(4,5) = \frac{1}{2} m^2_{3/2}\left[9 - 2x + x^2 \pm (3+x)\sqrt{9-10x+x^2}\right] \tag{50}$$

and two neutral (Majorana) fermions with masses

$$m^2_0(3) = m^2_{3/2} \quad , \quad m^2_0(4) = m^2_{3/2}(2-x)^2 \tag{51}$$

For the scalar quarks and leptons one has the general formula[21]

$$M^\pm_{sq(sl)} = \left| m_{3/2} \pm m_{q(\ell)} \right| \tag{52}$$

Without direct gaugino masses the photinos and gluinos are massless at the tree level. However, due to an explicit breaking of R-symmetry through the supergravity induced breaking terms, they get one-loop radiative masses $O(\alpha_{e.m.} m_{3/2})$ and $O(\alpha_s m_{3/2})$ respectively from the splitting in masses of charged and/or coloured heavy particles[22]. The light charged particles of Eqs. (47) and (48) also contribute to the gaugino masses, while the quark and lepton multiplets give a smaller contribution for $m_{q,\ell}/m_{3/2} \ll 1$. It is also possible to get a direct gaugino mass from non-minimal kinetic terms for the gauge sector of the theory[7]. In this case the natural order of magnitude of gaugino masses, irrespectively of their quantum numbers, would be $m_{3/2}$ [21,25].

Application of the present ideas to GUTs can be carried out by introducing a suitable superpotential for the matter fields y_i, which describes a unified theory of electroweak and strong interactions: $G_{GUT} \supset SU(3) \times SU(2) \times U(1)$. Unification brings into the theory a second scale M_X which may or may not be identified with M_P. If $M_X \ll M_P$ we can still apply the previous formalism to describe a GUT theory with a spontaneously broken local supersymmetry.

For example, an embedding of the minimal SU(5) supersymmetric model in an N = 1 supergravity coupled theory can be constructed with a tree level hierarchy[24]

$$\langle H_2 \rangle, \langle H_2^c \rangle, \langle Y \rangle \sim O(m_{3/2})$$

$$\langle \Delta_{24} \rangle = O(M_X) \quad , \quad \langle Z \rangle = O(M_P)$$

However, as recently pointed out[26], GUT models based on N = 1 supergravity, with $m_{3/2} = O(M_W)$ do not generally have a stable hierarchy in perturbation theory

$$M_W \ll M_X$$

This is due to the presence, in these theories, of light singlets, such as y, which couple to heavy fields in the superpotential, which in turn have large $(\text{mass})^2$ splittings $O(m_{3/2} M_X)$ coming from the ΔV term given by Eq. (37) in the effective scalar potential. These light singlets induce radiative square masses to the Weinberg-Salam doublets $O(m_{3/2} M_X)$ which destroy the tree-level hierarchy. This quite important phenomenon imposes severe constraints on GUTs based on the combined Higgs and superHiggs effect. It is possible to construct a class of models which do not have this instability in perturbation theory[22]. These models contain a more complicated set of SU(5) representations than the minimal supersymmetric SU(5) GUT.

A general feature of these models is that they contain 75, 50 and $\overline{50}$ SU(5) representations[27] through which the Higgs triplets of the 5 and $\bar{5}$ representations receive masses, while the Higgs doublet sector only fills the gravitino mass scale but not the GUT scale M_X.

To summarize, we have shown that the hypothesis of a spontaneously broken supersymmetric theory based on supergravity-induced Higgs and superHiggs phenomena can accommodate realistic theories for particle interactions. A gravitino mass scale O(1 TeV) can lead to a satisfactory description of a low-energy physics.

The major problems we have found with this approach, and for which an explanation is still lacking, is the unsatisfactory fine-tuning of the (nowadays) cosmological constant and the obscure origin of the Goldstino interaction with matter. Perhaps these questions will be answered by an understanding of the relation which might exist between this phenomenological N = 1 supergravity Lagrangian and a more general theory (N = 8 supergravity?)[28] from which the present approach should emerge.

REFERENCES

1. J. Wess and B. Zumino, Nucl. Phys. B70:39 (1974); Phys. Lett. 49B:52 (1974).
2. J. Wess and B. Zumino, Phys. Lett. 49B:52 (1974).
   J. Iliopoulos and B. Zumino, Nucl. Phys. B76:310 (1974.
   S. Ferrara, J. Iliopoulos and B. Zumino, Nucl. Phys. B77:413 (1977).
   S. Ferrara and O. Piguet, Nucl. Phys. B93:261 (1975).
   M. T. Grisaru, M. Rocek and W. Seigel, Nucl. Phys. B159:429 (1979).
3. E. Gildener, Phys. Rev. D14:1667 (1976).
   E. Gildener and S. Weinberg, Phys. Rev. D15:3333 (1976).
   L. Maiani, Proceedings of the Summer School of Gif-sur-Yvette p. 3 (1979).

E. Witten, Nucl. Phys. B185:153 (1981).

S. Dimopoulos and S. Raby, Nucl. Phys. B199:353 (1981).

4. J. Ellis, S. Ferrara and D. V. Nanopoulos, Phys. Lett. 114B:231 (1982).

5. R. Barbieri, S. Ferrara and D. V. Nanopoulos, Z. Phys. C13:267 (1982); Phys. Lett. 116B:6 (1982).

M. Dine and W. Fischler, Nucl. Phys. B204:346 (1982).

J. Ellis, L. Ibanez and G. Ross, Phys. Lett. 113B:283 (1982).

S. Dimopoulos and S. Raby, Los Alamos preprint LA-UR-82-1282 (1982).

S. Polchinski and L. Susskind, Phys. Rev. D26:3661 (1982).

6. D. V. Volkov and V. A. Soroka, JEZP Lett. 18:312 (1973).

S. Deser and B. Zumino, Phys. Rev. Lett. 38:1433 (1977).

E. Cremmer, B. Julia, J. Scherk, S. Ferrara, L. Girardello and P. van Nieuwenhuizen, Phys. Lett. 79B:231 (1978); Nucl. Phys. B147:1051 (1979).

7. E. Cremmer, S. Ferrara, L. Girardello and A. van Proeyen, Phys. Lett. 116B:231 (1982); Nucl. Phys. B212:413 (1983).

8. J. Bagger, Nucl. Phys. B211:302 (1983).

J. Bagger and E. Witten, Phys. Lett. 115B:202 (1982).

9. R. Barbieri, S. Ferrara and C. A. Savoy, Phys. Lett. 119B:343 (1982).

10. H. P. Nilles, M. Srednicki and D. Wyler, Phys. Lett. 120B:346 (1983).

11. D. Z. Freedman, P. van Nieuwenhuizen and S. Ferrara, Phys. Rev. D13:3214 (1976).

12. S. Ferrara and B. Zumino, Nucl. Phys. B79:413 (1974).
A. Salam and A. Strathdee, Phys. Lett. 51B:353 (1974).
B. de Wit and D. Z. Freedman, Phys. Rev. D12:2286 (1975).
13. R. Barbieri, S. Ferrara, D. V. Nanopoulos and K. S. Stelle, Phys. Lett. 113B:219 (1982).
14. B. A. Ovrut and J. Wess, Phys. Lett. 112B:347 (1982).
15. P. Fayet and J. Iliopoulos, Phys. Lett. 51B:461 (1974).
16. B. Zumino, Phys. Lett. 87B:203 (1979).
17. M. T. Grisaru, M. Rocek and A. Karlhede, CALTECH preprint CALT-68-949 (1982).
18. S. Ferrara, L. Girardello and F. Palumbo, Phys. Rev. D20:403 (1979).
19. E. Cremmer, P. Fayet and L. Girardello, Phys. Lett. 122B:41 (1983).
20. L. Girardello and M. T. Grisaru, Nucl. Phys. B194:65 (1982).
21. E. Cremmer, P. Fayet and L. Girardello, Ecole Normale Supérieure preprint LPTENS 82/30 (1982).
22. S. Ferrara, D. V. Nanopoulos and C. A. Savoy, CERN preprint TH.3442 (1982).
23. R. Barbieri and S. Cecotti, Scuolo Normale Superiore preprint SNS 9/1982.
24. R. Arnowitt, A. H. Chamseddine and P. Nath, Phys. Rev. Lett. 49:970 (1982).
L. Ibanez, CERN preprint TH.3374 (1982).
25. J. Ellis, D. V. Nanopoulos and K. Tamvakis, Phys. Lett. 121B:123 (1983).

26. H. P. Nilles, M. Srednicki and D. Wyler, CERN preprint TH.3461 (1982).

27. A. Masiero, D. V. Nanopoulos, K. Tamvakis and T. Yanagida, Phys. Lett. 115B:380 (1982).

28. E. Cremmer and B. Julia, Phys. Lett. 80B:48 (1978); Nucl. Phys. B159:141 (1979).

B. de Wit and H. Nicolai, Phys. Lett. 108B:285 (1981); Nucl. Phys. B208:323 (1982).

H. Nicolai, Ref. TH3455 CERN, to appear in "Supergravity 82" (Eds. S. Ferrara, J.G. Taylor, P. van Nieuwenhuizen), Proceedings of the Trieste September School on Supergravity and Supersymmetry, 1982 (World Scientific Publishing Co., Singapore).

D I S C U S S I O N S

CHAIRMAN: S. FERRARA

Scientific Secretaries: A. Albrecht, J. Cole, J. Lykken

DISCUSSION 1

- *KREMER:*

Why do you speak of the supersymmetry algebra rather than the supersymmetry group?

- *FERRARA:*

For me they are the same.

- *YANG:*

Why do you say that? They are very different!

- *FERRARA:*

They are very different in the sense that you have to exponentiate the supersymmetry algebra to get the supersymmetry group. Is that what you mean?

- *YANG:*

The exponentiated algebra is not a group.

- *FERRARA:*

It is a graded group.

- *YANG:*

Yes. And I never understood what is so good about a graded group.

- *FERRARA:*

Since a graded Lie group involves anti-commuting variables, I think it is a debated question what a supersymmetry group is, in fact - especially in the application to physics. However, in order to get, in a Lagrangian field theory all physical information, I don't need to carry out this exponentiation. The reason being that I can obtain all of the Ward identities needed to fix the couplings, and so on, just from the conservation of the supersymmetry current. From a more fundamental point of view I think one would also like to have the notion - and the physical relevance - of what a graded Lie group is, but this, as yet, is an unsolved question.

- *LYKKEN:*

I would like to understand the ultraviolet-finite model that you mentioned in your lecture. What is the divergence structure like, and are all cancellations of divergences due just to supersymmetry, or does more go into it?

- *FERRARA:*

Let me tell you what the theory is. The theory has N = 4, that is, it is a realization of the supersymmetry algebra with four spinorial charges. The particle content of this theory is: a gauge field in the adjoint representation of an arbitrary compact semi-simple group G, four Majorana fields in the adjoint representation of G, and six scalar fields also in the adjoint representation of the Yang-Mills group. This theory is the maximally extended theory which does not contain gravitation. It is described by a unique Langrangian, which is the usual Yang-Mills action together with some Yukawa couplings and ϕ^4 interactions, all defined in terms of single coupling constant which is the usual gauge coupling constant. As I explained this morning, in supersymmetric theories there are cancellations of ultraviolet divergences. This can be understood explicitly in N = 1 theories, showing the absence of quadratic divergences. What is new in this theory, which is not the case in N = 1 and N = 2 gauge theories, is that the β function, which is related to the charge renormalization, vanishes identically. This property cannot be understood as in the case of N = 1 supersymmetry; it occurs only for N = 4. There are more cancellations of ultraviolet divergences in N = 4 supersymmetry than for theories with 1 or 2 supersymmetry generators. If you use power counting, taking account of supersymmetry in this theory, and working in a particular gauge - the light cone gauge - as Mandelstam did, one can show that this β function vanishes to all orders of perturbation theory. I don't think there is any symmetry explanation other than this explicit calculation for the finiteness of this theory.

- *ALBRECHT:*

What good is an ultraviolet-finite theory with zero β function?

- *FERRARA:*

Computationally it is a simpler theory. But I don't think there is any physical justification unless there are people who are afraid of renormalization.

- *ALBRECHT:*

I have another question. You said that one nice thing about supersymmetry is that it gives you an excuse for the Higgs field as a partner of a fermion. But in the models invented so far the fermion partner of the Higgs is not a known particle. So it seems like you've just had to invent more particles?

- *FERRARA:*

Yes, in this $N = 1$ supersymmetry you are essentially duplicating every known species of particle. These new particles allow you to avoid ultraviolet quadratic divergences. So the reason for introducing them is just to get a theory which is more convergent in the ultraviolet region.

VAN BAAL:

One can transform chiral invariance in the fermion sector to some invariance in the Bose sector. Will this protect the scalars from acquiring mass through radiative corrections, if we throw away the fermion part of the Lagrangian - apart from the fact that this might give a nonrenormalizable theory?

- *FERRARA:*

No, you need supersymmetry because you need cancellations of boson diagrams with fermion diagrams.

- *GOITY:*

Can you have Majorana masses in $N > 1$ supersymmetry without breaking supersymmetry?

- *FERRARA:*

Well in N = 8 supersymmetry you have a case where the quarks and leptons must be considered as composite objects, not as fundamental particles. This theory, as a subconstituent theory, has a chiral symmetry coming from an SU(8) symmetry. So in N = 8 there is a chiral symmetry for these fermions.

- *KAPLUNOVSKY:*

Excuse me, what is wrong with a massive N = 8 supermultiplet?

- *FERRARA:*

The massive supermultiplet for N = 8 supersymmetry has particles with spin higher than two.

- *ZICHICHI:*

Is there any theoretical reason to fix the number N ?

- *FERRARA:*

If we are to have particles with no spins greater than one, which is the basic criterion for renormalizable theories, one must have no more than four supersymmetric charges.

- *VAN DE VAN:*

Is it still true that physicists have been unable to construct quantum field theories for spin larger than 2 ?

- *FERRARA:*

Yes, for interacting particles. For free particles one can of course construct Lagrangians for any spin.

- *VAN DE VEN:*

But is there proof that one cannot construct such theories ?

- *FERRARA:*

No, there is no proof. In fact it probably is not true. I think the reason you cannot construct theories with higher spin is related to the fact that when you have higher spin you must have gauge invariance which eliminates the unphysical degrees of freedom.

For spin 1 you have just the usual Yang-Mills invariance.

For spin $^3/_2$ this was solved with supergravity. Supergravity gives the first example of a consistent theory of spin $^3/_2$ massless particles because there is an associated gauge invariance.

For spin 2 it was solved by the gravitational interaction. All these particles are essentially gauge fields of associated gauge symmetries.

For spin higher than 2 no such symmetry is known. Perhaps there is some way of describing interacting particles with spins higher than 2 but nobody knows it.

- *VON BIJ:*

What if they are massive ?

- *FERRARA:*

You will have the same problem. In fact in the spin 2 Einstein theory the field cannot be massive. In spin $^3/_2$ and spin 1 non-Abelian gauge theories masses are only allowed through spontaneous breaking of the symmetry. The basic interaction must be gauge invariant. In the spin 1 Abelian case you can have an explicit mass term. For other theories I do not know how to incorporate mass.

- *VON BIJ:*

This is only if you insist on renormalizability I guess, but with gravity you don't have renormalizability anyway.

- *FERRARA:*

It is not only renormalizability, but causal propagation etc. which gives you problems when you try to construct a normal massive theory of, say, a spin $^3/_2$ particle.

- *MUKHI:*

If you have a world in which supersymmetry is exact would there be any experimental way to distinguish between bosons and fermions ?

- *FERRARA:*

That is a question which is interesting but I don't know how to answer it.

- *YANG:*

Merely by counting the number of states of a spin $^1/_2$ particle and a spin 1 particle you can tell which one is which.

- *MUKHI:*

I have another question : In the 1960's there was a "no-go" theorem restricting the possible symmetries of a relativistic S-matrix to those already known and supersymmetry violated that because of graded Lie algebras. Now we have a new no-go theorm including supersymmetry. Do you know any mathematical reason why we can or cannot have a new extension of graded Lie algebras?

- *FERRARA:*

This is related to the theorem of Haag et al. These people have only proven that in the context of a graded Lie algebra the spinorial charges must have spin $^1/_2$ and not spin $^3/_2$ for instance. However, you can imagine that there is some as yet undiscovered algebraic structure which allows you to have a more general framework in which you can evade this theorem.

- *MARTIN:*

How can Fermi and Bose amplitude be equal when they have different Lorentz indices?

- *FERRARA:*

They are not equal in supersymmetry, but they are related. They are equal up to trivial kinematical factors.

- *RICHARDS:*

Has supersymmetry been observed as a physical principle in other branches of particle physics?

- *FERRARA:*

Supersymmetry has been used in nuclear physics where symmetries between boson and fermion states have been observed. However, the formalism is very different and the connection with particle physics is not clear.

- *LOEWE:*

Is it possible to suggest a real experiment which would tell us that supersymmetry has anything to do with physics and is it possible to distinguish between N = 1, 2, 3, etc., supersymmetric theories.

- *FERRARA:*

The only supersymmetric theory which it is possible to test at low energies is N = 1 supersymmetry. The only test of this supersymmetry, since it is clearly a broken symmetry, is to find new particles, which are the partners of the quarks, the Higgses and the usual gauge bosons.

- *PERNICI:*

Has a supersymmetric regularization been found? If it has not, do you think that this is only a technical point, or is it a fundamental point?

- *FERRARA:*

For the N = 4 theory there is a supersymmetric regularization. It is a method which uses higher covariant derivative regulators by Stelle.

- *PERNICI:*

And in the other cases? Supergravity?

- *FERRARA:*

I don't think it can be used in supergravity.

- *PERNICI:*

I have a comment about an earlier question: "In an exactly supersymmetric world can you distinguish fermions and bosons?" Supersymmetry transformations do not change the number of states. For example a Majorana spinor has two states. If you make a supersymmetric rotation, the spinor is rotated into two scalars, which have two states again. So you can't distinguish bosons and fermions by the number of states. So, I want to ask again if, in an exactly supersymmetric world you can distinguish bosons and fermions.

- *FERRARA:*

I think this question has already been answered.

- *MUKHI:*

No, Professor Yang answered that the number of states is what you count and that they are different but Pernici is saying that the number of states is the same.

- *FERRARA:*

Bosons and fermions are not only distinguished by the number of states.

- *MUKHI:*

Well that was the only answer which I understood.

- *KAPLUNOVSKY:*

You rotate the apparatus by 360° and you will find the difference in sign to clearly distinguish fermions from bosons.

- *MUKHI:*

Remember that your apparatus is also supersymmetric, I don't think that this is a trivial question.

- *ZICHICHI:*

This question has to be clarified in another session.

- *DEVCHAND:*

Can one have a lattice version of a supersymmetric theory?

- *FERRARA:*

As far as I understand, yes, in a certain formulation of lattice gauge theories, for example, the T.D. Lee formulation with random lattices. Probably it is also possible in other formalisms.

- *KAPLUNOVSKY:*

You told us that you want to have non-vector like theories (i.e. parity non-conserving). We cannot do it with $N > 1$.

What about odd N, say N = 3. Then it would be just fine you can start with helicity +1 and finish with $-\frac{1}{2}$ and have a different number of left-handed and right-handed fermions.

- *FERRARA:*

The N = 3 theory turns out to be exactly equal to the N = 4 theory. If you look at the unique Lagrangian for the N = 3 theory it turns out to be the same as the N = 4 theory.

- *KAPLUNOVSKY:*

Is it due to the reality of the gauge boson?

- *FERRARA:*

Well, it is certainly due to the fact that all particles sit in the adjoint representation of the gauge group.

DISCUSSION 2 (Scientific Secretaries: L. Castellani, L. Petrillo, M. Pernici.)

- *CASTELLANI:*

You have shown that auxiliary fields are necessary when supersymmetry is broken. Are there other fundamental reasons for introducing them? For example, is it possible to quantize the theory without a closed gauge algebra?

- *FERRARA:*

In order to discuss in a model independent way the supersymmetry breaking it is necessary to introduce the auxiliary fields in order to make clear the analogy between the supersymmetry breaking and the usual breaking of a Yang-Mills group. As for quantization, I do not think that auxiliary fields are really needed.

- *CASTELLANI:*

Could you discuss in some more detail what you mean by supersymmetry breaking as a gravitational effect?

- *FERRARA:*

I have shown that when you couple a Y.M. theory to supergravity you get a modification of the mass relations and you get an additional contribution coming from the gravitino mass. It is a purely gravitational effect, and if the gravitino mass is of the order of 100 GeV, you get a feedback in the low energy theory, because this gravitino mass is going to shift the masses of all the particle fields, including the scalar partner of quarks and leptons.

- *PERNICI:*

In a spontaneous broken supersymmetry the mass formula gets modified by radiative corrections. We want the scalar partner of the lepton to remain light. How can we guarantee that it will happen and not the other way around?

- *FERRARA:*

This is the reason why I said that the way of getting the correct mass shifts is somehow artificial, because it is very model dependent. You have to construct a model in such a way that the scalar masses have the correct shift and do not receive big radiative corrections.

- *PERNICI:*

In supersymmetry it is convenient to introduce auxiliary fields; is the functional generator of Green function well defined in presence of a auxiliary field?

- *FERRARA:*

I think so.

- *BELLUCCI:*

As pointed out by Pagels and Primack and mostly by Weinberg, the gravitino must be so heavy that almost all gravitinos would have decayed before the time of the helium synthesis, or so light (less than about 1 keV) that they do not produce too large a cosmic deceleration. In the first case you get that symmetry breaking scales are bounded between 10^{11} and 10^{16} GeV. In the second case you have an upper bound of the order of 10^{6} GeV on the scale of spontaneous supersymmetry breaking. Can you exclude this second possibility?

- *FERRARA:*

I cannot exclude that. The only thing I said is that the

mechanism I discussed at the end would be totally irrelevant if the gravitino is very light. An almost massless gravitino (compared to the scale of the weak interactions) would never cure this wrong splitting between the lepton and quark masses and their scalar superpartners.

- *BELLUCCI:*

But it depends on the coupling constant of the goldstino in global supersymmetry.

- *FERRARA:*

You can put in an unnatural value for this coupling constant. The only object which enters in the formula is the gravitino mass. Thus the cosmological bounds are on the gravitino mass, which is approximately given by $\frac{M_{SUSY}^2}{M_{PLANCK}}$. A too light mass would produce completely negligible gravitational effects. These gravitational effects become relevant only if M_G is of the order of magnitude of the weak vector boson masses.

- *BELLUCCI:*

You mentioned that spin-0 leptons and quarks may acquire large masses from radiative corrections. What about the additional symmetry breaking scales which would be necessary in the theory in this case?

- *FERRARA:*

These models which have breaking through radiative corrections must be cooked up in such a way that both the shift on the masses and the usual SU(2)xU(1) breaking come from radiative corrections. Essentially the input of the theory is the supersymmetry breaking scale, which has to be put in the tree Lagrangian. The weak vector boson mass scale will be generated by radiative corrections and at the same time will trigger the SU(2)xU(1) breaking and the shift between scalars and fermions.

- *BELLUCCI:*

It is possible to cancel the anomalies produced by the axial coupling of the U-gauge boson to leptons and quarks, by introducing new heavy leptons and quarks, with axial coupling of the opposite sign? Would this impose restrictions on the scale of supersymmetry breaking?

- *FERRARA:*

I think that under certain reasonable assumptions Barbieri, Nanopoulos and myself have shown that this is impossible. You cannot get with this extra gauge interaction an anomaly free theory. If you want this extra interaction, the only thing you can do is to push the vector boson mass at very high energies in which the renormalizability criterion hopefully is no longer valid. In that case the supersymmetry breaking scale should be of the order I discussed in my lecture. Another possibility, discussed by Fayet, is to keep the U boson light but with an extremely small gauge coupling constant so that it almost decouples from matter.

- *NANOPOULOS:*

The point is that we can enforce the cancellation of anomalies, but then one gets only supersymmetric solutions.

- *FERRARA:*

Usually the cancellation of anomalies is in contradiction with the possibility of avoiding supersymmetric solutions. Since supersymmetric solutions are always global minima of the potential, you should avoid them. In general the fields that you have to introduce in order to cancel these anomalies tend to restore supersymmetric solutions. So there is no possibility to construct a reasonable model which is anomaly free and exclude supersymmetric minima.

- *KREMER:*

If a supersymmetric theory is spontaneously broken, do the UV divergences still cancel?

- *FERRARA:*

Yes, but you can generate scalar masses proportional to the supersymmetry breaking scale. This is why the effective supersymmetry breaking scale must be of the order of magnitude of the weak interaction scale. The cancellations are exact in an exactly supersymmetric theory but they have to go to zero with the supersymmetry breaking scale. This means that the scalar can get only masses which are of the order of the supersymmetry breaking scale.

This scale appears in the theory essentially through the vacuum energy in a spontaneously broken theory. This does not mean that this scale is the same which controls the split in the masses in supersymmetric multiplets. In fact, you can make theories in which the shift in the masses of the low energy sector of quarks and leptons is only of the order μ, if the supersymmetry breaking scale is given by $\sqrt{\mu M}$, where M is a big scale and μ is a small scale. This is what happens in all the models I was describing before, including the one where the breaking comes from gravitation.

- *KREMER:*

Is there a phenomenological difference between the Fayet-Iliopoulos type and the O'Raifertaigh type of symmetry breaking?

- *FERRARA:*

Yes, it produces a completely different spectrum of particles.

- *GOCKSH:*

You want to give the fermion masses through radiative corrections. What happens to chiral symmetry? I remember you saying that chiral symmetry prevents fermions from acquiring a mass.

- *FERRARA:*

In these theories, in order to generate fermion masses, you have to break supersymmetry at the tree level. The radiative corrections break SU(2) x U(1), and fermions acquire masses through the usual mechanism.

- *ZICHICHI:*

If you can have SU(2) x U(1) breaking via radiative effects, you should not need any Higgs fields. In other words, what do you mean by radiative corrections?

- *FERRARA:*

The Lagrangian at the tree level does not break SU(2) x U(1). But the effective potential, which includes radiative corrections, breaks SU(2) x U(1). That is, the loops give vacuum expectation values to the Higgs fields.

- *NANOPOULOS:*

In these models (where SU(2) x U(1) is broken by radiative corrections) the interesting thing is that one gets a negative mass for the Higgs bosons without introducing it by hand in the Lagrangian.

- *LOEWE:*

You discussed the mass relations at the tree level for a gauge theory. They were related to the absence of triangle anomalies ($\Sigma = 0$). In the supergravity case, the same mass relation is related to the mass of the gravitino. What happens with the anomalies?

- *FERRARA:*

Without gravity, the only way to avoid unwanted mass splittings is another gauge interaction which acts chirally on all quarks and leptons. In gravitational theories you can have an SU(3) x SU(2) x U(1) theory without introducing a new gauge interaction provided the gravitino mass is of the order of 100 - 1000 GeV.

- *LYKKEN:*

Don't you have problems with a large cosmological constant when you include the effects of gravity on supersymmetry?

- *FERRARA:*

In fact, the opposite happens. I use as basic ingredient the fact that the minimum of the potential occurs at zero. Otherwise I cannot have the super-Higgs effect in ordinary space-time. Of course one gets a one-loop cosmological constant which must be fine-tuned to zero. I don't know how to solve this new fine-tuning problem.

- *LYKKEN:*

I do not understand why in your scheme the mass splittings are of the order of the gravitino mass. I would have expected that they would be of the order of the supersymmetry breaking scale. Is this model dependent?

- *FERRARA:*

The supersymmetry breaking scale, defined as the coupling of the would-be Goldstino, or defined as the vacuum energy of the flat potential, would be $\sqrt{M_W M_{Planck}}$. However the gravitino mass is the square of this number divided by the Planck mass, so it is of the order of M_W. The gravitino mass is the relevant scale which can appear in the final Lagrangian, and this mass gives the supersymmetry scale.

- *VAN DE VEN:*

You showed us a supersymmetry breaking mechanism by coupling a supersymmetric theory to supergravity. Did you consider a non-perturbative breaking mechanism à la Witten?

- *FERRARA:*

Unfortunately Witten's criteria tell us that supersymmetry breaking cannot be produced nonperturbatively in a large class of theories.

- *ANTONELLI:*

If supersymmetry is broken at the tree level, can it be restored in higher orders of perturbation theory?

- *NANOPOULOS:*

Recently we have proven (with Howe, Lahanas and Nicolai) that if you have spontaneous symmetry breaking at the tree level, radiative corrections cannot restore symmetry. It is the converse of the theorem which says that supersymmetry, if unbroken at the tree level, cannot be broken in perturbation theory.

- *DOBREV:*

Is it true for supergravity?

- *FERRARA:*

In fact, the opposite happens. I use as basic ingredient the supersymmetry appears to be explicitly broken at low energies. Therefore there is no question whether it can be restored or not.

- *DOBREV:*

In your last considerations on the supersymmetry breaking scale you demanded the cosmological constant to be zero. Is this condition crucial?

- *FERRARA:*

The formulas I have written are valid only if you cancel the cosmological constant, in the sense that it is very small compared to the other parameters of the theory. If the cosmological constant is not very small, there is a problem in the interpretation of the terms which appear as mass terms in the Lagrangian, because they are no longer mass terms. So if the cosmological constant is huge,

I do not have a Minkovski space where the particles live, and the interpretation of masses is a problem.

- *ANTONELLI:*

What happens to axions in supersymmetric theories?

- *FERRARA:*

This problem will be considered by Professor Nanopoulos in his lectures.

- *RAUH:*

Is it possible that two already known particles are supersymmetric partners of each other, e.g. that the Higgs are the scalar partners of the quarks?

- *FERRARA:*

This cannot happen, because the scalar partner of the quark carries the same quantum numbers.

- *OGAWA:*

It seemed that you attached importance to the absolute value of the potential, while usually the potential can be changed by an additive constant. Could you explain why?

- *FERRARA:*

The additive constant is fixed by the constant term which appears in the transformation law of the Goldstone spinor, and measures the supersymmetric breaking. Since this object has dimension 3/2 and the supersymmetry parameter has dimension -1/2, you see that this constant must have dimension 2 (in energy units). This object is the vacuum expectation value of some of the auxiliary fields.

- *MUKHI:*

Is the fact that the Hamiltonian is positive semi-definite also true for supergravity theory?

- *FERRARA:*

No, it is not true. This is the reason why in supergravity you can cancel the cosmological constants.

- *MUKHI:*

Is there any known physical interpretation for a conserved supersymmetric charge?

- *FERRARA:*

It has no physical meaning because it is a fermion. However, it does relate couplings of bosons with couplings of fermions, so it has physical consequences in the sense that it connects matrix elements of different amplitudes.

- *HOFSÄSS:*

The auxiliary fields which you introduce are not dynamical. Is this necessary, in the sense that a dynamical field would lead to inconsistency?

- *FERRARA:*

They are nondynamical if I want a Lagrangian of dimension 4. I could, from a mathematical point of view, introduce kinetic terms for these fields, but they would correspond to ghosts.

- *DEVCHAND:*

Does the supersymmetric algebra close both on shell and off-shell with the same auxiliary fields?

- *FERRARA:*

Yes, but without auxiliary fields the supersymmetric algebra closes only on-shell.

- *MAINA:*

What is the situation as far as extended supergravities are concerned?

- *FERRARA:*

In extended supersymmetries, the main unwanted feature is that if you want to identify particles with the right quantum numbers, the multiplets would contain higher spin states.

- *WIGNER:*

I have two questions. The first one is based on your observation that supersymmetries are approximate. Is it possible that they conflict with some other symmetries as, for instance, the L-S symmetries or the SU(4) symmetry for nuclei conflicts with the independent particle model, based on a definite j value for each orbit? Is a similar conflict possible in the supersymmetries considered in particle physics, now or in the future?

- *FERRARA:*

In the case of supersymmetry, if one of them is approximate, all of them are approximate. That is, if one has many spinorial generators, breaking one of them implies breaking all the others.

- *WIGNER:*

My second question relates to the discussion of the existence of another queer kind of particle, the particle with mass zero but infinite spin. Has the possibility of the existence of such "particles" being discussed in the literature? I was opposed to believing in this possibility because the specific heat of vacuum would be infinite if they could be realized. But we discussed many exotic particles' possible existence.

- *FERRARA:*

I am not aware of the use of such particles. These states are related to the infinite dimensional representations of the Poincaré group. In principle these representations exist also in supersymmetric theories.

DISCUSSION 3 (Scientific Secretary: S. Bellucci)

- *CASTELLANI:*

I think for 6 space-time bosonic ordinary dimensions the maximum number of spinorial charges you can have is 4.

- *FERRARA:*

Yes, because the dimensionality of a spinor increases with the number of space-time coordinates. You cannot have more than 32 spinorial coordinates: this is a space-time independent statement.

- *CASTELLANI:*

You do not want it because you do not want particles with elicity higher than 2.

- *KREMER:*

In your first lecture you mentioned that quarks and leptons cannot be regarded as elementary in N = 8 supergravity, but you just showed a supersymmetry breaking pattern:

$$SU(8) \rightarrow SU(5) \rightarrow SU(3) \otimes SU(2) \otimes U(1)$$

So, what is wrong with elementary quarks and leptons?

- *FERRARA:*

For N = 1 supersymmetry quarks and leptons can be elementary. In the framework of N = 8 supergravity the effective N = 1 supersymmetry should be viewed as a theory for composite fields. You can get this composite fields by making bound states of the N = 8 supergravity with the same quantum numbers of quarks and leptons. This is what Ellis, Gaillard, Maiani and Zumino have proposed.

- *LOEWE:*

You said that at the scale of 10^{19} GeV all particles are massless. What happens with the mass of the Z^o, $W^{\pm}$ and the superheavy vector bosons?

- *FERRARA:*

You know that already in Grand Unified Theories these elementary fields are massless in the symmetry limit. This has nothing to do with supersymmetry.

- *LOEWE:*

What is the mathematical structure of the superspace? Is it a Cartesian product?

- *FERRARA:*

It is not a Cartesian product because the anticommutator of two spinorial charges gives rise to a space-time translation, so there is a connection between these coordinates: the Q's communicate with the space-time translation.

- *RICHARDS:*

You have introduced the fermionic coordinates θ_i. Could you specify what values these coordinates take?

- *FERRARA:*

You cannot observe a fermion field, you cannot observe a spinor. These θ's are anticommuting coordinates, they are not normal c-numbers but Grassman variables. You do not measure them. The only measurable consequence of this algebra is that you have relations between boson and fermion interactions.

- *PERNICI:*

Do you think it is possible to introduce supersymmetric theories without using the Grassman variables θ at a fundamental level? For example I think it was proven that the Wess-Zumino model can be derived imposing to a spin 0-spin 1/2 some non-renormalization theorems. Can every supersymmetric theory be derived in this way, without using anticommuting c-numbers at a fundamental level?

- *FERRARA:*

Yes. They can be derived looking for the spinorial currents J^μ_α (which do not contain Grassman variables) and then looking for all the theories for which $\partial_\mu J^\mu_\alpha = 0$ is satisfied, i.e. such that the supersymmetric charge $Q_\alpha = \int d^3 x\, J^o_\alpha(x)$ is conserved.

- *COLE:*

In N = 8 supersymmetry, starting from one graviton we presumably generate 8 types of gravitinos. Can we distinguish between these types in broken or unbroken supersymmetry theories?

- *FERRARA:*

If the supersymmetry is broken, and you have a hierarchy to N = 1 supersymmetry, (for instance in N = 8 supergravity,) you would expect that 7 gravitinos have a mass of the order of the Planck mass and that the residual gravitino has a mass of the order of 100 GeV. If the supersymmetry is unbroken, all the gravitinos are massless because the gravitino is a gauge field. If the supersymmetry is broken at very high energy, then the gravitino could acquire a large mass.

- *KAPLUNOVSKY:*

The spectrum of the matter gravity supermultiplet of N = 8 supergravity contains 28 gauge bosons which apparently close to SO(8) group. Could you clarify the way in which you enlarge this group to SU(8) ?

- *FERRARA:*

The SU(8) symmetry is there, but it is realized in a non-linear way. Only its SO(8) is realized linearly. This is similar to an U(1) example with the Lagrangian:

$$\mathcal{L} = \left[(\partial_\mu + ieA_\mu)\phi^*. \quad (\partial_\mu - ieA_\mu)\phi\right] - V(\phi^*\phi).$$

The U(1) local symmetry is transparent here. But there is no kinetic term for A_μ, so it could be integrated out and one remains with:

$$\mathcal{L} = |\partial_\mu\phi|^2 - \tfrac{1}{4}(\phi^*\overleftrightarrow{\partial_\mu}\phi)^2/\phi^*\phi - V(\phi^*\phi)$$

which describes complicated non-polynomial interactions. But U(1) local symmetry is still here. The same kind of thing happens to the SU(8) symmetry generators.

- *KLEINERT:*

Let me point out that there exists an application of supersymmetry to vulcanization processes by Parisi and Sourlas in which they discovered that every complex pair of fermionic variables has the effect of reducing the effective space dimension by one unit. This is most easily seen from the following calculation valid for a supersymmetric function:

$$\int d^D x d\bar\theta d\theta \phi(x,\bar\theta,\theta)$$

$$\sim \int dr^2 (r^2)^{\frac{D}{2}-1} d\bar\theta d\theta \underbrace{\phi(r^2 + \bar\theta\theta)}$$

$$\hookrightarrow \phi(r^2) + \frac{d}{dr^2}\phi(r^2)\bar\theta\theta + \ldots$$

notice: $\int d\bar\theta d\theta\bar\theta\theta = 1$ all others give zero

$$\sim \int dr^2 (r^2)^{\frac{D}{2}-1} \frac{d}{dr^2}\phi(r^2)$$

$$\sim -(\tfrac{D}{2}-1)\int dr^2 (r^2)^{\frac{D}{2}-2}\phi(r^2)$$

$$\sim \int d^{D-2}x\phi(x)$$

The upper line has D plus two fermion dimensions; the lower has D-2 dimensions. Having a theory with two fermionic degrees of freedom is the same as having a theory in a space with two dimensions less. This is the reason why supersymmetric theories are so convergent.

PHYSICAL CONSEQUENCES OF GLOBAL AND LOCAL SUPERSYMMETRY

D. V. Nanopoulos

Theory Division, CERN
Geneva, Switzerland

In these lectures I discuss in some detail the structure and physical consequences of global and local supersymmetric (SUSY) gauge theories. Section 1 contains motivations for SUSY theories, whilst Sections 2 and 3 explain what supersymmetry is, and what are its physical properties. The observable consequences of SUSY at low energies and super-high energies are discussed in Sections 4 and 5. The physical structure of simple (N = 1) local SUSY (≡ supergravity) is given in Section 6, whilst Section 7 contains the physics of simple supergravity both at super-high as well as at low energies. Amazingly enough, we find that gravitational effects, as contained in supergravity theories, may play a rather fundamental role at all energy scales. This strong interrelation between gravity and particle physics is unprecedented.

1. MOTIVATION(S) FOR SUPERSYMMETRY

Unification of all elementary particle forces has been the Holy Grail of theoretical physics. The first realistic step towards this end has been the highly successful unification between electromagnetic and weak interactions, now called electroweak interactions. The obvious (and natural) next step is then the amalgamation of electroweak and strong interactions, justifiably called Grand Unified Theories [1] (GUTs). The qualitative successes (e.g., charge quantization, equality of different coupling constants and equality of certain quark-lepton masses at super-high energies, natural understanding of quark and lepton quantum numbers, etc.), as well as the quantitative successes (e.g., disparity of coupling constants and of quark-lepton masses at low energies, determination of the electroweak mixing angle (θ_{e-w}), possible limit on the number of flavours, virtually massless neutrinos, etc.) are rather well-known [1]. It is also well-known that the grand unification scale M_X is rather large

$$M_X \sim 10^{15} \text{ GeV}. \tag{1}$$

GUTs contain baryon and lepton number violating interactions, and the presently observed proton stability ($\tau_p > 10^{31-32}$ years) immediately puts a lower bound on M_X which more or less is saturated by (1). The existence of two scales, the electroweak scale ($M_W \sim 100$ GeV) and the GUT scale M_X, so different

$$\frac{M_W}{M_X} \leqq 10^{-13} \tag{2}$$

creates a fundamental problem for GUTs.

The gauge hierarchy problem: how is it possible to keep these two scales separate, incommunicado? In ordinary field theories, even if we fix the parameters at the tree level to satisfy (2), radiative corrections will undo it and we will have to adjust it at each order in perturbation theory. This does not sound right if we claim to have a natural theory. The heart of the problem is the existence of scalar fields. The only way to break a gauge theory and keep renormalizability is through Spontaneous Breaking (SB). The simplest way to achieve SB is to have certain scalar fields get a vacuum expectation value (v.e.v.). In GUTs we need two sets of scalars, one set to break the big group G which supposedly contains all interactions down to SU(3) x SU(2) x U(1) at M_X, and thus the v.e.v. of these fields $V \sim M_X$, and another set to break SU(3) x SU(2) x U(1) down to SU(3) x $U(1)_{E-M}$, and thus they should get v.e.v.'s $v \sim M_W$.

Since the masses of these Higgs fields are proportional to their v.e.v.'s, we end up with light Higgs of masses $O(M_W)$ and heavy Higgs of masses $O(M_X)$. Again this can be arranged at the tree level. But, since light and heavy Higgs are both coupled to gauge bosons, fermions and scalars, already at the 1-loop level there is a communication between the light and heavy Higgs, through the above mentioned fields running around the loop. That immediately creates corrections to light scalar masses $O(M_X)$, while only corrections at most $O(M_W)$ are allowed, i.e., end of the gauge hierarchy!

This happens because there is no symmetry able to keep scalars massless (or virtually massless: $M_W \lll M_X$), in contrast to gauge or chiral symmetries which keep gauge bosons or fermions massless.

One way out would be to abandon completely the use of scalar fields as means of SB. Then we have to attempt dynamical SB. This approach has been tried (technicolor, extended technicolor, ...) but led to fatal flaws, so it became evident that technicolor has created more heat than light [2]. So back to scalars again. One may try now a different way. Is there any possibility that when one adds up all the 1-loop corrections to light scalar masses, they practically vanish, becoming at most of order M_W? We may exploit the fact that thanks to the "spin-statistics theorem and all that", boson and fermion loops differ by an all-over minus sign. Then if suitable relations exist between fermion and boson masses on the one hand and gauge, Yukawa and scalar self coupling constants on the other, the hope of cancellation between different 1-loop diagrams may be realized. Well, this is a very neat way to discover supersymmetry [3] (SUSY). Actually, one may rigorously prove that the only way to get the desired 1-loop cancellation is through supersymmetry [4]. As we will see later, a remarkable property of supersymmetry is that if the cancellation occurs at the 1-loop level, then it occurs automatically to all orders in perturbation theory. The technical aspect of the gauge hierarchy problems then has been solved. Incidentally, since supersymmetry has to be broken, the corrections to light Higgs masses will be $O(M_{LESB})$ and thus:

$$M_{LESB} \leqq O(M_W) \tag{3}$$

where M_{LESB} refers to the Low-Energy Supersymmetry Breaking, i.e., the SUSY breaking that the low-energy world suffers. Discussing about hierarchies, another serious problem, this time concerning quantum chromodynamics (QCD), naturally comes to mind. Non-perturbative effects in QCD have the disturbing feature of adding a term:

$$\varepsilon_{\mu\nu\rho\sigma} F^a_{\mu\nu} F^a_{\rho\sigma} \cdot \theta \tag{4}$$

where $F^a_{\mu\nu}$ is the gluon field strength and θ is a parameter constrained to be

$$\theta < 10^{-9} \tag{5}$$

from the present upper bound [5] on the Dipole Electric Moment of the Neutron (DEMON). This is the strong CP-hierarchy problem. Again supersymmetry solves the technical aspect of this problem [6]. Starting with $\theta = 0$, one proves [6] that in SUSY type theories, θ naturally lies below the limit posed by (5). The same type of miraculous cancellations, as mentioned above, occur again. Supersymmetric theories are very well behaved; they respect hierarchies.

We have by now enough physical motivation to have a close look at the structure of supersymmetric theories.

2. SUPERSYMMETRY (SUSY)

All kinds of symmetries that one normally uses in particle physics, global (like Isospin, Eight-foldway, ...) or gauge (SU(3) x x SU(2) x U(1), SU(5), O(10), E_6, ...) are always transform fermions to fermions and bosons to bosons. Supersymmetry, or Fermion-Boson Symmetry [3], tries to bypass this prejudice and aims to a theory in which a fermion-boson transformation will also be possible. Indeed, such theories have been constructed [3] and are in full accord with all the standard laws of quantum field theory. In its simplest form, one has to extend the usual Poincaré Algebra of the generators of space-time rotations and translations, $M^{\mu\nu}$ and P^{μ}, to contain a self-Conjugate (Majorana) spin 1/2 generator, Q_α, which turns boson fields to fermion fields and vice versa. Schematically,

$$Q_\alpha|\text{boson}> = |\text{fermion}>; \tag{6}$$

α = 1, 2, 3, 4 is the Dirac (spinor) index.

They satisfy the following (anti) commutation rules:

$$[Q_\alpha, M^{\mu\nu}] = i(\sigma^{\mu\nu}Q)_\alpha \tag{7}$$

$$[Q_\alpha, P^{\mu}] = 0 \tag{8}$$

$$\{Q_\alpha, \bar{Q}_\beta\} = -2(\gamma\mu)_{\alpha\beta}P^{\mu} \tag{9}$$

in which $\sigma^{\mu\nu} = \frac{1}{4}[\gamma^\mu, \gamma^\nu]$ and $\bar{Q} = Q^T\gamma^0$.

The Q_α are four hermitian operators and we use Majorana representation for Dirac Matrices. Because of the spinorial character of the generators Q_α, the extended Poincaré Algebra, called supersymmetry algebra, involves both commutation and anticommutation relations. It is not an ordinary Lie Algebra, but what the mathematicians call a Graded Lie Algebra (GLA). The spinorial generator Q_α is a grading representation of the Poincaré Lie Algebra. Supersymmetry extends this algebra in a rather non-trivial way; one can associate in irreducible representations a finite number of bosons and fermions. This fact is extremely important if one wants to construct conventional renormalizable quantum field theories invariant under supersymmetry and satisfying the usual Wightman axioms. Graded Lie Algebras play perhaps a unique role in particle physics, because they realize truly relativistic spin-containing symmetries in which particles of different spin belong to the same supermultiplet. It is remarkable that by making the spinorial generators transform as some representation of an internal symmetry group, the resulting algebra provides also a fusion between space-time and internal symmetry overcoming previous no-go theorems [7]. Irreducible multiplets combine in this case fermions and bosons with different internal quantum numbers.

The physical meaning of the (anti) commutation rules (7), (8) and (9) is rather apparent. Eq. (7) simply states that Q transforms as a spinor, while Eq. (8) states that the spinor charges are conserved and are translation invariant. Presumably the most important is Eq. (9), and it suggests the terminology that the supersymmetry charges are the "square root of translations". Clearly, SUSY involves the structure of space-time and it is this connection that is fully developed in local supersymmetry or supergravity [8].

It is very easy to show that SUSY implies a relation between particles of different spin. Apply the spinor charge Q_α to a particle state $|P,S\rangle$ of definite momentum and helicity:

$$Q_\alpha |P,S\rangle = a|P,S+1/2\rangle + b|P,S-1/2\rangle \; ; \qquad (10)$$

because of (8), the RHS is a superposition of particles of the same momentum and energy-thus <u>the same mass</u>. Furthermore, addition of angular momentum implies that these particles have helicities $S \pm 1/2$. Supersymmetry transformations connect states which differ by 1/2 unit of spin. These states fill in the so-called supermultiplets. They therefore relate bosons to fermions. One may easily generalize the (anti) commutation rules (7), (8) and (9) to involve more than one spinorial charge, say Q_α^N, $N = 1, 2, \ldots$ Then we talk about N-extended supersymmetry in contrast to $N = 1$ or simple supersymmetry. Since here we are not going to be involved with $N > 1$ supersymmetries we stick to the above given (anti) commutation rules.

When trying to mix SUSY with gauge theories we better keep in mind certain general features and constraints that more or less come out from first principles:

1) In global supersymmetry the supercharges (Q_α) always commute with gauge symmetries; the commutator, if not zero, would be a supersymmetry transformation depending on the infinite number of parameters of the gauge symmetry, so would have to be a local SUSY. Thus,

$$[Q_\alpha^N, G] = 0. \qquad (11)$$

There are two immediate consequences of this fact:

i) all members of a supermultiplet have the same internal quantum numbers. This means that
ii) only $N = 1$ (global) SUSY makes phenomenological sense.

$N \geqq 2$ (global) supersymmetric theories yield always fermions in real representations of SU(3) x SU(2) x U(1) in sharp contrast with what we observe experimentally at low energies (parity violation exists both in charged and neutral currents). This observation justifies our attitude of not considering seriously any $N \geqq 2$ (global) SUSY theories.

2) An immediate consequence of Eq. (9) is that in global SUSY theories with SB the Hamiltonian H is the sum of the squares of the supersymmetry charges

$$H = \sum_{\alpha=1}^{4} Q_\alpha^2. \tag{12}$$

Since H is the sum of squares of Hermitian operators, the energy of any state is positive or zero.

Clearly, if there exists a SUSY invariant state, that is a state annihilated by Q_α, then it is automatically the true vacuum state since it has zero energy, and any state that is not invariant under supersymmetry has positive energy. Thus, in contrast with ordinary gauge theories, if a SUSY state exists, it is the ground state and SUSY is not SB. Only if there does not exist a state invariant under SUSY, SUSY is SB. In this case the ground state energy is positive. Obviously, it is far more difficult to achieve SUSY SB than achieving gauge symmetry SB. The supersymmetric state would have to be ostracized from the physical Hilbert space.

It follows that in global SUSY theories, it is impossible to spontaneously break an N-extended SUSY to an $N'(N>N'\geq 1)$ SUSY, because all Q_α^N satisfy separately Eq. (12) and if one breaks down some N, then all of them break down as well. No step-wise extended SUSY SB in global supersymmetric theories is possible. Things are different though in supergravity.

We are ready now to move to the construction of N = 1 SUSY models.

3. PHYSICAL PROPERTIES OF SUPERSYMMETRY

From our previous discussion of the general features of supersymmetric theories, we recall the fact that only N = 1 global SUSY theories make phenomenological sense. Thus each supermultiplet contains only two kinds of particles, with identical internal quantum numbers but with a "spin-shift" of 1/2 unit.

Let us take an arbitrary gauge group, with gauge mesons A_μ^a (spin 1) and fermionic partners λ^a (spin 1/2) called gauginos, belonging to the adjoint representation of the gauge group. They consist of the vector multiplet. The gauginos have to have spin 1/2 and not 3/2 because of renormalizability. In addition, we may introduce left-handed fermions ψ_L^i in an arbitrary multiplet of the gauge group. They form supersymmetry multiplets

$$\begin{bmatrix} \psi_i^L \\ \phi_i \end{bmatrix}$$

with complex scalar bosons ϕ_i. They consist of the chiral multiplets. The right-handed fermion fields are the complex conjugates of the left-handed fields, $\psi_{jR} = (\psi^j_L)^*$ and their SUSY partners are the complex conjugates ϕ^*_j of the ϕ^i. We have to use scalar fields and not spin-1 fields to complete the "fermion" multiplet because of renormalizability; the only allowed spin-1 bosons are gauge bosons, but then the fermions should belong to the adjoint representation of the gauge group, in contradiction to what we see experimentally. It is very interesting that renormalizability plus "observation" define the superpartners uniquely. The superpartners of the observed fermions (quarks, leptons) are called sfermions (squarks, sleptons) and have spin 0, while the superpartners of Higgs are called Higgs-inos and have spin 1/2. We will see below why in N = 1 global SUSY, the usual Higgs fields cannot be identified with the superpartners of the observed quarks and leptons.

In addition, we introduce a function $f(\phi_i)$, which is known as the "superpotential". f must be an analytic function of the ϕ_i, i.e., a function of ϕ_i but not of their complex conjugates ϕ^*_j. For a renormalizable theory f should be at most cubic in the ϕ_i, otherwise f is restricted only by gauge invariance. The general form of f is $f(\phi_i) = C_i\phi^i + C_{ij}\phi^i\phi^j + C_{ijk}\phi^i\phi^j\phi^k$, where C_i, C_{ij}, C_{ijk} are gauge covariant tensors.

Notice that in SUSY theories the usual Higgs fields cannot be identified with the superpartners of the observed quarks and leptons (sfermions). In ordinary gauge theories we can use a Higgs doublet (H_2) to give masses to up quarks, while its charge conjugate (H^C_2) can provide masses to charged leptons and down quarks. In SUSY theories, since the superpotential f is function only of ϕ_i's and not of ϕ^*_j's, H_2 and H^C_2 should be chosen to be completely unrelated, different, fields. Thus, even if we identified the ($s\nu$,se) doublet with H^C_2, we are missing H_2, i.e., we are left with massless up quarks (u,c,t)!

Furthermore, in a GUT theory, quarks and leptons are sitting in the same multiplet; thus by identifying sleptons with "weak" Higgs fields, we have to interpret squarks as color Higgs fields which is catastrophic, since these color Higgs fields will make protons to decay instantly. For example, in SU(5), the Higgs doublet H_2 is sitting in the same multiplet, a 5-plet of SU(5), with a colored Higgs triplet, H_3. The color Higgs mediates proton decay and since it is coupled to quarks and leptons with the normal Yukawa couplings, it had better be superheavy ($\gtrsim 10^{10}$ GeV), otherwise, matter will desintegrate instantly! Thus, if we identified s-ups or s-downs with colored Higgs fields, we are in big trouble since s-ups and s-downs cannot weigh much more than M_W($\ll 10^{10}$ GeV) if we want to solve the gauge hierarchy problem. Finally, the existence of a pair of Higgs supermultiplets of opposite helicities is crucial in cancelling the Adler-Bell-Jackiw anomalies of the Higgsino sector.

Then, a SUSY Lagrangian can be written in the form

$$\mathcal{L} = \mathcal{L}_k + \mathcal{L}_Y + \mathcal{L}_S \tag{13}$$

where $\mathcal{L}_k$ is the standard kinetic energy term ..., while $\mathcal{L}_Y$ describes the Yukawa interactions

$$\mathcal{L}_Y = \frac{\partial^2 f}{\partial\phi_i \partial\phi_j} \psi_L^i \psi_L^j + \text{h.c.}, \tag{14}$$

and $\mathcal{L}_S$ is the scalar potential

$$-\mathcal{L}_S = V(\phi_i, \phi_i^*) = \sum_i \left|\frac{\partial f}{\partial\phi_i}\right|^2 + \frac{1}{2} \sum_a g_a^2 |(\phi^*, T^a\phi)|^2 \tag{15}$$

where the second sum runs over all generators a of the gauge group, the g_a are the gauge coupling constants and T^a are the generators of the gauge group acting on the representation of the group furnished by the ϕ_i. If the gauge group is not semi-simple but contains U(1) generators, say Y_ℓ with charge g_{Y_ℓ}, then its contribution to (15) becomes

$$\frac{1}{2} g_{Y_\ell}^2 |(\phi^*, Y_\ell\phi) + \xi_\ell|^2 \tag{16}$$

where ξ_ℓ are arbitrary constants of mass dimension 2. Usually one defines

$$F_i \equiv \frac{\partial f}{\partial\phi_i} \tag{17}$$

and $$D_a \equiv (\phi^*, T^a\phi) + \sigma a Y_\ell \cdot \xi_\ell \qquad (T^{Y_\ell} \equiv Y_\ell), \tag{18}$$

and thus $$V(\phi_i, \phi_i^*) = \sum_i |F_i|^2 + \frac{1}{2} \sum_a g_a^2 |D_a|^2. \tag{19}$$

In the classical approximation, the zero point energy of the fields may be neglected, and the energy of the ground state just equals the minimum of the potential $V(\phi_i, \phi_i^*)$. SUSY is unbroken if V = 0 for $\phi_i = \langle\phi_i\rangle$. As V is a sum of squares, V = 0 if each term separately vanishes. Thus the condition for unbroken SUSY is

$$F_i \equiv \frac{\partial f}{\partial\phi_i} = 0 \tag{20}$$

for each field ϕ_i, and that

$$D_a \equiv (\phi^*, T^a\phi) + \delta a Y_\ell \cdot \xi_\ell = 0 \tag{21}$$

for every generator T^a of the gauge group.

If (20) and (21) have a simultaneous solution, SUSY is unbroken at the tree level. Otherwise, SUSY is spontaneously broken. It is not difficult to show, that a necessary condition to have spontaneous SUSY breaking at the tree level in a supersymmetric gauge theory, is to have one of the following two conditions satisfied:

i) The group G should contain at least a neutral field X with linear terms in the superpotential f. This is called F-type breaking because (20) cannot be satisfied.

ii) The group G should contain at least an Abelian factor U(1) with a non-vanishing ξ in (18). This is called D-type breaking because (21) cannot be satisfied.

Similarly to the SB of global symmetry where there are Goldstone bosons, in the case of SB of global SUSY a Goldstone fermion, called goldstino, should be present. In general, up to a normalization factor the goldstino is given by:

$$\psi_g = \frac{1}{2} g_a D_a \lambda^a + \frac{\partial f}{\partial \phi_i} \psi_i . \tag{22}$$

Let us define the coupling M_S^2 of the supercurrent $S_{\mu\alpha}$ ($Q_\alpha = \int d^3 x\, S_{0\alpha}$) to the goldstino by:

$$<0|S_{\mu\alpha}|\psi_\beta> = M_S (\gamma_\mu)_{\alpha\beta} . \tag{23}$$

There is a simple and fundamental relation between the value of the potential at the minimum (vacuum energy) V_0 and M_S^2:

$$V_0 = (M_S^2)^2 = M_S^4 . \tag{24}$$

The physical meaning of M_S is apparent: M_S^2 is the "order parameter" of supersymmetry. A value of M_S different from zero implies that supersymmetry is spontaneously broken. In particular, if we denote by ε the coupling of the goldstino to a supermultiplet, then there is a mass splitting between the boson and the fermion inside the supermultiplet:

$$m_B^2 - m_f^2 = M_S^2 \varepsilon . \tag{25}$$

The from of the mass splitting (25) is very suggestive. If one wishes, one may arrange things in such a way that, by making the coupling of the goldstino to certain supermultiplets small ($\varepsilon \ll 1$), these supermultiplets suffer mass splittings, much smaller than the primordial SUSY breaking scale M_S. This simple mechanism, the SUSY DECoupling mechanism (SUDEC), has been discovered only recently [9].

Its importance in constructing realistic SUSY models is difficult [9,10] to overestimate. The reason is very simple. As we have seen before (see (3)), because of the gauge hierarchy problem, M_{LESB}, the mass splitting that the low-energy world supermultiplets suffer has to be of the order of M_W. Then if we identified M_S with M_{LESB}, no realistic model can be constructed [9,11]. It seems that we definitely need

$$M_S >> M_{LESB} \tag{26}$$

which, in turn, means that necessarily the SUDEC mechanism [9] has to be employed ($\varepsilon<<1$). Actually, thanks to the "magic" properties of SUSY theories (non-renormalization theorems), one can prove [9, 12] that (26) persists to all orders in perturbation theory, i.e., it is stable against large radiative corrections. Indeed, realistic models have been already constructed [9,10] where one usually finds that

$$M_S^2 \simeq M_W M_{Planck} \tag{27}$$

implying that, by using (3) and (25),

$$\varepsilon_{LOW} \simeq \frac{M_W}{M_{Pl.}}. \tag{28}$$

Things become very interesting because local SUSY or supergravity [8] cannot be neglected anymore [13].

In general, the goldstino becomes the missing longitudinal components of the spin 3/2 gravitino, the gauge fermion of local supersymmetry (the superpartner of the spin 2 graviton) through the superhiggs effect [14,15]. In analogy with ordinary gauge theories in which the gauge boson mass is given by

$$M \sim g\langle\phi\rangle, \tag{29}$$

where g is the gauge coupling constant and $\langle\phi\rangle$ is the v.e.v. of the some scalar field causing the breaking, i.e., the order parameter of the gauge symmetry, the mass of the gravitino is given by [15]:

$$m_{3/2} \sim \sqrt{G_{Newton}}\, M_S^2 = \frac{M_S^2}{M_{Pl.}} \tag{30}$$

in terms of the gravitational coupling constant $\sqrt{G_N}\left(\equiv \frac{1}{M_{Pl.}}\right)$ and M_S^2, the order parameter of supersymmetry. One then finds [16,17] extra contributions to the RHS of (25) proportional to $m_{3/2}^2$. By putting together (27) and (30) we find that

$$m_{3/2} \sim M_W, \tag{31}$$

implying that the extra gravitational corrections to (25) cannot be neglected anymore [13], as being of the same order of magnitude with the non-gravitational ones. One may even suspect that it is possible to create the whole M_{LESB} through gravitational effects. Indeed it is possible! Furthermore, if we do not want to upset the gauge hierarchy problem, we had better demand that [17]

$$m_{3/2} \leqq O(M_W) \tag{32}$$

or, equivalently, that

$$M_S^2 \leqq M_W M_{Pl.}. \tag{33}$$

Incidentally, there is another (non-spontaneous) way of breaking global supersymmetry: the soft way. It sounds arbitrary, ad-hoc and ugly, but it is allowed if it occurs through a certain type of breaking terms [18]. The ultraviolet properties of the theory, e.g., softening of divergences etc., hold true [18] in soft breaking as they do in SB, so soft breaking may be employed in physical applications as good as SB. Amazingly enough, recent progress has shown that it is almost impossible [10] to construct appealing SUSY models with spontaneous global SUSY breaking, while realistic SUSY models have appeared [19] satisfying all possible phenomenological constraints, with a very definite pattern of soft SUSY breaking emerging from the spontaneous breakdown of local supersymmetry [19]. Gravity seems to play a rather fundamental role here, as will be discussed later.

One of the central features of supersymmetric theories is that, thanks to their fermion-boson symmetry, there are a lot of cancellations between "badly" behaving graphs. This amounts to a much better behaved field theory at the ultraviolet, compared with the ordinary field theories, which make them very attractive. Actually, recently it has been proven that the N = 4 SUSY Yang-Mills field theory in real (4) space-time dimensions is FINITE [20]! A more quantitative statement on the ultraviolet behavior of SUSY theories is provided by the so-called "Non-Renormalization theorems" [21]. There is no renormalization, finite or infinite, for the mass terms, Yukawa couplings or Higgs self couplings independent of the gauge coupling renormalizations. In SUSY theories one can, if one wishes, impose arbitrary relations among mass and Yukawa parameters or set some of them equal to zero. No problem.

"SET IT AND FORGET IT". It is this property that solves [22] the technical aspect of the gauge hierarchy and strong CP-hierarchy [6] problems that we discussed in the beginning as motivation for SUSY theories. With the same token, one can prove that if SUSY is unbroken at the tree level, it remains unbroken to all orders in perturbation theory [21]. It sounds like a magic world full of miracles. But still, it is true!

4. "LOW ENERGY" PHYSICS AND SUSY

Indeed, if SUSY provides the solution [22] to the gauge hierarchy problem, then the phenomenological implications are tremendous. Firmly, each one of the "standard" particles of the low energy world (quarks, leptons, gauge bosons, Higgs) should have their superpartners in a mass range at most not far above M_W. This fact makes the situation very exciting because the hope exists to discover these particles in the not very far future. Present experimental limits from PETRA and PEP put the mass of any new charged particles approximately above 20 GeV, which puts a lower bound on the mass of charged SUSY particles. I will not discuss here how to find SUSY particles since it has been discussed in lengthy detail elsewhere [23].

Here I will discuss the constraints that well-established phenomenological facts like absence of flavor changing neutral currents (FCNC), absence of strong CP-violation, g-2, etc., impose on SUSY models. Clearly, the introduction of new particles which are coupled to the low-energy world with ordinary gauge or Yukawa couplings and of mass not far above M_W, sounds like trouble. For example, analogously to the gauge boson-fermion-fermion coupling there is a gaugino-sfermion-fermion coupling of comparable strength. Such kinds of couplings contribute to all kinds of rare processes, like Re and Im part of K_L-K_S, $K_L \to \mu\mu$, $\mu \to e\gamma$, etc.

FCNC are naturally suppressed in SUSY theories by a simple super-GIM mechanism. However, the extent of this suppression depends on the differences Δm^2_{sf} in $(\text{mass})^2$ between spin-zero superpartners (squarks and sleptons) of the quarks and leptons in different generations. We find [24] that $\Delta m^2_{sf}/m^2_{sf}$ must be much smaller than unity both for squarks and for sleptons. Our results [24] are set out in the Table, where it is seen that the best constraints on the first two super-generations come from

$$\mathrm{Im}(K_1 - K_2) \Rightarrow \frac{\Delta m^2_{sq}}{m^2_{sq}} < O(10^{-3})$$

and

$$\mu \to e\gamma \Rightarrow \frac{\Delta m^2_{sl}}{m^2_{sl}} < O(10^{-3}) \tag{34}$$

if the SUSY partners W, B of the SU(2) and U(1) gauge bosons have masses O(100) GeV, and the super-Cabibbo angles are not much smaller than the familiar Cabibbo angle. These results imply that the thresholds in e^+e^- annihilation for the pair production of different flavors of squark or slepton with identical charges must be almost degenerate. In particular, the su and sc (or sd and ss) thresholds may be closer together than the e^+e^- beam energy spread!

These limits [24,25] look very severe indeed and, at first

Table 1. FCNC Constraints on Broken SUSY Theories

Flavor-changing Transition	Constraint for arbitrary $m_{\tilde{W}}$	if $m_{\tilde{W}}$ or $m_{\tilde{B}}$ = 100 GeV
$\mathrm{Re}(K_1-K_2)$	$\frac{1}{m_{\tilde{W}}^2}\left(\frac{\Delta m_{sq}^2}{m_{sq}^2}\right)^2 < O(10^{-7})\ \mathrm{GeV}^{-2}$	$\left(\frac{\Delta m_{sq}^2}{m_{sq}^2}\right) < O(\frac{1}{30})$
$\mathrm{Im}(K_1-K_2)$	$\frac{1}{m_{\tilde{W}}^2}\left(\frac{\Delta m_{sq}^2}{m_{sq}^2}\right)^2 < O(10^{-10})\ \mathrm{GeV}^{-2}$	$\left(\frac{\Delta m_{sq}^2}{m_{sq}^2}\right) < O(10^{-3})$
(D_1-D_2)	$\frac{1}{m_{\tilde{W}}^2}\left(\frac{\Delta m_{sq}^2}{m_{sq}^2}\right)^2 < O(10^{-6})\ \mathrm{GeV}^{-2}$	$\left(\frac{\Delta m_{sq}^2}{m_{sq}^2}\right) < O(\frac{1}{10})$
$K_L^0 \to \mu\mu$	$\frac{1}{m_{\tilde{W}}^2}\left(\frac{\Delta m_{sq}^2}{m_{sq}^2}\right) < O(10^{-5})\ \mathrm{GeV}^{-2}$	$\left(\frac{\Delta m_{sq}^2}{m_{sq}^2}\right) < O(\frac{1}{10})$
$K_L^0 \to \mu e$	no useful constraint	
$\mu N \to eN$	$\frac{1}{m_{\tilde{W}}^2}\left(\frac{\Delta m_{s\ell}^2}{m_{s\ell}^2}\right) < O(10^{-5})\ \mathrm{GeV}^{-2}$	$\left(\frac{\Delta m_{s\ell}^2}{m_{s\ell}^2}\right) < O(\frac{1}{10})$
$\mu \to e\bar{e}e$	$\frac{1}{m_{\tilde{W}}^2}\left(\frac{\Delta m_{s\ell}^2}{m_{s\ell}^2}\right) < O(10^{-4})\ \mathrm{GeV}^{-2}$	$\left(\frac{\Delta m_{s\ell}^2}{m_{s\ell}^2}\right) < O(1)$
$\mu \to \gamma$	$\frac{1}{m_{\tilde{B}}^2}\left(\frac{\Delta m_{s\ell}^2}{m_{s\ell}^2}\right) < O(10^{-7})\ \mathrm{GeV}^{-2}$	$\left(\frac{\Delta m_{s\ell}^2}{m_{s\ell}^2}\right) < O(10^{-3})$

glance, one seems to need some kind of fine-tuning to satisfy them. Fortunately, things are different. The problem arises because the fermion and sfermion mass matrices are not in principle simultaneously diagonalizable. There is a contribution to the sfermion mass matrix, through the Yukawa couplings, identical to the fermion mass matrix, but the main contribution to the sfermion mass matrix has nothing to do with the fermion mass matrix. Thus, except the case where the main contribution to the sfermion mass matrix is proportional to the unit matrix, we are heading for a disaster. Lo and behold, a universal contribution to all sfermion masses does occur in most SB SUSY models of F- or D-type or in softly broken SUSY models where the soft breaking is provided from supergravity effects. Then, indeed, the sfermion and fermion mass matrices are almost simultaneously diagonalizable which leads to an almost complete evasion of the most stringent limits (34), as will be argued later.

Similar results are obtained from the analysis of g-2 [26] or of the absence of strong CP-violation [6]. Once again, we find that F- or D-type SB SUSY models, or supergravity induced softly broken SUSY models, easily satisfy very stringent types of constraints [26]. It is remarkable that realistic SUSY models satisfy automatically severe low energy phenomenological constraints [24-26], which have been the Nemesis of other alternatives like technicolor models [2].

Being very happy with the low energy front, let us move now to the GUT front.

5. GUTS VERSUS SUSY GUTS

Ordinary GUTs

GUTs explain naturally the charge quantization, <u>but</u>, at the same time, entail that α, the electromagnetic fine structure constant, should be constrained to [1]:

$$1/170 \leq \alpha \leq 1/120 \tag{35}$$

rather severe bounds, given that $\alpha \simeq 1/137$.

GUTs demand that at some superhigh energy limit all three interactions have more or less the same strength or, inversely, that at low energies the three interactions should have different strengths, as is observed experimentally. In such theories one finds that the electroweak mixing angle, (θ_{e-w}), as measured at present energies, is given by [1]:

$$\sin^2\theta_{e-w}(M_W) \simeq 0.214 \pm 0.002 \tag{36}$$

which compares most favorably with the radiatively corrected experimental average

$$\sin^2\theta_{e-w}(M_W) = 0.215 \pm 0.014. \tag{37}$$

GUTs not only explain naturally the similar behavior under electroweak interactions of quarks and leptons and their disparity under strong interactions, <u>but</u> entail quark-lepton mass relations like [1]:

$$m_b/m_\tau\big|_{\text{low energies}} \simeq 2.8 - 2.9$$

in full accordance with what is observed experimentally. Incidentally the spectacular agreement between Eq. (38) and its experimental value entails that there are at most six flavors and that the top quark mass has an upper bound

$$m_t < 155 \text{ GeV}. \tag{39}$$

Being on the mass front, it is worth recalling that in GUTs neutrinos are either massless or acquire a very tiny mass (<100 eV), in accordance with terrestrial and cosmological observations.

Certainly the most dramatic consequence of GUTs has to do with matter instability. As is well-known, there are grand unified interactions that violate baryon (B) number and lepton (L) number conservation, thus making the proton unstable. However, for $\Lambda_{ms} = 0.1$ to 0.2 GeV, minimal SU(5) GUTs predict [1]

$$M_X = (1 \text{ to } 2)10^{15}\, \Lambda_{ms} = (1 \text{ to } 4)10^{14} \text{ GeV}, \tag{40}$$

and thereby a proton lifetime

$$\tau_p \simeq 10^{29\pm2} \text{ years} \tag{41}$$

where the quoted uncertainty reflects due allowance for our ignorance about the baryon decay matrix elements. Most calculations expect $p \to e^+\pi^0$ to be the dominant decay mode with a BR of order of 30%. Thus the IMB lower limit [27]

$$\tau_p(p\to e^+\pi^0) > 10^{32} \text{ years} \tag{42}$$

is very bad news for minimal SU(5) [28]. Drastic modifications are needed and they are freely provided for in SUSY GUT models.

GUTs also play a very important role in the evolution of the very early universe. They not only contain all the highly desirable ingredients necessary for creating a baryon asymmetry, like B violation, CP-violation and C-violation, but in conjunction with an expanding universe, one finds [1]

$$n_B/n_\gamma \sim 10^{-6} \text{ to } 10^{-12} \tag{43}$$

certainly inside the observed number

$$n_B/n_\gamma \sim 10^{-8\pm2}. \tag{44}$$

Supersymmetric GUTs

The main reason for supersymmetrizing grand unified theories is of course the solution [22] of the cumbersome gauge hierarchy problem. We have seen that a proliferation of the "low energy" particle spectrum is then necessarily unavoidable. Every "known" particle, fermion, Higgs boson or gauge boson should have its corresponding superpartner with characteristic mass differences of order $O(M_W)$. Additional problems to the ones discussed in the previous section appear. The new "low energy" degrees of freedom will definitely modify the standard program of grand unification and, in general, there is the danger that the whole program will be mucked up. It is remarkable that in SUSY GUTs the standard success of ordinary GUTs remains more or less intact. So let us see how the unification program changes. Our SUSY GUT should contain at least the supersymmetrized SU(3) x SU(2) U(1) model. This piece of information is enough to give a kind of general analysis. It is clear from the beginning that the unification point is going to be raised. The new "light" degrees of freedom involve fermions and scalars, thus their contribution to the various β functions has the effect of delaying the change of the various coupling constants with energy. Notably, the strong coupling constants fall down with energy much smoother than before and so it will take "longer" for the different coupling constants to "meet". At the same time, one expects a larger unification coupling constant. More precisely, in "minimal" type SUSY GUTs [29] one finds, for the coefficients of the SU(3), SU(2) and U(1) β-functions,

$$\beta_3 = 9 - f$$

$$\beta_2 = 6 - f - \frac{h}{2} \qquad (45)$$

$$\beta_1 = - f - \frac{3h}{10}$$

where f represents the number of flavors ($f \gtrsim 6$) and h stands for the number of "light" Higgs doublets ($h \gtrsim 2$).

Concerning the coupling constants we get, using Eq. (45),

$$\frac{1}{\alpha_3(m)} = \frac{1}{\alpha_{SG}} - \frac{1}{2\pi}[9-f]\ell n\left(\frac{M_{SX}}{m}\right)$$

$$\frac{1}{\alpha_2(m)} = \frac{1}{\alpha_{SG}} - \frac{1}{2\pi}\left(6-f-\frac{h}{2}\right)\ell n\left(\frac{M_{SX}}{m}\right)$$

$$\frac{1}{\alpha_1(m)} = \frac{1}{\alpha_{SG}} - \frac{1}{2\pi}\left(-f-\frac{3h}{10}\right)\ell n\left(\frac{M_{SX}}{m}\right), \qquad (46)$$

where as usual $\alpha_i \equiv \frac{g_i^2}{4\pi}$ (i=1,2,3), α_{SG} is the SUSY GUT unification fine structure constant, M_{SX} is the SUSY GUT unification mass and m a "low energy" mass scale larger than or equal to $\sim O(M_W)$. We can recast Eqs. (46) in a more useful form

$$\ln \frac{M_{SX}}{M_W} = \frac{2\pi}{18+h} \left(\frac{1}{\alpha(M_W)} - \frac{8}{3}\frac{1}{\alpha_3(M_W)}\right) \tag{47}$$

$$\sin^2\theta_{e-w}(M_W) = \frac{(3+h/2)+(10-h/3)\alpha(M_W)/\alpha_3(M_W)}{18+h} \tag{48}$$

and

$$\frac{1}{\alpha_{SG}} = \frac{(9-f)1/\alpha(M_W)-(6-(8f/3)-h)1/\alpha_3(M_W)}{18+h} \tag{49}$$

where, for simplicity, we have identified the supersymmetry breaking scale (M_{LESB}) with M_W.

Using Eqs. (47) - (49), and taking into account higher order corrections, we get [30,31],

$$M_{SX} \simeq \begin{cases} 6.10^{16}\,\Lambda_{\overline{MS}} \text{ for } h = 2 \\ 3.10^{15}\,\Lambda_{\overline{MS}} \text{ for } h = 4 \end{cases} \tag{50}$$

where the present favorable value of $\Lambda_{\overline{MS}}$ (the QCD scale parameter evaluated in the modified minimal subtraction scheme with four flavors) is between 100 and 200 MeV. The electroweak angle is calculated to be [30,31]:

$$\sin^2\theta_{e-w}(M_W) = \begin{cases} 0.236 \pm 0.003 \text{ for } h = 2 \\ 0.259 \pm 0.003 \text{ for } h = 4 \end{cases} \tag{51}$$

while $\alpha_{SG} \simeq 1/24$ to $1/25$ for six flavors and two light Higgs doublets.

We move next to the m_b/m_τ ration in SUSY GUTs. Here we find [30,31]:

$$\frac{\left(\frac{m_b}{m_\tau}\right)_{SUSY}}{\left(\frac{m_b}{m_\tau}\right)_{ORD}} = \left[\frac{\alpha_3(M_W)}{\alpha_{SG}}\right]^{8/9} \Big/ \left[\frac{\alpha_3(M_W)}{\alpha_G}\right]^{4/7} \tag{52}$$

and substituting $\alpha_{SG} \simeq 1/24$, $\alpha_G \simeq 1/41$ and $\alpha_3(M_W) \simeq 0.12$, we get

$$\frac{\left(\frac{m_b}{m_\tau}\right)_{SUSY}}{\left(\frac{m_b}{m_\tau}\right)_{ORD}} = 1.0! \tag{53}$$

and thus, by using Eq. (38), declaring that $(m_b/m_\tau)_{SUSY}$ is in full accordance with its experimental value. We find this "coincidence" remarkable. The situation is rather clear. As was expected the unification scale moves upward and the unification coupling constant increases, as does the electroweak angle always compared to the ordinary GUTs results (cf. Eqs. (36) and (40)). The m_b/m_τ remains unchanged, a surprise at least to me! Concerning the value of $\sin^2\theta_{e-w}$, it seems to be a bit high for the case of two light Higgs doublets compared with the experimental value (see Eq. (37)). On the other hand, the increase of the grand unification scale by a factor of O(10) with respect to ordinary GUTs, suppresses the conventional (gauge-boson mediated) proton decay mode, $p \to e^+\pi^0$, by a factor of (10^4) compared to the ordinary GUTs value (see (41)), thus evading any conflict with the present experimental lower bound [27]. However, the show is not over! It has been remarked [32] that in a large class of SUSY grand unified theories, if there are no preventing symmetries, there are loop diagrams that may cause rapid proton decay. For example, by "dressing up" diagrams of the form

Figure 1

where s_f and H_{SX} represent the SUSY partners of "light" fermions (f) and "superheavy" coloured Higgs triplets respectively, one may get "looping" proton decay

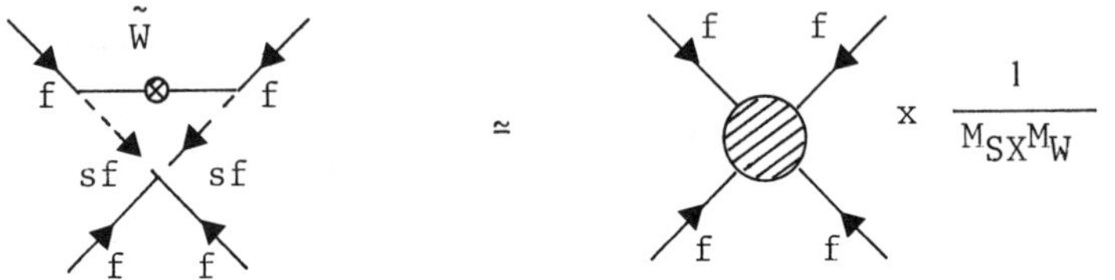

Figure 2

where again $\tilde{W}$ stand for the SUSY partners of the charged weak bosons. The bizarre thing here is that $\tau_p \propto M_{SX}^2 M_W^2$ and not $\tau_p \propto M_{SX}^4$.

One may then naively think that these kinds of SUSY theories are dead because they cause a too rapid proton decay [32]. A more careful analysis [31] showed though that things are different. Indeed, we have found [31] that in such theories the proton lifetime can easily be 10^{31} years or a bit longer (not much longer though), and with the very "peculiar" characteristic decay mode [31] $\bar{\nu}_\tau K^+$. Thus, in the so-called softly broken SUSY GUTs (without "preventing" symmetries) we find [31]

$$\tau_N \simeq O(10^{31\pm 2}) \text{ years}$$

$$\tau(N\to\bar{\nu}_\tau K) >> \Gamma(N\to\bar{\nu}_\mu K) >> \Gamma(N\to\bar{\nu}\pi, \mu^+K) >> \Gamma(N\to\mu^+\pi)$$

$$>> \Gamma(N\to e^+K) >> \Gamma(N\to e^+\pi). \quad (54)$$

But the surprises are not over. Very recently we have found [33] that SUSY GUTs may solve naturally the monopole problem. In doing so though, we may upset the standard solution of the baryon asymmetry problem. One way to reconcile this puzzle and keep both solutions intact [33] is the existence of "light" superheavy triplets, i.e., $M_{H_3} \sim 10^{10}$ GeV. Actually we find [34] in this case $\sin^2\theta_{e-w} \simeq 0.220$ much closer to the experimental value given by Eq. (36), than in other SUSY GUTs (see Eq. (51)). But it is well-known [35] that such Higgsons mediate proton decay with lifetime $\sim O(10^{31\pm 2})$ years, and we find that in SUSY GUTs the decay modes are given by [34],

$$\tau_N \simeq O(10^{31\pm 2}) \text{ years}$$

$$\Gamma(\bar{\nu}_\mu K^+, \mu^+K^0) : \Gamma(\bar{\nu}_e K^+, e^+K^0, \mu^+ "\pi^0") : \Gamma(e^+\pi^0, \bar{\nu}_e\pi^+)$$

$$\simeq 1 : \sin^2\theta_c : \sin^4\theta_c \quad (55)$$

$$(\theta_c \cong \text{Cabibbo angle}).$$

All these predictions have to be contrasted with the ordinary GUT predictions [1]:

$$\tau_N \simeq 10^{31\pm 2} \text{ years}$$

$$B(N\to e^+ \text{ non-strange}, \bar{\nu}_e \text{ non-strange}, \mu^+ \text{ or } \bar{\nu}_\mu \text{ strange}):$$

$$B(N\to e^+ \text{ strange}, \bar{\nu}_e \text{ strange}, \mu^+ \text{ or } \bar{\nu}_\mu \text{ non-strange}) =$$

$$= 1 : \sin^2\theta_c. \quad (56)$$

The contrast between Eqs. (54), (55) and (56) is rather dramatic. Apart from the case where the protons decay in the "conventional" way (Eq. (56)) but with $\tau_p \propto M_{SX}^4$ and M_{SX} as given by Eq. (50) (which will make life very, very difficult, if not impossible), all other possibilities are very interesting and hopefully not impossible to test experimentally. It should be emphasized once more that <u>proton decay</u> in <u>SUSY GUTs</u> at an <u>observable rate</u> always involves <u>strange particles</u> (K,...) in the <u>final state</u>. This striking difference in the proton decay modes between SUSY and ordinary GUTs is maintained also in supergravity models, as we will see later. Experiment will tell us!

I find it very remarkable that despite the proliferation of the "low energy" spectrum, SUSY GUTs have succeeded in passing the tests of $\sin^2\theta_{e-w}$, $\frac{m_b}{m_\tau}$, τ_p without much difficulty. Certainly the main advantage of SUSY GUTs is their capacity to provide a natural solution to the gauge hierarchy problem. It should be recalled that we would like to understand in a natural, satisfactory way:

1) Why $\frac{M_W}{M_X} \leqq O(10^{-14})$?

2) How to separate, at the tree-level the masses of the Higgs doublet and its GUT partner, the Coloured Higgs Triplet?

3) The absence of the Cosmological constant.

4) How to incorporate gravitational interactions, etc.

A possible answer to all these questions may be potentially found in the framework of local SUSY theories or supergravity [8], where we move next. It should be stressed that the move to supergravity theories is not only for aesthetical reasons but it is entailed by the structure of realistic SUSY models, as discussed above (see the remarks after (31)).

6. PHYSICAL STRUCTURE OF SIMPLE (N = 1) SUPERGRAVITY

We are then led to consider local SUSY gauge theories [16]. The effective theory below the Planck scale must be [36] N = 1 supergravity. The restriction to N = 1 follows from the apparent left-right asymmetry of the "known" gauge interactions. Since we are dealing with local SUSY, the breaking of SUSY must be spontaneous, not explicit, if Lorentz invariance or unitarity are not to be violated. It is remarkable that the effective theory below $M_{P\ell}$ has been uniquely determined [36] to be a spontaneously broken N = 1 local SUSY gauge theory [16].

We start with a reminder of the structure of N = 1 supergravity actions [16] containing gauge and matter fields (if not explicitly stated, we use natural units $k^2 \equiv 8\pi G_N = (8\pi/M_{P\ell}^2) \equiv 1/M^2 = 1$)

$$A = \int d^4x d^4\theta E\Big(\Phi(\phi,\bar{\phi}e^{2V}) + \mathrm{Re}[R^{-1}g(\phi)] + \mathrm{Re}[R^{-1}f_{\alpha\beta}(\phi)W_a^\alpha \varepsilon^{ab} W_b{}^\beta]\Big) \tag{57}$$

where E is the superspace determinant, Φ is an arbitrary real function of the chiral superfields ϕ and their complex conjugates $\bar{\phi}$, V is the gauge vector supermultiplet, R is the chiral scalar curvature superfield, g is the chiral superpotential, $f_{\alpha\beta}$ is another chiral function of the chiral superfields ϕ, and W_a^α is a gauge-covariant chiral superfield containing the gauge field strength. In addition, to all the obvious general co-ordinate transformations, local supersymmetry and gauge invariance, the action (57) is also invariant [16] under the transformations

$$
\begin{aligned}
J \equiv 3\ell n(-\tfrac{1}{3}\Phi) &\to J + K(\phi) + K^*(\bar{\phi}) \\
g(\phi) &\to e^{K(\phi)} g(\phi).
\end{aligned}
\tag{58}
$$

These transformations can be related to a description of the chiral superfields ϕ as co-ordinates on a Kähler manifold with Kähler potential Φ, and the transformations (58) are known as Kähler gauge transformations [37]. One particular manifestation of this Kähler gauge symmetry is in the effective scalar potential

$$
V = -\exp(-G)(3 + G'_i G''^{i-1}_j G'^j) + (\text{gauge terms}) \tag{59}
$$

where

$$
G \equiv J - \ell n(\tfrac{1}{4}\, |g|^2) \tag{60}
$$

which is clearly invariant under the transformations (58). In general, the action depends on a real function

$$
\hat{\Phi} \equiv \Phi \,/\, |g(\phi)|^{2/3} \tag{61}
$$

and on the chiral function $f_{\alpha\beta}(\phi)$. The most familiar forms of these functions are $J = -\phi\bar{\phi}/2$, giving canonical kinetic energy terms for the chiral superfields, $g(\phi)$ a cubic polynomial giving renormalizable matter interactions of dimension ≤ 4, and $f_{\alpha\beta} = \delta_{\alpha\beta}$. We expect that more complicated functions will contain terms $O(\phi/M_{P\ell})^n$ relative to these canonical leading terms.

Ellis, Tamvakis and myself have suggested [36] interpreting Eq. (57) as an effective action suitable for describing particle interactions at energies $\ll M_{P\ell}$ just as chiral SU(N) x SU(N) Lagrangians were suitable for describing hadronic interactions at energies $\ll$1 GeV. In much the same way as we know that physics gets complicated at E = 1 GeV, with many new hadronic degrees of freedom having masses of this order, we also expect many new "elementary particles" to exist with masses $O(M_{P\ell})$. It may well be that all the known light "elementary particles", as well as these heavy ones, are actually composite, and that at energies $\gg M_{P\ell}$ a simple preonic picture will emerge, analogously to the economical description of high-energy hadronic interactions in terms of quarks and gluons. It may even be that these preonic constituents are themselves ingredients in an extended supergravity theory [38]. But let us ignore these speculations for the moment and return to our pedestrian phenomenological interpretation of the action (57).

The well-known rules of phenomenological Lagrangians [39] are that one should write down all possible interactions consistent with the conjectured symmetries (e.g., chiral SU(2) x SU(2)), and only place absolute belief in predictions which are independent of the general form of the Lagrangian (e.g., $\pi\pi$ scattering lengths). These are the reliable results which could also be obtained using current algebra arguments. It does not make sense to calculate strong interaction radiative corrections (read: supergravity loop corrections) to these unimpeachable predictions: these are ambiguous until we know what happens at the 1 GeV scale (read: $M_{P\ell}$), and our ignorance can be subsumed in the general form of the phenomenological Lagrangian, in which any and all possible terms are present a priori (read: non-trivial J, non-polynomial g and $f_{\alpha\beta}$). On the other hand, non-strong interaction radiative corrections can often be computed meaningfully (e.g., the $\pi^+ - \pi^0$ mass difference, large numbers of pseudo-Goldstone boson masses in extended technicolor theories). Similarly, it makes sense to compute matter interaction (gauge, Yukawa, Higgs) corrections to the tree-level predictions of the effective action (57).

Since the supergravity action is non-renormalizable, and since both the Φ and $f_{\alpha\beta}$ terms in the action (57) have a $\int d^4\theta$ form, we expect general variants of them to be generated by loop corrections. Presumably, radiative corrections maintain the essential geometry of the Kähler manifold [37]. Therefore, we expect loop corrections to fall into the class of Kähler gauge transformations (58). The only analogous transformation allowed in a conventional renormalizable theory is K = constant, corresponding to a wave function renormalization. In our case, more general gauge functions $K(\phi)$ might appear.

In N = 1 SB local SUSY gauge models, called for abbreviation supergravity or SUGAR models, one usually distinguishes two sectors: (i) the "observable" sector containing quarks, leptons, Higgs and gauge bosons of electroweak and GUT types, as well as their SUSY partners; (ii) the "hidden" sector containing at least the goldstino and its SUSY associate. The "observable" and "hidden" sectors both couple to supergravity, but not directly to each other. In other words, ignoring supergravity, physics in the "observable" sector would be completely supersymmetric. A realization of this program occurs, for example, by splitting the superpotential of the theory into the sum of two terms

$$g(\phi_i) = h(Z_i) + f(\chi_i) \tag{62}$$

where χ_i are the "observable" fields and Z_i the "hidden" fields. The f-part of the superpotential has to do with the "observed" sector and contains most of the physics, while the h-part of the superpotential has to do with SB of local SUSY and the vanishing of the cosmological constant. The scalar fields of the hidden sector typically have v.e.v.'s of $M_{P\ell}$ which cause SUSY to be SB at a scale M_S determined by the parameters in the h-part. We may choose these parameters to be such that (27), or equivalently (31), are satisfied, as well as cancelling the cosmological constant at the same time "superHiggs effect") [16]. A celebrated example is the Polonyi potential [40]:

$$h(Z) = M_S^2(Z+B) \tag{63}$$

which, in the absence of other fields, and for $B = (2-\sqrt{3})M$ and $\langle Z\rangle = (\sqrt{3}-1)M$, ($M \equiv (M_{P\ell}/\sqrt{8\pi}) \simeq 2.5 \cdot 10^{18}$ GeV being the appropriate supergravity scale (superPlanck mass)), implies SB of local SUSY at M_S and vanishing cosmological constant. The communication between the two sectors is mediated by the auxiliary fields of the SUGAR multiplet and takes place at tree-level in a model-independent way. Exchange of gravitons or gravitinos plays no role at this level. The elimination of these auxiliary fields produces non-renormalizable interactions between the two sectors. These non-renormalizable interactions include the ones between the hidden fields Z_i, usually taken to be neutral under all gauge symmetries, and gauge fields.

All these effects have been summed into an effective scalar potential [16] (Eq. (59), but with canonical kinetic energy terms for the chiral superfields):

$$\begin{aligned} V(\phi) = \exp\left(\frac{1}{M^2}\sum_i \phi_i\bar{\phi}_i\right)\Big(\sum_i \left|\frac{\partial g(\phi)}{\partial \phi_i} + \frac{\bar{\phi}_i g(\phi)}{M^2}\right|^2 - \\ - \frac{3|g(\phi)|^2}{M^2}\Big) + \frac{1}{2}\sum_a D_a^2(\phi,\bar{\phi})g_a^2 \end{aligned} \tag{64}$$

This mess involves the scalar fields ϕ_i, the superpotential g and the D-terms, which have their usual global SUSY form (see (18)). Incidentally, by comparing (64) with the global SUSY potential (19)

$$V(\phi) = \sum_i \left|\frac{\partial g(\phi)}{\partial \phi_i}\right|^2 + \frac{1}{2}\sum_a D_a^2(\phi,\bar{\phi})g_a^2$$

we notice the richer structure of the SUGAR potential (64) and thus we anticipate that more physics will come out from it. When the scalar components of the hidden fields are replaced by their v.e.v.'s all superheavy fields are integrated out, and expanding in powers of 1/M, the resulting effective theory, just below M_X, for the light observable fields, will contain both the usual SUSY terms and a soft SUSY breaking piece [17,41-43]:

$$\begin{aligned} \mathcal{L}_{SOFT} = \sum_i m_i^2|\chi_i|^2 + m_{3/2}\sum_n (A_n f_n + h.c.) - \\ - \sum_{\alpha=1}^{3}\left(\frac{1}{2}M_\alpha\lambda_\alpha\lambda_\alpha + h.c.\right) \end{aligned} \tag{65}$$

where λ_α is a gauge fermion (gluino (α=3), wino (α=2) or bino (α=1)), f_n is any term in f and χ_i is any scalar (Higgs, squark or slepton). The A_n's [42] are expected to be of order one, while m_i and M_α

should be in general of order $m_{3/2}$ [17,36,44]. Actually, when corrections [43] from integrating out superheavy fields are ignored, the following relations hold [17]:

$$m_i^2 = m_{3/2}^2 \tag{66}$$

for every scalar field, and [42,43]

$$A_n = A - 3 + d_n \tag{67}$$

where A is a universal number [42] (assumed real) and d_n is the number of fields multiplied together in term n. If we imagine that the low energy theory is embedded in a GUT model at some GUT scale below the Planck mass, then all gaugino masses are equal at the GUT scale, so that only one single parameter M_0 is needed (M_0 <u>in principle</u> may be of $O(m_{3/2})$):

$$M_\alpha(M_X) = M_0, \quad \alpha=1,2,3 \tag{68}$$

while at lower energies M_α evolves in a manner identical for the gauge couplings:

$$\frac{M_\alpha(\mu)}{M_0} = \frac{\alpha_\alpha(\mu)}{\alpha_G}, \quad \alpha=1,2,3 \tag{69}$$

with α_α, α_G the usual SU(3), SU(2), U(1), GUT fine structure constants.

In the following, we shall assume that corrections [43] from integrating out superheavy fields are not very important and proceed with the simplified soft SUSY breaking piece, just below M_X,

$$\mathcal{L}_{SOFT} = m_{3/2}^2 \sum_i |\chi_i|^2 + m_{3/2} \sum_n (A-3+d_n) f_n + h.c. - \\ - \frac{1}{2} M_0 \sum_\alpha \lambda_\alpha \lambda_\alpha + h.c. \tag{70}$$

The soft operators in (65) or (70) are determined by the couplings of the hidden fields and by radiative corrections at the Planck scale, including those due to gravity. In spite of this, the very interesting thing is that sometimes it is possible to give the form of these operators without making detailed assumptions about either the hidden sector or the effects of gravitational radiative corrections. For example, in (70), the form of the hidden sector enters only through the three parameters $m_{3/2}$, A and M_0. The non-renormalizable interactions of the hidden fields with gauge fields, discussed before, will <u>generally</u> lead to soft Majorana masses, M_0, for the gauginos ($f_{\alpha\beta} \neq \delta_{\alpha\beta}$ in (57)), while the non-renormalizable interactions between the "observable" and "hidden" sectors, commented before, will produce the soft operators in the scalar potential (first two terms in (65) or (70)).

In a nutshell, the SB of SUSY in SUSY gauge theories coupled to N = 1 SUGAR leads to an effective theory below the Planck scale in which global SUSY is explicitly broken by a constrained set of soft operators, at an effective scale:

$$(M_S)\mathrm{eff} \simeq m_{3/2}. \tag{71}$$

As discussed before, physics constraints (see (3)) impose the condition $m_{3/2} \sim O(M_W)$ (see (31)), which creates a new hierarchy problem. Since we are dealing with gravitational phenomena, naively, the natural mass scale for the gravitino is $M_{P\ell}$ and not M_W. I call this the supergravity hierarchy problem, or the SUGAR hierarchy problem. On the other hand, it should be emphasized that the automatic soft breaking of global SUSY that is provided in the SUGAR framework not only splits "low energy" supermultiplets in the right way (all scalars and gauginos getting masses $O(m_{3/2})$), but it has the correct form to pass unscathed through all the traps set out by low energy phenomenology, discussed in Section 4.

In the physics applications which follow, we shall make extensive use of two main characteristics of the general framework discussed above. First, since we are dealing with an effective theory (the N = 1 SUGAR action is non-renormalizable), the superpotential g is not anymore necessarily constrained by renormalizability to be at most cubic, but it may contain any higher powers, suitably scaled, by inverse powers of $M_{P\ell}$, the natural cut-off of the theory [36]. Secondly, because of the non-renormalization theorems [21] of SUSY (SET IT AND FORGET IT principle), we may set, as we wish, certain parameters equal to zero, even if no symmetry implies that - a very different situation from ordinary gauge theories. Here, no apologies are needed. As explained in detail before, most of the physics is contained in the "observable" sector superpotential $f(\chi_i)$ (see (62)). Here we shall assume that, in one way or another, the "hidden" sector has played its role, as discussed previously, and we shall concentrate on the form of $f(\chi_i)$. We follow the natural (cosmic) evolution of things starting at energies below $M_{P\ell}$ and "coming down" to M_W. So we distinguish physics around the GUT scale (M_X) and physics around the electroweak (e-w) scale (M_W).

All physics from $M_{P\ell}$ down to (and including) low energies should emerge from such a program. We will show next that this is indeed possible.

7. PHYSICS WITH SIMPLE (N = 1) SUPERGRAVITY

A. Physics Around the GUT Scale (M_X)

The superhigh energy regime ($\sim 10^{16}$ GeV) is the theorists' paradise. There is a lot of freedom in building models, even though the constraints both from particle physics and cosmology become tighter and tighter. For definiteness, simplicity, and out of habit, we shall take as our prototype GUT an SU(5) type model [28, 29]. All GUT physics information will be contained in f_{GUT}, the GUT part of the "observable sector" superpotential. There is no consensus about the definite form of this superpotential, but it should unavoidably contain a piece (f_I) that breaks SU(5) down to SU(3) x SU(2) x U(1) and if possible, a piece (f_{II}), providing some explanation about the tree-level gauge hierarchy problem, so we write:

$$f_{GUT} = f_I + f_{II}. \tag{72}$$

For example, we may take [45]:

$$f_I = \frac{a_1}{M} X^4 + \frac{a_2}{M^2} X^2 \mathrm{T}\tau(\Sigma^3) \tag{73}$$

and [46]:

$$\begin{aligned} f_{II} &= \bar{\theta} H(\lambda_1 \frac{\Sigma^2}{M} + \lambda_2 \frac{\Sigma^3}{M^2} + \ldots) + \\ &+ \bar{H}\theta(\lambda_1' \frac{\Sigma^2}{M} + \lambda_2' \frac{\Sigma^3}{M^2} + \ldots) + M_\theta \bar{\theta}\theta \end{aligned} \tag{74}$$

where $X = \underline{1}$, $\Sigma = \underline{24}$, $\overset{(-)}{\theta} = \overset{(-)}{50}$, $\overset{(-)}{H} = \overset{(-)}{5}$ are chiral superfields of SU(5). The Higgs fields H and $\bar{H}$ couple to quark and lepton fields in the usual way. All components of θ and $\bar{\theta}$ have a bare mass M (which is taken to be of order M_X or larger), and so remain heavy after SU(5) breaks to SU(3) x SU(2) x U(1). After minimizing the potential, obtained by plunging into (22) the sum of f_I and f_{II} as given by (73) and (74), we get zero v.e.v.'s for $\overset{(-)}{H}$ and $\overset{(-)}{\theta}$ but non-zero ones for:

$$\begin{aligned} \langle X \rangle &= \left(\frac{m_{3/2}}{M}\right)^{3/8} M, \\ \langle \Sigma \rangle &= \left(\frac{m_{3/2}}{M}\right)^{1/4} M. \end{aligned} \tag{75}$$

Furthermore, we find [45] that the SU(3) x SU(2) x U(1) symmetric minimum is the lowest one for all values of a_1 and a_2, with a value [45]:

$$V_{eff} \simeq - \left(\frac{m_{3/2}}{M}\right)^{5/2} M^4. \tag{76}$$

What do these results mean? First, since the v.e.v. of Σ sets the scale of SU(5) breaking, we find that the GUT scale M_X satisfies [45]:

$$M_X^4 \simeq O(m_{3/2}M^3) \tag{77}$$

which is a highly successful relation. Using as an input the <u>non-hierarchical</u> and easy to explain ratio $(M_X/M) \sim 10^{-2} - 10^{-4}$, we obtain that $m_{3/2} \sim O(100\ \text{GeV})$! More generally, relations of the form $M_X{}^{2p-2} \simeq O(m_{3/2}M^{2p-3})$ with $p \geqq 3$, are also possible [45] by suitably modifying the exponents in (73). The supergravity hierarchy problem has been solved in a rather simple way.

Secondly, the SU(3) x SU(2) x U(1) symmetric minimum is lower in energy density than the SU(5) symmetric minimum $X = \Sigma = 0$ by an amount $(m_{3/2}/M)^{5/2}M^4$. Thirdly, the barrier between these two minima is never larger than $(m_{3/2}/M)^{5/2}M^4$, the same as the splitting between the states. Why this is so can be seen by noting that if we replace X by its v.e.v. (75) in (73), the effective renormalizable self-coupling of Σ is 10^{-12} tr(Σ^3). Thus we have generated [45] a small renormalizable coupling for Σ from our starting point of only non-renormalizable interactions among X and Σ. This small coupling suppresses the barrier between the SU(5) and the SU(3) x SU(2) x U(1) phases. The consequences of this suppression for supercosmology [33] are difficult to overestimate. Simply, it now makes possible the transition from the SU(5) to the SU(3) x SU(2) x U(1) phase at temperatures $T \sim 10^{10}$ GeV, which was previously blocked, since the barrier between the two phases was of the order of $(M_X)^4$. Incidentally, in this picture, the number density of GUT monopoles is naturally suppressed [33] below its present experimental upper bound.

It should be clear that the basic result - small renormalizable couplings arising from non-renormalizable ones suppressed only by inverse powers of M - is quite general and does not depend on the detailed form of the superpotential (f_I) [45]. The main characteristics of these types of models [45,47] are that they provide relations of the type (77); they make possible "delayed" SU(5) to SU(3) x SU(2) x U(1) phase transitions at $T \sim 10^{10}$ GeV, and they contain [45,47] more "light" particles than the ones in the minimal SUSY SU(3) x SU(2) x U(1) model. This last fact may sound dangerous when calculating M_X, $\sin^2\theta_{e-w}$ and m_b/m_τ, since in general an arbitrary increase of "light" stuff gives an out-of-hand increase [30, 31,48] and thus experimentally unacceptable values for the above-mentioned quantities. A more careful analysis [49] of these cosmological acceptable models (CAM) [49] shows that they make predictions as successful (for $\sin^2\theta_{e-w}$, m_b/m_τ, ...) as <u>at least</u> the ones [30,31] of the phenomenologically acceptable minimal type models (MIM). For a detailed, thorough phenomenological analysis of CAMs, see ref. [49].

Next, we discuss [46] physics related to f_{II} as given in (74). The v.e.v. of Σ does not only break SU(5) to SU(3) x SU(2) x U(1) but also provides a mass term which mixes the color triplets in H and $\bar{H}$ with those in θ and $\bar{\theta}$. However, there is no weak doublet in the 50, and so the weak doublets in H and $\bar{H}$ remain massless. The color triplets will have a mass matrix [46]:

$$\begin{pmatrix} 0 & \sim \dfrac{M_X^2}{M} \\ \sim \dfrac{M_X^2}{M} & M_\theta \end{pmatrix} \tag{78}$$

where M_θ should be of order M_X or larger ($\lesssim M$), to avoid having particles from θ and $\bar{\theta}$ influencing the renormalization group equations at scales below M_X (or even M). The eigenvalues of this mass matrix are $O(M_\theta)$ and $O(M_X^4/M_\theta M^2)$; this latter eigenvalue is about 10^{10} GeV for $M_X \sim O(10^{16}$ GeV) and $M_\theta \sim O(M)$. In this case, the Higgs color triplet can be used to generate [33,45] the baryon number of the universe after the SU(5) to SU(3) x SU(2) x U(1) transition which, as discussed earlier, occurs at temperatures $T \sim 10^{10}$ GeV in CAMs [49]. It is remarkable that $O(10^{10}$ GeV) is the lower bound [35] allowed for color triplet Higgs masses from present limits [27] on proton decay ($\tau_p > 10^{31-2}$ years). If indeed there are 10^{10} GeV Higgs triplets, then protons should decay predominantly [34] to $\bar{\nu}_\mu K, \mu K$ with a lifetime $\sim O(10^{31-2}$ years), see (55).

The role of supergravity in this natural explanation [46] of the Higgs triplet-doublet mass splitting ($\equiv$ tree-level gauge hierarchy problem) is fundamental, in several aspects. The same kind of explanation had been suggested before [50] in the framework of renormalizable global SUSY GUTs, where Σ^2 in (74) was replaced by a 75 of SU(5) and higher than two powers of Σ were absent [50]. Unfortunately, the use of 75 drastically conflicts with cosmological scenarios [33,45,47] based on SUSY GUTs. The barrier between the SU(5) and SU(3) x SU(2) x U(1) phases is impossible to overcome unless most of the 75 is very light ($\sim M_W$). But then, all hell breaks loose. A light 75 makes the gauge coupling in the SU(5) phase decrease at lower energies so there is no phase transition at all [33,45]. Furthermore, the presence of these new light particles in the SU(3) x SU(2) x U(1) phase changes the renormalization group equations, and prevents perturbative unification. On the contrary, in SUGAR theories, since we may use non-renormalizable terms, we may replace the fundamental 75 by an "effective" 75 contained in Σ^2. Unlike a light 75, a light 24 neither makes the SU(5) gauge coupling decrease at energies below M_X, nor upsets perturbative unification [46]. The previously mentioned cosmological scenarios [33,45,47] can proceed without modification. In addition, SUSY non-renormalization theorems [21] ensure the stability of the triplet-doublet splitting to all orders in perturbation theory. Since the only modifications of the theory are at the GUT scale M_X, it seems that we have got [46] a harmless and elegant solution of the tree-level, and for that matter, to all orders in perturbation theory, gauge hierarchy problem.

SUGAR models give good physics at the GUT scale - unique, cosmologically acceptable breaking of SU(5) to SU(3) x SU(2) x U(1), with an explanation of the smallness of the gravitino mass [45] ((77)-like relations), and a natural explanation [46] of the Higgs triplet-doublet splitting, cosmologically fitted and general enough. We believe that even if the very specific form of f_I in (73) may change, then f_{II} as given by (74) (or its obvious generalization to other GUT models) will be always a useful part of the f_{GUT}.

After finding plausible explanations for the SUGAR hierarchy problem (gravitino mass $\sim$ O(100 GeV)), the tree-level and higher orders gauge hierarchy problem (triplet-doublet Higgs splitting), it is time to explain the last gauge hierarchy problem (2), i.e., why does $M_W/M_X \leqq O(10^{-13})$? This problem brings us naturally to our next subject.

B. Physics Around the Electroweak (e-w) Scale (M_W)

Although there is no consensus on the best way to incorporate grand unification in SUGAR models, a unique minimal low energy model has recently emerged [36,51-53]. In this model, the physics of the TeV scale is described by an effective SU(3) x SU(2) x U(1) gauge theory, in which the breaking of weak interaction gauge symmetry is induced by renormalization group scaling of the Higgs $(\text{mass})^2$ operators [36]. Much of the attractiveness of this model stems from the fact that no gauge symmetries or fields beyond those required in any low energy SUSY theory are included. Sometimes, it may happen, as is the case of Cosmological Acceptable Models [49] (CAMs), that there are GUT relics which are light ($\sim M_W$), but they do not seem to play any fundamental role at low energies, so we may neglect them in our present discussion. Furthermore, adding random chiral superfields to the low energy theory may be problematic. For example, the presence of a gauge singlet superfield coupled to the Higgs doublets and added to trigger SU(2) x U(1) breaking [41-44,54], usually (but not always [55]) destroys [56] any hope of understanding the gauge hierarchy problem; the reason being [56] that in a GUT theory, the gauge singlet does not only couple to the Higgs doublets but also to their associate, superheavy color triplets. Then we have to try hard [55] to avoid 10^{10} GeV Higgs doublet masses, generated by [56] one-loop effects involving color triplets. Something smells fishy.

We focus then on the standard low energy SU(3) x SU(2) x U(1) gauge group, containing three generations of quarks and leptons, along with two Higgs doublets, as chiral superfields. The low energy effective superpotential (f_{LES}) of the model consists only of the usual Yukawa couplings of quark and lepton superfields to the Higgs superfields, along, in general, with a mass term coupling the two Higgs doublets, H_1 and H_2. Explicitly, in a standard notation:

$$
\begin{aligned}
f_{LES} &= h_{ij} U_i^c Q_j H_2 + \tilde{h}_{ij} D_i^c Q_j H_1 + \\
&+ f_{ij} L_i E_j^c H_1 + m_4 H_1 H_2
\end{aligned} \tag{79}
$$

where a summation over generation indices (i,j) is understood and $Q(U^c)$ denote generically quark doublets (charge −2/3 antiquark singlet) superfields, while $L(E^c)$ refers to lepton doublets (charge −1 antilepton singlet) superfields. With the exceptions of the top quark Yukawa coupling and the mass parameter m_4, which in principle may be of order $O(m_{3/2})$, all other parameters appearing in (79) contribute to the masses of the observed quarks and leptons and are known to be small. Neglecting these small couplings, the effective Low Energy Potential (V_{LEP}) can be written as (see (64), (65) and (70):

$$
\begin{aligned}
V_{LEP} = & \sum_{i=1}^{3} \{ m_{L_i}^2 |L_i|^2 + m_{E_i}^2 |E_i^c|^2 + m_{Q_i}^2 |Q_i|^2 + \\
& + m_{U_i}^2 |U_i^c|^2 + m_{D_i}^2 |D_i^c|^2 \} + \\
& + m_1^2 |H_1|^2 + m_2^2 |H_2|^2 + A h_t m_{3/2} (U_3^c Q_3 H_2 + \text{h.c.}) + \\
& + B m_{3/2} m_4 (H_1 H_2 + \text{h.c.}) + h_t m_4 (H_1^+ Q_3 U_3^c + \text{h.c.}) + \\
& + h_t^2 (|Q_3|^2 |U_3^c|^2 + |Q_3|^2 |H_2|^2 + |U_3^c|^2 |H_2|^2) + \\
& + \text{"D-terms"}.
\end{aligned} \tag{80}
$$

The effective parameters appearing in (80) take, at large scales ($\sim M_X$), the values:

$$
\begin{aligned}
& m_1^2(M_X) = m_2^2(M_X) = m_{3/2}^2 + m_4^2(M_X) \\
& m_{Q_i}^2(M_X) = m_{U_i}^2(M_X) = m_{D_i}^2(M_X) = m_{L_i}^2(M_X) = m_{E_i}^2(M_X) = m_{3/2}^2 \\
& A(M_X) = A; \; B(M_X) = A-1; \; (i=1,2,3)
\end{aligned} \tag{81}
$$

as dictated by (70). It should be stressed once more that the boundary conditions (81) are exact, if we only neglect corrections at the Planck scale, ignore the scaling of parameters from M to M_X, and pay no attention to corrections [43] at the GUT scale. All these effects are expected to be small and it is assumed that they do not seriously disturb (81) and the picture hereafter.

It is apparent from (80) that SUGAR models can easily succeed in giving weak interaction scale masses ($m_{3/2} \sim M_W$) to squarks, sleptons and gauginos (see (70)). Alas, SUGAR models also give large positive (mass)^2 to the Higgs doublets, thus making the breaking of SU(2) x U(1) difficult. One way to overcome this difficulty is the introduction [41-44,54] of a gauge singlet coupled to H_1 and H_2, but, as mentioned above, with disastrous effects [56] for the gauge hierarchy. A particularly simple solution to the SU(2) x U(1) breaking relies upon the fact that the boundary conditions (81) need be satisfied only at M_X (or M), and that large renormalization group scaling effects can produce a negative value for m_H^2 at low energies [36]. The full set of renormalization group equations for the parameters in V_{LEP} (80) has been written elsewhere [57]. Here we concentrate on the most interesting equation, the one for the mass-squared of the Higgs (m_2^2), which gives mass to the top quark:

$$\mu \frac{\partial}{\partial \mu} \begin{bmatrix} m_2^2 \\ m_{U_3}^2 \\ m_{Q_3}^2 \end{bmatrix} = \frac{h_t^2}{8\pi^2} \begin{bmatrix} 3 & 3 & 3 \\ 2 & 2 & 2 \\ 1 & 1 & 1 \end{bmatrix} \begin{bmatrix} m_2^2 \\ m_{U_3}^2 \\ m_{Q_3}^2 \end{bmatrix} + \frac{|A|^2 h_t^2 m_{3/2}^2}{8\pi^2} \begin{bmatrix} 3 \\ 2 \\ 1 \end{bmatrix} - \frac{h_t^2 m_4^2}{8\pi^2} \begin{bmatrix} 3 \\ 2 \\ 1 \end{bmatrix} - \frac{8\alpha_3}{3\pi} M_3^2 \begin{bmatrix} 0 \\ 1 \\ 1 \end{bmatrix} \tag{82}$$

where we have neglected gauge couplings other than the "colored" one, $\alpha_3 (\equiv g_3^2/4\pi)$, M_3 is the gluino mass (see (69) and (70)), and Yukawa couplings other than h_t, for the top, have been dropped. The physics content of (82) is apparent. Since μ is decreasing (we come from high energies down to low energies), the sign of the first two terms in (82) is such as to make all m_2^2, U_3, Q_3 smaller at low energy with the decrease of m_2^2 becoming more pronounced because of the 3:2:1 weighting. On the other hand, the sign of the last two terms in (82) is such as to make m_{U3}^2 and m_{Q3}^2 (the squark masses) larger at low energy, but have no direct effect on m_2^2 (notice the "zeros" in the corresponding matrices in (82)). Indirectly though, the net effect on m_2^2 of the last two terms in (82) is to enhance further its decrease at low energies, by increasing m_{U3}^2 and m_{Q3}^2, which then drive down m_2^2 via the first two terms of (82). This is exactly what we are after! We want large ($\sim M_W^2$) and positive squarks and slepton (masses)^2, <u>but</u> negative Higgs (mass)^2 to trigger SU(2) x U(1) breaking. The ways of obtaining negative Higgs (mass)^2 now become clear (see (82)). We have to use either a large top Yukawa coupling (h_t), <u>or</u> large A, <u>or</u> large m_4, <u>or</u> a fourth generation to provide large Yukawa couplings, <u>or</u> some suitable, physically plausible combination of the above possibilities. There are pros and cons for every one of the above situations. In the case of large h_t, a lower bound on the mass of the top quark is set [36,51-53]:

$$m_t > O(60\ \text{GeV}) \tag{83}$$

which some people may find uncomfortable. We may avoid a large h_t by moving it into the large A(>3) regime [51,58]. The price though is high. The phenomenologically acceptable vacuum becomes unstable against tunnelling into a vacuum in which all gauge symmetries, including color and electromagnetism, are broken. We must [51,58] then arrange things in such a way that the lifetime for this vacuum decay process is greater than the age of the universe. Some people, not without reason, may find this possibility dreadful. We may avoid large h_t and/or large A by using [49] non-vanishing $m_4(\sim m_{3/2})$ where a rather satisfactory picture then emerges [49]. Some people may object here to the basic assumption of large $m_4(\sim m_{3/2})$, since in the case of natural triplet-doublet Higgs splitting-type models [46] (see (74)), m_4 has a tendency to be small, if not zero, even though other sources of m_4 may be available.

Finally, we come to the possibility of a fourth generation which, suitably weighed, may help us to avoid large h_t, A, or m_4. The problem here is that low energy phenomenology (evolution of coupling constants, m_b/m_τ, ...) [49] as well as firm cosmological results like nucleosynthesis (especially ^{4}He abundance) [59], may suffer almost unacceptable modifications. Furthermore, one has to watch out for the mass of the fourth generation charged lepton, since it is going to behave like m_2^2 in (82), and thus $m^2_{L^4,E^4}$ may easily go negative, breaking electromagnetic gauge invariance.

Whatever mechanism (if any) turns out to be correct, it is rather remarkable that in SUGAR-type models, there is a simple explanation of the <u>breaking</u> of SU(2) x U(1) and of the <u>non-breaking</u> of SU(3) x $U(1)_{E-M}$. Furthermore, for the first time, we have a simple explanation of why $M_W \lll M_X$ (or M), i.e., a <u>simple solution</u> of the cumbersome <u>gauge hierarchy problem</u>. Starting with a positive Higgs (mass)^2, of order $m^2_{3/2}$ at M_X, and noticing that (see (82)) the evolution with μ^2 of the Higgs (mass)^2 is very slow (logarithmic), it is not surprising that we have to come down a long way in the energy scale, before the Higgs (mass)^2 turns negative and is thus able to trigger SU(2) x U(1) breaking. For example, in a class of models [51] characterized by "small" gravitino masses (<< $O(M_W)$) and by a Coleman-Weinberg-type [60] radiative SU(2) x U(1) breaking, occurring naturally, we get [61], by dimensional transmutation

$$M_W \simeq \Lambda_{QCD} \cdot \frac{g_2(M_W)}{24} \exp\Big(\frac{2\pi}{\alpha_G} F[\frac{h_t^2(M_X)}{\alpha_G}; A]\Big) \tag{84}$$

where g_2 denotes the SU(2) gauge coupling constant, α_G is the GUT fine structure constant and F is a rather involved function of its indicated variables. Using standard values for $g_2(M_W)$ ($\sim$0.67), α_G($\sim$1/25) and reasonable values for $h_t(M_X)$ (0.2 − 0.3), $A(M_X)$ (2-3), (84) gives [61]:

$$M_W \simeq (300-600)\Lambda_{QCD}, \tag{85}$$

a rather remarkable equation from many points of view. It does not only give $M_W <<< M_X$ as it was required by the gauge hierarchy, but it also provides a new and successful relation between the scale of electroweak unification M_W($\sim$80 GeV) and the fundamental scale of strong interactions, Λ_{QCD} ($\sim$0.15 − 0.3 GeV) in terms of dimensionless parameters. Clearly, this occurs because the rapid final stages of evolution of m_2^2 are driven by the increases in t-quark Yukawa coupling, and more importantly, in the squark masses which occur when $g_3^2/4\pi$ becomes large. It is only in the SUSY Coleman-Weinberg scenario [49,51] that the weak interaction scale is related to that of strong interactions. This contrasts with what usually happens in weak gauge symmetry breaking in SUGAR models [52-53] where M_W is connected to $m_{3/2}$, but it is not directly related to the strong interaction scale.

Another very amazing fact is that the values of the parameters of the low energy world seem to co-operate with us. Since quarks are feeling strong interactions, (82) tells us that quarks may enjoy large masses (Yukawa couplings) without making squark (masses)^2 negative, because of the last term ($\sim \alpha_3$), which easily balances off large Yukawa couplings, without any sweat. On the other hand, since leptons are not feeling strong interactions, the balance-off between the weak gauge couplings and large Yukawa couplings becomes extremely delicate and could be problematic. How nice that for all three generations, leptons and down quarks weigh less than 5 GeV and especially for the third generation that the top quark (t) is heavier than the bottom quark (b). An inverse situation would be disastrous, because in any reasonable GUT, a very heavy b quark would mean a very heavy τ lepton, thus making electromagnetic gauge invariance tremble in such SUGAR-type schemes. I will not go any further into the esoterics of this type of SU(2) x U(1) breaking models, since a rather thorough and detailed exposé of these types of theories and of their phenomenological consequences is now available [49]. It should be stressed that things are now very constrained, as we see from the Table taken from ref. [49], where the whole low energy spectrum is worked out in terms of very few parameters, $m_{3/2}$, $A(M_X)$, $\xi \equiv M_0/m_{3/2}$ (see (68)) and $m_4(M_X)$. Eventually, with more theoretical insight, we hope to determine even these very few parameters, thus predicting uniquely the low energy spectrum. For example, we have already discussed ways of determining $m_{3/2}$ (see (77)), while some people may favor $A(M_X) = 3$ as a natural solution [62] to the absence of the cosmological constant problem, etc. Among other interesting things contained in the Table, the existence of a very light ($\sim$[3-6) GeV] neutral Higgs, with the usual Yukawa couplings to matter,

Table 2. Particle Spectrum [49]*

		CAM	MIM	CAM	MIM	CAM	MIM
$A(M_X)$		3	3	2.8	2.8	2.0	1.6
$m_{3/2}$		15	15	15	15	15	15
ξ		2.8	2.2	3.2	3.1	3.5	1.9
$m_4(M_X)$		15	17	16	18	11	7
top		25	25	35	35	50	50
All families	$(\text{sleptons})_L$	29	27	32	36	35	25
	$(\text{sleptons})_R$	21	20	22	23	23	19
1st and 2nd families	$(\text{squark})_L$	58	77	66	108	72	67
	$(\text{squark})_R$	54	74	61	104	67	65
		54	74	60	103	66	65
3rd family	$(\text{sbottom})_L$	58	76	64	106	68	66
	$(\text{sbottom})_R$	54	74	60	104	66	65
	$(\text{stop})_L$	81	96	95	132	112	106
	$(\text{stop})_R$	26	54	23	78	21	37
	Charged Higgses	96	93	95	94	88	83
	Neutral Higgses	106	104	105	105	100	95
		3	3	4.4	5.3	6	5
	"Axion"	51	46	49	48	35	19
	Gluinos	42	84	47	118	52	72
	Photino	11	4.6	9.4	7.3	4	3
	HW, WH-inos	89	87	87	94	84	90
		78	82	79	79	82	75
	HZ, HZ-inos	99	108	99	116	101	106
		92	85	91	80	88	83
	Axino	26	23	24	24	17	9

* Physical mass spectrum of the cosmologically acceptable model (CAM) and the minimal model (MIM) corresponding to the same gravitino mass $m_{3/2} \simeq 15$ GeV for top quark masses equal to 25 GeV, 35 GeV and 50 GeV respectively. ξ denotes the ratio of the gaugino to the gravitino mass at M_X. All masses are in GeV units. The light neutral Higgs gets its mass via radiative corrections.

should not escape our attention. Since such a particle is a common feature of a large class of models [49,51], a search in the $Y \to H^0 + \gamma$ channel, which is expected to be a few per cent of the $Y \to \mu^+\mu^-$ decay, may turn out to be very fruitful.

There are other, very interesting features concerning low energy phenomenology stemming out from the general form of V_{LEP} (see (80) and (81)) in SUGAR models. Very tight constraints coming from natural suppression of flavor changing neutral currents [24-25] (FCNC), absence of large corrections to (g-2) [26] and ρ [52,63] ($\equiv (M_W/M_Z\cos\theta_{e-W})^2$) as well as to θ_{QCD} [6], which have been the nemesis of SUSY models with arbitrary and explicit soft SUSY breaking, are satisfied in SUGAR models. The highly-constrained set of soft SUSY operators (70) in SUGAR models fits the bill [36,52]. Concerning FCNC, (81) guarantees the super-GIM mechanism, since the mass matrices for the quarks and leptons are diagonalized by the same transformation that renders the mass matrices for their scalar partners and gluino couplings generation diagonal. Despite the fact that this property does not survive, in general, after renormalization, it has been shown [64,65] that these effects are controllable. Furthermore, the Buras stringent upper bound [66] on the top quark mass (<O(40 GeV)), coming from kaon phenomenology (K_L-K_S and $K_L \to \mu^+\mu^-$ systems), is avoided [64] in SUGAR models. There are a lot of cancellations between ordinary and SUSY contributions in K processes [67], such that the top quark mass may be stretched up to 100 GeV without problem [64]. That sounds very satisfactory, especially for SUGAR models [51-53] that do need a large top quark mass for SU(2) x U(1) breaking. It looks like a self-service situation. Similar comments apply in the case of $(g-2)_\mu$ or ρ, where it has been shown that SUGAR model contributions are acceptable [49,52, 63]. Typical values for SUGAR contributions are [49,68] $|\Delta(g-2)_\mu| \lesssim (3 \cdot 10^{-9})$ and [52,63] $\Delta\rho \leqq 0.01$, which compare favorably with the present experimental upper bounds of $(4 \cdot 10^{-8})$ and (0.03) respectively, but are large enough to be interesting. Better experimental bounds, especially on $\Delta\rho$, could be revealing.

Concerning θ_{QCD}, it has been shown [36,69] that in SUGAR-type models, we not only understand the non-renormalization [6] of θ, but we also understand [36,69] why θ is zero or small ($<10^{-9}$) to start with. This fact is related with our freedom, discussed before, to use non-minimal ($\neq \delta_{\alpha\beta}$) $f_{\alpha\beta}$ in (57). It has already been observed [16] that the gauge kinetic term in (57), as well as giving rise to the canonical

$$-\frac{1}{4} F^\alpha_{\mu\nu} F^\beta_{\mu\nu} (\mathrm{Re}\ f_{\alpha\beta}), \tag{86}$$

could also yield the CP-violating θ vacuum term

$$\varepsilon^{\mu\nu\rho\sigma} F^\alpha_{\mu\nu} F^\beta_{\rho\sigma} (\mathrm{Im} f_{\alpha\beta}). \tag{87}$$

We know that θ_{QCD} is $<O(10^{-9})$ experimentally [5], and that θ is not renormalized in a supersymmetric theory [6]. It is finitely renormalized when supersymmetry is broken, but [6] this is plausibly only by an amount $\delta\theta = O(10^{-16})$ in the popular Kobayashi-Maskawa model [70]. Thus we see that θ should be less than $O(10^{-9})$ in a supersymmetric GUT and may be very small. An attractive hypothesis is that $f_{\alpha\beta}$ is a function with only real coefficients as found in extended supergravities [71]. In this case, $\text{Im}f_{\alpha\beta} = 0$ when $\langle 0|\phi|0\rangle =$ $= 0$, and the theory is CP-invariant in the gauge sector. If some of the ϕ then acquire complex vacuum expectation values, they will induce a non-zero value of $\text{Im}f_{\alpha\beta}$ and hence violate CP spontaneously in the gauge sector, which is a new twist on an old proposal [72]. If the moduli of some of these complex $\langle 0|\phi|0\rangle$ were $O(M_{P\ell})$, then the effective θ parameter would be $O(1)$ which is phenomenologically unacceptable. However, it is easy to imagine scenarios where θ is much smaller. For example, if $f_{\alpha\beta} \doteq \delta_{\alpha\beta} + O(\phi^2/m_{P\ell}^2)$ and the culprit $\langle 0|\phi|0\rangle = O(m_X)$, then

$$\theta = O\left(\left[\frac{M_X}{M_{P\ell}}\right]^2 \text{ x some small (?) angle}\right) \lessapprox O(10^{-7}) \tag{88}$$

and the phenomenological constraint on θ_{QCD} could easily be respected. If the only complex $\langle 0|\phi|0\rangle$ were $O(m_W)$, or if all the $\langle 0|\phi|0\rangle$ were real as in all supersymmetric GUTs proposed to date, then the bare $\theta = 0$. Hence supergravity offers the other half of an answer to the θ vacuum problem. It should be stressed that low energy supergravity models have new sources [52,69,73,74] of CP-violation beyond the standard model. Unfortunately, they do not shed more light on the smallness of the observed CP-violation in the K system [73]. On the other hand, potential large contributions [69,73] to ε', θ_{QCD} and to the Dipole Electric Moment Of the Neutron (DEMON) put, in general, severe constraints [69,73] on these new possible CP-violating phases. Actually, it seems almost unavoidable [6,52,69,74] that in SUGAR models the DEMON should be near, but not above, the present experimental upper bound [5] of $6 \cdot 10^{-25}$ e.cm, a rather drastic and experimentally testable prediction, in sharp contrast with the standard model prediction [75] $O(10^{-30}$ e.cm). We may know soon.

Turning now to baryon decay, an interaction of the form [36]

$$f \ni \frac{\lambda}{M} \bar{F}TTT \tag{89}$$

where $\bar{F}$ is a $\bar{5}$ of matter (quark + lepton) chiral superfields in SU(5), T is a $\underline{10}$ of matter superfields and λ is some generic Yukawa coupling, could replace the Higgs exchange in the Weinberg-Sakai-Yanagida [32] loop diagram for baryon decay. The magnitude of the diagram with (89) relative to the conventional Higgs diagram (see Figure 2) is:

$$\left[\frac{\lambda}{M}\right] / \left[\frac{\lambda^2}{M_{H_3}}\right] \simeq \left[\frac{M_{H_3}}{\lambda M}\right]. \tag{90}$$

The ratio (90) could easily be >1, making a non-renormalizable superpotential interaction the dominant contribution to proton decay. A careful analysis of SUGAR-induced baryon decay shows [76], surprisingly enough, that the expected hierarchy of decay modes is similar to that [31-34] coming from conventional minimal SUSY GUTs as given by (54) and (55). One might have wrongly expected that no hard and fast predictions could be made about gravitationally-induced baryon decay modes. Anyway, this mechanism could give observable baryon decay even if the GUT mass $M_X \simeq M$.

Incidentally, similar terms like (89) have been considered [77] in efforts to explain the "lightness" of the first two generations of quarks and leptons. One replaces [77] direct Yukawa couplings for the first two generations with (very schematically):

$$\begin{aligned} f \ni \frac{\tilde{\lambda}_2'}{M} \bar{H}\Sigma T_2 \bar{F}_2 + \frac{\tilde{\lambda}_2'}{M} H T_2 T_2 \Sigma + \\ + \frac{\tilde{\lambda}_1}{M^2} \bar{H}\Sigma^2 T_1 \bar{F}_1 + \frac{\tilde{\lambda}_1'}{M^2} H T_1 T_1 \Sigma^2 + \ldots \end{aligned} \tag{91}$$

which not only repairs [36,77] wrong relations like $m_d(M_X) \simeq m_e(M_X)$, very difficult to correct [78] in conventional SUSY GUTs, <u>but</u> also provides reasonable masses for the first two generations. Indeed, it follows from (91) that the second generation is getting masses $(M_X/M)M_W \sim$ (0.1-1 GeV), while the first generation masses are $(M_X/ /M)^2 M_W \sim$ (1-10 MeV), exactly what was ordered. It is amazing that in SUGAR models, by increasing $M_X(\sim 10^{16}$ GeV), relative to its ordinary GUT value (10^{14} GeV), and by decreasing $M_{P\ell}$, what is relevant is the superPlanck scale $M(\sim 10^{18}$ GeV), the highly-desired ratio $(M_X/M) \sim 10^{-2}$, which appears naturally. It seems now, for the first time, that gravitational interactions may be responsible for the masses of at least the first two generations. Once more, non-renormalizable interactions contained in SUGAR models provide a simple solution [77] to another hierarchy problem, the <u>fermion mass hierarchy problem</u>.

We have shown that gravitational effects, as contained in SUGAR theories, cannot be neglected anymore in the regime of particle physics. On the contrary, it may be that supergravitational effects are really responsible: <u>for</u> the SU(5) breaking at M_X with an automatic triplet-doublet Higgs splitting; <u>for</u> the SU(2) x U(1) breaking (and SU(3) x $U(1)_{e-m}$ <u>non-breaking</u>) at M_W, naturally exquisitely smaller than M_X; <u>for</u> the "constrained" soft SUSY breaking at $m_{3/2}$, hierarchically smaller than M in a natural way; <u>for</u> definite, at present experimentally acceptable depatures from the "standard" low energy phenomenology (like the DEMON, or $\Delta\rho$, with values below, but not far from, their present experimental upper bounds, or the existence of very light (<O(10 GeV)) neutral Higgs bosons), as well as a rather well-defined low energy SUSY spectrum; <u>for</u> observable baryon decay even if $M_X \simeq M$; and for the light fermion masses of the first two generations. Furthermore, supergravity theories may provide, for the first time, a problem-free cosmological scenario, from primordial inflation [79] through GUT phase transitions [45,47] to baryon and nucleosynthesis, ostracizing troublesome particles such as GUT monopoles [45,47], gravitinos [80], Polonyi fields [81] or other SUSY relics [82].

Putting the whole thing together, it becomes apparent that spontaneously broken N = 1 local SUSY gauge theories, with their prosperous and appropriate structure, may well serve as an effective theory describing all physics from $M_{P\ell}$ down to (and including) low energies, with well-defined and rich experimental consequences. What's next then? Well, we really have to understand where this highly successful theory comes from. There are reasons to believe [38] that N = 8 extended SUPERGRAVITY, suitably broken [83] down to N = 1 supergravity, may provide the fundamental theory. But this next move asks for a deep understanding of physics at Planck energies, which is as exciting as it is difficult, taking into account that even QUANTUM MECHANICS may need modification [84], if quantum gravitational effects have to be considered seriously.

ACKNOWLEDGMENTS

I would like to thank Nino Zichichi for organizing once more such a stimulating and exciting Summer School.

REFERENCES

1. For reviews, see:
   D. V. Nanopoulos, Ecole d'Eté de Physique des Particules, Gif-sur-Yvette, (IN2P3, Paris, 1980), p. 1 (1980);
   J. Ellis, in: "Gauge Theories and Experiments at High Energies", K. C. Bowler and D. G. Sutherland, eds., (Scottish Universities Summer School in Physics, Edinburgh 1981), p. 201; and CERN préprint TH.3174 published in the Proceedings of the 1981 Les Houches Summer School (1981);
   P. Langacker, Phys. Rep., 72C:185 (1981), and Proceedings of the 1981 International Symposium on Lepton and Photon Interactions at High Energies, W. Pfeil, ed., Bonn University, p. 823 (1981).
2. For a review, see:
   E. Fahri and L. Susskind, Phys. Rep., 74C:277 (1981).
3. Y. A. Gol'fand and E. P. Likhtman, Pis'ma Zh. Eksp. Teor. Fiz., 13:323 (1971).
   D. Volkov and V. P. Akulov, Phys. Lett., 46B:109 (1973).
   J. Wess and B. Zumino, Nucl. Phys., B70:39 (1974).
4. M. Vetman, Acta Phys. Pol., B12:437 (1981).
   T. Inami, H. Nishino and S. Watamura, Phys. Lett., 117B:197 (1983).

N. G. Deshpande, R. J. Johnson, E. Ma, Phys. Lett., 130B:61 (1981).
5. I. S. Altarev et al., Phys. Lett., 102B:13 (1981);
W. B. Dress et al., Phys. Rev., D15:9 (1977);
N. F. Ramsey, Phys. Rep., 43:409 (1978).
6. J. Ellis, S. Ferrara and D. V. Nanopoulos, Phys. Lett., 114B: 231 (1982).
7. S. Coleman and J. Mandula, Phys. Rev., 159:1251 (1967).
8. For a review, see:
P. Van Nieuwenhuisen, Phys. Rep., 68C:189 (1981).
9. R. Barbieri, S. Ferrara and D. V. Nanopoulos, Zeit. für Phys., C13:267 (1982); Phys. Lett., 116B:16 (1982).
10. J. Ellis, L. Ibañez and G. G. Ross, Phys. Lett., 113B:983 (1982); Nucl. Phys., B221:29 (1983).
11. G. R. Farrar and S. Weinberg, Phys. Rev., D27:2732 (1983).
12. J. Polchinsky and L. Susskind, Phys. Rev., D26:3661 (1982).
13. R. Barbieri, S. Ferrara, D. V. Nanopoulos and K. Stelle, Phys. Lett., 113B:219 (1982).
14. D. V. Volkov and V. A. Soroka, JETP Lett., 18:312 (1973).
15. S. Deser and B. Zumino, Phys. Rev. Lett., 38:1433 (1977).
16. E. Cremmer et al., Phys. Lett., 79B:931 (1978); Nucl. Phys., B147:1051 (1979); Phys. Lett., 116B:231 (1982); Nucl. Phys., B212:413 (1983).
17. J. Ellis and D. V. Nanopoulos, Phys. Lett., 116B:133 (1982).
18. L. Girardello and M. T. Grisaru, Nucl. Phys., B194:65 (1982).
19. For recent reviews, see:
R. Barbieri and S. Ferrara, CERN preprint TH-3547 (1983);
J. Polchinski, Harvard preprint HUTP-83/A036 (1983);
J. Ellis, CERN preprint TH-3718 (1983).
20. S. Mandelstam, Phys. Lett., 121B:30 (1983); Nucl. Phys., B213: 149 (1983); P. S. Howe, K. S. Stelle and P. K. Townsend, Imperial College preprint ICTP/82-83:20.
21. J. Wess and B. Zumino, Phys. Lett., 49B:52 (1974);
J. Iliopoulos and B. Zumino, Nucl. Phys., B76:310 (1974);
S. Ferrara, J. Iliopoulos and B. Zumino, Nucl. Phys., B77:413 (1974);
S. Ferrara and O. Piquet, Nucl. Phys., B93:261 (1975);
M. T. Grizaru, W. Siegel and M. Rocek, Nucl. Phys., B159:420 (1979).
22. L. Maiani, Proceedings of the Summer School of Gif-sur-Yvette, p. 3 (1979);
E. Witten, Nucl. Phys., B185:153 (1981);
R. K. Kaul, Phys. Lett., 109B:19 (1982).
23. Proceedings of the "Supersymmetry versus Experiment", workshop edited by D. V. Nanopoulos, A. Savoy-Navarro and Ch. Tao, CERN preprint TH-3311/EP-82:63 (1982), to appear as a Physics Report.
24. J. Ellis and D. V. Nanopoulos, Phys. Lett., 110B:44 (1982).
25. R. Barbieri and R. Gatto, Phys. Lett., 110B:211 (1982).

26. J. A. Grifols and A. Méndez, Phys. Rev., D26:1809 (1982);
J. Ellis, J. Hagelin and D. V. Nanopoulos, Phys. Lett., 116B: 283 (1982);
R. Barbieri and L. Maiani, Phys. Lett., 117B:203 (1982).
27. R. M. Bionta et al., Phys. Rev. Lett., 51:97 (1983).
28. H. Georgi and S. L. Glashow, Phys. Rev. Lett., 32:438 (1974).
29. S. Dimopoulos and H. Georgi, Nucl. Phys., B193:150 (1981).
N. Sakai, Zeit. für Phys., C11:153 (1981).
30. M. B. Einhorn and D. R. T. Jones, Nucl. Phys., B196:475 (1982).
31. J. Ellis, D. V. Nanopoulos and S. Rudaz, Nucl. Phys., B202:43 (1982).
32. S. Weinberg, Phys. Rev., D26:187 (1982);
N. Sakai and T. Yanagida, Nucl. Phys., B197:533 (1982).
33. D. V. Nanopoulos and K. Tamvakis, Phys. Lett., 110B:449 (1982);
M. Srednicki, Nucl. Phys., B202:327 (1982); ibid., B206:139 (1982);
D. V. Nanopoulos, K. A. Olive and K. Tamvakis, Phys. Lett., 115B:15 (1982).
34. D. V. Nanopoulos and K. Tamvakis, in Ref. 33; Phys. Lett., 113B: 151 (1982); Phys. Lett., 114B:235 (1982).
35. J. Ellis, M. K. Gaillard and D. V. Nanopoulos, Phys. Lett., 80B:360 (1979).
36. J. Ellis, D. V. Nanopoulos and K. Tamvakis, Phys. Lett., 121B: 123 (1983).
37. B. Zumino, Phys. Lett., 87B:203 (1979).
38. E. Cremmer and B. Julia, Nucl. Phys., B159:141 (1979);
J. Ellis, M. K. Gaillard and B. Zumino, Phys. Lett., 94B:343 (1980).
39. S. Weinberg, Phys. Rev., 166:1568 (1968);
S. Coleman, J. Wess and B. Zumino, Phys. Rev., 177:2239 (1968);
C. Callan, S. Coleman, J. Wess and B. Zumino, Phys. Rev., 177: 2247 (1968).
40. J. Polonyi, Budapest preprint KFKI-1977:93 (1977).
41. R. Barbieri, S. Ferrara and C. A. Savoy, Phys. Lett., 119B:343 (1982).
42. H. P. Nilles, M. Srednicki and D. Wyler, Phys. Lett., 120B:346 (1982).
43. L. Hall, J. Lykken and S. Weinberg, Phys. Rev., D27:2359 (1983).
44. E. Cremmer, P. Fayet and L. Girardello, Phys. Lett., 122B:41 (1983).
45. D. V. Nanopoulos, K. A. Olive, M. Srednicki and K. Tamvakis, Phys. Lett., 124B:171 (1983).
46. C. Kounnas, D. V. Nanopoulos, M. Srednicki and M. Quiros, Phys. Lett., 127B:82 (1983).
47. C. Kounnas, J. Leon and M. Quiros, Phys. Lett., 129B:67 (1983);
C. Kounnas, D. V. Nanopoulos and M. Quiros, Phys. Lett., 129B: 223 (1983).
48. D. V. Nanopoulos and D. A. Ross, Phys. Lett., 118B:99 (1982).
49. C. Kounnas, A. B. Lahanas, D. V. Nanopoulos and M. Quiros, Phys.Lett., 132B:95 (1983) and CERN prepriut TH-3657 (1983).

50. A. Masiero, D. V. Nanopoulos, K. Tamvakis and T. Yanagida, Phys. Lett., 115B:380 (1982);
B. Grinstein, Nucl. Phys., B206:387 (1982).
51. J. Ellis, J. Hagelin, D. V. Nanopoulos and K. Tamvakis, Phys. Lett., 125B:275 (1983).
52. L. Alvarez-Gaumé, J. Polchinski and M. Wise, Nucl. Phys., B221: 495 (1983).
53. L. Ibañez and C. Lopez, Phys. Lett., 126B:54 (1983).
54. A. H. Chamseddine, R. Arnowitt and P. Nath, Phys. Rev. Lett., 49:970 (1982).
55. S. Ferrara, D. V. Nanopoulos and C. A. Savoy, Phys. Lett., 123B:214 (1983).
56. H. P. Nilles, M. Srednicki and D. Wyler, Phys. Lett., 124B:337 (1983);
A. B. Lahanas, Phys. Lett., 124B:341 (1983).
57. K. Inoue, A. Kakuto, H. Komatsu and S. Takeshita, Prog. Th. Phys., 68:927 (1982).
58. M. Claudson, L. J. Hall and I. Hinchliffe, LBL preprint 15948 (1983).
59. K. A. Olive, D. N. Schramm, G. Steigman and J. Yang, Ap. J., 246:557 (1981).
60. S. Coleman and E. Weinberg, Phys. Rev., D7:1888 (1973).
61. J. Ellis, J. Hagelin, D. V. Nanopoulos and K. Tamvakis, to be published.
62. E. Cremmer, S. Ferrara, C. Kounnas and D. V. Nanopoulos, Phys.Lett., 133B:61 (1983).
63. R. Barbieri and L. Maiani, Rome preprint (1983);
C. S. Lim, T. Inami and N. Sakai, INS Rep., 480 (1983).
64. A. B. Lahanas and D. V. Nanopoulos, Phys. Lett., 129B:461 (1983).
65. J. F. Donoghue, H. P. Nilles and D. Wyler, Phys. Lett., 128B: 55 (1983).
66. A. J. Buras, Phys. Rev. Lett., 46:1354 (1981).
67. T. Inami and C. S. Lim, Nucl. Phys., B207:533 (1982).
68. D. A. Kosower, L. M. Krauss and N. Sakai, Harvard preprint HUTP-83:A056 (1983).
69. D. V. Nanopoulos and M. Srednicki, Phys. Lett., 128B:61 (1983).
70. J. Ellis and M. K. Gaillard, Nucl. Phys., B150:141 (1979).
71. M. T. Grisaru, M. Roček and A. Karlhede, Phys. Lett., 120B:110 (1983).
72. T. D. Lee, Phys. Rep., 9:143 (1973).
73. F. del Aguila, J. A. Grifols, A. Mendez, D. V. Nanopoulos and M. Srednicki, Phys. Lett., 129B:77 (1983).
74. W. Büchmuller and D. Wyler, Phys. Lett., 121B:321 (1983);
F. del Aguila, M. B. Gavela, J. A. Grifols and A. Mendez, Phys. Lett., 126B:71 (1983);
J. Polchinski and M. Wise, Phys. Lett., 125B:393 (1983).
75. For a recent review, see:
S. Pakvasa, J. Ellis and D. V. Nanopoulos, CERN preprint TH-3464 (1983).
76. J. Ellis, J. Hagelin, D. V. Nanopoulos and K. Tamvakis, Phys. Lett., 124B:484 (1983).

77. D. V. Nanopoulos and M. Srednicki, Phys. Lett., 124B:37 (1983).
78. L. Ibañez, Phys. Lett., 117B:403 (1982);
A. Masiero, D. V. Nanopoulos and K. Tamvakis, Phys. Lett., 126B:337 (1983).
79. J. Ellis, D. V. Nanopoulos, K. A. Olive and K. Tamvakis, Nucl. Phys., B221:524 (1983);
D. V. Nanopoulos, K. A. Olive, M. Srednicki and K. Tamvakis, Phys. Lett., 123B:41 (1983);
D. V. Nanopoulos, K. A. Olive and M. Srednicki, Phys. Lett., 127B:30 (1983);
G. Gelmini, D. V. Nanopoulos and K. A. Olive, Phys.Lett., 131B:53 (1983).
G. Gelmini, C. Kounnas and D. V. Nanopoulos, CERN prepriut TH-3777 (1983).
80. J. Ellis, A. D. Linde and D. V. Nanopoulos, Phys. Lett., 118B: 59 (1982).
81. D. V. Nanopoulos and M. Srednicki, Phys.Lett., 133B:287 (1983).
82. J. Ellis, J. Hagelin, D. V. Nanopoulos, K. A. Olive and M. Srednicki, SLAC-PUB-3171 (1983).
83. R. Barbieri, S. Ferrara and D. V. Nanopoulos, Phys. Lett., 107B:275 (1981);
J. Ellis, M. K. Gaillard and B. Zumino, Acta Phys. Pol., B13: 253 (1982).
84. J. Ellis, J. Hagelin, D. V. Nanopoulos and M. Srednicki, CERN preprint TH-3619/SLAC-PUB-3134 (1983).

DISCUSSIONS

CHAIRMAN: D.V. NANOPOULOS

Scientific Secretaries: A. Gocksch, V. Kaplunovsky, N. Turok

DISCUSSION I

- *KAPLUNOVSKY:*

Suppose, as you have said, gluini are not very heavy. What would be the experimental signal for them?

- *NANOPOULOS:*

Gluini would manifest themselves as R-hadrons.

- *KAPLUNOVSKY:*

What are R-hadrons?

- *NANOPOULOS:*

Ordinary hadrons are $\bar{q}q$ mesons and qqq baryons. Of course, one may combine a $\bar{q}q$ pair into a colour octet instead of a colour singlet, supplement them with a gluon and obtain a colour singlet meson of type $\bar{q}qg$. If one has a gluino $\tilde{g}$ which is also in a colour octet it could do the same. So one would obtain hadrons of type $\bar{q}q\tilde{g}$ and $qqq\tilde{g}$. However they are not ordinary mesons and baryons since they obey the opposite statistics.

These hadrons are called R-hadrons since they have negative R-parity. What we call (after Fayet) R-parity is basically

$$R = (-)^F \cdot (-)^{3B + L}$$

All ordinary particles (quarks, leptons, gauge and Higgs bosons, etc.) are R-positive while their supersymmetry partners are R-negative. But it is not the nomenclature that matters here!

These R-hadrons behave like ordinary ones. You may produce them in a beam dump experiment $p + N \to R + \bar{R} + X$ just like you produce e.g. charmed particles.

You may also look for R-hadrons with similar methods. If you do not see anything you obtain an upper bound on their production cross-section and thus you can find a lower limit on their masses.

This limit on R-hadron masses you translate into a limit on the gluino mass. One should stress that this naive procedure which takes the gluino as one of the constituents may be wrong! It may be that the masses of R-hadrons bear no relation to the gluino mass which may even be zero. But basically what most people do is to take the gluino as a constituent.

- *KAPLUNOVSKY:*

What would be the experimental signal for such a creature? Should it be stable?

- *NANOPOULOS:*

It is not clear. If they are stable then one has limits from FNAL etc. But they may also decay. For example, the gluino may recombine with one of the quarks to form a squark which in turn decays into a quark and goldstino or photino.

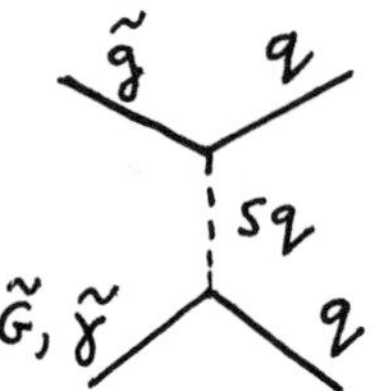

In effect one has R-hadron decays into photinos and ordinary hadrons e.g. pions. As for the experimental signal for such a process you may look for short tracks or for rescattering of the photino which may produce some funny events, or for missing energy in the final state etc. So one would be able to find such R-hadrons when they become accessible.

- *GOITI:*

Do you expect some -inos to be stable.

- *NANOPOULOS:*

The answer is model-dependent. In many models one expects the photino to be stable. Indeed the lightest R-odd particle should be stable in most of the models. In some models the photino has very low mass so it would be stable but would not create any cosmological problems.

- *VAN BAAL:*

Isn't there a discrepancy between your M_{SUSY}, around M_W, and the supersymmetry breaking scale Prof. Ferrara used $(10^{10} - 10^{11}$ GeV)?

- *NANOPOULOS:*

No. You see there are two worlds. For the low energy world mass relations like $M_{sq}^2 - M_q^2 \sim M_W^2$ hold. However, in the superheavy world one can have $M_{sh}^2 - M_{ssh}^2$ equals whatever one chooses for this, the primordial scale. (sh = superheavy, ssh = supersymmetric partner of sh.)

- *LAU:*

What is the mass limit on the supersymmetric partner of the electron obtained from measurement of g - 2?

- *NANOPOULOS:*

For the supersymmetric partner of the muon, the smuon, you get a limit of 15 GeV from g - 2. However for the selectron there are suppression factors and the limit is not so good. But in this case there is a limit from the PETRA experiments of 16-17 GeV or so.

- *LYKKEN:*

You mentioned that, supersymmetric theories which do not employ U(1) type breaking have a hard time satisfying the condition coming from the absence of FCNC on the squark $(M_{SC} - M_{SU})$ mass differences. Would you therefore consider SUSY schemes other than the U(1) scheme to be excluded?

- *NANOPOULOS:*

No, the F-type supersymmetry breaking is also fine. It is only the soft explicit breaking of supersymmetry that presents problems in obtaining these mass differences - you really have to stand on your nose.

- *CASTELLANI:*

Supersymmetry solves the hierarchy problem in so far as it prevents the masses of the scalars being lifted to the GUT scale. However, can supersymmetric models explain the ratio $M_{GUT}/M_W \sim 10^{13}$?

- *NANOPOULOS:*

No, there are models in which this ratio can be obtained but we certainly do not have as straightforward an understanding as for the scalar mass problem.

- *ALBRECHT:*

One of the problems of particle physics has been to explain charge quantization. GUT's offer an answer to this question, but by introducing a new U(1) to break the supersymmetry aren't you introducing the same problem all over again?

- *NANOPOULOS:*

Your question touches a sensitive point so let me answer in some detail. It seems virtually impossible to create a $SU(3) \times SU(2) \times U(1) \times \widetilde{U(1)}$ model with an anomaly-free, traceless $\widetilde{U(1)}$ which cannot reach a supersymmetric minimum, if the $\widetilde{U(1)}$ is broken at the scale M_W. Barbieri, Ferrara and myself have constructed a realistic model in which the $\widetilde{U(1)}$ is broken at $M_p \sim 10^{19}$ GeV, which has anomalies. Of course these are irrelevant - who cares about anomalies at 10^{19} GeV! Furthermore, when we constructed a GUT theory, the easiest way to implement the same mechanism was to use say $SU(5) \times \widetilde{U(1)}$. In such models charge quantisation and the successful neutral current phenomenology are automatic i.e. there is no mixing between $SU(2) \times U(1)$ and $\widetilde{U(1)}$.

- *TUROK:*

How does supersymmetry help solve the 'monopole problem'?

- *NANOPOULOS:*

In a large class of supersymmetric GUT models there is a natural delay of the phase transition, $G \to SU(3) \times SU(2) \times U(1)$ where G is the GUT group. This was shown by K. Tamvakis and myself, and independently by M. Srednicki. The phase transition occurs around $T_c \sim 10^{10}$GeV and thus using the simple formula

$$\frac{n_{monopole}}{n_\gamma} \sim \left(\frac{T_c}{m_{pl}}\right)^3$$

one finds $n_{monopole}/n_\gamma < 10^{-27}$ which is experimentally acceptable.

DISCUSSION II (Scientific Secretaries: A Gocksch, V. Kaplunovsky, N. Turok).

- *KREMER:*

As I understand from Professor Ferrara's discussion session, as soon as supersymmetry is spontaneously broken, the ultraviolet divergences no longer cancel. What does this imply for your explanation of CP violation?

- *NANOPOULOS:*

The ultraviolet divergences do still cancel but one obtains a small finite part proportional to the scale of supersymmetry breaking.

- *RAUH:*

I did not understand your remarks on the inflationary universe. It seems to involve a faster expansion rate than the standard model. How is this state entered and which problems does it solve?

- *NANOPOULOS:*

The expansion rate of the universe is given by the Hubble constant $H = \dot{R}/R \sim \frac{\sqrt{\rho}}{m_{pl}}$. In a radiation dominated universe $\rho \sim T^4$ and thus $H \sim T^2/m_{pl} \sim 1/t$. In inflationary models the density contains a constant term from the vacuum energy which, when it is dominant yields $H \sim$ const. and thus R increases exponentially. Inflation occurs during a first order phase transition when the universe is trapped in a metastable state. This scenario solves many cosmological problems - the horizon problem, the homogeneity problems etc. and has been extensively discussed by Lindé, Albrecht and Steinhardt and others.

- *LYKKEN:*

I did not understand why in some supersymmetric models, protons decay preferentially via kaon modes.

- *NANOPOULOS:*

The coupling of the superheavy Higgs boson to the decay product quark in proton decay

(see figure 1 in the text)

is proportional to the quark mass, so the decay tends to go into the heaviest quark around, energetically permissible.

- *CASTELLANI:*

What happens to the neutrino mass in SO(10) or E_6 when you introduce supersymmetry?

- *NANOPOULOS:*

It depends on how you break the supersymmetry. For instance if you use a Fayet-Illiopoulos U(1), down at low energies, then you have to be careful with the quantum numbers of the neutrino. If the Majorana mass term violates the U(1) it is not allowed. However you don't necessarily have to use a U(1) or to break U(1) at low energies.

- *ITOYAMA:*

I have a question about the electric dipole moment of the neutron in broken SUSY GUT's. In addition to the θ term there is another contribution coming from CP violation in the weak interactions. Which effect is dominant?

- *NANOPOULOS:*

If you do an honest to god calculation of the dipole moment of the neutron from weak CP violating effects you obtain 10^{-30} e cm or so. If θ is not much less than 10^{-9} as indicated in my lecture ($d \sim \theta \cdot 10^{-16}$e cm) the contribution from SUSY breaking is 10^{-26}e cm or so. Thus the SUSY effect is dominant.

TESTING QCD IN HADRONIC PROCESSES

P.V. Landshoff

DAMTP, University of Cambridge
Cambridge, England

1. INTRODUCTION

The widespread belief among particle physicists that QCD is the correct theory of the strong interactions is supported by a wide range of tests. Because of calculational difficulties, these tests are qualitative rather than quantitative, but they are impressive for their variety[1]. For both experimental and theoretical reasons, most of them concern processes that involve leptons in either the initial or the final state, but my brief at this School is to discuss purely hadronic reactions.

Calculations of scattering processes rely on perturbation theory. The coupling α_s of QCD becomes small at short distance, which is necessary in order that a perturbation expansion may have a chance of being valid. In the vast majority of hadron-hadron collisions, there is no short-distance interaction and so there is no reason to believe that a perturbation-theory calculation is applicable. The problem, then, is to pick out the small fraction of events in which short-distance effects are dominant. We have no precise way of doing this, but the folklore is that we recognise such events by looking for an untypically large momentum transfer t. It is certainly plausible that large t is generated only in those rare events where the two initial hadrons collide head-on, rather than peripherally as in the more typical collisions.

There are two kinds of large-t event, exclusive reactions such as wide-angle elastic scattering, and inclusive reactions in which one or more particles carry off large transverse momentum. I shall discuss both of these in my two lectures.

Even if we can correctly identify the collisions in which the short-distance force dominates, there are severe calculational difficulties, which are encountered in all applications of QCD[1]. At least at present, and probably for the forseeable future, precise calculations are possible only up to the leading power of the large variable t. It is far from sure that non-leading powers of t, called higher twists in the jargon, are numerically insignificant for those values of t that are experimentally accessible. But if one simply assumes that all is well and that the leading-power (leading twist) perturbation expansion is good enough, there is the further problem that calculations usually find that it is very slow to converge. Much has been written about this problem[1], which is associated with the renormalisation-scheme dependence of the expansion, and I do not have time to discuss it here.

The conclusion must be that if experimental data seem to agree with predictions based on simple applications of perturbative QCD, we should be pleased, but if things are more complicated we should not be too surprised.

2. ELASTIC SCATTERING AT LARGE t

Elastic scattering at small t is characterised by a very sharp peak in the forward direction. At sufficiently large energy, a dip is found, around $t = -1.4\ \text{GeV}^2$ in the case of pp

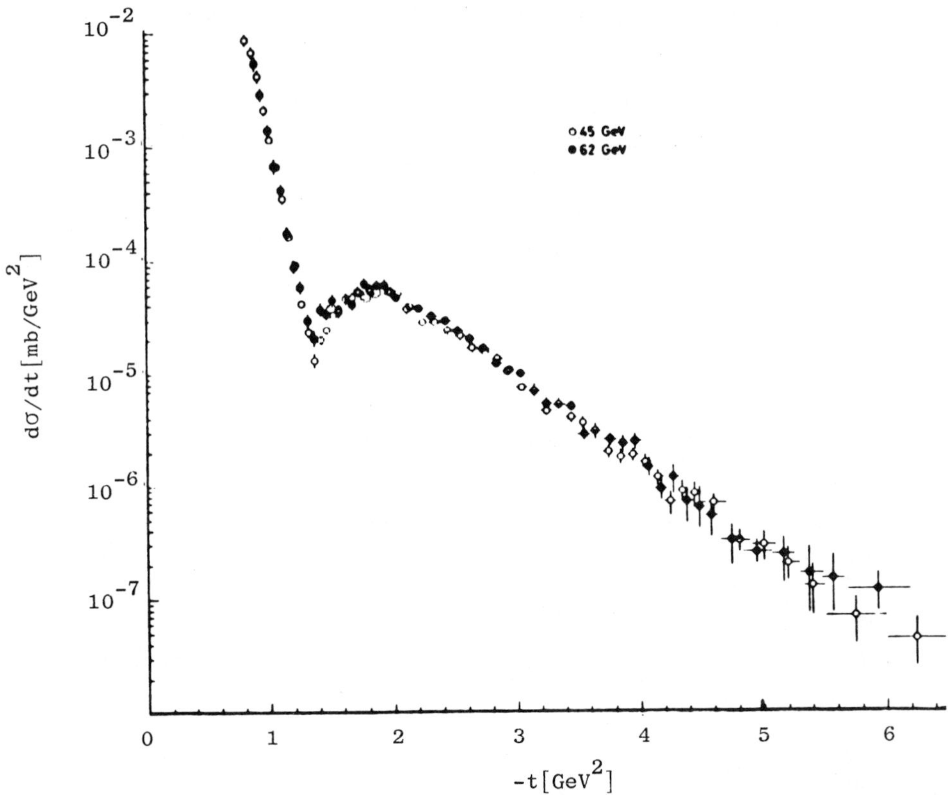

Fig. 1. pp elastic scattering data (CERN-Hamburg-Heidelberg-Annecy-Vienna).

scattering (figure 1) and around -4 GeV^2 for πp scattering. The differential cross section out to values of t beyond the dip may be understood either from a geometrical picture or in terms of exchange of Regge poles and Regge cuts[2]. The only thing I need say about this is that perturbative QCD is almost certainly not relevant. Beyond the dip, dσ/dt falls much less slowly with increasing $|t|$, and it is possible that in this region the data may be described by perturbative QCD.

Although there are no data for it, think first about high-energy ππ elastic scattering at large t. An obvious diagram to

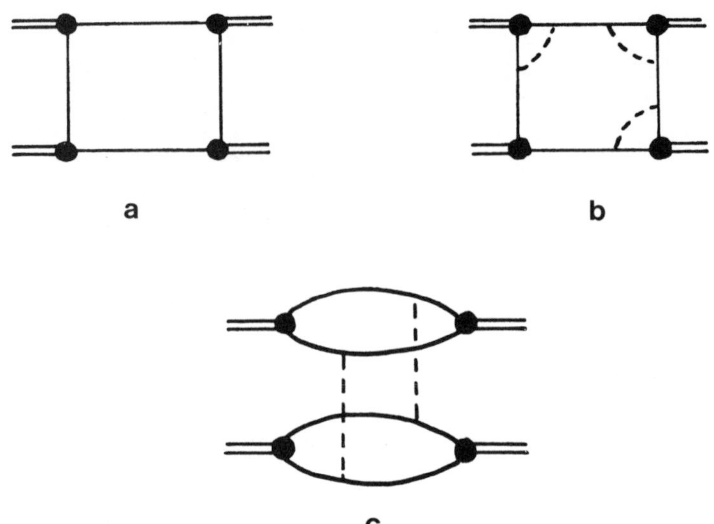

Fig. 2. Diagrams for ππ elastic scattering at large t; the blobs are pion vertex functions. Diagram (a) would need the tails of the vertex functions. (b) is an example of a CIM diagram and (c) is the double-scattering diagram.

try is that of figure 2a, where the blobs represent the vertex function that couples the pion to its constituent quark and antiquark. Calculation finds that when s and t are both large, for at least three of the four vertex functions it is their tails that matter, where the attached quarks are far off shell. It is now widely accepted that the tail of the vertex function is determined by gluon exchange between the quark legs, so that one arrives, for example, at figure 2b.

This is an example of a class of diagrams known as <u>constituent interchange</u> (CIM) diagrams[3]. These diagrams satisfy a rule that was first formulated independently of QCD, the <u>dimensional counting rule</u>[4], which says that for the large-t scattering AB → CD,

$$\frac{d\sigma}{dt} \sim t^{-m} F(\theta_{CM}) \tag{2.1}$$

with

$$m = n_A + n_B + n_C + n_D \tag{2.2}$$

where n_H is the number of valence quarks in hadron H. For $\pi\pi \to \pi\pi$ this gives $m = 6$.

However, there is a rival mechanism[5], which violates the dimensional counting rule. This is the <u>double-scattering</u> diagram of figure 3c; here none of the quarks is far off shell and only the two exchanged gluons carry large momentum transfer. This diagram gives $m = 5$.

For pp elastic scattering, there are several thousand diagrams of the CIM type. They all correspond to $m = 10$, with $F(\theta)$ a function that is too complicated to calculate. For triple scattering, which is the pp analogue of figure 1c, $m = 8$. Furthermore, when t is large but much less than s, it turns out that triple scattering yields an $F(\theta)$ that is simply a constant, so that

$$\frac{d\sigma}{dt} \sim t^{-8} \tag{2.3}$$

independent of energy.

It is really only pp that so far has extensive high energy data at large t. The data seem already to be interesting in the PS energy range. When s and t are both large, $t/s \sim -\sin^2\frac{1}{2}\theta$, so that (2.1) is equivalent to

$$\frac{d\sigma}{dt} \sim s^{-m}\, G(\theta) \tag{2.4}$$

with $G(\theta) = F(\theta)(-\sin^2\frac{1}{2}\theta)^{-m}$. Figure 3 shows data[6] at various fixed values of θ; it is seen that (2.4) is well satisfied with the CIM value $m = 10$. It is found that at each angle θ the lowest value of $|t|$ for which the points fall on the straight

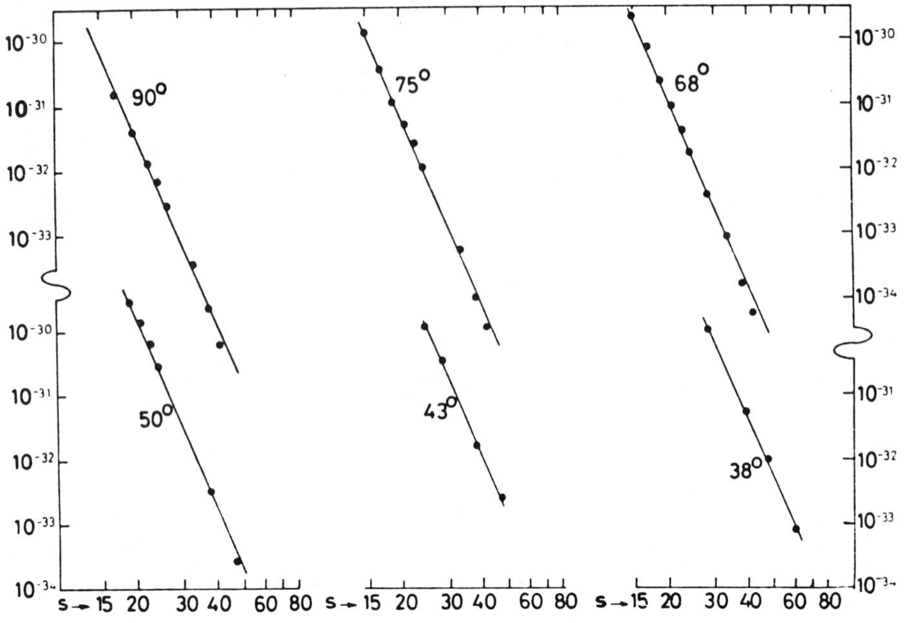

Fig. 3. pp elastic scattering data at PS energies. The straight lines have slope 9.7.

lines is about 2.5 $(\text{GeV})^2$.

It is found also that the function $G(\theta)$ is such that, if instead one plots the data at fixed t they fall very steeply with increasing s in the PS energy range. However, at higher energies this decrease disappears, until at ISR energies the large-t data appear to be energy-independent (though it is possible to believe[7] that there is still a very slow decrease if one has reason to want it). See figure 4. At energies high enough for the energy dependence to have disappeared, the data fit very well[8] to t^{-8}; this is shown in figure 5.

It is tempting, then, to suppose that at PS energies the CIM mechanism dominates because its coefficient function $G(\theta)$ is larger than the corresponding triple-scattering coefficient function. However, because m is smaller for triple scattering, at any fixed θ its fall-off with energy is slower, and that is

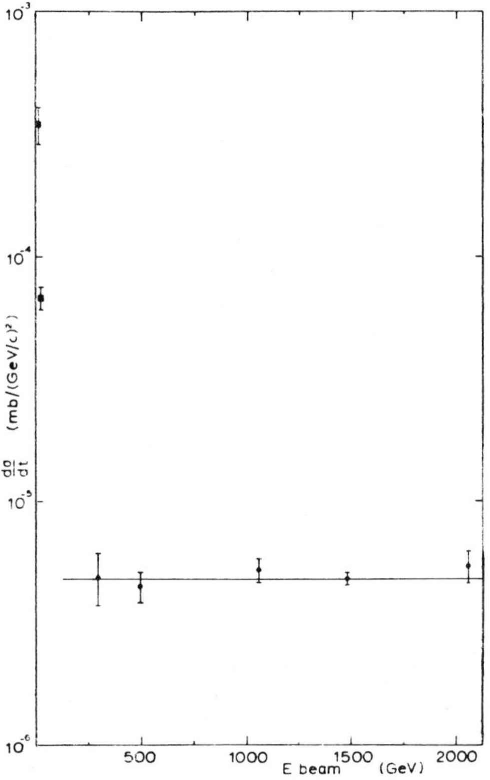

Fig. 4. Energy dependence of pp elastic scattering at $t = -3.35\ (\mathrm{GeV})^2$.

why triple scattering appears to be dominant at ISR energies.

This simple picture receives some support from $\bar{p}p$ elastic scattering data. Estimates of the CIM coefficient function $\bar{G}(\theta)$ for $\bar{p}p$ find that $\bar{G}(\theta) \ll G(\theta)$. On the other hand, the triple-scattering mechanism gives equal, energy-independent contributions to $\bar{p}p$ and pp (for $|t| \ll s$). So one might expect that $\bar{p}p$ might become dominated by triple scattering at a lower energy than pp, and that at this energy its large t differential cross section is already equal to that of pp at higher energy. This seems to be the case, as is seen in figure 6.

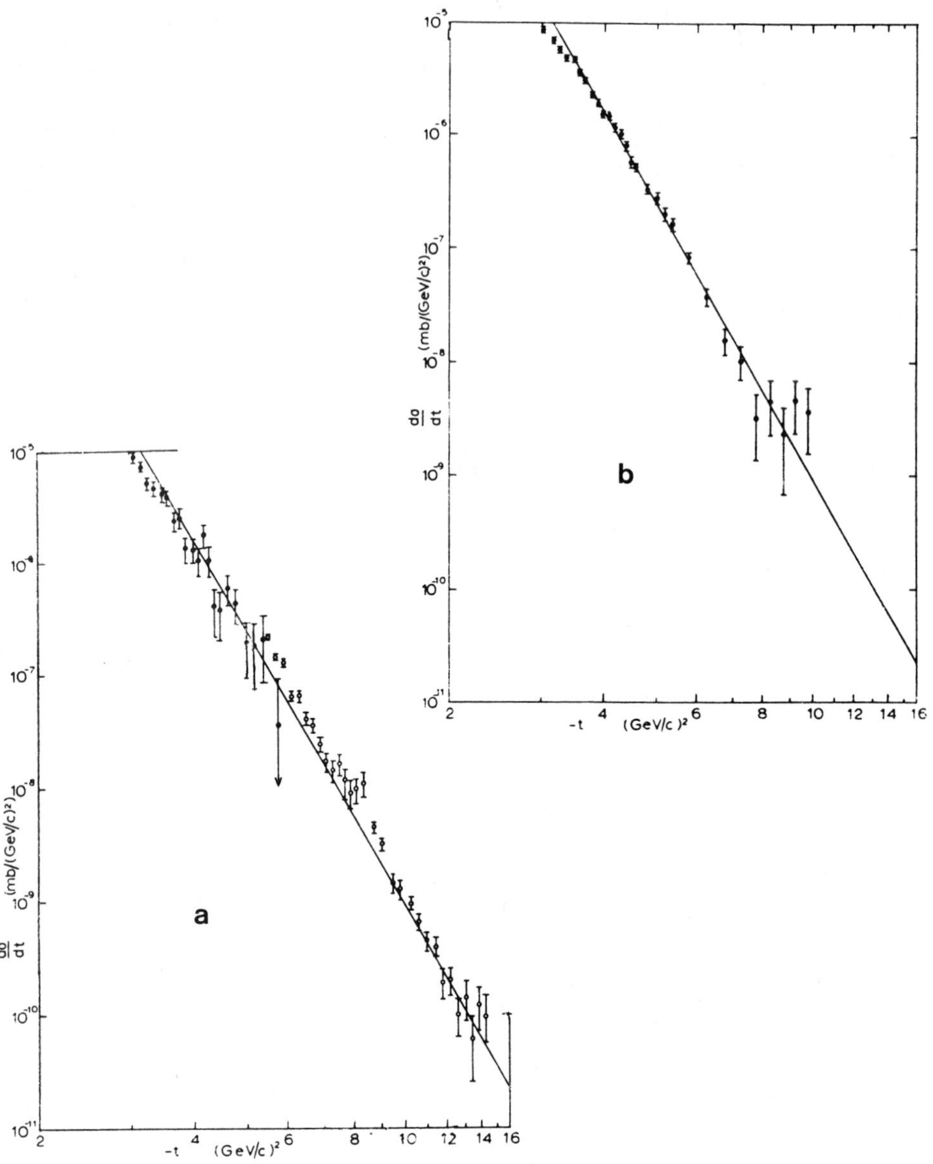

Fig. 5. t dependence of data, (a) from FNAL (400 GeV/c) and the ISR (494 GeV/c) and (b) from the ISR (1482 GeV/c). In each case the line is $0.09\,t^{-8}$.

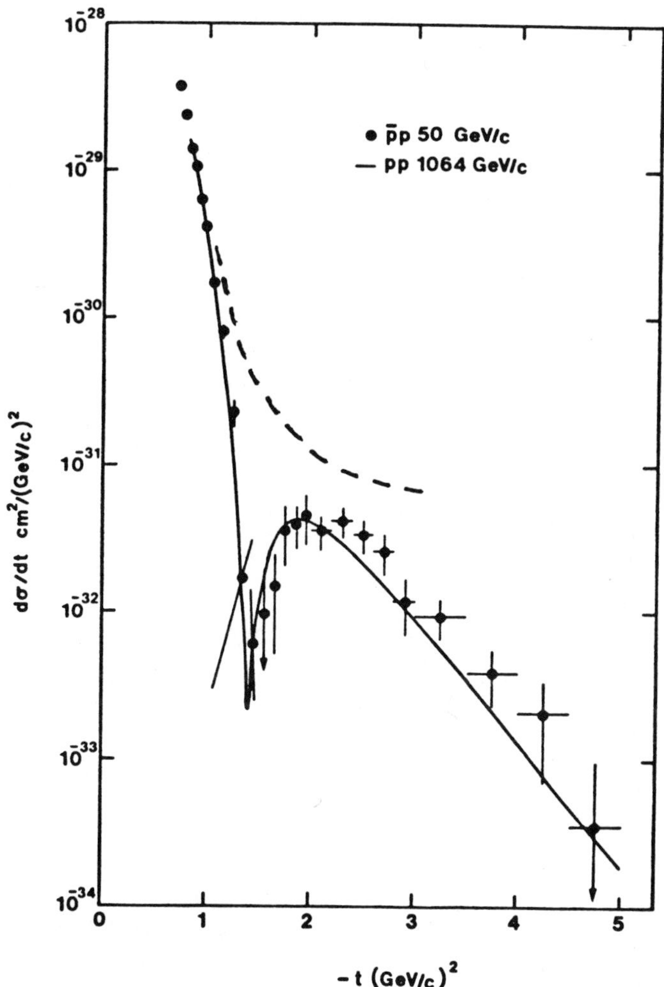

Fig. 6. $\bar{p}p$ elastic scattering at 50 GeV/c (Annecy-CERN-Copenhagen-Genova-Oslo-UCL). The upper curve corresponds to pp at the same energy, and the lower curve to pp at ISR energies.

Difficulties with QCD Models

Attractive though the simple picture is, it has a number of problems. Firstly, Collins and Gault[7] argue that at ISR energies the Regge-pole-plus-Regge-cut picture, which presumably applies for smaller values of t and in particular gives the dip, also describes the data out to values of t well beyond the dip. This may be true, but that picture does have a large number of parameters that need adjustment, while the triple-scattering model has only one, the constant that multiplies t^{-8}. (Collins and Gault go further, and argue that triple scattering is actually excluded by the data, but I do not agree that it is necessarily possible to draw such a conclusion[9].)

The other problems concern the effects of inserting additional gluons in the CIM and multiple scattering diagrams. For pp scattering, the lowest-order CIM diagram gives a contribution to $d\sigma/dt$ that is proportional to g^{20}, where g is the QCD coupling. For triple scattering, the lowest order contribution is proportional to g^{12}. One might guess that putting in extra gluons changes this fixed coupling to a running coupling giving respectively $[\alpha_s(Q^2)]^{10}$ and $[\alpha_s(Q^2)]^6$. Without an explicit calculation, which has not been done, one cannot be sure whether in fact this is correct, nor can one know what is the appropriate argument Q^2 for the coupling. However, there is a serious danger that such high powers of α_s will appreciably change the $s^{-10}G(\theta)$ behaviour of CIM, or the t^{-8} behaviour of triple scattering, so removing the good agreement with experiment. One possible way out is to suppose that the effective value of the QCD scale Λ is rather small, at most a few tens of MeV, so that α_s is very nearly constant.

Another problem associated with the insertion of additional gluons concerns so-called Sudakov factors. In a diagram where two near-shell quarks are in short-distance interaction, the insertion of an additional gluon connecting the two quarks introduces[10] a large $(\log)^2$ factor. In a CIM diagram such as figure 2b, such an additional gluon may be inserted in several ways, connecting different pairs of quark lines, and it is found that the corresponding $(\log)^2$ factors cancel. However, a similar cancellation does not occur in the case of the triple scattering diagram. There, it is found that the net effect of adding to the original diagram all the various one-gluon insertions is to multiply (2.3) by the Sudakov factor[11]

$$1 - b_1 \log s \log t + b_2 (\log t)^2 \tag{2.5}$$

where

$$b_1 = 18 \frac{g^2}{8\pi^2}, \quad b_2 = 10 \frac{g^2}{8\pi^2} \tag{2.6}$$

Adding in further gluons in the same diagram gives higher powers of $(\log)^2$. It is believed that if one sums all these higher powers, the result is to convert (2.5) into an exponential:

$$\exp[-b_1 \log s \log t \ + \ b_2 (\log t)^2] \tag{2.7}$$

This exponentiation has been checked[11] up to the $(\log)^4$ term in the series expansion, that is up to two-gluon insertions in the original diagram.

The appearance of such an exponential factor would completely remove the agreement between (2.3) and experiment. However, the result (2.7) assumes that it is valid to calculate each term correct only to leading power of $(\log)^2$ and then to sum these leading terms. Such a procedure is valid in some cases but not, as here, when the resulting sum turns out to be very small. It can well be that the non-leading logarithmic powers sum to something larger than (2.7). Mueller[12] concludes that indeed this happens, with the result that the corrected triple-scattering diagrams sum to a form that closely approximates to (2.4), with $m \approx 9.9$. That is, asymptotically, CIM and triple scattering correspond to almost the same inverse power of s, with triple scattering dominating (just!).

My guess is that none of this analysis is relevant to

existing experimental data, for at least two reasons. First, as I understand it Mueller's analysis assumes that t and s are of the same order of magnitude, while the FNAL/ISR data correspond to $|t|$ large but $\ll$ s. More important, all the estimates of the Sudakov and related effects are asymptotic estimates, valid only for very large s and t. Involving, as they do, contributions from very large numbers of gluons, these asymptotic forms are very probably not valid for presently accessible values of s and t.

It has been pointed out recently[13] that the Sudakov-related effects change also the phase of the triple-scattering amplitude. Pire and Ralston argue that this may explain the oscillation about the simple form $s^{-10}G(\theta)$ that may just be seen in the plot of the PS-energy-range data of figure 3. However, they do note that this explanation should be treated with caution, for the reason that I have given in the second part of the previous paragraph.

Conclusions

The conclusion, then, about elastic scattering is that lowest-order QCD seems to give a very good description of the experimental data. However, higher-order terms in the perturbation expansion may spoil the agreement with experiment. More theoretical work is needed on this, and it would also be very useful to have as much data as possible for a variety of elastic processes at high energy and high momentum transfer. For example, there is a prediction[8] that for πp elastic scattering at high enough energy and beyond any dips etc,

$$\frac{d\sigma}{dt} \sim t^{-7} \qquad (2.8)$$

The energy at which this behaviour might set in cannot be predicted, but some data at 50 GeV/c are not unencouraging: see figure 7.

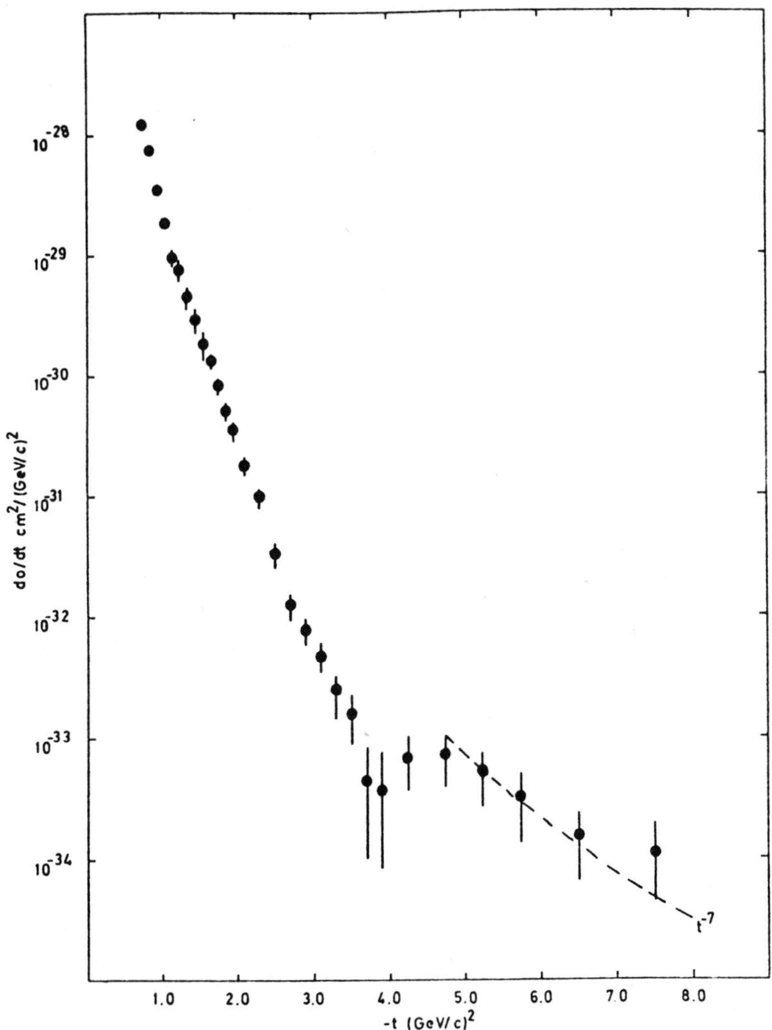

Fig. 7. π^+p elastic scattering data at 50 GeV/c (Annecy-CERN-Copenhagen-Oslo-UCL).

3. INCLUSIVE PROCESSES

I explained in §1 that it is interesting to trigger the detection apparatus with a large-transverse-momentum particle, in the hope that this will select events in which a short-distance interaction has occurred. Inclusive events of this type are commonly supposed to be described by a hard-scattering

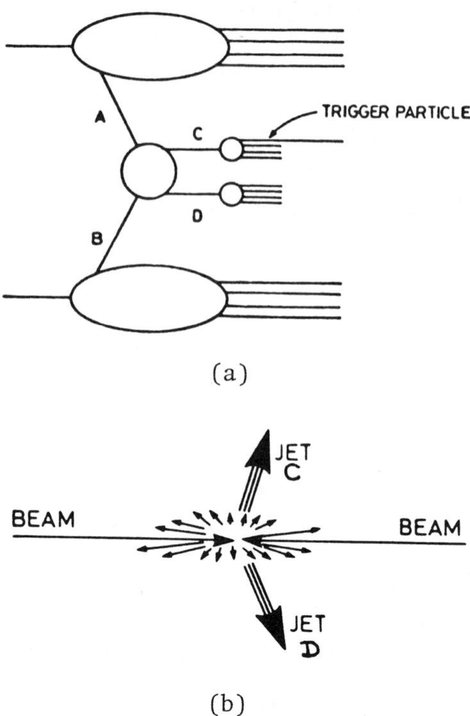

Fig. 8. (a) Hard-scattering model for large p_T inclusive processes, which results (b) in two transverse jets C and D, in addition to a pair of longitudinal jets composed of the residual beam fragments.

mechanism, figure 8a. Here, some constituent A of one of the initial hadrons scatters at wide angle on a constituent B of the other initial hadron. This results in two objects C and D that emerge sideways as jets, and one of these jets includes the large-p_T trigger particle. Figure 8b, which is equivalent to figure 8a, shows the two transverse jets, and in addition the longitudinal jets which are what is left of the initial hadrons after the constituents A and B have been removed from them.

In spite of a great deal of experimental and theoretical activity over the past decade, the identity of the objects A, B, C and D has proved hard to establish. The existence of a jet

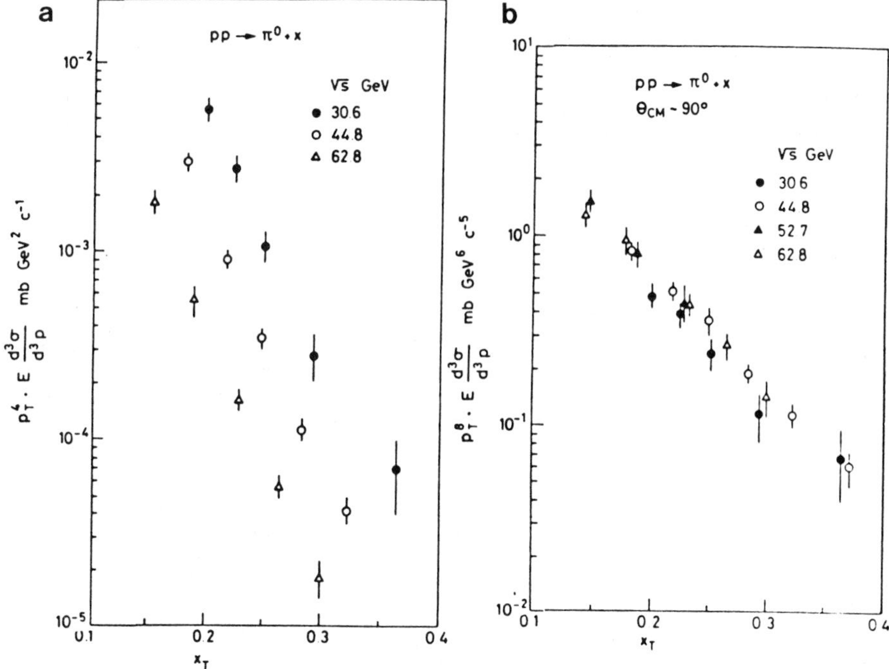

Fig. 9. ISR data for π^0 production at 90° (CERN-Columbia-Oxford-Rockefeller). Plots of (a) $p_T^4 \cdot E\, d\sigma/d^3p$ and (b) $p_T^8 \cdot E\, d\sigma/d^3p$ against $x_T = 2p_T/\sqrt{s}$.

structure such as is shown in figure 8b was established[14] in the mid-1970's by several experiments at the CERN ISR, though at ISR energies it is not easy to separate the transverse jets from the longitudinal ones.

If one makes the simplest assumptions, a calculation of the hard-scattering mechanism of figure 8a yields for the inclusive spectrum of particles produced with high p_T at 90° an expression of the form

$$E \frac{d\sigma}{d^3p} = p_T^{-n} F(x_T) \tag{3.1}$$

where

$$x_T = 2p_T/\sqrt{s} \tag{3.2}$$

In perturbative QCD, apart from logarithmic effects associated with the introduction of the Λ parameter through renormalisation, the dynamics are scale-free. Therefore if p_T is large enough for perturbative QCD to give a valid description, $F(x_T)$ cannot contain any fixed dimensional parameter and so must be a dimensionless function. Hence p_T^{-n} must have the same dimensions as $E\, d\sigma/d^3p$, that is $n = 4$.

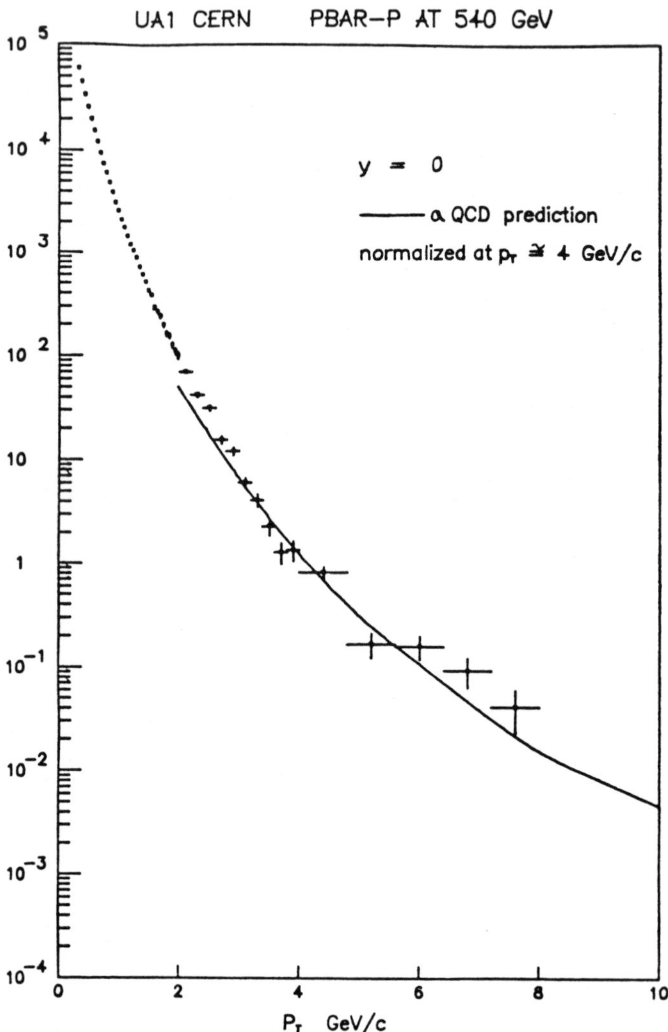

Fig. 10. Large p_T production at the SPS Collider.

As is seen in figure 9, data from the ISR do not fit n = 4, but rather n = 8. Different explanations for this may be found in the literature; see references 14, 15 and 16 for some of the relevant papers. The matter is highly controversial, and my own impression is that there is no one model that is capable of explaining _all_ the available data. This is sad after so much hard work, but preliminary results from the much higher energy available at the SPS Collider offer a glimmer of hope that at last perturbative QCD is beginning to be apparent. This is seen in the data of figure 10, which contains also a QCD prediction. Notice that the latter is normalised so as to agree with the data at p_T = 4 GeV/c. The absolute normalisation of the QCD prediction is subject to considerable uncertainty, from for example the effects of scale-breaking and more particularly of intrinsic transverse momentum of the scattering constituents within their parent hadrons[16], and also from higher-order QCD corrections[17]. I estimate that, altogether, the uncertainty is something like an order of magnitude.

Transverse-Energy Triggers

Figure 11 shows an interesting discovery made by the NA5 Collaboration at CERN. They have measured the transverse energy deposited in a calorimeter that surrounds the beam. The upper set of points correspond to using the whole calorimeter, which extends over 2π in azimuthal angle, while the lower set of points correspond to using only two portions of the calorimeter, each extending over $\pi/2$ in azimuthal angle and on opposite sides of the beam. The huge difference, of some two orders of magnitude, between the two sets of data points shows clearly that they cannot be explained by supposing that large-transverse-energy events can be explained solely in terms of a two-transverse-jet structure. This is confirmed by the distribution of particle tracks in the $\Delta\phi = 2\pi$ case; there is no sign of a two-jet

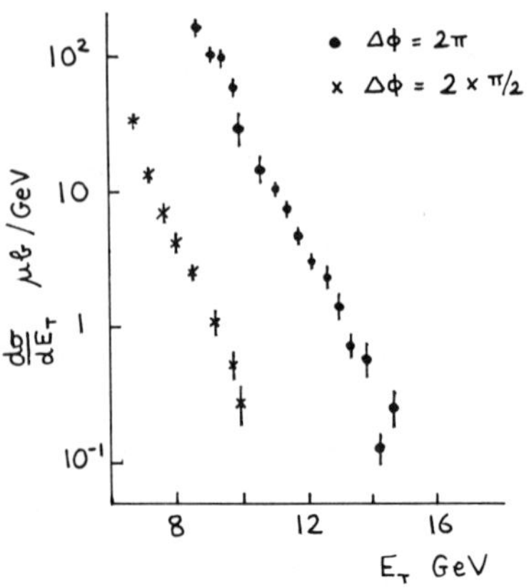

Fig. 11. Transverse energy distributions in 300 GeV/c pp collisions (Bari-Krakow-Liverpool-Munich-Nijmegen).

structure and also the multiplicity is found to be surprisingly high.

It has sometimes been said that this experiment proves that two-transverse-jet events do not exist. However, it does not show this. What it does show is that there is another mechanism for producing large transverse energy, which also produces large multiplicity. There is no reason why this mechanism, whatever it may be, should not co-exist with the two-transverse-jet process. Which of the two mechanisms dominates in a particular experiment depends on the triggering conditions. Reducing the azimuthal acceptance of the triggering calorimeter, or using a single-particle trigger, is more likely to pick out the two-transverse-jet process, while a large-azimuthal-angle calorimeter

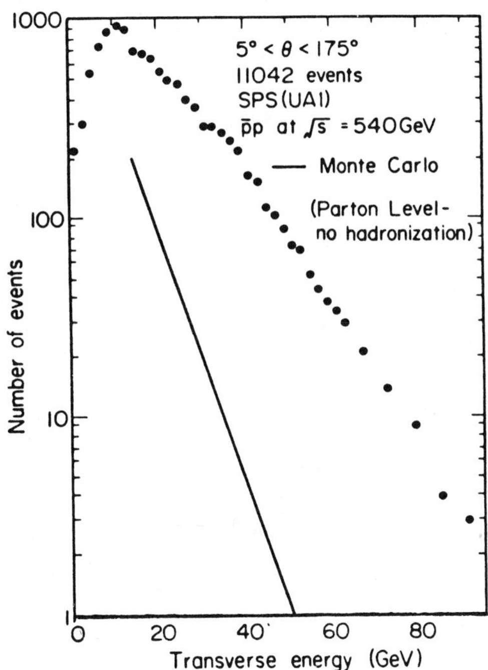

Fig. 12. Transverse energy distribution at the SPS Collider, with a calculated curve[18] (see text).

is much more likely to select the large-multiplicity mechanism.

It remains to explain the dynamics of the large-multiplicity mechanism. A very interesting recent calculation[18] shows how perturbative QCD can yield high multiplicity; it assumes the hard-scattering model of figure 8a with A, B, C, D all quarks, and in addition allows the quarks to radiate gluons. However, the calculations do not succeed in explaining the transverse energy distributions that have recently been measured at the SPS Collider (figure 12). A possible conclusion is that perturbative QCD is not applicable: it may be that, instead of a single short-distance collision, a large number of long-distance collisions are involved.

Heavy-Quark Production

Most of the particles produced with large p_T or large E_T

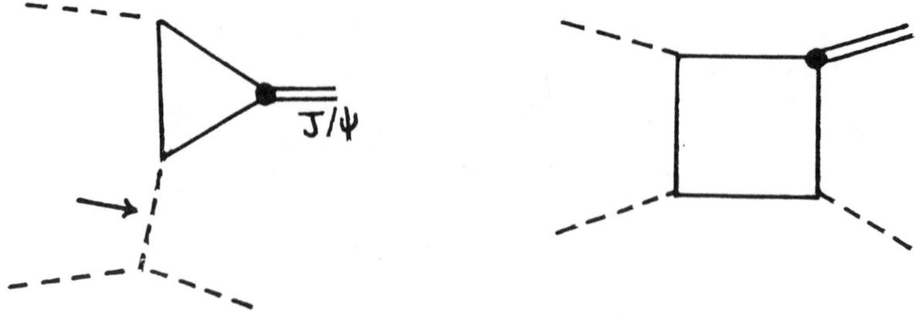

Fig. 13. Typical diagrams for large-p_T J/ψ production; these diagrams involve the interaction of a pair of gluons.

are pions. It is interesting to consider also the production of heavy particles at large p_T, for example J/ψ. The calculation of the total cross-section for heavy particle production is subject to considerable uncertainty (this is discussed in Professor Odorico's lectures at this School), but large p_T cross-sections are theoretically much cleaner. Typical diagrams are shown[19] in figure 13. In these diagrams, a gluon constituent is used from each of the initial hadrons. The gluon marked with an arrow carries large momentum transfer in the case of large p_T production; if one were calculating the total cross section, the kinematics would allow it to carry small momentum transfer, so making perturbative QCD invalid (unless some rather arbitrary cut-off is introduced).

Fig. 14. Diagram for $J/\psi \to e^+e^-$ via a virtual photon, or for weak decay of a charged meson via a virtual W.

The diagrams of figure 13 each involve a vertex function that couples the J/ψ to charmed quarks. Standard assumptions are made about the form of its coupling, in particular its normalisation is determined from the decay $J/\psi \to e^+e^-$, which is calculated from the diagram of figure 14. Calculations[19] of large-p_T J/ψ production, together with data, are shown in figure 15. Agreement is good, though it should be remembered that the calculations involve the gluon distribution in the proton, which is not well known. Also, they involve the QCD coupling $\alpha_s(Q^2)$, and it is not clear what one should choose for Q^2 or Λ; this is connected[1]

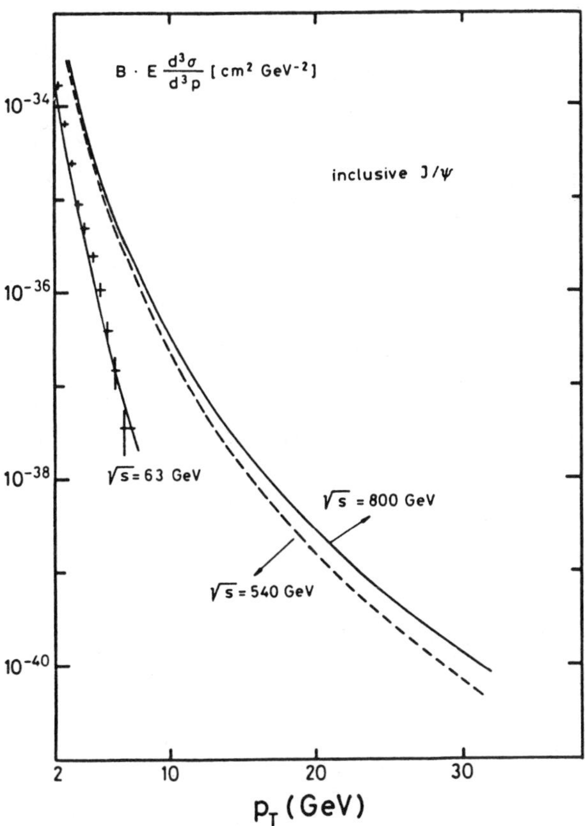

Fig. 15. Calculations[19] of large p_T J/ψ production, with data from the Athens-Brookhaven-CERN-Stony Brook-Yale Collaboration.

with the uncertainties about renormalisation scheme dependence and higher-order contributions, which I mentioned in §1.

Following the discovery of surprisingly copious production of naked-charm particles at the ISR, apparently by some diffractive mechanism[20], thoughts have naturally turned to possible naked-top production at the SPS Collider. According to folklore, diffractive production varies with the mass M of the particle and the incident energy as

$$\frac{1}{M^2} \log s \tag{3.3}$$

If M is in the range 20 to 50 GeV, this would predict a cross-section ranging between 10^{-3} and 10^{-4} mb, but a QCD mechanism has recently been suggested[21], which could give about 10^{-2} mb. This mechanism corresponds to a set of diagrams of which figure 16a is an example, and gives the production of a pair of mesons $T = (t\bar{u})$ or $(t\bar{d})$. Again there is need to decide the normalisation of the vertex function that couples the meson to its constituent quarks, and of course in the case of the T this is not known. This normalisation may be measured in terms of a constant f_T, which is the analogue of the constant f_π that determines the decay $\pi \to \mu\nu$ (f_π = 93 MeV). In a potential-theory model for the T, f_T would be related to the size of the Schrödinger

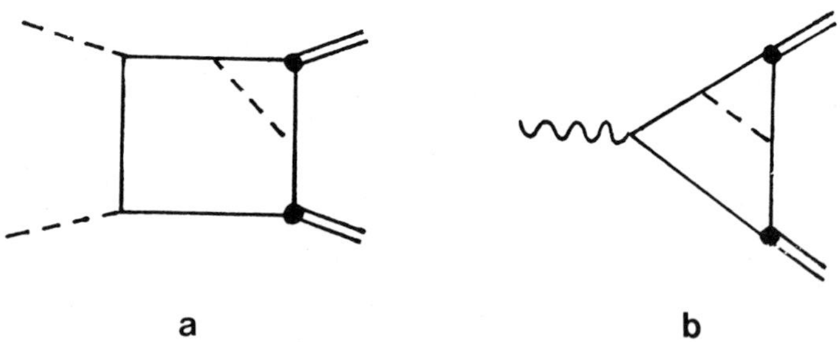

Fig. 16. Diagrams for $T\bar{T}$ production (a) by gluon-gluon interaction in $p\bar{p}$ collisions and (b) in Z^0 decay.

wave function at the origin, though it is most unlikely[21] that potential theory gives a valid description of the T. For neutral mesons whose predominant decay is electromagnetic instead of weak, an analogous constant may be defined; in either case the decay corresponds to a diagram such as figure 14, with the wavy line either a virtual γ or a W. The value of the constant has been measured for π, K, ρ, ω, ϕ, ψ and Υ. Except for the case of the pion, which is so often anomalous, the values are all consistent with the empirical rule

$$f^2/M = 25 \text{ MeV} \tag{3.4}$$

(for the pion the value is 60 MeV). In the result that I quoted above, this rule was assumed to apply also to the T. Note, though, that the cross-section is very sensitive to the value of f_T: it is proportional to f_T^4.

A similar calculation[22] of Z^o decay, based on the diagram of figure 16b, would give a surprisingly large width for the Z^o. The width is again proportional to f^4, and with the assumption (3.4) the calculation gives <u>25 GeV</u> if the t-quark mass is 40 GeV. An answer as large as this must, I think, imply that the perturbation theory has broken down, but it does raise the possibility that the Z^o width is rather greater than has previously been supposed. Of course, the rule (3.4) may overestimate f_T by an order of magnitude or more; at present we do not know.

Final-State Interactions and K-Factors

I have mentioned already that the higher-order QCD corrections to the hard-scattering model of figure 8 seem to be very large[17] — at rather more than the 100% level, though there is some doubt as to just how they should be calculated. The theory of this is somewhat complicated, so here I shall discuss the slightly simpler problem of the Drell-Yan process.

In this process the central hard scattering of figure 8a is replaced by a simple hard process, the point coupling of a $q\bar{q}$ pair to a massive virtual photon (or W or Z).

In his lectures at this School, Professor Eisele has explained that the standard Drell-Yan formula fits the available data very well, except that it has to be multiplied by a constant, the so-called K-factor, whose value is about 2.2.

This K-factor is believed[23] to have its origin in QCD corrections to the standard Drell-Yan formula. QCD corrections result in the formula

$$q^2 \frac{d\sigma}{dq^2} = \frac{4\pi\alpha^2}{3q^2} \sum_{i=u,d,\ldots} e_i^{-2} \int dx_1 dx_2 dz \; \delta(x_1 x_2 z - q^2/s)$$

$$F_i(x_1,q^2)\, F_{\bar{i}}(x_2,q^2)\, [\delta(1-z) + \frac{\alpha_s}{\pi} f(z) + O(\alpha_s^2)] \tag{3.5}$$

Here, $\sqrt{q^2}$ is the mass of the virtual photon, and F_i is the contribution to the deep-inelastic structure function F_2 from the quark of flavour i, including scale-breaking effects from gluon insertions. The $\delta(1-z)$ in the square bracket corresponds to the standard Drell-Yan formula, while the next term represents one-gluon corrections. It turns out that f(z) itself contains a $\delta(1-z)$ term:

$$f(z) = N\delta(1-z) + \ldots. \tag{3.6}$$

and this is the part of f(z) that, numerically, is the most significant. So if one neglects the rest of f(z), one retrieves the standard Drell-Yan formula, but with the additional factor

$$K = 1 + \frac{\alpha_s}{\pi} N + O(\alpha_s^2) \tag{3.7}$$

It is widely believed that this is an exponential series, so that

$$K = \exp(\frac{\alpha_s}{\pi}N).$$

Explicit calculation finds that

$$N = \tfrac{1}{2}C_F\,[1 + \pi^2/3 + C\pi^2] \qquad (3.8)$$

where C_F is a colour factor, $C_F = 4/3$. The value of C seems to depend on the calculational procedure, C = 1 or 2.

This apparent ambiguity in the value of C is the so-called π^2-problem. There cannot really be an ambiguity, because the formula (3.5) relates quantities that are measurable and unambiguously defined. It has been found recently[24] that to solve this problem one has to consider carefully final-state interactions among the hadrons that are produced together with the virtual photon.

The first work on these final-state interactions was done several years ago[25]. In those days, it was natural to consider such interactions in terms of pomeron exchange; nowadays one would say that this describes just the long-range part of the force. It was found that the sum of all final-state interactions just cancels with the effects of initial state interactions. That is, if one is interested just in the virtual-photon cross-section $d\sigma/dq^2$ one can forget such interactions, though they do rearrange the final-state hadrons. Presumably this theorem explains why experiments find that on nuclear targets the Drell-Yan cross-section varies with mass number as A, instead of the $A^{2/3}$ that is found for the total hadronic cross section. (The $A^{2/3}$ is explained by saying that initial and final-state interactions make it hard for the beam to penetrate a nucleus, and even if it does any particles produced inside find it hard to escape, so that only the surface nucleons are important in the target. In the Drell-Yan process, on the other hand, the theorem says that the initial and final-state interactions can be ignored.)

To prove the theorem, a technical assumption is needed[25]: the coupling of the pomeron to a quark becomes very small if the quark goes far off shell. Nowadays, we ascribe the short-range part of the force to gluon exchange, and gluon exchange does not satisfy this assumption: the gluon/quark coupling is a constant, at least to lowest order. So the theorem does not immediately apply to the short-range part of the force.

There has recently been a suggestion[26] that two-gluon final-state interactions completely wreck the Drell-Yan formula (3.5), but this is controversial[27]. I shall discuss only one-gluon effects, where the formula is not in question.

The origin of the term involving C in (3.8) is as follows[23,28]. Part of it comes from diagrams where the gluon is attached just to "active" quarks, that is the quarks that fuse to form the virtual photon. See figure 17. If one calculates in leading log approximation, one finds that the virtual gluon insertion of figure 17a gives a factor

$$\mathrm{Re}\left[-\frac{\alpha_s}{2\pi}C_F\, C\log^2(-q^2)\right] \qquad (3.9)$$

This is another example of the Sudakov factors that I discussed in §2. Notice that the argument of the logarithm is $-q^2$; this is because[29] the amplitude for a vertex correction is real for

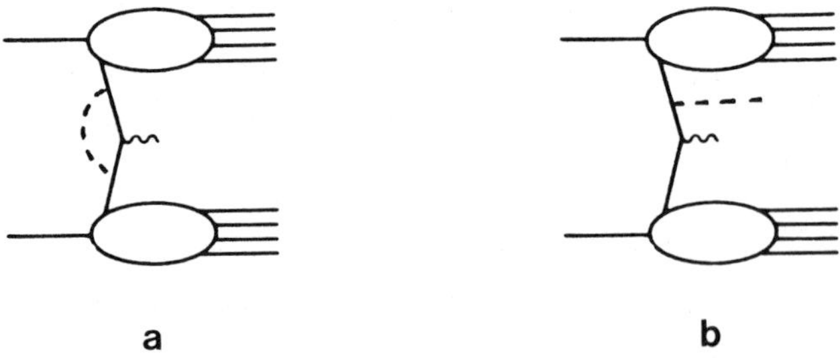

Fig. 17. One-gluon corrections to the Drell-Yan process: (a) virtual gluon, (b) real gluon.

negative q^2. For the positive values of q^2 that we want, we must write

$$\log(-q^2) = \log q^2 - i\pi \tag{3.10}$$

The real-gluon correction of figure 17b gives, again in leading log approximation, the factor

$$\mathrm{Re}\left[\frac{\alpha_s}{2\pi} C_F \, C \log^2 q^2 \right] \tag{3.11}$$

If we add (3.9) and (3.11), using (3.10), we obtain

$$\frac{\alpha_s}{2\pi} C_F \, C \, \pi^2 \tag{3.12}$$

There are two ways to calculate the value of C. The first, and possibly the more popular, is to forget the incoming hadrons and the final-state spectator hadrons, and to pretend that the two active quarks are the incoming particles. In this case, the two quarks are on shell, with the result that both the diagrams of figure 17 are infrared divergent. A regulator is introduced to remove these divergences, for example the gluon is given a small mass. Then the diagrams are calculated and added together, and the regulator can be removed at the end of the calculation (the infrared divergences cancel between the two diagrams). This procedure gives C = 1.

If, however, one takes seriously the fact that the active quarks are not free particles, but are constituents of the incoming hadrons, then they should be taken off shell. In this case the diagrams of figure 17 have no infrared divergence and no regulator is needed. Calculation then finds[28] that the value of C is independent of how far the quarks are off-shell, and gives C = 2. However, with this way of calculating one should include also the possibility that the gluon attaches not to an

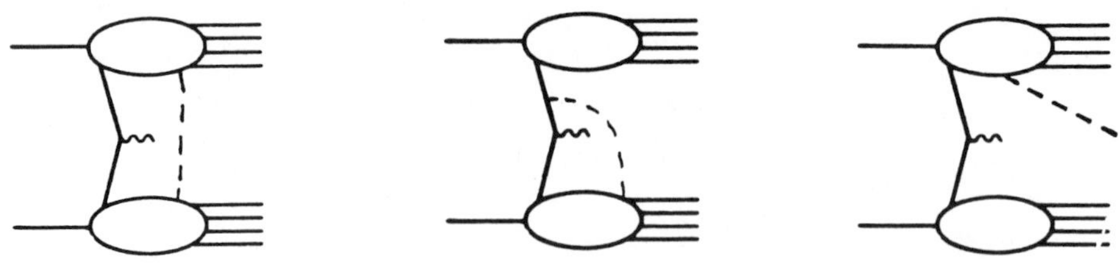

Fig. 18. Spectator-interaction corrections to the Drell-Yan process.

active quark, but to a spectator. Calculation[24] of the sum of diagrams shown in figure 18 finds an additional contribution -1 to C, so that the value C = 1 is again retrieved.

Conclusions

As in the case of elastic scattering, comparison between theory and experiment has some encouraging features. There is a real hope that the high energy available at the SPS Collider will make much cleaner tests of QCD, but at present we are very far from firmly establishing that QCD can correctly describe purely hadronic reactions. Further theoretical work needs to be done, in particular the analysis of final-state interactions and K-factors has not been done properly for purely hadronic processes. As is true also for semileptonic processes, the problem of renormalisation-scheme dependence is a very real one.

References

1. For reviews, see for example P.V. Landshoff, rapporteur talk at EPS Lisbon Conference (1981); A. Buras, rapporteur talk at Bonn Lepton-Photon Conference (1981); I. Hinchliffe, talk at APS Santa Cruz Conference (1981).
2. T.T. Chou and C.N. Yang, Phys. Rev. D19, 3268 (1979); P.D.B. Collins and F.D. Gault, Phys. Lett. 73B, 330 and 74B, 432 (1978).
3. S.J. Brodsky and G.P. Lepage, Phys. Rev. D22, 2157 (1980).
4. V.A. Matveev, R.M. Muradyan and A.N. Tavkelidze, Lett. Nuov. Cim. 7, 719 (1973); S.J. Brodsky and G.R. Farrar, Phys. Rev. Lett. 31, 1153 (1973).

5. P.V. Landshoff, Phys. Rev. D3, 1024 (1973).
6. P.V. Landshoff and J.C. Polkinghorne, Phys. Lett. 44B, 293 (1973).
7. P.D.B. Collins and F.D. Gault, Phys. Lett. 112B, 255 (1982).
8. A. Donnachie and P.V. Landshoff, Z. Physik C2, 55 (1979).
9. This will be discussed in detail in a forthcoming paper by A. Donnachie and P.V. Landshoff.
10. J.M. Cornwall and G. Tiktopoulos, Phys. Rev. D13, 3370 (1976).
11. P.V. Landshoff and D.J. Pritchard, Z. Physik C6, 69 (1980).
12. A. Mueller, Physics Reports 73, no.4 (1981).
13. B. Pire and J.P. Ralston, McGill preprints.
14. For reviews see, for example, M. Jacob and P.V. Landshoff, Physics Reports 48, no. 4 (1978); N.A. McCubbin, Rutherford Appleton report RL-81-041.
15. D.I. Sivers, R. Blankenbecler and S.J. Brodsky, Physics Reports 23C, no. 1 (1976).
16. R.P. Feynman, R.D. Field and G.C. Fox, Nuclear Physics B128, 1 (1977); R.R. Horgan and P. Scharbach, Nuclear Physics B181, 421 (1981); N. Fleishon and W.J. Stirling, Nuclear Physics B188, 205 (1981).
17. R.K. Ellis, M.A. Furman, H.E. Haber and I. Hinchliffe, Nuclear Physics B173, 397 (1980).
18. G.C. Fox and R.L. Kelly, Berkeley preprint LBL-13985.
19. R. Baier and R. Rückl, Munich preprint MPI-PEA/PTh 16/82. See also J. Finjord, G. Girardi, P. Mery and P. Sorba, Phys. Lett. 112B, 489 (1982).
20. For a review, see for example B.L. Combridge, talk at Moriond Workshop on New Flavours, Rutherford Appleton report RL-82-027.
21. R.E. Ecclestone and D.M. Scott, Cambridge preprint DAMTP 82/17.
22. R.R. Horgan, P.V. Landshoff and D.M. Scott, Phys. Lett. 110B, 443 (1982).
23. G. Altarelli, R.K. Ellis and G. Martinelli, Nuclear Physics B143, 521 and B146, 544 (1978).
24. P.V. Landshoff and W.J. Stirling, Cambridge preprint DAMTP 82/3.
25. J.L. Cardy and G.A. Winbow, Phys. Lett. 52B, 95 (1974); C.E. DeTar, S.D. Ellis and P.V. Landshoff, Nuclear Physics B87, 176 (1975).
26. G.T. Bodwin, S.J. Brodsky and G.P. Lepage, preprint SLAC-PUB-2787.
27. D.G. Ross and C.T. Sachrajda, paper submitted to 1982 Paris Conference.
28. F. Khalafi, Cambridge preprint DAMTP 82/2.
29. R.J. Eden, P.V. Landshoff, D.I. Olive and J.C. Polkinghorne, The Analytic S-Matrix, §2.3 (Cambridge University Press, 1966).

D I S C U S S I O N S

CHAIRMAN: P. Landshoff

Scientific Secretaries: S. Bellucci and Y. Leblanc

DISCUSSION 1

- *YANG:*

In the calculation you reported with sums of leading logarithms, in defining the leading terms, you must have made assumptions about either you are considering 1) fixed θ_{CM}, large s, or 2) fixed and large t, large s. Which of these two cases was considered?

- *LANDSHOFF:*

Calculation has been done in both regimes. As I said in the lecture, the most complete calculation is Mueller's; I believe that this was only done for fixed θ_{CM}.

- *YANG:*

I have a remark. You mentioned that there is a geometrical picture which is really a diffraction picture of the elastic scattering. That picture, in my opinion, has been extremely successful especially in predicting the existence and the position of these dips. In fact, T.T. Chou and I are fully convinced that at higher energies there will be increasingly many dips, and the reason for that is very obvious in the diffraction picture, when the hadrons become very black. The dip movement with the blackness increase is predicted by this geometrical picture. It is very interesting that at the collider there are very high cross sections (something like 66 mb). With this large blackness such parameters like b would be very interesting to watch, and there are definite predictions for them from the geometrical picture. They are listed in a paper by Chou and me in 1979 in the Physical Review.

- *MUKHI:*

I have a comment on Sudakov form factors in QCD. The elementary quark-quark-gluon form factor in QCD has been summed to all orders in leading logs, and shown to exponentiate in all leading and non-leading logs, contrary to your statement that the result has been shown only in QED and only in leading logs. What is more, the sub-leading logs have been found not to dominate.

- *LANDSHOFF:*

I think that you are talking about two things: one is just looking at a simple vertex where you have a gluon and then you put in extra-gluons; but then there is the question of what happens when you put in some more complicated hard scattering structure than a simple gluon vertex, and that is a much more difficult problem.

- *ETIM:*

I would like to make a comment about what is here referred as the Sudakov effect. This effect is essentially nothing more than the effect of infrared radiative corrections. We are familiar with these in QED where we know they exponentiate, and know enough about a few perturbative calculations of them in QCD, but unfortunately do not know if they will exponentiate.

- *LANDSHOFF:*

I agree, partly.

- *DOBREV:*

Why do you take three gluon corrections in the calculation of the $\pi\pi$ scattering, in the Constituent Interchange Model?

- *LANDSHOFF:*

Momentum analysis shows that one must take at least three vertices with at least one gluon correction. Of course, there are many more diagrams with three gluon lines and that makes the calculation very difficult.

- *VAN BAAL:*

You showed the plot with coinciding data for $p\bar{p}$ at 50 GeV and pp at approximately 1000 GeV; this could be understood in the perturbative region. But the coincidence takes place over the full

$p_\perp$ range. Might there be some deep principle behind this?

- *LANDSHOFF:*

One can understand it in the region of low p_T just by Regge theory looking at what trajectories contribute. It turns out that the contribution of the ones with intercept ½ in the region of the dip cancels in $\bar{p}p$ and not in pp.

- *VAN BAAL:*

I have another question. You made a remark that the exponential term of Regge poles could also be understood in terms of geometrical models. I did not understand your reference to Prof. Yang's lecture.

- *LANDSHOFF:*

I said that some people talk about the small t region in terms of pomeron and multipomeron exchange; Prof. Yang has a geometrical model and I suspect there must be a connection between the two.

- *YANG:*

About the connection between the geometrical picture and the Regge picture, the one emphasizes the s channel and diffraction, and the other emphasizes the t channel. From the geometrical point of view the close identity of the $\bar{p}p$ and the pp data is not surprising at all. The total cross sections of $p\bar{p}$ and pp at 50 GeV and 2000 GeV respectively happen to be approximately the same (about 43-44 mb). In the geometrical picture once the shape of the hadrons and the total cross section are determined to be identical, the differential cross sections are exactly the same. (This is not to say that one cannot view that from the Regge viewpoint.) Next let me make a remark about what happens in the geometrical point of viewpoint to the differential cross section, between 2 and 6 in t values. The calculations with the geometrical picture always showed a second dip, somewhere around $|t| = 4$ to 6, so far not experimentally found. I repeat what I said earlier: with increasing blackness you will have infinitely many dips, eventually, and the question of this second dip is mainly a question of energy not high enough.

- *HOFSÄSS:*

What do theory and experiment say about the existence of diquarks? Are they ruled out because they would lead to a different power behaviour?

- *LANDSHOFF:*

You can incorporate diquarks into the picture; you could, for example, think that the proton is made up of 2 diquarks plus a quark. In this picture you get higher twist that is powers of t which are larger and more negative than the ones that I have been discussing. So they may be present but we hope they are smaller.

- *HOFSÄSS:*

There must have been attempts to obtain the normalization of pp scattering without calculating the 1000 diagrams of the CIM, using instead a simple representation for the bound state wave-functions and the box-like diagram of the CIM. Can you comment on such attempts?

- *LANDSHOFF:*

If you are talking about the region where maybe the CIM model applies, an accurate calculation is very hard because there are several thousand diagrams. At first sight the cross section is surprisingly large and one can explain that by saying that there are so many diagrams. In the region where triple scattering maybe is dominant, this calculation has not been done, but you can do a very rough order of magnitude calculation, which corresponds to taking the three valence quarks cored in the proton having a radius something less than 1 fermi (but this is not controversial and I think everybody believes that is true).

- *KAPLUNOVSKY:*

What are power-counting rules for multiple exchange diagrams and how to obtain them?

- *LANDSHOFF:*

There is a simple rule for CIM, but there is not such one for multiple scattering. You must pick up the diagram and calculate it.

- *EISLER:*

You have mentioned that problems arise unless Λ is very small (of order 10 MeV). Could you explain this?

- *LANDSHOFF:*

If you calculate any physical quantity P whose first

non-vanishing term in perturbation theory is $O(\alpha_s^n)$:

$$P \propto \alpha_s^n(\ldots)\left[1 + A\frac{\alpha_s}{\pi} + \ldots\right]$$

and if you just calculate the leading power α_s^n, then you have no idea of what you should take as the argument for α_s. If you calculate the next term $A\frac{\alpha_s}{\pi}$, then the value that you get for the constant A depends on what you choose as the argument for α_s. If you change the argument of α_s or, alternatively, if you change the renormalization scheme, then you have a new α_s which is related to the other one by the equation:

$$\alpha_s \to \alpha_s\left[1 + B\frac{\alpha_s}{\pi} + \ldots\right]$$

This changes the value of the coefficient A of the expansion of P: $A \to A + nB = A'$. You want the perturbation series to converge as rapidly as possible. Because you have not calculated the other terms of the expansion, you do not know what choice of the argument of the α_s or of the renormalization scheme is the best one. You can make your choice in such a way that the A' term is very small but, at this stage, you do not know anything about the order of magnitude of the next term in the series.

- *D'ALI'*:

The $\bar{p}p$ $d\sigma/dt$ distribution at PS energies compares very well with the same distribution worked out for pp interactions at much higher energies; the agreement is good in particular in the low t region. How can this be related to the fact that the perturbative QCD calculation you presented predicts that in $\bar{p}p$ the triple scattering behaviour should indeed begin at much lower energy than in pp; in other words, can this comparison be extrapolated in some way to the high t region?

- *LANDSHOFF*:

No, there is no reason for which the extrapolation could be done. I cannot predict the exact energy where the T.S. mechanism should dominate the high t behaviour of $d\sigma/dt$ distribution in pp. I only notice that the good agreement of $\bar{p}p$ data at PS energy and pp $d\sigma/dt$ behaviour at ISR energy is a hint that already at this energy it could occur, and I am surprised about that.

DISCUSSION 2

- *KAPLUNOVSKY:*

Are ϕ-isotropic events planar (i.e. planarity of outcoming particles disregarding the initial beam axes)? If they are, it suggests a smeared 3-jet structure. If they are spherical blobs they could be an indication of a new kind of particles.

- *ZICHICHI:*

Could you specify the definition of ϕ-isotropy? Is it based on the trace distribution or is it based on energy? In the last case I don't understand how from an isotropic distribution you can get jets back to back unless there are substatistical fluctuations.

- *LANDSHOFF:*

As far as I know, a typical event looks with no systematic trace of any clustering... Obviously there will be fluctuations so that you will see clusters in some of the events but there is nothing systematic about it...

- *YANG:*

I don't understand the meaning of isotropy in ϕ in your CERN 300 GeV/c $p\bar{p}$ data E_T plot. I think the definition of isotropy should be re-examined because if you plot the ellipsoid of the momentum distribution in 2 dimensions, the major and minor axes must be on the average quite different otherwise you cannot have the red plot you showed this morning.

- *ANTIC:*

First these events don't look isotropic. It is actually a very rare situation but the point is that they don't look as "jetty" as in the case of e^+e^- experiments. Also we have investigated the events structure in term of planarity and in the limited solid angle the events don't look as "jetty" as in e^+e^- experiments.

- *ODORICO:*

Concerning the NA5 "puzzle" I would like to point out that in a recent paper that I presented at the Paris conference, it is shown that when triggering on the total transverse energy E_T, results largely depend on details of beam jet dynamics and in particular on the slope of the tail of the multiplicity distribution. If a Field-Feynman model is chosen for beam jets (as in the analysis of the NA5 collaboration), the multiplicity distribution is much narrower than the experimental one and one obtains that the (probabilistically) preferential way of obtaining large E_T is by means of not too large p_T (because of QCD hard scattering), although the contribution to E_T from small p_T particles remains sizable (~ 50%). But if the beam jet model has a KNO multiplicity distribution, the sizable KNO tail at large multiplicities makes it probabilistically more convenient to get large E_T with large multiplicity events, all the particles carrying $p_T \sim \langle P_T \rangle \simeq 350$ MeV. This means that with such a beam jet model by triggering on large E_T, paradoxically one anti-selects hard scattering events. Therefore, there is no problem in reproducing the NA5 data or other similar data with large acceptance calorimeters with a QCD model, but the results depend almost exclusively on a fine tune-up of the beam jet model. Triggering on E_T, it appears, one learns very little about QCD hard dynamics. Fox has recently suggested that initial gluon radiation may also have a part in that. I have calculated it but I have found that while it is substantial at SPS $p\bar{p}$ collider energies (corresponding to ~ 30% of the total E_T at moderate and large E_T) it gives too small effects ($\lesssim$ 10% of total E_T) at the energy of the NA5 experiment (300 GeV).

- *LANDSHOFF:*

When one calculates the cross section for 2 jets production using QCD diagrams like diagrams of quark-quark scattering, do the data exclude jets at the level that QCD predicts?

- *ANTIC*

No, they don't exclude them.

- *LANDSHOFF:*

So the message is that there are possible jets but there is also a lot else which may be what is usually "log s physics", that is nothing with any hard interaction in it. So jets could also be present in addition to Odorico's mechanism.

- *ANTIC:*

Let me make the following comment on the NA5 results. In this experiment, events triggered where large transverse energy is deposited in the calorimeter covering the full 2π azimuthal acceptance. Events structure doesn't look like expected from QCD-4-jet model based on perturbative QCD calculations of hard constituent scattering + fragmentation parametrized by e^+e^- jets, but it also cannot be explained by the fluctuations of the isotropic events (as one gets from low p_T cluster model).

- *LANDSHOFF:*

The conclusion is that we have apparently a mixture of hard scattering QCD and beam jet events which have no hard scattering and which you find depending on how you trigger. The way you trigger is quite delicate.

- *KAPLUNOVSKY:*

Can you decipher KNO?

- *ODORICO:*

KNO stands for Kobes, Nielsen and Olesen. It is basically a scaling. They considered multiscale processes and fragmentation. The result is higher multiplicity than in the FF case, but less hard scattering.

- *HUSTON:*

I have a question and a comment. First, one of your transparencies shows the single particle inclusive p_T spectrum seen at the SPS $p\bar{p}$ collider and a comparison to ISR results. The differential cross section seems to fall less steeply with p_T in the SPS results. Is QCD shining through at last or is it due to dealing in a different area of x_T?

My comment is the following. In one of your transparencies you showed $d\sigma/dE_T$ versus E_T for SPS $p\bar{p}$ results. Also shown on the transparency was a comparison to a Monte Carlo calculation by Fox and Kelly. The shape of the differential cross sections agreed but the prediction falls far below the data. Recently Fox has extended this calculation including the effect of hadronization and has found a much more reasonable agreement between theory and experiment.

- *LANDSHOFF:*

Inclusive particle cross sections are usually parametrized in the form $p_T^{-n}F(x_T,\theta)$. It is hard to extract a power of n without performing the experiment at different energies. Assuming $F(x_T,\theta)$ to be constant, perhaps a reasonable assumption, a value of n~5 is extracted. This is closer to the naive QCD value of 4.

- *HUSTON:*

My understanding of the way the value of n evolves from 4 to 8 in QCD is that effects such as the running of the running coupling constant, scaling violations in the distribution and fragmentation functions, and intrinsic k_T effects can provide this drop. In $p\bar{p}$ data, α_S still runs with p_T^2 being the relevant variable and k_T smearing is still present. The different power of p_T may be due to the low x region of the parton distribution being probed.

- *LANDSHOFF:*

The only question is if the overall normalization of the data looks reasonable. Somebody suggested this morning that the normalization of the data was rather large compared to what you would expect from QCD. But I don't know whether this is true or not.

- *SHERMAN:*

Is there any theoretical justification for the constancy of the ratio of the meson decay constant and the square of its mass (F_X/M_X^2) considering the fact that the Υ and ψ are considered non-relativistic systems while the π,ϕ,K, etc. are highly relativistic?

- *LANDSHOFF:*

I consider it as an accident of numerics.

- *LOEWE:*

What about a cut that only slices partially the pomeron in the one pomeron exchange correction to Drell-Yan processes?

- *LANDSHOFF:*

This cut is not dominant and can be neglected.

- *DEVCHAND:*

We have heard a lot today about experimentally testing QCD in particular you mentioned some ISR and UA1 collider data which seem to disagree with perturbative QCD predictions. Could you summarize for us your view on how useful perturbative QCD is for describing hadronic physics?

- *LANDSHOFF:*

There is yet no contradiction with perturbative QCD. However the perturbative calculation could be hidden by non-short distance effects.

QCD PREDICTIONS FOR HEAVY FLAVOUR PRODUCTION

R. Odorico

Istituto di Fisica dell'Università and I.N.F.N.

Bologna

ABSTRACT

Predictions of Quantum Chromodynamics for the hadronic production of heavy flavours are reviewed, with special attention to the progress made in the last year. The contribution of flavour excitation diagrams is discussed in detail and a procedure is given which allows to cure their divergences and to stabilize their calculation. A comparison of the results with charm cross-section data is made, showing that the level and the energy behaviour of the charm cross-section are thus correctly reproduced. The observed production of diffractive Λ_c is shown to be naturally understood by means of a simple recombination model. Expectations for the general features of charm production at collider energies are worked out and in particular it is shown that i) $\sigma(c\bar{c}) \simeq 10$ mb at the CERN $\bar{p}p$ collider, which entails $e/\pi \approx 5\ 10^{-3}$ near 90°; ii) charm events should have relatively high multiplicities. Problems arising in the theoretical prediction of bottom and top particle production are outlined.

INTRODUCTION

Recent experimental surveys of hadronic production of charm can be found in the talks by Treille[1] at the Bonn Symposium (1981) and by Fisher[2] at the Paris Conference (1982). The theoretical understanding of the data has been reviewed by Phillips[3] at the Madison Workshop (1981) and more recently by Halzen[4] at the Paris Conference (1982). About one year ago the main conclusions about the conventional treatment of charm hadroproduction in terms of QCD fusion diagrams (Fig. 1) alone were[3]:

i) production of hidden charm is adequately reproduced;

ii) cross sections for open charm are too low when compared to data;

iii) calculated longitudinal spectra of charmed hadrons are too soft and insufficient to explain diffractive production of Λ_c.

Because of this unsatisfactory situation a number of alternatives to perturbative QCD models were proposed. Among them diffraction excitation[5], intrinsic charm[6] and diquark recombination[7].

In the last year there has been considerable progress in the ability to make a more complete calculation of what should actually be expected from perturabtive QCD in this process[4]. Specifically, ways have been found to determine the contribution of the hard to calculate flavour excitation diagrams[8]. As a result of that, the seeming difficulties met by perturbative QCD in describing hadronic production of open charm have disappeared.

Flavour excitation diagrams were often neglected before because of poor knowledge of the charm distribution, $c(x,Q^2)$, which is

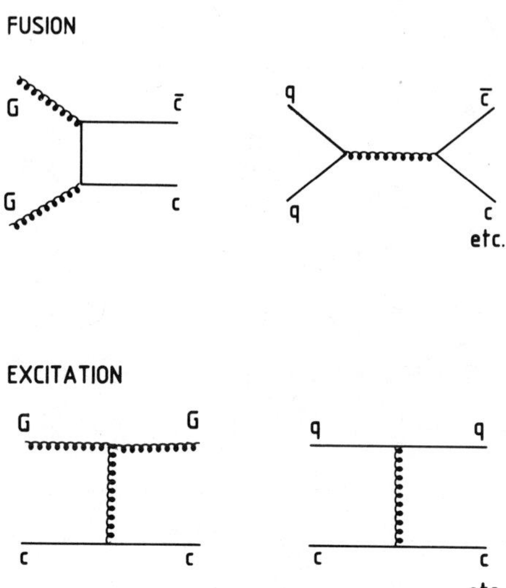

Fig. 1. Typical fusion and flavour excitation diagrams contributing to heavy flavour production.

required in their calculation. Combridge[9] was the first to give serious consideration to them. To estimate their contribution he took $c(x,Q^2)$ as given by the Buras and Gaemers[10] parametrization. In [10] $c(x,Q^2)$ is evoluted as normal (light) sea apart from a delay in the evolution, which starts at $Q^2 = Q_o^2$ (taken equal to 1.8 $(GeV/c)^2$ from fits to the SLAC-MIT electroproduction data). No phase space effects associated with the massiveness of the charm quarks are taken into account. In part because of this unconstrained evolution of $c(x,Q^2)$, and in part because Combridge assumes $Q^2 = \hat{s}$, the c.m. energy square in the subprocess, as the evolution scale a huge $\sigma_{c\bar{c}}$(excitation) is found, of the order of several mb at collider energies. But recent experimental determinations[11,12] of the charm structure function, F_2(charm), show that the Buras and Gaemers parametrization is not able to account, even qualitatively, for the experimental behaviour of F_2(charm).

A paper by Barger, Halzen and Keung[13] (BHK) has revived interest in these diagrams. They assume that a properly QCD evoluted $c(x,Q^2)$ should have a shape $\sim x^{m-1}(1-x)^m$ at $Q^2 \simeq 4m_c^2$, which they take as the appropriate evolution scale (calculations are presented for $m = 1$). Further assuming that the nucleon momentum fraction carried by charmed quarks is $\simeq 0.5\,\%$, they find a $\sigma_{c\bar{c}}$(excitation) much higher than $\sigma_{c\bar{c}}$(fusion) at accelerator and ISR energies. Because of the shape they take for $c(x,Q^2)$ near $x \simeq 0$ the two become comparable, however, at collider energies. With a very simple but rather extreme model for the charm hadron originated by the spectator c quark (the hadron is assumed to carry all the nucleon momentum left by the active quark), they are also able to reproduce the diffractive spectrum observed for Λ_c production. At the Q^2 typical of deep inelastic scattering it is argued that the ordinary QCD degradation brings the charm distribution back to the normal shape expected for a sea component, i.e. peaked at $x \simeq 0$, and as a consequence that there is no difficulty in meeting the experimental upper bounds from deep inelastic multi-lepton data.

These estimates make it clear that $\sigma_{c\bar{c}}$(excitation) can play a central role in the perturbative QCD approach. That is essentially due to the fact that at large $\hat{s}$:

$$\begin{aligned}
&\hat{\sigma}(Gc \rightarrow Gc) \propto \text{constant}\\
&\hat{\sigma}(qc \rightarrow qc) \propto \text{constant}\\
&\hat{\sigma}(GG \rightarrow c\bar{c}) \propto \frac{1}{\hat{s}} \log \hat{s}\\
&\hat{\sigma}(q\bar{q} \rightarrow c\bar{c}) \propto 1/\hat{s}
\end{aligned} \qquad (1)$$

and to the $d\hat{\sigma}/d\hat{t}^2 \propto 1/\hat{t}^2$ behaviour exhibited by the excitation contribution. If the non-perturbative cutoff, $\hat{t}_{min}$, is sufficiently small, $\sigma_{c\bar{c}}$(excitation) can turn out to be very large. A proper calculation of these diagrams, however, requires:

i) a correct procedure to fix $\hat{t}_{min}$ (BHK take $-\hat{t}_{min} \sim m_c^2$, where $m_c = 1.5$ GeV is the charm quark mass, but they recognize that an ambiguity by a factor $2 \div 4$ remains; Combridge simply assumes $-\hat{t}_{min} = Q_o^2 = 1.8\ (\text{GeV}/c)^2$);

ii) a reliable perturbative QCD calculation of the evoluted $c(x,Q^2)$ and, to discuss hadron distributions, of the spectator charm quark distribution.

In order to comply with ii) one must absolutely use a calculational framework in which phase-space constraints associated with the massiveness of charm quarks are duly taken into account. This rules out the inclusive treatment offered by the conventional QCD evolution equations and calls necessarily for an <u>exclusive</u> treatment. (This becomes even more necessary if one wants to calculate the spectator charm distribution). At present the only known way to do that is the QCD Monte Carlo.

CHARM SEA EVOLUTION BY QCD MONTE CARLO

The QCD Monte Carlo technique has been originally developed for parton jet calculation in electron-positron annihilation[14]. More recently it has been extended to calculate the evolution of structure functions in the LLA[15,16]. By appropriate modifications of splitting probability functions one can include in the calculation also next-to-leading order corrections[17]. There are no limitations in the precision attainable with the technique. Fig. 2 shows a comparison of next-to-leading order corrections to the QCD evolution of non-singlet structure function moments as calculated by the Monte Carlo with the corresponding analytic results[17]. Of course the QCD Monte Carlo gives much more information than an analytic calculation can ever provide, since it calculates the full final state, including longitudinal and transverse distributions and all conceivable parton correlations. For instance it is easy to calculate with it the LLA evolution of the transverse momentum distribution of Drell-Yan pairs including also non-abelian effects[18] (see also[19]), which is practically impossible to do analytically. FORTRAN programs for the simulation of parton jets in electron-positron annihilation[20] and for the QCD evolution of structure functions, including heavy quark effects[16], are available.

The only dynamical input in this mathematical technique is the elementary emission probability:

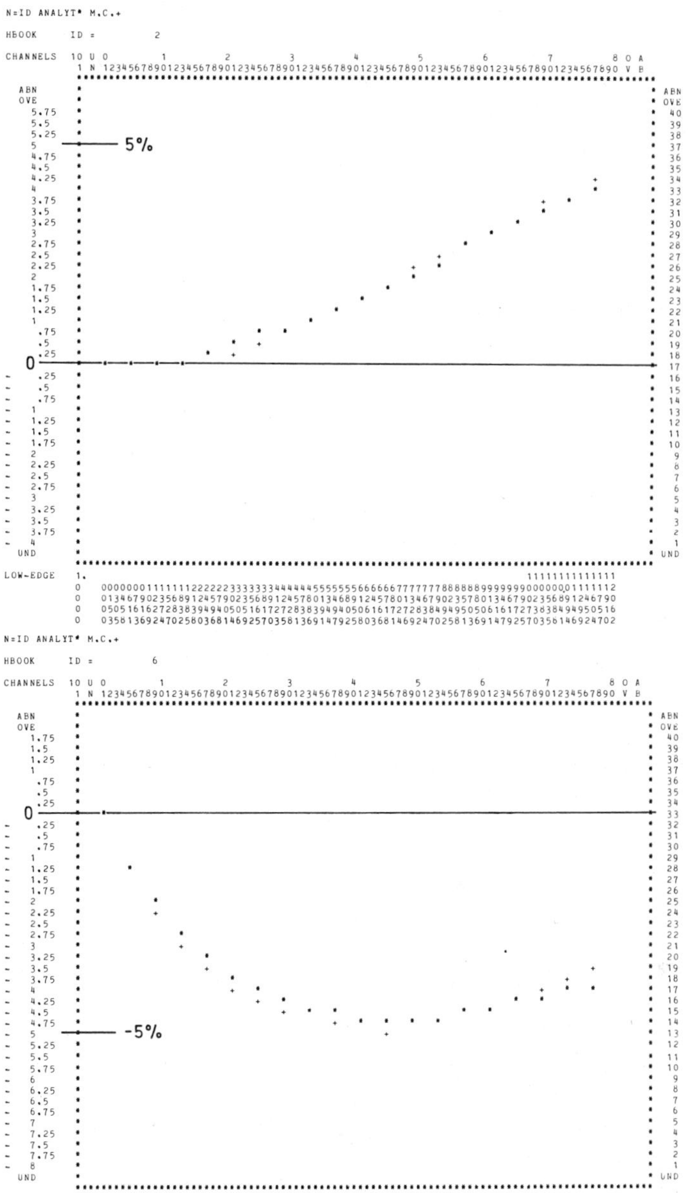

Fig. 2. $(M_n^{NXT} - M_n^{LLA})/M_n^{LLA}$ versus $\log(\log(Q^2/\Lambda^2)/\log(Q_o^2/\Lambda^2))$, where M_n^{LLA} and M_n^{NXT} are the nth moments of the non-singlet structure function calculated in the LLA and including next-to-leading corrections, respectively (Λ = 0.3 GeV, Q_o^2 = 1 $(GeV/c)^2$, Q^2_{MAX} = 200 $(GeV/c)^2$). Stars: analytic calculation. Crosses: Monte Carlo results. n = 2 above, n = 6 below[17].

$$d\,P_E = \frac{\alpha_s(K^2)}{2\pi}\,\frac{d\,K^2}{K^2}\int_{\varepsilon_1(K^2)}^{1-\varepsilon_2(K^2)} P(z) \tag{2}$$

where $P(z)$ is one of the appropriate splitting probability functions ($q \to qG$, $G \to GG$, $G \to q\bar{q}$). $\varepsilon_1(K^2)$ and $\varepsilon_2(K^2)$ embody phase-space bounds. For a $G \to c + \bar{c}$ branching in a spacelike evolution the bounds can be derived from the positivity condition on the transverse momentum square K_T^2 generated in the branching[8,21]:

$$K_T^2 = z(1-z)\left(K^2 - \frac{K_{act}^2}{z} - \frac{K_{spect}^2}{1-z}\right) \geq 0 \tag{3}$$

where K^2 (<0), K_{act}^2 and K_{spect}^2 are the 4-momentum squares of the parent, the spacelike secondary and the timelike secondary, respectively, and z is the fraction of the momentum of the parent carried by the spacelike secondary. Since $K_{spect}^2 \geq m_c^2$, where m_c is the mass of the c quark, and $K_{act}^2 \geq -Q^2$, where Q^2 (>0) is the evolution scale ($|K_{act}^2| \gg |K^2|$ for the LLA to be valid), one obtains immediately after the branching[8,21]:

$$x_{spect} \geq x_{act}\,\frac{m_c^2}{Q^2} \tag{4}$$

For $Q^2 \gg m_c^2$ (e.g. the deep inelastic scattering regime) the condition is ineffective. But for $Q^2 \approx m_c^2$ it clearly favours the flow of momentum to the timelike quark. For gluon initiated cascades with $x \equiv 1$, one obtains the results of Fig. 3 for small Q^2 ($Q^2 = m_c^2/4$, $m_c = 1.5$ GeV, and K^2 replaced by $K^2 + 1.5\,m_c^2$ in eq. 2, see below). The $x \simeq 0$ peak in the spectator distribution gives no contribution in the final calculation, because of the threshold condition to pass. At larger Q^2 the $x \simeq 1$ peak gradually goes away. In an actual calculation these distributions

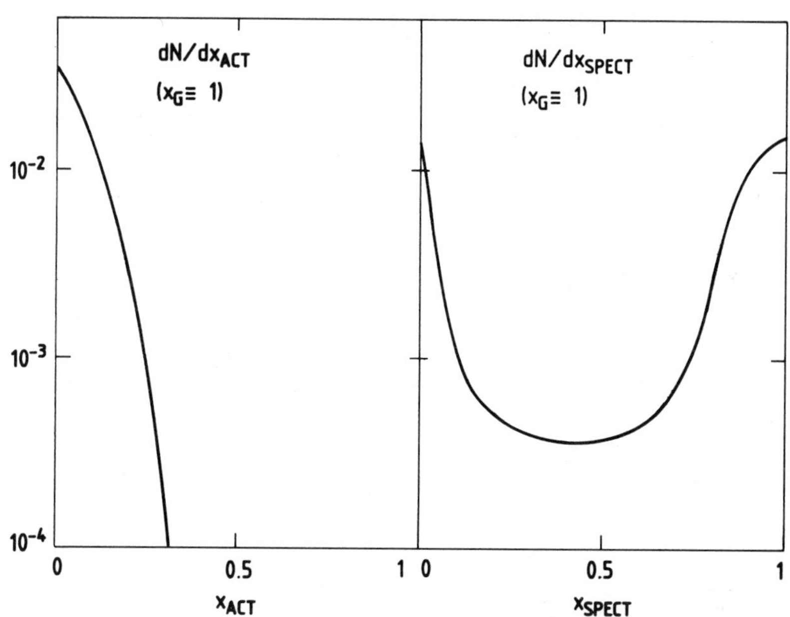

Fig. 3. x distributions of active (spacelike) and spectator (timelike) charm quarks resulting from gluon initiated QCD cascades as calculated by the QCD Monte Carlo ($x_G \equiv 1$ for the initial gluon, $\alpha_s = \alpha_s(Q^2 + 1.5\ m_c^2)$, $m_c = 1.5$ GeV, $Q^2_{evol} = 0.25\ m_c^2$).

are to be convoluted with the initial gluon distribution to obtain the final answer. The result of Fig. 3 is of course of crucial importance when discussing diffractive production of charm. It simply follows from taking kinematics thoroughly into account, which is something that LLA analytic calculations are not able to do.

One should be aware that the calculation depends on the treatment of the $K^2 \simeq 0$ divergence present in eq.2. It is clear that it is beyond the possibilities of perturabtive QCD to cure it, and that (in lack of a non-perturbative calculational technique) one must adopt an effective procedure using experimental and/or phenomenological input. For this sake one can use the recent experimental measurements[11,12] of F_2(charm). In photon-gluon fusion model calculations covering these data one obtains satisfactory fits by assuming[3] $\alpha_s = \alpha_s(Q^2 + m(c\bar{c})^2)$, where $m(c\bar{c})$ is the invariant mass of the $c\bar{c}$ system. Mimicking this successful recipe we replace $K^2 \rightarrow K^2 + a\, m_c^2$ in eq. 2, where a is a free parameter to be fixed by a best fit to the data. For $a = 1.5$ one obtains the curves of Fig. 4 which provide a fit of the same quality as that of the photon-gluon fusion model (data compilation from[22]; $x\,G(x) = 0.5\,(1+\eta)\,(1-x)^{\eta}$ with $\eta = 5$ has been assumed for the initial gluon distribution; initial quarks give a contribution $\lesssim$ 1/10 of the total).

CALCULATION OF FLAVOUR EXCITATION

The basic formula for the charm cross section is

$$\sigma_{c\bar{c}} = \iint_{\hat{s}_{th}/s} dx_1 dx_2\, f_1(x_1,Q^2)\, f_2(x_2,Q^2)\, \hat{\sigma}(\hat{s}) \tag{5}$$

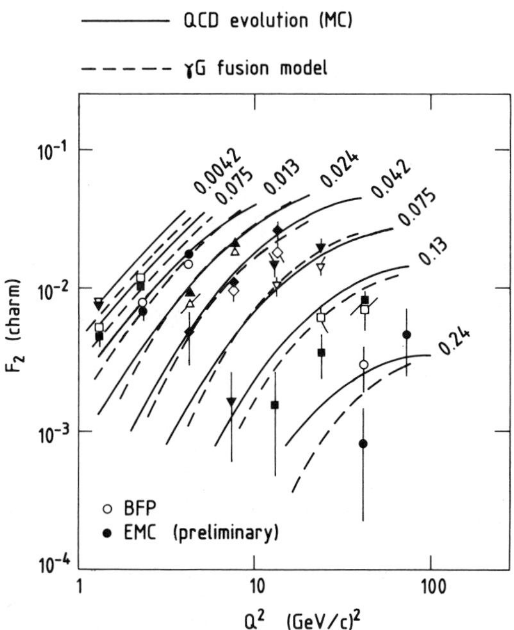

Fig. 4. Comparison of F_2(charm) calculated by the QCD Monte Carlo ($\alpha_s = \alpha_s(Q^2 + 1.5\ m_c^2)$) with experimental data[11,12,22]. Dashed lines represent the photon-gluon fusion model results[22].

where $\hat{\sigma}$ is the cross section for the parton subprocess, f_1 and f_2 are the appropriate parton density functions, s and $\hat{s}$ are the c.m. energy squares of the hadron and parton processes, respectively, $\hat{s}_{th}$ defines the threshold of the subprocess. When s increases, smaller x's become accessible. Because of the $1/x$ behaviour of the parton densities this gives an extra $\log^2 s$ factor besides the energy dependence implied by $\hat{\sigma}(\hat{s})$. The calculation of the cross section results from the combination of an essentially small factor, the charm sea, and a large one, the excitation cross section (see Fig. 5). The calculation is therefore potentially unstable, and special care is required when fixing the intervening parameters.

In an actual quantitative calculation one has to include in eq. 5 convolutions over the transverse momenta of the colliding partons. For flavour excitation, the transverse momentum of the charm quark includes the effects of parton emission in the QCD evolution yielding the charm quark, which turns out to be an important effect.

$\hat{\sigma}(\hat{s})$ for excitation is dominated by the $d\hat{\sigma}/d\hat{t} \propto 1/\hat{t}^2$ singularity, which must be regularized. Adopting an obvious analogy with deep inelastic scattering, the virtual "hard" gluon replacing the virtual photon, it appears sensible to take $Q^2_{evol} = -\hat{t}$, where Q^2_{evol} is the scale up to which the charm sea is evoluted. Since events accumulate at $\hat{t} \simeq \hat{t}_{min}$, to avoid inessential complications in the calculation one can simply take $Q^2_{evol} = -\hat{t}_{min}$. Thus increasing Q^2_{evol} the charm sea is increased, but $\hat{\sigma}$ is reduced. One expects $Q^2_{evol} \approx m_c^2$, but the resulting cross section $\sigma_{c\bar{c}}$ varies considerably for Q^2_{evol} moving around this value (Fig. 6).

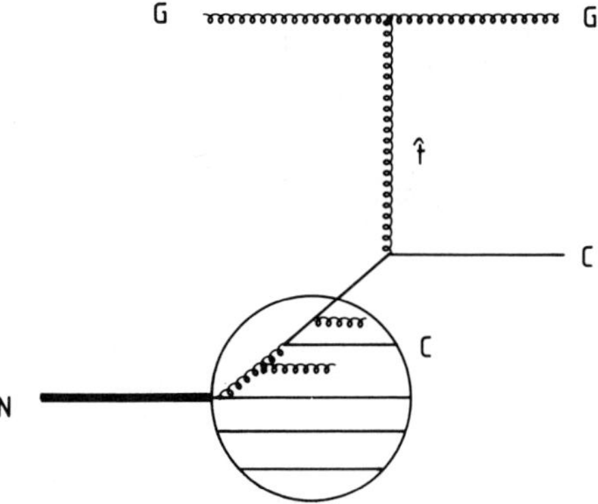

Fig. 5. Basic scheme of the flavour excitation calculation.

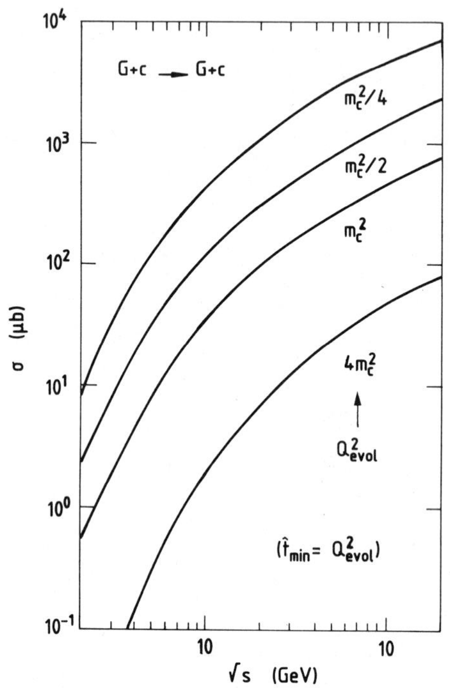

Fig. 6 . Dependence of the G+c → G+c contribution to σ(charm) on Q^2_{evol}.

In practice, what is left undefined because of the divergence is the radius of the interaction. In the calculation of ref. 8 this radius has been fixed using existing data on the p_T distributions of charm hadrons. Experimentally at ISR[23] $dN/dp_T^2 \propto \exp(-b\, p_T)$ with $b = 2.5 \pm 0.4\ (\mathrm{GeV}/c)^{-1}$ for Λ_c, and $b = 2.35 \pm 0.47\ (\mathrm{GeV}/c)^{-1}$ for D^o and D^+, which corresponds to $\langle p_T \rangle \simeq 0.8$ GeV/c. The transverse momentum of the spectator charm quark is originated in the evolution process. I calculate it with the QCD Monte Carlo, assuming an intrinsic transverse momentum for the initiating gluon $\langle K_T^2 \rangle_{intrinsic} = 0.4\ (\mathrm{GeV}/c)^2$ obtained from an LLA analysis[18] of Drell-Yan data (only a fraction $\sim x_c^2$ of it goes to the charm quark). The resulting p_T grows fast with Q^2_{evol} (Fig. 7). The struck charm quark gets its transverse momentum also from the collision process. For near collinear kinematics: $(p_T^2)_{interaction} \simeq -\hat{t}$. Taking into account that some extra transverse momentum is generated in the hadronization process one thus obtains approximately Q^2_{evol}, $-\hat{t}_{min} \lesssim m_c^2/4$, with $m_c = 1.5$ GeV. In ref. 8 it is taken $Q^2_{evol} = -\hat{t}_{min} = 0.25\ m_c^2$, as obtained from fits to the p_T data and assuming a recombination model for the hadronization (see below). Results, however, negligibly depend on the hadronization picture used. The fact that the effective scale $\sim m_c^2/4$ obtained with this procedure is only a fraction of what a priori one would consider as the natural scale for the process should be compared with results obtained in QCD calculations for large p_T processes. There one observes that the reabsorption of higher order corrections into a redefinition of the scale appearing in α_s leads to a reduction of the "natural" scale by a substantial factor ($\sim 1/7$)[24].

Other less important sources of ambiguity in the calculation

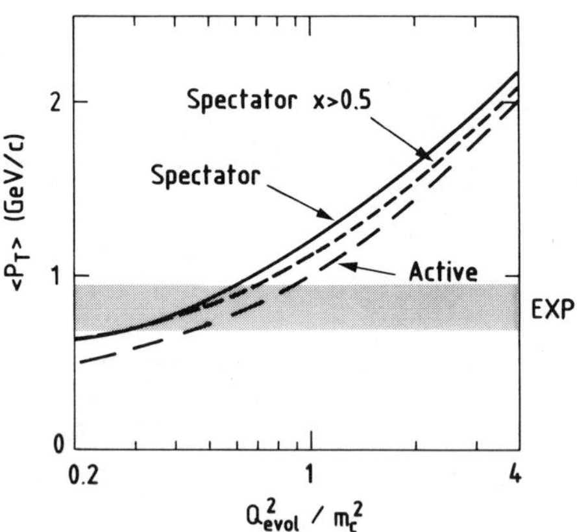

Fig. 7. Dependence of $\langle p_T \rangle$ for the spectator and active charm quarks, immediately after the evolution, on the evolution scale. The dotted region corresponds to the $\langle p_T \rangle$ measurement for charmed hadrons of the BCF experiment[23].

are associated with the choice of the argument of α_s and the off-shellness of the initial (spacelike) charm leg. The results of ref. 8 are obtained for $\alpha_s = \alpha_s(-\hat{t}_{min})$ ($\Lambda = 0.5$ GeV), and neglecting the off-shell mass of the initial charm quark. However, full account is taken of the transverse momentum of colliding partons when dealing with the threshold condition.

Figs. 8, 9 and 10 contain the results for $\sigma_{c\bar{c}}(s)$ and a comparison with data. $\sigma_{c\bar{c}}$(excitation) turns out to be considerably larger than $\sigma_{c\bar{c}}$(fusion) and the relative ratio increases with energy. The experimental level for the cross section and the fast rise between accelerator and ISR energies are correctly reproduced, and thus the difficulty met when considering fusion diagrams alone is finally eliminated.

The huge excitation cross section found does not alter the already satisfactory situation for hidden charm cross section[3], since $\sigma_{c\bar{c}}$(excitation) does not contribute to the $c\bar{c}$ mass interval of interest (Fig. 11).

CHARM HADRON DISTRIBUTIONS

Before claiming that flavour excitation is the main mechanism responsible for the observed production of charm, one must verify that it also accounts for the experimental diffractive production of Λ_c. In order to discuss that a hadronization model is needed. The conventional Field and Feynman[25] quark fragmentation model looks inadequate in this case. The model was developed to describe quark fragmentation in deep inelastic scattering, electron-positron annihilation, and large p_T processes. These are all processes in

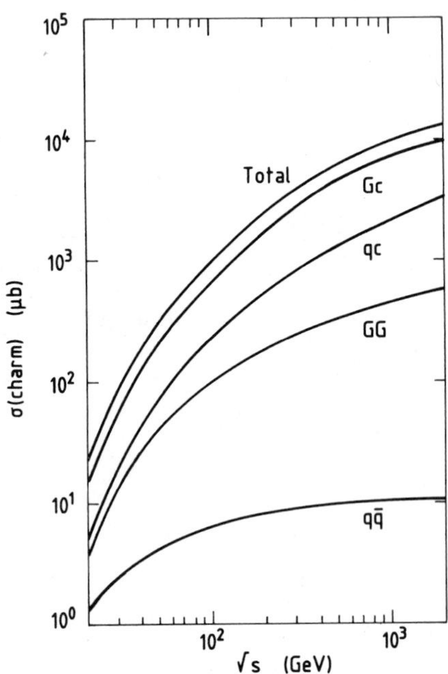

Fig. 8. Calculated charm cross sections for $\bar{p}p$ collisions.

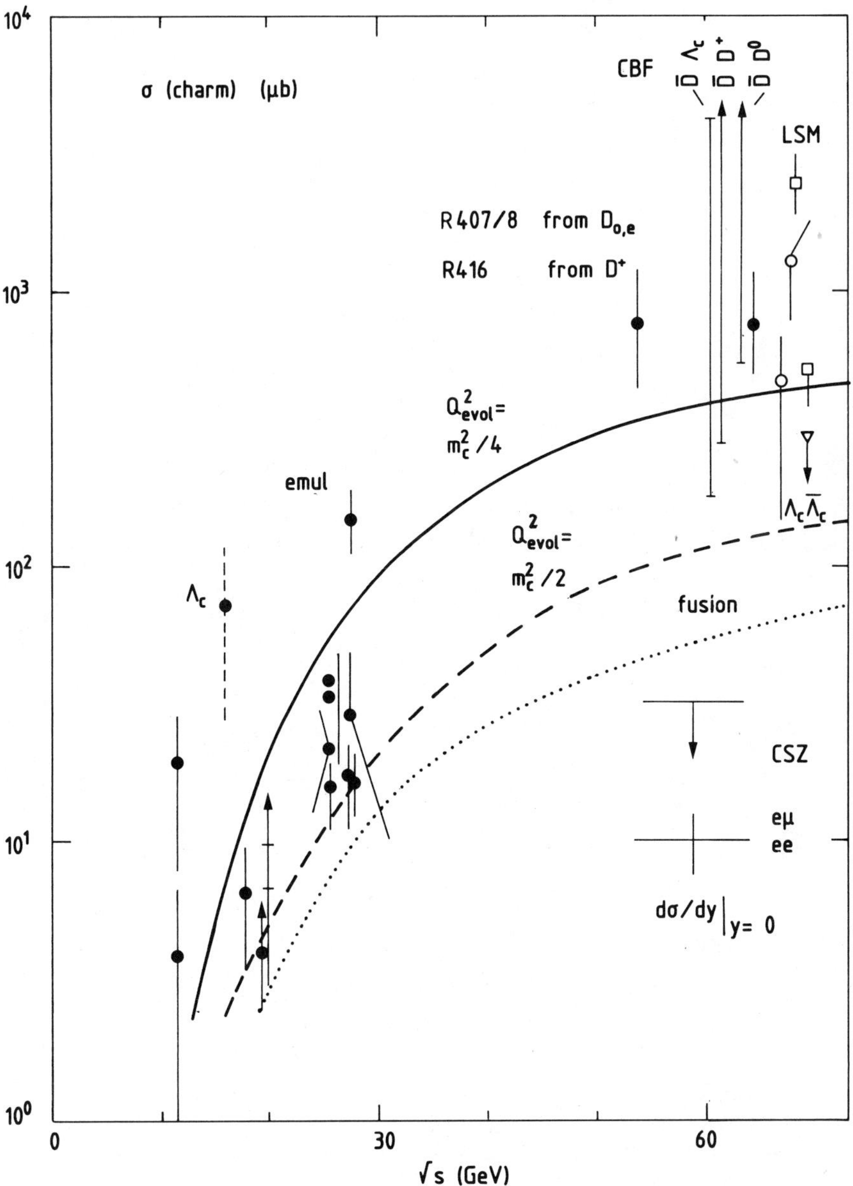

Fig. 9. Comparison of the calculated charm cross section with experimental data (data compilation from [1]).

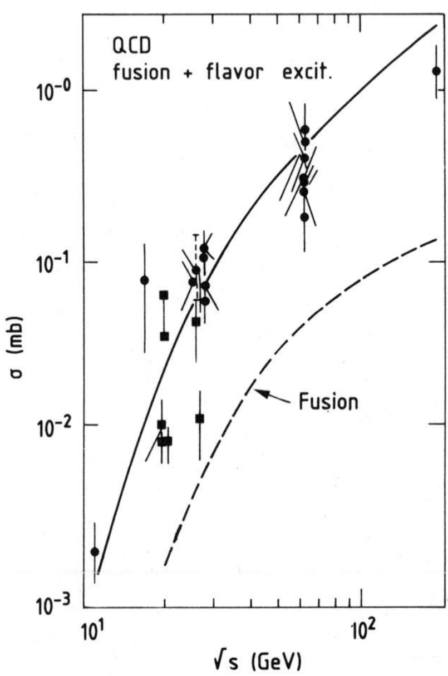

Fig. 10. Comparison of the calculated charm cross section with experimental data (data compilation and figure from [4]).

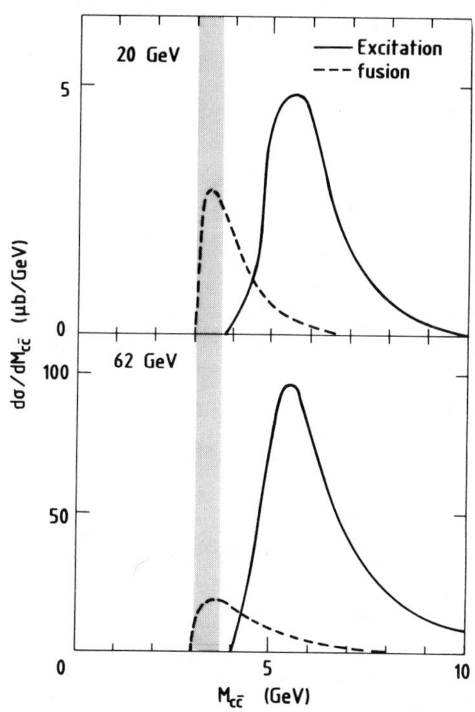

Fig. 11. Invariant $c\bar{c}$ mass distribution resulting from fusion and excitation contributions. The dotted region corresponds to the mass interval attributed to hidden charm production.

which the fragmenting quark is far in phase space from other fast quarks and recombinations among them are therefore unlikely. But in our case it is not so. There are other fast spectator quarks around which may recombine with the charm quarks into (low mass) hadrons which as a result move faster than the original charm quark (oppositely to what one would expect from a naïve application of the Field and Feynman model).

The recombination model[26] has some degree of success in explaining longitudinal distributions of conventional hadrons. Criticism against its use in connection with charm production has been raised[3] in two respects: i) fast D^+ $(c\bar{d})$ are observed in pp collisions[27], whereas from the recombination model one would expect a much softer distribution since recombination with a sea quark is involved; ii) by Bjorken's argument[28] (requiring closeness in rapidity for the recombining partons) the c quark should be fast in order to have a fast charmed hadron, and thus the momentum contribution of the uncharmed quarks should not affect the hadron spectrum sizeably anyway.

With a Monte Carlo calculation one can easily verify whether the second type of criticism has reason to exist. Indeed one can directly impose a mass cut on the invariant mass of the recombining quarks as the condition to have a hadron recombination:

$$m_c \leq m_{recombination} \leq m_{hadron} + h \qquad (6)$$

where the term h is meant to take into account the effects of soft interaction dynamics (h = 0.5 GeV in the calculation). To use eq. 6 one must give masses and transverse momenta also to the uncharmed quarks (m_q = 0.350 GeV and $\langle K_T^2 \rangle_q = 0.05\ (\mathrm{GeV}/c)^2$, so that $\langle K_T^2 \rangle_\pi = 0.1\ (\mathrm{GeV}/c)^2$, have been used). As to the

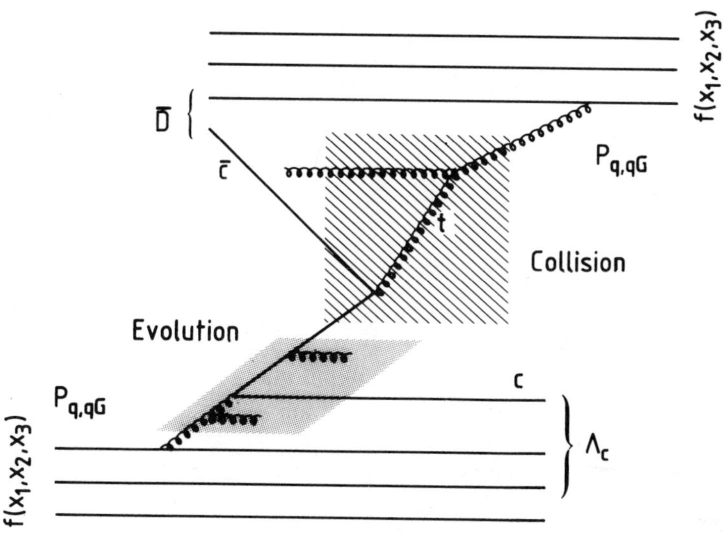

Fig. 12. Basic scheme of the calculation involving the recombination process.

longitudinal momenta of the uncharmed quarks, one must realize that in generating them it is important to keep track of the momentum flow in all the steps of the calculation: initial parton distributions, QCD evolution, interaction, and recombination (Fig. 12). For instance, if the gluon originating the $c\bar{c}$ pair is more energetic, thus leading to a faster $c_{spectator}$, the valence quarks with which $c_{spectator}$ will recombine are slower on average, and thus the effect on the charm hadron momentum is in part compensated. Therefore one must start from a correlated distribution for the initial partons and then go on. It is clear that in order to keep the calculation within manageable proportions only few-parton configurations can be treated. But according to the recombination model these are just the ones mainly responsible for the production of <u>fast</u> hadrons ($x_F \gtrsim 0.5$).

Let us assume a correlated valence quark distribution[29]:

$$f(x_1,x_2,x_3) \propto x_1\, x_2\, x_3\, \delta(1-x_1-x_2-x_3) \tag{7}$$

which complies with the counting rule result[30]:

$$f_q(x_1) = \iint dx_2\, dx_3\, x_1\, x_2\, x_3\, \delta(1-x_1-x_2-x_3) = \frac{1}{6}\, x_1\, (1-x_1)^3 \tag{8}$$

The vanishing at $x_1 \simeq 0$ is of course due to the neglect of multiparton configurations, and therefoere is not physical.

Let us further assume that gluons are generated off valence quarks with a distribution $dN = P_{q,Gq}(z)\, dz \propto dz/z$, where z is the momentum fraction of the parent valence quark taken by the gluon. The resulting gluon distribution is $f_G(x) = (5-4(1-x))\,(1-x)^4/x$.

To treat recombinations of the spectator charm quark (into fast hadrons) one must: i) generate the momentum fractions of the valence quarks and of the gluon in the charm originating nucleon; ii) evolute the gluon; iii) look for recombinations of the charm spectator with the three valence quarks after the interaction has occurred.

For recombinations of the struck charm quark one can adopt a similar procedure. Indeed one finds that fast struck c quarks (present only at small and moderate c.m. energies, Fig. 13) all move in the direction opposite to that of the charm originating nucleon. Therefore they will recombine with the quarks of the nucleon, and in order for them to be fast the other colliding parton must be energetic. For recombinations of the struck c quark one can thus generate the parton configuration of the other nucleon in the same way as above.

Results for nucleon-nucleon collisions at the energies of the CERN $\bar{p}p$ collider ($\sqrt{s}$ = 540 GeV) and ISR (62 GeV) are shown in Figs. 14-17. Although results are reported for the full range of x_F, it is understood that they apply only to the large x_F region, $x_F \gtrsim 0.5$.

At collider energies one observes for baryon recombinations (3 quarks) a clear leading particle effect which is absent in meson recombinations (quark-antiquark). The difference is simply due to the fact that for baryons one adds up 3 quark momenta, whereas for mesons only 2. An interesting (theoretical) feature is that fast baryons are almost exclusively generated by the charm spectator, in agreement with the suggestion of [13]. This is less true for fast mesons.

At ISR energies the overall results remain the same, but the theoretical picture behind them appears distorted with respect to the simple situation observed at higher energies. Now the effects

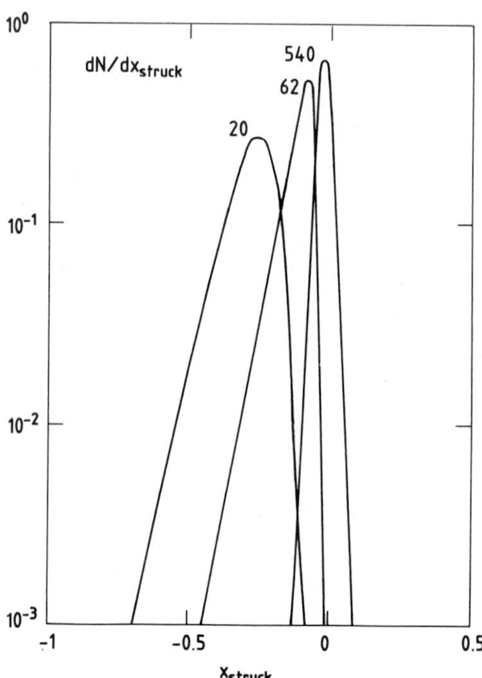

Fig. 13. x distributions of the struck charm quark after interaction at $\sqrt{s}$ = 20, 62 and 540 GeV.

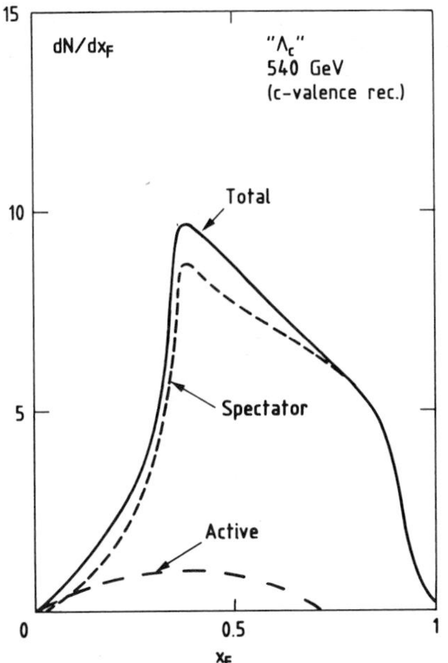

Fig. 14. x_F distribution of fast charmed baryons at $\sqrt{s}$ = 540 GeV.

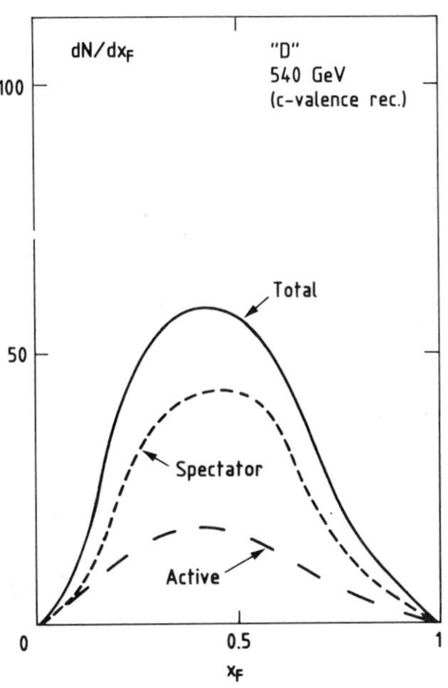

Fig. 15. x_F distribution of fast charmed mesons at $\sqrt{s}$ = 540 GeV.

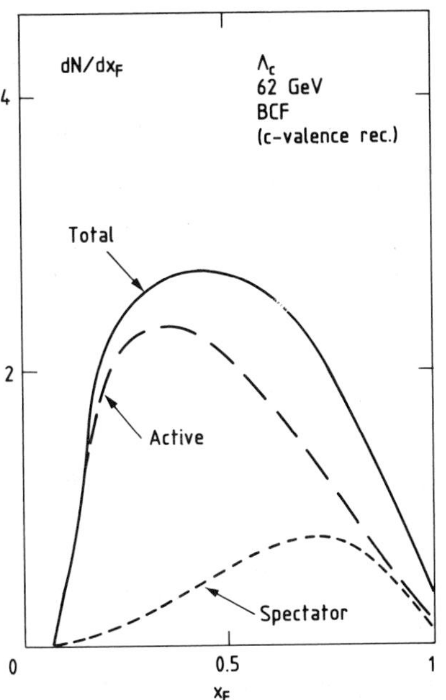

Fig. 16. x_F distribution of fast charmed baryons at $\sqrt{s}$ = 62 GeV.

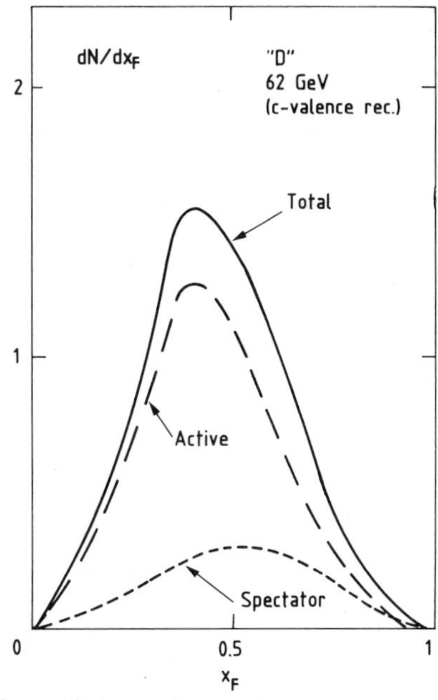

Fig. 17. x_F distribution of fast charmed mesons at $\sqrt{s}$ = 62 GeV.

of the threshold condition become more severe, and in order to pass it the momentum fraction of c_{active} must be favoured at the expenses of $c_{spectator}$. For the same reason the other colliding parton is more energetic on average, which results in a harder spectrum for c_{struck}. As a consequence the role of $c_{spectator}$ in fast hadron recombinations is reduced and that of c_{struck} increased. The theoretical curve for baryons nicely agrees with the Λ_c data of the BCF experiment[31] (the relative normalization is free, Fig. 18). A similar comparison is not possible for D mesons, since existing data do not cover the $x_F \gtrsim 0.5$ region.

Fig. 19 shows a comparison of calculated p_T distributions with the BCF data[23] for Λ_c and D^o. The mean p_T of these data has been used to fix Q^2_{evol}.

It is interesting to look at the x distribution of spectator c quarks entering baryon recombinations (Fig. 20). One learns that once all effects are quantitatively taken into account (including transverse momenta, and thus the m_T fluctuations which alter rapidities) $c_{spectator}$ must not be necessarily very fast, but just moderately fast, in order to recombine with valence quarks.

The results of Figs. 14-17 for the spectator recombinations can be qualitatively reproduced with a simple, although approximate, analytic treatment. If we introduce a recombination probability $R = (x_c/x_{hadron})^n$, which penalizes recombinations of slow charm quarks, and we further approximate $x_c \simeq x_G$, where x_G is the momentum fraction of the gluon initiating the evolution cascade, by making simple analytic integrals one obtains at large x $(n = 1)$: $f_{baryon}(x) \propto (1+24(1-x))$ and $f_{meson}(x) \propto (1-x)^2$, in qualitative agreement with the Monte Carlo results.

The meson distributions which have been obtained are of course

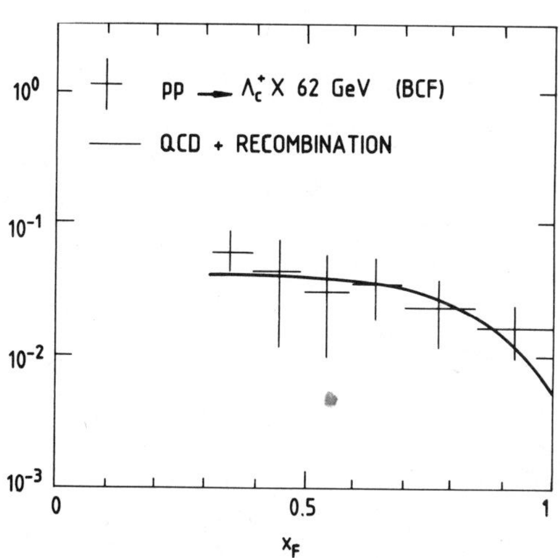

Fig. 18. Calculated x_F distribution for fast charmed baryons at $\sqrt{s}$ = 62 GeV compared with BCF data[31] for Λ_c (the relative normalization is free).

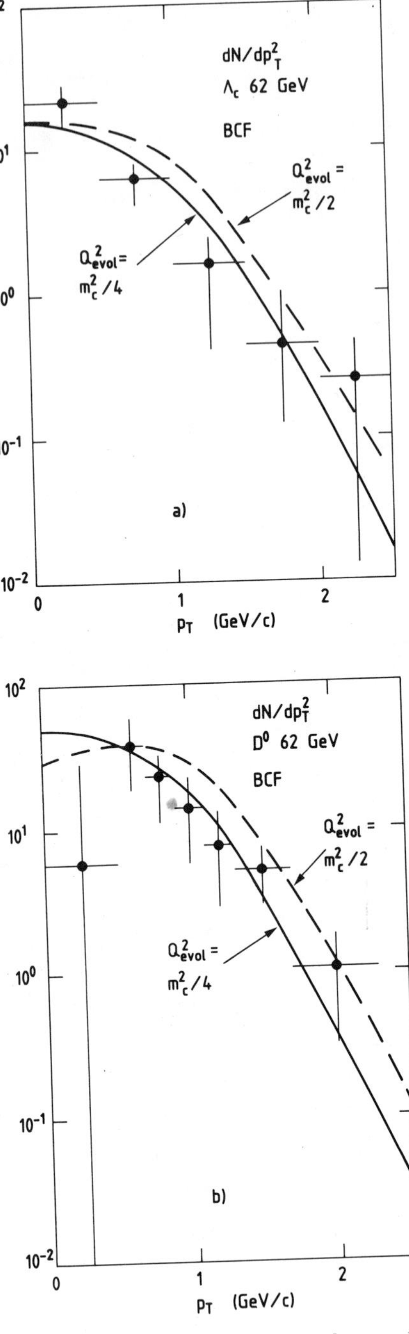

Fig. 19. Calculated p_T distribution of charmed hadrons at $\sqrt{s}$ = 62 GeV compared with experimental data[23] (the relátive normalization is free). a) Λ_c ; b) D^o .

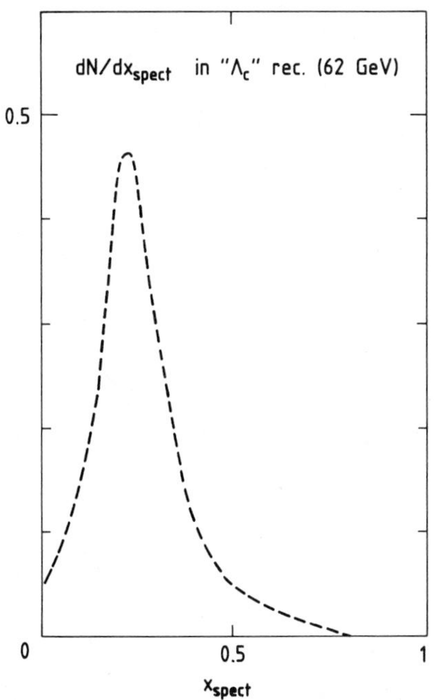

Fig. 20. x distribution of spectator charm quarks entering baryon recombinations with valence quarks ($\sqrt{s}$ = 62 GeV).

meant for charm mesons resulting from recombinations with valence quarks. If a recombination with a sea quark is involved one expects a much softer spectrum. This appears in contrast with the observation of fast D^+ $(c\bar{d})$ in pp collisions. I believe, though, that this does not disprove the direct recombination model, but rather clarifies its limitations. In fact, one may first have a recombination into a high mass excited baryon, e.g. Λ_c^*, which subsequently decays, e.g. $\Lambda_c^* \longrightarrow D^+ + n$. If Λ_c^* is produced with a $\sim(1-x)$ distribution, two body kinematics for the decay gives $\sim(1-x)^2$ for the D^+. In other words, if a channel appears depleted for direct recombinations, it may be easily contaminated or even refilled by multi-step processes. Therefore care must be used in interpreting the predictions of the model in such cases.

SOFT CHARMED MESONS

While fast charmed hadrons can be discussed by a simple, handy model involving recombinations with only the few valence quarks, for slower charmed hadrons one should correspondingly consider recombinations with sea quarks. No viable way of doing that is in view, however, at least at present. Existing data, though, allow us to make some observations of phenomenological interest on the production of soft charmed mesons.

Fig. 21 shows the Feynman x_F distributions at various c.m. energies of the charm quark emerging from the interaction in the charm excitation mechanism. Superimposed on them there are data for the x_F distributions of D mesons at accelerator[32] and ISR[23] energies. One realizes that a simple hadronization model based on a

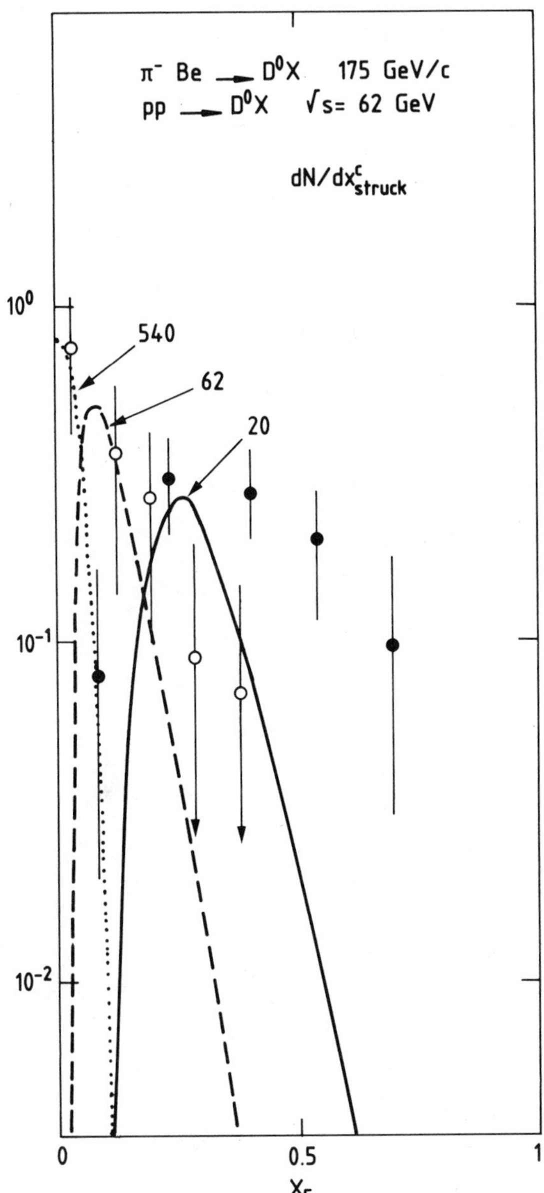

Fig. 21. x_F distributions of the struck charm quark at $\sqrt{s}$ = 20, 62 and 540 GeV[33]. Superimposed on them there are NA11 data[32] for $\pi^- Be \rightarrow D^{*\pm}X$, $D^* \rightarrow D^o\pi$ at 175 GeV/c ($\sqrt{s}$ $\simeq$18 GeV), and pp $\rightarrow$ D^oX data[23] at $\sqrt{s}$ = 62 GeV.

small, fluctuating exchange of momentum between the charm quark and the surrounding spectator sea quarks is enough to bridge the gap between the charm quark spectra and the experimental D meson distributions. In fact, the basic shapes of the D meson distributions (in particular the minimum at $x_F \simeq 0$ at accelerator energies[32]) are already qualitatively approximated by the calculated charm quark spectra, and such a hadronization model would have the effect of smearing the latter so as to reproduce the former.

The results of Fig. 21 for the struck charm quark are easy to understand. For flavour excitation, e.g. $G+c \rightarrow G+c$, the threshold condition for the parton subprocess reads $x_G x_c s \geq \hat{s}_{th}$ (neglecting transverse momenta), where s is the c.m. energy square of the hadron collision, x_G and x_c are the momentum fractions of the colliding gluon and charm quark, respectively, and $\hat{s}_{th}$ represents the threshold for the subprocess. At small s the product $x_G x_c$ must be large. On the other hand x_c is typically small. Therefore x_G must be large, and as a consequence $x_G - x_c$ is large too. After boosting from the c.m. system of the parton subprocess to the laboratory system the final x_F of the struck charm quark is some finite fraction of $\simeq x_G - x_c$, and therefore is large.

Besides the struck charm quark one must also consider the contribution to D meson production from the spectator charm quark, generated in the QCD evolution of the hadrons. Even in this case one should expect a progressive softening of the resulting D meson distribution as energy increases. This time, however, not because of the production characteristics at the parton level, but because of the hadronization process. Within the hadron in which the $c\bar{c}$ pair is originated, after the active charm quark has been emitted to give rise to the interaction, all the remaining spectator partons, including the charm spectator, must recombine into a system of hadrons of a certain invariant mass M_x. The value of M_x considerably depends, at low energy, on the recombination process undergone by the charm spectator. If $c_{spectator}$ recombines with other two spectator quarks into a baryon (e.g. Λ_c) M_x can be relatively low. But if $c_{spectator}$ recombines into a D meson and among the fragmentation products there is also a leading nucleon, M_x may be rather large when the D meson is produced centrally. For $y_D \simeq 0$ and $y_N \simeq (y_N)_{max}$ one obtains

$$\frac{M_x}{\sqrt{s}} \gtrsim \sqrt{\frac{m_D}{\sqrt{s}}} \simeq \begin{cases} 0.30 & \sqrt{s} = 20 \text{ GeV} \\ 0.17 & \quad '' \quad 60 \quad '' \\ 0.06 & \quad '' \quad 540 \quad '' \end{cases}$$

(If the nucleon carries only $\frac{1}{2}$ of its maximum momentum these numbers should be divided by $\sqrt{2}$). At low energy, therefore, the production of central D mesons on the part of $c_{spectator}$ represents an expensive proposition when viewed within the framework of global kinematics, with which the hadronization mechanism must of course comply. The price to pay becomes progressively lower as energy increases, and also in this case, therefore, one should expect a corresponding softening of the D meson distribution.

CHARM PRODUCTION AT COLLIDER ENERGIES

In order to study the implications of a very soft charm meson distribution at collider energies I have incorporated it in a Monte

Carlo simulation of the events[33]. Ordinary events, containing only light hadrons, are treated according to a longitudinal phase-space model including leading baryon effects. Charm events are simulated, for the sake of our considerations, by generating a pair of D mesons according to the charm quark distribution of Fig. 21 at $\sqrt{s}$ = 540 GeV, and degrading the rest of the energy into light hadrons as for ordinary events. Admittedly such a simulation is rather crude, but it should be enough to clarify the effects of a soft D meson distribution as that suggested by Fig. 21. We have assumed a ratio of events with charm to events without charm of 1/3 , roughly corresponding to a charm cross section $\sigma(c\bar{c}) \simeq$ 10 mb, expected from the perturbative QCD model, a total cross section $\sigma_T \simeq$ 50-60 mb, and a fraction of elastic events and single diffraction dissociation events (to which the UA5 experiment is not sensitive) of about 18% each[34].

Fig. 22 shows a comparison of the Monte Carlo simulation with data from the UA5[35] and UA1[36] experiments. Basically, charm decay products are concentrated in the central region, and give rise there to an increase of the visible transverse energy by about $\simeq$ 0.5 GeV per unit of pseudorapidity, mainly because of the kaon content. Of course, similar effects can also be obtained by increasing $\langle p_T \rangle$ or, as far as the η distribution is concerned, by increasing the mean energy fraction carried away by the leading baryons. The comparison of Fig. 22 is only meant to show that our expectations are compatible with existing data.

The most notable effect to be caused by the softening of the D meson distribution at collider energies is the considerable enrichment of the charm event fraction in high multiplicity events (Fig. 23a). In spite of the fact that the global event ratio (charm)

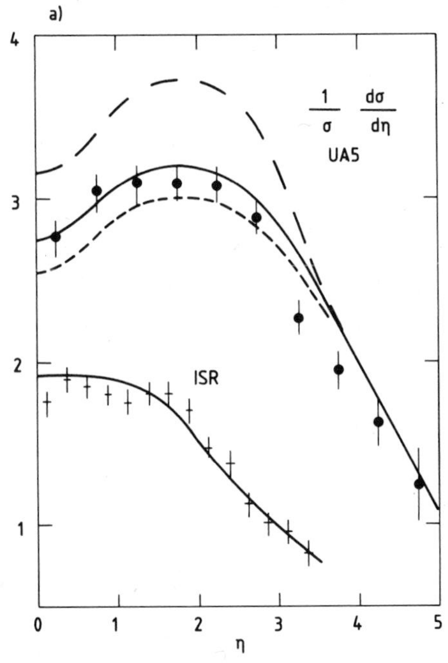

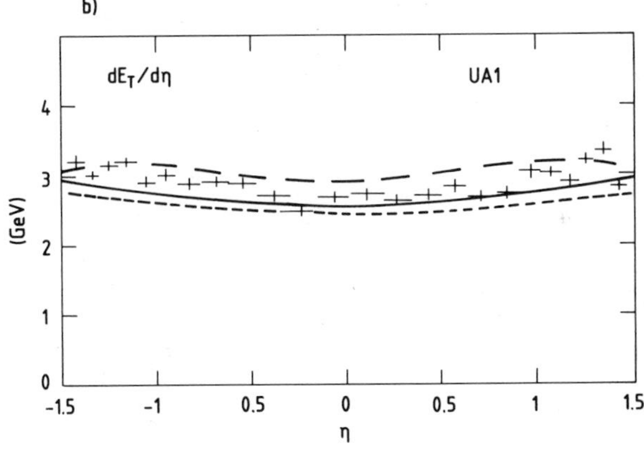

Fig. 22. a) Comparison of Monte Carlo results (solid curves) with data for the pseudorapidity distribution of the UA5 collaboration[35] at the CERN collider ($\sqrt{s} = 540$ GeV) and ISR ($\sqrt{s} = 53$ GeV). Long and short dashed curves correspond to the same quantity for events with charm ($\sigma_C^{-1} d\sigma_C / d\eta$) and without charm ($\sigma_{NC}^{-1} d\sigma_{NC}/d\eta$), respectively.
b) Comparison of Monte Carlo results (solid line) with data for the visible transverse energy per unit of pseudorapidity measured in the UA1 experiment[36]. Long and short dashed lines correspond to the same quantity for events with charm and without charm, respectively.

/(no charm) has been taken equal to 1/3, for 60 $\lesssim n_{ch} \lesssim$ 90 the ratio between the two becomes almost one to one. This is a consequence of the fact that soft D mesons give an extra contribution to the multiplicity without spending much of the available energy. A practical implication is that the search for charm mesons should be considerably more efficient when restricted to high multiplicity events. (Incidentally, in Fig. 23a a better fit to the data could be achieved increasing the fraction of the charm component). The association of charm events (or charm event candidates) with high multiplicities has already been noted earlier in cosmic ray data[37].

From the Monte Carlo one can calculate the electron/pion ratio. In Fig. 23b e/π is plotted at $\theta \simeq 90°$ as a function of p_T. Clearly, a measured $e/\pi \simeq 10^{-2} - 10^{-3}$ would provide an immediate indication that $\sigma(c\bar{c})$ has increased up to several millibarns at the CERN collider.

BOTTOM AND TOP PRODUCTION

The stability of the calculation for the charm excitation cross section has largely depended on the use of experimental input to fix the evolution of the charm sea (F_2(charm) data) and of the interaction radius (p_T distribution of charmed hadrons). For higher flavours crucial information for the heavy sea is missing at present. This is a pity, because of the overwhelming interest in having reliable predictions for the yields of heavy flavour particles. E.g., in connection with the design of experiments devoted to the identification of such particles and with future machine projects. More theoretical effort should go into that, but at the moment there are no firm results available.

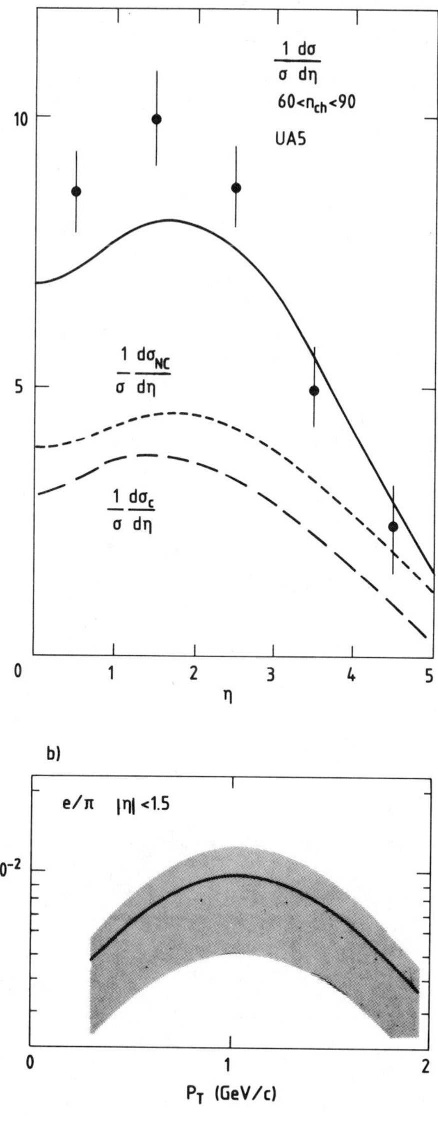

Fig. 23. a) Pseudorapidity distribution for $60 \lesssim n_{ch} \lesssim 90$. Data points from [35]. The solid line represents the global result of the Monte Carlo. Long and short dashed lines represent the contributions to the global result from charm events ($\sigma^{-1}\, d\sigma_C/d\eta$) and events without charm ($\sigma^{-1}\, d\sigma_{NC}/d\eta$), respectively. b) e/π ratio near 90° versus p_T. The shaded area corresponds to an estimate of the error deriving from the uncertainties on $\sigma(c\bar{c})$ and σ(inelastic).

CONCLUSIONS

The difficulties met by perturbative QCD models in explaining the data for hadroproduction of open charm appear to go away after a proper determination of the flavour excitation contribution. The calculation for the charm cross section that has been discussed in some detail[8] does not suffer from the ambiguities encountered in previous attempts because: i) an appropriate technique, taking into account phase-space bounds, is used to calculate the evolution of the charm sea; ii) the remaining uncertainty for the charm sea is eliminated using the recent F_2(charm) data; iii) the radius of the interaction, left theoretically undefined because of the perturbative $\propto 1/\hat{t}^2$ divergence, is fixed using recent measurements of the p_T distribution of charmed hadrons. As a result, the level and the energy behaviour observed for the charm cross section are correctly reproduced.

The leading particle effect observed in Λ_c production appears to be naturally understood with a simple recombination model.

Using guidance from existing data for the D meson longitudinal distribution, the perturbative QCD model suggests a progressive softening of such a distribution with increasing energy, and in particular at collider energies. As a consequence, with a charm cross section $\sigma(c\bar{c}) \simeq 10$ mb expected at the CERN $\bar{p}p$ collider (which should bring the e/π ratio up to a value $\sim 5\ 10^{-3}$ near 90°) almost one every two events with high n_{ch} ($\gtrsim 60$) should contain charm mesons[33].

Unfortunately, predictions for higher flavours are unreliable at present because of lack of the necessary experimental input to stabilize the calculation.

REFERENCES

1. D. Treille, Proc. Int. Symp. on Lepton and Photon Interactions at High Energies, Bonn (1981), p. 750.
2. C. Fisher, Proc. XXIth Int. Conf. on High Energy Physics, Paris (1982).
3. R.J.N. Phillips, Proc. Int. Conf. on High Energy Physics, Madison (1980), p. 1470.
4. F. Halzen, Proc. XXIth Int. Conf. on High Energy Physics, Paris (1982).
5. G. Gustafson and C. Peterson, Phys. Lett. 67 B, 81 (1977); C. Peterson, Talk at the Topical Workshop on Forward Production of High-Mass Flavours at Collider Energies, Paris (1979), Nordita preprint 80-2 (1980).
6. S.J. Brodsky, P. Hoyer, C. Peterson and N. Sakai, Phys. Lett. 93 B, 451 (1980);
   S.J. Brodsky, C. Peterson and N. Sakai, Phys. Rev. D 23, 2745 (1981);
   D.P. Roy, Phys. Rev. Lett. 47, 213 (1981);
   R.V. Gavai and D.P. Roy, Tata preprint, TIFR/TH/81-7 (1981).
7. R. Horgan and M. Jacob, CERN-TH 2824 (1981).
8. R. Odorico, University of Bologna preprint IFUB 82/3 (1982), to be published in Nucl. Phys. B; Proc. of the Workshop on Forward Collider Physics, Madison, Wisconsin (1981), p. 100 ;
   Talk at the Moriond Meeting on Heavy Flavours, Les Arcs (1982).

9. B.L. Combridge, Nucl. Phys. B 151, 429 (1979).

10. A.J. Buras and K.J.F. Gaemers, Nucl. Phys. B 132, 249 (1978).

11. A.R. Clark et al. (BFP Collaboration), Phys. Rev. Lett. 45, 1465 (1980).

12. C.H. Best (EMC Collaboration), Talk at the XVI Rencontre de Moriond, Les Arcs (1981), Rutherford preprint RL-81-044 (1981); G. Coignet (EMC Collaboration), Talk at the EPS Int. Conf., Lisbon (1981).

13. V. Barger, F. Halzen and W.Y. Keung, Phys. Rev. D 25, 112 (1982).

14. R. Odorico, Nucl. Phys. B 172, 157 (1980);
P. Mazzanti and R. Odorico, Phys. Lett. 95 B, 133 (1980);
Z. Physik C 7, 61 (1980);
R. Odorico, Phys. Lett. 103 B, 465 (1981);
P. Mazzanti, R. Odorico and V. Roberto, Nucl. Phys. B 193, 541 (1981);
G.C. Fox and S. Wolfram, Nucl. Phys. B 168, 285 (1980);
S. Wolfram, Talk at the XIV Rencontre de Moriond, Les Arcs (1980);
R.D. Field, Talk at the Conference on Perturbative QCD, Tallahassee (1981).

15. R. Odorico, Phys. Lett. 102 B, 341 (1981).

16. R. Odorico, Computer Physics Communications 25, 253 (1982).

17. A. Sansoni, Thesis, University of Bologna (1981).

18. R. Odorico, B 199, 189 (1982).

19. G.C. Fox, Lectures at the SLAC Summer Institute (1981);
G.C. Fox and R.L. Kelly, Proc. of the Workshop on Forward Collider Physics, Madison, Wisconsin (1981), p.435.

20. R. Odorico, Computer Physics Communcations 24, 73 (1981).

21. R. Odorico, Phys. Lett. 107 B, 231 (1981).

22. M. Strovink, Proc. Int. Symp. on Lepton and Photon Interactions

at High Energies, Bonn (1981), p. 594.

23. M. Basile, CERN-EP/81-23, 127 (1981), submitted to Lett. Nuovo Cimento.

24. R.K. Ellis, M.A. Furman, H.E. Haber and I. Hinchliffe, Nucl. Phys. B 173, 397 (1980);
W. Furmanski and W. Slominski, Cracow preprint, TPJU-11/81 (1981);
W. Celmaster and D. Sivers, Argonne preprint, ANL-HEP-PR-80-61 (1980);
J.F. Owens, Talk at the XIII Int. Symp. on Multiparticle Dynamics, Notre Dame (1981).

25. R.D. Field and R.P. Feynman, Nucl. Phys. B 136, 1 (1978).

26. K.P. Das and R. Hwa, Phys. Lett. 68 B, 459 (1977);
L. Van Hove, CERN-TH 2628 (1979).

27. D. Drijard et al., Phys. Lett. 81 B, 250 (1979); 85 B, 452 (1979).

28. J.D. Bjorken, Phys. Rev. D 17, 171 (1978).

29. J.F. Gunion, Phys. Rev. D 12, 3469 (1975);
S.J. Brodsky and P. Lepage, SLAC-PUB-2478 (1980);
J.F. Gunion, Talk at the XI Int. Symp. on Multiparticle Dynamics, Bruges (1980), SLAC-PUB-2607 (1980).

30. S.J. Brodsky and G.R. Farrar, Phys. Rev. Lett. 31, 1153 (1973).

31. M. Basile et al., CERN-EP/81-22 (1981), submitted to Lett. Nuovo Cimento.

32. P. Weilhammer (NA11 Collaboration), Talk at the Moriond Meeting on Heavy Flavours, Les Arcs (January, 1982).

33. R. Odorico, University of Bologna preprint IFUB 82/10 (1982), to be published in Phys. Lett. B.

34. G. Alberi and G. Goggi, Phys. Rep. 74, 1 (1981).

35. K. Alpgard et al. (UA5 Collaboration) CERN/EP 81-152, 153 (1981);
R. Meinke (UA5 Collaboration), Proc. of the Workshop on Forward

Collider Physics, Madison, Wisconsin (1981), p. 347.

36. G. Arnison et al. (UA1 Collaboration), CERN preprint (1981); A. Kernan (UA1 Collaboration), Proc. of the Workshop on Forward Collider Physics, Madison, Wisconsin (1981), p. 325.

37. T.K. Gaisser and F. Halzen, Phys. Rev. D 14, 3153 (1976); F. Halzen, Talk at the VII Int. Coll. on Multiparticle Reactions, Tutzing, Germany (1976).

D I S C U S S I O N

CHAIRMAN: R. Odorico

Scientific Secretaries: Z. Dubničková and J.L. Goity

DISCUSSION

- *LOEWE:*

Can you explain in a simple way why the flavour excitation mechanism dominates over the fusion mechanism?

- *ODORICO:*

Since $\sigma_{fusion} \propto \frac{1}{\hat{s}} \log \hat{s}$ and $\sigma_{excitation} \propto$ const., above some energy $\sigma_{excitation} > \sigma_{fusion}$. The different energy behaviours are due to having exchanged in the $\hat{t}$ channel a quark (spin ½) in the case of fusion, and a gluon (spin 1) in the case of excitation.

- *LOEWE:*

You did a Monte Carlo analysis for your leading log approximation. Why did you not do it analytically using, for example, the KUV (Konishi, Ukawa, Veneziano) model?

- *ODORICO:*

Analytically one can only calculate inclusive quantities (double inclusive at most). In our case one needs full (exclusive) information about the longitudinal and transverse momenta of the active and spectator charm quarks and of the correlations between them.

- *LOEWE:*

Can you consider some new x-variable (for example Nachtman's x) in order to describe mass effects?

- *ODORICO:*

In our case mass and threshold effects cannot be reproduced (not even qualitatively) by introducing some new effective variable.

- *RAUH:*

How do you explain constant or even rising cross section within asymptotic freedom?

- *ODORICO:*

The relevant scale in our case is not the energy but the heavy quark mass. Therefore by going to higher energies one does not probe the theory at shorter distances.

- *ERREDE:*

You discussed at length hadronic production of charm. Are the characteristics of beauty production similar? For example, you predicted a 10 mb cross-section for charm production at the SPS collider. Would you therefore expect a 1 mb cross-section for beauty at the SPS in pp interactions simply from scaling the squares of the ratio of heavy quark masses?

- *ODORICO:*

In calculating charm production the model is constrained by the existing data for the charm structure function. Similar data are not available for beauty at present.

- *DOBREV:*

Calculating the charm quark interaction, did you use the constituent interchange model or the triple scattering model or what?

- *ODORICO:*

I used only perturbative QCD.

- *GOCKSCH:*

dP_E is proportional to $\alpha_s(k^2)$. Do you take the standard asymptotic freedom $\alpha_s(k^2)$? If so, is your fit to the total cross-section including fusion and flavour excitation good enough to say anything about Λ_{QCD}?

- *ODORICO:*

I took Λ_{QCD} = 0.5 GeV in the calculation. By changing it within reasonable limits the results are only slightly modified. Because of other sources of uncertainty, however, this is certainly not the best way to determine Λ_{QCD}.

- *HUSTON:*

Could you repeat the motivations behind Halzen applying the $A^{2/3}$ correction to the charm cross-sections on nuclear targets?

- *ODORICO:*

They are essentially empirical motivations. I refer you for that to Halzen's talk at the Paris Conference (1982).

EXACT RESULTS FOR pp and p$\bar{p}$ DIFFRACTION SCATTERING AT HIGH ENERGIES

A. Martin

CERN, Geneva, Switzerland

INTRODUCTION

The CERN colliders offer a unique opportunity to test general consequences of analyticity and unitarity of scattering amplitudes obtained during the period 1960 to 1975, and considered at the time as very remote from experiment. The ISR, allowing a comparison of pp and p$\bar{p}$ amplitudes and total cross-sections at energies from 12 + 12 to 31 + 31 GeV makes it possible to test the Pomeranchuck theorem and its generalizations. The SPS collider makes it possible to explore a new range of energies, 540 GeV c.m. (equivalent to 150 000 GeV lab energy) and test asymptotic bounds and high energy models. In fact there is some possibility to run at 400 + 400 GeV equivalent to 340 000 GeV lab energy.

POMERANCHUCK THEOREMS AND ISR EXPERIMENTS

Notations

E = lab energy = s/2M (in GeV) $F^+ = F_{pp} + F_{p\bar{p}}$, even signature amplitude, $F^- = F_{p\bar{p}} - F_{pp}$, odd signature amplitude.

The Pomeranchuck theorem says:

$$\sigma_{Tot}^{\bar{p}p} - \sigma_{Tot}^{pp} \to 0 \quad \text{if} \quad F^-/(E\log E) \to 0 \text{ for } E \to \infty \; ^{1)}$$

or $\mathrm{Re}F^-/(\mathrm{Im}F^- \log E) \to 0$ [2] or $\mathrm{Re}F^- \times \mathrm{Im}F^- > 0$ for $E > E_0$.[3]

F^-, up to recently, was not directly accessible to experiment and one could prefer to the weaker statement[4]:

$$\sigma_{tot}^{p\bar{p}} / \sigma_{tot}^{pp} \to 1 \quad \text{if either} \quad \sigma_{tot}^{p\bar{p}} \to \infty \quad \text{or} \quad \sigma_{tot}^{pp} \to \infty$$

Analysis of previous experiments[5], with the best fit

$$\sigma_{tot}^{p\bar{p}/pp} = 42\,E^{-0.37} \pm 24\,E^{-0.55} + 27 + 0.17\,(\log 2E)^{2.1}\ \text{mb}$$

suggest that indeed $\sigma_T \to \infty$.

Finally, there is a version of the Pomeranchuck theorem for elastic scattering[5]: $(d\sigma^{p\bar{p}}/dt)(d\sigma^{pp}/dt) \to 1$ inside the diffraction peak. In particular $b_{p\bar{p}}/b_{pp} \to 1$, where $b(s,t) = d/dt\,[\log(d\sigma/d\omega)(s,t)]$, is the slope parameter.

Two experiments at the ISR test these properties, R 210 and R 211. R 210[6] measures σ_T and $b(s,t)$ for $0.02 < |t| < 0.7\ (\text{GeV})^2$. They find that $\sigma_T^{p\bar{p}} > \sigma_T^{pp}$. Their best fit is

$$\sigma_{tot}^{pp} = 38.3 + 0.5\,\log^2\left(\frac{E}{62}\right)\ \text{mb}$$

$$\Delta\sigma = \sigma_{tot}^{p\bar{p}} - \sigma_{tot}^{pp} = 79\,(2E)^{-0.58}\ \text{mb}$$

In particular

$$\Delta\sigma = 1.85 \pm 0.8 \quad \text{at} \quad p_{lab} = 500\ \text{GeV}$$
$$\Delta\sigma = 1.5 \pm 0.55 \quad \text{at} \quad p_{lab} = 1500\ \text{GeV}$$
$$\Delta\sigma = 0.85 \pm 0.45 \quad \text{at} \quad p_{lab} = 2000\ \text{GeV}.$$

So not only $\sigma_T^{p\bar{p}}/\sigma_T^{pp} \to 1$ but $\sigma_T^{p\bar{p}} - \sigma_T^{pp} \to 0$. Concerning slopes they get, at $E_{CM} = 53$ GeV, $b_{p\bar{p}} = 14 \pm 2.6\ \text{GeV}^{-2}$, $b_{pp} = 12.6 \pm 1.4\ \text{GeV}^{-2}$ for $t < 0.1\ (\text{GeV})^2$ consistent with $b_{p\bar{p}}/b_{pp} = 1$, $b_{p\bar{p}} = 10.4 \pm$ $\pm\ 0.6$, $b_{pp} = 10.4 \pm 0.4$ for $t > 0.1\ \text{GeV}^2$. The theorem on the slopes is therefore satisfied. In fact, for $<|t|> \sim 0.2\ \text{GeV}^2$, we have,

$$b_{p\bar{p}}/b_{pp} = 1.12 \quad \text{at} \quad E_{lab} = 30\ \text{GeV}$$
$$1.05 \quad \text{at} \quad E_{lab} = 100\ \text{GeV}$$
$$1.00 \pm 0.06 \quad \text{at} \quad E_{lab} = 1500\ \text{GeV}$$

R 211[7] measures differential cross-sections at very small t, $10^{-3} < |t| < 60 \times 10^{-3}$ GeV^2, and sees the Coulomb peak and the Coulomb nuclear interference. Hence they measure $\rho = \text{ReF}/\text{ImF}$, σ_T and $b(s,t \cong 0)$, both for pp and $p\bar{p}$ scattering. Their result are at

$E_{CM} = 30.4$ GeV	$E_{CM} = 52.8$ GeV
$\sigma_{Tot}^{pp} = 41.7 \pm 0.5$ mb	$\sigma_{tot}^{p\bar{p}} = 43.68 \pm 0.2$ mb
$\Delta\sigma = 1.5 \pm 0.5$ mb	$\Delta\sigma = 0.98 \pm 0.36$ mb
$\rho_{pp} = 0.040 \pm 0.005$	$\rho_{pp} = 0.060 \pm 0.006$
$\rho_{p\bar{p}} = 0.031 \pm 0.021$	$\rho_{p\bar{p}} = 0.101 \pm 0.018$
	$b_{pp} = 12.85 \pm 0.12$ GeV^{-2}
	$b_{p\bar{p}} = 13.36 \pm 0.53$ GeV^{-2}

(Here b is postulated to be 12.2 GeV^{-2} for pp and 12.6 GeV^{-2} for $p\bar{p}$).

It is remarkable that here, for the first time, both the real and imaginary parts of F^{pp} and $F^{p\bar{p}}$, and therefore of F^+ and F^- are obtained. These measurements:

1) favour, like R 210, $\Delta\sigma \to 0$;
2) show that $|\text{ReF}^-/\text{ImF}^-|$ is not large and also that $\text{ReF}^- \times \text{ImF}^- > 0$ and hint that two sufficient conditions for the validity of the Pomeranchuck theorem might be satisfied;
3) make very unlikely unorthodox possibilities such as the presence of "odderons".

"Odderons" are contributing to the odd signature amplitude which, to most phenomenologists, look rather odd.

Odderon 1 behaves like a Froissarton $Cs(\log s - i(\pi/2))^2$, C real. As shown by Łukaszuk and Nicolescu it is not incompatible with s channel unitarity if the even signature amplitude saturates the Froissart bound. The difference between σ_T^{pp} and $\sigma_T^{p\bar{p}}$ goes to $-\infty$ but the ratio of the cross-sections approaches 1 as it should as we first said. The data showing a steady decreases of $\Delta\sigma$ and consistency with extrapolations from low energy seem to exclude this possibility.

Odderon 2 is milder: $C\,s(\log s - i(\pi/2))$, C real. It produces a non-zero asymptotic value for $\Delta\sigma$. For the same reasons as the previous one it is essentially excluded.

Odderon 3 is just a real constant. To exclude it one has to check the over-all consistency of an unsubtracted dispersion relation for the odd signature amplitude. Also it would destroy the relation

$$\mathrm{Re}F^-/\mathrm{Im}F^- = \cot\alpha \quad \text{if} \quad \Delta\sigma \sim s^{-\alpha}$$

The errors on ρ_{pp}, $\rho_{p\bar{p}}$ and $\Delta\sigma$ at 52 GeV c.m. are too large to make a serious test of this relation but the dispersion relation at low energies indicates that the contribution of odderon 3, if any, has to be very small.

On the other hand, the ratio of the slopes, at E_{CM} = 53 GeV, is compatible with unity in agreement with our theorem.

The data at E_{CM} = 62 GeV are not yet analyzed.

From the existing data a fit of σ_T has been proposed, compatible with a $(\log s)^2$ increase, which predicts[8)] σ_T = 71 mb at E_{CM} = 540 GeV or 66 mb if $\sigma_T(\infty)$ = 130 mb while R 210 predicts σ_T = 69 mb at E_{CM} = 540 GeV and Ref. 5) predicts σ_T = 63 mb at E_{CM} = 540 GeV.

A fit of the slope[9)] based on R 211 indicates that a linear fit in log E is unacceptable and that a quadratic fit $b = A + B \log E + (\log E)^2$ is needed.

ASYMPTOTIC BOUNDS AND BEHAVIOUR, AND SPS COLLIDER EXPERIMENTS

The fastest growth of the total cross-section allowed by field theory[10)] is given by the "Froissart bound" first derived from Mandelstam representation[11)]

$$\sigma_{Tot} < \mathrm{Const} \times (\log E)^2 .$$

If

$$\sigma_{Tot}/(\log E)^2 \rightarrow \mathrm{Const} \neq 0 ,$$

a situation which, to the best of our knowledge, is compatible with analyticity and unitarity in all channels[12)] we have[13)]

$$\sigma_{elastic}/\sigma_{Tot} \rightarrow \mathrm{Const} \neq 0 ,$$

$$b(s, t=0) \sim (\log E)^2 ,$$

$$\frac{d\sigma}{dt} = \frac{d\sigma}{dt}(t=0) \times F\left(t(\log E)^2\right),$$

F(z) = entire function of order 1/2 = analytic function of z, bounded by exp $C\sqrt{|z|}$ in the whole complex plane.

Another possible asymptotic behaviour is that of the critical Pomeron[14)], for which

$$\sigma_{Tot} \sim (\log E)^{0.26}$$

$$\sigma_{el}/\sigma_{Tot} \sim (\log E)^{-0.86}$$

$$b \sim (\log E)^{1.13}$$

and

$$\frac{d\sigma}{dt} = \frac{d\sigma}{dt}(t=0)\; G\left(t(\log E)^{1.13}\right)$$

Data at energy $\leq$ ISR energies are compatible with a qualitative saturation of the Froissart bound [the best fit of Ref. 5) gives $\sigma_T \sim (\log E)^{2.1\pm0.1}$], but this is based on a 10% increase of σ_T. The great interest of the SPS collider is that it allows to observe a much more impressive effect. The UA4 collaboration has measured at 540 GeV c.m. the slope of the diffraction peak for 0.05 < < $|t|$ < 0.18 GeV^2 [15)]; it is

$$b = 17.2 \pm 1.0 \; GeV^{-2}$$

By extrapolating to t = 0, they get L × dσ/dt (0) = const × L × × $(\sigma_T)^2(1+\rho^2)$. L luminosity can be eliminated by measuring the total rate of events, L σ_T, and one gets (with $\rho = 0$; remember that $\rho \leq 0.15$)

$$\sigma_{Tot} = 66 \pm 7 \pm 3 \text{ mb}.$$

This measurement is compatible with the various extrapolations from ISR and slightly favours a model saturating the Froissart bound.

The slope, thought first to be anomalously large, is in fact compatible with a model saturating the Froissart bound but also with the critical Pomeron.

The UA1 group, on the other hand, has made measurements of b at larger t $0.15 < |t| < 0.26\ GeV^2$ and found a much smaller value[16])

$$b = 13.3 \pm 1.5\ GeV^{-2}$$

The tendency of the slope to decrease with $|t|$ is seen at lower energies but seems to be more marked here. This is not incompatible with theory because, if $\sigma_T \sim (\log E)^2$ we expect a non-uniform behaviour of $b(E,t)$ as a function of t: $b(s,t=0) \sim (\log s)^2$ while $b(s,t)_{t\ fixed < 0} \sim \log s$.

In this short review[17]) we have tried to show the great potential of the SPS collider elastic scattering experiments. It is only in one or two years'time that, with increased accuracy, we shall be able to draw firmer conclusions. If real part measurements were also made at these energies they would allow to guess what happens to σ_{total} at much higher energies (as was done at the ISR) and this would be a most valuable information.

REFERENCES

1. A. Martin, Nuovo Cimento 39:704 (1965).
2. N.N. Khuri and T. Kinoshita, Phys. Rev. B140:707 (1965).
3. J. Fischer, Phys. Reports 76:157 (1981).
4. R.J. Eden, Phys. Rev. Lett. 16:39 (1966);
   T. Kinoshita, in Perspectives in Modern Physics, Ed. R.E. Marshak, Wyley and Sons, New York (1966), p. 211.
5. U. Amaldi et al., Phys. Lett. 62B:460 (1976).
6. G. Carboni et al., Phys. Lett. 108B:87 (1982);
   M. Ambrosio et al., Phys. Lett. 115B:495 (1982).
7. D.Favart et al., Phys. Rev. Lett. 44:1191 (1981) and Communication by N. Amos et al., to the XXI International Conference on High Energy Physics, Paris 1982.
8. M. Block and R. Cahn, CERN preprint TH.3307 (1982).
9. M. Block and R. Cahn, CERN preprint TH.3342 (1982).
10. A. Martin, Nuovo Cimento 42:930 (1966).
11. M. Froissart, Phys. Rev. 123:1053 (1961).
12. J. Kupsch, CERN preprint TH.3282 (1982), to appear in Nuovo Cimento.

13. G. Auberson, T. Kinoshita and A. Martin, Phys. Rev. D3:3185 (1971).
14. See for instance,
M. Moshe, Nucl. Phys. B198:13 (1982).
15. R. Battiston et al., Phys. Lett. 115B:333 (1982) and EP 82-111 (1982), to appear in Phys. Lett.
16. G. Arnison et al., Communication to the XXI International Conference on High Energy Physics, Paris 1982.
17. For more theoretical details, see
A. Martin, Zeitschrift für Physik C15:185 (1982).

DISCUSSION

CHAIRMAN: A. Martin

Scientific Secretaries: A. van de Ven and M. Kremer

DISCUSSION

- *KREMER:*

It is certainly not possible to measure physical quantities up to infinity. If the assumptions of the Pomeranchuk theorems are only satisfied in a certain large but finite region only, is it then possible to obtain results which are similar to the Pomeranchuk theorems, and which maybe also only hold in a certain finite region?

- *MARTIN:*

I think, one has to have an up-to-infinity-condition. I will give you one example. You have to know something about what is happening at infinity, but not very much. In pion - nucleon - scattering there is the Goldberger sum rule which says that the $I = \frac{3}{2}$ scattering length minus the $I = \frac{1}{2}$ scattering length is equal to something like

$$\int_{m_\pi}^{\infty} \frac{\sigma^+(E) - \sigma^-(E)}{E - m_\pi} dE$$

($\sigma^\pm = \pi^\pm$ p cross section)

(this formula holds if there are no subtractions in the sum rule and if the integral converges).

The scattering length is measured in low energy πN scattering. If you cut off the integral at some finite energy E_{max}, you have still to make the assumption that beyond E_{max}, $\sigma^+ - \sigma^-$ has a definite sign, in order to know the sign of the error you are making. If you know this sign, you obtain inequalities like

$$a_{\frac{3}{2}} - a_{\frac{1}{2}} \geqslant \int_{m_\pi}^{E_{max}} \frac{\sigma^+ - \sigma^-}{E - m_\pi} dE$$

- *VAN DE VEN:*

You mentioned that the experimentally apparent saturation of the Froissart bound implies a geometrical picture for pp and $p\bar{p}$ scattering. Can you explain what this geometrical picture is?

- *MARTIN:*

If σ_{tot} satisfies the Froissart bound i.e.

$$\sigma_{tot} \sim c(\log s)^2$$

then one can prove that the elastic cross section σ_{el} behaves like $(\log s)^2$. I will give you a proof for the special case when the scattering amplitude is dominantly imaginary (this assumption is not really necessary). There is a bound on the logarithmic derivative of the absorptive part of the amplitude

$$\frac{d}{dt} \log A(s,t)\Big|_{t=o} > C_1 \frac{\sigma_{tot}}{4\pi}$$

$$\frac{d}{dt} \log A(s,t)\Big|_{t=o} > C_2 \frac{(\sigma_{tot})^2}{4\pi\sigma_{el}}$$

which can be derived from unitarity.

Because $A(s,\cos\theta) = \sum_\ell (2\ell + 1) \, \mathrm{Im}\, f_\ell \, P_\ell(\cos\theta)$, unitarity yields

$$0 < \mathrm{Im}\, f_\ell < 1.$$

From this one can derive two bounds for σ_{tot} and for $\sigma_{tot}^2/\sigma_{el}$ (Martin et al.).

$$\frac{d}{dt} \log A(s,t) \bigg|_{t=o} = \frac{b(s,o)}{2}$$

(because $\sigma = |A|^2$).

If $\sigma_{tot} \sim (\log s)^2$, then also $b \sim (\log s)^2$ and $\frac{\sigma_{tot}}{\sigma_{el}} \to$ const.

Proof : from the bound on $\frac{\sigma_{tot}^2}{\sigma_{el}}$ we see that if $\sigma_{tot}/\sigma_{el} \to \infty$, one would get that b grows faster than $(\log s)^2$, because then

$$\frac{\sigma_{tot}^2}{\sigma_{el}} \to \infty \times (\log s)^2. \qquad (*)$$

But this is impossible because of the analyticity properties of the scattering amplitude.

The absorptive part of the scattering amplitude is defined in the physical region $-1 \leqslant \cos\theta \leqslant +1$. It can be continued inside an ellipse with boundary $\cos\theta = 1 + \frac{4\,m^2}{2\,k^2}$ where k is the c.m. momentum. In the t variable the region is bounded by $t = 4\,m^2$.

It can be proved (Y.S.Jin and A. Martin) that inside this ellipse the absorptive part of the scattering amplitude is bounded by s^2. This implies

$$b(s,0) < (\log s)^2.$$

Together with (*) this yields

$$b(s,0) \simeq (\log s)^2.$$

But then the bound on $\frac{\sigma_{tot}^2}{\sigma_{el}}$ tells us that $\frac{\sigma_{tot}}{\sigma_{el}}$ is bounded.

So we have the following asymptotic behaviour :

$$\sigma_{tot} \simeq (\log s)^2$$

$$b \simeq (\log s)^2$$

$$\frac{\sigma_{tot}}{\sigma_{el}} \simeq \text{const}$$

- *CASTELLANI:*

Could you briefly sketch the proof of one of the Pomeranchuk - type theorems? The result seems to be intuitive ($\sigma^{tot}_{pp} - \sigma^{tot}_{p\bar{p}} \underset{s \to \infty}{\to} 0$), but I would like to see where the hypothesis of the theorem comes into play.

- *MARTIN:*

Let us prove the theorem 2 with assumption 1, i.e.

$$\frac{F^-}{E \log E} \to 0$$

Take the odd signature amplitude (which controls the Pomeranchuk theorem) :

$$F^- = F^{\text{particle}} - F^{\text{antiparticle}}.$$

The full scattering amplitude is analytic in the cut E - plane.

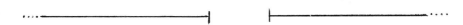

It has two cuts. Approaching the right hand cut from above yields the particle - particle amplitude, whereas approaching the left-hand cut from below yields the particle - antiparticle amplitude.

$$f(z) := \frac{F^-}{E}$$

defines an even analytic function of $z = E^2$ with a single cut. $f(z)$ is bounded by $\sqrt{z}(\log z)^2$ (Froissart bound). So one can write a dispersion relation for $f(z)$:

$$f(z) = A + \frac{z}{\pi} \int_{z_o}^{\infty} \frac{\operatorname{Im} f(x)\, dx}{x(x - z)}$$

According to our assumption

$$\frac{f(z)}{\log z} \to 0 \qquad \text{along the physical cut.}$$

Now we use a mathematical theorem :

If f is an analytic function in the cut complex plane and if

$$\lim_{|z| \to \infty} \frac{f(z)}{e^{\sqrt{|z|}}} = 0 \qquad \text{in all complex directions}$$

and $\quad f(z) \to 0 \qquad$ along the cut

then $\quad f(z) \to 0 \qquad$ in all directions.

In particular we have then

$$\lim_{z \to -\infty} \frac{f(z)}{\log z} = 0 \qquad \text{along the real axis.}$$

Now $\operatorname{Im} f(x)$ is proportional to $\sigma^P - \sigma^A$. Assume that for $t > x_o$

$$\sigma^P(t) - \sigma^A(t) > \varepsilon > 0$$

i.e. $\operatorname{Im} f(x) > \varepsilon$ for $x > x_o$. Putting $z = -t$ with t real and positive we obtain from the dispersion relation

$$f(-t) > A - \frac{t}{\pi} \int_{z_o}^{x_o} \frac{\mathrm{Im}\ f(x)dx}{x(t+)} - \frac{t\varepsilon}{\pi} \int_{x_o}^{\infty} \frac{dx}{x(t+x)}$$

The last term behaves like log t, so now we have a contradiction, since we get

$$\frac{f(t)}{\log t} \nrightarrow 0.$$

So either $\sigma^P - \sigma^A \to 0$, or there is still the other possibility that Im f(z) is an oscillating function. But it is certain that

$$\int_{z_o}^{\infty} \frac{\sigma^P - \sigma^A}{E}\, dE \qquad \text{converges.}$$

- *HOFSÄSS :*

I wonder whether QCD has anything to say about the results on pp and $p\bar{p}$, which you presented. In particular I have in mind the Regge calculations which have been done in QCD. Have they been pushed to a level at which one can compare them to these experiments?

- *MARTIN:*

There is a claim by A. White that QCD has much to do with the behaviour of these total cross sections.

Second, if QCD enters, it is via a model of the pomeron. In a very simple model the pomeron consists of a chain of pairs of gluons in QCD. So it is in this kind of hidden way that QCD contributes to these questions.

- *ETIM:*

I do not think that A. White believes that QCD says something precise about hadron - hadron scattering. He showed that one can have Regge behaviour in a non Abelian gauge theory. He also tried to incorporate hadrons, but he did not succeed in this point; but the main emphasis of his work was on quarks and gluons.

- *A.N:*

If the total cross section rises as $(\log s)^2$, does that mean that the Pomeranchuk singularity is a branch cut?

- *MARTIN:*

It can only be a triple pole or a branch cut.

- *LOEWE:*

Can one say that the present experimental data exclude the critical pomeron?

- *MARTIN:*

No. First, one should wait for better data. Second, if the cross section becomes of the order of 66 mb, then the specific model of Moshe-Moshe could be excluded, but it is always possible to construct a model for critical behaviour of the pomeron, which agrees with present experimental data and reaches the asymptotic regime only for extreme energies.

- *YANG:*

I will make some remarks concerning the very exciting fact that is verified experimentally at the collider and the ISR, namely, that pp and $p\bar{p}$ total cross sections σ_t are increasing at higher energies. Prof. Martin just told us about the theoretical work concerning this fact, utilizing the concept of the analyticity of scattering amplitudes. His work and that of others in this area have been most interesting. Another approach was mentioned by Prof. Landshoff yesterday, the Regge approach, which emphasizes the analyticity in the t-plane. I would like to discuss a third approach which I believe has been very fruitful and will be increasingly so in the future. This is the geometrical approach which emphasizes the impact parameter b. It should be stated that these different approaches are complementary to each other, in that each looks at hadron-hadron collisions from a different view-point. (It is unfortunate that some physicists seem to take the view that they are exclusive of or contradictory to each other.)

The quantities

σ_t = total cross section

σ_{el} = total elastic cross section

and $$b = \frac{d}{dt}(\ln \frac{d\sigma}{dt})\Big|_{t=0}$$

all have the dimension of an area, probably all $\to \infty$ as $E_{CM} \to \infty$ in pp and $p\bar{p}$ collisions.

This was discussed in a 1979 Phys. Rev. paper by T.T. Chou and me. Experimentally $z \equiv \sigma_{el}/\sigma_t$ is 1 for E_{CM} below pion threshold in pp collision, decreases rapidly after that, reaching a minimum of about .17 to .18 at ISR energies. Chou and I believe that at the collider energies z will increase to about .22 and would eventually approach ½ as $E_{CM} \to \infty$. That is the value of z for scattering from an infinitely black disc.

If we write the S-matrix, for angular momentum L, as $S_L = S(b)$, where b is defined by L = k.b and k = CM momentum of each incoming particle, then

$$\sigma_t = \iint (2 - S - S^*) d^2b$$

$$\left(\frac{d\sigma}{dt}\right)_{el} = \frac{1}{4\pi}\left|\iint (1 - S) \quad e^{i\vec{k}\cdot\vec{b}} \, d^2b \right|^2 \qquad (\vec{k}^2 = -t)$$

$$\sigma_{el} = \iint |1 - S|^2 \, d^2b$$

$$\sigma_{inel} = \iint (1 - SS^*) d^2b \qquad (1)$$

Chou and I believe that as $E_{CM} \to \infty$, S becomes approximately a step function so that S(b) = 0 for small b, and S(b) = 1 for large b, with a transition region that is not too large in Δb. A different possibility, also in the geometrical approach, is to have S(b) for small b non-vanishing :

$$S(b) \cong S_o > 0 \text{ for small b.}$$

Such is a semi-transparent model. In such a case, according to equation (1) above,

$$\frac{\sigma_{el}}{\sigma_t} \to \tfrac{1}{2}\,(1-S_o) \text{ as } E_{CM} \to \infty.$$

Thus the limit of (σ_{el}/σ_t) is related to whether the hadron-hadron collision becomes totally black, or semi-transparent, which is a very important physical difference.

Related to this is the presence of dips in the elastic angular distribution. With a black disc there should be infinitely many dips, a point that Chou and I have emphasized. Experimentally one dip in pp and $p\bar{p}$ has already been found at $-t = 1.2$ to 1.4 (GeV/c^2), where we predicted it to be. At collider energies maybe a second dip, or a "shoulder", the precursor of a dip, will be found.

- *WIGNER:*

The increase of the cross sections, both elastic and inelastic to infinity when the energy also increases to infinity is surprising. Could you describe it in terms of the collision's angular momenta? It surely means that the cross sections become high even at very high angular momenta when the energy increases but a description of the nature of this increase would be very interesting.

- *YANG:*

The S matrix S_L is found to depend mainly on the impact parameter b defined by $L = k.b.$ Thus $S_L = S(b)$. As the energy increases, the curve S(b) vs. b becomes deeper, and also pushes outwards, as shown.

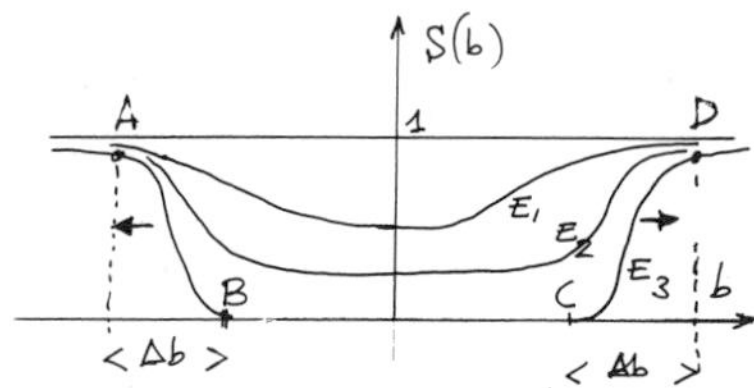

I believe at very high energies the curve assumes the shape of a basin, with the bottom at S = 0. Provided the part A-B is not too wide in Δb, the scattering would be like that from an infinitely absorptive disc, and we should have $\sigma_{el}/\sigma_t \sim \frac{1}{2}$. Such a case is what I called the "black centre" case. The mere requirement that $\sigma_t \to \infty$ does not require that in the centre B-C, S = 0. It could be that there $S = S_o > 0$. Such is what I called a "semi-transparent centre" case. In such a case σ_{el}/σ_t remains less than $\frac{1}{2}$.

- *WIGNER:*

But is it not true that if the cross section increases indefinitely, the length BC over which S(b) = 0 will have to increase indefinitely also? The concern I have expressed is based on a somewhat non quantum mechanical, that is somewhat classical picture, as is the picture referring to S(b). But it would be good, least in order to project an elementary picture, to clear this up.

- *MORSE:*

If the pp total cross section approaches infinity as the energy goes to infinity, the S matrix predicts the strong interaction becomes long range. Is there a simple way to understand this?

- *YANG:*

Yes. The Yukawa tail of the proton extends to infinity. As the energy approaches infinity, new channels are opened up and the range of the strong interaction goes to infinity.

- *HEUSCH:*

I would like to pick up where a previous question left off. It bothers me that here we have two beautifully self consistent presentations by A. Martin and C.N. Yang; in the framework of analyticity and of a geometrical picture. Then, one of our students here, to whom our present orthodoxy teaches that there is a (probably) correct theory of the strong interactions, called QCD, asks for an interpretation of these lecturer's messages in terms of this very QCD theory. The answer he receives, briefly, is that there is no overlap between these approaches.

Now, conceptually, the standard answer to the question of "soft" phenomena (σ_{el}, σ_{diff}, ...) in QCD is: they are dictated by multiple gluon exchange. We resort to such mysterious concepts as automatic color neutralization, a fair amount of general handwaving and I see very few attempts to do anything quantitative. Odorico reasonably argued that for heavy quarks (q = c, b, ..) their mass alone may entitle them to permit a perturbation description of processes involving them. That leaves the largest fraction of such well observed processes as pp $\rightarrow$ X out of the picture.

It must surely appear odd to the student - and, I surmise to all of us - that our most promising theory of the Strong Interaction has little or nothing to say about such a large fraction of the observables of particle dynamics. Is it really true that the elegant treatments of our present speakers have nothing to say about basic hadronic sub-processes?

Landshoff showed us what contortions it takes to calculate certain "simple" vertex corrections where added gluon exchange modifies some basic Pomeron concept. Does it look hopeless that extrapolation and summation concepts can be developed that bridge the gap between basic QCD graphs and global properties of the scattering amplitudes discussed today?

- *MARTIN:*

We have no such connection to date.

- *ZICHICHI:*

The geometrical model described by Prof. Yang gives an energy dependence of σ_{el}/σ_t. The asymptotic limit is $\sigma_{el}/\sigma_t \sim \frac{1}{2}$. At present energies however $\sigma_{el}/\sigma_t \sim 0.2$. Your theorem says that σ_{el}/σ_t should be constant, if the basic three quanties (σ_t, b, σ_{el}) grow with energy as fast as allowed. How do we reconcile the two predictions with each other?

- *MARTIN:*

The theorem only says that σ_{el}/σ_t approaches a constant, different from zero at high energies. The question is, have we already reached the asymptotic regime or not? If Chou and Yang are right this is not the case yet, and there is no real contradiction.

- *ODORICO:*

This is the answer to a student's question about Prof. Yang's model. The question was : "How does asymptotic freedom relate to the geometrical model that you presented?" My answer : In perturbative QCD models calculating heavy flavor production in hadronic interactions one finds contributions to σ(charm), σ_t(bottom), etc., which increase as $\log^2 s$. This implies an infinite total cross section at infinite energy, as in the geometrical picture of Prof. Yang. Application of perturbative QCD is justified in this case by the "large" mass of the heavy quarks.

- *VAN DER SPUY:*

In the geometrical picture of high energy scattering what carries the long range interaction when the radius of the black disc (~ ln s) is greater than 10^{-13}cm?

- *YANG:*

The proton has hadronic interactions around it. If you merely take the tail of the Yukawa potential, this is very weak, of course, at a distance of, let's say, 2 f., but if a proton passes by, although there is not very much of the target proton left at that point, at very high energies there are enormously many channels opened up, and it does interact. It does interact because we know that the cross-section can become quite large. So the effectiveness of the tail is what gives rise to the increasing cross-section. There is always an exponential tail. Now if you say that E^a times e^{-Kr} is of the order of E incident, then you can see that r is proportional to log E. And that says that if effectiveness increases by some power of E then the effective size r would increase logarithmically with E. This is only a heuristic argument.

- *VAN DER SPUY:*

If quarks are confined does that exponential law somewhat change?

- *YANG:*

Confinement does not say that the proton does not have a form factor, and I am saying that the form factor extends over a certain distance and the tail becomes more and more effective as the energy becomes big because of the opening of channels. That is presumably what makes the proton effective cross-section increase as $(\log s)^2$.

- *VAN DER SPUY:*

Does that agree with the bag model?

- *YANG:*

I cannot comment on that.

PHOTON SCATTERING AT VERY HIGH ENERGIES --
OR: HOW DOES THE PHOTON EVOLVE?

Clemens A. Heusch

Institute for Particle Physics
University of California
Santa Cruz, California 95064

INTRODUCTION

In the framework of this Course on the Gauge Interaction, it is appropriate that we take a close look at the quantum of that basic gauge field which we believe to have a solid grip on, at the photon. The carrier of the electromagnetic field has, of course, been shown to be precisely described by Quantum Electrodynamics (QED) in its interactions with point-like leptons. The question of interest here is this: what have we gleaned from the manifestations of the photon in kinematical regimes where its coupling to heavier particles becomes intimately entangled with observables?

The photon "evolves" as energies increase, as transverse momenta in the interaction increase, as its "mass-squared" moves from the light cone into the time-like or space-like regimes. The particle with the quantum numbers of the photon which we observe in hadronic interaction, the "hadronic photon", may well be seen as a source of information on hadronic features that become accessible through its unique couplings. It is particularly instructive to observe parallels and disparities between the "evolution" of the photon and that of the gauge particle of the strong interaction, the gluon.

In this lecture, we will deal exclusively with those features which manifest themselves in the framework of photon-hadron scattering, with $Q^2 = 0$ or $Q^2 > 0$, but small,[1] i.e., of real or slightly space-like photons. What happens as they acquire high energy, as they interact with large transverse momenta, as they move slightly off the mass shell?

The basic graph of photon-hadron scattering is seen in Fig. 1. The photon beam, massless (for real photons) or massive (for the space-like case), interacts with the nucleon or with one of its sub-constituents. The final state will contain a system with baryon number B = 0, reflecting the hadronization of the photon in "soft" processes, and a B = 1 system indicative of the target structure (Fig. 1a). In harder processes, characterized by larger momentum transfers, the photon is known to act more like a point-like probe of the nucleon, so that even the B = 0 component contains information about target "partons" rather than about the hadronic photon components (Fig. 1b).

In an increasingly informative set of experiments that have become possible at e^+e^- machines, these two aspects of photon interactions can be combined: Fig. 2 illustrates that, for the case where one photon has large $Q^2 > 0$ and the other is essentially on the mass shell, $Q^2 \approx 0$, one acts as the point-like probe of the hadronic structure of the other. These promising investigations, realizable in the reaction

$$e^+e^- \to e^+e^-(\gamma\gamma) \to e^+e^- + \text{hadrons}$$

combine the aspects of soft and of hard photon interactions. They are treated in detail elsewhere in these proceedings.[2]

WHAT IS THE NATURE OF THE PHOTON?

At low energy E_γ, "mass" $\sqrt{Q^2}$, and transverse momentum $k_\perp$, we well understand the particulate photon as our only stable vector "particle" with the quantum numbers

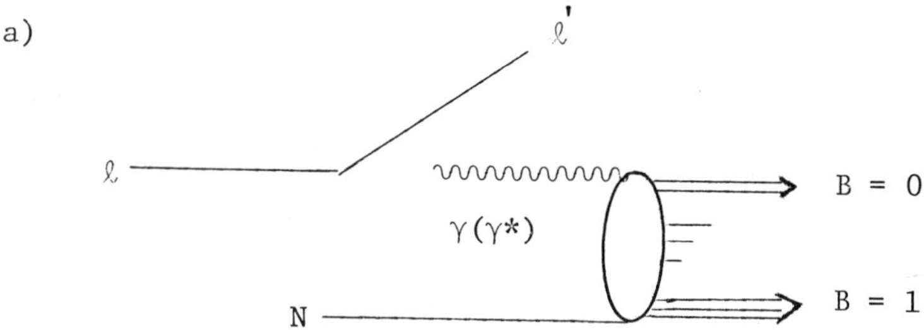

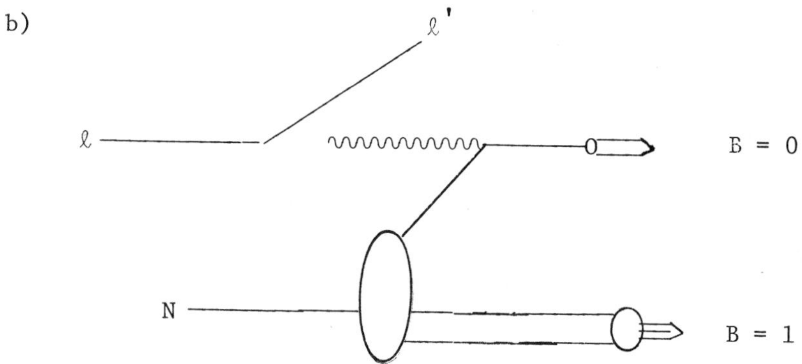

Fig. 1 Basic photoproduction process: a) The real (or spacelike) photon produces a mesonic system *via* diffractive or peripheral interaction with the target nucleon: "soft interaction". b) The photon locally couples to the charge of one of the target constituent quarks, producing a "current fragmentation" B = 0 system indicative of target structure: "hard interaction".

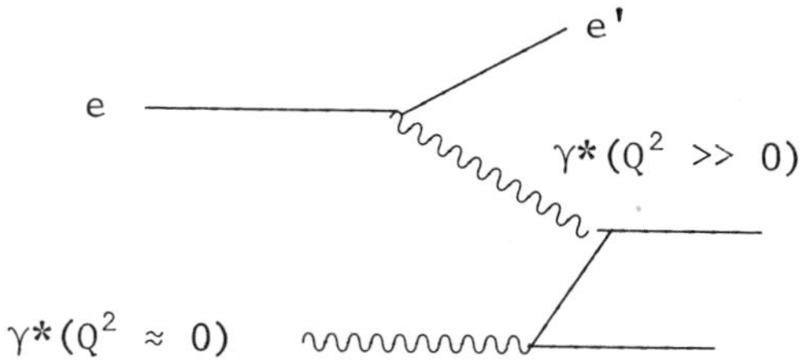

Fig. 2 Graph for the process $e^+e^- \to e^+e^- + \gamma\gamma \to e^+e^-$ + hadrons: the two photons assume the roles of the hard-scatter and soft-scatter aspects of Figs. 1a,b for a probe of photon structure.

$J^{PC} = 1^{--}$, U spin = 0.

All interactions described by QED are well-defined, and add to our confidence in QED. The candidate theory for the strong interaction, Quantum Chromodynamics (QCD), should take over at higher values for $\sqrt{Q^2}$ and $k_\perp$, but QCD is <u>not</u> an established and understood theory. We will explore to what extent the basically understood photon couplings can shed light on otherwise ill-defined concepts as the realm of kinematically permitted photon interactions evolves from the genteel area of QED into the hurly-burly of QCD territory. This is the peculiar interest in photon-hadron evolution.

In diagrammatic form, we can denote the photon, including its vacuum polarization effects, in the following way:

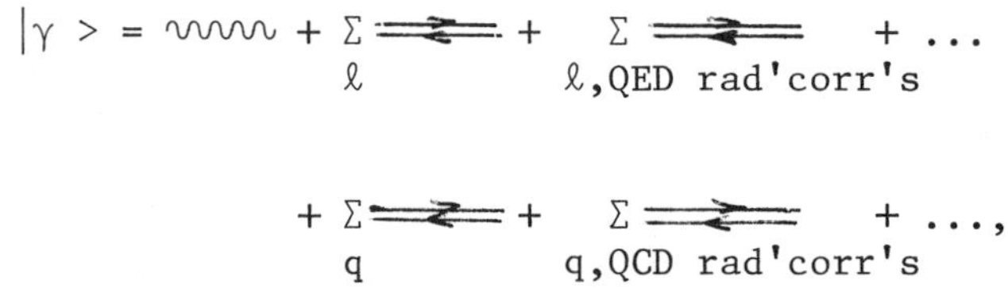

$$|\gamma > = \sim\sim\sim + \sum_{\ell} \rightleftharpoons + \sum_{\ell,\text{QED rad'corr's}} \rightleftharpoons + \ldots$$

$$+ \sum_{q} \rightleftharpoons + \sum_{q,\text{QCD rad'corr's}} \rightleftharpoons + \ldots,$$

with the sums extending over all accessible lepton (ℓ) and quark (q) flavors. This is written with the understanding that we have knowledge of the appropriate couplings

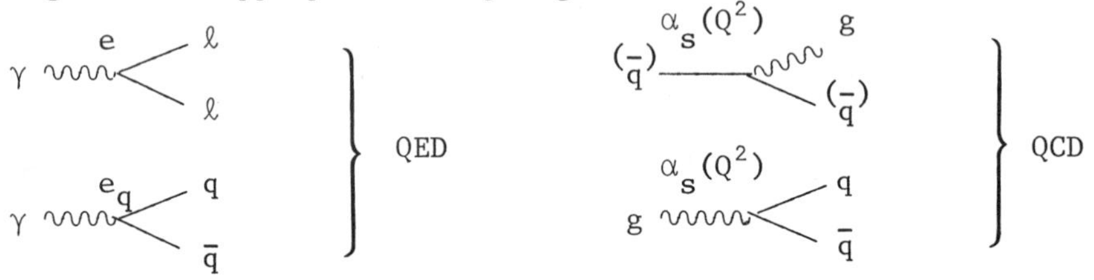

$\gamma \sim\sim\sim<$ (coupling e) ℓ, ℓ	QED	$\overset{(-)}{q}$ —— (coupling $\alpha_s(Q^2)$) g, $\overset{(-)}{q}$	QCD
$\gamma \sim\sim\sim<$ (coupling e_q) q, $\bar{q}$		g $\sim\sim\sim<$ (coupling $\alpha_s(Q^2)$) q, $\bar{q}$	

The QED couplings are known to depend on the charge alone; the QCD coupling is assumed to be independent of flavor, but does depend on the masses (or Q^2 values) involved in the process. Phenomenology of hadronic processes has convinced us that we do have a fair understanding of <u>hard</u> processes involving localizable interactions of point-like objects. It has also consistently been proven to be, at best, at the handwaving stage for all <u>soft</u> processes: our understanding of diffractive processes, of vacuum ("Pomeron") exchange, of singularities in the scattering plane, in QCD terms, is at best very loose and qualitative.

EVOLUTION OF THE HADRONIC PHOTON

Let's write down a wave function for the photon and follow its opening-up with kinematics. We choose light-cone variables: for fractional longitudinal momenta x, 1-x:

$$\gamma \to q\,(x,k_\perp) \quad \bar{q}\,(1-x,\, k_\perp)$$

and transverse momenta $k_\perp$, we can then write[3] the "Born term" for the lowest-order hadronization

$$\psi_\gamma^{(0)} = \frac{e_q}{k_\perp^2 + m_q^2}\; x(1-x)\; \frac{\langle \bar{q}_\uparrow(x,k_\perp)\,|\vec{\gamma}\cdot\vec{\varepsilon}\,|\,q_\downarrow(1-x,k_\perp)\rangle}{\sqrt{x}\;\sqrt{1-x}} \tag{1}$$

Since $\lambda_q = -\lambda_{\bar{q}}$, we can write this

$$\psi_\gamma^{(0)} = \frac{e_q}{k_\perp^2 + m_q^2}\; x(1-x) \begin{cases} i\sqrt{2}\,\dfrac{k_\perp}{x} & \text{for } \varepsilon = \dfrac{x+iy}{\sqrt{2}} \\[2ex] \sqrt{2}\,\dfrac{k_\perp}{(1-x)} & \text{for } \varepsilon = \dfrac{x-iy}{\sqrt{2}} \end{cases} \tag{2}$$

This implies a verifiable result:

$$\psi_\gamma^{(0)} \sim \frac{1}{k_\perp}\,. \tag{3}$$

We can derive from this the zeroeth-order QCD structure function of the photon by squaring, averaging over ε, summing over quark helicities λ_q: it defines the quark content of the photon

$$q_\gamma^{\,0}\,(x,k_\perp) = |\psi^{(0)}_{\gamma\to q\bar{q}}|^2 = \frac{2\,e_q^2}{k_\perp^2 + m_q^2}\left[(1-x)^2 + x^2\right]. \tag{4}$$

During evolution, Q^2 sets the scale for $k_\perp$ values available:

$$q_\gamma^0(x,Q^2) = \int_{k_\perp^2<Q^2} \frac{d^2k_\perp}{16\pi^3} q_\gamma^0(x,k_\perp) \tag{5}$$

$$= \frac{\alpha e_q^2}{2} \ln \frac{Q^2}{m_q^2} \left[(1-x)^2 + x^2\right] \tag{6}$$

This can be written in the form of an evolution equation:

$$\frac{d}{d \ln Q^2/Q_0^2} q_\gamma^{(0)}(x,Q^2) = \frac{\alpha e_q^2}{2\pi} \left[(1-x)^2 + x^2\right] . \tag{7}$$

This remarkably simple expression applies for real as well as almost-real spacelike photons, where we have limited ourselves to transverse polarization. The next step clearly has to include strong couplings. The Altarelli-Parisi[4] equations apply: The quark evolution proceeds according to

$$\frac{d}{d \ln Q^2/Q_0^2} q_i(y,Q^2) = \frac{\alpha_s(Q^2)}{2\pi} \int_x^1 \frac{dy}{y} \left[q_i(y,Q^2)P_{qq}(\tfrac{x}{y}) + g(y,Q^2)P_{qg}(\tfrac{x}{y})\right] \tag{8}$$

whereas gluons evolve after a comparable law

$$\frac{d}{d \ln Q^2/Q_0^2} g(x,Q) = = \frac{\alpha_s(Q^2)}{2\pi} \int_x^1 \frac{dy}{y} \left[\sum_i q_i(y_i Q^2)P_{gq_i}(\tfrac{x}{y}) + g(y,Q^2)P_{gg}(\tfrac{x}{y})\right] \tag{9}$$

These expressions give quark distributions of flavor i, q_i, and gluon distributions, g, in terms of an assumedly flavor-independent set of quark and gluon degradation functions P_{qq}, P_{qg}, P_{gq}, P_{gg}. Here, the second index refers to the parent, the first to the newly contributed q or g in the graphs:

P_{qq}, P_{gq} , P_{qg} , P_{gg} .

The y are appropriate fractional longitudinal momenta.

The resulting $q_i(x)$, $g(x)$ show characteristically different behavior for photon and for hadron evolution: the Born term acts as a "driving term" for the photon case, and results in a much flatter x dependence (Fig. 3a) for $q_i(x)$ and $g(x)$, than what we are expecting in the hadron case. The latter is well parameterized in terms of a distribution (Fig. 3b) resulting from the usual

$$q_i(x),\ g(x) \sim x^m(1-x)^n. \tag{10}$$

All the above is due to perturbative approaches. But most of what happens is not accessible to perturbative treatment. In that domain, let us assume an exponential fall-off of the photon wave function with the available "transverse mass"

$$\psi_\gamma(x_i, k_{\perp_i}) \propto e^{b\varepsilon_\gamma} . \tag{11}$$

This ansatz, due to Brodsky et al.,[5] defines a "binding energy" parameter

$$\varepsilon_\gamma = Q^2 - \sum_i \left(\frac{k_\perp{}^2 + m^2}{x} \right)_i \quad ; \quad \varepsilon_\gamma < 0 \tag{12}$$

for the hadronic constituents i of the photon in its vacuum-polarized state. Typically, then, for fixed Q^2 (e.g., $Q^2 = 0$ for real photons), Eq. 12 tells us how sensitive photon interactions are to hadronic wave function components. For $k_\perp \gg m_i$, this should reduce to the $\psi \propto k_\perp{}^{-1}$ behavior described above.

The most successful realization of a specific picture for this regime is the vector dominance model (VDM),[6] which postulates the saturation of the "soft" photon component by a sum over vector hadron states with appropriate couplings

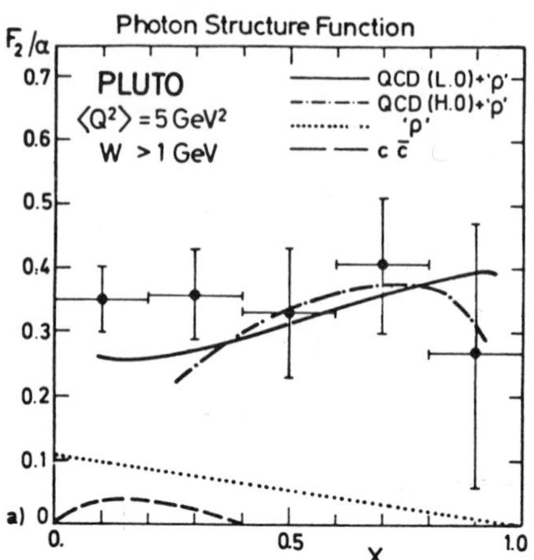

Fig. 3a The photon structure function F_2 as measured by the PLUTO Collaboration, showing various fits. The Born term (Eq.6) is responsible for the forward peaking feature. It is not suppressed by subsequent evolution.

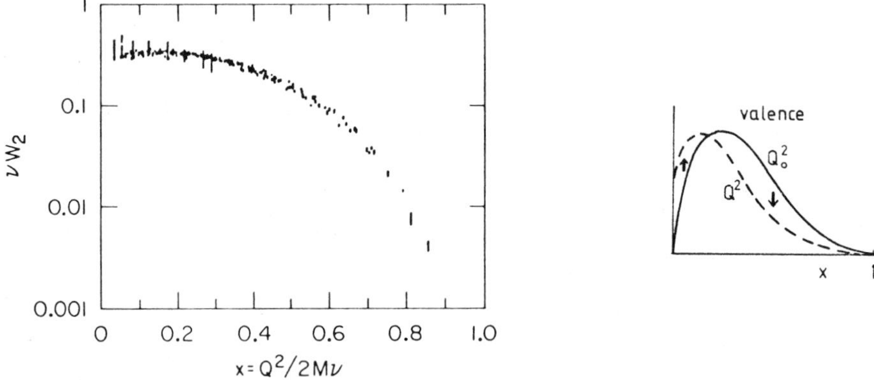

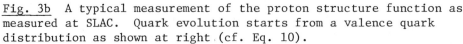

Fig. 3b A typical measurement of the proton structure function as measured at SLAC. Quark evolution starts from a valence quark distribution as shown at right (cf. Eq. 10).

$$|\gamma\rangle = \sum_i \frac{e}{\gamma_{V_i}} |V_i{}^0\rangle \; . \tag{13}$$

In recent years it has become experimentally feasible to look at thresholds where such quasi-valence quark contents would include charmed and bottom quarks: for Q^2 (or $k_\perp{}^2$) > 4 $m_c{}^2$, e.g., we might then search for these valence charmed quarks as well as for the q_c component of the regular photon-quark evolution.

EXPERIMENTAL DEFINITION OF PHOTON EVOLUTION

These considerations should convince us that the photon's evolution as evidenced by its coupling to heavy-quark states can be studied in terms of two components: the perturbative component of QCD evolution, calculable through its point-like couplings and the prescriptions of QCD, will be sensitive to high-$k_\perp$ (or-$\sqrt{Q^2}$) interactions: although it makes up but a small fraction of the total hadronic photon cross-section, it is distinctive by its $k_\perp{}^{-2}$ dependence.

The "soft" component, not accessible to perturbative treatment, will dominate the total cross-section. Any valence heavy-quark component, or long-lived component of vacuum-polarization origin, will be able to emerge. The approximate $\exp(-k_\perp{}^2)$ dependence will make this quasi-mesonic photon component lose its preeminence at larger transverse momenta.

Experimentally, photon evolution studies do well to concentrate on photon couplings to specific, recognizable quarks or hadronic final states:

Real photon beams are available, at high-energy proton accelerators, with energies up to some 200 GeV^2. Their energies can be "tagged" with an accuracy of ~ 1%, and even their polarization state can be defined under certain conditions. We then expect real photons to undergo quantum fluctuations into hadronic states with lifetimes

$$\tau_{q\bar{q}} \sim 0\;(E_\gamma\, m_{q\bar{q}}{}^{-2}) \tag{14a}$$

or, in terms of interaction-defined parameters

$$\tau_{q\bar{q}} \sim 0\ (E_\gamma\ k_\perp^{-2})\ . \qquad (14b)$$

Similarly, space-like photons γ^* are available by the well-known techniques of inelastic muon scattering; their energies are defined by the energy loss of the muons:

$$E_{\gamma^*} = E_\mu - E_{\mu'} \equiv \nu$$

Energies available in the laboratory match those for real photons; polarizations are defined by ν and scattering angles, and can include longitudinal components. Lifetimes of quantum fluctuations now have to compete with Q^{-2}.

How do these evolved photons scatter? Do the final states give us a hint of the true nature of the photon? Of the target? We observe evolution in terms of the parameters E_γ, Q^2, and $k_\perp$; the first two are defined by the beams of real or space-like photons, the latter by the interaction as revealed in the final-state. We will look at three kinematical regimes: we define them with reference to the scale parameter of QCD, $\Lambda(\approx 0.1 - 0.3$ GeV), and to appropriate quark masses m_q.

EXPERIMENTAL OBSERVATIONS - I.

1. For $k_\perp \lesssim m_q$ (and Λ) and $k_\perp << E_\gamma$, τ is long: we expect quasi-mesonic behavior, and we expect that VDM will provide a good description. Diffractive processes will yield the most telling examples, and will permit a probing of the hadronic photon structure by current fragmentation.

But let's be aware of the fact that, for $m_q >> \Lambda$ (i.e., for $m_q \geq m_c$), a quasi-perturbative treatment may be justified notwithstanding the condition $k_\perp < m_q$! We may test this notion by looking at the diffractive photoproduction of vector states for q = u, d, s, c. Fig. 4 illustrates our notion of how the process occurs: the real photon couples to a vector meson, which then scatters elastically off a nucleon by vacuum exchange:

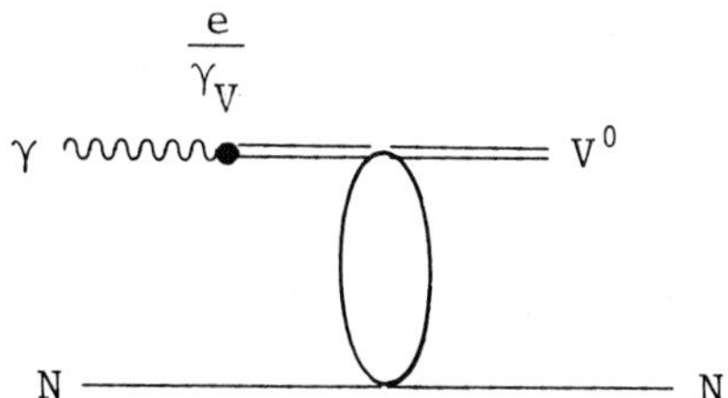

Fig. 4 Quasi-elastic vector meson production by real or space-like photons.

The coupling ratios for the 4 vector mesons ρ^0, ω^0, ϕ^0, ψ are defined in unbroken SU4 symmetry as

$$\gamma_\rho^{-2}:\gamma_\omega^{-2}:\gamma_\psi^{-2}:\gamma_\psi^{-2} = 9:1:2:8 \ , \tag{15}$$

and can be derived simply from the quark charges: this ratio is modified by the "size" of the $q\bar{q}$ object at the coupling, $\psi(\dot{0})$; this quantity can be extracted from $e^+e^- \to V^0$ measurements and changes the above ratios to

$$\gamma_{eff,\rho}^{-2}:\gamma_{eff,\omega}^{-2}:\gamma_{eff,\phi}^{-2}:\gamma_{eff,\psi}^{-2} = 9:1:1.3:1.5 \ . \tag{16}$$

The subsequent elastic scattering of $V^0N \to V^0N$ has been independently measured. Factorization then lets us expect the ratios of cross-sections for V^0 photoproduction

$$\frac{d\sigma}{dt}(\gamma p \to V^0 p) = \left(\frac{\alpha\pi}{\gamma_{V^0}{}^2}\right) \frac{d\sigma}{dt} \quad (V^0 p \to V^0 p) \tag{17}$$

to be reproduced by appropriately weighted elastic scattering ratios. Table 1 indeed shows that the ratios are compatible with direct experimental evidence on the photoproduction of the vector mesons mentioned.

We conclude that, for $k_\perp \lesssim m_c$, the lifetime of the quasi-hadron is long enough to have a normal elastic scattering process proceed. We illustrate this point by mentioning two precision experiments recently done by the Santa Cruz group at Fermilab. Their joint statement is that the vector dominance assumptions are valid indeed even as $E_\gamma > 100$ GeV, as long as $k_\perp \lesssim m_c$: simultaneous measurement of elastic Compton scattering on the proton[7a]

$$\gamma p \to \gamma p$$

and quasi-elastic vector meson production[7b]

$$\gamma p \to \omega^0 p \quad (\omega^0 \to \pi^0\gamma)$$

test the ideas of VDM and the additive quark model in a stringent way: these, together with the optical theorem, provide the relations between the differential cross-section for the above process with elastic $\pi^\pm p$ scattering (cf. Appendix):

$$\frac{d\sigma}{dt}(\gamma p \to \omega p) = \left(\frac{\alpha\pi}{\gamma_\omega^2}\right) \frac{d\sigma}{dt}(\omega p \to \omega p) \tag{18a}$$

$$= \left(\frac{\alpha\pi}{\gamma_\omega^2}\right) \left[\frac{1}{2}\left(\frac{d\sigma}{dt}(\pi^+ p)\right)^{\frac{1}{2}} + \frac{1}{2}\left(\frac{d\sigma}{dt}\right)(\pi^- p)^{\frac{1}{2}}\right]^2 . \tag{18b}$$

For the slopes of a momentum transfer dependence defined by

$$\frac{d\sigma}{dt} \sim e^{b(t)t} , \tag{19}$$

this model predicts $b_\omega = \frac{1}{2}(b_{\pi-} + b_{\pi+})$. (20)

The experiment, performed in the tagged photon beam at the Fermi National Accelerator Laboratory, measured only radiatively decaying final states in the forward direction with precision, thus restricting effective vector meson measurement to the ω^0 with its large decay width into $\pi^0\gamma$ ($\to 3\gamma$). But since it also had a drift chamber array that defined recoiling target particle trajectories accurately, the full kinematics of such final states as γp, $\omega^0 p$ could be reconstructed, and inelastic admixtures were excluded. Fig. 5 shows a mass plot for $\pi^0\gamma$ combinations with a clean ω^0 signal. Fig. 6 gives the differential cross-section for 3 different energy bins spanning photon energies from 30 to 120 GeV.

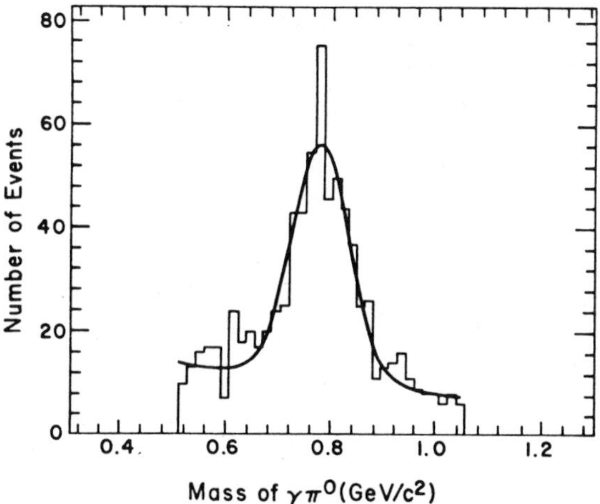

Fig. 5 Invariant mass spectrum for 3-photon systems detected by forward shower detectors in a photoproduction experiment at Fermilab. The ω^0 signal ($\omega^0 \rightarrow \pi^0\gamma \rightarrow 3\gamma$) stands out.[7a]

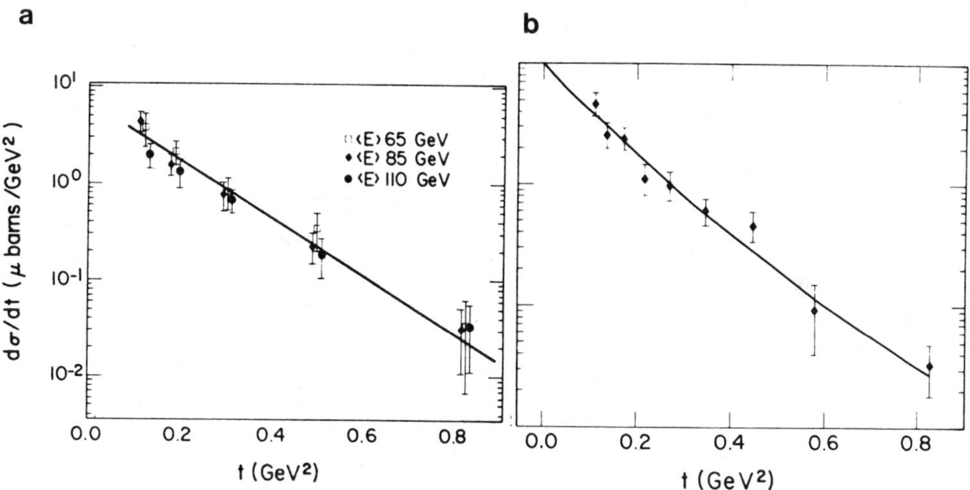

Fig. 6 Results from ω^0 photoproduction experiment at Fermilab;[7b] a) t dependence for 3 different energy bins shows no shrinking of a diffraction peak. b) combining all energies for increased statistical weight, a good fit is provided by a double exponential with slope parameters taken from $\pi^{\pm}p$ elastic scattering.

The information to be gleaned from Fig. 6 is:

- there is no obvious shrinkage of the diffraction peak over the energy range covered: one slope parameter will fit all energies (Fig. 6a),
- an excellent fit to the data, combined into one energy bin for greatest statistical accuracy, is provided by a double exponential form

$$\frac{d\sigma}{dt} (\gamma p \to \omega^0 p) = A \exp (8.9t + 2.2t^2),$$

where the exponential slope parameters are taken directly from $\pi^{\pm}p$ measurement[8] via Eq. 20.

This is the only experiment in high-energy photoproduction where the full final state was measured, excluding inelastic contributions; it reinforces our confidence in the quark evolution of the photon as parameterized, for low $k_{\perp}$ values, in terms of a vector meson dominance picture.

The same approach is taken one step further in the observation of elastic photon scattering, in the same experiment: if this process is convincingly parameterized in terms of meson-nucleon scattering, then an interpretation in terms of the quark content of the photon at the appropriate level of $k_{\perp} < m_c$ is compelling. The appropriate VDM inspired relation, illustrated in Fig. 7, is, in analogy with equations 13 and 17,

$$\frac{d\sigma}{dt} (\gamma p \to \gamma p) \to \sum_i \left(\frac{\alpha\pi}{\gamma_{V_i}^2}\right) \frac{d\sigma}{dt} (\gamma p \to V_i^0 p). \tag{21}$$

We limit ourselves to the simple VDM, excluding its various upgrades. It is experimentally most straightforward to compare slope parameters; evolution involving heavier quarks will be mirrored by the admixture of a telling flatter component to the diffraction peak: it is well known that simple-exponential fits yield[9] $b_{\psi} \simeq 2$, whereas $b_{\rho} \simeq b_{\omega} \simeq 6\text{-}8\ (\text{GeV/c})^{-2}$. Fig. 8a shows the Santa Cruz data, for three energy bins: again, as in the ω production case, there is no

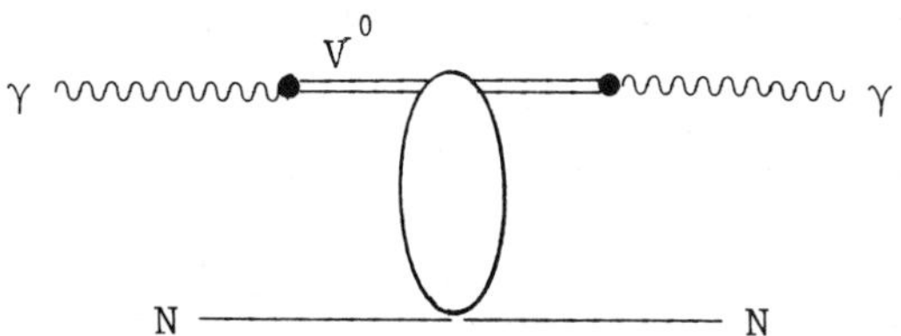

Fig. 7 Vector meson dominance diagram for elastic Compton scattering.

change of slope as a function of energy from 30 to 130 GeV within the accuracy of the experiment. The procedure of including all data in one energy bin can then be repeated as in the ω^0 case. Figure 8b shows the resulting cross-section plot. Trying the same approach of fitting the data within the slope parameters known from elastic $\pi^{\pm}p$ scattering,[8] we arrive at an excellent representation of the data in the $|t|$ range covered by the experiment (the fit involves only the normalization. We conclude that:

- Real photons scatter elastically off protons just like the vector mesons composed of the quarks contained in the evolution of the appropriate $k_\perp$ range.
- Heavy quark components do not contribute at a noticeable level to elastic scattering: otherwise, their characteristic small slope parameter (b_ψ = 1.5-2) should be in evidence. We might expect this situation to change at higher $\sqrt{t}$ ($k_\perp$) values $\gtrsim m_c$.
- Elastic Compton Scattering at low to moderate t can therefore be interpreted as a mirror of photon evolution in the t (or $k_\perp$) range covered by the experiment.

There are other points of interest: (real) photon beams are the only (massless) vector beams available to us: Parameterization of

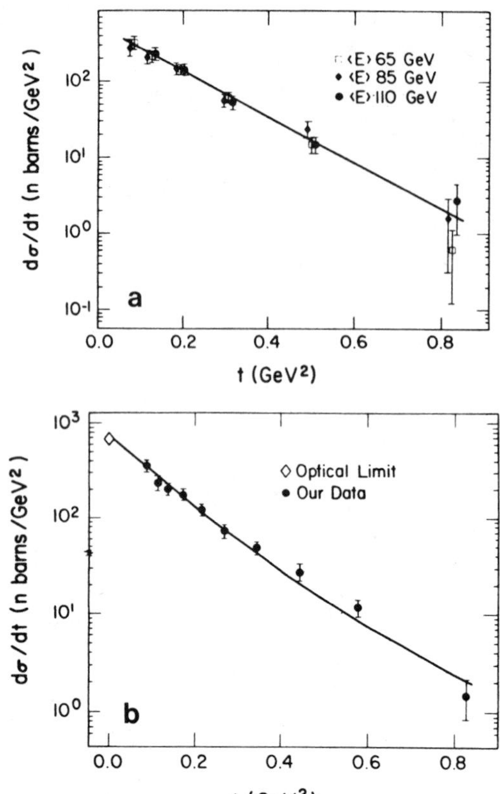

Fig. 8 Elastic photon scattering. Results of the Santa Cruz group at Fermilab[7a] show:

a) the absence of noticeable shrinking of the diffraction peak

b) a successful fit to the t dependence of the combined data using slope parameters measured in elastic $\pi^{\pm}p$ scattering.

elastic scattering terms of Pomeron (P) exchange then runs into trouble: for $t \to 0$, $J(P) = 1$; but the triple vector vertex (photon-photon-Pomeron) is forbidden owing to space-time arguments.[10] The unbroken diffraction peak down to small t values argues against this Pomeron decoupling, and for the introduction of a compensating $J = 1$ fixed pole.[11] In QCD language, quantitatively unproven at small $k_{\perp}$, an explanation in terms of double (or multiple) gluon exchange appears more naturally capable of removing this embarrassment.

Moreover, elastic photon scattering permits a test of the original Gell-Mann-Goldberger-Thirring dispersion relation.[12] Using the optical theorem, we can incorporate it in the following expression:

$$\frac{d\sigma}{dt}(t=0) = \frac{\pi}{\nu^2}|f(\nu)|^2$$

$$= \frac{\sigma_{tot}(\nu)^2}{16\pi} + \frac{\pi}{\nu^2}|\mathrm{Re}\ f_1(\nu)| + \frac{\pi}{\nu^2}|f_2(\nu)|^2 , \qquad (22)$$

where f_1 describes the spin-averaged term, f_2 is the spin-perpendicular amplitude. We use the fit to the diffraction peak in Fig. 8b

$$\frac{d\sigma}{dt}(\gamma p \to \gamma p) = (726 \pm 38)e^{8.9t + 2.2t^2} \qquad (23)$$

to determine, under the assumption that the 2nd and 3rd terms on the RHS are negligible, a value for σ_{tot} ($\gamma p \to$ hadrons) = 116 μb. This is entirely compatible with the only measurement to date of σ_{tot}[13], and leaves little room for (Re f_1) and (f_2).

What of heavier quarks and vector dominance? A fair amount of information has become available on the process

$$\gamma N \to \psi + \ldots$$

both from photoproduction and from muoproduction experiments at small Q^2. The latter have the disadvantage of not being able to see the full final state, and for frequently being done on heavy targets. Together with older photoproduction data, an energy dependence for photoproduction of ψ in γp (or γN, from heavier targets) collisions can be inferred to be a strong function of E_γ at lower E_γ values, increasing more gently above $E_\gamma \simeq 50$ GeV. While this energy dependence is well described by the lowest order QCD graph (cf. below), that model has nothing to say about t dependence. The original Fermilab data,[9] (Fig. 9a) had indicated a value of

$b_{\psi} \simeq 2\ (GeV/c)^{-2}$, and muoproduction results are compatible with this number.[14] A recent photoproduction experiment by Binkley et al.,[15] (Fig. 9b) does not confirm this picture, but is sensitive to inelastic admixtures. A small b value reflects a small object diffracting off the nucleon. Detailed recent calculations[16] involving lowest-order QCD attempt to interpret these data in terms of the photon-gluon fusion graph (Fig. 11 below), but account for only a fraction of the reported cross-section. The energy trend of the photon-gluon fusion process,[17] however, is that observed in the data, so that the approximate quantitative agreement of the vector dominance relations. Eq. 17 may be a sign of overlapping applicability of that phenomenological approach and a perturbative treatment due to the small distance scale m_c^{-1}: the condition $k_{\perp} \lesssim m_c$ no longer

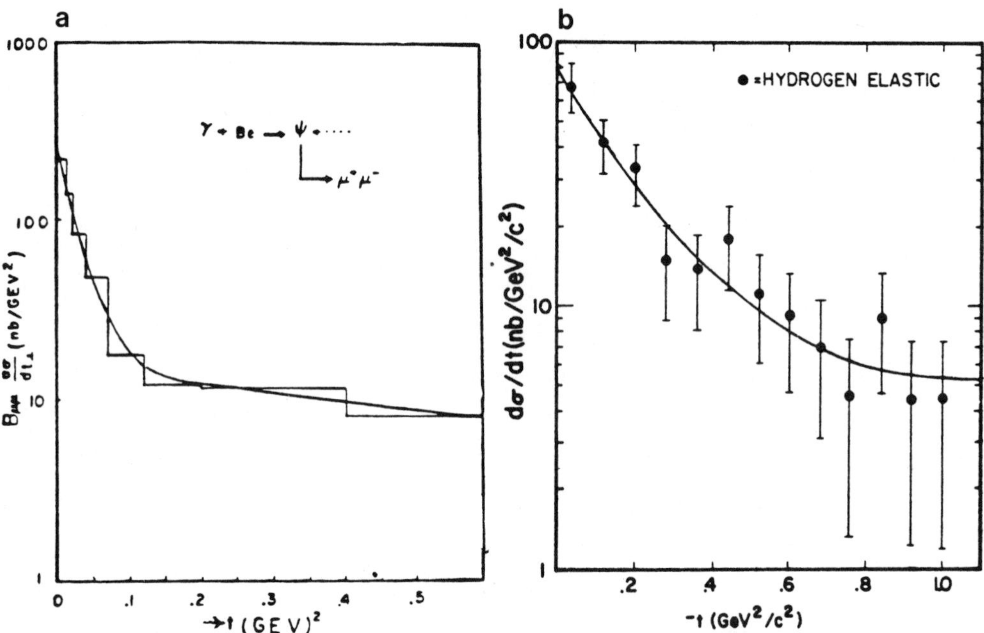

Fig. 9 Elastic photoproduction of ψ. Early results (a) show[9] a coherent peak due to production off Be nuclei and a subsequent slope of $b \approx 2(GeV/c)^{-2}$. Recent results off H_2 (b) give[15] no clear single-exponential fit.

implies $k_\perp \lesssim \Lambda$, as for u, d, s. The lifetime τ, for given E_γ, decreases as $m_{q\bar{q}}^{-2}$!

EXPERIMENTAL OBSERVATIONS - II.

$k_\perp > \Lambda, > m_q$ but still $<< E_\gamma$; $\tau << \Lambda^{-1}$.

The transverse momentum is now such that it lets separately evolving $q,\bar{q}$ beams develop and hadronize, as schematically shown in Fig. 10. In particular, the photon-nucleon interaction now proceeds

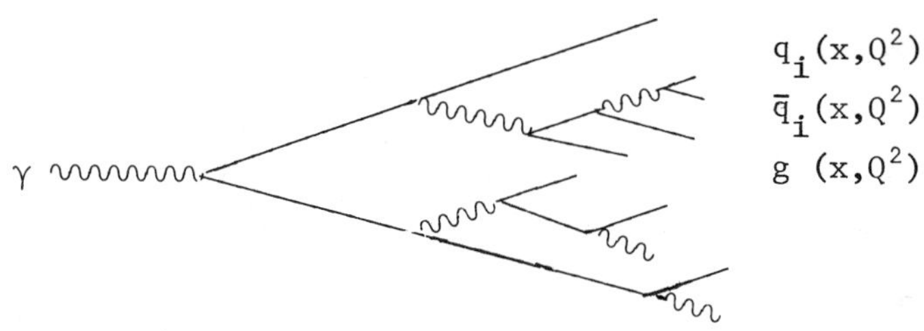

Fig. 10 For $k_\perp > m_q$, quark beams hadronize.

via exchange of a gluon by one of the $q(\bar{q})$ with the nucleon; this process, coupling the evolution graph to the nucleon via its gluon distribution, is therefore equivalent to the lowest-order QCD process, and is straightforwardly calculable.[17] (cf. Fig. 11).

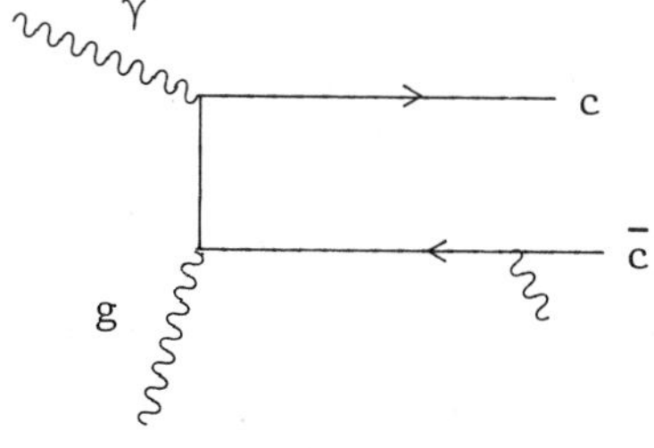

Fig. 11 For this regime ($k_\perp > m_q$), the photon couples to the target gluon field via quark formation. Evolution follows.

This "photon-gluon fusion" process, driven by the basic Born term, no longer permits quasi-mesonic behavior for the scattering photon. On the other hand, it does not probe the valence structure of the nucleon, but rather the gluon distribution inside the nucleon (or, more precisely, a convolution of gluon distributions in the hadronic photon and in the nucleon: but the small lifetime $\tau(\sim k_\perp^{-2})$ makes the influence of the former minimal).

Experimentally, things become less tangible: for the region $k_\perp > m_c$, open-charm production defines the process of interest. A total cross-section experiment[13] indicated a potential open-charm cross-section of up to 4 μb. But the search for individual charmed hadron production channels has been agonizingly difficult. With the possible exception of the F meson, not a single charmed hadron state has been discovered by either photo- or hadro-production. Final-state multiplicities are frequently high, making background subtractions difficult; moreover, experiments cover various fractions of phase space, and production process modelling is needed before a cross-section can be extracted; such models may involve questionable assumptions.[18]

We give only one example here to illustrate the point: the Omega Spectrometer Collaboration at CERN[19] uses their elaborate set of apparatus in the tagged photon beam at the CERN SPS to find the charmed-strange F meson via its decays into states containing η (and thus $s\bar{s}$) particles. The observed modes, illustrated in Fig. 12, do not give dramatic signals (and have shown up only in channels $F \to \eta + n\pi$ with n = odd). They occur at the 4 S.D. level at best, and add up to a total of ~ 0.1 μb.

Still, it is worthwhile to note that, in this intermediate region of photon evolution into individual quark hadronization processes, the overall picture is adequately parameterized in terms of the lowest-order QCD graph; the critical ingredient to such calculations is the gluon distribution inside the nucleon, which follows Eq. 10 with m = - 1, n = 5. If we sum over all observed

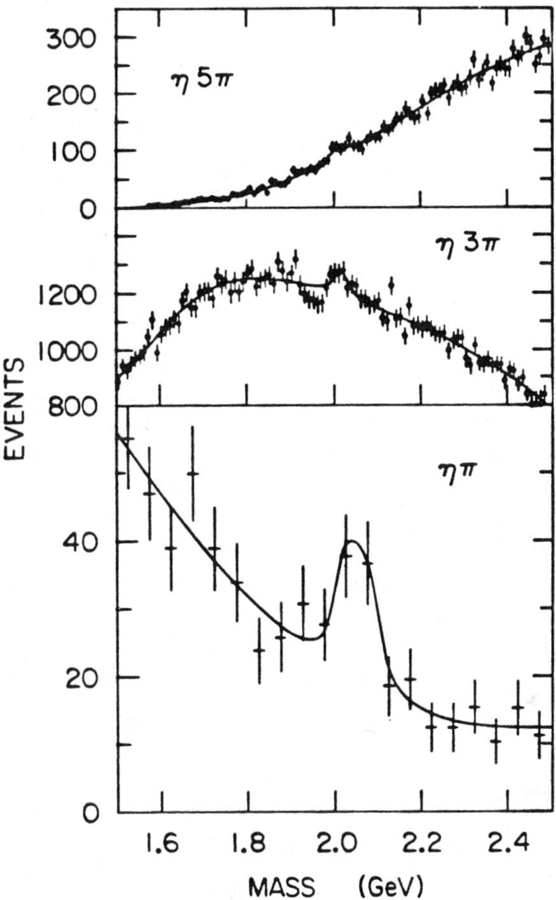

Fig. 12 Photoproduction of strange-charmed F mesons in the Omega Spectrometer at CERN:[19] No hints are obtained in final states involving an η and even numbers of pions. The F meson has not been definitively seen in e^+e^- annihilation.

modes of charmed hadron photoproduction, including mainly D,D*, and F mesons as well as Λ_c baryons,[18] the cross-sections loosely add up to

$$\sigma_{cc}(\gamma N) \simeq 1\text{-}2\ \mu b.$$

EXPERIMENTAL OBSERVATION - III.

$$k_\perp >> \Lambda,\ m_q;\ k_\perp \sim O(E_\gamma);\ \tau \sim E_\gamma^{-1}$$

This is basically the point-like photon, as in QED. It has almost no hadronic structure (in addition to renormalization of the point vertex

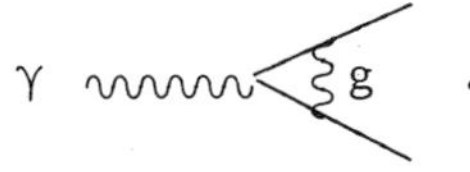

The photon acts as an elementary quantum, and therefore as a local probe with known couplings (QED), i.e., to the quark charges. This has a number of important manifestations:

a) Every photon has a "hard" component, even elastic Compton scattering; for available luminosities in real photon beams, these are mostly too small to be of practical importance.

b) The large $k_\perp$ values can become available from larger Q^2 values for space-like photons. This aspect of the $k_\perp >> m_q$ condition (for light quarks) has been at the basis of the enormously informative electro- and muo-production experiments that have probed nucleon structure.

c) This condition is also at the basis of the immense success of time-like photons (from e^+e^- annihilations) as sources of new flavors, and of our understanding of their spectroscopy: In this regime, the condition $k_\perp \sim O(E_\gamma)$ assures that the parent quarks separate sufficiently in phase space before hadronizing so as to make background subtractions less forbidding than in the previous section.

We have excluded from this lecture time-like and deeply space-like processes. In the real (or slightly space-like) region, however, processes with $k_\perp \sim O(E_\gamma$ or $\nu)$ are too rare to have been observed through their quark hadronization. Instead, the most ambitious part of real-photon experimentation attempts to identify high-$k_\perp$ events in the process

$$\gamma p \to \gamma + \text{hadrons}$$

that are due to the so-called inelastic Compton graph (Fig. 13): if all available $k_\perp$ from the incident photon is re-emitted via a final-state photon, then the small lifetime between absorption and reemission of the photon, $\tau \sim k_\perp^{-1}$, will not permit communication

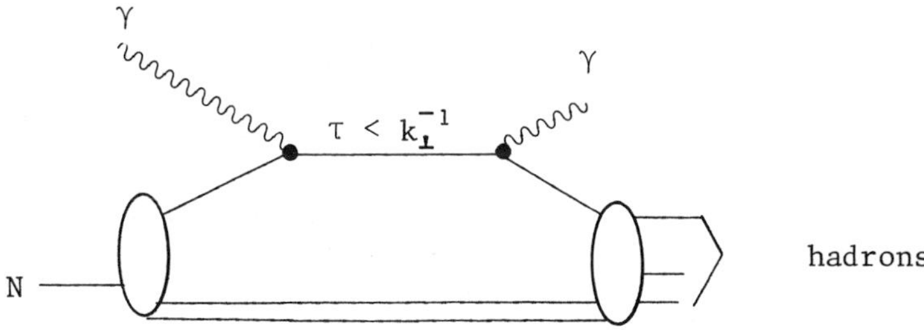

Fig. 13 The inelastic Compton graph emits a high-$p_\perp$ photon before the absorbed transverse momentum can be shared.

between the parton line struck inside the nucleon and the other constituents: there is no gluon exchange prior to re-emission of the photon. The photon does not have any time to "evolve", but may provide vital information on the nucleon's structure *via* its known couplings to parton charges. As early as 1969, Bjorken and Paschos[20] had pointed out the usefulness of this process for a "measurement of parton charges": if scaling is observed in this process, the structure function

$$\nu W_2^{\gamma N} = \sum_n P(n)\ f_n(x)\ \langle \sum_i^n Q_i^{\ 4} \rangle_N \tag{24}$$

(with P(n) the likelihood of finding n partons in the nucleon, $f_n(x)$ the x distribution of these n partons, $x = \frac{Q^2}{2m\nu}$; and Q_i = parton charges) can be used for a comparison with well-known electroproduction (or muoproduction) structure functions:

$$\nu W_2^{eN,\mu N} = \sum_n P(n)\ f_n(x)\ \langle \sum_i^n Q_i^{\ 2} \rangle_N \tag{25}$$

The relation of measured cross-sections

$$\frac{d^2\sigma}{d\Omega dk}\ (\gamma N) = \frac{\nu^2}{kk'}\ \frac{d^2\sigma}{d\Omega dk}\ (e/\mu N) \tag{26}$$

thus permits the determination of the ratio

$$R = \frac{\langle \Sigma_i Q_i^{\ 4} \rangle_{p,n}}{\langle \Sigma_i Q_i^{\ 2} \rangle_{p,n}}\ . \tag{27}$$

For integrally charged partons, R is clearly predicted to be 1; for fractionally charged nucleon constituents, $\frac{1}{3} < R < \frac{5}{9}$.

Experimentally, $k_\perp > m_c$, or even $k_\perp = O(E_\gamma)$, means small cross-sections at large $|t|$ values. As a consequence, neither the Santa Cruz experiment[7] nor a more recent CERN effort (NA 14 collaboration[21]) have been able to isolate signals that are clearly due to the local Compton graph. In Fig. 14 we show that the principal backgrounds to such a determination, inclusive $\pi^0(\to 2\gamma)$ and $\eta^0(\to 2\gamma)$ production, are reasonably in hand, but that there is no discernible evidence for the presence (or absence) fo a direct-photon component at high $|t|$ or $k_\perp$.

Note, however, that we have strayed from our concept of photon evolution with E_γ, $k_\perp$: the large $k_\perp$ (or $|t|$) values of this section were utilized, in the lightlike or spacelike photon case we are considering, only to define the strictly local character of this photon component: we have thus expressly excluded photon structure and evolution in this regime.

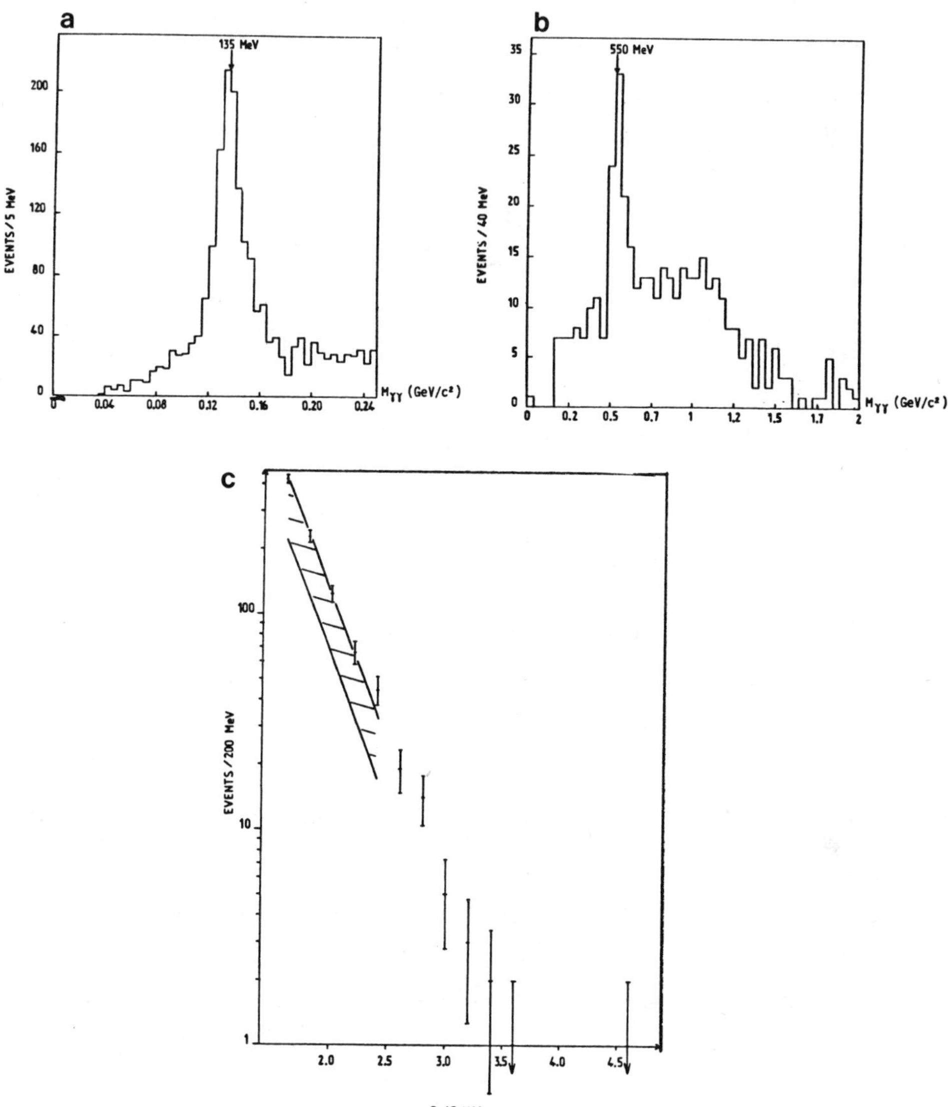

<u>Fig. 14</u> The NA14 Collaboration at CERN[21] was able to give well-identified π^0 (a) and η^0 (b) signals from reactions $\gamma N \to \gamma\gamma + \ldots$; isolation of the Compton graph (Fig. 13) would imply an excess of photon production above the shaded area in (c); no such signal in inclusive photon production can be claimed.

WHAT HAPPENS AT (SMALL) SPACELIKE $Q^2 > 0$?

As we progressed through the three different $k_\perp$ regimes of the previous section, we were able to carve out features that appeared to show mesonic behavior, quark/gluon evolution of the $\gamma q\bar{q}$ Born term, and, finally, the "bare photon" characteristic of a highly local probe. A difficulty in the categorization has been the appropriate scale for these phenomena -- is it Λ^{-1} or m_q^{-1}?

Quark mass thresholds between m and m_c make it difficult to follow $k_\perp$ evolution in detail; but when considering spacelike photons, it becomes possible to map out the regime of small Q^2 through hadronization channels: what is the transition from mesonic to quark/gluon behavior?

The standard way in which VDM enthusiasts have described the Q^2 dependence of a photon-initiated process is the introduction of a "transverse V^0 propagator term" to connect photo- with $Q^2 > 0$ electroproduction:[22]

$$\frac{d\sigma}{dt}(Q^2) = \frac{m_V^{\;2}}{m_V^{\;2} + Q^2} \frac{d\sigma}{dt}(Q^2=0)$$

(This expression neglects the influence of longitudinally polarized photons).

This ansatz, while giving the proper trend, is quantitatively incompatible with the best data available. In fact, $\sigma(Q^2)/\sigma(Q^2=0)$ falls precipitously at small values $Q^2 > 0$, then appears to settle into a gentler slope. Remarkably, this appears to be true not only for ρ and ω (there are not enough data to check ϕ production), but possibly also for ψ production. Fig. 15 illustrates the point, and shows the need for a steeper fall-off at $Q^2 < m_q^{\;2}$.

Similar surprises, unexplained by standard arguments of photon phenomenology, are visible in general inclusive hadron production data:[24] fractional prong cross-sections are essentially independent of Q^2 once the Λ^2 (or $m_q^{\;2}$) threshold has been passed; but at <u>small</u> values, the one-prong and three-prong cross-sections change more

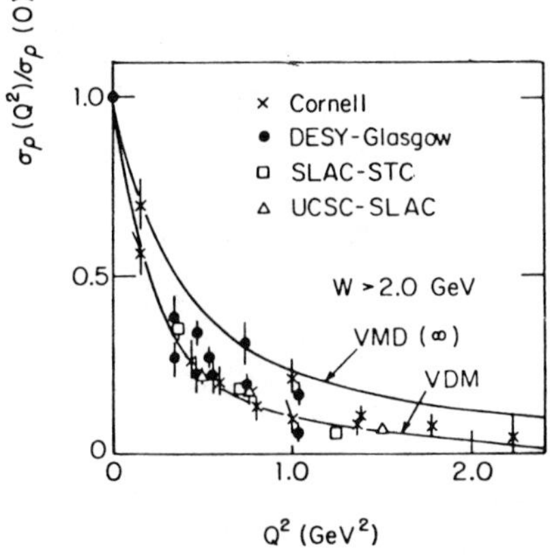

Fig. 15a ρ^0 photo- and electro-production permit a mapping of the Q^2 region $0 \leq Q^2 \leq 1$ (GeV/c)^2; the transverse VDM propagator term cannot quantitatively describe ρ^0 (or ω^0) production at low Q^2 unless forcibly adjusted (lower curve), cf. data mentioned in Ref. 23.

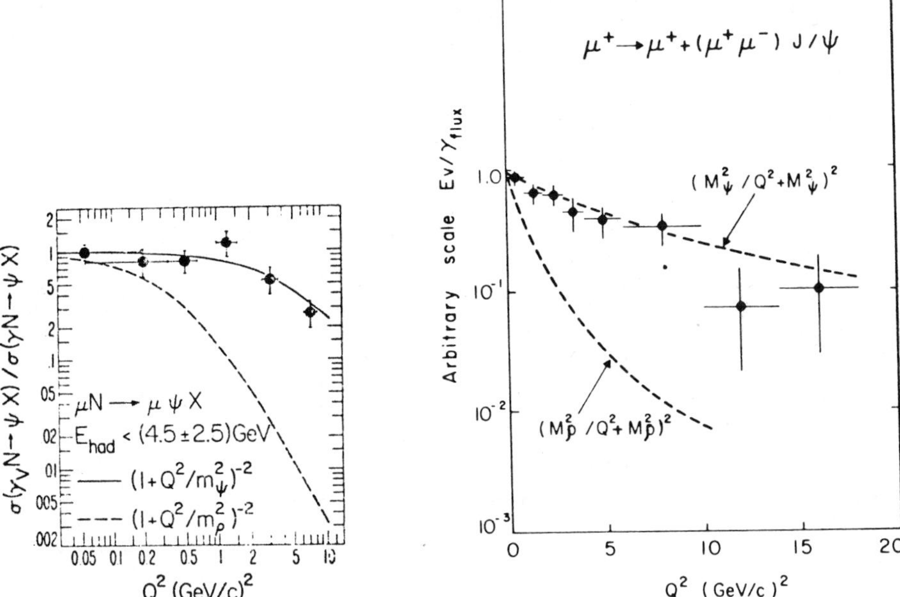

Fig. 15b Muoproduction of ψ's at Fermilab[14] (left) and CERN-EMC (right) shows that, while the Q^2 trend is well represented by a transverse VDM propagator, the dropoff at small Q^2 again appears to fall noticeably below the propagator.

rapidly than can be explained by the Q behavior of their known constituent cross-sections (Fig. 16).[25]

It is tempting to assume that the Q^2 dependence for the almost real photon is again governed by the ansatz Eq. 11, which will prescribe a steep Q^2 dependence to processes sensitive to the light-quark VDM component of the photon (ρ,ω production, prong cross-sections), while relaxing into a gentler Q^2 (or $k_\perp{}^2$) dependence beyond that limit, in accordance with the Born-term dependence of Eq. 3.

The breaks between the two regimes would again be either at $Q^2 \sim \Lambda^2$, or at $Q^2 \sim m_q{}^2$ (where we would choose constituent quark masses for the light quarks). The latter would give qualitative credence to the picture in ψ muoproduction (note that a transverse vector meson propagator with the proper quark masses can not fit

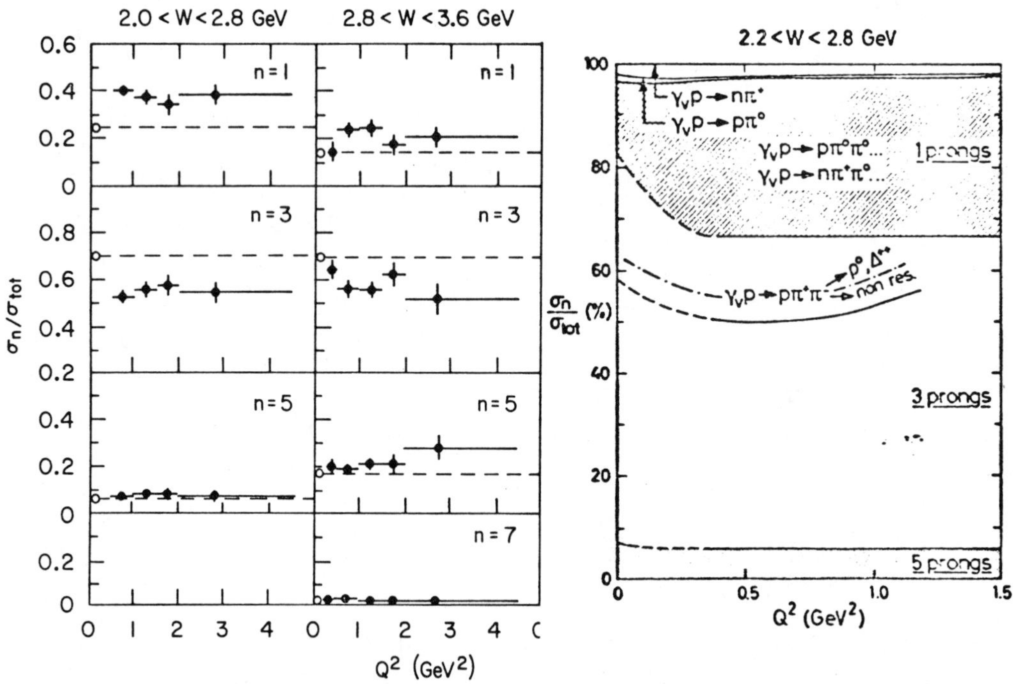

Fig. 16 Inclusive particle production observed by the Santa Cruz/ SLAC Collaboration[24] a) and DESY/Cornell[25] b): For n = 1 and n = 3, there are strong changes between $Q^2 = 0$ and $Q^2 = 0.3$.

the data!).[14] The message emerging for $Q^2 > 0$ photoproduction then is as follows: For Q^2 values permitting $k_\perp < \Lambda$ or m_q only, the phenomenological ansatz of the vector-dominated hadronic photon wave function qualitatively describes hadron production; at $Q^2 > \Lambda^2$ or m_q^2, perturbative photon/quark/gluon evolution takes over. But the details of the transition remain to be investigated.

ACKNOWLEDGMENT

I thank P. Zerwas and F.M. Renard for helpful discussions, and A. Zichichi for the excellent and stimulating atmosphere at the 1982 Erice School.

TABLE I

	ρ	ω	ϕ	ψ
$\Gamma\left(V^o \to e^+e^-\right)$ [keV]	6.8	0.76	1.27	4.4
$\left(\frac{1}{\gamma^2}\right)_{eff} \times 10^2$	3.9	0.44	0.56	0.64
$\sigma_{tot}\left(V^o p\right)$ [mb]	27	27	10	2
$\frac{\alpha}{4}\left(\frac{1}{\gamma^2}\right)_{eff} \sigma_{tot}^2\left(V^o p\right)$ [μb GeV^{-2}]	129	14	3.1	0.12
$\left.\frac{d\sigma}{dt}\right\|_{t=0}\left(\gamma p \to p V^o\right)$ [μb GeV^{-2}]	100	11	2.6	0.1

REFERENCES

1. We will use the customary convention $Q^2 \equiv -q^2$, where q is the photon momentum four-vector.
2. C. Berger, these proceedings.
3. S.J. Brodsky, SLAC-PUB-2747 (1981); Proc. 4th Int. Coll. on Photon-Photon Interactions, Paris (1981).
4. G. Altarelli, G. Parisi, Nuc. Phys. B126, 298 (1977).
5. S.J. Brodsky, T. Huang, G.P. Lepage, SLAC-PUB 2868 (1982).
6. J.J. Sakurai, lectures at this School (1973); published in: Laws of Hadronic Matter (A. Zichichi ed.), Academic Press (1975).

7a. A.M. Breakstone et al., Phys. Rev. Lett. 47, 1778 (1981).

b. A.M. Breakstone et al., Phys. Rev. Lett. 47, 1782 (1981).

8. D.S. Ayres et al., Phys. Rev. D15, 3105 (1977).
9. W. Lee, Proc. 1975 Int. Symposium on Electron and Photon Interactions, Stanford (1975).
10. C.N. Yang, Phys. Rev. 77, 242 (1950).
11. P.V. Landshoff, J.C. Polkinghorne, Phys. Rev. D5, 2056 (1977).
12. M. Gell-Mann, M.L. Goldberger, W. Thirring, Phys. Rev. 95, 1612 (1954).
13. D.O. Caldwell et al., Phys. Rev. Lett. 40, 1222 (1978).
14. M. Strovink, Proc. 1981 Int. Symposium on Lepton and Photon Interactions at High Energies (W. Pfeil, ed.) Bonn (1981).
15. M. Binkley et al., Phys. Rev. Lett. 48, 73 (1982).
16. R. Beier, R. Rückl, MPI Preprint PAE-PTH58-82 (1982).
17. This is done, e.g., in J.P. Léveillé, T. Weiler, Nuc. Phys. B147, 147 (1979).
18. For a review, see C.A. Heusch, Proc. of 1981 SLAC Summer Institute (A. Mosher, ed.), Stanford (1982).
19. D. Aston et al., Phys. Lett. 100B, 91 (1981); Nuc. Phys. B189, 205 (1981);
    M. Atkinson et al., CERN-EP/82-128 (to be published).
20. J.D. Bjorken, E.A. Paschos, Phys. Rev. 185, 1975 (1969).
21. P. Petroff, Proc. XVIIth Rencontre de Moriond (1982).
22. See, e.g., the review article of A. Donnachie and G. Shaw in: Electromagnetic Interactions of Hadrons, vol. 2. Plenum Press, N.Y. (1978).
23. See the compilation in: T.H. Bauer et al., Revs. Mod. Phys. 50, 261 (1978); also C. del Papa et al., Phys. Rev. D19, 1303 (1979);
    L.A. Ahrens et al., Phys. Rev. Lett. 42, 208 (1979).
24. K. Bunnell et al., Phys. Rev. D17, 2847 (1978).
25. H. Spitzer in: Leptons and Multi-Leptons (Moriond 1977), J. Tran Thanh Van ed., Paris (1977).

APPENDIX

Two approximate photon-hadron cross section relations inspired by the Vector Meson Dominance Model (VDM) are open to experimental checks:

$$\sigma\left(\gamma p \to V^o p\right) = \frac{4\pi\alpha}{\gamma_V^2}\left(\sigma_{el}\ V^o p\right) \tag{A-1}$$

For the LHS, good data exist; for the RHS, good direct data are available for $V^o = \rho$ at best. We can instead use the quark-parton model to reexpress

$$\sigma_{el}(\rho p) \doteq \sigma_{el}(\omega p) = \left\{\frac{1}{2}\left[\sigma_{el}\left(\pi^+ p\right)\right]^{\frac{1}{2}} + \frac{1}{2}\left[\sigma_{el}\left(\pi^- p\right)^{\frac{1}{2}}\right]\right\}^2$$
$$\sigma_{el}(\phi p) = \left\{\left[\sigma_{el}\left(K^+ p\right)\right]^{\frac{1}{2}} + \left[\sigma_{el}\left(K^- p\right)\right]^{\frac{1}{2}} - \left[\sigma_{el}\left(\pi^- p\right)\right]^{\frac{1}{2}}\right\}^2 \tag{A-2}$$

The data for evaluation of the RHS are available and lead to a successful fit for ρ,ω photoproduction, but to a discrepancy of a factor -2 for ϕ photoproduction at low energies (see Bauer et al., Ref. 23). For $V^o = \psi$, the RHS of eq. (A-2) would involve unknown Dp cross sections.

For the forward diffractive cross-section, we can use the optical theorem to write, using (A-1)

$$\left.\frac{d\sigma}{dt}\right|_{t=0}\left(\gamma p \to V^o p\right) = \frac{\alpha}{4}\ \frac{1}{\gamma_V^2}\ \sigma^2_{tot}\left(V^o p\right) \tag{A-3}$$

where all quantities are open to experiment at this time. Note that this expression neglects the real part of the $V^o p$ scattering amplitude. The effective couplings $(1/\gamma_V)$ are obtained from

$$\Gamma\left(V^o \to e^+ e^-\right) = \frac{4\pi\alpha^2}{3}\ \frac{m_V}{\gamma_V^2} \tag{A-4}$$

The unbroken SU_4 ratio for the ρ,ω,Φ,Ψ couplings of 9:1:2:8 [eq. (15)] is then modified to approximately 9:1:1.3:1.5 [eq. (16)], as evident from line 2 of Table I, above. Using these couplings and accepted values for the total cross sections for $V^o p$ interactions (line 3), we obtain an evaluation of the RHS of eq. (A-3) in line 4. It compares acceptably with directly measured values of the forward diffractive production cross section as quoted in line 5 (where we do not quote error bars, but note that all values are due to an extrapolation which can account, by itself, for the difference between lines 4 and 5).

D I S C U S S I O N

CHAIRMAN: C.A. HEUSCH

Scientific Secretaries: W. Janke and M.L. Schäfer

DISCUSSION

- *LOEWE:*

You said that the one-Pomeron exchange is not sufficient for the small t-region. Do you think that in the forward region it is necessary to incorporate in your description more Pomeron exchanges and also triple-P vertices ?

- *HEUSCH:*

If you want to use the simple-minded interpretation of Pomeron exchange by a massless vector beam scattered in the forward direction, you need some coupling of two vectors (photons) to another spin-1 object with the quantum numbers of the vacuum. For space-time reasons, this is not a possible vertex. So you have to put in a vertex function which contains a fixed pole, for example.

The main point of my lecture was to demonstrate that we now have for the first time a clear, simple-minded case of the possibility to look at a spin 1 - spin 1 - spin 1 coupling. The absence of an observation of a forward dip, which we expected, then confirms that we still need a more complicated description.

- *ETIM:*

I would like to make a short comment on the vector meson-proton total cross sections, expecially in connection with the large yield of charmed states. Experimentally, the situation is the following: (see A. Silverman's lecture in the Proceedings of the 1975 International Symposium on Lepton and Photon Interactions at High Energies (Stanford, California), edited by W.T. Kirk (SLAC Stanford 1976) as well as a summary report by D. Hithin at the 1976 SLAC Summer School)

$$\sigma_{\rho p}(\simeq \sigma_{\omega p}) : \sigma_{\phi p} : \sigma_{\psi p} = 8 : 4\text{-}5 : 1$$

These ratios are in disagreement with the mass scaling law

$$m_V^2 \, \sigma_{Vp} = \text{const.}$$

where m_V is the vector meson mass and the constant is independent of c.m. energy and of m_V.

The following modified scaling law

$$m_V^2 \, \sigma_{Vp} = \frac{\ln(m_V^2/\Lambda_o^2)}{1 + \frac{\alpha_s}{\pi}(m_V^2/\Lambda_o^2)}$$

fits the data. α_s is the QCD coupling constant. This equation plays an important role in the application of QCD to hadron-hadron scattering. This result will soon be published.

- *HUSTON:*

I would like to make a comment on the branching ratio for $\rho^o \to \pi^o\gamma$. The most recent experiments using the Primakoff-effect give a value twice the old one.

- *HEUSCH:*

Thank you for bringing the new value of your collaboration to my attention. You realize, however, that $\rho^o \to \pi^o\gamma$ decay is still not a promising final state for measuring ρ^o production - but it will have to be taken into account during data analysis for the $\omega^o\gamma$ final state.

- *TUROK:*

Can one understand the factor $\exp(b\varepsilon_\gamma)$ in the photon wave-function physically?

- *HEUSCH:*

This factor is simply an ansatz based on the concept of this 'binding energy' ε_γ of virtual quark constituents in the photon. You can look at it in the context of the usual interpretation of diffraction peaks. You will see more in the literature (Brodsky's papers, e.g. SLAC-PUB 2868, 1982).

- *OGAWA:*

Could you please explain the connection between vector meson photoproduction and pion cross section in a little more detail?

- *HEUSCH:*

A diagram for photoproduction of vector mesons in which Pomeron exchange is the scattering mechanism can be related to the total elastic scattering of scalar mesons with the same flavour content. Since the constituents of the vector mesons ω/ρ are u, $\bar{u}$, d, $\bar{d}$, ω/ρ photoproduction can be related to $\pi^{\pm}p$ elastic scattering. On the other hand, the constituents of the ϕ are s, $\bar{s}$, so ϕ photoproduction is related to the $\pi^{\pm}p$ and $K^{\pm}p$ elastic cross sections.

The relationship employs the optical theorem, the simple quark model for the valence constituents of the mesons, and a direct coupling constant of photons to vector mesons, γ_V.

- *RAUH:*

Is the photon beam really a bremsstrahlung beam?

- *HEUSCH:*

Yes, the γ's from the π^0 decay are first converted into e^+e^- pairs. Then the electrons pass through a radiator, where they are reconverted into γ's via bremsstrahlung. The photon momentum can be determined by measuring the electron momentum before and behind the last radiator.

- *VAN DE VEN:*

I think you mentioned that photon scattering at high energies is a nice way to test QCD in the soft (nonperturbative) regime. That confuses me, because QCD doesn't predict anything in this regime.

- *HEUSCH:*

I did not say that one can understand QCD from low - $k_\perp$ processes - there is no theory in the QCD framework of low - $k_\perp$ processes with multiple gluon exchanges in a nonperturbative way. The data can only be analysed by means of parametrizations that make sense in the general framework described here.

What I did say, however, is that we can reasonably characterize the different kinematical regimes of photon scattering in terms of comparisons of E_γ and $k_\perp$ with basic QCD parameters (like the much abused Λ cutoff) - and that therefore there should be hope that, ultimately, a successful theoretical link will be found that will permit a QCD calculation of low - $k_{\perp}$ processes.

- *RAUH*:

You showed us the F - meson peak in the $F \to \eta\pi^{\pm}$ channel. Do these data agree with the e^+e^- and emulsion events?

- *HEUSCH*:

As far as I know, the F has not yet been uniquely identified in e^+e^- or emulsion events.

HIGGS PARTICLES FROM PURE GAUGE FIELDS

Hagen Kleinert

Physics Department
Free University Berlin
1000 Berlin 33

I Introduction

One of the important outstanding questions in unified gauge theories of weak and electromagnetic interactions is the physical nature of the Higgs particles required by renormalization. Until recently, ideas have been guided by the first historical appearance of such fields in the context of superconductivity. In 1934, Gorter and Casimir[1] first proposed the use of a space time independent order parameter for the description of the temperature behavior of the specific heat in the superconductive phase transition. In 1937, Landau[2] made the order parameter a local order field by introducing gradient terms permitting the study of spatial fluctuations. This theory has the characteristic that as the temperature of the system passes below a certain temperature T_c, the mass term which stabilizes fluctuations changes sign, thereby leading to a non-vanishing expectation value $\langle\varphi\rangle \equiv \varphi_0 \neq 0$ of the order field. There is a second order phase transition to an ordered phase.

If the order field $\varphi(\underset{\sim}{x})$ is complex and the energy depends only on $\varphi^+\varphi$, there are two degenerate fluctuations above T_c. Below T_c there is the phenomenon of spontaneous symmetry breakdown: The size of the order parameter has massive, the phase, however, has long-range massless fluctuations which join smoothly to symmetry transformations in the limit $\underset{\sim}{k} \to 0$. These are called Nambu-Goldstone bosons. In 1950, finally, Ginzburg and Landau[3] took this theory and coupled it to a vector potential in a gauge invariant way. In this way they arrived at the prototype of present models of unified gauge theories. These are characterized by two properties. First, due to the covariant coupling

$$|(\underset{\sim}{\partial} - ie\underset{\sim}{A})\varphi|^2 \tag{1}$$

the ordered phase generates a photon mass term which leads to a finite magnetic penetration depth (Meissner effect[4]). Simultaneously, the gauge invariance of the coupling allows for the removal of the phase fluctuations by a gauge transformation and absorb them completely in the vector field A. There they become part of the massive gauge field fluctuations required by rotational symmetry (Higgs effect)[5]. The Meissner-Higgs effect has turned out to be crucial for generating a short range weak interaction without destroying the renormalizability of long-range gauge theories[6].

The remaining size fluctuations of the order field are presently referred to as Higgs particles. Nine years after Ginzburg and Landau, Gorkov[7] was able to explain the order field microscopically as a field of bound states of pairs of electrons (Cooper pairs). Inspired by this it is widely believed that the Higgs particles in unified gauge theories might be bound states of some more fundamental underlying microscopic constituents (e.g. technicolor quarks).

It is the purpose of this lecture to draw attention to another type of Higgs particles which exist in a variety of many-body systems and which could, in fact, represent the correct explanation for the origin of these fields in the context of unified theories of weak and electromagnetic interactions. These Higgs fields appear naturally in many non-linear problems in which the non-linearity leads to line-like singular, or almost singular, field configurations. They have physical properties opposite to the order field and are therefore called disorder fields. In contrast to order fields, they acquire non-vanishing expectation as the temperature passes above a certain critical value. We are far from being able to describe their specific role in the context of non-abelian gauge theories. In abelian gauge theories, however, they emerge quite naturally as a way of parametrizing global group excitations. Since such excitations are known to be of extreme importance also in non-abelian lattice gauge theories it is suggested that the Higgs fields appearing in unified theories are nothing but a convenient way of parametrizing non-linear line-like singular field configurations of a pure gauge theory. The development presented in this chapter will be completely parallel to recent discussions for crystals[8], pion condensates[9], and magnetic superconductors[10]. In all systems, line-like, almost singular field configurations are known under the name of defect lines. They play an important role in understanding fluctuation properties and phase transitions.

II Lattice Gauge Theory

Let us recall Wilson's definition[11] of a gauge theory on a simple, cubic lattice. If $\underset{\sim}{x}$ denotes the discrete lattice sizes and $\underset{\sim}{i}$ the oriented link going from $\underset{\sim}{x}$ to the next neighbor along the possible $\hat{x}_i$ axes, the partition function is given by

$$Z = \prod_{\underset{\sim}{x},i} \int dU_i(\underset{\sim}{x})\, e^{\frac{\beta}{2d_c} \operatorname{tr} \sum_{\underset{\sim}{x},i \neq j} \left(\operatorname{Re} U_{ij}^{\square} - 1 \right)} \tag{2}$$

where dU_i is the invariant group measure and $U_{ij}^{\square}$ a product of four group elements, one for every link on a plaquette formed by $\underset{\sim}{x}$, $\underset{\sim}{x}+\underset{\sim}{i}$, $\underset{\sim}{x}+\underset{\sim}{i}+\underset{\sim}{j}$, $\underset{\sim}{x}+\underset{\sim}{j}$:

$$U_{ij}^{\square} \equiv U_i(\underset{\sim}{x})\, U_j(\underset{\sim}{x}+\underset{\sim}{i})\, U_i^{+}(\underset{\sim}{x}+\underset{\sim}{j})\, U_j^{+}(\underset{\sim}{x}) \tag{3}$$

The number d_c is a conventional factor denoting the dimensionality of the U_i matrices. For small field fluctuations, $U_i \sim 1$ can be parametrized in terms of $d_c \times d_c$ matrices A_i as

$$U_i(\underset{\sim}{x}) = e^{iagA_i(\underset{\sim}{x})} = 1 + iagA_i(\underset{\sim}{x}) + \tfrac{1}{2}(iag)^2 A_i(\underset{\sim}{x})^2 + \ldots$$

and one finds the limit $A_i \to 0$

$$U_i(\underset{\sim}{x})U_j(\underset{\sim}{x}+\underset{\sim}{i}) = 1 + iag\,(A_i + A_j) + \tfrac{1}{2}(iag)^2(A_i^2 + A_j^2) + ia^2g\,\partial_i A_j + (iag)^2 A_i A_j + \ldots$$

such that

$$\begin{aligned} \mathrm{Re}\, U_{ij}^{\Box} - 1 &= -\tfrac{1}{2}\left(U_i(\underset{\sim}{x})\,U_j(\underset{\sim}{x}+\underset{\sim}{i}) - U_j(\underset{\sim}{x})\,U_i(\underset{\sim}{x}+\underset{\sim}{j})\right) \\ &\quad\cdot\left(U_j^{+}(\underset{\sim}{x}+\underset{\sim}{i})U_i^{+}(\underset{\sim}{x}) - U_i^{+}(\underset{\sim}{x}+\underset{\sim}{j})U_j^{+}(\underset{\sim}{x})\right) \\ &\to -\tfrac{1}{2}a^4g^2F_{ij}^2 \equiv -\tfrac{1}{2}a^4g^2\{\partial_i A_j - \partial_j A_i + ig[A_i, A_j]\}^2 \end{aligned} \tag{4}$$

thus arriving at the conventional continuum theory in D dimensions

$$Z = \prod_{\underset{\sim}{x}, i} \int_{-\infty}^{\infty} dA_i(\underset{\sim}{x})\, e^{-\frac{1}{4}\int d^Dx\, F_{ij}^2} \tag{5}$$

if one identifies

$$\beta \equiv \frac{d_c}{a^{4-D}}\,\frac{1}{g^2} \tag{6}$$

and a as the lattice spacing. In other words, the temperature $T \equiv \beta^{-1}$ in such a description is proportional to the square of the coupling constant. In the case of an abelian theory $U_i = e^{iagA_i}$

with A_i=real and the partition function becomes

$$Z_1 = \prod_{\underset{\sim}{x},i} \int_{-\pi}^{\pi} \frac{dA_i(\underset{\sim}{x})}{2\pi} e^{-\beta \sum_{\underset{\sim}{x},i<j} (1-\cos(\nabla_i A_j - \nabla_j A_i))} \tag{7}$$

where

$$\nabla_i A_j(\underset{\sim}{x}) \equiv A_j(\underset{\sim}{x}+\underset{\sim}{i}) - A_j(\underset{\sim}{x}) \tag{8}$$

is defined as the lattice derivative.

The important feature which distinguishes this lattice gauge theory from the small field limit (5) is the physical spectrum. In (5), there are only non-interacting photon fields A_i. In (7) there are, in addition, global group excitations which in the following will be called defects. They arise from the possibility that if the difference of two adjacent A_i fields $\nabla_i A_j(\underset{\sim}{x}) = A_j(\underset{\sim}{x}+\underset{\sim}{i}) - A_j(\underset{\sim}{x})$ changes by an integer multiple of 2π, the cosine function is invariant and the energy is unchanged.

Defects have an important impact upon the statistical mechanics of the system due to their high entropy. This is why they lead to a drastic change of thermodynamic properties at a certain critical value β_c. For small β, i.e. in the hot phase, they proliferate and cause a phase transition due to the well-known relation

$$F = E - TS$$

which shows the importance of configurations of high entropy at high temperature even if the energy is large.

The point we would now like to make is that starting from a pure gauge theory, the proliferation of these defects can most easily be classified by rewriting the theory as another gauge theory involving the <u>dual</u> field strength coupled minimally to complex

Higgs fields. The Higgs field accounts for the random ensemble of closed defect lines. For $T > T_c$, it takes a non-vanishing expectation. Thus it is properly a disorder field in the sense defined above.

Since the field energy is dual to the original one, the ensemble of defect lines is nothing but the world lines of gas of magnetic monopoles. It is the magnetic analogue of Debye screening that in this gas color electric flux lines are compressed into small tubes thereby leading to a linear potential between color electric quarks. In order to develop this Higgs field theory, it is most convenient to approximate the periodic energy $1-\cos(\nabla_i A_j - \nabla_j A_i)$ by another expression which has the same periodicity and shape around the minima of the energy, where the configurations are most probable, and introduces an error only where the energy is large, i.e. for improbable configurations. This approximation was invented by Villain and is excellent for not too small values of β [12]. He observed that for any configuration $A_i(\underset{\sim}{x})$ near the minimum of the energy there is always a configuration of integer numbers $n_{ij}(\underset{\sim}{x})$ such that

$$1-\cos\left(\nabla_i A_j(\underset{\sim}{x}) - \nabla_j A_i(\underset{\sim}{x})\right) \sim \frac{1}{2}\left(\nabla_i A_j(\underset{\sim}{x}) - \nabla_j A_i(\underset{\sim}{x}) - 2\pi n_{ij}(\underset{\sim}{x})\right)^2 \tag{9}$$

Thus, if we perform this replacement in the partition function and sum over all $n_{ij}(\underset{\sim}{x})$ we introduce little error since the wrong values of $n_{ij}(\underset{\sim}{x})$ are strongly suppressed by a high energy. In this way, Villain arrived at a partition function of the form

$$Z_1 \sim Z_{1V} = \prod_{\underset{\sim}{x},i} \int_{-\pi}^{\pi} \frac{dA_i(\underset{\sim}{x})}{2\pi} \sum_{\{n_{ij}(\underset{\sim}{x})\}} e^{-\frac{\beta}{2}\sum_{\underset{\sim}{x},i<j}\left(\nabla_i A_j - \nabla_j A_i - 2\pi n_{ij}(\underset{\sim}{x})\right)^2} \tag{10}$$

This has the great advantage that the exponent is quadratic in all fluctuating variables. Therefore, we can introduce an antisymmetric auxiliary field tensor F_{ij} and rewrite the partition function in the form[13,14]

$$Z_V = \prod_{\underset{\sim}{x},i} \int_{-\pi}^{\pi} \frac{dA_i(\underset{\sim}{x})}{2\pi} \sum_{\{n_{ij}(\underset{\sim}{x})\}} \prod_{\underset{\sim}{x},i<j} \int_{-\infty}^{\infty} \frac{dF_{ij}(\underset{\sim}{x})}{\sqrt{2\pi\beta}} e^{-\frac{1}{2\beta}\sum_{\underset{\sim}{x},i<j} F_{ij}^2 + i\sum_{\underset{\sim}{x},i<j} F_{ij}(\nabla_i A_j - \nabla_j A_i - 2\pi n_{ij})} \tag{11}$$

The sum over $n_{ij}(\underset{\sim}{x})$ can be performed using Poisson's formula

$$\sum_{n=-\infty}^{\infty} e^{-2\pi i F n} = \sum_{f=-\infty}^{\infty} \delta(F-f) \tag{12}$$

which squeezes the field F_{ij} onto the integers f_{ij}. If we now execute the A_i integrals, the integer numbers f_{ij} are seen to have no lattice divergence. Hence

$$\begin{aligned} &\prod_{\underset{\sim}{x},i} \int_{-\pi}^{\pi} \frac{dA_i(\underset{\sim}{x})}{2\pi} \sum_{\{n_{ij}(\underset{\sim}{x})\}} e^{i\sum_{\underset{\sim}{x},i<j} F_{ij}(\nabla_i A_j - \nabla_j A_i - 2\pi n_{ij})} \\ &= \sum_{\{f_{ij}(\underset{\sim}{x})\}} \delta_{\overset{*}{\nabla}_i f_{ij},0} \, \delta(F_{ij} - f_{ij}) \end{aligned} \tag{13}$$

In order to arrive at this result we have used the lattice analogue of partial integration, $\sum_{\underset{\sim}{x}} g^*(\underset{\sim}{x}) \nabla_i h(\underset{\sim}{x}) = -\sum_{\underset{\sim}{x}} (\overset{*}{\nabla}_i g(\underset{\sim}{x}))^* h(\underset{\sim}{x})$, where

$$\overset{*}{\nabla}_i g(\underset{\sim}{x}) = g(\underset{\sim}{x}) - g(\underset{\sim}{x} - \underset{\sim}{i}) \tag{14}$$

Because of the lattice divergence, we can rewrite f_{ij} as a curl of an integer valued lattice vector potential which is dual to A_i

$$f_{ij}(\underset{\sim}{x}) = \varepsilon_{ijkl} \overset{*}{\nabla}_k \hat{a}_l(\underset{\sim}{x} - \underset{\sim}{l}) \tag{15}$$

a decomposition which is gauge invariant under $\hat{a}_l(\underset{\sim}{x}) \to \hat{a}_l(\underset{\sim}{x}) + \nabla_l \hat{\Lambda}(\underset{\sim}{x})$ In this way we arrive at

$$Z = (2\pi\beta)^{-3N} \sum_{\{\hat{a}_l(\underset{\sim}{x})\}} \delta_{a_1,0}\, e^{-\frac{1}{2\beta}\sum_{\underset{\sim}{x},i<j} \tilde{f}_{ij}^2} \tag{16}$$

where $\delta_{a_1,0}$ is some gauge fixing which is compatible with the integer valuedness of $\hat{a}_l$, N is the total number of lattice sites, and

$$\tilde{f}_{ij} \equiv \nabla_i \hat{a}_j - \nabla_j \hat{a}_i = \varepsilon_{ijkl} f_{kl} \tag{17}$$

is the dual field strength of f_{ij}.

Since we are not used to thinking in terms of integer fields it is advantageous to allow $\hat{a}(\underset{\sim}{x})$ to be continuous variables but enforce their integer values via one more application of Poisson's formula (12). Then Z_1 can be written as

$$Z_1 = (2\pi\beta)^{-N} \prod_{\underset{\sim}{x},i} \int_{-\infty}^{\infty} \frac{d\hat{A}_i(\underset{\sim}{x})}{\sqrt{2\pi\beta}} \, \delta(\hat{A}_1) \, e^{-\frac{1}{2\beta}\sum_{\underset{\sim}{x},i<j} \tilde{F}_{ij}^2} \tag{18}$$

$$\cdot \sum_{\{l_i(\underset{\sim}{x})\}} \delta_{\overset{*}{\nabla}_i l_i,0} \, e^{2\pi i \sum_{\underset{\sim}{x},i} l_i \hat{A}_i}$$

The $\delta_{\overset{*}{\nabla}_i l_i,0}$ restriction has to be inserted to account for the gauge invariance of $\hat{F}_{ij}$. This restriction implies that the sum over integer $l_i(\underset{\sim}{x})$ configurations

$$\sum_{\{l_i(\underset{\sim}{x})\}} \delta_{\overset{*}{\nabla}_i l_i,0} \, e^{2\pi i \sum_{\underset{\sim}{x},i} l_i \hat{A}_i} \tag{19}$$

can be decomposed into the sum over closed loops in which each line carries the number $l_i(\underset{\sim}{x})=1$. It is this sum which represents the additional degree of freedom due to the line-like defects in the lattice gauge theory. The coupling (19) is the same as the minimal coupling of an electric current. Here, however, the field A_i is dual such that the currents are the world lines of magnetic monopoles with respect to the A_i fields.

If the A_i fields are integrated out, the partition function becomes[13,14]

$$Z_1 = (2\pi\beta)^{-3N/2} \sum_{\{l_i(\underset{\sim}{x})\}} \delta_{\overset{*}{\nabla}_i l_i,0} \, e^{-\frac{\beta}{2}4\pi^2 \sum_{\underset{\sim}{x},\underset{\sim}{x}'} l_i(\underset{\sim}{x})\, v(\underset{\sim}{x}-\underset{\sim}{x}')\, l_i(\underset{\sim}{x}')} \tag{20}$$

where

$$v(\underset{\sim}{x}) \equiv \sum_{\underset{\sim}{k}} e^{i\underset{\sim}{k}\underset{\sim}{x}} \frac{1}{2\sum_{i=1}^{4}(1-\cos k_i a)} \equiv \sum_{\underset{\sim}{k}} e^{i\underset{\sim}{k}\underset{\sim}{x}} \frac{1}{|K|^2} \tag{21}$$

is the lattice version of the Coulomb potential. Eq. (20) is recognized as the sum over Biot-Savart type of energies of the magnetic current loops. It is convenient to remove the large value at $\underset{\sim}{x}$=o and rewrite

$$v(\underset{\sim}{x}) \equiv v'(\underset{\sim}{x}) + v(\underset{\sim}{0})\,\delta_{\underset{\sim}{x},\underset{\sim}{0}} \tag{22}$$

where $v(\underset{\sim}{0}) \simeq .155$ and

$$v'(\underset{\sim}{x}) \equiv \sum_{\underset{\sim}{k}} e^{i\underset{\sim}{k}\underset{\sim}{x}} \left(\frac{1}{|K|^2} - \frac{1}{N}\, v(\underset{\sim}{0})\right) \tag{23}$$

is a potential which vanishes at $\underset{\sim}{x}$=o. Then (20) becomes

$$Z = (2\pi\beta)^{-3N/2} \sum_{\{l_i(\underset{\sim}{x})\}} \delta_{\bar{\nabla}_i l_i, 0} \cdot e^{-\frac{\beta}{2} 4\pi^2 \sum_{\underset{\sim}{x},\underset{\sim}{x}'} l_i(\underset{\sim}{x})\, v'(\underset{\sim}{x}-\underset{\sim}{x}')\, l_i(\underset{\sim}{x}') - \frac{\beta}{2} 4\pi^2 v(\underset{\sim}{0}) \sum_{\underset{\sim}{x},i} l_i^2(\underset{\sim}{x})} \tag{24}$$

The first term can again be rewritten in the form (18) and

becomes

$$Z_1 = (2\pi\beta)^{-N} \prod_{\underset{\sim}{x},i} \int_{-\infty}^{\infty} \frac{d\hat{A}_i(\underset{\sim}{x})}{\sqrt{2\pi\beta}} \, \delta(\bar{\nabla}_i \tilde{A}_i) \, e^{-\frac{1}{2\beta}\sum_{\underset{\sim}{x},i<j} F_{ij}'^{\,2}}$$

$$\cdot \sum_{\{l_i(\underset{\sim}{x})\}} \delta_{\bar{\nabla}_i l_i,0} \, e^{-\frac{\beta}{2}4\pi^2 v(0)\sum_{\underset{\sim}{x},i} l_i^2 + 2\pi i \sum_{\underset{\sim}{x},i} l_i \tilde{A}_i} \tag{25}$$

The prime of the field strength denotes the fact that F_{ij}^2 in momentum space is multiplied by $\left(1-\frac{v(0)}{N}|K|^2\right)^{-1}$ corresponding to the replacement $v(\underset{\sim}{x}) \to v'(\underset{\sim}{x})$. This expression has the advantage that it displays the suppression of magnetic current loops of higher current strength due to the self-interaction. Since the vector potential $\hat{A}(\underset{\sim}{x})$ propagates with (23), the residual interactions have no self-energy due to

$$\langle \tilde{\hat{A}}_i(\underset{\sim}{x})^2 \rangle = 0 \tag{26}$$

We are now ready to show that (25) can be rewritten as a Higgs like disorder field theory representing the grand-canonical ensemble in the original pure gauge theory!

III Higgs Disorder Fields

The derivation proceeds in two steps. First, we follow Peskin[14] and observe that the sum over closed l_i loops in (25) is equivalent to an XY model in an external vector potential A_i, by which we mean the following partition function[14]

$$Z_{\text{loops}}^{\hat{A}_i} = \prod_{\underset{\sim}{x}} \int_{-\pi}^{\pi} \frac{d\gamma(\underset{\sim}{x})}{2\pi} \, e^{\frac{1}{t}\sum_{\underset{\sim}{x},i} \cos(\nabla_i \gamma - 2\pi \hat{A}_i)} \tag{27}$$

The auxiliary temperature t of this model is to be identified with the factor of the $\frac{1}{2}\sum_{\underset{\sim}{x},i} \ell_i^2$ term as follows

$$t = \beta\, 4\pi^2 v(0) \tag{28}$$

The proof is quite simple: Using Villain's approximation, (27) becomes

$$Z_{loops}^{\hat{A}_i} \sim \prod_{\underset{\sim}{x}} \int_{-\pi}^{\pi} \frac{d\gamma(\underset{\sim}{x})}{2\pi} \sum_{\{m_i(\underset{\sim}{x})\}} e^{-\frac{1}{2t}\sum_{\underset{\sim}{x},i}\left(\nabla_i\gamma - 2\pi\hat{A}_i - 2\pi m_i\right)^2} \tag{29}$$

Following the same arguments as after (10) we can rewrite this as

$$Z_{loops}^{\hat{A}_i} \sim \prod_{\underset{\sim}{x},i} \int_{-\infty}^{\infty} \frac{dL_i(\underset{\sim}{x})}{\sqrt{2\pi/t}} \prod_{\underset{\sim}{x}} \int_{-\pi}^{\pi} \frac{d\gamma(\underset{\sim}{x})}{2\pi} \sum_{\{m_i(\underset{\sim}{x})\}} \exp\left\{-\frac{t}{2}\sum_{\underset{\sim}{x},i} L_i^2 - i\sum_{\underset{\sim}{x},i} L_i\left(\nabla_i\gamma - 2\pi\hat{A}_i - 2\pi m_i\right)\right\} \tag{30}$$

Summing over $m_i(\underset{\sim}{x})$ forces L_i to be integers ℓ_i, due to (12), and integrating out $\gamma(\underset{\sim}{x})$ we find $\delta_{\bar{\nabla}_i \ell_i, 0}$, just as in (13). Thus we arrive at

$$Z_{loops}^{\hat{A}_i} \sim \left(\frac{2\pi}{t}\right)^{-3N/2} \sum_{\{\ell_i(\underset{\sim}{x})\}} \delta_{\bar{\nabla}_i \ell_i, 0}\, e^{-\frac{t}{2}\sum_{\underset{\sim}{x},i} \ell_i^2 + 2\pi i \sum_{\underset{\sim}{x},i} \ell_i \hat{A}_i} \tag{31}$$

which is indeed the loop sum in (25), after identifying (28). But there is no problem in turning (27) into a complex Higgs like field theory. For $\tilde{A}_i = o$ we can write

$$\sum_{\underset{\sim}{x},i} \cos \nabla_i \gamma = \sum_{\underset{\sim}{x},i} \cos \left(\gamma(\underset{\sim}{x}+\underset{\sim}{i}) - \gamma(\underset{\sim}{x})\right) \tag{32}$$

$$= \sum_{\underset{\sim}{x},i} \left[\cos\gamma(\underset{\sim}{x}+\underset{\sim}{i}) \cos\gamma(\underset{\sim}{x}) + \sin\gamma(\underset{\sim}{x}+\underset{\sim}{i}) \sin\gamma(\underset{\sim}{x}) \right]$$

$$= \sum_{\underset{\sim}{x},i,a} S_a(\underset{\sim}{x}) \left(1+\nabla_i\right) S_a(\underset{\sim}{x})$$

where

$$\underset{\sim}{S}(\underset{\sim}{x}) = \left(\cos\gamma(\underset{\sim}{x}),\ \sin\gamma(\underset{\sim}{x})\right)$$

Moreover, using

$$\sum_{\underset{\sim}{x},i} S_a(\underset{\sim}{x}) \nabla_i S_a(\underset{\sim}{x}) = \frac{1}{2} \sum_{\underset{\sim}{x},i} S_a(\underset{\sim}{x}) \nabla_i S_a(\underset{\sim}{x}) - \overset{*}{\nabla}_i S_a(\underset{\sim}{x})\, S_a(\underset{\sim}{x})$$

$$= \frac{1}{2} \sum_{\underset{\sim}{x},i} S_a(\underset{\sim}{x}) (\nabla_i - \overset{*}{\nabla}_i) S_a(\underset{\sim}{x}) = \frac{1}{2} \sum_{\underset{\sim}{x},i} S_a(\underset{\sim}{x}) \left(S_a(\underset{\sim}{x}+\underset{\sim}{i}) - S_a(\underset{\sim}{x}) + S_a(\underset{\sim}{x}-\underset{\sim}{i}) - S_a(\underset{\sim}{x})\right) \tag{33}$$

$$= \frac{1}{2} \sum_{\underset{\sim}{x},i} S_a(\underset{\sim}{x})\, \overset{*}{\nabla}_i \nabla_i\, S_a(\underset{\sim}{x})$$

we can rewrite (27) as

$$Z_{\text{loops}}^{\hat{A}_i=0} \sim \prod_{\underset{\sim}{x}} \int_{-\pi}^{\pi} \frac{d\gamma(\underset{\sim}{x})}{2\pi}\, e^{\frac{D}{T} \sum_{\underset{\sim}{x},a} S_a(\underset{\sim}{x}) \left(1+\sum_i \overset{*}{\nabla}_i \nabla_i / 2D\right) S_a(\underset{\sim}{x})} \tag{34}$$

where D=4 is the space dimensionality.

Now we introduce a two component real auxiliary field φ_a and rewrite

$$Z_{loops}^{\hat{A}_i=0} \sim \det\Big(1+\sum_i \overset{*}{\nabla}_i\nabla_i/2D\Big)^{-1} \prod_x \int_{-\infty}^{\infty} \frac{d\varphi_1 d\varphi_2}{4\pi D/t} \prod_x \int_{-\pi}^{\pi} \frac{d\gamma(\underset{\sim}{x})}{2\pi} \tag{35}$$

$$\exp\Big\{-\frac{t}{4D}\sum_{\underset{\sim}{x},\underset{\sim}{x}',a} \varphi_a(\underset{\sim}{x})\Big(1+\sum_i \overset{*}{\nabla}_i\nabla_i/2D\Big)^{-1}(\underset{\sim}{x},\underset{\sim}{x}')\varphi_a(\underset{\sim}{x}') + \sum_{\underset{\sim}{x},a} S_a\varphi_a\Big\}$$

The integration over $\gamma(\underset{\sim}{x})$ gives $I_0(|\varphi|^2)$. In order to remove the determinantal factor it is useful to go to fields

$$\Psi_a(\underset{\sim}{x}) \equiv \sum_{\underset{\sim}{x}'} \Big(1+\sum_i \overset{*}{\nabla}_i\nabla_i/2D\Big)^{-\frac{1}{2}}(\underset{\sim}{x},\underset{\sim}{x}')\,\varphi(\underset{\sim}{x}') \tag{36}$$

such that

$$Z_{loops}^{\hat{A}_i=0} \sim \int \frac{d\Psi d\Psi^+}{4\pi D/t} \tag{37}$$

$$\cdot \exp\Big\{-\frac{t}{4D}\sum_{\underset{\sim}{x}}|\Psi(\underset{\sim}{x})|^2 + \sum_{\underset{\sim}{x}} \log I_0\Big(\Big|\Big(1+\sum_i \overset{*}{\nabla}_i\nabla_i/2D\Big)^{\frac{1}{2}}\Psi\Big|\Big)\Big\}$$

where we have gone to complex field notation. The field $\Psi(\underset{\sim}{x})$ may be considered as the second quantized field version of the random loops represented in (31) (at $\tilde{A}_i$=o)(in the euclidean version).

Close to a phase transition, we can expand the exponent in powers of the field Ψ and include only the lowest gradient terms

$$\sum_{\underset{\sim}{x}} \left\{ -\frac{t}{4D} |\psi|^2 + \frac{t}{4} |\psi|^2 - \frac{1}{8D} |\nabla_i \psi|^2 - \frac{1}{64} |\psi|^4 \right\} + \ldots \tag{38}$$

We can now include the coupling to the gauge field $\tilde{A}_i$ by simply replacing the derivatives by their covariant versions

$$\begin{aligned} \nabla_i \psi(\underset{\sim}{x}) &\to D_i \psi(\underset{\sim}{x}) \equiv e^{-2\pi i A_i(\underset{\sim}{x})} \psi(\underset{\sim}{x}+\underset{\sim}{i}) - \psi(\underset{\sim}{x}) \\ \overset{*}{\nabla}_i \psi(\underset{\sim}{x}) &\to \overset{*}{D}_i \psi(\underset{\sim}{x}) \equiv \psi(\underset{\sim}{x}) - e^{2\pi i A_i(\underset{\sim}{x}-\underset{\sim}{i})} \psi(\underset{\sim}{x}-\underset{\sim}{i}) \end{aligned} \tag{39}$$

As a result we find for the original <u>pure electric</u> $U(1)$ lattice gauge theory a <u>magnetic</u> $U(1)$ lattice gauge theory <u>with Higgs fields.</u>

$$Z \sim \prod_{\underset{\sim}{x},i} \int_{-\infty}^{\infty} \frac{d\tilde{A}_i}{\sqrt{2\pi\beta}} \prod_{\underset{\sim}{x}} \int \frac{d\psi\, d\psi^+}{4\pi D/t} \tag{40}$$

$$\exp\left\{ -\frac{1}{4\beta} \sum_{\underset{\sim}{x},i,j} \tilde{F}_{ij}^{\,2} - \sum_{\underset{\sim}{x}} \left(\frac{1}{8D} |D_i \psi|^2 + \frac{t-D}{4D} |\psi|^2 + \frac{1}{64} |\psi|^4 \right) \right\}$$

It is now straightforward to see that there is a phase transition at

$$t = D \tag{41}$$

For $t < D$, i.e. $T > T_c \equiv \frac{D}{4\pi^2 v(0)}$, which corresponds to strong coupling, the monopole disorder proliferates leading to a non-vanishing $\psi_0 \neq 0$. As a consequence of the minimal coupling, which in the long wavelength limit reduces to the form (1), the dual vector potential becomes massive and the color electric flux is screened, implying linear potentials and confinement. For $t > D$ i.e. $T < T_c$ (weak coupling) there are no monopoles and the forces are Coulomb like. It is gratifying to note that this temperature will not be shifted by seagull diagrams

$$\langle \tilde{A}_i^2 \rangle \psi^+ \psi \quad = \qquad\qquad (42)$$

since the propagator $v'(\underline{x})$ ensures the absence of closed photon loops (recall (26)).

Up to now, all operations have been performed on a D dimensional lattice whose spacing a was normalized to unity. In order to extract the consequences for a field theory on a four dimensional continuous space-time we have to reintroduce the parameter a at the appropriate places and take the limit $a \to 0$. Then lattice derivatives go over into proper derivatives as $\nabla_i \psi \to a \partial_i \psi$. In order to arrive at the conventional gradient term $\int d^4x |\partial_i \psi|^2$ we have to renormalize the field by a factor a i.e. we have to replace $\psi \to \sqrt{24}\, a \psi$ Then $\nabla_i \psi \to \sqrt{24}\, a^2 \partial_i \psi$ and the expansion (38) becomes in four dimensions

$$- \int d^4x \left\{ \frac{2}{a^2}(t-D)|\psi|^2 + |\underline{\partial}\psi|^2 + 9|\psi|^4 \right\} + \ldots \qquad (43)$$

If the fields are to have a finite mass, say M^2, we have to require that $t-D$ approaches zero with a as

$$t - D = \frac{a^2}{2} M^2 \to 0 \qquad (44)$$

Thus the continuum limit lies automatically in the critical regime.

Physically, this situation may be described as follows: Critical fluctuations are characterized by a very large coherence length as compared with the lattice spacing. But then, in the limit of zero lattice spacing a, any finite wavelength M^{-1} is extremely long as compared with respect to a such that it is critical in the underlying infinitesimally fine lattice. We can convince ourselves that indeed all higher terms which are present in the exponent of (37) disappear in this continuum limit: Every higher power of ψ carries an extra factor a and so does every further gradient ∂_i.

Thus as announced in the introduction, starting from a pure gauge theory we have indeed arrived, by purely formal manipulation, at another dual gauge theory coupled to Higgs fields. The conclusions of this observation may be far reaching:

IV Conclusion

The discussion in this work was based on abelian gauge theory on the lattice. The question arises as to what we can learn from this as far as non-abelian theories are concerned. The basic lesson to be extracted from the many Monte Carlo computer calculations is that abelian and non-abelian theories are both characterized by the proliferation of monopoles above a certain value of the coupling. This is a consequence of global group excitations which are present in the compact abelian theory as well as the non-abelian theory. The difference between the two arises only in the weak coupling limit. While in the $U(1)$ theory for $D > 3$, there is a value β_c above which there are no more monopoles, and therefore no confinement, the non-abelian version has no phase transition and remains confining for arbitrarily large β. We therefore are led to conjecture that whatever we derive in the abelian theory for strong coupling should be valid in the non-abelian theory also for weak coupling. In fact, this is necessary if we want to use information from lattice calculations and extract from them statements about the continuum limit of the theory. Assuming that this can be done we arrive at the following exciting hypothesis[15] concerning the structure of strong and weak and electromagnetic interactions. The strong interactions are a pure gauge theory involving 8 colored gluons with a Lagrangian $\mathcal{L} = \frac{1}{4g_s^2}\sum_{ij,a} F_{ij}^{a\,2}$. The weak interactions are also a pure gauge theory with 3 flavored vector mesons W_i : $\mathcal{L} = \frac{1}{4g_W^2}\sum_{ij,\alpha} \tilde{F}_{ij}^{\alpha\,2}$. Contrary to the strong interactions, the action involves the dual tensor $\tilde{F}_{ij}^{\alpha}$ of the fields W_i. The coupling g_W is much larger than g_s, opposite to what one would naively think. Due to the weakness of g_s, there are very few color magnetic monopoles in the ground state which lead to color electric confinement but allow for a large energy range where perturbation theory is applicable.

The weak interactions, on the other hand, are characterized by a strong g_W. Therefore there exists a high "monopole" density of the dual theory which are flavor electric particles. They are most conveniently described by Higgs fields. These have a non-zero vacuum expectation such that the W_i fields are massive. Due to the largeness of g_W, the world lies deeply in the Higgs phase and perturbation theory with massless W's is impossible up to very high energies. This is observed as the short range property of weak interactions. At the same time, the flavor electric interaction between Higgs particles is very weak. This picture is quite different from current hypotheses of unifying weak, electromagnetic, and strong interactions via a single, multicomponent non-abelian gauge theory. In our case the unified theory has the generic form

$$\mathcal{L} = \frac{1}{4g_s^2}\sum_{i,j,a} F_{ij}^{a\,2} + \frac{1}{4g_W^2}\sum_{i,j,\alpha} \tilde{F}_{ij}^{\alpha\,2}$$

with small g_s and large g_w. The quarks are electric sources for A_i^a and magnetic ones for $\tilde{W}_i^\alpha$. Certainly, many details remain to be investigated.

It goes without saying that the field theory on the lattice with the partition function (11) can also form the basis for studying real magnetic monopoles, if the single event discovered recently by Cabrera finds successors. In this case the field A_i is the usual electromagnetic one which is coupled immediately to electrons. The dual version (40) can be seen as a possible generalization of quantum electrodynamics which naturally gives rise to Dirac monopoles.

Let us recall that Dirac[17] derived his quantization condition by postulating the invisibility of the backflow of magnetic field lines with respect to electrons. In the present formulation, this is automatically ensured if the electron is coupled minimally via the covariant derivative $D_i\Psi(\underset{\sim}{x}) = e^{-2\pi i A_i}\Psi(\underset{\sim}{x}+\underset{\sim}{i}) - \Psi(\underset{\sim}{x})$, since this derivative is indifferent to jumps of A_i by 2π. There was, however, one problem which Dirac was not able to solve within continuum electrodynamics: Even though he had succeeded in making the string invisible to electrons, it still carried an infinite density. In order to circumvent this problem, many tricky procedures were invented and you have heard about them in Prof. Yang's lecture. It is gratifying to note that the present modification (7) of the electromagnetic action completely avoids these energetic problems without involving additional mathematical structures. The scalar fields, on the other hand, describe spin zero particles which couple locally to the dual electromagnetic potential A_i. Thus, with respect to the original potential A_i, they correspond to magnetic monopoles. Their magnetic charge is $4\pi\sqrt{\beta} = \frac{4\pi}{g}$ as we can read off equ. (40). This is precisely Dirac's value for magnetic monopoles which has magnetic charge= 4π /electric charge (in our units). Thus the partition function (7), (10) has a chance of being the correct extension of quantum electrodynamics into the short-distance regime and the monopoles would be scalar particles with well-defined complicated repulsive self-interactions.

References

1) C.J. Gorter and H.B. Casimir, Phys. Z. 35, 963(1934).

2) L.D. Landau, Phys. Z. Sovietunion 11, 129,545(1937), reprinted in The Collected Papers of L.D. Landau, ed. by ter Haar, Gordon and Breach-Pergamon, New York, 1965, p.193.

3) V.L. Ginzburg and L.D. Landau, Zh. Eksp. i. Teor. Fiz. 20, 1064(1950).

4) W. Meissner and R. Ochsenfeld, Naturwiss. 21, 787(1933).

5) F. Englert and R. Brout, Phys. Rev. Lett. 13, 321(1964).
P.W. Higgs, Phys. Lett. 12, 132(1964), 13, 508(1964).
G.S. Guralnik, C.R. Hagen, and T.W.B. Kibble, Phys. Rev. Lett. 13, 585(164).

6) C. Itzykson, and J.-B. Zuber, Quantum Field Theory, McGraw-Hill, New York, 1980.

7) L.P. Gorkov, Zh. Eksp. Teor. Fiz. 36, 1918, 37, 1407(1959), (Sov. Phys.) ETP 9, 1364(1959), 10, 998(1960).

8) H. Kleinert, Phys. Lett. A89, 295(1982).

9) H. Kleinert, Lett. Nuovo Cimento 34, 103(1982).

10) H. Kleinert, Phys. Lett. A90, 259(1982).

11) F.G. Wilson, Phys. Rev. D10, 2445(1974).

12) J. Villain, H. Phys. (Paris) 36, 581(1975).

13) T. Banks, R. Myerson, and J. Kogut, Nucl. Phys. B129, 493(1977).

14) M.E. Peskin, Ann. Phys. (New York) 113, 122(1978).

15) In a less definite way, this hypothesis was advanced in H. Kleinert, Lett. Nuovo Cimento 34, 209(1982).

16) B. Cabrera, Phys. Rev. Lett. 48, 1378(1982)

17) P. Dirac, Proc. Roy. Soc. A133, 60(1931)

D I S C U S S I O N

CHAIRMAN: H. KLEINERT

Scientific Secretaries: T. Hofsäss and P. van Baal

DISCUSSION

- *GOCKSCH:*

How is your action related to the standard Wilson action? In particular, can one explain the appearance of the $n_{ij}(x)$ through this action?

- *KLEINERT:*

I did not have time to explain this during the lecture, but you will find it in the lecture notes, paragraph II up to equation (9).

- *KREMER:*

I am afraid that the monopoles in your theory might form dipoles or some condensate. How can you be sure that the monopoles exist individually?

- *KLEINERT:*

Indeed the monopoles do form a condensate if the coupling is so strong such that the system is inside the Higgs phase. In the weak coupling phase, monopoles and antimonopoles can form dipoles through the electromagnetic interaction. These dipoles can be calculated from my final action. All forces are known: the long range ones from the electromagnetic coupling $g = 2\pi/e$, the short range ones from the interaction terms in the Higgs field. Notice that these terms are unknown in Prof. Yang's theory of monopoles while here they arise from the fact that the sum:

$$\sum_{\ell_i(x)\in\mathbb{Z}} \delta_{\nabla_i \ell_i(x),0}$$

(cf. equation (25)) does not involve all random loops but that backtracking is forbidden.

Examples:

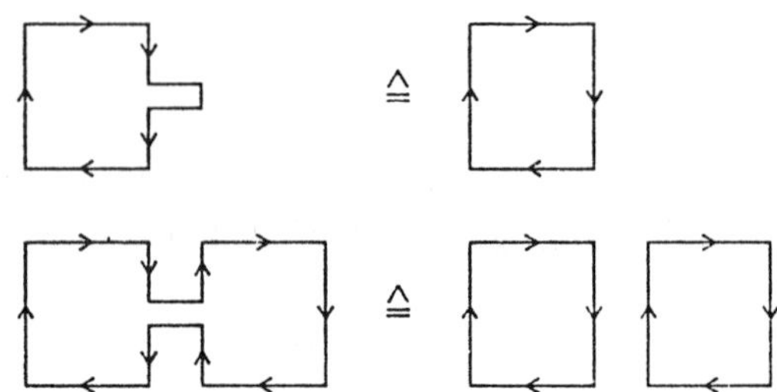

The loops on the left-hand side are absent, since they would double-count the equivalent loops on the right-hand side.

These restrictions are automatically ensured by the short range self-interactions of the ψ field shown in equation (37).

- *SCHÄFER:*

What will happen to your construction of the Higgs field if you make the same analysis for a non-Abelian theory?

- *KLEINERT:*

In detail, I cannot say. But it appears as if Abelian and non-Abelian theories are quite similar in the strong-coupling phase. The principal difference between the two seems to lie in the fact that in four dimensions the physics in the Abelian theory changes abruptly at a critical coupling, where the spaghetti of monopoles disappears, while in non-Abelian theories the strong-coupling physics can be extrapolated without discontinuity into the weak coupling regime (with only a roughening transition wiping out the memory of the lattice). By transitivity, we might expect that the strong-coupling Abelian theory can be used to learn something about the weak-coupling non-Abelian theory.

- *LYKKEN:*

Can you determine the parameters of the Higgs potential in your scheme?

- *KLEINERT:*

Yes, because the coefficients of the $|\psi|^{2n}$ expansion are completely determined by the theory, compare equation (37).

- *VAN DER BIJ:*

Can you tell us what the masses of the Higgs and the gauge particles are?

- *KLEINERT:*

Yes, from equation (40) you see that the Higgs mass is given by (t<D):

$$m_H^2 = 4D(D-t)$$

and the mass of the neutral vector meson is:

$$m_Z^2 = \frac{(2\pi)^2\beta}{4D}\,|\langle\psi\rangle|^2$$

inserting $|\langle\psi\rangle|^2 = 8(D-t)$ and $\beta = e^2/(2\pi)^2$ we find:

$$m_H/m_Z = \sqrt{2}\,D/e \simeq 20$$

which brings m_H into the TeV range. Before we compare with experiment, however, we should not forget this is not the full Georgi-Weinberg-Salam theory but just a model.

- *TUROK:*

But how do we obtain in this lattice theory the weak interaction scale of approximately 100 GeV?

- *KLEINERT:*

In every renormalizable theory you have an observable mass, which is given by:

$$m = a^{-1}f(g) \quad (*)$$

It is characteristic of renormalizability that m (and any other observable parameter) is invariant under a change of the scale of a as long as the coupling is appropriately adjusted as a function of a. Thus the relation (*) does not really determine the mass, but everytime you choose a new a you have to adjust g such that m is the mass you want it to be. After all, the mass is the proper dimensionally transmuted parameter to characterize the theory (which replaces the pair (a,g(a)) to be used otherwise).

- *MUKHI:*

Can you distinguish your Higgs particle from those of other theories?

- *KLEINERT:*

By more detailed dynamical consequences of the theory. Since all parameters are fixed, these are now calculable. For example, my Higgs particles will never split into smaller fermionic substructures (such as technicolor quarks).

- *DOBREV:*

You were stressing that you do not need to have sections. In the continuum they are needed to have a mathematically well defined description of monopoles. Of course, on the lattice you do not need sections, for topological reasons.

- *KLEINERT:*

It is not the topology which makes the strings invisible in my lattice formulation but the specific form of the energy. Had I added a term $-n_{ij}^2$ in the exponent, strings would have shown up.

My reservation against the section way of avoiding a string energy derives from the impossibility of making a quantum field theory in which a quantum field $\psi(x)$ is supposed to know which is the upper and which the lower section of the many-monopole states it can describe. Maybe Prof. Yang can explain that.

- *YANG:*

How to go from a finite number of particles to a field theory is a similar problem as in usual QED, and there is no difficulty. The difficulty that there is no Feynman rule derives from the fact that there is no interaction representation.

- *KLEINERT:*

Exactly, and this in turn is due to the sections.

- *CASTELLANI:*

Are there other ways to generate Higgs fields without using the lattice formulation?

- *KLEINERT:*

Most non-linear systems manufacture their own defects which have finite energy and therefore can appear in the laboratory (solitons, vortex lines in superfluid ^{4}He, and disclinations in liquid crystals). All these can in principle be represented by a Higgs field through a lattice formulation, and you can find this in recent Phys. Lett.A and Lettere Nuovo Cimento of mine.

What I presented can be seen as an idealization of this type of non-linear systems, in which smooth defects are sharpened to be local, which can be described by a local field theory, and in which the residual interactions among the defects are completely linearized. In this idealization, the lattice is convenient in preventing energetic infinities to appear. Also, it simplifies the construction of the Higgs field.

- *CASTELLANI:*

I was thinking more of recent appearances of Higgs fields in dimensional reduction schemes.

- *KLEINERT:*

How those Higgs fields are related to my defect Higgs fields, I do not know, but would be interested in finding out.

- *HOFSÄSS:*

I want to point out, that if one uses your procedure to generate the Higgs field which is needed to unify electromagnetic and weak interactions, one has a very restrictive scheme: as all the parameters for the Higgs field are fixed, the scheme could predict the mass of the electron.

- *PERNICI:*

Do you have an analogue of the Dirac quantization condition?

- *KLEINERT:*

Yes. My monopole satisfies $eg = 2\pi$ which, by the way is the same as Dirac's $2eg = 1$, since he used the convention:

$\mathcal{L} = \frac{1}{16\pi}F_{\mu\nu}^2$ instead of $\frac{1}{4}F_{\mu\nu}^2$

- *TUROK:*

What is the spin of your monopole? Don't you need spin 1/2 for the full electromagnetic symmetry?

- *KLEINERT:*

My monopole's spin is zero.

- *ZICHICHI:*

Spin zero is just garbage.

- *KAPLUNOVSKY:*

Why for heaven's sake should it have spin 1/2?

- *KLEINERT:*

Don't you know. If Prof. Zichichi wants it to have spin 1/2 it must have spin 1/2?

DETECTORS PROPOSED FOR LEP

Günter Wolf

Deutsches Elektronen-Synchrotron, DESY
Hamburg, Germany

INTRODUCTION

In January of this year seven large groups[1-7] submitted letters of intent expressing the desire to do experiments at the new large electron positron storage ring LEP which is under construction at CERN. Six of these groups presented proposals for new detectors. These detectors are considerably larger and more ambitious than those operating presently at e^+e^- storage rings.
In order to understand better their design it is helpful to have first a brief look at the expected properties of LEP and the physics one hopes to do with it.

THE LEP MACHINE

Fig. 1 shows the location of LEP with respect to the present installations of CERN. Table 1 lists some of the design parameters of the machine.

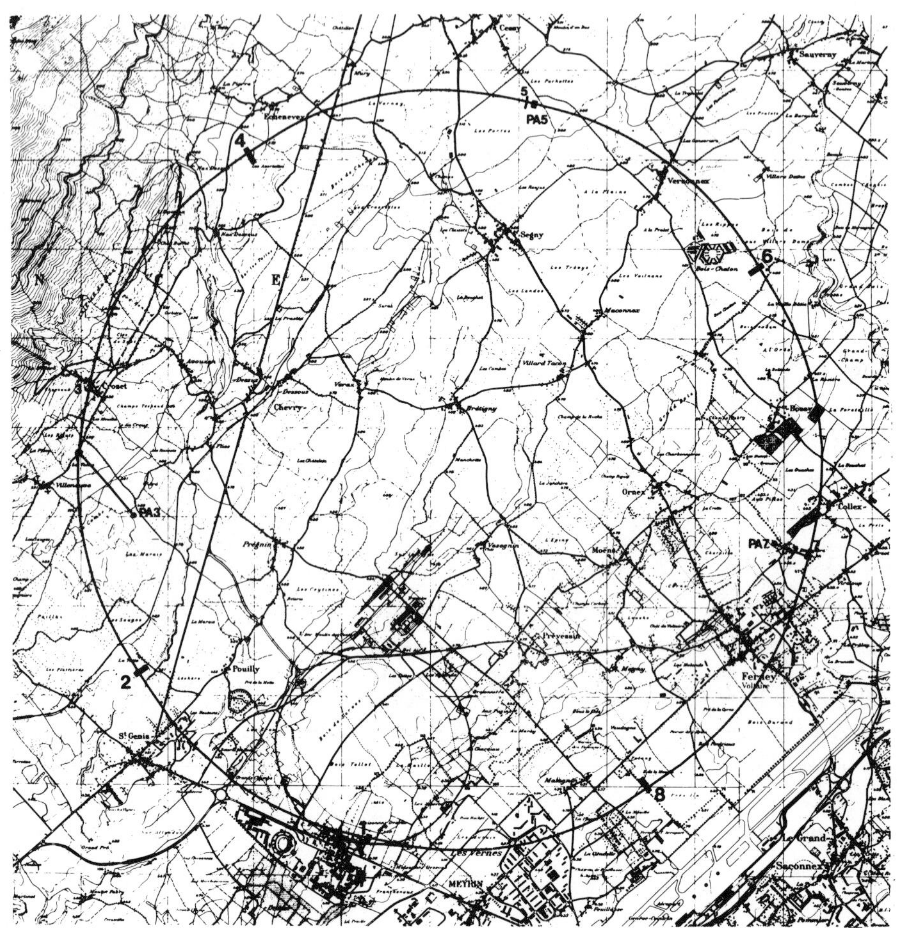

Fig. 1 Site of the LEP ring.

Table 1. Design Parameters of LEP

Circumference	26.66 km
maximum beam energy	60 GeV (phase I)
	110 GeV (with superconducting cavities)
maximum luminosity	10^{31} $cm^{-2}s^{-1}$
number of interaction points	4 (can be increased to 8)
first injection of beams	1987

The geological section given in Fig. 2 shows how LEP is embedded in the terrain near CERN. Note that the ring plane has a 1.5 % slope.

SOME PHYSICS TOPICS AT LEP ENERGIES

The first phase of LEP is optimized for c.m.s. energies near 100 GeV and centers around the study of the (still hypothetical) Z^o boson. At the Z^o pole the cross section and consequently the event rates are large. This will allow a thorough probing of the electroweak interactions and will produce a wealth of information on couplings, mixing angles and possibly on new particles. At a later stage when LEP is running at c.m. energies well above 200 GeV the major topics could be the search for and study of scalars and a first look into the "near" desert. Here the cross sections are small and the number of events expected per day only a hand full.

A detailed account of the physics expected at and above the Z^o can be found e.g. in the various studies for LEP[8)] and SLC[9)]. Here only a few of the basic processes will be mentioned. The cross sections given are based on the standard electroweak theory[10)].

$e^+e^- \rightarrow \mu^+\mu^-$

The sum of the contributions from photon and Z^o exchange (Fig. 3) lead to the following

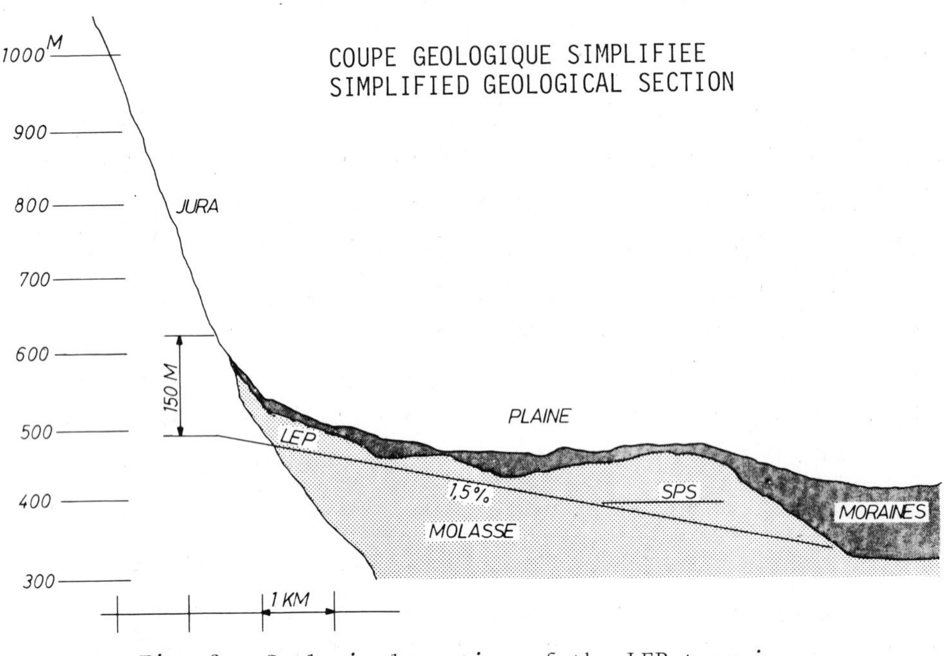

Fig. 2 Geological section of the LEP terrain.

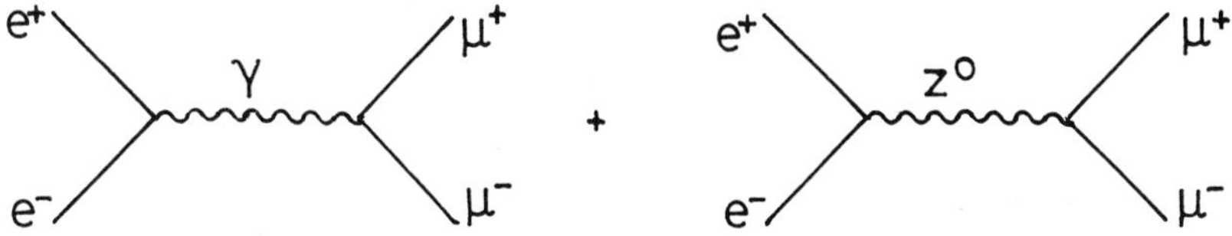

Fig. 3 The electromagnetic and weak contributions to μ pair production

expression for the cross section:

$$\sigma(e^+e^- \to \mu^+\mu^-) = \sigma^o_{\mu\mu}\left\{1 - \frac{2s\, v^2 g}{(\frac{s}{m_Z^2} - 1) + \Gamma_Z^2/(s - m_Z^2)} + \frac{s^2(v^2 + a^2)^2 g^2}{(\frac{s}{m_Z^2} - 1)^2 + \Gamma_Z^2/m_Z^2}\right\}$$

where $s = W^2$ = square of the c.m. energy

$$\sigma^o_{\mu\mu} = \frac{4\pi\alpha^2}{3s} \approx \frac{87.6 \text{ nb}}{s}, \quad s \text{ in GeV}^2$$

m_Z, Γ_Z = Z^o mass and width

$$m_Z = \frac{37.4 \text{ GeV}}{\sin\Theta_W \cos\Theta_W} \approx 93 \text{ GeV}$$

The total and leptonic widths of the Z^o are

$\Gamma_Z \simeq 2.6$ GeV, $\Gamma_{Z \to e^+e^-} \approx 0.1$ GeV

$v \equiv 2\, g_V$, $a \equiv 2g_A$, g_V, g_A are the vector and axial vector couplings

$v = -1 + 4\sin^2\Theta_W \approx -0.08$

$a = -1$

$g = G_F/(8\sqrt{2}\pi\alpha) = 4.5 \cdot 10^{-5}$ GeV^{-2}.

The μ pair cross section in the neighbourhood of the Z^o is shown in Fig. 4 in terms of $R = \sigma(e^+e^- \to \mu^+\mu^-)/\sigma^o_{\mu\mu}$. It reaches a peak

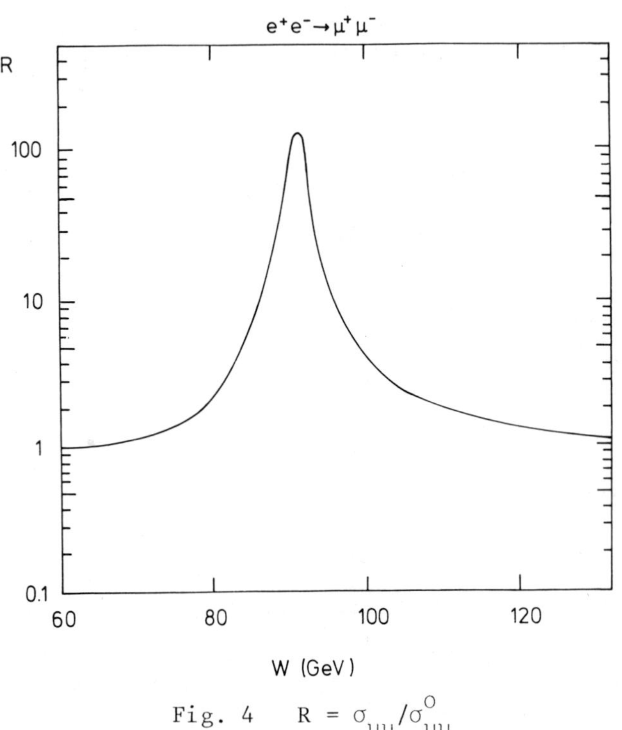

Fig. 4 $R = \sigma_{\mu\mu}/\sigma^0_{\mu\mu}$

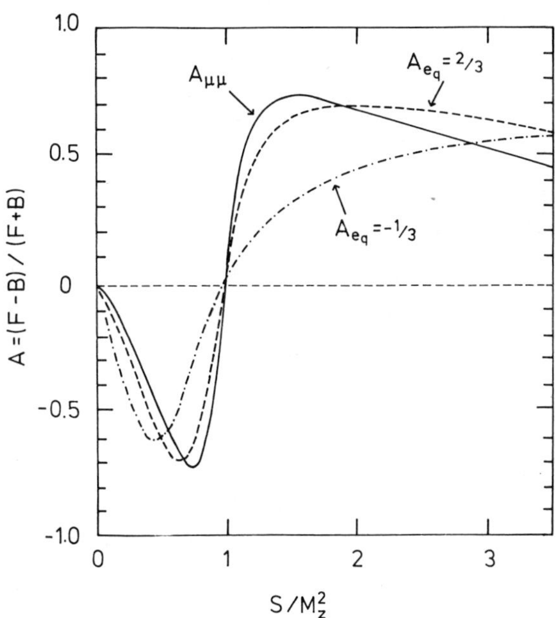

Fig. 5 The forward backward asymmetry for $e^+e^- \to \mu^+\mu^-$ and for $e^+e^- \to q\bar{q}$ with quark charge e_q = 1/3 and 2/3, respectively.

value of R ≈ 140 corresponding to a cross section of 1.4 nb. Hence an average luminosity of 500 nb^{-1}/day produces 700 μ pair events per day.

The muon angular distribution with respect to the beam axis is given by

$$\frac{d\sigma}{d\Omega} = \frac{\alpha^2}{4s}\left\{(1 + \cos^2\Theta)(1 + 2v^2 D + (v^2 + a^2)^2 D^2) + 4\cos\Theta(a^2 D + 2v^2 a^2 D^2)\right\}$$

where $D = g \cdot s/(s/m_Z^2 - 1)$. The interference between γ and Z^o exchange leads to a forward-backward asymmetry:

$$A = \frac{F - B}{F + B} = \frac{3}{2} \quad \frac{a^2 D + 2v^2 a^2 D^2}{1 + 2v^2 D + (v^2 + a^2)^2 D^2}$$

The s dependence of A is shown in Fig. 5. A varies rapidly near the Z^o mass. Precise measurements of the energy variation of the cross-section and the asymmetry near the Z^o will permit an accurate determination of the moduli of g_V and g_A.

$e^+e^- \rightarrow$ hadrons

The cross section for hadron production is calculated in an analogous manner replacing the muons by quarks (see Fig. 6) and summing over all possible quark flavours:

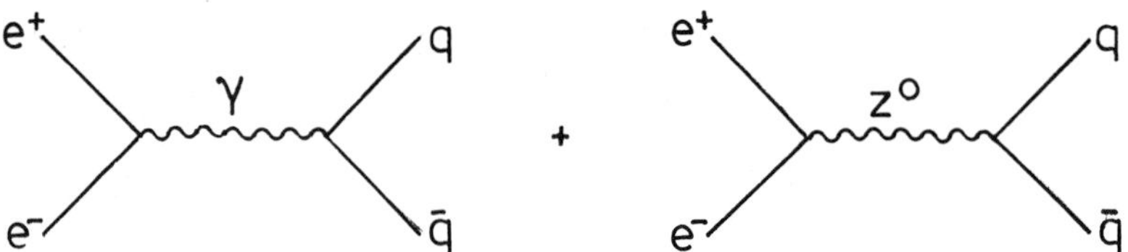

Fig. 6 The electromagnetic and weak contributions to hadron production

$$R \equiv \frac{\sigma_{tot}\ (e^+e^- \to \text{hadrons})}{\sigma^o_{\mu\mu}}$$

$$= 3 \sum_q \left\{ e_q^2 - \frac{2s\ e_q\ vv_q \cdot g}{(\frac{s}{m_Z^2} - 1) + \Gamma_Z^2/(s - m_Z^2)} + \frac{s^2(v^2 + a^2)(v_q^2 + a_q^2)g^2}{(\frac{s}{m_Z^2} - 1)^2 + \Gamma_Z^2/m_Z^2} \right\}$$

where e_q = quark charge. The predictions of the standard theory for the coupling constants are

for u,c,t: $v_q = 1 - 8/3\ \sin^2\Theta_W \approx 0.39, \quad a_q = 1$

for d,s,b: $v_q = -1 + 4/3\ \sin^2\Theta_W \simeq -0.69, \quad a_q = -1$

The ratio R reaches a peak value of ~4000 at the Z^o (Fig. 7). This corresponds to a cross section of 40 nb which for 500 nb^{-1}/day will lead to roughly 20 000 hadronic events per day - an enormous rate for e^+e^- experiments.

The forward-backward asymmetry is shown in Fig. 5. It has different s dependences for $e_q = -1/3$ and $e_q = 2/3$ quarks.

The determination of the weak coupling constants for a given quark q will require an identification of the hadronic events from this particular quark. (Note: In order to measure the asymmetry, one has to be able to distinguish between quark jet and antiquark jet.) This appears to be possible for charm quarks via the detection of D or D^* mesons and via the semileptonic decays $c \to \ell^+X$, $\ell = e,\mu$; for bottom and top quarks via their semileptonic decays, $b,t \to \ell X$. Consequently, in the design of the LEP detectors lepton identification plays a key role.

The average charged particle and photon multiplicities expected from hadronic decays of the Z^o are $\langle n_{CH}\rangle = 22$, $\langle n_\gamma\rangle = 20$. These particles are emitted in narrow jet cones as illustrated by Fig. 8 which displays an event of the type $e^+e^- \to b\bar{b}g$. The large multiplicities and the strong collimation require a high segmentation of the charged particle tracking system and of the shower detectors.

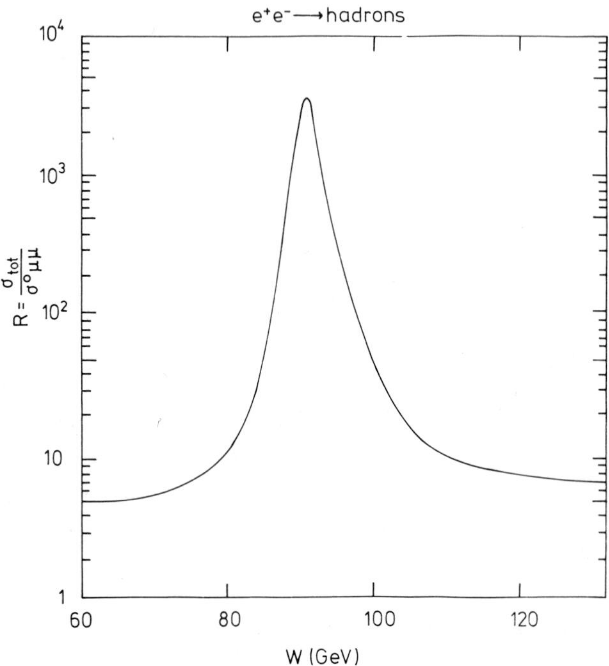

Fig. 7 $R = \sigma_{tot}(e^+e^- \to \text{hadrons})/(\sigma^o_{\mu\mu}$

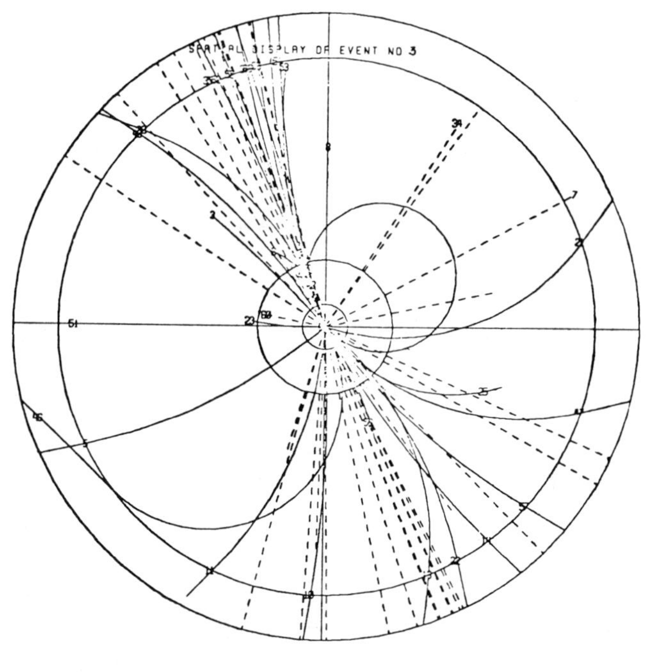

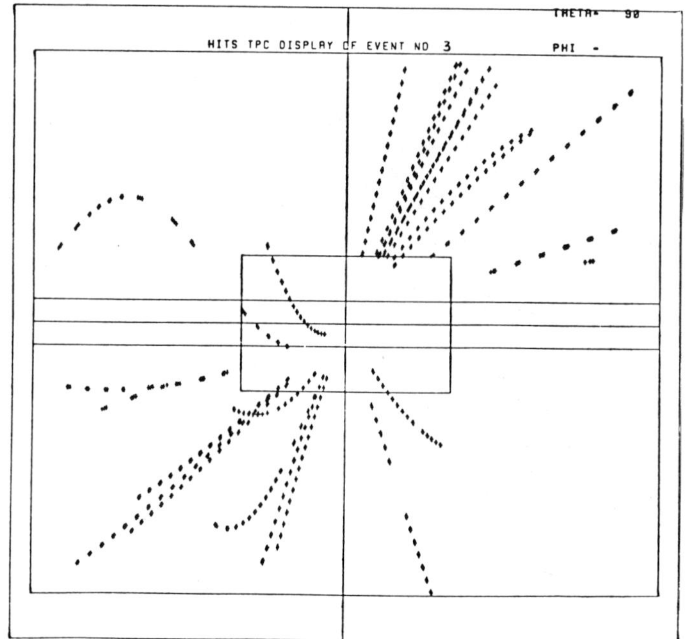

Fig. 8a A Monte Carlo event in the DELPHI detector.

$b\bar{b}g \rightarrow (bd)^{\circ}\rho^{-}\rho^{+}\pi^{-}\Omega\rho^{\circ}\Omega K^{+}\rho^{\circ}\eta'\rho^{-}(\bar{b}s)^{*}\eta'\rho^{+}\pi^{\circ}\pi^{\circ}$

$\pi^{\circ}\pi^{-}\rho^{-}\Omega^{+}D^{*+} \qquad (\bar{b}s)\gamma$

$D^{\circ}\pi^{+} \qquad e^{+}F^{-}\rightarrow F^{-}\gamma$

$K^{-}\pi^{+} \qquad K^{-}K^{\circ}$

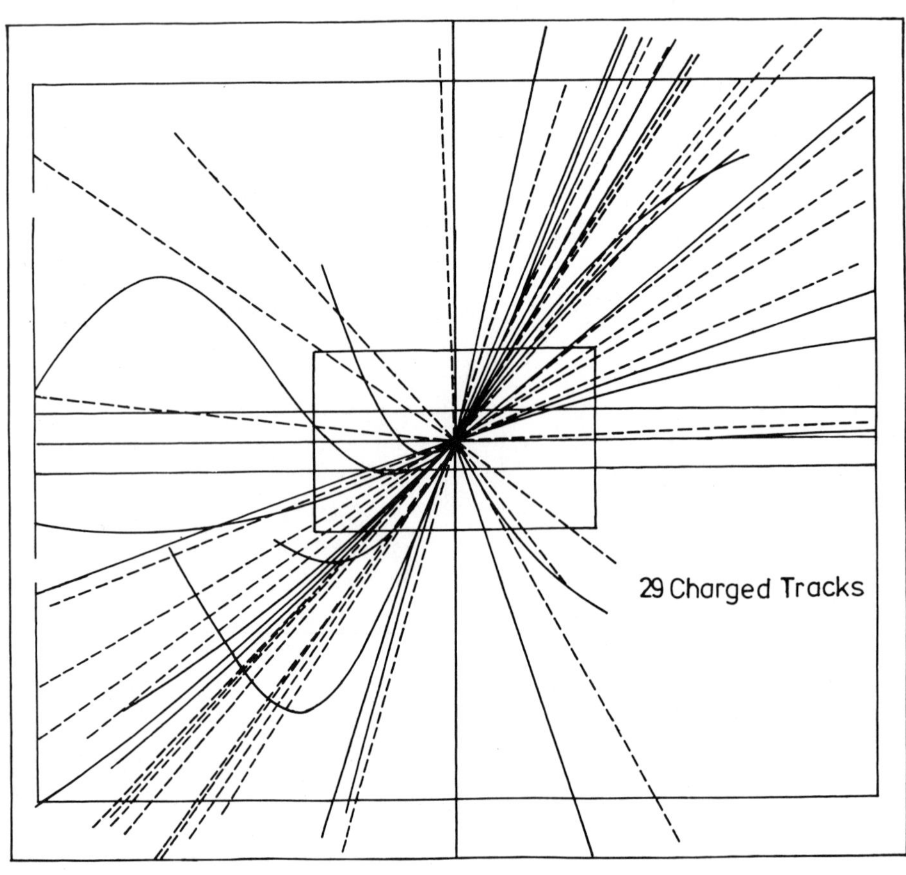

Fig. 8b A Monte Carlo event in the LEP detector.

Higgs Production

As mentioned before the search for heavy scalars - neutral or charged - will be play a central role at LEP. We shall consider as an example the search for Higgs[11].

a) $m_H < m_Z$

If the Higgs is lighter than the Z^o it can be produced by Z^o decay,

$$e^+e^- \to Z^o \to e^+e^- H^o$$

The decay rate is estimated to be

$$\frac{\Gamma_{Z \to e^+e^- H^o}}{\Gamma_{Z \to e^+e^-}} \approx 2.5 \cdot 10^{-3}$$

which leads to R_{peak} $(e^+e^- H^o) \approx 0.5$. The Higgs can be observed as a peak in the recoil spectrum of the e^+e^- system.

b) $m_H > m_Z$

If the Higgs is heavier than the Z^o a possible production mechanism is

$$e^+e^- \to Z^o H^o$$

which has an R value of

$$R(Z^o H^o) \approx 0.15 \quad \text{for} \quad \sqrt{s} >> m_Z + m_H.$$

This corresponds to a cross section of ~0.3 pb at $\sqrt{s}$ = 200 GeV yielding 0.15 events/day for 500 nb^{-1}/day. The Higgs may be detected by studying the momentum spectrum of the e^+e^- system resulting from the decay of the Z^o,

$$e^+e^- \to Z^o \; H^o, \quad Z^o \to e^+e^-$$

Unfortunately the event rate is very small, 0.015 events/day. For this reason it seems better to look for Higgs decays into two quark jets,

$$H^o \to q\bar{q}$$

which is the dominating decay channel and to observe the Higgs as a peak in the jet invariant mass distribution,

$$e^+e^- \to Z^o \, H^o$$
$$\quad\quad\quad\quad\quad\; \lfloor\!\to q\bar{q}$$

The expected event rate is 0.1 events/day. This makes the search just barely feasible provided the background is small. The background can be suppressed by using a well designed hadron calorimeter which provides a good jet jet mass resolution.

THE DETECTORS

In this section a brief description of the six detectors proposed for LEP is given.

ALEPH[1)]

Fig. 9 shows a side view of the ALEPH detector. It uses a superconducting coil, 2.5 m in radius, 7 m long with a magnetic field of 1.5 T. The coil is a so called thick wall with 6 r.l. wall thickness.

Charged particle tracking is achieved with a small drift chamber close to the beam and a large TPC (time projection chamber) with 4.4 m length, 0.2 and 1.8 m inner and outer radii, respectively and with a pressure of 1 atm. Its volume is roughly four times larger than that of the TPC built by the Lawrence Berkeley group[12)]. The TPC provides space points and dE/dx measurements. Fig. 10 shows the cathode pad arrangement on the end flanges. In total 21 (r,φ) and 400 (z, dE/dx) measurements are made with an expected accuracy of $\sigma_{r\varphi} \approx 300\ \mu$, $\sigma_Z \approx 1.5$ mm, $\sigma(dE/dx)/(dE/dx) \simeq 4.7\ \%$. The momentum resolution is calculated to be $\sigma_p/p \approx 2\cdot 10^{-3}\cdot p$, p in GeV/c.

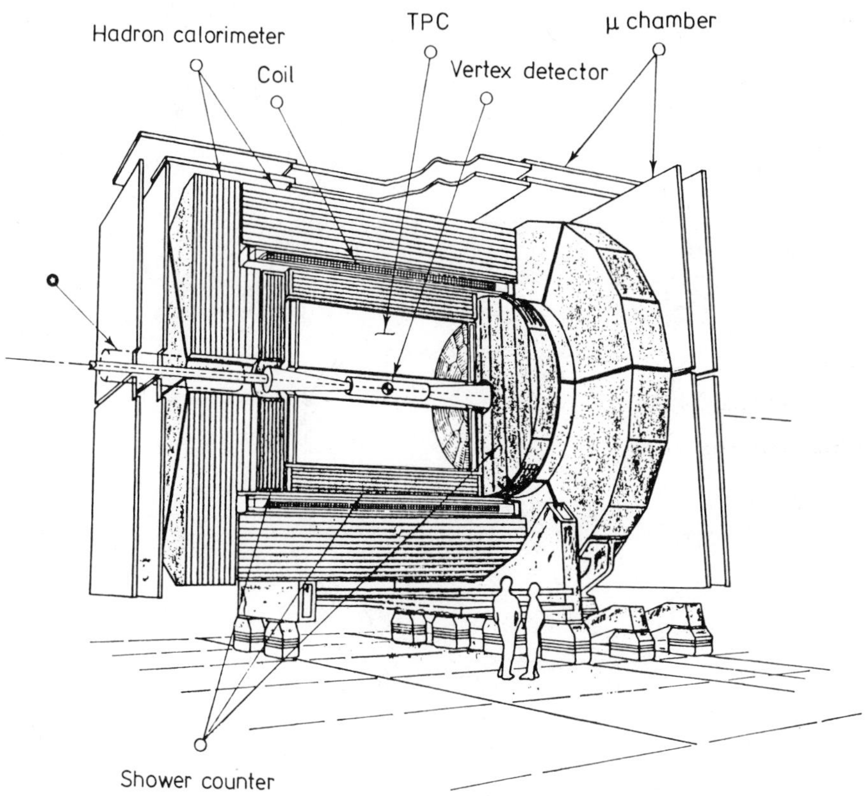

Fig. 9 The ALEPH detector.

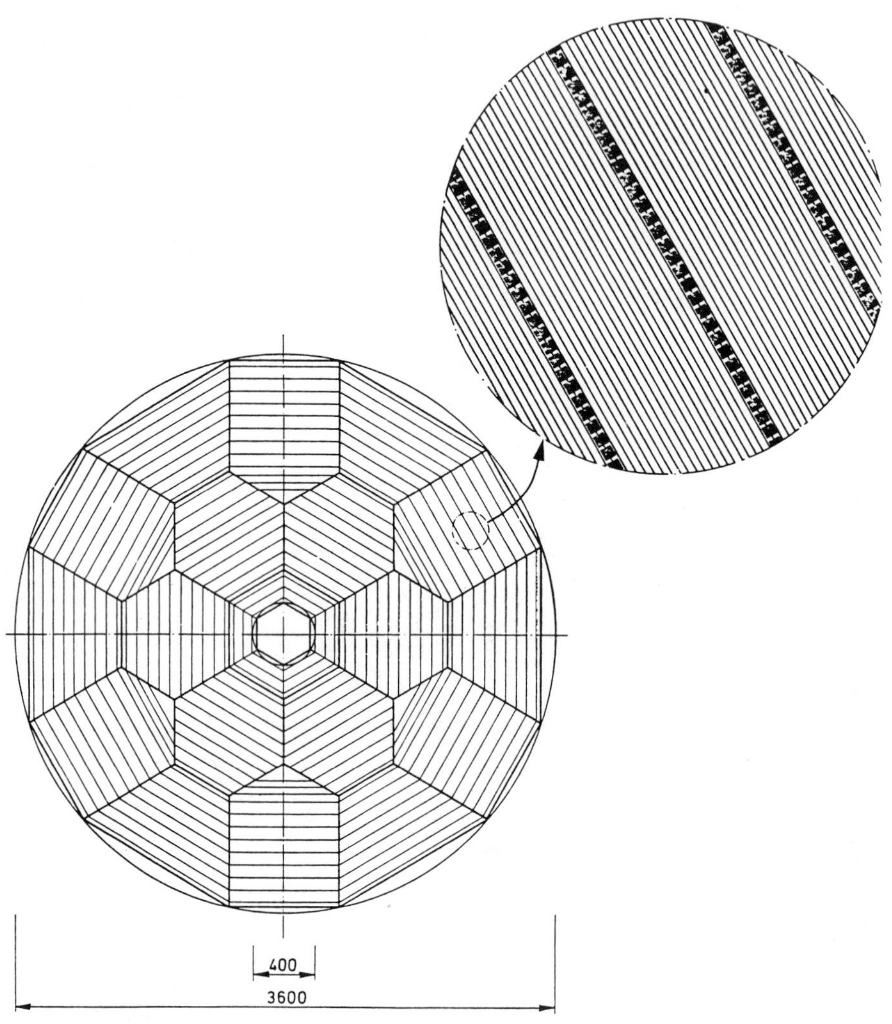

Fig. 10 ALEPH: The TPC endflange geometry.

The TPC is followed by an electromagnetic shower counter (still inside the magnetic coil) which covers also the endcap area. It consists of proportional tube chambers sandwiched between lead sheets (2.4 mm thick). Charged particles traversing the tubes produce signals on the proportional wires and induce signals in cathode pads at the surface of the chamber. The cathode pads in successive layers form $1^o \times 1^o$ towers which are directed towards the interaction point. The shower energy is determined in the towers. The expected photon energy resolution is $\Delta E/E \approx 16\ \%/\sqrt{E} + 1\ \%$. The localization of the shower is made with the help of the towers and the proportional wires.

The solenoid is surrounded by a hadron calorimeter followed by two layers of muon chambers. The hadron calorimeter consists of a sandwich of iron sheets (4 mm thick) and streamer tube chambers read out by cathode pads.

OPAL

In Fig. 11 a vertical cross section of the OPAL detector is shown. It uses a thin superconducting coil of 7.6 m length, 2.0 m radius, 1.1 r.l. wall thickness and 0.5 - 1.0 T field.

Charged particle tracking is done in a vertex detector close to the beam and a jet chamber with 0.1 and 1.8 m inner and outer radii, respectively, and 4.5 m length. The chamber pressure is 2 atm. It will provide 160 $r\varphi$ and dE/dx measurements along the radius. In addition by charge division readout a coarse measurement of the z coordinates is obtained. Fig. 12 shows the arrangement of the anode wires. The expected accuracy is $\sigma_{r\varphi} = 150\ \mu$, $\sigma_z \approx 3$ cm, $\sigma(dE/dx)/(dE/dx) \approx 3.5\ \%$ and $\sigma_p/p \approx 2\cdot 10^{-3}\cdot p$. The design of the jet chamber is similar to that used by the JADE group at PETRA. The jet chamber is followed by a z chamber which provides an accurate measurement of the z coordinate at this radius.

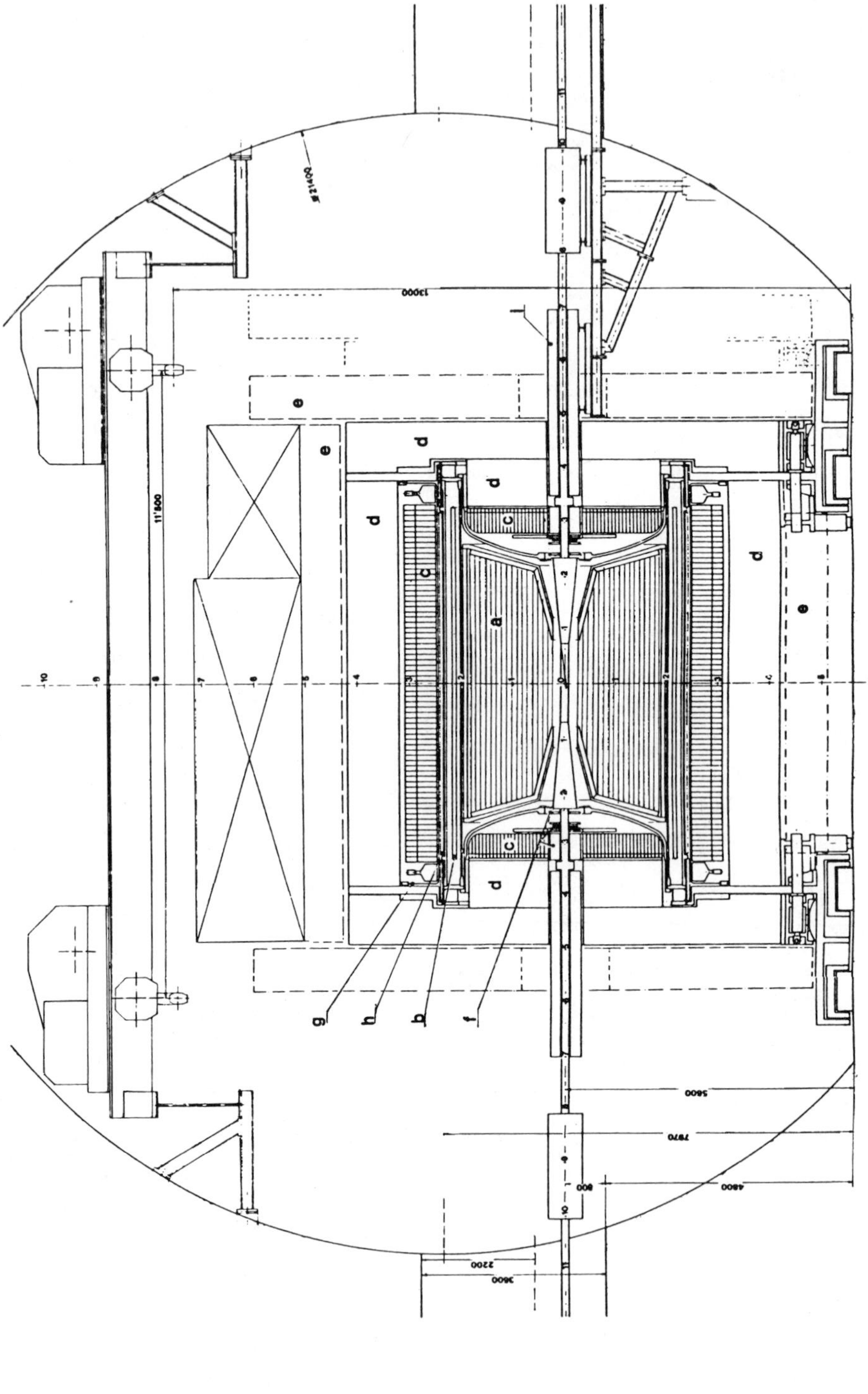

Fig. 11 The OPAL detector.

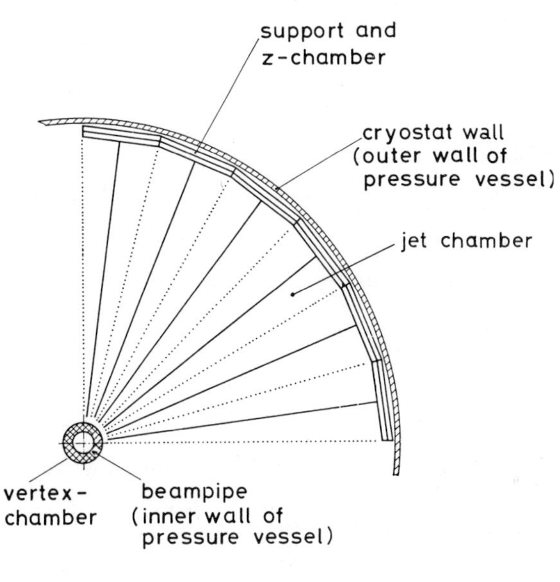

Fig. 12 OPAL: The cell structure of the jet chamber.

The shower counter system consists of two parts, the barrel and the endcap counters. The barrel counter sits behind the coil. It contains 11 000 lead glass blocks with cross sections of $10 \times 10\ cm^2$. The lead glass is read out by photomultipliers. The lead glass counter is preceded by the presampler, a proportional chamber which measures the position of showers that have started in the coil. The energy resolution expected for the shower counter is

$$\sigma_E/E = \sqrt{\left(\frac{4\%}{\sqrt{E}}\right)^2 + \left(\frac{8\%}{E}\right)^2} + 0.1\ \%$$

The endcap shower counter consists of 1500 lead glass blocks which are read out by photodiodes.

The shower counter is followed by the hadron calorimeter which consists of a sandwich of 16 steel plates (5 cm thick) and scintillator plates with BBQ read out. The hadron energy is collected in $50 \times 50\ cm^2$ towers. The hadron calorimeter is surrounded by additional drift chambers to detect muons.

The L3 Detector

The layout of the L3 detector is given in Fig. 13. A large normal conducting coil provides a field of 0.45 T over a volume of 12 m in diameter and 12 m in length.

Charged particles are tracked in a small drift chamber (see Fig. 14) with a novel read out (time expansion chamber). The chamber has an outer radius of 50 cm and is about 1 m long. The expected position resolution is $\sigma_{r\varphi} \approx 50\ \mu$ providing a momentum resolution of $\Delta p/p = 1.4\ \%\cdot p$.

The drift chamber is surrounded by a BGO shower counter consisting of 12 000 elements each subtending $2^o \times 2^o$. The BGO or bismuth-germanate ($Bi_4\ Ge_3\ O_{12}$) crystals are a new type of shower material which in many respects is superior to NaJ or lead glass.

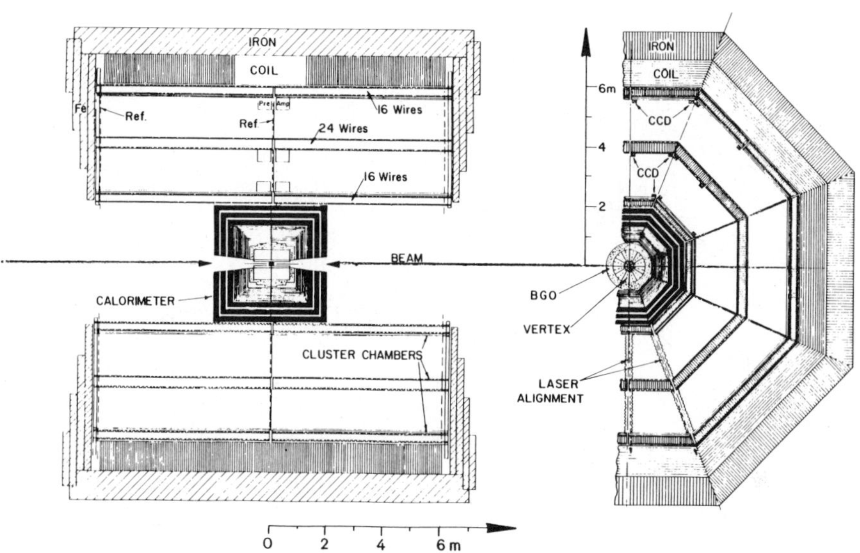

Fig. 13 The L3 detector.

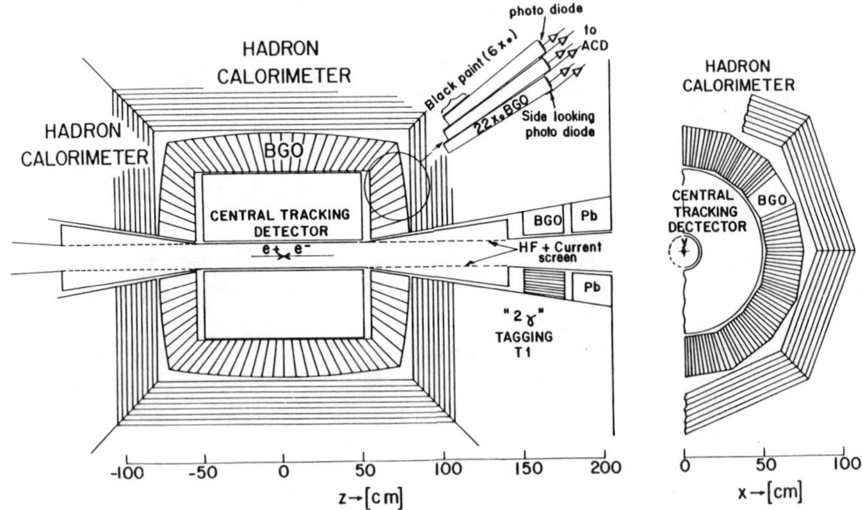

Fig. 14 The inner part of the L3 detector.

Due to its high density ($\rho = 7.1\ g/cm^3$) it has a short radiation length (1 r.l. = 1.12 cm) and a small shower radius. The light output is smaller than that of NaJ but larger than of lead glass. It can be read out by photo diodes which can operate in a magnetic field and which seem to have a very stable gain factor. The energy resolution of the shower counter is estimated to be

$$\sigma/E \approx 0.5\ \%/\sqrt{E} + 0.3\ \%.$$

The shower counter is followed by a hadron calorimeter which starts at a radius of ~80 cm. Since the inner radius is so small a calorimeter with good energy resolution appears to be affordable. The design is not yet finalized.
The drift space between calorimeter and coil is reserved for muon detection. Muon tracks are measured in 3 layers of large drift chambers (with length 6 m) with a planned position resolution of 180 μ per wire. A large effort is needed to achieve a comparable accuracy in the alignment of the chambers.

ELECTRA[5)]

The ELECTRA detector is shown in Fig. 15. A thin superconducting coil, 2.25 m in radius, 7.4 m long, 0.8 r.l. wall thickness provides a field of 1 T.

A high precision drift chamber detects short lived particles close to the beam. Next follows a drift chamber with 1.7 m outer radius and 4.7 m length (Fig. 16). The wire geometry is similar to that used by MARK II at PEP or TASSO at PETRA. The $r\varphi$ position is measured by 31 layers of wires which run parallel to the chamber axis; the z coordinates are determined by 9 layers of stereo wires. A position resolution of $\sigma_{r\varphi} = 200\ \mu$ and $\sigma_Z = 6$ mm is expected. The momentum resolution is calculated to be $\sigma_p/p = 1.5 \cdot 10^{-3} \cdot p$.

The drift chamber is surrounded by a transition radiation detector (TRD) designed to recognize electrons. The TRD consists of 5 layers of carbon wool which acts as radiator (Fig. 17). The

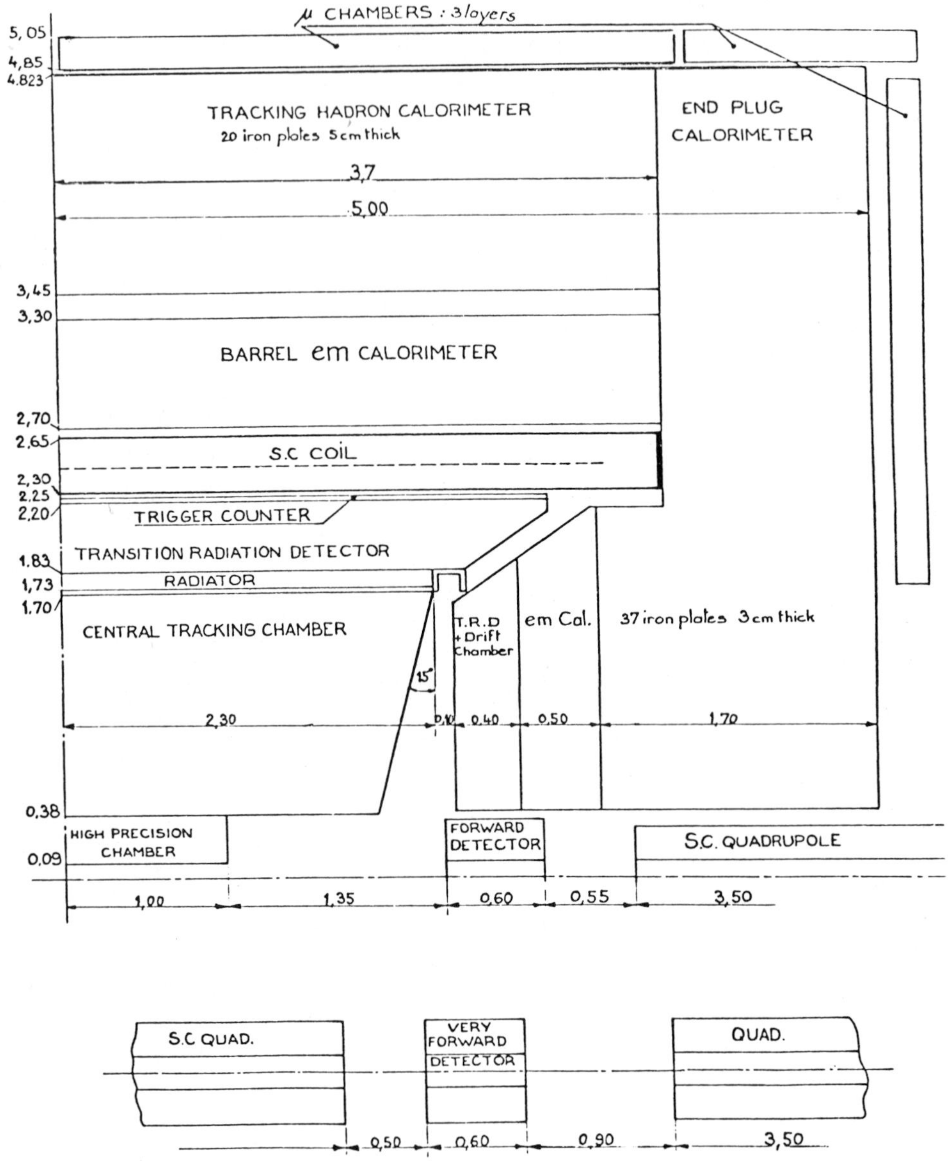

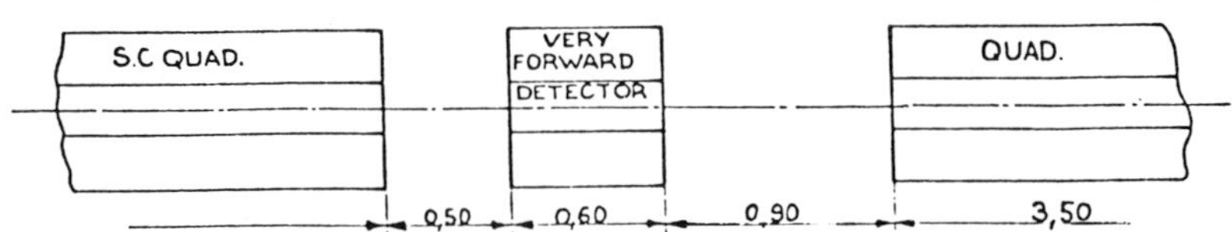

A cross-section showing one quadrant of the ELECTRA detector.

Fig. 15 The ELECTRA detector.

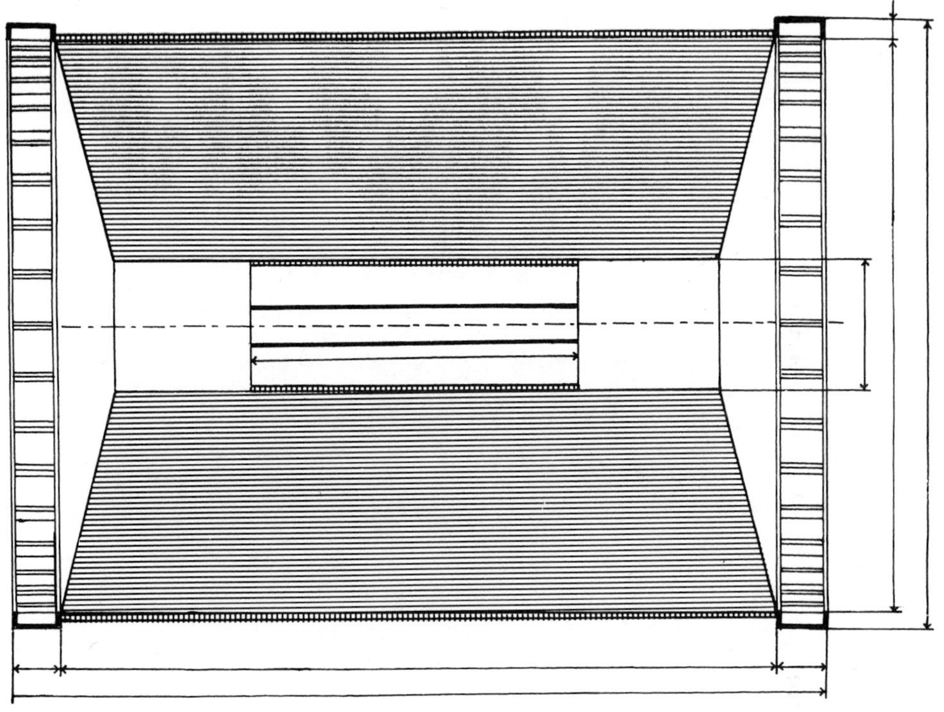

Fig. 16 ELECTRA: Layout of the central drift chamber.

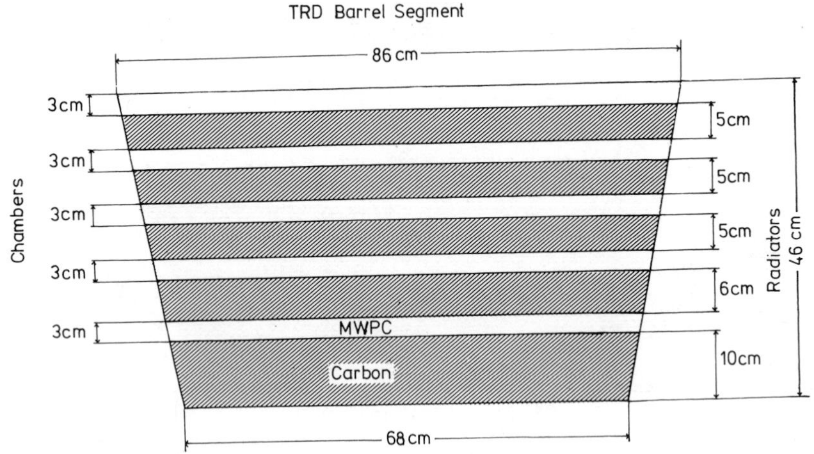

Fig. 17 ELECTRA: The transition radiation detector.

transition radiation is detected in 5 multiwire proportional chambers. The total thickness of the TRD is 46 cm. The probability for misidentifying pions as electrons is $\varepsilon_{\pi\to e} \approx 1\cdot 10^{-2}$ for pion momenta above 1.5 GeV/c.

The shower counter is situated behind the coil and in the endcap area. It is designed as a lead-liquid argon calorimeter with towers pointing to the interaction point for the energy measurement and strips for an accurate position measurement. The expected energy resolution is $\sigma/E \approx \sqrt{(10\ \%/\sqrt{E})^2 + (8\ \%/E)^2}$.

The shower counter is followed by an ironplate calorimeter with streamer tubes for readout. The design energy resolution is $\sigma/E \approx 70\ \%/\sqrt{E}$.

DELPHI[6)]

The DELPHI detector uses a superconducting coil (2.55 m radius, 7.5 m length) with a field of 1.2 T (Fig. 18). Charged particle tracking is accomplished with a vertex detector, a TPC and an outer detector. The TPC has roughly the dimensions of the TPC used by the Berkeley group: outer radius 1.26 m and length 2.9 m. It is operated at normal pressure.

The outer detector consists of 3 layers of drift tubes and provides additional track points behind the Cerenkov counter which leads to a considerable improvement of the momentum resolution. The combined information from TPC and outer detector leads to $\sigma_p/p = 1.8\cdot 10^{-3}\cdot p$.

The special feature of DELPHI is the use of ring imaging Cerenkov counters (RICH) for charged particle identification. The principle of operation of a RICH counter is sketched in Fig. 19. A charged particle produces Cerenkov light in the radiator. The Cerenkov cone opens in the drift region. The Cerenkov photons enter a TPC like chamber fill with the photo ionizing gas TMAE (tetrakis dimethylamine ethylene) which has a mean free path for

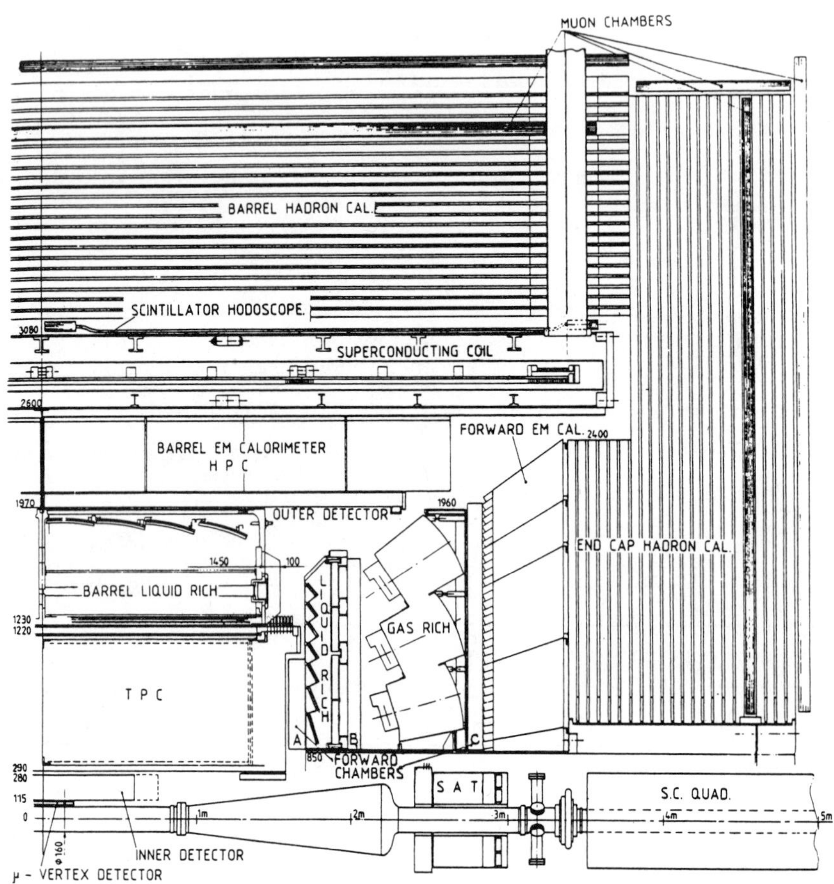

Fig. 18 The DELPHI detector.

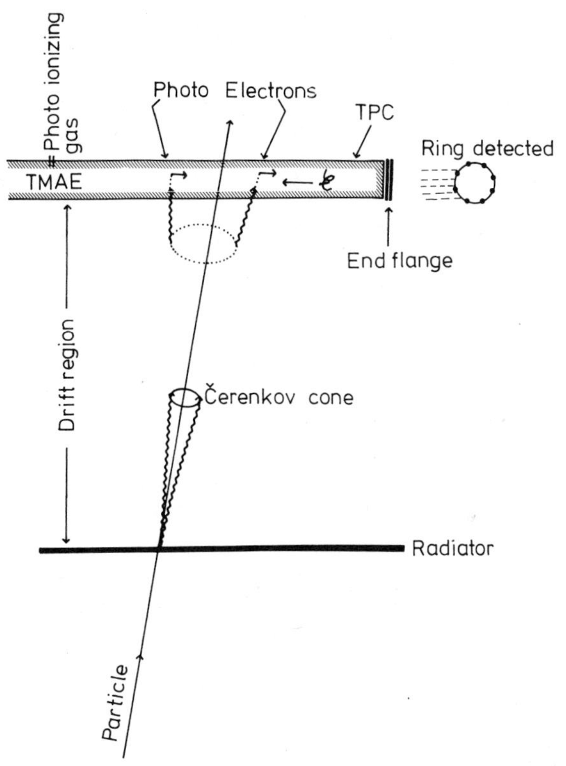

Fig. 19 Principle of operation of a RICH counter.

photons above 5.4 eV of about 1 cm. The photoelectrons produced by the Cerenkov photons drift in the TPC along the electrical field lines and are detected at the endflange. Fig. 20 shows a Monte Carlo simulation for an event with 23 charged particles traversing the barrel RICH. An average of 30 photoelectrons per particle was assumed which is probably on the optimistic side. The evaluation of the hit pattern is not as difficult as it may seem. It will e.g. not require a pattern recognition analysis. The path of the charged particle and therefore the positions of the Cerenkov rings for the π, K and p hypotheses are known. A chi-square comparison between the predicted rings and the observed hits will allow a decision between the different particle hypotheses. The barrel RICH should allow π/K separation up to 8 GeV/c and K/p separation up to 11 GeV/c (Fig. 21).

The shower counter is located inside the coil. Its design is not yet finalized. Following the coil the detector is surrounded by a hadron calorimeter, made of iron plates and layers of streamer tubes. The streamer tube readout is highly segmented such as to allow also the detection of muon tracks.

LOGIC[7)]

The geometry of the LOGIC detector is markedly different from that of the other detectors (see Fig. 22). A thin superconducting coil (0.8 m radius, 0.2 m thick) together with iron cones under an angle of $\Theta \approx 20^{o}$ with respect to the beam provide a roughly longitudinal field of 1 T in the centre. The important feature of this geometry is that apart from the coil region near $\Theta \simeq 90^{o}$ and the iron cones the solid angle is unobstructed by heavy material.

Charged particles are detected in the central tracker before they enter two RICH counter systems. Note that due to the open geometry, to provide sufficient space for the RICH counters is no

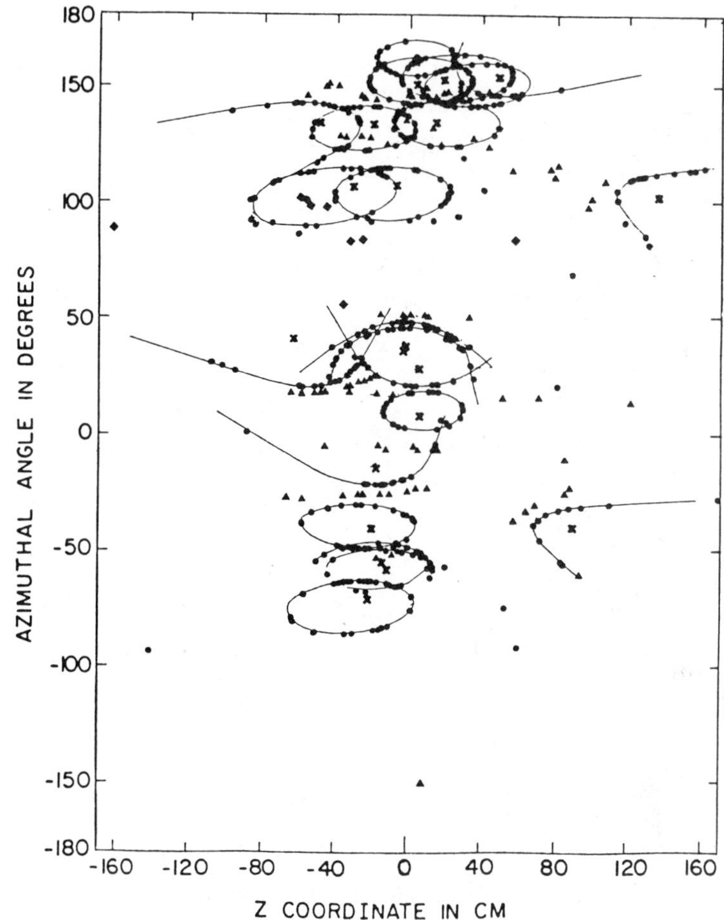

Fig. 20 DELPHI: Monte Carlo simulation of the RICH response for a typical hadronic event.

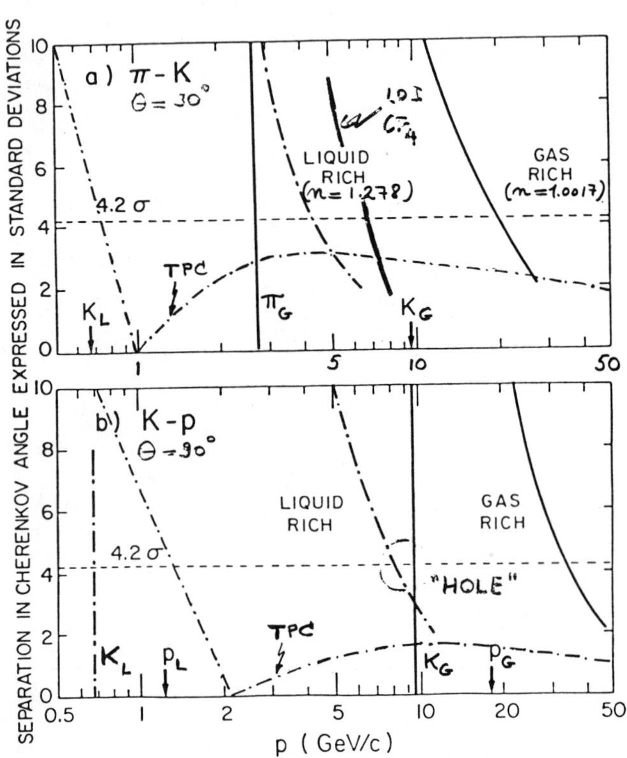

Fig. 21 DELPHI: Particle identification with the barrel RICH

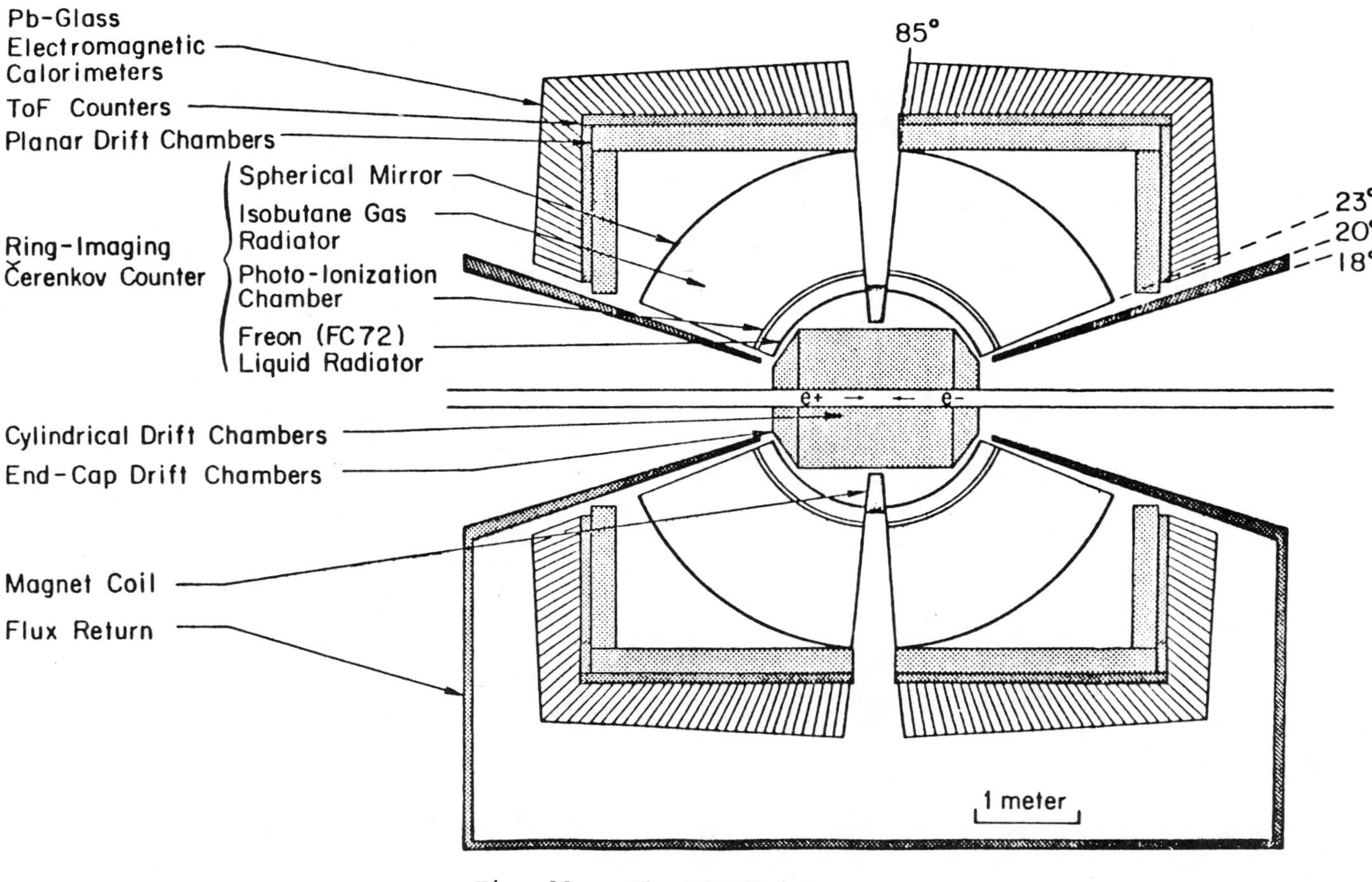

Pb-Glass
Electromagnetic
Calorimeters
ToF Counters
Planar Drift Chambers
Spherical Mirror
Isobutane Gas
Radiator
Ring-Imaging
Čerenkov Counter
Photo-Ionization
Chamber
Freon (FC72)
Liquid Radiator
Cylindrical Drift Chambers
End-Cap Drift Chambers
Magnet Coil
Flux Return
85°
23°
20°
18°
e+
e-
1 meter

Fig. 22a The LOGIC detector

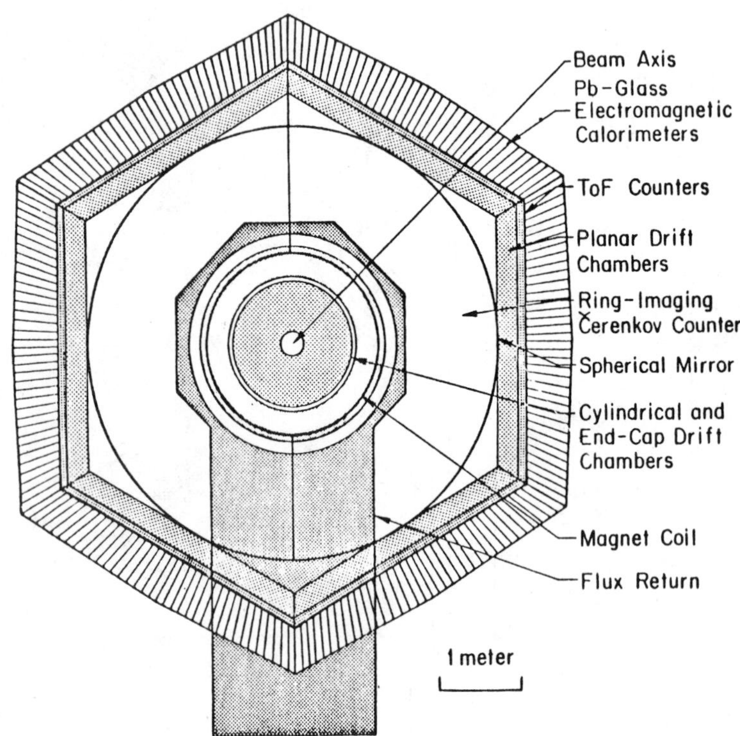
Beam Axis
Pb-Glass
Electromagnetic
Calorimeters
ToF Counters
Planar Drift
Chambers
Ring-Imaging
Čerenkov Counter
Spherical Mirror
Cylindrical and
End-Cap Drift
Chambers
Magnet Coil
Flux Return
1 meter

Fig. 22 b

problem. Particle separation is possible for π/K up to 52 GeV/c and K/p up to 70 GeV/c.

The RICH counters are surrounded by planar drift chambers in order to obtain an additional point on particle tracks. The momentum resolution expected is $\sigma_p/p = (4 - 8\cdot 10^{-3})\cdot p$.

The planar drift chambers are followed by the shower counter which consists of 18 000 lead glass blocks directed onto the interaction point and read out by photo tubes. Since the total material in front of the shower counter is only 0.2 r.l. a good energy resolution is expected also at low photon energies:

$$\sigma/E = \sqrt{(4\%/\sqrt{E})^2 + (0.6\%)^2}$$

The detector has no hadron calorimeter, no muon detector and no shower counter coverage of the small angle region, $\Theta < 20^{o}$.

REFERENCES

1. Letter of intent to study e^+e^- annihilation phenomena, Bari et al.
2. Letter of intent for LEP, OPAL collaboration, Birmingham et al.
3. Letter of intent, University of Lund et al.
4. LEP letter of intent from BCF collaboration, Bologna, CERN, Frascati
5. Letter of intent to study e^+e^- annihilations at LEP, ELECTRA collaboration, Aachen et al.
6. DELPHI, a detector with lepton, photon and hadron identification Collège de France et al.
7. LOGIC detector, Cal. Tech. et al.
8. Physics with very high energy e^+e^- colliding beams, CERN Yellow Rep. 76-18 (1976); Proc.of the LEP summer study, CERN Yellow Rep. 79-01 (1979)
9. Proc. SLC workshop on experimental use of the SLAC linear collider, SLAC Rep. 247 (1982).
10. S. Weinberg, Phys.Rev.Lett. 19 (1967) 1264; A. Salam, Proc. 8th Nobel Symp., ed. N. Svartholm, (Almquist and Wiksell, Stockholm 1968), p.367.
11. See e.g. J. Ellis, M.K. Gaillard, G. Girardi, P. Sorba, LBL-13856, LAPP-TH-52 (1982) and references given therein.
12. PEP-4 TPC collaboration, H. Aihara et al., Proc. 1982 IEEE Nucl. Sci. Symp.

DISCUSSION

CHAIRMAN: G. Wolf

Scientific Secretaries: J. Huston, A. Ogawa, L. Petrillo

DISCUSSION

- *HUSTON:*

I have a question in regard to the transition radiation detector (TRD) system in the ELEKTRA spectrometer. There appear to be only 5 or so transition boundaries in the system. Most of the TRD's I have seen used very thin foild and one or two thousand boundaries in order to get a large probability of having transition radiation emitted. How does one get enough transition radiation to get adequate particle identification?

- *WOLF:*

The trick is to use wool which provides many surfaces over a short distance.

- *LOEWE:*

What, actually, are the new aspects of the detectors at LEP compared to the present detectors at PETRA. What are the improvements?

- *WOLF:*

First: size. You may not call that an improvement, but it certainly increases the momentum resolution by a factor of 10. Secondly, there is a new type of chamber, the TPC (Time Projection Chamber), which, if it works will provide accurate $r\varphi$, z and dE/dx measurements. Thirdly, there is a new shower counter material, the BGO which promises a much better photon energy resolution. Fourthly, there is the Ring Imaging Čerenkov (RICH) counter technique. With RICH, one has a hope to identify all charged particles in a jet.

- *LAU:*

What is the typical total energy resolution of these detectors?

- *WOLF:*

The total energy is measured in hadron calorimeters. Since their size in general is enormous it is very difficult (due to costs) to build a good calorimeter. One cannot afford, for example, a uranium or copper calorimeter. So the average resolution hoped for, but not really tested and seen is $\sigma_E/E = 100\ \%/\sqrt{E}$ (in GeV), i.e., a total energy of 100 GeV is measured with 10 % resolution. There is one detector that has a fairly concise central detector, where it might be possible to do better, perhaps $\sigma_E/E \approx 50 - 70\ \%/\sqrt{E}$.

- *LAU:*

My second question is: what is the typical absorption length of these detectors?

- *WOLF:*

The detectors are typically 5 - 10 absorption lengths thick.

- *LAU:*

So there may be some leakage of the hadronic energy.

- *WOLF:*

Yes.

- *LAU:*

Can one sufficiently cut down the punch-through at high energy, so as to obtain muon discrimination?

- *WOLF:*

No, the punch-through is a problem. And above, say, 5 - 10 GeV/c hadron energy the punch-through is of the order of several percent. What people hope to do is to spot $\pi \to \mu$ or $K \to \mu$ decays in the tracking chamber and/or to measure the particle's scattering angle in the muon filter. By measuring the track before and after, one can try to reduce the hadronic background (by requiring the muon candidate to have only a small scattering angle).

- *HOFSÄSS:*

You quoted event rates at the Z^0 peak. These of course depend on the spread of the beam momentum. Are you very sure about how this will be?

- *WOLF:*

Yes. The energy spread of the beam is much smaller than the width of the Z^0. The beam spread depends a little on how you run the machine, but will be of the order 50-80 MeV. The width of the Z^0 is ~2.6 GeV. So there is no problem.

- *D'ALI:*

Could you tell us if there are different physics aims of the experiments you've shown us?

- *WOLF:*

That's difficult to say. The big detectors, OPAL, ALEPH, ELEKTRA and, to a large extent DELPHI aim, as a first goal, at the hadronic states: analyzing jet patterns, getting at quark flavours, measuring the weak coupling constants. On the other hand L3, for example, leans more towards leptonic particles, and, in particular the search for scalars. But in the end, everybody can do everything, apart from particle identification.

- *SAKURAI:*

I have a crucial question. You have four intersection points and six detectors. Does this mean two of the detectors will be eliminated.

- *WOLF:*

Yes.

- *NEUMANN:*

What is the status of approval?

- *WOLF:*

I will summarize what the committee has done and what it's

going to do. We first had an open presentation of the proposals. Then, at the next meeting, we looked in great detail at the technical feasibility of these detectors. Then the committee asked each group to answer physics questions of the type: "Can you detect free quarks? How would you detect the bottom quark through its leptonic decay? How would you determine Higgs?" The answers to these questions, which all required elaborate, enormous Monte Carlo work, showed in a quite convincing way the potential of the detectors. This allowed us to evaluate them and weigh one against the other. This has been done.

Having received all the answers, the committee made the recommendation that there should be no more than four experiments. The committee could have recommended six, because there are eight interaction points, and from the machine point of view there would be no restriction. Most members feel that one could do all the physics known at the moment with three detectors. That is why the number is less or equal to four.

Then in a meeting in July first indications were given as to which detectors should be preferred.

That has to be followed up now by discussion on the risky components, like "Can you build TPC's that size? Can you get BGO and can you pay for it?". Things of that sort have to be looked into.

Final recommendations by the committee are planned for November.

- *EISELE:*

It seems to me that one of the major problems of these huge detectors is mechanical alignment. It especially seems to be true for L3 where one wants to measure muon trajectories out to very large distances. What are the technical means and what are the possibilities to get alignment to a fraction of a millimetre?

- *WOLF:*

L3 has invested an enormous amount of effort investigating this problem, and the referees who specifically looked into this were convinced that you can align these chambers, which are each 6 m long and 2 m apart, to within 100 micron.

Lasers and photosensors are permanently mounted on the chambers to sense position shifts. A temperature change of 2 to 3 degrees can cause a position shift of several hundred microsn. If you don't want to correct for it later on in the analysis, remote control movement of the chambers is necessary. In terms of alignment precision, L3 is the most sensitive detector.

There are also problems in other detectors, though. In the DELPHI case there is a TPC in front of the RICH, and other chambers behind. One hopes to align these two with respect to each other to within a few hundred microns. If the temperature can be kept stable enough, the relative position of the two chambers can be determined with cosmic rays to this precision.

- *OGAWA:*

I'm interested in what distorts the circles in the Ring Imaging Cerenkov counter into what appear to be parabolas. Are these highly inclined tracks?

- *WOLF:*

These were tracks going at an angle. They were not at normal incidence but at 40-50 degrees. The light is emitted in a cone and the detector cuts the cone under an angle.

- *TRIPICCIONE:*

I would like to ask about the physics motivation of LEP. While my question may appear to be a little provocative, I assure you it is not. I understand the physics motivation of LEP phase I, but from your sketchy discussion the physics motivation for extending LEP to 130 GeV per beam is not clear to me. It seems too high for the standard model and too low for supersymmetric particles and things like that.

- *WOLF:*

Firstly, it is not true that the standard model does not profit from the increase of the LEP beam energy from 60 to 130 GeV. For example, for the pair production of W's, $e^+e^- \rightarrow W^+W^-$, the beam energy probably has to be above ~90 GeV. Secondly, the increase in energy extends the range over which one can search for any pointlike charge particle to masses as large as ~130 GeV. Candidates one will look for are of course the Higgs, other charged scalars, heavy leptons and heavy quarks, and the whole zoo of supersymmetric particles for which virtually no mass predictions exist.

- *D'ALI:*

Could one experiment do everything?

- *WOLF:*

The answer is no for the following reasons:

1. The only known way to identify $\pi^{\pm}$, $K^{\pm}$, p,$\bar{p}$ up to high momenta in jets is to use Ring Imaging Čerenkov counters, RICH. RICH needs a lot of space (70 - 100 cm in radial direction) which costs momentum resolution and a loss in the performance of the shower counter.

2. Good momentum resolution on the other hand which is important for charm particle studies and for searches for new particles requires large field strength B and a large track length in the tracking chamber ($\Delta p \sim p^2B^{-1}L^{-2}$).

3. The best energy resolution for photons and electrons is offered by BGO. Because of the large costs (at present 12 $ per ccm) one can at best afford a 4π BGO shower counter which starts at a radius of 50 cm from the interaction point. Hence, if one goes for optimal γ,e resolution the charged particle tracking will suffer. Similarly in order to be able to afford a good hadron calorimeter the central detector (charged particle tracking and shower counter) has to be small.

Hence it is clear that particle identification, optimal charged and neutral particle resolution and good hadron calorimetry are contradicting requirements. In order to have all features at LEP one needs at least three dectors.

There is another point of consideration which favours in fact four detectors. There should at least be one of the detectors ready to take data as soon as LEP delivers luminosity. Hence one of the detectors should be a conventional one, i.e., one which employs only well proven techniques.

REVIEW OF HIGH ENERGY e^+e^- PHYSICS

James G. Branson

Massachusetts Institute of Technology
Cambridge/Massachusetts 02139/USA and

Deutsches Elektronen-Synchrotron DESY
Hamburg/Fed. Rep. of Germany

ABSTRACT

I report on results from high energy electron positron collisions at PETRA and PEP. I will describe the basic physical processes which are important in e^+e^- then review the present status of the data on the subjects of hadron production, QCD measurements of electroweak effects, and study of B meson decays. Since many experiments have produced quantitative results on similar topics, I will show several comparisons to check the experimental accuracy of the measurements. However, in many cases where the results are of a more qualitative nature, only one example of the data will be given.

CONTENTS

1.0 SIMPLE ELECTRON POSITRON INTERACTIONS

2.0 HADRON PRODUCTION

2.1 Total Cross Section

2.2 Hadronic Jets

3.0 STUDY OF QCD IN HADRON PRODUCTION
3.1 Testing the QCD Differential Cross Section
3.2 The Strong Interaction Coupling Constant
3.3 Quark and Gluon Fragmentation
3.4 Characteristics of the Final State Hadrons
4.0 ELECTROWEAK INTERACTIONS
4.1 Bhabha Scattering
4.2 Muon and Tau Pair Production
4.3 Charge Asymmetry
4.4 Interpretation of Leptonic Data
4.5 Electroweak Reactions of Quarks
4.6 B Meson Lifetime Limit
4.7 Production of Leptons in Hadronic Events
4.8 Search for Structure in the Fermions
4.9 Search for Symmetry Breaking Scalars.

1.0 SIMPLE ELECTRON POSITRON INTERACTION

At high energies, the dominant processes in electron positron collisions are particularly simple. Most of the interactions which we measure are fermion pair production, calculable using the Feynman diagram below.

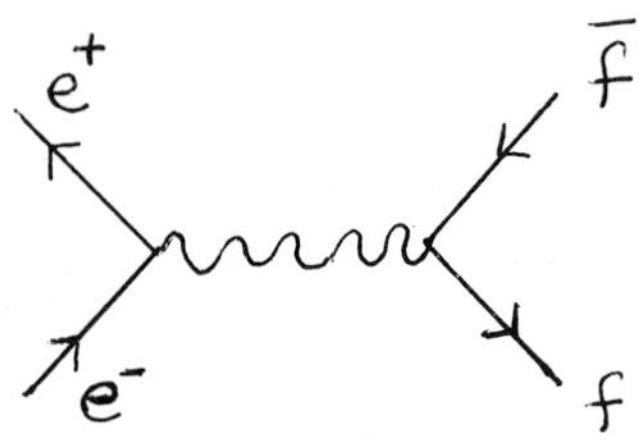

The electron and positron annihilate forming a virtual photon which has a mass equal to the center of mass energy. This photon may then decay into any pair of charged fermions that is energetically allowed. The processes of this sort which have been observed at PETRA are

$$e^+e^- \rightarrow \mu^+\mu^-$$

$$e^+e^- \rightarrow \tau^+\tau^-$$

$$e^+e^- \rightarrow q\,\bar{q} \quad \text{(u,d,s,c and b quarks seen)}$$

In lowest order Quantum Electrodynamics, the total cross section for muon pair production far above threshold is

$$\sigma_{\mu\mu} = 87.6/s \text{ nanobarns}$$

where s is equal to the center of mass energy squared in GeV^2. In a colliding beams machine this gives

$$s = (2\,E_{BEAM})^2$$

The production of spin 1/2 particle pairs has a $(1+\cos^2\theta)$ angular distribution with respect to the beam direction in pure QED. The total cross section for pair production of other fermions is similar to that for muon pairs.

$$\sigma_{f\bar{f}} = Q_f^2\,\sigma_{\mu\mu}$$

Q_f is the charge of the fermion in question. This factor arises because the coupling of any particle to the photon is proportional to its charge. From the above relations, we then know

$$\sigma_{\tau\tau} = \sigma_{\mu\mu}$$

$$\sigma_{u\bar{u}} = \sigma_{c\bar{c}} = \frac{4}{3}\,\sigma_{\mu\mu}$$

$$\sigma_{d\bar{d}} = \sigma_{s\bar{s}} = \sigma_{b\bar{b}} = \frac{1}{3}\,\sigma_{\mu\mu}$$

where a factor of 3 increase in the quark pair production cross sections comes from the fact that there are 3 colors of quarks. This of course leads to the famous expression for the ratio R

$$R = \frac{\sigma(e^+e^- \rightarrow \text{HADRONS})}{\sigma_{\mu\mu}} \quad 3\sum_f Q_f^2$$

where it is assumed that all hadron production is by the mechanism of quark pair production. At PETRA energies this is a very good approximation since the quarks are pointlike at our energy scale, $q^2 \simeq 1000$ GeV^2, while charged hadrons show structure at q^2 less than 1 GeV^2. Of course charged particles with spins other than 1/2 can also be in e^+e^-. For example scalar particle pairs would be produced with 1/4 the cross section of spin 1/2 pairs.

The simplicity of the production mechanism in e^+e^- interactions allows us not only to search for the production of new particles but also to put limits on their masses if none are found. This is particularly clear for the case of a sequential heavy lepton, a heavier version of the muon or the τ, and for a new quark flavor since these production cross sections will be exactly the same as those of their lighter predecessors except for threshold effects which are calculable.

A number of important processes remain which are not solely due to annihilation through one photon. Bhabha scattering, $e^+e^- \rightarrow e^+e^-$, is of course quite different from muon pair production because an additional diagram where a photon is exchanged between the electron and positron is involved. The total cross section for this process is infinite as in coulomb scattering. The differential cross section $d\sigma/d\Omega$ can be easily calculated in QED. In fact, the small scattering angle part of this cross section is dominantly due to low q^2 photon exchange. Since QED is well tested at low q^2, we can use this small angle part of Bhabha scattering to monitor the luminosity of the machine without worrying about the effects of new physics at high q^2. We may then use this luminosity to calculate cross sections for all the other processes we measure.

$$\text{(number of events)} = \text{(cross section)(luminosity)}.$$

The final order α^2 QED process is photon pair production, $e^+e^- \rightarrow \gamma\,\gamma$.

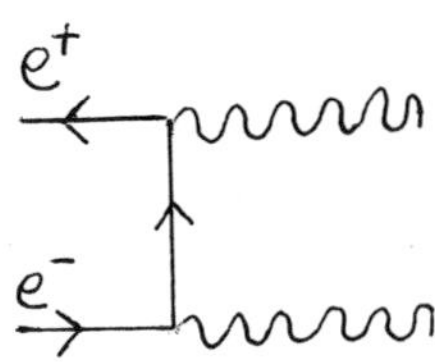

Finally, the so called two photon processes $e^+e^- \to e^+e^- f\bar{f}$, are calculable in order α^4 QED.

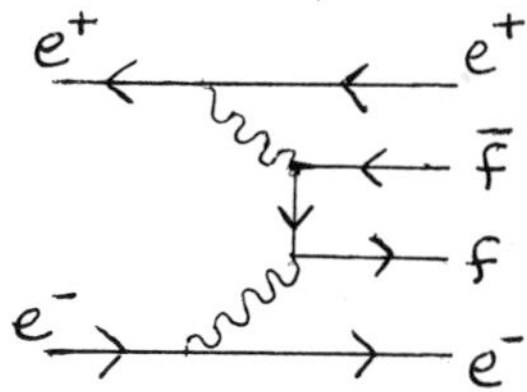

Although this diagram is of higher order than the annihilation processes, the total two photon cross sections are larger because the intermediate photons can be much closer to their mass shell.

Of course lowest order QED calculations may not be accurate enough to compare to data. In general, we find that processes like muon pair production or Bhabha scattering, which have leading terms of order α^2, have order α^3 terms which are quire important at our level of accuracy.

Estimates of the next order corrections, however, indicate that they will not be important in most cases. An example of an order α^3 effect is shown in Fig. 1 which gives acollinearity distribution for the final state in Bhabha scattering. This acollinearity is due to photon emission.

$$e^+e^- \to e^+e^-\gamma.$$

One can see that only a small fraction of the data has large acollinearity and that the order α^3 QED calculation fits well. I will discuss radiative corrections, which are simply corrections to the data for higher order QED effects so that they can be compared to the lowest order QED formula, along with most of the physics subjects in this paper.

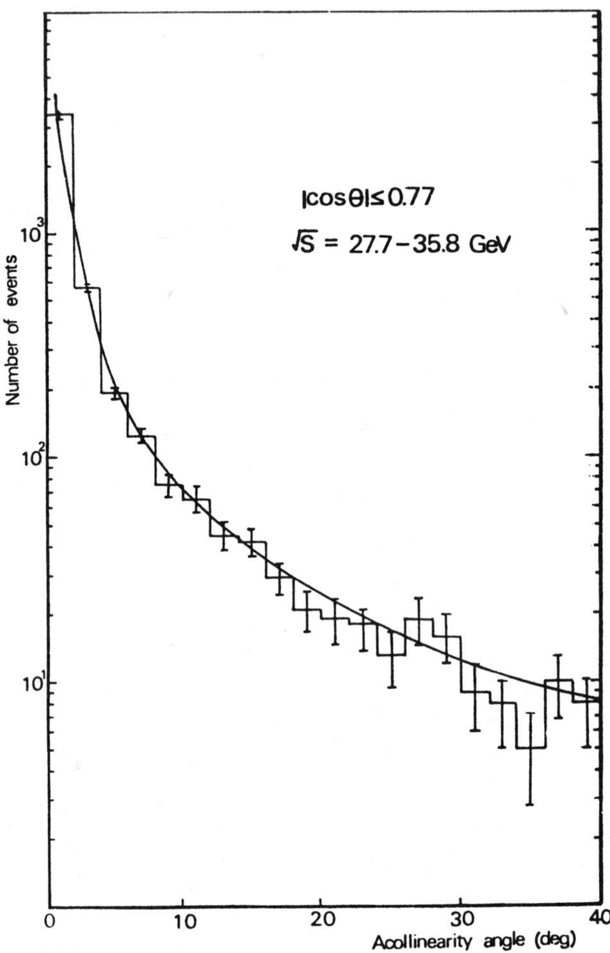

Fig. 1 Acollinearity distribution for Bhabha scattering from JADE.

2.0 HADRON PRODUCTION

2.1 Total Cross Section

The total cross section for $e^+e^- \to$ hadrons has been measured over a wide range of center of mass energies. Its ratio, R, to the μ pair production cross section is shown in Fig. 2

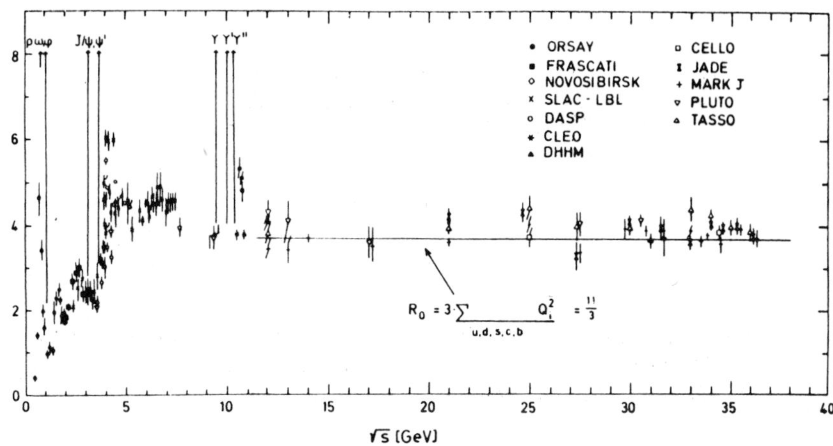

Fig. 2 Summary of Measurements of R.

The gross features of this plot can be easily understood in terms of the quark parton model in which R is equal to the sum of the squares of the quark charges with the sum running over all quark flavors which are energetically allowed. That means that, when a threshold opens up for the production of a new flavor of quark, we expect a step in the value of R. Such a step is quite apparent in the 3 to 4 GeV region in the figure where the charm threshold is crossed. This step is expected to be 4/3 units of R. Another step of 1/3 units of R should lie around 10 GeV where the b flavor threshold opens up. Above 10 GeV, R is quite consistent with a constant indicating that no new flavor thresholds have been crossed. In this plot of R, one can also see the quarkonium resonances, that is bound states of a quark and antiquark, that occur just below a new flavor threshold. At PETRA, a fine energy scan has been made at the highest machine energies to search for a new resonance. The combined results of the scan are shown in Fig.3.

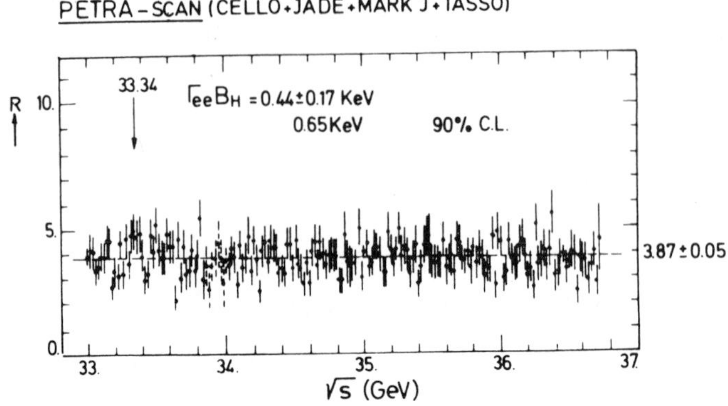

Fig. 3 High energy fine scan of R combined from data of four PETRA experiments places the most stringent lower limit on the top quark mass.

Again R is constant within statistics and no indication of a narrow resonance is seen. What we would expect for a charge 2/3 and for a charge 1/3 quarkonium resonance is shown in Fig. 4 along with some scan data from MARK-J in the region of the maximum sized signal for a resonance found.

A fit to constant plus a gaussian with a width equal to the machine energy resolution shows that a new charge 2/3 quarkonium resonance is ruled out by the MARK-J data and that a charge 1/3 resonance is disfavored. If the combined PETRA data are used, the charge 1/3 resonance can also be excluded. The qualitative behavior of R is in excellent agreement with the quark parton model prediction and is probably the best evidence for the quark substructure of hadrons. The accurate measurement of R is also of great importance to out understanding of elementary particles and their interactions.

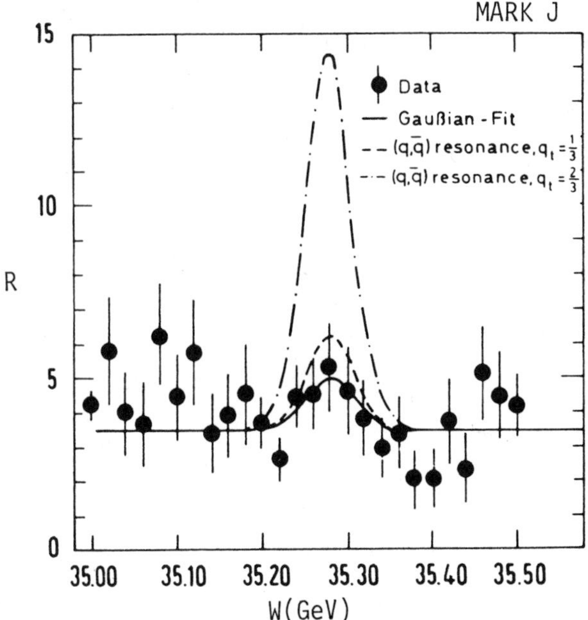

Fig. 4 Largest signal for a resonance in the scan data of MARK-J. The solid line is the best fit to a constant plus a gaussian with a width equal to the machine energy resolution. The dashed line and the dash-dot line show the expectation for the first quarkonium state for charge 1/3 and 2/3 quarks.

As well as depending on the number of quark flavors, R also depends slightly on the value of the running coupling constant in QCD(1) and on weak neutral current effects(2). If we include these effects in our calculation, new Feynman diagrams interfere with the first order QED production of quarks.

We then find:

$$R = 3\sum_f \left| Q_f^2 + F(\sin^2\theta_W) \right| \left| 1 + \alpha_s/\pi + O(\alpha_s^2) + \ldots \right|$$

The weak effects, represented by $F(\sin^2\theta_W)$, give about a 1% correction at 35 GeV with $\sin^2\theta_W = 0.23$. The correction is typically larger for other values of $\sin^2\theta_W$ in the electroweak section of this report. The strong interaction correction is around 5% in $O(\alpha_s)$ and the second order correction is only 0.3%. This is then a very clear prediction of perturbative QCD. The series converges very nicely. The effect, however, is rather small and is therefore experimentally very difficult to measure. We must make an effort to make the systematic error on R as small as possible since this is probably the best way to measure α_s.

Events for the process $e^+e^- \to$ hadrons are selected by making cuts which vary from experiment to experiment. The basis for most of these cuts, however, is that the amount of energy measured should be within errors of the center of mass energy, that the energy should be balanced since the sum of momenta should be zero, and that the charged multiplicity be large. The systematics of these cuts have been investigated. For example, JADE finds that by varying the energy cut between 1.0 and 1.6 times the beam energy, the computed value of R varies by less than 1%. To measure R accurately, one must make a careful calculation of backgrounds, acceptance, and radiative corrections. The important backgrounds to $e^+e^- \to$ hadrons come from

$$e^+e^- \to \tau^+\tau^-$$
$$e^+e^- \to e^+e^- + \text{hadrons}$$
$$e^+e^- \to e^+e^-\tau^+\tau^-$$

Typically, these backgrounds total a few tenths of a unit of R. Other sources of background are found to be negligible. Monte Carlo models are used to make this small subtraction.

Acceptances are also calculated with a Monte Carlo model of hadron events in which an initial pair of quarks fragment to give final state hadrons. QCD effects are also included. The parameters of the model are tuned up to fit the measured data distributions quite well. With the rather loose cuts applied to events, it is expected that the model dependence in the R measurement is quite small. More details on our model of hadron events will be presented when we compare QCD predictions to the data.

The order α^3 radiative corrections to R are quite important. Radiation in the final state should be negligible compared to the strong fragmentation effects. Initial state radiation is included in our Monte Carlo model. This along with vacuum polarization substantially raises the hadron production cross section above the lowest order cross section. With the photon energy cut offs in our model, this is about a 35% increase. Many of these events with hard photons radiated, however, are not accepted since the photon with a substantial fraction of the center of mass energy escapes down the beam pipe. Overall, we find that the radiative correction to the cross section is around 15% and depends on detector configuration and cuts. Because of this dependence, the data are corrected for these radiative effects so that they should be compared with the lowest cross section.

When R is calculated, we do not use the measured muon pair cross section because radiative corrections and acceptance differ for the two processes and because this would increase the statistical error on R. Rather, we compare the corrected hadron cross section to the theoretical lowest order cross section for muon pair production. The high statistics luminosity measurement from Bahbha scattering is of course used to compute the hadron cross section.

A summary of the high energy measurements of R is shown in Table I. The experiments agree within errors with the average being 3.92±0.07. The χ^2 associated with this average is 0.45 for 4 degrees of freedom when systematic errors are included. From this average we may determine α_s.

$$\alpha_s = 0.18 \pm 0.06$$

This is in agreement with measurements from 3 jet events which I will discuss later in this section. The error, however, is large and will be difficult to improve because it is limited by systematics.

2.2 Hadronic Jets

When a pair of quarks are produced, they must somehow fragment into hadrons so that only color singlet particles are observed in our detectors. One of the predictions of the quark parton model is that the hadrons from the quark fragmentation should appear in "jets" along the initial quark direction. This has been observed previously but becomes more and more evident as the beam energy is increased. At high energy, hadronic events typically have the topology of two back to back jets of hadrons. Several variables have been devised to show how two jetlike an event is. Probably the most widely used is thrust.

$$T = \text{Max}\ \{\Sigma|\vec{p}_i \cdot \hat{t}|\}\ /\ \Sigma|p_i| \simeq (\Sigma_p\ {}^{||}\ \text{to quark})/p_{TOT}$$

where the maximization is done by choosing the thrust axis, $\hat{t}$, which

TABLE I

MEASURED VALUES R AT HIGH ENERGY

Group	R ± Statistical ± Systematic
JADE	3.93 ± 0.03 ± 0.09
MARK-J	3.84 ± 0.03 ± 0.22
TASSO	4.01 ± 0.03 ± 0.20
MAC	3.87 ± 0.05 ± 0.14
MARK II	3.90 ± 0.05 ± 0.25

gives the largest total momentum parallel to it. The value of the thrust, T, varies from 0.5 for a perfectly spherical event, to 1.0 for an event that is perfectly two jetlike. In Fig. 5, the average value of <1-T> is plotted vs. center of mass energy.

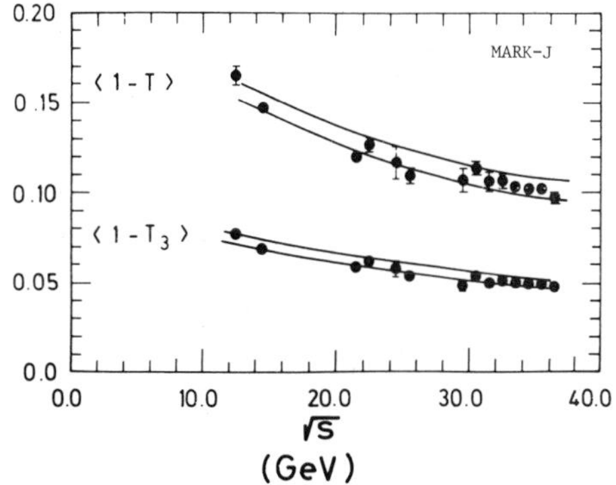

Fig. 5 Average value of thrust and triplicity versus energy.

As the energy is increased, the average value of thrust increases. At the highest PETRA energies, an average value of 0.9 is attained. The average thrust seems to be approaching a constant less than one however. This is due to the appearance of three jet events which I will discuss shortly.

Another simple prediction of the quark parton model is that the quarks should have a $(1+\cos^2\theta)$ angular distribution as do all spin 1/2 pairs produced in e^+e^-. Since the hadronic jets go along the initial quark direction, we may look at the angular distribution of the thrust axis of the events. In Fig. 6 the measured thrust axis angular distribution is shown in comparison to what is expected after the original $(1+\cos^2\theta)$ distribution is corrected for radiation of photons by the initial state electrons and detector acceptance. It is clearly seen that the thrust axis follows the $(1+\cos^2\theta)$ distribution well.

One important use of the thrust is made in the search for the production of new massive flavors of quarks. Since the masses of the five known quark flavors are all small compared to the beam energy, the thrust distribution is peaked near 1.0. If a new quark flavor with a mass only slightly less than a beam energy were produced, it would give rise to "spherical" events with an average

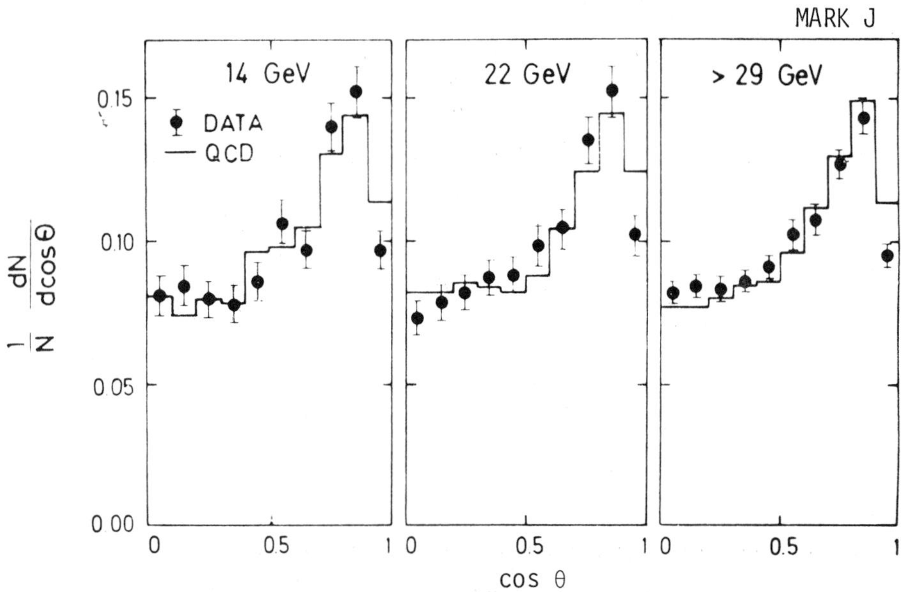

Fig. 6 Polar angular distribution of the event thrust axis with respect to the beam direction.

value of thrust which is much lower. This is because the produced quark is not highly relativistic. The high multiplicity of decay products from the weak decay of the massive quark come out fairly uniformly in the quark rest frame and are not boosted into a narrow jet.

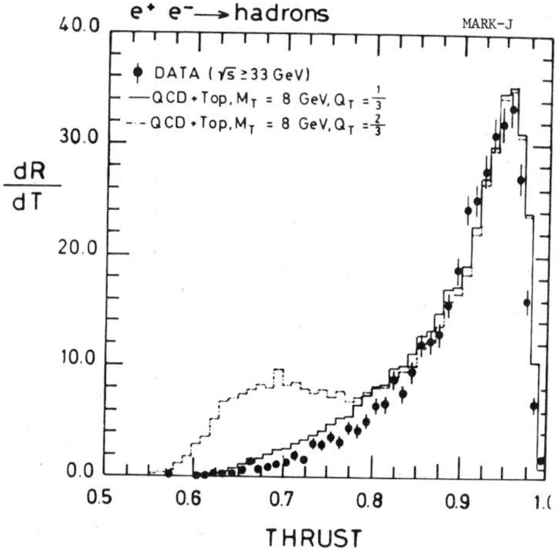

Fig. 7 Thrust distribution from data compared to that for charge 1/3(solid line) and charge 2/3(dashed line) quarks.

In Fig. 7 the thrust distributions expected for new quark flavors of charge 2/3 and 1/3 are compared to data. Both predictions are excluded statistically but, systematically, one worries whether the Monte Carlo model used was tuned too carefully to fit the data. This systematic error is of particular importance in the charge 1/3 case, where the effect of the new flavor is not a qualitative change in the shape of the distribution.

One way to check this is to look only at the small fraction of events containing muons. Because the massive quarks produce muons in their weak decay, we can enhance the signal by requiring a muon. Thrust distributions for events containing muons are shown in Fig. 8

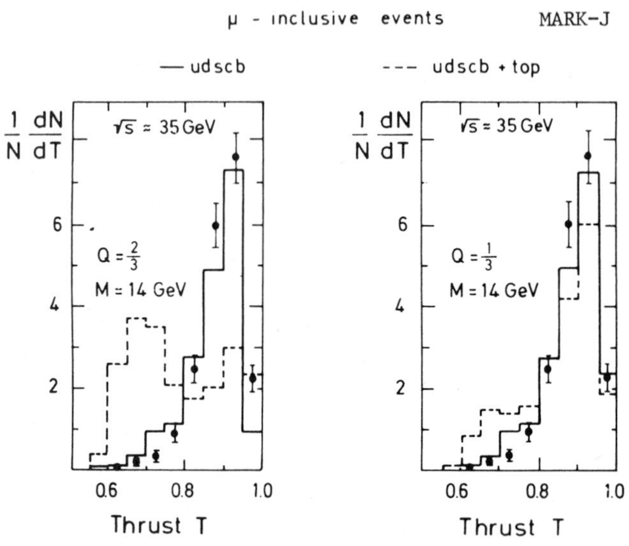

Fig. 8 Thrust distribution for hadron events containing muons versus that expected with a new heavy quark of charge 1/3 or 2/3.

3.0 STUDY OF QCD IN HADRON PRODUCTION

As quark jets become narrower when the beam energy is increased it becomes possible to see the effects of QCD in jet formation. In particular QCD predicts the emission of hard gluons which are not along the same direction as the quarks in much the same way as QED allows us to calculate the rate of photon bremsstrahlung.

$$e^+e^- \to q\,\bar{q}\,g$$

These hard noncollinear gluons will also fragment into hadrons giving a third jet. By now the existence of three jet events has been very clearly established(3). This was done by observing a large excess of planar events over what is expected from $q\bar{q}$ final states and by seeing that the particles in the planar events lie along three jet axes.

3.1 Testing the QCD Differential Cross Section

In Monte Carlo studies of the $q\bar{q}g$ events, it has been found that the 3 jet directions and energies that are reconstructed quite accurately correspond to the parton directions and energies. We can then use these reconstructed jet energies to test whether the distribution of the jets agrees with the prediction of QCD. In order to quantify how well we have tested QCD, it is useful to compare the data with another model giving 3 jet final states. For this purpose we have taken a model with "scalar gluons" which has gluon bremsstrahlung but with a different distribution than in QCD.

Since the production of 3 jets has the same kinematics as a 3 body decay of a particle, the total information about the decay distribution is contained in a Dalitz plot of the three jet energies. In Fig. 9 such a Dalitz plot is shown for the data and for a QCD Monte Carlo. The three axes correspond to normalized jet energies.

$$x_i = 2\,E_{JET}/E_{VIS}$$

The limits on the plot are set by the energy conservation condition

$$x_1 + x_2 + x_3 = 2$$

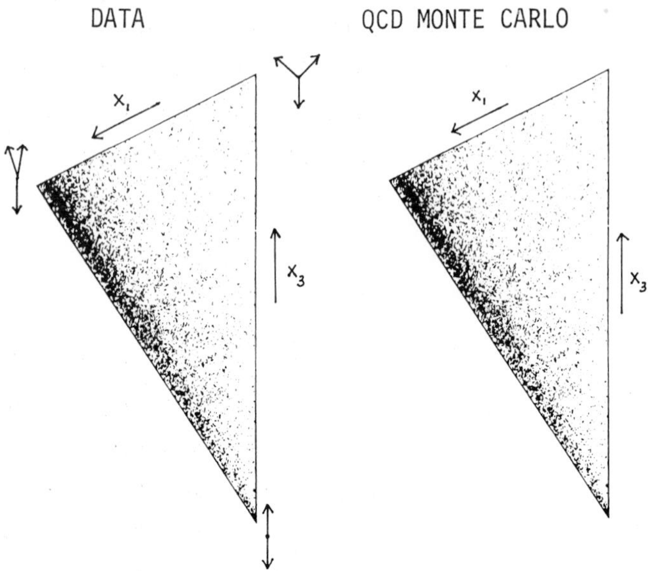

Fig. 9 Dalitz plot of 3 jet energies for data and QCD Monte Carlo from the MARK-J collaboration.

and by an ordering of jet energies

$$x_1 > x_2 > x_3$$

which we must use because we cannot distinguish between quark, antiquark and gluon jets. In this Dalitz plot, two jet events cluster along the lower left edge of the plot, where either x_3 is small or jet 2 and 3 are collinear, while events with three equal energy jets will appear in the upper right hand corner. These two dimensional plots appear to agree with each other and, in fact, a comparison of the data with a higher statistics Monte Carlo result gives a χ^2 of 59 for 72 degrees of freedom. Our scalar gluon model on the other hand, gives 110 for 72 degrees of freedom. We, therefore, can see a very clear difference in the quality of the fit between vector and scalar gluons.

The difference between vector and scalar can be most clearly demonstrated in one dimensional distributions. In Fig. 10 the distribution of the data from TASSO[(4)] in the Ellis-Karliner[(5)] angle is shown.

$$\cos(\tilde{\theta}) = \frac{x_2 - x_3}{x_1}$$

In Fig. 11 MARK-J data are shown as a function of a variable, S/V, which is simply the ratio of the expected scalar gluon cross section to the expected vector gluon cross section computed for each measured event configuration.

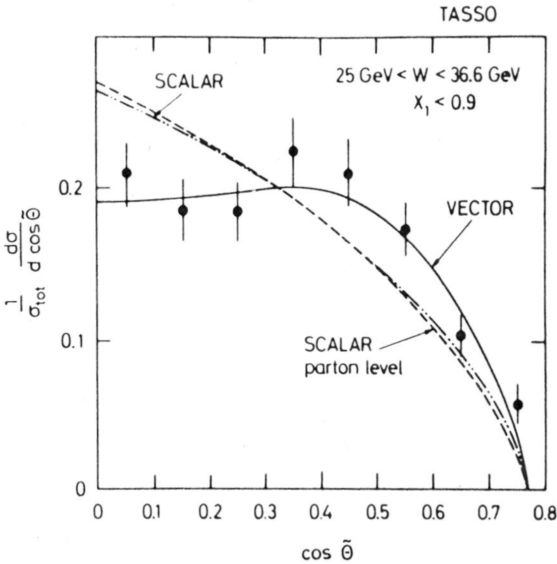

Fig. 10 Histogram of events vs. the Ellis-Karliner angle compared to predictions for vector and scalar gluon theories.

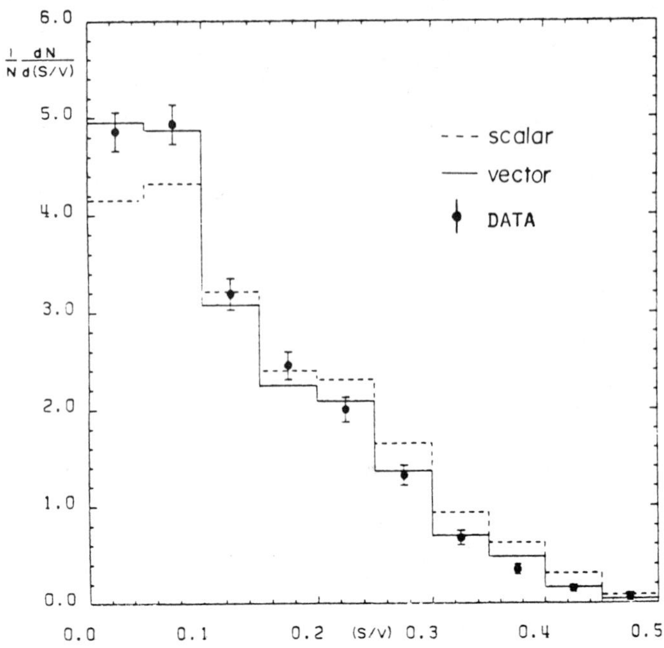

Fig. 11 Histogram of the ratio of the scalar gluon to the vector gluon cross sections, S/V, from MARK-J.

$$d\sigma_{VECTOR} \propto \frac{x_1^2 + x_2^2}{(1-x_1)\ (1-x_2)}$$

$$d\sigma_{SCALAR} \propto \frac{x_3^2}{(1-x_1)\ (1-x_2)}$$

$$S/V \equiv \frac{x_3^3}{x_1^2 + x_2^2}$$

The main difference between these two cross sections is that the vector cross section diverges if the gluon is soft ($x_3 \to 0$) or collinear ($x_1 \to 1$) while the scalar cross section diverges only in the collinear case. Therefore the main difference is in the region where the gluon is soft. In the S/V distribution of Fig. 11, the vector theory gives a χ^2 of 10.5 for 9 degrees of freedom while the scalar gluon theory gives 98.1 for 9 degrees of freedom. This corresponds to about a 9 standard devistion difference between the two theories.

The conclusion is that the data are in very good agreement with QCD and are sufficiently accurate to distinguish between QCD and another gluon bremsstrahlung theory at a very high degree of confidence. Therefore, QCD is quite well tested in the distribution of jets in three jet events.

3.2 The Strong Interaction Coupling Constant

The PETRA experiments have determined the value of the strong interaction coupling α_s in the past by comparing their data on three jet production to a first order perturbative QCD calculation[6] or to a similar calculation[7] which includes four parton final states. Because we do not measure the quarks and gluons directly, we must check that our determination does not depend on parton fragmentation models. The methods used to do this can be simply illustrated in Fig. 12. The Feynman-Field fragmentation model parameter with the largest effect on a determination of α_s is the transverse momentum, σ_q, imparted to a quark pulled out of the sea in fragmentation. The figure shows how our results can be made insensitive to variations of this parameter. What is plotted is the best fit value of α_s as a function of the assumed value of σ_q. If we choose a variable like the broad jet oblateness O_B, which is the difference between the transverse momentum in the event plane and the transverse momentum perpendicular to the event plane, then the fit value of α_s depends only very weakly on σ_q. On the other hand, a variable like the thrust of the jet in the narrower hemisphere, T_N, can be used to fix the value of σ_q since it depends very strongly on it. Because we must fit both distributions with unique values of the two parameters, we take the intersection of the two lines in the figure. The small remaining error on σ_q does not contribute significantly to the overall error on α_s determined from the O_B distribution.

It would be a great improvement over these previous determinations of α_s if a second order perturbative QCD calculation could be used. This of course has the advantage of being more accurate and giving an idea of how well the perturbation series is converging but most important, by going to second order, we obtain a relationship between α_s and $\Lambda_{\overline{MS}}$.

Three groups of theorists[(8)] have made the QCD calculation of the three parton and four parton final state cross sections to second order in α_s. It has been found that the three calculations agree in detail. One of the groups, however, has made some different assumptions about how to apply their result to get a thrust distribution. They apply Sterman-Weinberg cuts to their final states to, for example, remove four jet events where two of the

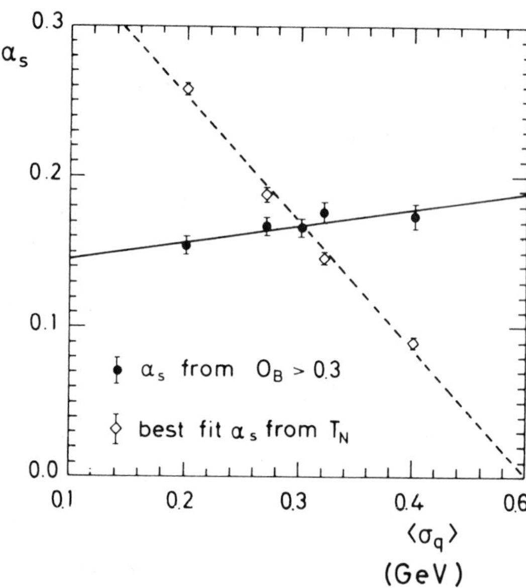

Fig. 12 The best fit value of α_s from MARK-J is plotted versus the assumed average transverse momentum (σ_q) in fragmentation for two jet variables.

jets will not be physically distinguishable from one jet. These are then moved into the three jet category under the assumption that the fragmentation model already includes the effect of soft or collinear gluon radiation. Further study has shown that the size of the second order correction to the thrust distribution depends critically on the values chosen for ε and δ, the Sterman-Weinberg cut parameters. This means that the QCD calculation leaves a reasonably large amibiguity in a determination of α_s from the thrust distribution of distributions of a similar nature.

The JADE collaboration[9] has presented a determination of α_s, using the second order calculation of Fabricius et al. in which the cuts are similar to those used in a cluster analysis of the data. They find in second order:

$$\alpha_s = 0.16 \pm 0.015 \pm 0.03$$

This may be compared with their value in first order of 0.20.

It may be possible, however, to find distributions that do not depend as strongly on cuts as does thrust and therefore remove the amibiguity in the QCD calculation. One such distribution is the cross section asymmetry in energy-energy correlations. The energy energy correlation can be computed simply from the energy distribution in a hadron event. No minimization, maximization nor determination of an event axis is needed. In its most often used form, the energy-energy correlation cross section is a function of one angle, χ, which is the angle between a pair of final state particles, i and j.

$$\frac{1}{\sigma_{TOT}} \frac{d\Sigma(\chi)}{d(\cos\chi)} = \frac{1}{\Delta\cos\chi} \frac{1}{N_{EVENT}} \sum_{evts} \sum_{i,j} \frac{E_i E_j}{E_{VIS}^2}$$

The asymmetry in this cross section around π is an interesting quantity since it is nearly independent of fragmentation effects and, therefore, perhaps also less sensitive to Sterman-Weinberg cuts in the QCD calculation.

$$\frac{d\sigma_{ASY}}{d(\cos\chi)} = \frac{d\Sigma(\chi)}{d(\cos\chi)} - \frac{d\Sigma(\pi-\chi)}{d(\cos\chi)}$$

A study of the cross section asymmetry, as a function of different ε cuts, has shown that it is independent of the value of the cut. The angle χ is equivalent to a δ cut, so that by excluding the region near $\chi=\pi$, one can also be independent of the value of δ. In addition it is found(10) that the second order corrections to the cross section asymmetry are very small as is shown in Fig. 13.

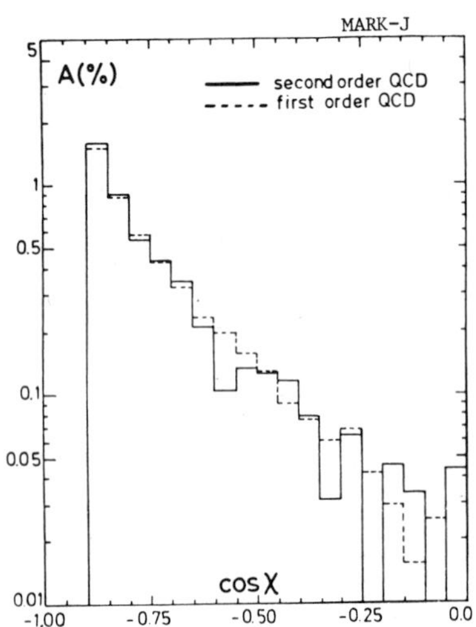

Fig. 13 QCD calculation of the cross section asymmetry in the energy-energy correlations at the parton level for first and second order using the same value of Λ.

In Fig. 14 the measured cross section asymmetry is compared to the prediction of first order QCD modified by fragmentation and detector acceptance. Both of the QCD curves are for the best fit

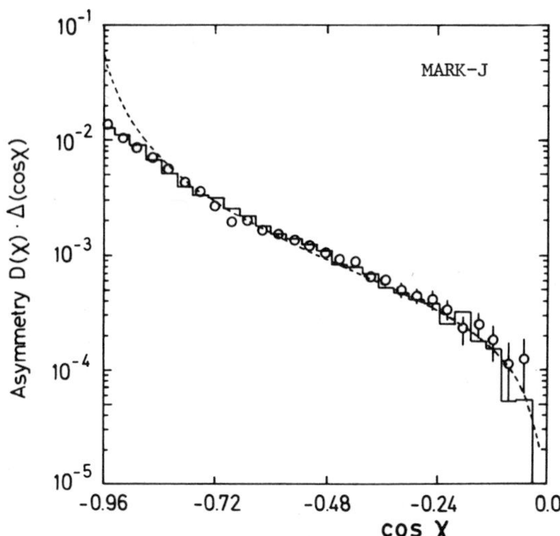

Fig. 14 The cross section asymmetry from the uncorrected data compared to a first order QCD Monte Carlo with α_s=0.16.

value of α_s of 0.16. If a second order calculation is used, α_s is found to be 0.14±0.01±0.02 which corresponds to $\Lambda_{\overline{MS}} \simeq$ 200 MeV. Although this result is still preliminary, it appears hopeful that a method like this will allow us to determine α_s in second order with the ambiguties inherent in QCD minimized.

3.3 Quark and Gluon Fragmentation

Several models of quark and gluon fragmentation have been made to parameterize the nonperturbative of confinement part of QCD. At the present time, it is not possible to calculate, directly from the theory, what the details of the final state hadrons should be. These models of fragmentation we use are therefore only motivated by the theory and must be tuned up to agree as well as possible with data at various center of mass energies. When one attemps to measure α_s, or test QCD, any dependence of the result on the fragmentation model is then part of the uncertainty in the measurement.

The most widely used model of fragmentation is that of Feynman and Field[12]. This has proven adequate to describe a large amount of experimental data with reasonable accuracy. In this model, each of the partons fragment independently. The final hadrons are distributed in momentum along the parton direction according to a fragmentation function, f(z), which is free to be changed to best fit the data. The transverse momentum distribution of the hadrons is gaussian with an RMS value of $\sqrt{2}\,\sigma_q$, where σ_q is another parameter of the model.

While this model has worked well for two jet events, its direct extension to events with three of more jets involves additional assumptions. In the particular models we use, Hoyer et al.[6] and Ali et al.[7], all three jets are fragmented independently in the center of mass frame of the event. These models have the property that the energies and momenta of the particles in a jet will sum to give the energy and direction of the parton which produced that jet. This means that, for sufficiently well separated jets, we can reconstruct the parton momenta very well.

This is not true in the other model of fragmentation which has been tried. This is known as the Lund model[11]. In this model, fragmentation occurs along the color strings connecting partons. For a two jet event, the two fragmentation models are equivalent. For a three jet event, the Lund model assumes that two color strings exist, one along the line between the antiquark and the gluon and the other along the line between the quark and the gluon. Because the line between the quark and the gluon does not coincide with the quark or gluon direction, this model produces jets which do not reconstruct to give the parton momenta. The Lund model has the advantage that it is Lorentz invariant. One example of how the

the Lund model works is useful. Take an event from the process $e^+e^- \to q\,\bar{q}\,\gamma$:

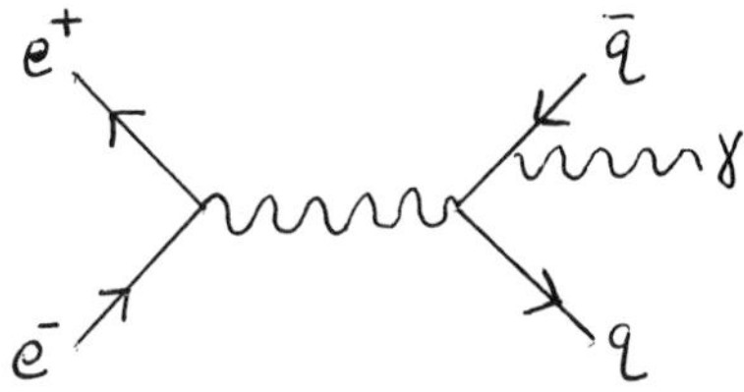

In this case the Lund model has a color string connecting the quark and the antiquark. Fragmentation in any frame gives the same result. One can get the equivalent result from Feynman-Field fragmentation by boosting to the center of mass frame of the $q\bar{q}$ system, fragmenting, and then boosting back to the event center of mass system. This procedure obviously is the correct extension of Feynman-Field fragmentation of simple $q\bar{q}$ events. It gives the same result as Lund: that the fragmentation has changed the directions of the jets from the original parton directions. This is essentially due to the boost of the hadrons, which are not at $\beta=1$, back to the event center of mass system. When we go to $q\bar{q}g$ case, the choice of method is not so clear cut because in this case, we have three colored partons and we do not know how the color strings, if indeed there are any, arrange themselves when soft gluon effects are taken into account. It is in this sense that the Hoyer and Lund models are extreme cases. Hoyer assumes that the parton directions are not at all affected by color strings, while Lund assumes that the two original color strings are preserved throughout the fragmentation process, thus changing the direction of the jets maximally.

These two models are very different from each other and in that sense give a good indication of the model dependence of a result. It would of course be useful to try many possible fragmentation models, but the time required to tune each model to fit the data well has so far proved prohibitive.

CELLO has found[13] that both the Lund and Feynman Field models of fragmentation, in conjunction with a first order QCD calculation, give a good description of their data. They find, however, that the best fits are with values of α_s that differ by 28% to 67% between the two models with that determined using the Lund model being larger. Other experiments find similar, but somewhat smaller variations of the value of α_s using first order QCD. This fragmentation dependence needs to be explored in more detail in the future. Presumably, methods can be found with the least dependence on fragmentation possible. Also, we might hope that the fragmentation model dependence will be reduced as we go to higher orders in the QCD calculation.

It should be possible to determine which of these two fragmentation models more closely fits the data. This probably does not remove the model dependence of various measurements since other fragmentation schemes could be envisioned. However, it is interesting to see which scheme is closer to the truth to learn about the fragmentation process. JADE has found[14] the multiplicity of low momentum particles between jets agrees with the Lund model but not with models using Feynman-Field fragmentation. This is probably too small a detail to differentiate between models, since the amount of energy flowing between jets is nearly the same in the two models. A more fundamental prediction of the Lund model, which is also expected based on QCD, is that the gluon jet should be broader than the quark jets. This can be tested because the gluon tends to be the lowest energy jet and we can therefore look for differences in breadth between this jet and the other two jets. JADE[15] has used the average P_t of a jet plotted against jet energy to look for this effect. They find that the lowest energy jet is indeed broader than the other jets in the data as shown in Fig. 15.

This is also found in the Lund model but not in the Hoyer Monte Carlo where the gluon fragments in the same way as a quark.

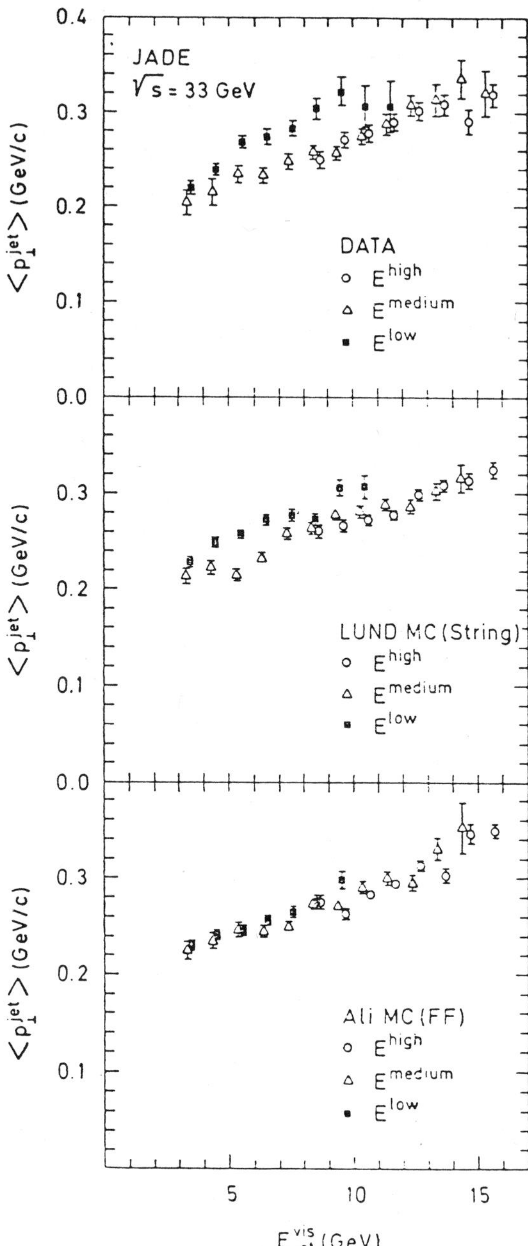

Fig. 15 Average transverse momentum in jets for 3 jet events. The jets are energy ordered with jet 1 having the largest energy.

This result has not yet been verified by another group. Of course a broader gluon jet can also easily fit within the framework of Feynman-Field fragmentation by simply assigning a different transverse momentum distribution for the gluon fragmentation.

3.4 Characteristics of the Final State Hadrons

A large amount of information has been gathered about the details of the distributions of those hadrons that we measure in $e^+e^- \rightarrow$ hadrons. This is of course useful in building our fragmentation models and in making projections for what kind of detectors are needed for the future. Because jets can be so clearly identified in e^+e^- reactions, the measurements we make can be used to make projections not only for higher energy e^+e^- machines but also, for example, for hard scattering of partons in protron-proton collisions.

One of the most fundamental distributions is the multiplicity distribution. Fig. 16 shows data combined from several experiments.

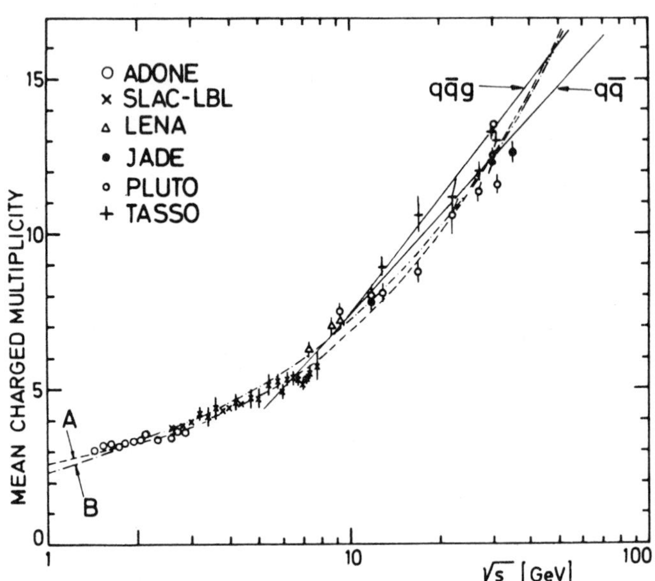

Fig. 16 The mean charged particle multiplicity in hadron events versus center of mass energy.

The mean charged multiplicity is plotted versus center of mass energy. As expected, the average multiplicity rises with energy. One may attempt to look at the detailed rate of increase to see if it is consistent with our understanding of hadron production. The dash dotted curve is that for the function

$$\langle N_{ch} \rangle = 2.3 \sqrt[4]{s} \qquad (s \text{ in GeV})$$

which gives the form of increase expected from phase space. This curve is in fair agreement with all of the data. The expected increases from b quark production and from QCD effects are also consistent with the data. The distribution of multiplicities, however, can not be predicted by simple phase space considerations. In Fig. 17 the charged multiplicity distribution is plotted as

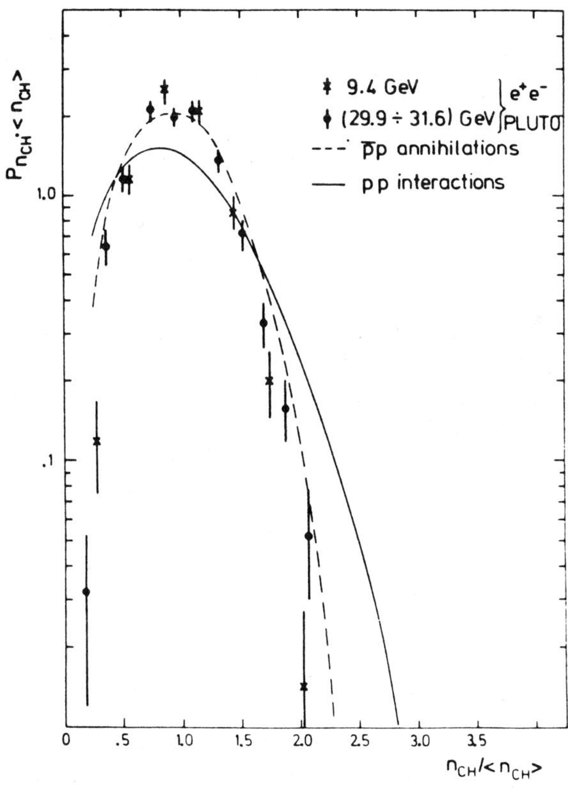

Fig. 17 KNO plot of charged multiplicity distribution for hadron events compared to what is found in ee and $p\bar{p}$ collisions.

$P(N_{ch}) \langle N_{ch}\rangle$ versus $N_{ch}/\langle N_{ch}\rangle$ where $P(N_{ch})$ is the probability of finding charged multiplicity N_{ch} in a given hadron event. On this type of plot, data from different center of mass energies should fall on top of each other if KNO[(16)] scaling is correct. One sees in the figure that data from 9.4 GeV and ∿ 30 GeV are quite consistent with each other. The width of the two distributions is the same, which is not what is expected from phase space where this type of graph would get narrower when the center of mass energy is increased. The data are also consistent with what is found in $p\bar{p}$ collisions but different from pp collision data.

In Fig. 18 the fraction of the measured energy in hadron events which is carried by photons and that by all neutral particles

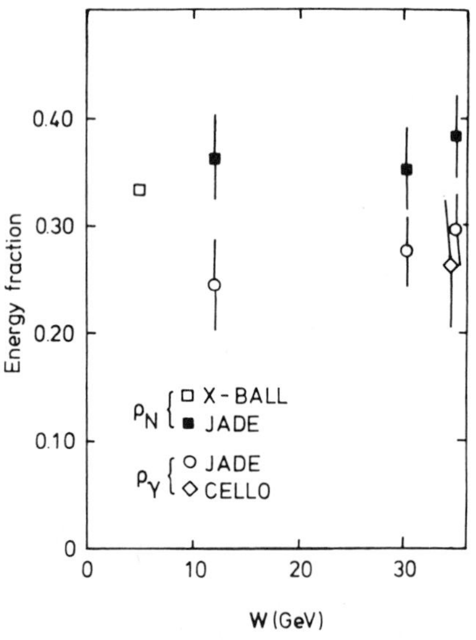

Fig. 18 Fraction of energy in hadron events carried by photons and by all neutral particles versus center of mass energy.

is shown. Within the errors, these fractions are independent of energy. The photon energy fraction is consistent with the dominant source being π^0's.

Many methods of particle identification have been used at PETRA and PEP to get information about the fragmentation of partons into pions, kaons and baryons. A summary of some of this information is shown in Fig. 19.

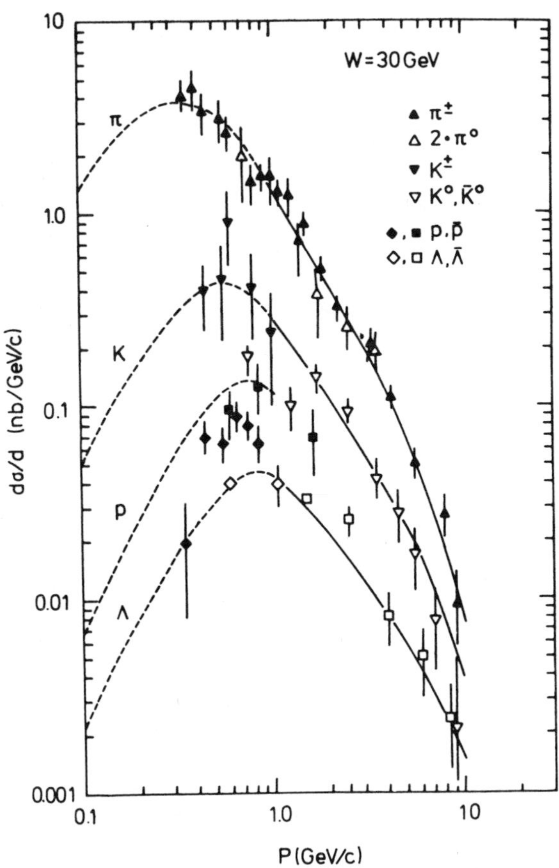

Fig. 19 Momentum distribution of various species of particles identified in hadron events.

All particle species have qualitatively similar momentum distributions. The rise at low momentum is due to phase space effects. At all momenta the dominant particle species is pion.

However, at large momentum, this dominance is reduced. In Fig. 20 sources of kaons are compared to the rate of kaon production measured. Although some are from primary s quark production or from primary b or c quark decay to an s quark, most kaons come from s quarks which are pulled out of the sea in fragmentation.

One very important question about fragmentation is whether the leading particles in a jet can be used to identify the original parton that produced the jet. One test of this has been done by the TASSO group. The produced quark and antiquark have opposite charges. If the leading particles in each jet have some probability to "remember" the parton charge, the charges computed from the two jets will tend to be opposite more often than like. Feynman and Field proposed to use a momentum weighted jet charge:

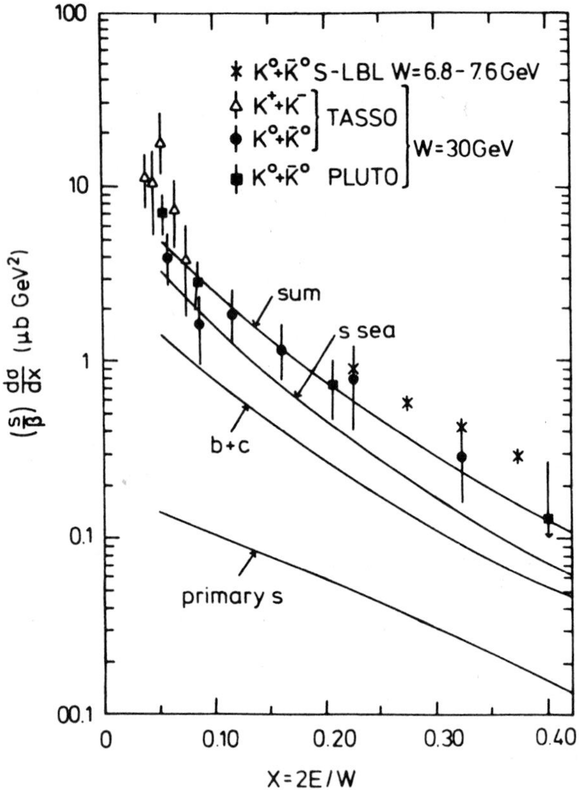

Fig. 20 Sources of K mesons versus momentum.

$$q_{JET}(\gamma) = \Sigma\, e_i\, x_i^{\gamma}$$

where γ is a parameter which should be optimized to best determine the charge. TASSO has taken out some of the γ dependence by redefining

$$q_{JET}(\gamma) \rightarrow q'_{JET}(\gamma) = (\Sigma\, x_i^{\gamma} / \Sigma\, x_i^{2\gamma})\, q_{JET}(\gamma)$$

In Fig. 21, the average product of the charges

$$P'(\gamma) = < q_1'(\gamma)\, q_2'(\gamma) >$$

is plotted versus γ for events where each jet has at least one particle with $x > 0.35$. $P'(\gamma)$ peaks near $\gamma = 1$. To test whether this indead indicates that leading particles remember the parton charge, the charges of particles were randomized within each jet. The resulting expectation for $P'(\gamma)$ is shown as the solid line in the figure. Obviously there is indeed some memory of the parton charge.

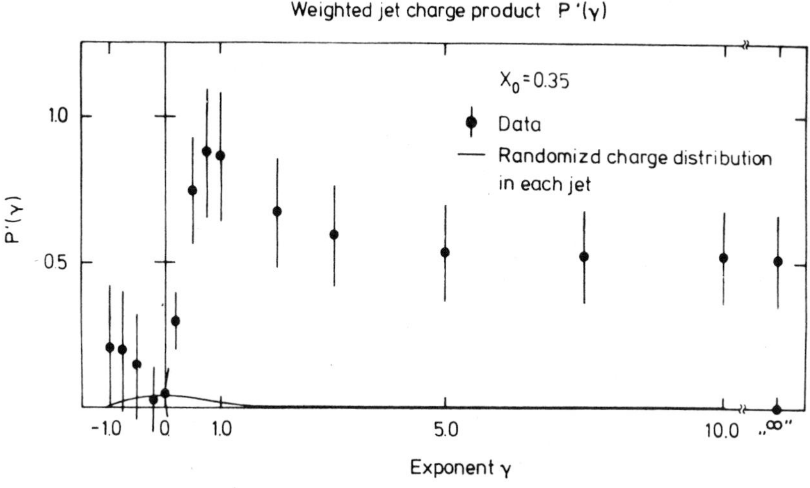

Fig. 21 P' versus γ for events which have one particle in each jet having x greater than 0.35.

4.0 ELECTROWEAK INTERACTIONS

The standard electroweak interaction model[17], based on broken SU(2) x U(1) symmetry has been very successful in fitting all of the presently available data. Indeed this model even predicted many of the data's features. It predicted the existence of a $\Delta S=0$ neutral current even though strangeness changing neutral currents were limited by experiment to be very small. Since the weak neutral current was discovered[18] in neutrino scattering, very detailed measurements[19] have been made of its structure, particularly by scattering neutrinos from quarks. All of the data from these experiments can be fit to the original model with its one free parameter, $\sin^2\theta_w$. Data from the purely leptonic process, neutrino electron scattering[20], also are fit by the model with the same value of $\sin^2\theta_w$. Finally, a very high precision measurement[21] of parity violation in electron deuteron scattering also was in good agreement with the model and determined the SU(2) multiplet structure of the fermions to be as originally expected.

The low energy limit of the model has therefore been tested very stringently by the combination of all of these experiments. However, some of this model's most interesting features, such as spontaneous symmetry breaking and the existence of intermediate vector bosons at predicted masses, have not yet been tested. As was first pointed out by Bjorken[22], all that is needed to fit the data is a global SU(2) symmetry, weak electromagnetic mixing and universality. The new measurements needed to pin down these important predictions of the Glashow-Weinberg-Salam model are to find the scalar particles responsible for symmetry breaking and to make measurements at high q^2 where the boson structure of the interaction may be determined.

On the surface it seems that a theory that made so many predictions which were subsequently verified, would be very likely to be correct. However, the large number of parameters in the Higgs sector has seemed unnatural to many and has spawned further searches for models that do not contain so many parameters yet naturally explain the low energy data. These models include grand unified theories (GUTs), technicolor models, supersymmetric GUTs and constituent models. Many of these models yield a different high energy behaviour than GWS.

To be more specific, several authors[(23)] have proposed alterations to the standard model which would give a different boson structure. The least radical change among these has first been proposed by Georgi and Weinberg. They showed that by extending the symmetry group from SU(2) x U(1) to SU(2) x U(1) x G, which may be embedded in a larger symmetry group in a GUTs model, all of the low energy predictions of the model remain unchangend. At high energy, however, the extended models have a richer boson structure. A more radical model is one in which the W's, Z's and fermions are composed of constituents and, at energies of the order of 1 TeV, a continuum of weak interactions occurs. At low energies in these models, weak interactions may be dominated by the lowest lying resonance, a situation analogous to vector dominance in the strong interactions of the photon where the ρ meson is a bound state of constituent quarks.

At low energy, all of the models, including the standard one, can be described by an effective neutral current Hamiltonian

$$2H_{NC} = \frac{-e^2}{q^2}\, j_{EM}{}^2 + \frac{8G_F}{\sqrt{s}}\left[(j^{(3)} - \sin^2\theta_W \cdot j_{EM})^2 + C \cdot j_{EM}{}^2\right]$$

where $j^{(3)}$ is the third component of the weak isospin current.

$$\vec{J}_\mu = \Sigma\, \bar{\psi}_i \gamma_\mu\, (1-{}_5)/2\, \vec{\tau}/2\, \psi_i$$

with the sum running over all weak fermion doublets. In the standard model, as in any model with a single Z boson, the constant C is equal to zero. In theories with more than one Z boson, C will be greater than zero.

Gounaris and Schildknecht[24] have given a nice interpretation of the parameter C in terms of a deviation from the standard model. They found

$$16\ C = \frac{\left[\int ds\ \frac{1}{s}\cdot\sigma(e^+e^-\to \text{ALL})\right]_{\text{ACTUAL}} - \left[\int ds\ \frac{1}{s}\cdot\sigma(e^+e^-\to \text{ALL})\right]_{\text{GWS}}}{\left[\int ds\ \frac{1}{s}\cdot\sigma(e^+e^-\to \text{ALL})\right]_{\text{GWS}}}$$

so that the quantity 16 C measures the deviation of the total e^+e^- cross section from the standard model, integrated over all energy with the weighting factor (1/s). C will therefore be a parameter of general interest to weak interaction model builders.

Throughout the article, I will describe the strength of the axial vector and vector couplings of the weak neutral current in terms of the dimensionless coupling constants g_A and g_V. In the standard model we have for the left handed fermion doublets under weak SU(2)

$$g_A = T_3$$
$$g_V = T_3 - 2\cdot q\cdot \sin^2\theta_w$$

where q is the charge of the particle and T_3 is the z component of its weak isospin. The reactions

$$e^+e^- \to e^+e^- \quad \text{Bhabha Scattering}$$
$$e^+e^- \to \mu^+\mu^- \quad \text{Muon Pair Production}$$
$$e^+e^- \to \tau^+\tau^- \quad \text{Tau Pair Production}$$

have been studied[25] at PETRA in terms of electroweak models. Although interference effects between the neutral weak boson and the photon mediated graphs could be measured in any of these interactions, the cleanest measurement of weak effects can be made in the forward backward charge asymmetry in muon and tau pair production. Besides being sensitive to weak effects, the measurements of the three reactions above are also uniquely sensitive to non-pointlike structure in the leptons since they are made at the highest available q^2.

For all of the measured processes, order α^3 QED calculations are necessary to test weak interaction effects because the order α^3 radiative corrections are generally about the same size as the weak effects. These calculations have been made by Berends, Gastmans, and Kleiss[(26)]. Monte Carlo event generators with the order α^3 matrix elements have been supplied to us by Berends and Kleiss. With these generators, we are able to pass the events through our detector simulation and analysis programs so that the effects of resolution and experimental cuts on the radiative corrections can be accurately represented. Of particular note is the inclusion of hadronic vacuum polarization effects in a form slightly modified from that originally published by Berends and Komen[(27)]. Some effects of radiation of photons can be measured as a partial check of the calculations. In Fig. 22, the measured accollinearity distribution for muon pair production is compared to the expectation from Monte Carlo. Good agreement is found.

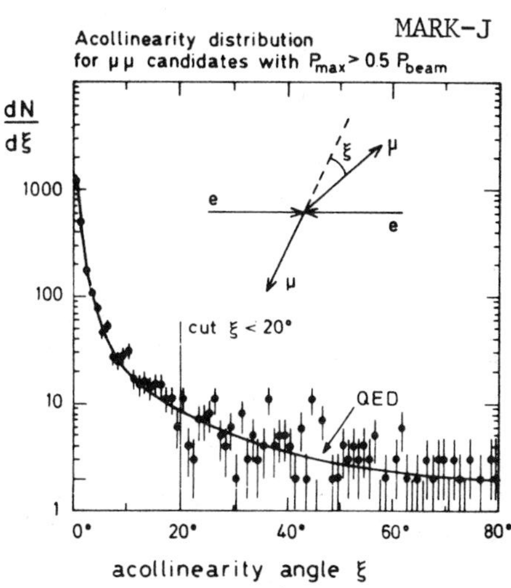

Fig. 22 Acollinearity distribution for muon pair production.

4.1 Bhabha Scattering

The data from Bhabha scattering that are of relevance to our tests of the electroweak interaction are the angular distributions of the final state electron and positron. In the absence of beam polarization, the distribution is symmetric in the azimuthal angle ϕ, but has a very strong dependence in the polar angle θ from the beam axis.

Fig. 23 is a graph of $s(d\sigma/d\cos\theta)$. Data, with error bars invisible on this scale, are compared with the Monte Carlo calculation of order α^3 QED. As the cross section is steeply falling, it is difficult to see the details of the match between measurements and theory. A more clear exposition of this may be obtained by plotting the fractional difference between data and theory:

$$\delta = \frac{N_{DATA} - N_{QED}}{N_{QED}}$$

In Fig. 24, the δ distribution from three PETRA groups is

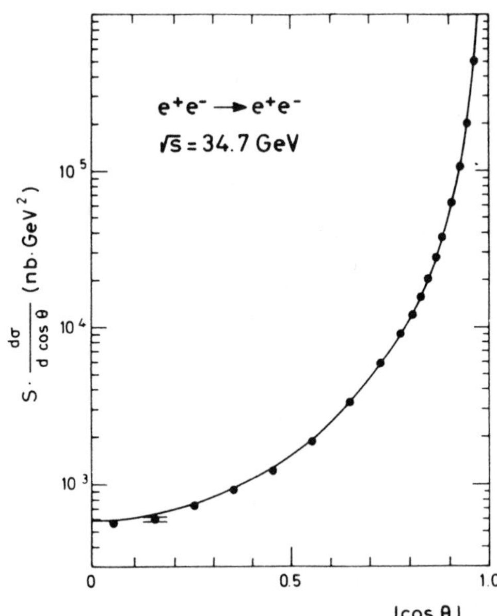

Fig. 23 Bhabha scattering angular distributions from MARK-J compared to pure QED.

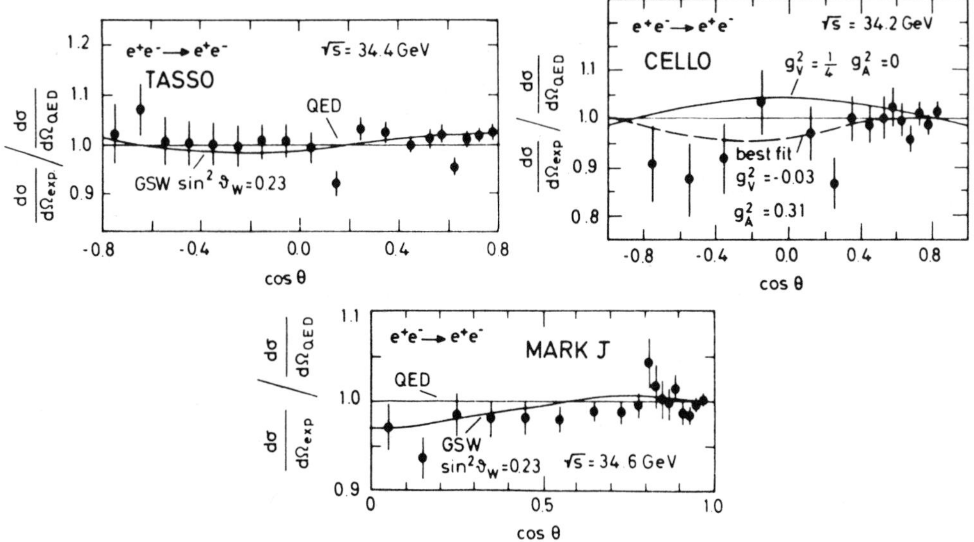

Fig. 24 The difference between data and QED is compared to the prediction of the GWS model with $\sin^2\theta_W = 0.23$. In the CELLO graph, curves are shown for different values of g_A and g_V.

shown vs. $\cos\theta$. If the data agreed exactly with QED, the points should lie on the horizontal line at 0. The solid line drawn in the figure is the expectation for the standard model of Glashow, Weinberg and Salam, with $\sin^2\theta_W = 0.23$. The data at high energy favor GWS over pure QED but are not conclusive. Although the weak effects are expected to be small in the standard model, variations from this model will in general yield larger effects.

4.2 Muon and Tau Pair Production

In muon pair production, the total cross section and the polar angular distribution are sensitive to weak effects. The weak effect is expected to be small for the total cross section in the case of the standard model. However, as with the results for Bhabha scattering, if one goes outside the standard model, the effects on the total cross section may be quite large. On the other hand, the weak effect in the angular distribution, and in particular, in the forward-backward charge asymmetry, is of measurable size in the standard model and generally larger in other models. We will, therefore, look at both the total cross section and the forward backward asymmetry, but expect to find our most interesting results in the charge asymmetry where we can clearly observe weak effects.

The radiative correction to $d\sigma/d\Omega$ for muon pair production is shown in Fig. 25 as a function of the polar angle θ. This correction is for the particular cuts on muon pair events, that each muon have at least 50% of a beam energy and that the two muons be collinear within 20°. The correction is not symmetric about 90°, indicating that there is a pure QED forward-backward charge asymmetry, which must be taken into account in order to measure weak effects. I will discuss this correction again when I come to the charge asymmetry.

The data on the total muon pair production cross section are summarized in Fig. 26. $R_{\mu\mu}$, the ratio of the measured muon pair production cross section to the cross section expected in QED, is

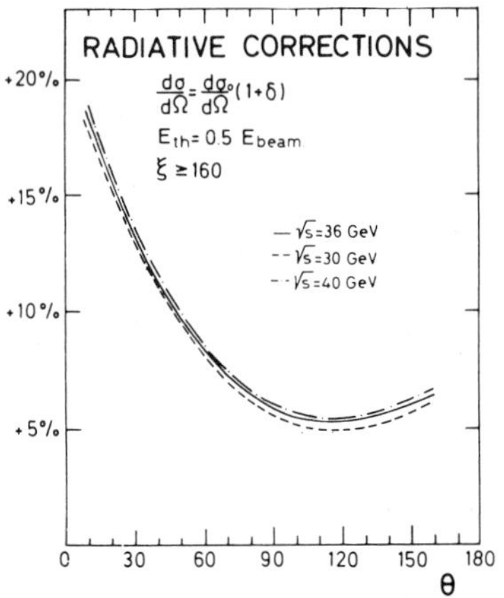

Fig. 25 Radiative correction to muon pair production as a function of polar angle.

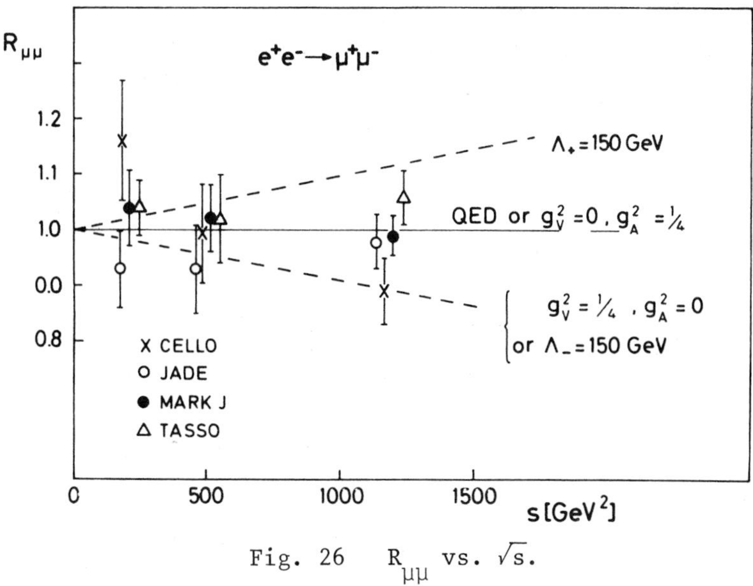

Fig. 26 $R_{\mu\mu}$ vs. $\sqrt{s}$.

shown as a function of $\sqrt{s}$ for all five PETRA experiments. The data, which have been corrected for radiative effects, are compared to models with nonpointlike behavior for muons which may be parameterized by an energy scale Λ. I will discuss these models in more detail later. Neither large disagreement nor unusual structure are found.

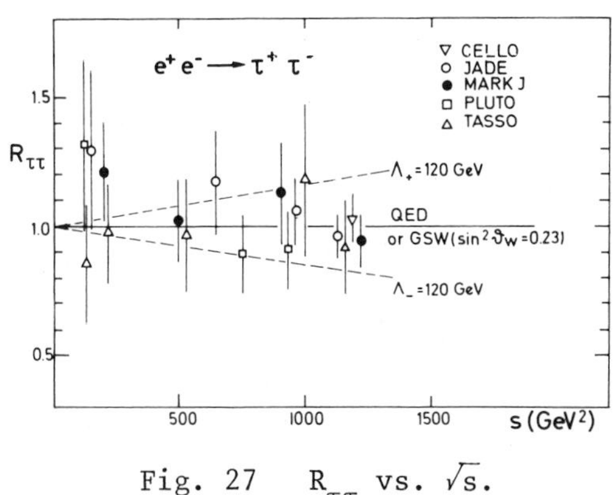

Fig. 27 $R_{\tau\tau}$ vs. $\sqrt{s}$.

Fig. 27 is similar to Fig. 26 but the tau pair production cross section is plotted rather than that for muons. Again, no unusual structure is seen.

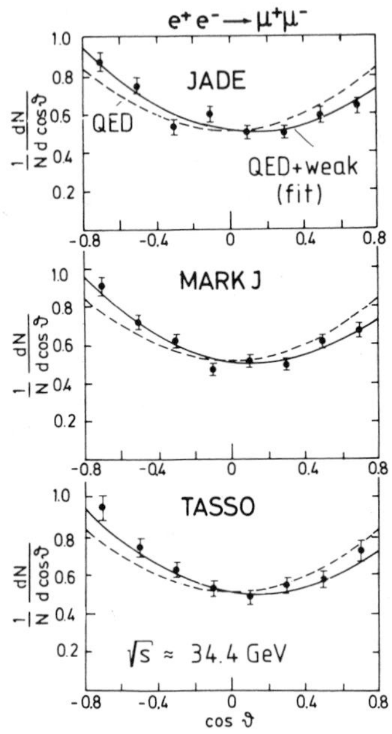

Fig. 28 Angular distribution for μ pair production from three groups.

The angular sitributions from three individual experiments are shown in Fig. 28. A large forward-backward charge asymmetry is already apparent in these distributions. The data are all quite consistent with the solid lines which are the expectation from the standard electroweak model. From these graphs it is clear that all three experiments are seeing a charge asymmetry of similar magnitude and that the angular distribution of this asymmetry is, within the statistical errors, that expected from weak effects. In Fig. 29, the tau pair angular distributions from four PETRA groups are shown.

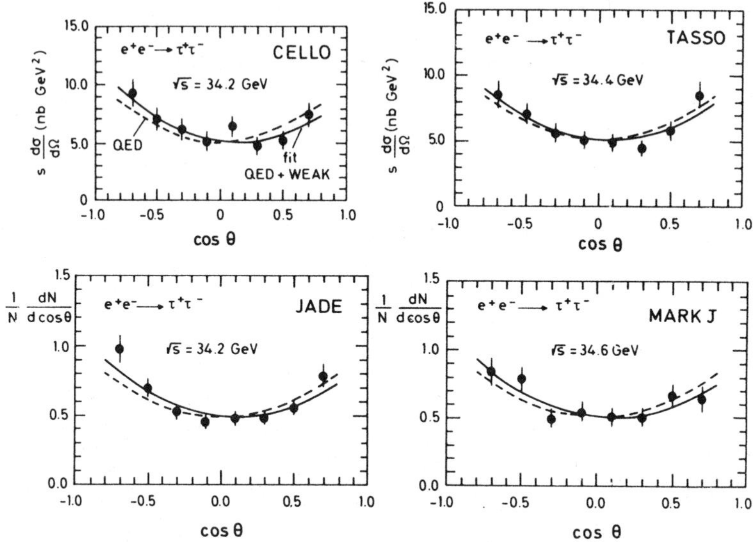

Fig. 29 Angular distribution for tau pair production from four groups.

4.3 Charge Asymmetry

The polar angle θ is defined as the angle between the initial electron direction and the final μ^- or τ^- direction. The forward hemisphere is then defined as $0^o < \theta < 90^o$ and the backward hemisphere as $90^o < \theta < 180^o$. The asymmetry as a function of θ is defined to be

$$A_{\mu\mu}(\theta) = \frac{N_{\mu^-}(\theta) - N_{\mu^-}(\pi-\theta)}{N_{\mu^-}(\theta) + N_{\mu^-}(\pi-\theta)}$$

Since statistics are still low, the number usually quoted is integrated over angle

$$A_{\mu\mu} = \frac{N_{FORWARD} - N_{BACKWARD}}{N_{FORWARD} + N_{BACKWARD}}$$

We can compute the asymmetry in lowest order using just the two diagrams

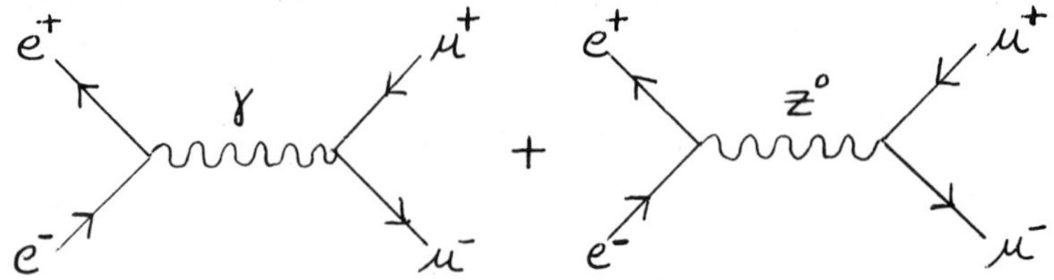

Using this lowest order computation, the differential muon pair cross section is

$$d\sigma/d\Omega \;=\; \alpha^2/4s\left[F_1(1+\cos^2\theta) + F_2\cos\theta\right]$$

where

$$F_1 = 1 + 8sgg_V^2\left(\frac{M_Z^2}{s-M_Z^2}\right) + 16s^2g^2(g_A^2+g_V^2)^2\left(\frac{M_Z^2}{s-M_Z^2}\right)^2$$

$$F_2 = 16sgg_A^2\left(\frac{M_Z^2}{s-M_Z^2}\right) + 128s^2g^2g_A^2g_V^2\left(\frac{M_Z^2}{s-M_Z^2}\right)$$

$$g = G_F/8\sqrt{s}\,\pi\alpha = 4.49\cdot10^{-5}\ \text{GeV}^{-2}$$

and g_A and g_V are the previously defined axial vector and vector coupling constants for the charged leptons. The asymmetric part of the cross section is the term proportional to $\cos\theta$. By putting in some numbers one can see that, at PETRA energies, F_2 is small compared to F_1 and that, in both F_1 and F_2, the first terms dominate. The first term in F_1 is the purely electromagnetic term and the first term in F_2 is due to weak electromagnetic interference. We may then write down an approximate formula for the asymmetry to investigate its dependence on the lepton coupling constants, the mass of the Z^o, the center of mass energy and the polar angle.

$$A_{\mu\mu} \simeq 7\cdot10^{-4}s\,g_A^2\left(\frac{M_Z^2}{s-M_Z^2}\right)\frac{\cos\theta}{1+\cos^2\theta}$$

This formula is graphed in Fig. 30.

The asymmetry in this approximation depends only on the axial vector coupling of the electron and muon. It grows as the center of mass energy squared for $s << M_Z^2$ and is negative in this region. As s approaches M_Z^2, the purely weak terms of course must also be included. Since the effect increases linearly with s, the square of the center of mass energy, measurements at the highest center of mass energy become most valuable. For example, we can see from the figure, that at 40 GeV, we get nearly twice the asymmetry as at 30 GeV. Finally, from the angular dependence shown in the figure, we should note that the maximum effect is at small angle to the beam, thus it is important to measure over a large acceptance to get the most precise measurement of weak effects.

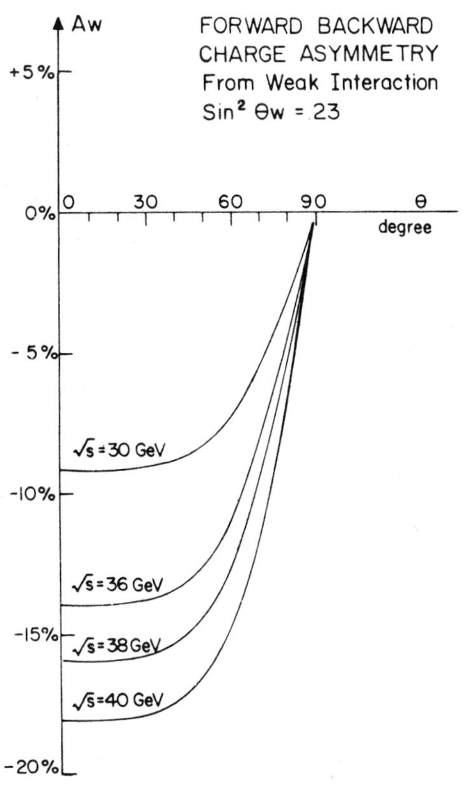

Fig. 30 Expected asymmetry vs. θ.

Of course to do the calculation properly, we must look at order α^3 effects from QED. These are included in the Monte Carlo generator and the QED asymmetry due to them is shown in Fig. 31. This asymmetry is somewhat smaller than that we expect from weak effects, but is of opposite sign. It increases as $\theta \to 0$ but the experiments do not measure for $|\cos\theta| > 0.8$ so that the maximum effect is around 3% and the angular averaged effect is about 1.5%. This indicates that higher order QED effects may be neglected. The data from each experiment are corrected for this pure QED radiative effect. In addition to the pure QED correction to the expected asymmetry, there is a radiative correction to diagrams containing a Z^o which will of course depend on the particular weak interaction model. One obvious effect is a reduction of the magnitude of an expected asymmetry due to the effective lowering of s by hard initital state photon emission. We use a Monte Carlo[(28)] to calculate the expected asymmetry and find that it is reduced by 0.8% ± 0.2% from the simple calculation at $\sqrt{s}$ = 34.6 GeV. Since this is model dependent, we do not correct the data for this but rather quote a corrected value of the expected asymmetry from the GWS model.

Experimentally, an asymmetry of this sort is a particularly easy thing to measure accurately. To make a systematic error, the

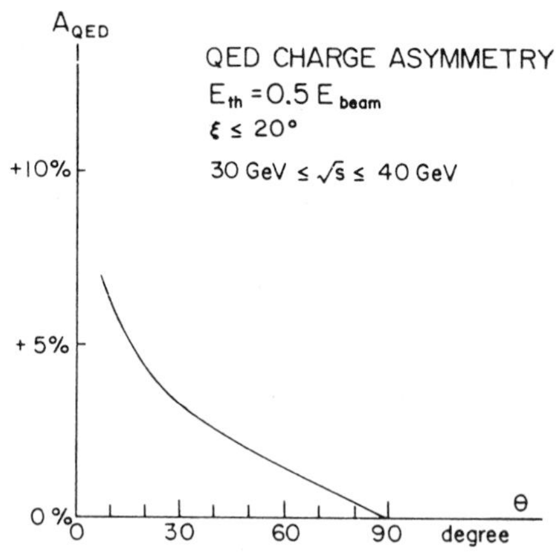

Fig. 31 Asymmetry due to QED vs. θ.

detector must have an unexpected difference in acceptance between high momentum μ^+'s and μ^-'s. Since the only differences must be die to the small muon track curvature in the magnetic field, one would expect these effects to be small. No absolute knowledge of acceptance or luminosity is needed.

Significant measurements of the charge asymmetry have been presented by four experiments at PETRA. All of these groups estimate that their systematic error in determining the charge asymmetry is very small, that is, approximately 1%. The experiments have made studies of their systematic errors on the charge asymmetry. Each experiment finds that the probability of double charge confusion is very small by comparing the number of μ pairs measured to have like charge to the number measured to have opposite charge. As an example, the data from MARK-J are shown in Fig. 32. Here the signed momentum of one muon is plotted vs. that for the other. One can see directly that the probability for charge confusion is small. Typically, the fraction of like charge is about 1% implying a very small amount of double charge confusion.

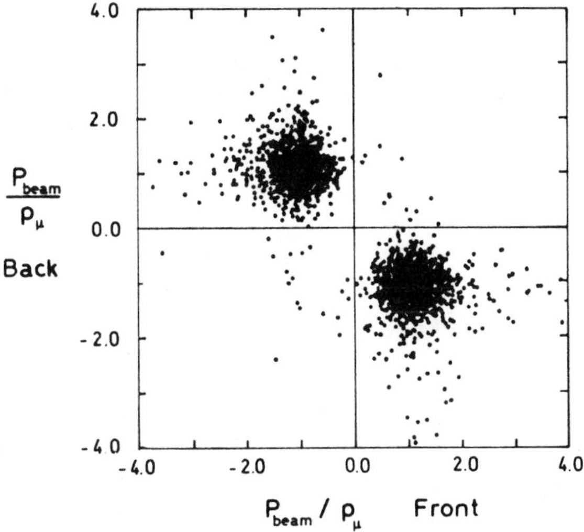

Fig. 32 Charge momentum for forward muon vs. charge momentum for backward muon.

Asymmetric backgrounds, such as cosmic rays of e^+e^- production, have been shown to be very small.

To determine whether the detector has a charge asymmetric acceptance, TASSO has studied the charge asymmetry in hadron production and found it to be very small. In order to more closely duplicate the topology and curvature of the μ pairs, JADE and MARK-J have measured the detector asymmetry using high momentum cosmic rays and each has found it to be $\lesssim 1\%$. Fig. 33 shows the MARK-J measurement of the detector asymmetry binned in cosθ. This was determined by measuring the cosmic ray asymmetry on each magnet polarity. One half the difference is the detector systematic error. MARK-J finds after integrating over mu-pair production distributions:

$$A_{DETECTOR} = \frac{1}{2}\,(A_{POL+} - A_{POL-}) \;=\; 0.6 \pm 0.9\%.$$

This proves that the inherrent detector asymmetry is less than a few percent.

In addition, MARK-J has the ability to change the polarity of its magnetic field at regular intervals during data taking without affecting the beams. A positive muon, bending under one polarity

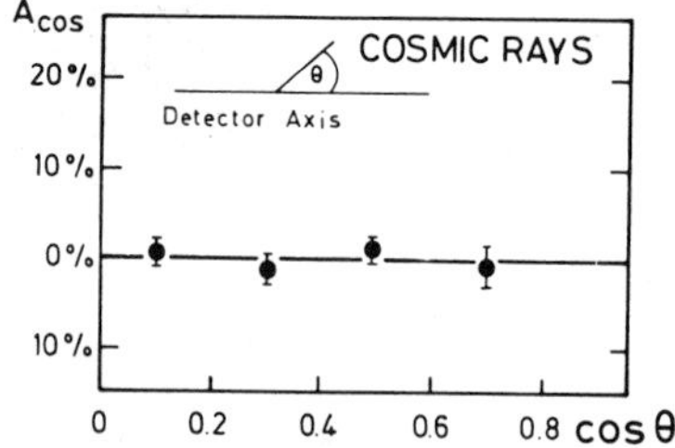

Fig. 33 Asymmetry measured for high momentum cosmic rays vs. cos(θ).

magnetic field, is identical in the detector to a negative muon bending under the opposite polarity. Therefore, by taking equal amounts of data at each polarity, MARK-J can cancel any detector asymmetry. This is particularly true since the polarity is changed often to assure cancellation of any time dependent effects. Therefore this maximum, of a few percent, detector asymmetry will be cancelled yielding an overall systematic error on the measurement of the charge asymmetry which is certainly less than 1%.

The measured asymmetries are given in Table II. A fit is made to the values of F_1 and F_2. The asymmetry extrapolated to 4π solid angle coverage is $3F_2/8F_1$. This number is quoted to facilitate averaging of experiments and as a number which can be compared to new theories, since it does not depend on the details of the detector.

The average value is of particular interest. Since the systematic errors are small, compared even to the combined statistical error, averaging greatly improves the precision of the measurement. A systematic error of 1% on each measurement has been included in calculating the overall error. The four data points give a χ^2 of 0.4 for three degrees of freedom.

The result, as shown in the table, is A = -10.4%±1.4%

TABLE II

MEASURED VALUES OF ASYMMETRIES

GROUP	$A_{\mu\mu}$	$A_{\tau\tau}$
CELLO	- 6.4 ± 6.4	-10.3 ± 5.2
JADE	-10.8 ± 2.2	7.9 ± 3.9
MARK-J	-10.4 ± 2.1	9.5 ± 5.9
TASSO	-10.4 ± 2.3	5.6 ± 4.4
COMBINED	-10.4 ± 1.4	-8.1 ± 2.4
EXPECTED	- 8.6 ± 0.2	-8.2 ± 0.2

expecting -8.6% in the standard model. Thus weak effects have been observed at PETRA in this seven standard deviation effect. This is also the highest q^2 at which weak effects have been measured, the average q^2 being around 1200 GeV^2.

From the measurement, we can determine the axial vector coupling constant for the charged leptons assuming only universality and that we are far from the Z^o pole(s):

$$g_{A_\mu} g_{A_e} = 0.30 \pm 0.04$$

At large enough q^2, the asymmetry depends upon the Z^o mass due to the propogator term $(M_Z^2/(s-M_Z^2))$.At the present value of q^2, this contributes 1.5% to the asymmetry if M_Z= 90 GeV. We can therefore place limits on the Z^o mass which do not depend on the GWS model. The only assumptions needed are that there is a single Z^o and that $g_A = \pm 0.5$ for electrons and muons as is expected in any model with the now well established SU(2) symmetry. This is a very important measurement since we have the opportunity to see the effects of bosons in the weak interaction for the first time by limiting the Z^o mass to be less than infinity. An infinite mass Z^o corresponds to the theory in which four fermions interact at a point as in the original theory of Fermi. Taking the data at face value we find

$$M_Z = 63 \, {}^{+17}_{-\ 7} \text{ GeV},$$

or at the 95% confidence level

$$52 \text{ GeV} < M_Z < 162 \text{ GeV}$$

A word of caution is in order however. The 95% confidence level corresponds roughly to only two standard deviations. The two standard deviation effect we see comes from the average of four experiments so that no single group is responsible for it. A four fermion interaction is not ruled out at the three standard deviation level because the effect is a highly nonlinear function of M_Z. I would therefore like to see this measurement improved since the result is of great importance.

This improvement should come in the near future as the PETRA energy is increased, first by four GeV this October, then by four GeV again next year. Since the difference between GWS and $M_Z = \propto$ is proportional to the center of mass energy to the fourth power, our separation will greatly improve. To see the size of the effect we can look at Fig. 34 where it is shown how the asymmetry depends on the Z^o mass and on s. The asymmetry measurements made to data are also shown. The most significant points are those from PETRA at $s = 1200\ \text{GeV}^2$. One can see that, with the present level of accuracy, the difference between the standard model and $M_Z = \infty$ is only about one standard deviation when the high energy PETRA data are combined. To compare to the graph, we will be getting data points at $s = 1600\ \text{GeV}^2$ this year and $s = 1900\ \text{GeV}^2$ next year.

The tau pair charge asymmetry is also given in Table II. Here none of the groups see a clear weak effect but the average of four groups shows an effect at greater than the three standard deviation level. Therefore, we do observe weak effect in tau pair production

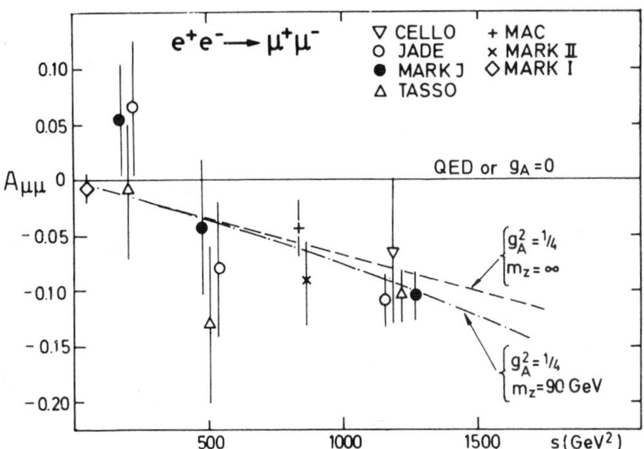

Fig. 34 The expected asymmetry is plotted cs. s for various values of Z^o mass and compared to data points from all energies.

and the data are in quite good agreement with what we expect if the tau axial vector coupling is the same as that of the muon and electron.

4.4 Interpretation of Leptonic Data

The simplest interpretation of the data is a measurement of $\sin^2\theta_W$ in the standard model. Table III give the results from the five PETRA experiments for $\sin^2\theta_W$ determined from purely leptonic processes.

The value of 0.25 for $\sin^2\theta_W$ is a very likely one because this type of experiment is not sensitive to the signs of g_A and g_V but only to g_A^2 and to g_V^2. If the term proportional to g_V^2 is measured to be less than zero, then the best that can be done within the standard model is to make $g_V^2 = 0$, in which case $\sin^2\theta_W = 0.25$. On the other hand, if g_V^2 is measured to be greater than zero, this can be produced by two values of $\sin^2\theta_W$ symmetric about 0.25. This means that the central value is likely to be near to 0.25 unless g_V^2 is actually limited to be greater than zero in which case the two solutions separate. Since g_V is indeed expected to be very small compared to the accuracy with which it is measured here, the experiments have a very good chance of measuring $\sin^2\theta_W = 0.25$. Note, however, that because M_Z^2 depends on $\sin^2\theta_W$ in the standard model, values other than $\sin^2\theta_W = 0.25$ can give the best fit, due primarily to propagator effects in the asymmstry, even if g_V^2 comes out less than 0.

TABLE III

$\sin^2\theta_W$ FROM PURELY LEPTONIC REACTIONS

CELLO	0.21 ± 0.12
JADE	0.25 ± 0.15
MARK-J	0.26 ± 0.09
PLUTO	0.23 ± 0.17
TASSO	0.27 ± 0.07

It should be pointed out again that no disagreement with the standard model has been found. So our result, in terms of the standard model, is that $\sin^2\theta_W \simeq 0.25 \pm 0.07$.

Outside the standard model, we may determine g_A and g_V for charged leptons. This will also show quantitatively whether the data are in agreement with the standard model. Fig. 35 shows the measured value of g_A^2 versus the measured value of g_V^2 for five PETRA experiments with error bars denoting the one sigma error. The points cluster together quite well in a region near $g_A^2 \approx 0.0$ and $g_A^2 \approx 0.3$. For $\sin^2\theta_W = 0.25$, g_V^2 is zero. The model predicts $g_A^2 = 0.25$ independent of $\sin^2\theta_W$. Most of the data points are within one sigma of the standard model with $\sin^2\theta_W = 0.23$ as measured in neutrino scattering from hadrons.

Alternatively, we may compare and combine our results with those of another purely leptonic process, neutrino electron scattering. In Fig. 36, the results of determination of g_A and g_V for the charged leptons are shown for ν-e scattering, and for the

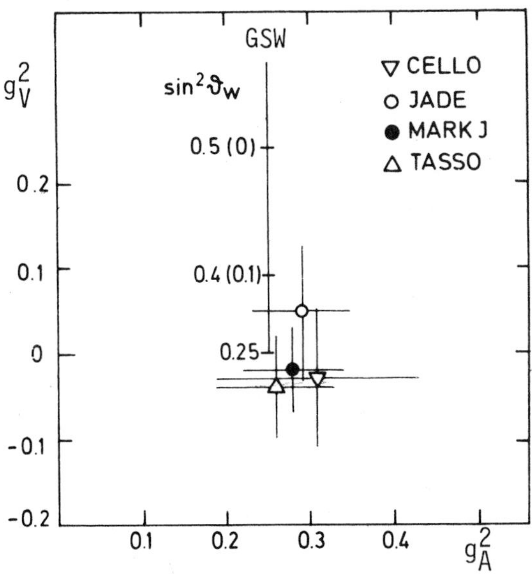

Fig. 35 Measured values g_A^2 and g_V^2.

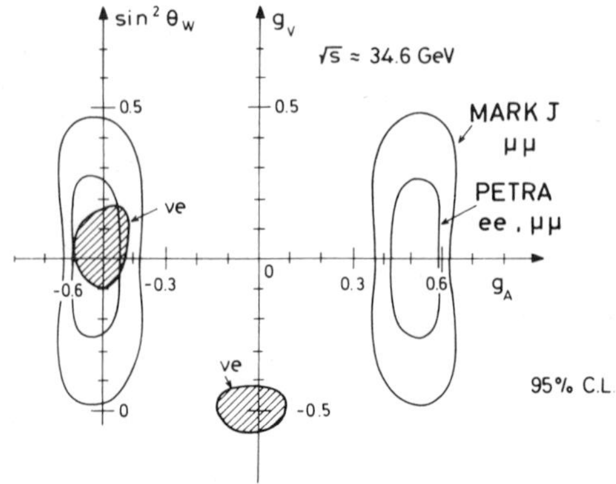

Fig. 36 Allowed regions in the $g_A - g_V$ plane from three neutrino electron scattering experiments and from PETRA.

combined ee and μμ data from PETRA. The ν-e scattering 95% confidence level limits are the two shaded regions combining three different kinds of neutrino experiments. We can discriminate between the two regions by using the data from PETRA. The two unshaded areas are the 95% confidence level contours from the e^+e^- data. These areas are symmetric about the lines $g_A = 0$ because only g_A^2 and g_V^2 have any significant effect on the e^+e^- data. These data nicely eliminate one of the two solutions, leaving only the one near $g_A = 0.5$ and $g_V = 0.0$. So combining all of the data, one can limit g_A and g_V to be a small region which again is quite consistent with the standard model shown by the line. This is true even if factorization is not assumed. These data, however, are from different q^2 regions, the ν-e data coming from very low q^2 and the e^+e^- data from $q^2 \sim 0.15\ M_Z^2$.

In the introduction, I defined the parameter C, which is the coefficient of a term in the weak neutral current Hamiltonian proportional to the elecfromagnetic current squared. This kind of term is expected in multiboson weak models and also in more radical models where there is a continuum of weak interactions perhaps due to a constituent nature of the weak bosons. With multiple bosons, the model can no longer be parameterized in terms of g_A and g_V only. C is expected to be greater or equal to zero in all models.

Limits on C, which are applicable to any model of the weak neutral current, have been obtained by all of the PETRA experiments and are shown in Table IV. All the results are consistent with C = 0.

If the take the tightest of these limits, we may put a limit on the normalized difference between the actual theory and the standard model, which is equal to 16 C. More specifically

$$\frac{\left[\int ds \frac{1}{s}\cdot\sigma(e^+e^- \to ALL)\right]_{ACTUAL} - \left[\int ds \frac{1}{s}\cdot\sigma(e^+e^- \to ALL)\right]_{GWS}}{\left[\int ds \frac{1}{s}\cdot\sigma(e^+e^- \to ALL)\right]_{GWS}} < 0.29$$

at the 95% confidence level. Therefore, in terms of this weighted integral over all energy, we have determined that the actual theory of weak interactions is within 30% of the standard model.

TABLE IV

95% CONFIDENCE UPPER LIMITS ON C

CELLO	0.031
JADE	0.039
MARK-J	0.021
PLUTO	0.060
TASSO	0.018

4.5 Electroweak Reactions of Quarks

The data in e^+e^- on production of hadrons can also be used to test the electroweak interaction. In particular, the total hadronic cross section is sensitive to weak effects. The ratio of the total hadronic cross section, corrected for QED radiative effects, to the lowest order QED cross section for the production of muon pairs is known as R. R is just the sum over flavors and colors of R for each quark species.

$$R = \sum_{\substack{\text{flavors} \\ \text{colors}}} R_q$$

To lowest order, R_q is[2]

$$R_q = Q_q^2 - 8sgQ_q g_{V_e} g_{V_q} \left(\frac{M_Z^2}{s-M_Z^2} \right) +$$

$$16s^2g^2(g_{V_e}^2 + g_{A_q}^2) \cdot (g_{V_q}^2 + g_{A_q}^2) \left(\frac{M_Z^2}{s-M_Z^2} \right)^2$$

where $g_{A_q} = \pm 0.5$ and g_{V_q} is, in the standard model, approximately 0.19 for a charge 2/3 quark and -0.35 for charge 1/3. This is in contrast to the charged leptons, where $g_V \tilde{\sim} -0.08$. For $\sin^2\theta_W$ = 0.23 the weak correction for R is only about 1%. Other values of $\sin^2\theta_W$ will, however, give measureably large changes in R. Of course, there is also a well known QCD correction[1] to R as described in section 2.1. This correction is taken account of in the analysis but the results are insensitive to the value of α_s. Table I in section 2.1 lists the values of R measured by 5 experiments from PETRA and PEP.

In Fig. 37, measurements of R from three PETRA experiments, at a wide range of center of mass energies, are compared to the

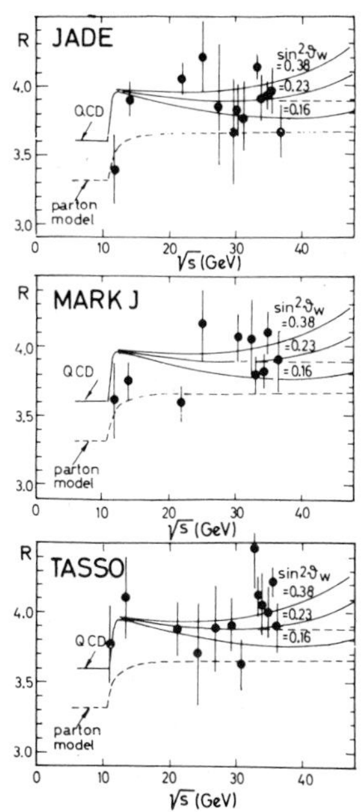

Fig. 37 R vs. $\sqrt{s}$ from three experiments compared to predictions for various values of $\sin^2\theta_W$.

predictions for different values of $\sin^2\theta_W$. Even though the systematic error on the measurement of R is large, the point to point error is estimated to be quite small, so that by measuring the energy dependence of R, we can make a good determination of weak effects[29] to R. PETRA has recently run at center of mass energies of 14 and 22 GeV, yielding high statistics measurements of R at those energies. From these data, together with the high energy points, JADE gets a value of

$$\sin^2\theta_W = 0.25 \pm 0.05.$$

This is now a very accurate determination of $\sin^2\theta_W$ at high q^2, $q^2 \sim 1300\ \text{GeV}^2$. Using a similar analysis, MARK-J has obtained a value from their combined data of $\sin^2\theta_W = 0.30 \pm 0.06$.

4.6 B Meson Lifetime Limit

By putting an upper limit on the B particle lifetime, measured in events containing muons, JADE has deduced[30] a limit on weak charged current mixing angles. These angles are analogous to the Cabibbo angle but, in a model with six quark flavors, there would be three free angles and one free phase factor. The definitions of the angles are not unique. In the formalism used by JADE, decays of a b quark to a u quark are proportional to $\sin(\beta)$ and decays to a c quark are proportional to $\sin(\gamma)\cos(\beta)$. The lifetime of the particle is then influenced by these angles, being longer if both $\sin(\gamma)$ and $\sin(\beta)$ are small. Fig. 38 shows their bounds on the angles derived from their determination that the lifetime is less than 1.4×10^{-12} sec. The upper limit on the angles comes from high accuracy measurements of the Cabibbo angle in different processes. The region above the dashed line is eliminated by the results from CESR which indicate that b quarks decay mainly into c quarks. Since the figure is drawn on a log scale, it should be noted that there is still substantial freedom for these angles.

4.7 Production of Leptons in Hadronic Events

It has been found that approximately 5% of hadronic events contain identifiable prompt muons and a similar 5% contain

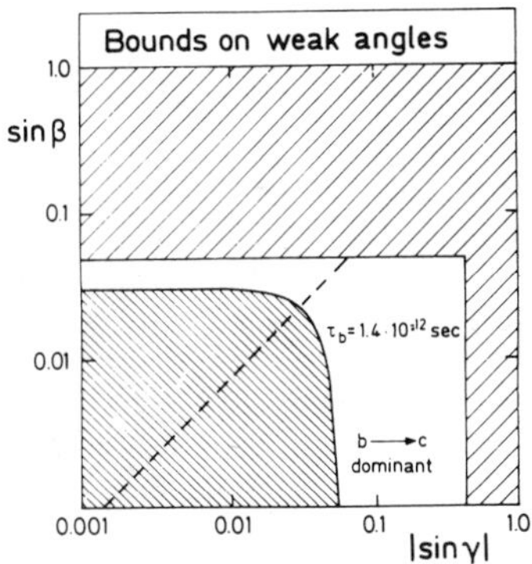

Fig. 38 Bounds placed on Miani angles by JADE's limit on the B particle lifetime.

electrons. The source of most of these leptons appears to be conventional, that is, they come from the semileptonic decay of mesons containing b or c quarks. Fig. 39 compares several jet properties of hadronic events containing muons to those for all hadronic events. No significant difference is seen. This is consistent with all of these leptons coming either from heavy meson decay or from standard background processes where hadrons or their decay products are misidentified as prompt leptons.

The leptons in hadronic events can give us much information about the electroweak production of heavy quarks and about their weak decays. An experimental program to understand these leptons is under way at PETRA. The primary goal of this is to measure the forward-backward charge asymmetry in the production of heavy quark

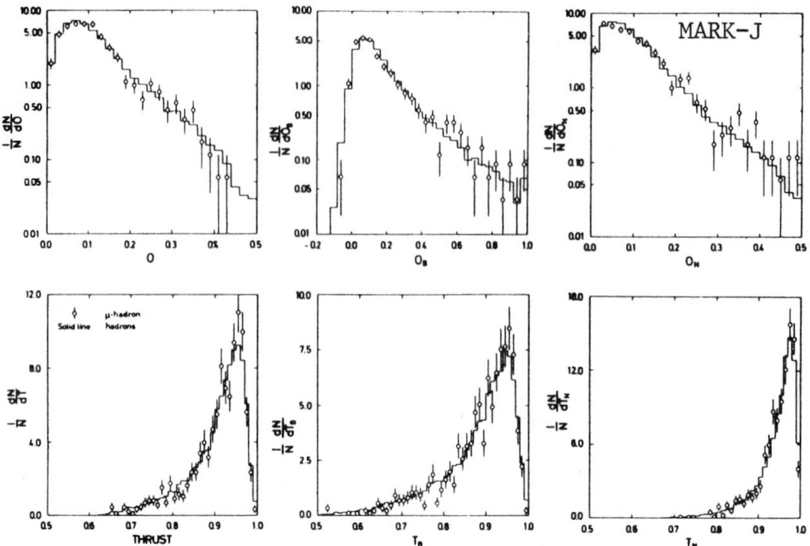

Fig. 39 Several properties of hadron events containing muons (points) compared to all hadron events (line).

pairs. Since this asymmetry is inversely proportional to the quark charge, the asymmetry expected for charmed quark production is 50% larger than for muon pairs and the asymmetry for bottom is a factor of three larger. The asymmetry in the production of heavy quarks is transmitted into the charge asymmetry in their decay muons.

Of course the leptons from heavy quark decay may allow us to measure several other quantities of interest, such as the charm and bottom fragmentation functions and the semileptonic decay branching ratios for charm and for bottom.

To make a partial separation of the muons from bottom decay, we look at the transverse momentum of the muons with respect to the jet axis of the jet containing the muon. Since transverse momentum due to fragmentation is small, the muon transverse momentum will largely come from the kick received in the decay of the massive parent particle. For a B particle decay, the average transverse momentum is considerable, around 1 GeV. In Fig. 40 the

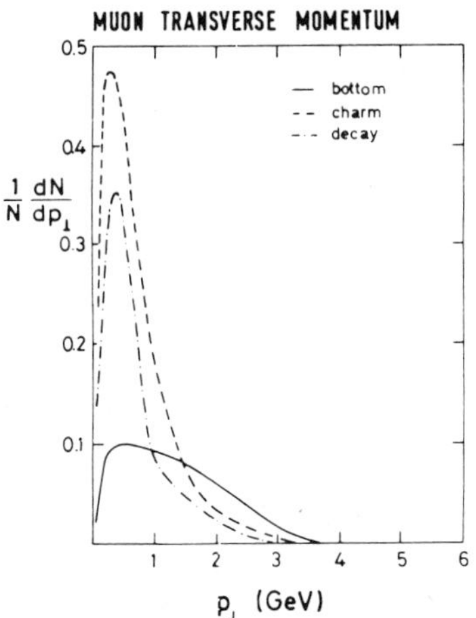

Fig. 40 Transverse momentum distribution from muons in hadron events according to Monte Carlo from charm decay, bottom decay and background.

Monte Carlo P_t distribution of muons from three sources is shown. Although the B decay events are only a small part of the inclusive muon event sample, when a cut of $P_t > 1.2$ GeV is applied, we find that we can enrich the fraction from bottom to 45% of the total. This means we can use the rate of events with P_t greater than 1.2 GeV to measure the bottom semileptonic branching ratio. In Fig. 41 the measured P_t distribution is compared to what we expect according to the Monte Carlo assuming a $B \rightarrow \mu + X$ branching ratio of 8%. The agreement over the entire distribution is quite good.

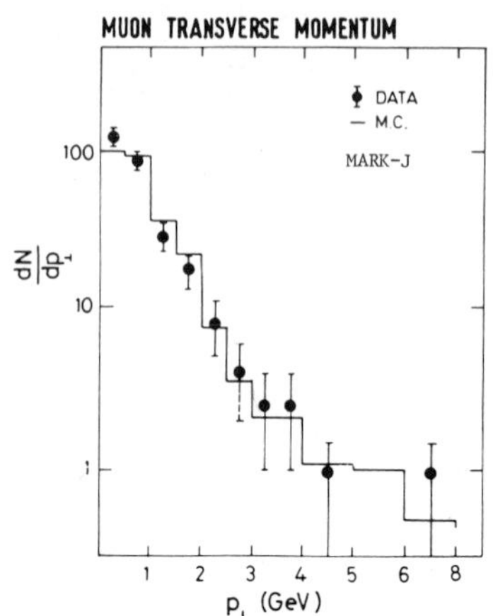

Fig. 41 Transverse momentum distribution from muons in hadron events. Monte Carlo expectation is shown as the solid line.

Using the rate for $P_t > 1.2$ GeV, we find

MARK-J

Br $(B \rightarrow \mu + X)$	= 9.3%	± 2.9%	statistical
		± 2.0%	systematic
Br $(C \rightarrow \mu + X)$	= 9.8%	± 1.1%	statistical
		± 2.0%	systematic

TASSO

Br $(B \rightarrow \mu + X)$	= 15.0%	± 3.5%	statistical
		± 3.5%	systematic
Br $(B \rightarrow e + X)$	= 13.6%	± 4.9%	statistical
		± 4.0%	systematic

MARK-II

Br (B → e + X) = 11.0% ± 3.0% statistical
± 2.0% systematic

Br (C → e + X) = 7.0% ± 2.0% statistical
± 2 0% systematic

where this branching ratio refers only to the primary decay of a B particle to a muon, not to muons from the cascade decay through charm.

With this separation between b and c quark, we can try to measure the heavy quark production asymmetry. This is hampered by the fact that the muons from c decay are expected to have an asymmetry of opposite sign to those from b decay. Thus, unless the separation can be made very clearly, the large expected asymmetry will be watered down by background. The situation for the b asymmetry is even slightly worse than that because the muons from the cascade decay $b \to c \to \mu$ also tend to cancel the asymmetry in $b \to \mu$. Present results on asymmetry measurements from MARK-J are shown in Table V.

It can be seen that better flavor identification methods must be found in order to accurately measure the heavy quark couplings. One way to do this may be through use of a high precision vertex chamber to identify charm decay vertices. Another method to measure the asymmetry in c quark production has been employed by TASSO[31]. They reconstruct 47 D^* mesons, expecting 11 events background, by requiring a good fit to the D^* mass in a $K\pi\pi$ channel and a good fit to the difference between that D^* mass and a D meson

TABLE V

MEASURED ASYMMETRIES IN B AND C SAMPLES

q	A measured	A expected	A original
b	−29 ± 10%	−6%	−23%
c	+ 7 ± 5%	+4%	+13%

mass from the kaon and one of the π's. Since it is believed that by far the dominant source of D mesons will be from pair production of charmed quarks, we may look at the D^* charge asymmetry to get the c quark asymmetry. Their result for the charm quark asymmetry is -35%±14% expecting -11% in the standard model. The statistical error bar is still quite large but this may improve in the future if more D decay channels can be used.

Finally, hadronic events with oppositely charged lepton pairs in the same jet may be a signal for flavor changing neutral currents in b quark decay. In models with no top quark, the b quark is in an SU(2) singlet and can only decay by a flavor changing neutral current. In this case the branching ratio for $b \to \mu\mu + X$ would be about 1%. MARK-J has searched for opposite sign dimuons with opening angles less than 15° in a sample of 28,400 hadron events. They find one candidate expecting one from background sources. Their acceptance for the flavor changing neutral current process is computed to be 7.1%, giving then a limit

$$Br(b \to \mu\mu + X) < 0.6\% \qquad 90\% \text{ C.L.}$$

A study from CLEO concluded that

$$Br(b \to \mu\mu + X) < 0.9\% \qquad 90\% \text{ C.L.}$$

Thus this kind of topless model is very unlikely.

4.8 Search for Structure in the Fermions

The data presented previously, on production of charged leptons, can be used to search for structure in the fermions by looking for a q^2 dependence to the cross sections that is not expected. In this measurement, we assume that the standard weak model is correct or at least that weak effects are nearly as small as in the standard model. Here, the measurements of R can also be used to look for structure in the quarks. A breakdown of the pointlike behavior of any of the fermions is parameterized in terms of a form factor with one parameter, Λ[32].

$$F_{\pm} = 1 \mp \frac{q^2}{q^2 - \Lambda_{\pm}^2}$$

The + and - refer to two different form factors, one of which increases the cross sections and one of which decreases them. Table VI lists the 95% confidence level lower limits for the Λ parameters. None of the fermions show evidence of structure up to energy scales of around 150 GeV.

4.9 Search for Symmetry Breaking Scalars

Of course an imporant aspect of the present weak interaction theory is spontaneous symmetry breaking by the Higgs mechanism. The standard model predicts that there should be an observable neutral scalar particle. As yet, no such particle has been found nor have any stringent limits been placed on its mass. This may

TABLE VI

95% C.L. LOWER LIMITS OF $\Lambda_{\pm}$

	e		μ		τ		q	
	Λ_+	Λ_-	Λ_+	Λ_-	Λ_+	Λ_-	Λ_+	Λ_-
CELLO	83	155	186	101	139	120		
JADE	112	106	142	126	111	93		
MARK-J	128	161	194	153	126	116	190	285
PLUTO	80	234	107	101	79	63		
TASSO	140	296	136	281	124	104		186

become possible if the top threshold is reached and the Higgs particle can be searched for in production by the Wilczek mechanism[33].

The method of symmetry breaking in elementary particle physics remains a mystery. Symmetry breaking is thought to be responsible for the generation of all fundamental particle maases, the weak mixing angles among quark species, CP violation and perhaps for the large parity violation seen in weak interactions. Among proposed forms of symmetry breaking, one prediction is universal, that is, the existence of new scalar particles, be they Higgs particles, technipions or supersymmetric scalar partners of fermions.

In the standard model, Higgs couplings to each fermion are independently free parameters which must be set to generate the fermion masses. More ambitious theories have been put forward to explain the spontaneous symmetry breaking of SU(2) x U(1) and the fermion mass generation. In grand unified theories, symmetry breaking is performed by a large number of Higgs particles some of which are charged. Although many of the Higgs particles are superheavy, some may have masses on the order of 10 GeV. Technicolor [34],[35] models, in which the symmetry breaking scalars are composed of constituents confined by a new strong force (technicolor), predict the existence of reasonably light charged particles called technipions. Since these are charged, they will be produced in e^+e^- annihilations through one photon. This implies $\Delta R = 1/4$ when well above threshold. In fact there are some specific technicolor models[36] which predict that the technipion mass lies between 5 and 14 GeV, which is the range that we can most easily explore at PETRA.

Fig. 42 shows a search, by the MARK-J group, for technipion or charged Higgs production. Since both of these particles should be produced at the same rate and both decay primarily into the

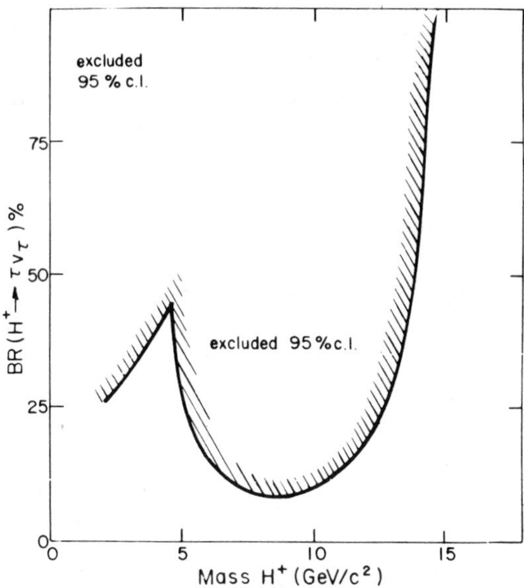

Fig. 42 Limit on mass and branching ratio into $\tau\nu$ for technipion or charged Higgs from MARK-J.

heaviest fermion antifermion pairs allowed by energy conservation, they may be studied simultaneously. In the mass range explored, the decay modes will be mainly $\bar{\tau}\nu$ and $c\bar{s}$. The branching ratio between these two modes depends on the specific theory. We have therefore left it, along with the technipion or Higgs mass, as a free parameter. Decays to a $c\bar{b}$ pair are found to give tighter limits than the $c\bar{s}$ decay. In the figure, the regions in the plane defined by the charged scalar mass and its branching ratio into the $\bar{\tau}\nu$ mode which are excluded at the 95% confidence level are displayed. Except for very small branching fractions to $\bar{\tau}\nu_\tau$, these scalars are excluded in the 4 to 15 GeV mass range.

TASSO[31] has looked particularly at the region where a technipion or charged Higgs particle decays only hadronically. By searching for 4 jet event shapes that could result from these hadronic decays, they have ruled out the remaining region where the branching ratio to $\tau\nu$ is small. This result is summarized in Fig. 43.

In supersymmetric grand unified theories, there is a symmetry between bosons and fermions that requires that each fermion have two scalar partners that are symmetrically associated with it. In our search for these scalars, we assume[37] that they decay rapidly into their fermion partner plus unobserved photinos or goldstinos. Mass limits[38] for scalar partners of the electron, muon and tau have been obtained by PETRA experiments and are shown in Table VII.

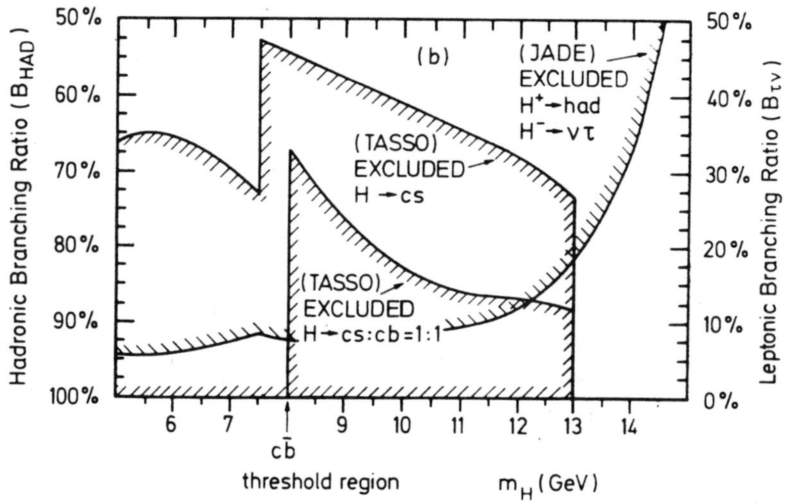

Fig. 43 Limit on mass and branching ratio into $\tau\nu$ for technipion or charged Higgs from combined TASSO and JADE results.

TABLE VII

LIMITS ON SUPERSYMMETRIC PARTNERS OF LEPTONS

s_e	m > 16 GeV
s_μ	m > 16 GeV
s_τ	m > 14 GeV

REFERENCES

(1) K.G. Chetyrkin et al., Phys. Lett. 85B (1979) 277.
M. Dine and J Sapierstein, Phys. Lett. 42 (1979) 668.
W. Celmaster and R. Gonsalves, Phys. Lett. 44 (1980) 560.

(2) R. Budny, Phys. Lett. 55B (1975) 227.
J. Ellis and M.K. Gaillard, "Physics with Very High Energy e^+e^- Colliding Beams", CERN 76-18 (1976) 21.

(3) D.P. Barber et al., Phys. Rev. Lett 43, (1979) 830.
R. Brandelik et al., Phys. Lett. 86B (1979) 243.
C. Berger et al., Phys. Lett. 86B (1979) 418.
W. Bartel et al., Phys. Lett. 91B (1980) 142.
P. Söding, DESY-Report 81-070 (1981).

(4) R. Brandelik et al., Phys. Lett. 92B (1980) 453.

(5) J. Ellis and I. Karliner, Nucl. Phys. B148 (1979) 141.

(6) P. Hoyer et al., Nucl. Phys. B161 (1979) 349.

(7) A. Ali et al., Phys. Lett. 93B (1980) 155 and Nucl. Phys. B168 (1980) 490.

(8) R.K. Ellis, D.A. Ross and A.G. Terrano, Phys. Rev. Lett. 45, (1980) 1226, and Nucl. Phys. B178 (1981) 421.
K. Fabricius, I. Schmitt, G. Schierholz and G. Kramer, Phys. Lett. 97B (1980) 431.
J.A.M. Vermaseren, K.J.F. Gaemers and S.J. Oldham, Nucl. Phys. B187 (1981) 301.

(9) W. Bartel et al., DESY-Report 82-060 (1982).

(10) A. Ali and F. Barreiro, DESY-Report 82-033 (1982).

(11) B. Andersson, G. Gustafson, and T. Sjostrand, Zeitschr. f. Physik C6 (1980) 235, Nucl. Phys. B197 (1982) 45.

(12) R.D. Feynman and R.P. Field, Nucl. Phys. B136 (1978) 1.

(13) H.J. Behrend et al., DESY-Report 82-061 (1982).

(14) W. Bartel et al., Phys. Lett. B101 (1981) 129.

(15) G. Heinzelmann, XXI International Conference on High Energy Physics, Paris (1982).

(16) Z. Koba, H.B. Nielsen, P. Olesen, Nucl. Phys. B40 (1972) 317.

(17) S.L. Glashow, NP 22 (1961) 579; Rev. Mod. Phys. 52 (1980) 539.
A. Salam, Phys. Rev. 127 (1962) 331; Rev. Mod. Phys. 52 (1980) 525.
S. Weinberg, Phys. Rev. Lett. 19 (1967) 1264; Rev. Mod. Phys. 52 (1980) 515.

(18) F.J. Hasert et al., Phys. Lett. 42B (1973) 121.

(19) K. Winter, Proc. of the 1979 International Symposium on Lepton and Photon Interactions, Fermilab (1979) p. 258;
P. Langacker et al., Proc. of Neutrino 79, Bergen (1979)p.276.

(20) H. Faissner, New Phenomena in Lepton-Hadron Physics (1979), ed. D.E. Fries and J. Wess (Plenum Publishing Corp., New York) p. 371.
F.W. Bullock, Proc. of Neutrino 79, Bergen 1979, p. 398.
R.H. Heisterberg et al., Phys. Rev. Lett. 44 (1980) 635.
L.W. Mo, Constribution to Neutrino 80, Erice (1980).
M. Roos and I. Liede, Phys. Lett. 82B (1979) 89, and references therein.
F.W. Büsser, Proc. of Neutrino 81, Wailea, Hawaii (1981).

(21) C.Y. Prescott et al., Phys. Lett. 77B (1978) 347.

(22) J.D. Bjorken, Unififation of Elementary Forces and Gravitation (Hartwood Academic Publishers, London 1978) p. 701; Proc. of the 13th Rencontre de Moriond, ed. Tran Thanh Van, p. 191; Phys. Rev. D19 (1979) 335; compare also the recent review, FERMILAB-Conf-80/86 THY (1980).
P.Q. Hung and J.J. Sakurai, Nucl. Phys. B143 (1978) 81.

(23) H. Georgi and S. Weinberg, Phys. Rev. D17 (1978) 275.
V. Barger, W.Y. Keung, and E. Ma, Phys. Rev. Lett. 44 (1980) 1169.
E.D. deGroot, D. Schildknecht and G.J. Gounaries, Phys. Lett. 90B (1980) 427.
E.H. deGroot and D. Schildknecht, Phys. Lett. 95B (1980) 149.

(24) G.J. Gounaris and D. Schildknecht, BI-TP 81/09 (1981).
D. Schildknecht, BI-TP 81/12 (1981)

(25) CELLO-Collaboration: H.J. Behrendt et al., Phys. Lett.103B (1981) 148; DESY-Report 82-019 (1981); DESY-Report 82-020 (1982).
JADE-Collaboration: W. Bartel et al., Phys. Lett. 88B (1979) 171; Phys. Lett. 92B (1980) 206 and Phys. Lett. 99B (1981) 281; Phys. Lett. 108B (1982) 140.
MARK-J Collboration: D.P. Barber et al., Phys. Rev. Lett. 43 (1979) 1915; Phys. Lett. 95B (1980) 149, Phys. Rev. Lett.

48 (1982) 1701.
PLUTO Collaboration: Ch. Berger et al., Zeitschr. f. Physik C1 (1979) 343; Zeitschr. f. Phys. C7 (1981) 289; Phys. Lett. 99B (1981) 489.
TASSO Collaboration: R. Brandelik et al., Phys. Lett. 92B (1980) 199; Phys. Lett. 94B (1980) 259; Phys. Lett. 110B (1982) 173; DESY-Report 82-032 (1982).

(26) F.A. Berends, K.F.J. Gaemers, and R. Gastmans, NP B63 (1973) 381.
F.A. Berends, K.F.L. Gaemers, and R. Gastmans, NP B68 (1974) 541.
F.A. Berends and R. Kleiss, DESY-Report 80-66 (1980).

(27) F.A. Berends and G.J. Komen, Phys. Lett. 63B (1976) 432.

(28) F. Berends, R. Kleiss and S. Jadach, private communication.

(29) D. P. Barber et al., (MARK-J Collaboration) Phys. Rev. Lett. 46 (1981) 1663.
W. Bartel et al. (JADE-Collaboration), Phys. Lett. 101B (1981) 361.

(30) W. Bartel et al., DESY-Report 82-014 (1982).

(31) D. Lüke, DESY-Report 82-073 (1982).

(32) S.D. Drell, Ann. Phys. 4 (1958) 75.

(33) F. Wilczek, Phys. Rev. Lett. 39 (1977) 1304.

(34) J. Schwinger, Phys. Rev. 125 (1962) 397, 128 (1969) 2425.
R. Jackiw and K. Johnson, Phys. Rev. D8 (1973) 2386.
J.M. Cornwall and R.E. Norton, Phys. Rev. D8 (1973) 3338.
M.A.B Bég and A. Sirlin, Ann. Rev. Nucl. Sci. 24 (1974)379.
S. Weinberg, Phys. Rev. D13 (1976) 974, D19 (1979) 1277.
L. Susskind, Phys. Rev. D20 (1979) 2619.

(35) For review, see K.D. Lane and M.E. Peskin, Rencontre de Moriond Lectures, NORDITA preprint 80/33 (1980).
E. Farhi and L. Susskind, CERN preprint TH 2975 (1980).
P. Sikivie, Rencontre de Moriond Lecture, CERN preprint TH 3083 (1981), and contributions to Proc. Cornell Z^o Theory Workshop, Eds.
M.E. Peskin and S.H.H. Tye, CLNS 81-485 by S.H.H. Tye, p.411; E. Eichten, p. 421; A. Ali, DESY-Report 81-032 (1981).
M.A.B. Bég, Rockefeller University preprint RU81/B/9 presented at the Lisbon International Conf. on High Energy Physics (1981).

(36) S. Dimopoulos, S. Raby, and G.L. Kane, Nucl. Phys. B182 (1981) 77.
S. Chadha and M.E. Peskin, Nucl. Phys. B185 (1981) 61.
S. Chadha and M.E. Peskin, Nucl. Phys. B187 (1981) 541.

(37) P. Fayet, CERN-Report TH-2864 (1980).

(38) J. Bürger, Proc. Int. Symp. Lepton and Photon Interactions at High Energy (Bonn, 1981).
D.P. Barber et al., Phys. Rev. Lett. 45 (1980) 1904.
D.P. Barber et al., MIT/LNS-125 (1982).

DISCUSSIONS

CHAIRMAN: J.G. Branson

Scientific Secretaries: K.-G. Rauh and D. Richards

DISCUSSION 1

- *MUKHI:*

Is the asymmetry in the energy-energy correlation more fragmentation model independent than the Sterman-Weinberg cross section?

- *BRANSON:*

In the case of a pure $q - \bar{q}$ event the partons fragment equally in the θ and $\pi - \theta$ directions, so that fragmentation effects cancel in the asymmetry. There is a contribution from the fragmentation of emitted gluons but this is much smaller. In the case of the energy-energy correlation the second order calculation has been shown to be independent of the Sterman-Weinberg cuts. This has not yet been shown for the Sterman-Weinberg cross section.

- *DUBNICKOVA:*

In which sense is R sensitive to new flavours of quarks?

- *BRANSON:*

As the energy threshold for the production of a new quark flavour is crossed, the value of R increases by $\frac{4}{9}$ or $\frac{1}{9}$ for quarks of charges $\frac{2}{3}$ and $\frac{1}{3}$ respectively.

- *OGAWA:*

Why is the value of R predicted by QCD usually higher than the value predicted by the quark-parton model?

- *BRANSON:*

In QCD the quark parton model prediction is modified by the factor $(1 + \frac{\alpha_s}{\pi} + O(\alpha_s^2))$ which numerically increases R by about 5%. There are further corrections to R from weak interactions which are dependent on the energy and $\sin^2\theta_W$.

- *ODORICO:*

This is a comment about QCD Monte Carlo (MC) calculations. The leading log approximation (LLA) MC, which is able to calculate multijet configurations with no limitations on jet multiplicity, does not have a larger number of free parameters than in $O(\alpha_s)$, $O(\alpha_s^2)$, etc. MCs of Ali et al. and Hoyer et al., but actually much less. In $O(\alpha_s^n)$ MCs one must fix at each energy arbitrary cutoffs on thrust, aplanarity etc. to cure divergences. If one freezes these cutoffs with respect to energy one gets absurd results. For instance the rate of 3-jet and 2-jet events would decrease with energy. Because of that, the multiplicity distribution of the quanta is completely arbitrary in this approach, while the theory predicts it, once the minimum invariant mass of the quanta is fixed. In the LLA MC there is only one free parameter (besides α_s) as far as the QCD calculation is concerned. This is the minimum invariant mass of the quanta, which represents a non-perturbative cutoff independent of energy and to be fixed once for all by a fit to the data. About the accuracy of the LLA calculation, I point out that for 3-jet events, for instance, the LLA matrix element is practically indistinguishable from the exact one.

- *PERNICI:*

How precise is the measurement of α_s obtained by comparing 3-jet and 2-jet events?

- *BRANSON:*

The statistical errors are of the order of 3%. The systematic errors are due to fragmentation, and within the Feynman-Field model these are small. However, incorporating more complicated fragmentation models, for instance an exponential p_T-distribution or a Lund Monte Carlo, leads to an increase in the systematic error.

- *LOEWE:*

Is it possible, by measuring 4-jet events, to distinguish between the prediction of Ellis, Ross, and Terrano and that of the Hamburg group?

- *BRANSON*:

The two calculations differ in the way they combine the divergent part of the 4-jet cross-section with the radiative corrections to the 3-jet cross-section. This corresponds to the case where 4-jet events mimic 3-jet events, and so cannot be resolved experimentally.

- *GRZADKOWSKI*:

When higher order QCD effects are included, the results depend on the renormalization prescription chosen. What choice has been used in the above calculations?

- *BRANSON*:

The calculations are done in the $\overline{MS}$ scheme.

- *TUROK*:

If a collision is simulated such that all outgoing particle momenta are in a purely random direction, what percentage of events would be called "4-jet" events?

- *BRANSON*:

At present it is not possible to distinguish between 4-jet events and phase-space events. Two- and 3-jet events alone are unable to explain the data, and a contribution from 4-jet events or phase-space events is needed.

- *RICHARDS*:

The CELLO collaboration fitted the lowest order QCD calculation to both the energy-energy correlation data and the asymmetry and obtained different values of α_s. However, when the second order QCD calculation is included the values of α_s are in much closer agreement. Is this a general feature of the PETRA data?

- *BRANSON*:

A somewhat higher value of α_s is obtained from the energy-energy correlation data but it does not improve that much when second order effects are included.

- *RAUH:*

Can you give an intuitive reason why QCD effects change the energy flow pattern in the way indicated in the figure you showed?

- *BRANSON:*

The models are completely different but QCD gives a definite 3-jet structure whereas phase-space does not.

- *OGAWA:*

Are you testing QCD with your data, or only fitting parameters into a phenomenological model?

- *BRANSON:*

Firstly we have clear evidence for 3-jet events. These cannot be explained by pure fragmentation, and suggest the presence of a third parton which is naturally identified with the gluon.

QCD has only one parameter, namely α_s, whereas fragmentation models have many parameters. We, therefore, try to look in a region where fragmentation effects are minimal. Here QCD fits the data well, and whereas minor modifications to the theory cannot be ruled out, the presence of a scalar gluon is certainly excluded.

DISCUSSION 2

- *PERNICI:*

Is it possible to determine the flavour of the quark producing a jet and if so how?

- *BRANSON:*

It looks hopeful that we may be able to determine the flavour of the quarks producing some jets. For example, for heavy quarks we can look for high p_T -decays when we may be able to determine that a jet has come from a bottom quark, and generally leptons in the final state indicate the presence of heavy quarks. By reconstructing D*'s we can also identify charm jets. We do see that the charge of the leading particles gives a good measure of the sign of the quark charge. However, no study has yet been done on flavour correlation in jets, though it is reasonable that if the leading particle in a jet remembers the charge of the fragmented quark then it may well remember the flavour.

- *TUROK:*

In your figure of the weighted jet charge product $P'(\gamma)$ the data points were again rising as γ became negative. This suggests that the slow particles in each jet also remember the charge of the parton producing the jet. Would you comment on this?

- *BRANSON:*

The data points are rising but by less than one standard deviation.

- *HOFSÄSS:*

I would like to make two comments on the results for $P'(\gamma)$. Firstly it is not the charge of the quark producing the jet that is being measured but its sign. Secondly as γ becomes increasingly large and negative $P'(\gamma)$ is weighting two adjacent particles at small x. Thus we are measuring short range correlation effects, which come from resonance decays, causing $P'(\gamma)$ to rise.

Secondly you stated that in $e^+e^- \to \mu^+\mu^-$ there is a contribution to the asymmetry which other groups have not taken into account and which puts upper and lower bounds on the mass of the Z_0. Would you describe this additional contribution?

- *BRANSON:*

The asymmetry increases with s, and the emission of an initial state photon reduces the effective s flowing through the γ or Z_0 propagator, which reduces the asymmetry. This correction has only been performed by the Mark J group, and when their correction is applied to the other data the expected asymmetry is changed from -9.3% to -8.2%, which is a change of almost one standard deviation. Hence if the mass of the Z_0 is set to infinity the asymmetry is more than two standard deviations away from the measurement. This means that we are effectively seeing the propagator mass, so that the mass of the Z_0 is not infinity. This is why we may give an upper bound for the mass of the Z_0.

- *ETIM:*

Have you also corrected for final state photon emission?

- *BRANSON:*

Yes, but the effect is very much greater from the initial state emission.

- *MUKHI:*

The ratio of the values of α_s obtained using the Lund Monte Carlo to that obtained using the Feynman-Field fragmentation model is highest in the case of the energy-energy correlation. Does this not contradict your statement that the energy-energy correlation is less fragmentation model dependent than other quantities?

- *BRANSON:*

I am unable to account for this other than to say that the more we look at only hard 3-jet events the smaller the difference between the two fragmentation models appears to be. In the case of the energy-energy correlation we measure all 3-jet events, not only the hardest.

- *RICHARDS:*

Using different fragmentation models different values of α_s are obtained. Is there any evidence that we can separate measurements into perturbative and fragmentation effects?

- *BRANSON:*

At the present time not enough work has been done on this problem. The difficulty arises because the Lund fragmentation model actually changes the direction of the jets.

- *DE BOER:*

I would like to make two comments. The first concerns the previous question about the asymmetry of the energy-energy correlation which has been widely advertised to be fragmentation independent. This is only true under the assumption that the final state particles follow the original parton directions. However, this is not true in a string-like fragmentation model, such as the Lund model.

In this model the Lorentz boost from the centre-of-mass frame of the string to the laboratory frame makes the final state look more 2-jet like than the initial state. Therefore, the asymmetry in the energy-energy correlation of the partons in the Lund model is roughly 50% higher than the asymmetry of the final state particles. In a model where the partons fragment incoherently, the asymmetries of initial and final states are equal. This is the main reason why the CELLO collaboration finds α_s to be 0.25 with the Lund model (LM) and 0.15 with the Hoyer model (HM), in which each parton fragments incoherently according to the Feynman-Field prescription.

The string picture is not the only difference between the HM and LM. Other differences are : a) different gluon fragmentation, b) different heavy quark fragmentation and decay (the c and b quarks make up 45% of the total hadronic cross-section), c) different cuts on the matrix elements to avoid divergences due to collinear and soft gluon emission, and d) different treatment of events with initial state radiation.

From a detailed Monte Carlo study the CELLO collaboration finds the fraction of 3-jet events generated for $\alpha_s = 0.2$ to be about 10% for the LM, about 12% for the LM without string kinematics, about 13% for the LM without string kinematics but with the gluon fragmenting like a quark, and about 14% for the HM. Here the 3-jet events were defined by topological cuts: sphericity greater than 0.25 and aplanarity less than 0.1. Therefore, for a given number of 3-jet events, α_s varies by 40% for the various choices of models. Note that this Monte Carlo study is independent of any experimental data.

As long as the experimental data is not precise enough to distinguish between the different models, one is forced to include a systematic uncertainty of the order of 40% in the value of α_s.

- *ERREDE:*

If we accept the possibility of a more complicated gauge theory of electroweak interactions, how does the presence of more than one Z_o affect the asymmetry parameter $A_{\mu\mu}$, and the limits on the Z masses?

- *BRANSON:*

In nonstandard electroweak interaction models with two Z_o's it is usually expected that one Z will have a mass smaller than the Z_{GWS} while the other will have a mass much greater than the Z_{GWS}. Thus the second Z_o is unlikely to have any direct observable effect at presently measurable energies. If it eventually turns out that the asymmetry is consistently higher than in the GWS model then a light Z_0 would be a possible explanation.

- *LAU:*

You stated that a radiative correction should be made to μ-pair asymmetry due to electroweak interference. Should a similar radiative correction be made to the QED asymmetry?

- *BRANSON:*

The data have been corrected by the QED asymmetry. The radiative correction to this is $O(\alpha^4)$ and has not yet been calculated. However history suggests that such a higher order QED correction will be very small.

- *LAU:*

In your measurement of the branching ratio of the production of charmed particles to muons have you taken an average of charged and neutral charmed particles, and, if so, is it possible to determine the relative number of charged and neutral particles produced, say, by dimuon events?

- *BRANSON:*

We have used an average of charged and neutral particles, but we do not have a sufficient number of dimuon events to determine the relative numbers produced.

- *ALBRECHT:*

In order to distinguish between the standard $SU(2)\otimes U(1)$ model for electroweak interactions and multiboson models with a $SU(2)\otimes U(1)\otimes G$ gauge group, you have introduced a quantity C. Is your definition of the effective Hamiltonian the most general?

- *BRANSON:*

At low energies I believe it is.

- *TUROK:*

The expression is only correct in the case where the electromagnetic current receives an additional contribution from the group G.

- *SHERMAN:*

What fragmentation function did you use to get the μ-spectrum from heavy quark decay?

- *BRANSON:*

We assumed a flat distribution in x with a cutoff due to mass. The error in the data includes differences due to varying the fragmentation function.

- *CASTELLANI:*

You stated that to fit the present data it is only necessary to have a global SU(2) symmetry, universality and electromagnetism. How can the asymmetry in $e^+e^- \to \mu^+\mu^-$ be explained without the inclusion of the Z_o?

- *BRANSON:*

Low energy data without precise measurements of the asymmetry can be fitted as you have described. However at higher energies the existence of the asymmetry requires the inclusion of the Z_o.

- *DEVCHAND:*

Is there any experimental evidence for spontaneous symmetry breaking in e^+e^--collisions?

- *BRANSON:*

There is no evidence as far as I know.

- *OGAWA:*

What is the source of the "kinematic reflection" seen in the K invariant mass, and can this affect measurements such as the D^0 lifetime?

- *BRANSON:*

The kinematic reflection is due to the three-body decay of the D^0 via a resonance. Such cases will give a lower result for the mass of the K -system. To determine the D^0 lifetime we only use events from the D^0-peak.

- *OGAWA:*

You stated that some corrections have to be applied to heavy quark asymmetries. Do they explain the difference in sign between the asymmetry for b and c quarks?

- *BRANSON:*

The difference in sign is just due to the differing signs of the quark charge. There are some corrections due to quark mass effects but these are small.

- *KAPLUNOVSKY:*

Is there any correlation between the production of baryons and their antiparticles in opposing jets? For example if a $\bar{\Lambda}$ is observed in one jet would the opposing jet be expected to contain a Λ or just p and K^- say?

- *BRANSON:*

Some experiments have been made but the data is insufficient to determine whether there are any such correlations.

- *KAPLUNOVSKY:*

Why should the quantity C be independent of Q^2?

- *BRANSON:*

The Q^2 dependence in C is integrated over and in the low energy approximation C is only dependent on boson masses.

- *D'ALI:*

You compared the muon transverse momentum data with a Monte Carlo simulation. Which matrix-element did you use in the Monte Carlo?

- *BRANSON:*

We used the spectator model for the matrix element where the b quark decays independently of the spectator quark and produces jets.

- *SUSINNO:*

What is the relative importance of heavy flavour production and perturbative QCD in the broadening of jets?

- *BRANSON:*

At the highest attainable energies the weak decay of the b-quark has an effect that is almost measureable, but it is much less than the α_s-corrections or even fragmentation effects.

- *GRZADKOVSKY:*

Do your measurements of B-meson lifetimes distinguish between the neutral and charged ones?

- *BRANSON:*

The result is only an upper limit without distinction between different types of B-mesons.

STATUS OF AND SEARCH FOR NEW LEPTONS AT PETRA

P. Duinker

National Institute for Nuclear and High Energy Physics
Amsterdam, The Netherlands

I. INTRODUCTION

This article reviews the status of leptons and the searches for new leptons at PETRA, the electron-positron colliding beam storage ring at DESY in Hamburg, Germany. Due to the excellent performance of the PETRA machine in 1981 and the first half of 1982 the number of events for the process $e^+e^- \rightarrow \mu^+\mu^-$ has increased in such a way that the weak effects in this reaction could be established beyond any doubt. A review of this situation is given in Section II.

A short review is given of the present status of tau-leptons in Section III. The existing data on the spin assignment, branching ratios for the various decay channels and the tau lifetime all lead to the conclusion that this particle is a sequential lepton.

The weak effects in Bhabha scattering $e^+e^- \rightarrow e^+e^-$ and the extraction of the parameters $\sin^2\Theta_W$ and the axial and vector coupling constants for the weak neutral currents are given in Section IV. Finally the searches for new leptons like excited states of muons and electrons, new heavy sequential leptons, heavy stable leptons, neutral heavy electrons, scalar leptons, charged Higgs particles and hyperpions are presented in Section V.

II. THE PROCESS $e^+e^- \to \mu^+\mu^-$

A. Introduction

One of the most important advances in physics in the last two decades is the development of the electroweak theory[1]. Neutrino experiments[2] in the last decade have shown agreement with this theory. These experiments are characterized by the experimental conditions: (1) the relatively low c.m. energy, namely $s \approx 200\ \mathrm{GeV}^2$; (2) the relatively low momentum transfer $Q^2 \approx -100\ \mathrm{GeV}^2$ in the spacelike region; (3) the use of nuclear targets.

The reaction

$$e^+e^- \to \mu^+\mu^- \tag{1}$$

proceeds via timelike momentum transfers, and at high energies $Q^2 = s \approx 1200\ \mathrm{GeV}^2$ offers the possibility of observing directly the effect of the neutral vector boson $Z^0 \to \mu^+\mu^-$. According to current[1] electroweak theory, reaction (1) proceeds through both photon and Z^0 intermediate states. The resulting differential cross section is[3]

$$\frac{d\sigma}{d\Omega} = \frac{\alpha^2}{4s}\{C_1(1 + \cos^2\theta) + C_2\cos\} \tag{2}$$

where

$$C_1 = 1 + 8g_V^2\, s\rho + 16(g_V^2 + g_A^2)^2 s^2\rho^2,$$

$$C_2 = 16g_A^2\, s\rho + 128\, g_V^2 g_A^2\, s^2\rho^2,$$

$$\rho = \frac{G_F}{8\sqrt{2}\pi\alpha}\cdot\left(\frac{1}{s/m_Z^2 - 1}\right),$$

and G_F is the Fermi coupling constant, m_Z the mass of the Z^0, g_V and g_A are the vector and axial couplings of the Z^0 to the lepton fields, and it is assumed that $\Gamma_Z \ll m_Z$. By integrating expression (2) one obtains the total cross section for $\mu^+\mu^-$ production,

$$\sigma_{\mu^+\mu^-} = \frac{4\pi\alpha^2}{3s} \cdot C_1 \tag{3}$$

As C_1 deviates from 1 only by a small amount, the weak effect in the total $\mu^+\mu^-$ cross section at the highest PETRA energy is very small, a precise measurement however can give a constraint on g_V, the vector coupling constant of the weak neutral current. The term $C_2 \cos\Theta$ in expression (2) produces a measurable forward-backward charge asymmetry defined as:

$$A_{\mu\mu} = \frac{N(\Theta<\pi/2) - N(\Theta>\pi/2)}{N(\Theta<\pi/2) + N(\Theta>\pi/2)} \tag{4}$$

where N is the number of μ^- with a polar angle Θ in the indicated range measured with respect to the e^- beam direction. Expression (2) leads to a neutral current prediction for the charged forward-backward asymmetry of

$$A^W = \frac{3}{8} \cdot \frac{C_2}{C_1} \tag{5}$$

In order to compare the measured quantity $A_{\mu\mu}$ from (4) with expression (5) the data has to be corrected for the small positive QED asymmetry A^{QED} resulting from α^3 contributions. The measured asymmetry can then be compared with the prediction from electro-weak theory and the QED corrections according to

$$A_{\mu\mu} = A^W + A^{QED} \tag{6}$$

B. QED Contributions to $A_{\mu\mu}$[4].

The contributions to the $\mu^+\mu^-$ scattering cross section from QED are evaluated to order α^3 and the relevant Feynman diagrams are shown in Fig. 1. Writing the QED cross section as:

$$\frac{d\sigma}{d\Omega} = \frac{d\sigma_0}{d\Omega}(1 + \delta) \tag{7}$$

where δ represents the radiative correction to order α^3 and where $d\sigma_0/d\Omega$ is the lowest order (α^2) contribution and equal to:

$$\frac{d\sigma_0}{d\Omega} = \frac{\alpha^2}{4s}(1 + \cos^2\Theta) \tag{8}$$

The real bremsstrahlung receives contributions from the four lowest diagrams of Fig. 1 where photons are emitted from the initial state electrons or positrons and the final state muons. The interference between the lowest order diagram and the diagrams in which closed loops occur introduces the QED asymmetry.

The phase space for real bremsstrahlung can be experimentally characterized by a simple boundary as shown in Fig. 2 by the contour ABC (solid line). By applying the cuts E_{th}, an energy threshold for the outgoing μ^- and μ^+ and ξ_{th}, a limit on the angle in space between the μ^- direction and the direction 180° to the μ^+-direction (acollinearity cut) a subarea in the available phase space (hatched area in Fig. 2) is defined. For a certain integrated luminosity the number of events produced according to the diagrams in Fig. 1 and contributing to the total area and the subarea can be calculated using the Monte-Carlo program of Behrends and Kleiss[5]. The ratio of the two contributions determines a phase space correction which has to be applied to $d\sigma/d\Omega$. The radiative correction δ can now be calculated and Fig. 3 shows δ for a acollinearity cut of $\xi_{th}<20°$ and $E_{th} = 0.5\ E_{beam}$ as function of the scattering angle Θ for three different total energies $\sqrt{s}$.

An experimental check of the radiative correction calculations can be made by comparing the distribution of the measured angle ξ with the QED prediction. In Fig. 4 the acollinearity angle ξ for $\mu^+\mu^-$ pair candidates as measured by the MARK-J group is shown. The agreement between the data and the calculations is excellent.

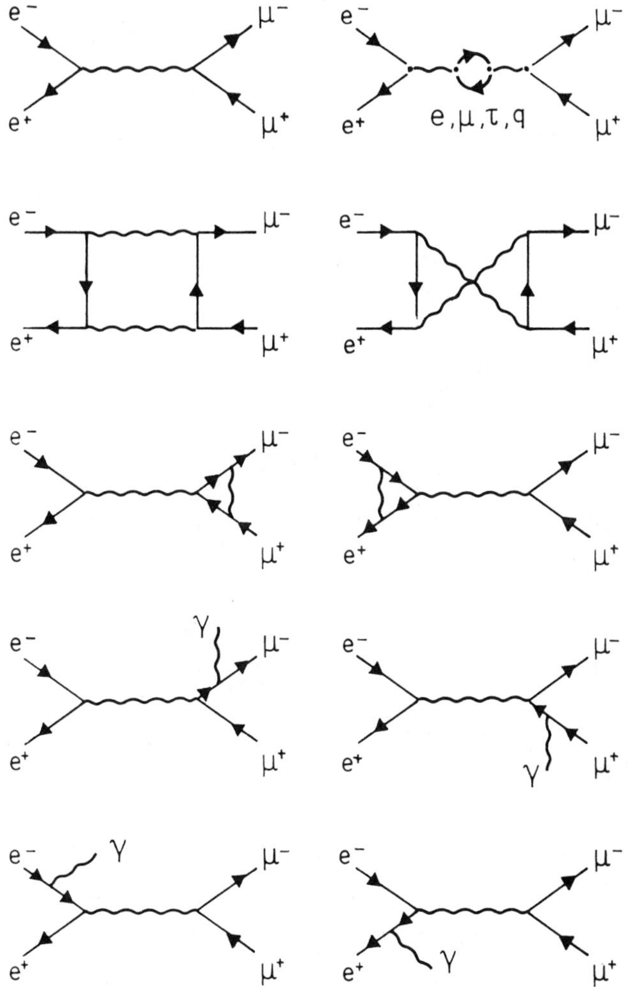

Fig. 1. Feynman diagrams used for the QED calculation and radiative corrections for the process $e^+e^- \to \mu^+\mu^-$.

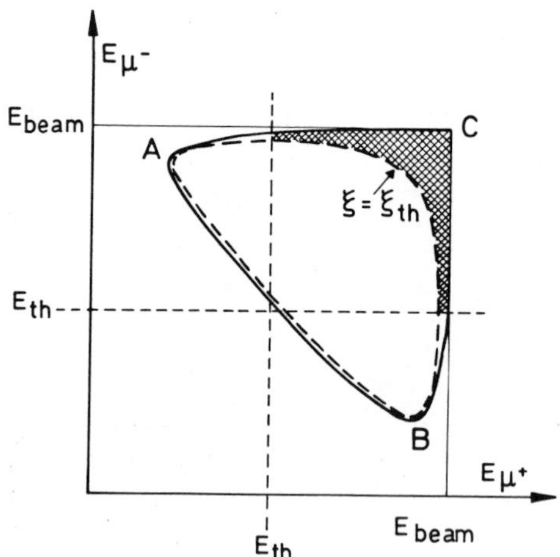

Fig. 2. Two-dimensional phase space diagram for the process $e^+e^- \rightarrow \mu^+\mu^-\gamma$. The energies of the outgoing μ^+ and μ^- are plotted along the x- and y-axis. The solid contour ABC is the outside boundary of the area permitted by energy and momentum conservation. The hatched area contains only those events which fulfil an acollinearity cut of $\xi<\xi_{th}$ and E_{μ^+} and $E_{\mu^-} > E_{th}$.

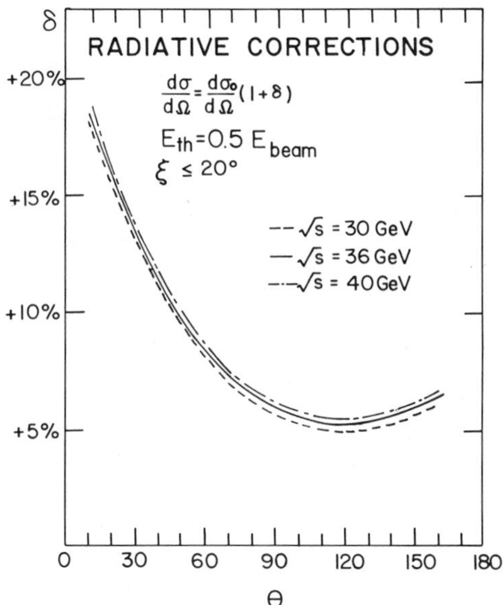

Fig. 3. The radiative correction δ as function of the scattering angle Θ for the process $e^+e^- \to \mu^+\mu^-$ in the MARK-J detector. The cuts and energies are as indicated.

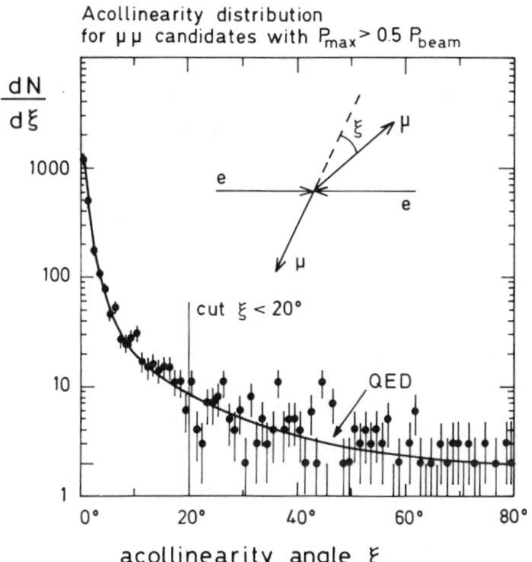

Fig. 4. Measured acollinearity distribution for the process $e^+e^- \to \mu^+\mu^-$ obtained by the MARK-J Collaboration. The measured momentum has to exceed 50% of the beam energy. The solid line is the QED prediction.

C. The measurements

Measurements of the process $e^+e^- \rightarrow \mu^+\mu^-$ at PETRA in the energy range 14 to 36.7 GeV have been reported earlier[6]. For some of the experimental details I will use the data of the MARK-J detector[7]. Their selection criteria for muon pair events are as follows.

1. Two minimum ionizing particles emerging within a cylindrical region of 20 centimeters length and 5 cm of radius centered at the beam crossing point, penetrating 90 to a 160 cm of magnetized iron, and reaching the outside driftchambers.
2. A time coincidence in the muon trigger counters to better than 5 nsec.
3. An acollinearity angle between the two muon tracks of less than 20°.
4. The measured muon momentum $p_\mu > \sqrt{s}/4$ for at least one muon.

The necessity of this last criterium is illustrated in the distribution of Fig. 5, where the maximum of the momenta of the μ^+ and μ^- tracks divided by the beam momentum, p_{max}/p_{beam}, is plotted. The overall distribution contains the contributions from the two-photon process:

$$e^+e^- \rightarrow e^+e^-\mu^+\mu^-$$

for which only the outgoing muons are detected and from the single photon process

$$e^+e^- \rightarrow \mu^+\mu^-$$

The muons at low values of p_{max}/p_{beam} are due to the two-photon process and they can be readily separated by accepting only events with $p_{max}/p_{beam} > 0.5$. The hatched area in Fig. 5 indicates the events with an acollinearity of less than 20°.

To measure the charge asymmetry precisely it is important to understand and minimize the systematic errors. Much effort has been spent

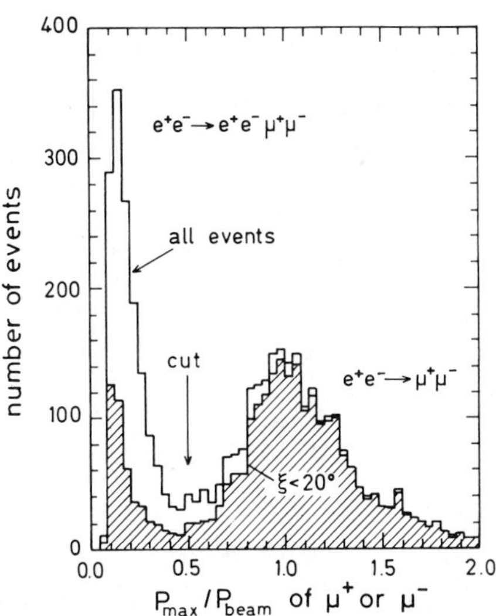

Fig. 5. The distribution of the quantity p_{max}/p_{beam} where p_{max} is the larger one of the two momenta of the μ^+ or μ^-. All candidate events with two muons in the MARK-J detector are included in this plot. The hatched area are those events for which the acollinearity angle is smaller than 20°.

to understand the systematic bias and to keep it below 1%. Some examples as given by the MARK-J group[8] are:

1. The uncertainty in the momentum measurement:

   This uncertainty affects the determination of the muon charge. Fig. 6 shows the p_{beam}/p_{μ} distribution measured in the back and front of the MARK-J detector for μ^+ and μ^- tracks separately. The inverse momentum resolution $k = 1/p_{\mu}$ which can be deduced from Fig. 6 is $\sigma(k)/k = 30\%$ for $p_{\mu} = 17.5$ GeV/c. With this resolution there are 50 events out of a sample of 2435 muon pair events where both muons are measured to have the same charge (2.1% of the sample), which are of course removed. This suggests that in less than 0.1% of the events, the wrong charge assignment was made for both muons.

2. The uncertainty in detector acceptance:

   This may cause a given solid angle for positive muons to be different for negative muons. The systematic error due to detector acceptance can be significantly decreased by alternating the polarity of the detector magnetic field. Any acceptance asymmetry will produce effects which are equal and opposite for positive and negative magnet polarities. Therefore, the difference between the measured asymmetries with each field polarity displays twice this systematic bias. To measure the detector asymmetry a sample of 20,000 cosmic ray muons with momentum greater than 10 GeV/c was collected. Most of these cosmic ray data were collected during normal data-taking periods, thus appropriately reflecting any possible time-dependent changes in detector response. After correction for the different time of flight and energy loss, the detector response to a cosmic ray muon going through the interaction region is equivalent to that for a collinear muon pair of the same energy. From a detailed study of these cosmic ray muon data[9], the detector asymmetry as function of the polar angle Θ could be calculated and the detector contribution to the charge asymmetry was determined to be less than 1%. Furthermore, this detector asymmetry is eliminated by collecting data in equal amounts with both magnet polarities, thereby cancelling to the first order all systematic errors which relate to the charge measurement of muons.

3. Trigger inefficiency:

From the measurement of muon yields from redundant triggers a <0.5% uncertainty is observed in the total muon pair rate. This implies that the uncertainty in muon asymmetry will be <0.1%.

4. Tau pairs:

The contamination from events of the process $e^+e^- \to \tau^+\tau^- \to \mu^+\mu^-\nu_\mu\bar{\nu}_\mu\nu_\tau\bar{\nu}_\tau$ is about 1% and is expected to yield a similar angular distribution to that from muon pairs.

5. Two photon process:

Muon pair events from the two-photon process $e^+e^- \to e^+e^-\mu^+\mu^-$ are characterized by lower muon momenta and by muon acollinearity angles that are roughly uniform up to 90°[10]. The rate of such events satisfying the above criteria was calculated to be less than 0.2% of the muon pair rate.

6. Cosmic rays:

Cosmic ray muons traversing the entire detector and going through the interaction region are recognized by the difference in times recorded in the two opposite trigger counters. The time difference distribution for muon pair events is centered at zero with a width of 0.8 nsec, whereas the flight time of a cosmic ray muon yields time differences greater than 9 nsec. One estimates a cosmic ray contamination to the muon pair sample of less than 0.4%.

From these studies the conclusion is drawn that by taking equal amounts of data at both magnetic polarities the total systematic error is <1%. As a consistency check on the systematic errors, Figures 7a and 7b display the distribution of muon pair events from the MARK-J group as a function of the scattering angle for $\sqrt{s}$ = 14 and 22 GeV. These distributions are corrected for the small (positive) asymmetry resulting from α^3 contributions to the cross sections, thus displaying the effects of the weak interaction alone. As can be seen, the deviations from the lowest order QED

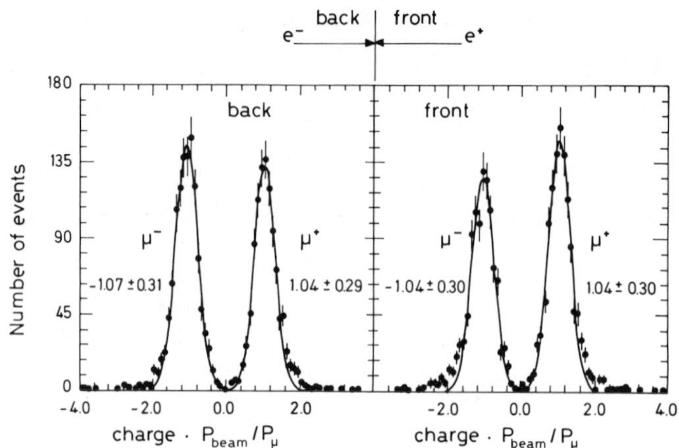

Fig. 6. Measured momenta distributions for the μ^+ and μ^- in the front and the back of the MARK-J detector.

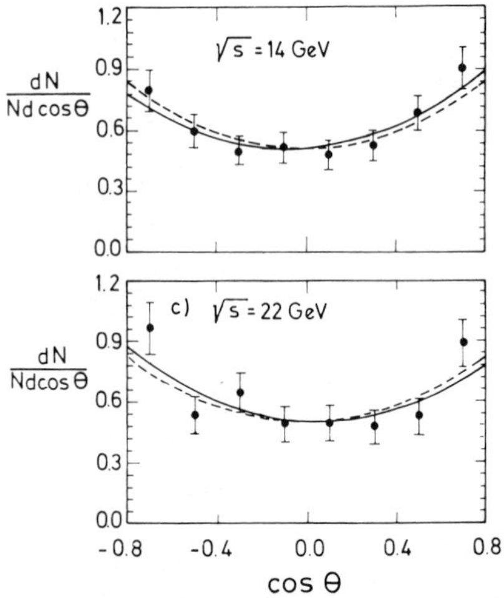

Fig. 7. The cosΘ distributions for muon pair events as measured by the MARK-J Collaboration at a) $\sqrt{s}$ = 14 and b) $\sqrt{s}$ = 22 GeV. The dashed lines are the QED predictions and the solid lines are a fit to the data including weak interaction effects.

angular distributions (the dashed curves) are small at $\sqrt{s}$ = 14 and 22 GeV; the asymmetry values are $A_{\mu\mu}$ = (+5.3 ± 5.0)% and $A_{\mu\mu}$ = (−4.3 ± 6.1)% respectively. This is consistent with the expected s-dependence of the asymmetry.

D. Results

The total cross sections for $e^+e^- \to \mu^+\mu^-$ at $\sqrt{s}$ = 14, 22 and 34.4 GeV normalized to the expectations from QED (after radiative corrections to order α^3) were determined by the four PETRA groups. The normalized cross sections $R_{\mu\mu}$ are shown in Fig. 8 as function of the centre of mass energy squared s. The systematic errors range from 3 to 5% for the different groups and are primarily due to the uncertainty in the luminosity measurements. No deviations from pure QED are seen in $R_{\mu\mu}$ and this is expressed in the cut-off parameters $\Lambda_\pm$ [11], which are determined to be greater than 150 to 200 GeV at 95% C.L. for the different groups. $R_{\mu\mu}$ can be related in a simple way to $\Lambda_\pm$ according to:

$$R_{\mu\mu} = 1 \pm \frac{s}{\Lambda_\pm^2}$$

and the dashed lines in Fig. 8 indicate the expectation for a modified QED with $\Lambda_\pm$ = 150 GeV. The possible weak effects in the quantity $R_{\mu\mu}$ are also indicated. The solution $g_V^2 = 0$ and $g_A^2 = \frac{1}{4}$ for the vector and axial vector coupling constants of the weak neutral current coincides with pure QED in the parameter $R_{\mu\mu}$ and the solution $g_V^2 = \frac{1}{4}$ and $g_A^2 = 0$ coincides with Λ_- = 150 GeV.

The distributions of the muon pair events as a function of the scattering angle cosΘ for $\sqrt{s} \approx$ 34.4 GeV are shown[12] in Fig. 9. The dashed lines in these plots show the $1 + \cos^2\Theta$ behaviour for zero order QED, the solid lines show the fit to the data of the electro-weak theory. The three sets of data show a distinct asymmetry. The asymmetry values obtained from the fit and extrapolated over the complete cosΘ range, $-1 < \cos\Theta < 1$ are listed in Table I for the two PEP groups, MAC and MARK-II, and the four

Table I. The measured muon asymmetry values for the various e^+e^- experiments. Also indicated are the expectations from the standard G.W.S. model with $\sin^2\Theta_W = 0.23$.

Experiment	$\sqrt{s}$ (GeV)	$A_{\mu\mu}$ (%)	Expected in G.W.S. with $\sin^2\Theta_W \approx 0.23$
MAC	29	-4.4±2.4	-6.3
MARK II	29	-9.6±4.5	-6.3
Combined	29	-5.6±2.1	-6.3
CELLO	34.2	-6.4±6.4	-9.1
JADE	34.2	-10.8±2.2	-9.2
MARK-J	34.6	-10.4±2.1	-9.4
TASSO	34.4	-10.4±2.3	-9.3
Combined	34.4	-10.4±1.2	-9.3

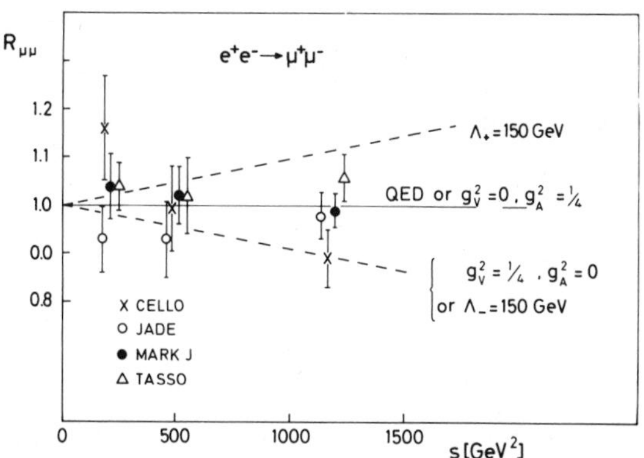

Fig. 8. The measurements of $R_{\mu\mu}$ for the four PETRA groups as function of s, the square of the centre of mass energy. The solid line at $R_{\mu\mu} = 1$ is the QED expectation coinciding with the solution for electro-weak theory with $g_V^2 = 0$ and $g_A^2 = ¼$. The dashed lines are the expectations for a modified QED with cut-off parameters $\Lambda_\pm = 150$ GeV. The $\Lambda_- = 150$ GeV coincides with an electro-weak theory solution with $g_V^2 = ¼$ and $g_A^2 = 0$.

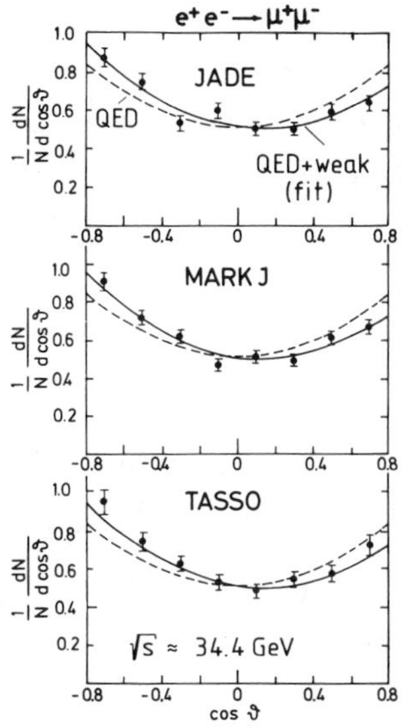

Fig. 9. The cosΘ distributions for the process $e^+e^- \to \mu^+\mu^-$ at $\sqrt{s} \approx 34.4$ GeV for the JADE, MARK-J and TASSO groups. The dashed lines are the QED expectations and the solid lines the fit to the data of the electro-weak theory.

PETRA groups, CELLO, JADE, MARK-J and TASSO. The average value for the asymmetry for the PEP groups is $A_{\mu\mu} = (-5.6 \pm 2.1)\%$ at $\sqrt{s} = 29$ GeV with an expected value of -6.3% from the standard model assuming $\sin^2\Theta_W = 0.23$. For the PETRA groups an average value of $A_{\mu\mu} = (-10.4 \pm 1.2)\%$ is found with an expected value of -9.3% at $\sqrt{s} \approx 34.4$ GeV.

In Fig. 10 all the available data for $A_{\mu\mu}$ is plotted as function of s. The solid line at $A_{\mu\mu} = 0$ is the QED expectation. The dashed and dashed-dotted lines show the predictions of the standard model for $g_A^2 = \frac{1}{4}$ and $m_Z = \infty$ and $m_Z = 90$ GeV/c^2 respectively. The standard model predicts the energy behaviour and the magnitude of the asymmetry at the highest PETRA energies in an impressive way. A lower limit on the Z^0 mass can be derived from the measurements and is found to be $m_Z > 56$ GeV/c^2 at 95% C.L.

III. THE τ-LEPTON

A. Introduction

The first experiment at a e^+e^- colliding beam machine to search for leptons with masses heavier than the muon was proposed in 1967 by A. Zichichi and his collaborators at ADONE[13]. Their search concentrated on the reaction

$$e^+e^- \rightarrow e^\pm + \mu^\mp + \text{anything},$$

and their result could be expressed as a lower limit for the mass of a heavy lepton of $m_{HL} > 1.45$ GeV/c^2 with 95% C.L.[14]. In 1975 the τ-lepton was discovered at SPEAR by M. Perl and the SLAC-LBL group[15]. They found 64 events of the type

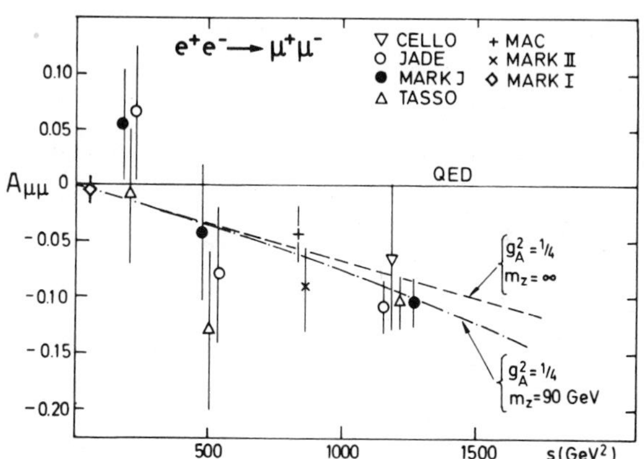

Fig. 10. The measurements of the muon asymmetry as function of s. The solid line of $A_{\mu\mu} = 0$ represents the QED expectation or the electro-weak theory solution with $g_A = 0$. The dashed and dashed-dotted lines are the electro-weak solutions for $g_A^2 = \frac{1}{4}$ and $m_Z = \infty$ and $g_A^2 = \frac{1}{4}$ and $m_Z = 90$ GeV.

in which no other charged particles and photons were detected. Confirmation of this discovery came subsequently from DESY[16] and the spin and mass were most accurately determined by the DELCO-group[17] at SPEAR. By collecting data in the centre of mass range of 3.5 to 4.4 GeV they determined from the cross section behaviour near threshold the mass to be $m_\tau = 1.782\,^{+0.002}_{-0.007}$ GeV/c^2 and the spin to be ½. By measuring the shape of the energy spectrum of the electrons originating from the decay

$$\tau^- \to e^- \bar{\nu}_e \nu_\tau$$

they were able to measure the Michel parameter $\rho = 0.72 \pm 0.15$. This measurement is in good agreement with the expectation of $\rho = 0.75$ for a charged weak V-A current[18]. For detailed information on the properties of the τ-lepton and their decay modes, see the reviews of M.L. Perl, G. Fluegge, S.C.C. Ting and G.J. Feldman[19].

In the next paragraphs I will give some of the results on a recent spin and mass determination by the MARK-II group, the measurements of decay modes into hadrons, and the determination of the tau lifetime. Cross section and asymmetry measurements at the highest PEP and PETRA energies will conclude Section III.

B. Spin and Mass Determination of the τ-Lepton

As mentioned before the first convincing determination of the spin and mass of the tau was given by the DELCO-group[17]. The cross section for the production of pointlike particles in e^+e^- annihilation for different spins is:

spin ½:
$$\sigma_{\tau\tau} = \frac{4\pi\alpha^2}{3s}\,\beta\,\left(1 + \frac{1-\beta^2}{2}\right) ,$$

where $\beta = v/c$,

spin 0:
$$\sigma_{\tau\tau} = \frac{1}{4}\,\frac{4\pi\alpha^2}{3s}\,\beta^3\,|F|^2 ,$$

$e^+e^- \to e^\pm + \mu^\pm$ + missing energy,

where F is the τ form factor,

spin 1: $$\sigma_{\tau\tau} = \frac{4\pi\alpha^2}{3s}\beta^3 \{(\frac{s}{4m_\tau^2})^2 + 5.\frac{s}{4m_\tau^2} + \tfrac{3}{4}\} .$$

By measuring the cross section dependence near threshold for a certain tau decay mode the different spin possibilities can be disentangled. The mass is a parameter in the determination of the spin due to the relation

$$\beta^2 = 1 - 4m_\tau^2/s ,$$

where m_τ is the mass of the tau.

The MARK-II group at SPEAR[20] based their measurements of the spin and the mass of the tau on the decay mode $\tau^- \to \pi^-\nu_\tau$. Events were selected with a charged pion and exactly one other particle. The total number of $\pi^\pm X^\mp$ events was 1138 ± 46 ± 174 (statistical and systematic errors respectively). The branching ratio for $\tau^- \to \pi^-\nu_\tau$ was found to be $B_\pi(\tau^- \to \pi^-\nu_\tau) = 0.117 \pm 0.004 \pm 0.018$. In Fig. 11 the quantity $\sigma_{\tau\tau}\cdot B_\pi / \sigma_{\mu\mu}$, the product of the $\tau^+\tau^-$ production cross section and B_π normalized to the pointlike muon production cross section, is plotted as function of the centre of mass energy. The spin 0 and 1 hypotheses for the τ-leptons are eliminated and a good description of the data is obtained for the spin ½ case with a value for the mass of $m_\tau = 1.803 {}^{+0.010}_{-0.018}$ GeV/c^2. The same data was used to determine an upper limit on the τ-neutrino mass. For the two-body decay $\tau^- \to \pi^-\nu_\tau$ the pion energy spectrum will be flat for mono-energetic taus. As a consequence the endpoints of these energy spectra are a function of the masses of the π, τ and ν_τ. In Fig. 12 the pion energy spectra are shown measured at six different energy settings. A fit to these spectra yields an upper limit of $m_{\nu_\tau} < 0.25$ GeV/c^2 at 95% C.L.[20].

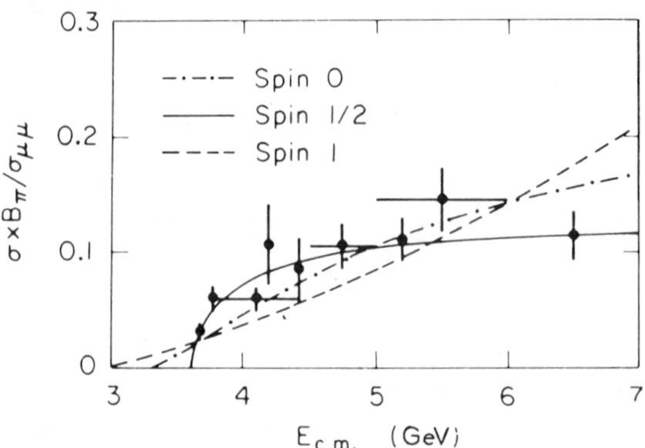

Fig. 11. Tau-pair production cross section as function of the centre of mass energy $E_{C.M.}$ as measured by the MARK-II Collaboration from the $\tau^- \to \pi^- \nu_\tau$ decay mode. The curves are fits for the different tau-spin assignments. The errors include an estimated point to point systematic uncertainty from background subtractions.

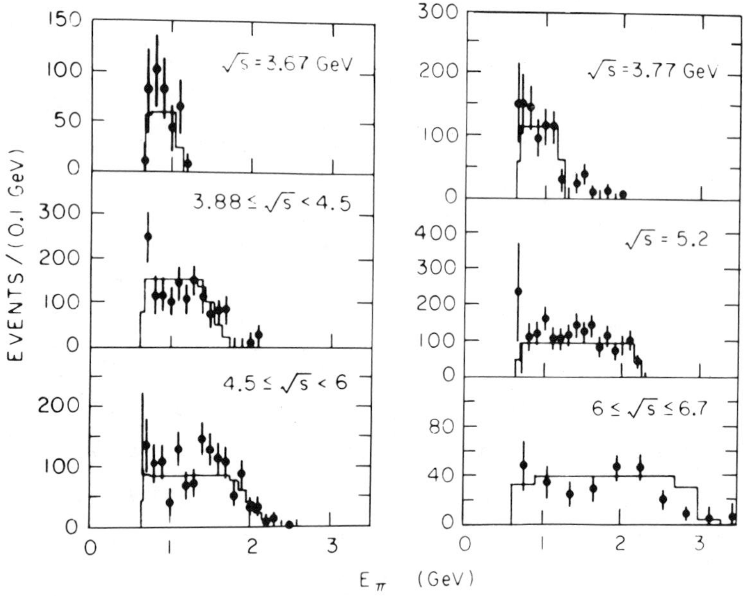

Fig. 12. The pion energy spectra for events where one of the taus decays according to $\tau \to \pi\, \nu_\tau$ at various energies as measured by the MARK-II detector. The background is subtracted bin-by-bin and efficiency corrections are applied. The curves are the expected spectra for m_τ = 1.782 GeV/c^2, m_ν = 0 and B_π = 0.117.

C. Decay Modes of the τ-Lepton

Within the framework of the standard model the possible decay modes of the τ-lepton mediated by the charged weak V-A current can be described. It is assumed that the tau is a sequential lepton with its own neutrino ν_τ and that we have e-μ-τ universality. The leptonic width for the decays $\tau^- \to e^- \bar{\nu}_e \nu_\tau$ and $\tau^- \to \mu^- \bar{\nu}_\mu \nu_\tau$ are given by

$$\Gamma_e \equiv \Gamma\left(\tau^- \to e^- \bar{\nu}_e \nu_\tau\right) = \frac{G_F m_\tau^5}{192\pi^3}$$

and

$$\Gamma_\mu \equiv \Gamma\left(\tau^- \to \mu^- \bar{\nu}_\mu \nu_\tau\right) = \Gamma_e \cdot F(y)$$

where $G_F = 1.02 / m_p^2 \cdot 10^{-5}$ (GeV^{-2}) and $F(y) = 0.973$ is a function depending on the masses of the muon and the tau.

The decay of the weak charged current into an $u\bar{d}$ pair gives rise to a hadronic system in the final state like a π, ρ, A_1 etcetera. Table II gives a recent compilation of the measured branching fractions (%) for the various decay channels taken from reference 21 and references therein. The decay modes are grouped according to the number of charged particles in the final state, one prong (B1) and three prong (B3) events. Recently direct measurements have become available for B1, B3 and B5 (decays of the τ into five charged hadrons) at the highest e^+e^- centre of mass energies from the MARK-II[21], TASSO[22], and CELLO groups[23]. The MARK-II group measured the charged particle multiplicity distributions at PEP to be:

$$B1 = 1 - B3 = (86 \pm 2 \pm 1)\% \text{ and } B5 < 0.5\% \text{ at } 95\% \text{ C.L.}$$

The TASSO group finds at PETRA,

$$B1 = (76 \pm 6)\%,\ B3 = (24 \pm 6)\% \text{ and } B5 < 6\% \text{ at } 95\% \text{ C.L.},$$

and CELLO obtained

Table II. Various decay modes of the τ-lepton with the experimental and theoretical values (in %).

Decay mode	Branching fraction (%)	
	Experimental	Theoretical
One prong:		
$\tau \to e\nu\nu$	17.6±1.1	17.6
$\to \mu\nu\nu$	17.1±1.1	17.1
$\to \pi\nu$	11.5±1.8	10.4
$\to K\nu$	1.3±0.5	0.7
$\to \rho\nu$	21.5±3.6	21.3
$\to K^{*}\nu$	1.7±0.7	1.4
$\to \pi^{\pm}2\pi^{0}\nu$	5.0±2.1	4.6
$\to \pi^{\pm}3\pi^{0}\nu$	-	2.0
	75.7±4.7	75.1
Three prongs:		
$\tau \to \pi^{\pm}\pi^{+}\pi^{-}\nu$	5.0±2.1	4.6
$\to \pi^{\pm}\pi^{+}\pi^{-}\pi^{0}\nu$	11.0±7.0	8.1
	16.0±7.0	12.7

$$B1 = (84 \pm 2)\%, \ B3 = (15 \pm 2)\% \text{ and } B5 = (1 \pm 0.4)\%,$$

the only result with a positive outcome for the five charged particle decay mode. It has to be noted that the high energy experiments find a considerably higher value for B1 than the lower energy experiments at DORIS and SPEAR where the world average for B1 was found to be $B1 = (68 \pm 10)\%$.

D. τ-Lifetime Measurements

Assuming τ-μ universality the lifetime of the tau is given by [24],

$$\tau_\tau = (m_\mu/m_\tau)^5 \cdot \tau_\mu \cdot B_e = (2.8 \pm 0.2) \times 10^{-13} \text{ sec},$$

where B_e is the branching fraction for $\tau \to e\nu\nu$, τ_μ the muon lifetime and m_μ and m_τ the masses of the muon and tau respectively. The uncertainty in the prediction comes from the error in the determination of B_e which was taken to be $B_e = 0.176 \pm 0.016$, a value measured by the MARK-II group[20]. Several groups at PEP and PETRA have measured τ_τ, the first one being the MARK-II group at PEP. As they have obtained the best value so far I will concentrate on their result[24].

Their data sample contains approximately 1.500 $\tau^+\tau^-$ pairs produced at a centre of mass energy of 29 GeV. From this data sample only those events are selected which have at least one jet with three charged particles and a charge of ±1. The visible invariant mass of each jet has to be less than 1.6 GeV/c^2 for the three charged particles and less than 1.8 GeV/c^2 for all detected particles including photons from decaying π^0's. Applying various other cuts to reduce possible backgrounds from $e^+e^- \to e^+e^-\tau^+\tau^-$, $e^+e^- \to$ hadrons and $e^+e^- \to e^+e^-\gamma$, there remain 284 events with 306 three prong τ-decays. Track fitting and additional criteria to obtain a common vertex for the three charged particles reduce the sample even further to 126 events. For these events the uncertainty in the vertex position (σ) is plotted in Fig. 13. The decay lengths for the events with $\sigma < 8$ mm and $\sigma < 4$ mm are shown in Fig. 14a and Fig. 14b respectively. An excess of events is observed with a decay length greater than zero. The measurement error is larger than the decay length and one assumes that the observed distribution is the convolution of an exponential decay curve with the experimental resolution function. For fake τ's from hadronic events the distribution of the decay length centers around zero as is shown in Fig. 14c. Knowing the resolution function the average decay length is measured to be 1.07 ± 0.37 mm, which leads to a tau lifetime of

$$\tau_\tau = (4.6 \pm 1.9) \times 10^{-13} \text{ sec.}$$

At the Paris conference of July 1982 the MARK-II group[25] reported the result obtained with a specially designed vertex detector. The value for the lifetime was found to be

$$\tau_\tau = (3.31 \pm 0.57 \pm 0.60) \times 10^{-13} \text{ sec.},$$

in good agreement with the abovementioned theoretical value.

Other measurements for the lifetime obtained with similar methods were reported by the MAC group[26], $\tau_\tau = (4.9 \pm 2.0) \times 10^{-13}$ sec., the TASSO group[27], $\tau_\tau = (0.8 \pm 2.2) \times 10^{-13}$ sec. and the CELLO group[28] with $\tau_\tau = \left(4.7 {}^{+3.9}_{-2.7}\right) \times 10^{-13}$ sec.

E. Cross section and Asymmetry Measurements of $e^+e^- \to \tau^+\tau^-$ at PEP and PETRA energies

The total cross sections for the process $e^+e^- \to \tau^+\tau^-$ at the highest e^+e^- beam energies have been measured by various groups. As only a fraction of the τ-events is accepted, depending on the decay modes studied, the number of obtained events is for most of the experiments smaller than in the case of $\mu^+\mu^-$ pair production. The systematic errors are higher than in the case of $\mu^+\mu^-$ pair cross section determinations due to the uncertainties in the decay modes of the τ-lepton and they range from 10 to 20%. In Fig.

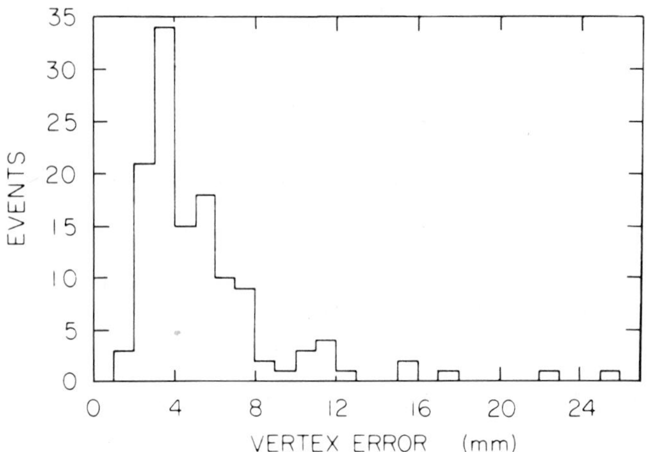

Fig. 13. Distribution of vertex uncertainties in the direction of the τ flight path as measured by the MARK-II Collaboration.

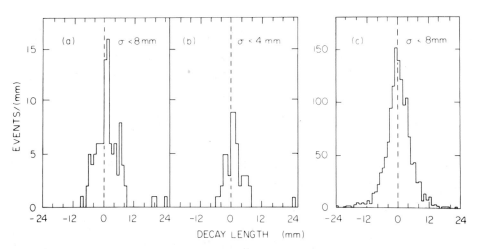

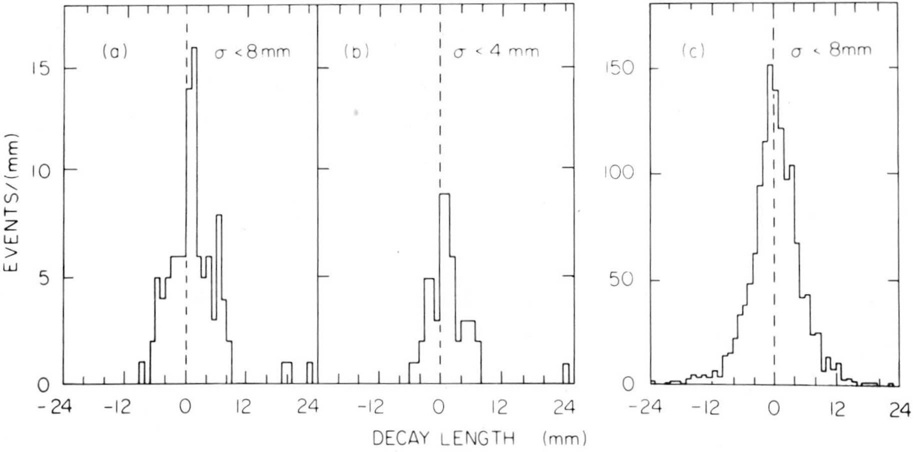

Fig. 14. Flight distance distributions for τ events with vertex uncertainties less than (a) 8 mm (b) 4 mm, and (c) for fake τ's made from hadronic events (MARK-II).

15 the total cross sections at various PETRA energies, normalized to the expectations from QED, $R_{\tau\tau}$, are shown. No deviations from pure QED are seen in $R_{\tau\tau}$ and this is expressed in the cut-off parameters $\Lambda_{\pm}$ [11]. $R_{\tau\tau}$ can be related in a simple way to $\Lambda_{\pm}$ according to:

$$R_{\tau\tau} = 1 \pm s/\Lambda_{\pm}^{2} ,$$

and the dashed lines in Fig. 15 indicate the expectations for a modified QED with $\Lambda_{\pm}$ = 120 GeV. Pure QED and the prediction of the standard electro-weak theory with $\sin^2\Theta_W = 0.23$ coincide and are indicated by the solid line at $R_{\tau\tau} = 1$.

In Fig. 16 the distributions of the tau pair events as function of the scattering angle $\cos\Theta$ at the highest PETRA energies are shown[29]. The dashed lines in these four distributions show the $1 + \cos^2\Theta$ behaviour for zero-order QED, the solid lines show the fit to the data of the standard electro-weak theory. In Fig. 17 the $\cos\Theta$ distributions from the PEP experiments is shown. The results for the tau asymmetry are listed in Table III. The combined PEP result from the MAC and MARK-II experiments at $\sqrt{s}$ = 29 GeV is $A_{\tau\tau} = (-1.7 \pm 2.5)\%$ while -6.3% is expected. The four PETRA groups, CELLO, JADE (preliminary result), MARK-J and TASSO obtain $A_{\tau\tau} = (-7.9 \pm 2.2)\%$ while - 9.3% is expected from the standard model.

IV. TESTS OF THE STANDARD ELECTRO-WEAK INTERACTION MODEL

A. Measurements of $\sin^2\Theta_W$

The theory of Glashow-Weinberg-Salam (GWS) [30] is characterized by a parameter denoted $\sin^2\Theta_W$ and neutrino-nucleon scattering experiments[31] yield a value of $\sin^2\Theta_W = 0.234 \pm 0.011$.

At e^+e^- machines complementary experiments have been performed measuring electro-weak parameters from the reactions:

$$e^+e^- \to (\gamma + Z^0) \to \ell^+\ell^- \quad (\ell = e, \mu, \tau) \tag{9}$$

Table III. The measured tau asymmetry values for the various e^+e^- experiments. Also indicated are the expectations from the standard G.W.S. model with $\sin^2\Theta_W = 0.23$.

Experiment	$\sqrt{s}$ (GeV)	$A_{\tau\tau}$ (%)	Expected in G.W.S. with $\sin^2\Theta_W = 0.23$
MAC	29	-1.3±2.9	-6.3
MARK II	29	-3.2±5.0	-5.0
Combined	29	-1.7±2.5	-6.0
CELLO	34.2	-10.3±5.2	-9.2
JADE (prelim.)	34.2	-7.9±3.9	-9.2
MARK-J	34.6	-8.4±4.4	-9.4
TASSO	34.4	-5.4±4.5	-9.3
Combined	34.4	-7.9±2.2	-9.3

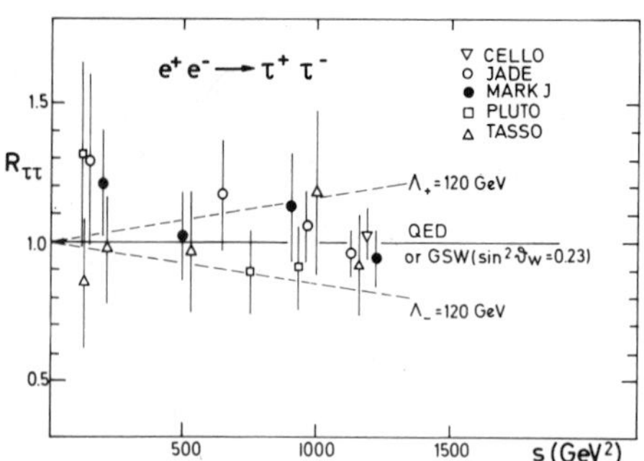

Fig. 15. The measurements of $R_{\tau\tau}$ for the four PETRA groups as function of s, the square of the centre of mass energy. The solid line at $R_{\tau\tau} = 1$ is the QED expectation coinciding with the solution for electro-weak theory with $\sin^2\Theta_W = 0.23$. The dashed lines are the expectations for a modified QED with cut-off parameters $\Lambda_\pm = 120$ GeV.

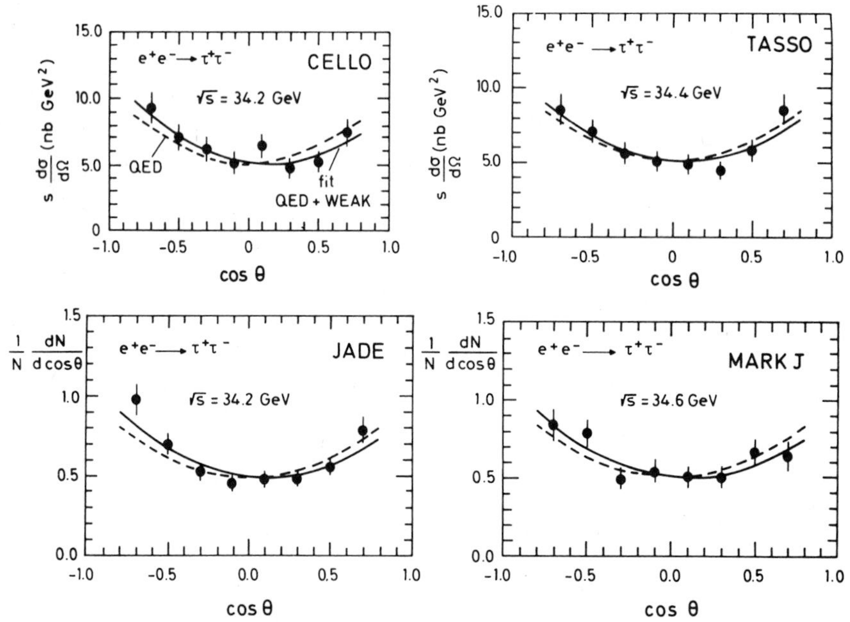

Fig. 16. The cosΘ distributions for the process $e^+e^- \to \tau^+\tau^-$ at the highest PETRA energy. The dashed lines are the QED expectations and the solid lines the fit to the data of the electro-weak theory.

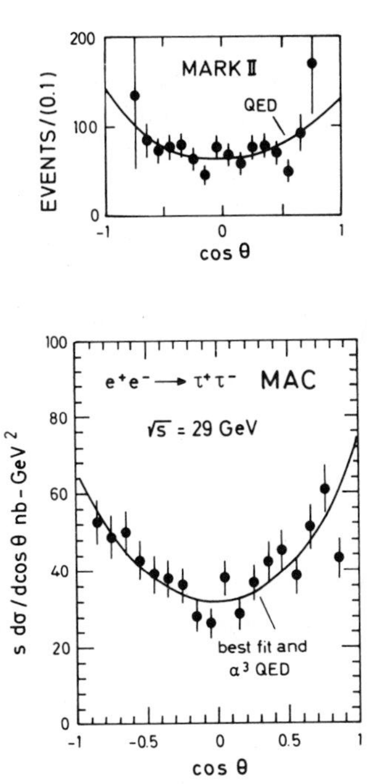

Fig. 17. The cosΘ distributions for the process $e^+e^- \rightarrow \tau^+\tau^-$ for two PEP experiments.

where the production of final state leptons is via both virtual photons (γ) and neutral intermediate vector bosons (Z^0).

The measurements consist of the already discussed data on muon and tau pair production. In addition the measurements of the angular distribution of the Bhabha scattering events are used. In Fig. 18 some of the available $e^+e^- \rightarrow e^+e^-$ data[29] at the highest PETRA energies are shown. The differential cross section $d\sigma/d\Omega$ is normalized with the first order QED cross section $d\sigma_{QED}/d\Omega$ and is plotted as function of $\cos\Theta$. The luminosity is measured by Bhabha scattering in the central detectors and independently in monitoring stations centered at small scattering angles (~30 mrad). The two luminosity measurements generally agree within 3% once radiative corrections have been taken into account[32,33] and one can take 3% as a conservative estimate of the error in the luminosity for reaction (9).

The results of the determination of $\sin^2\Theta_W$ are summarized in Table IV. The values obtained, assuming e-μ-τ universality range from $\sin^2\Theta_W = 0.21$ to 0.27. The average value for the e^+e^- experiments is found to be $\sin^2\Theta_W = 0.27 \pm 0.07$, which compares well with the world average of the νN and ed experiments of $\sin^2\Theta_W = 0.23 \pm 0.01$ and the value measured by the νe elastic scattering experiments of $\sin^2\Theta_W = 0.27 \pm 0.03$. In Fig. 19 the values for $\sin^2\Theta_W$ measured by the various types of experiments are plotted as function of Q^2, the momentum transfer, and one observes an impressive agreement.

B. Determination of g_A^2 and g_V^2

In comparing the data on $e^+e^- \rightarrow \ell^+\ell^-$ (ℓ = e, μ, τ) with the predictions of electro-weak theories containing one Z^0, the yields for reaction (9) are calculated according to QED using Monte Carlo programs which include order α^3 radiative corrections. The sole assumptions required for the determination of g_A^2 and g_V^2 are

1. e, μ, τ universality,
2. an effective interaction Hamiltonian which is a sum of products of

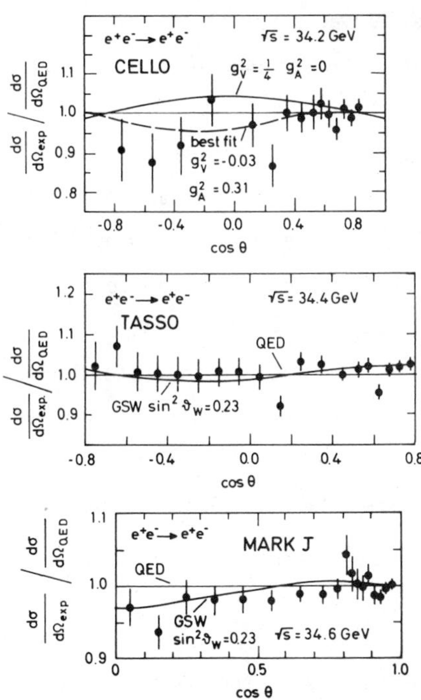

Fig. 18. The cosΘ distributions for the process $e^+e^- \to e^+e^-$ for three PETRA groups. The measured points are normalized with the QED expectations. The various curves are predictions of the electroweak models under assumptions as indicated.

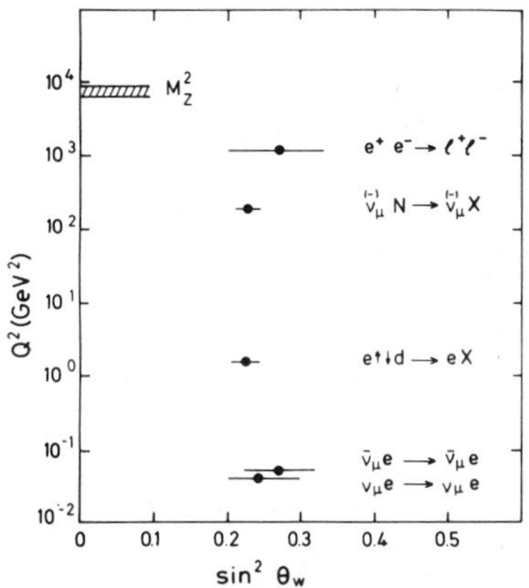

Fig. 19. The measured values of $\sin^2\Theta_W$ by various types of experiments as function of Q^2.

vector and axial-vector currents and,

3. the point-like nature of the leptons.

The results for g_V^2 and g_A^2 are summarized in Table IV and plotted in Fig. 20 in the $g_V^2 - g_A^2$ plane. The combined result of the JADE, MARK-J and TASSO groups of their ee and $\mu\mu$ data is $g_V^2 = 0.01 \pm 0.04$ and $g_A^2 = 0.28 \pm 0.04$, while 0.02 and 0.25 are expected from the GWS model with $\sin^2\Theta_W = 0.23$ respectively. In the $g_V^2 - g_A^2$ plane of Fig. 20 the GWS model is represented by a line parallel to the g_V^2 axis and intersecting the g_A^2 axis at 0.25. These results can be converted including correlations into an allowed region in the $g_V - g_A$ plane as shown in Fig. 21. Also shown in the figure are the combined results of the neutrino electron scattering data $\nu_\mu e^- \to \nu_\mu e^-$, $\bar{\nu}_\mu e^- \to \bar{\nu}_\mu e^-$, $\bar{\nu}_e e^- \to \bar{\nu}_e e^-$, which limit the possible value of g_A and g_V to two regions in the $g_V - g_A$ plane around ($g_V = 0$, $g_A = -\frac{1}{2}$) and ($g_V = -\frac{1}{2}$, $g_A = 0$)[34]. Combining the e^+e^- data and the neutrino scattering results the second solution is ruled out with more than 95% confidence. This confirms the conclusion drawn on the basis of deep inelastic neutrino nuclear scattering and polarized electron deuterium scattering data[35].

V. SEARCHES FOR NEW LEPTONS[36]

A. Search for μ^*

If a μ^* exists with a relatively small mass ($M < 10$ GeV/c^2), a sample of $e^+e^- \to \mu^+\mu^-$ events from one photon annihilation would be contaminated by events from $e^+e^- \to \mu^{*+}\mu^{*-}$. The number of events predicted from the latter process is shown as a function of mass M in Fig. 22a, along with the 95% C.L. upper limit which results from the MARK-J measurement of $\sigma(e^+e^- \to \mu^+\mu^-) = \sigma_{\mu\mu}$. As demonstrated by the figure, a μ^* with $M < 10$ GeV/c^2 is excluded.

A higher mass μ^* would lead to $\mu\mu\gamma$ final states from $e^+e^- \to \mu^*\mu$. The differential cross section for this process is

Table IV. The values obtained for $\sin^2\Theta_w$ and g_V^2 and g_A^2, the vector and axial coupling constants squared of the weak neutral current, by the various e^+e^- experiments.

Experiment	Reaction	$\sin^2\Theta_w$	g_V^2	g_A^2
MAC	ee, μμ	0.24±0.10	-	0.18±0.10
MARK II	ee, μμ	-	0.05±0.10	0.24±0.16
CELLO	ee, μμ, ττ	$0.21^{+0.14}_{-0.09}$	0.03±0.08	0.31±0.12
JADE	μμ		0.05±0.08	0.29±0.06
MARK-J	ee, μμ, ττ	0.26±0.09	-0.02±0.05	0.28±0.06
TASSO	ee, μμ, ττ	0.27±0.07	-0.04±0.06	0.26±0.07
Combined (JADE, TASSO MARK-J)	ee, μμ	0.27±0.07	0.01±0.04	0.28±0.04

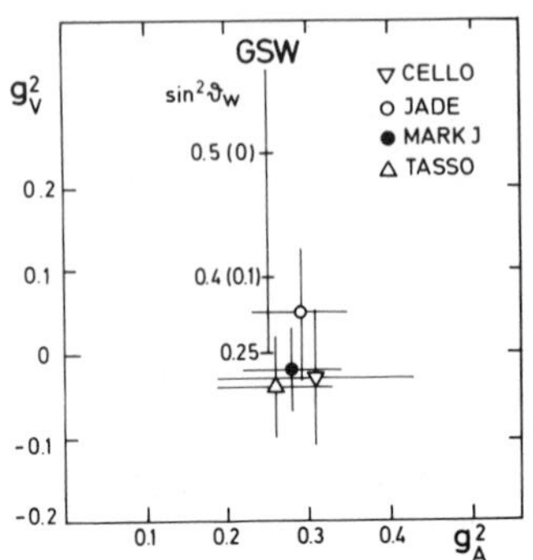

Fig. 20. Best fit values for g_V^2 and g_A^2 and their uncorrelated errors from the PETRA experiments.

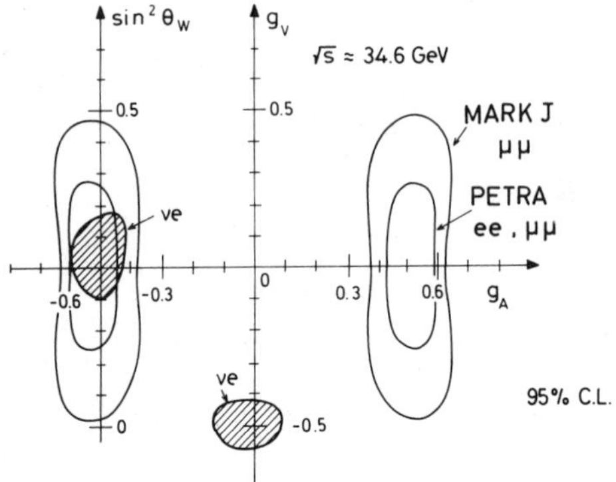

Fig. 21. Results obtained from neutrino experiments and the PETRA experiments expressed in terms of 95% confidence limits on g_V and g_A. The hatched areas are the elastic neutrino electron scattering results. The 95% C.L. contours from the PETRA experiments are indicated by solid lines.

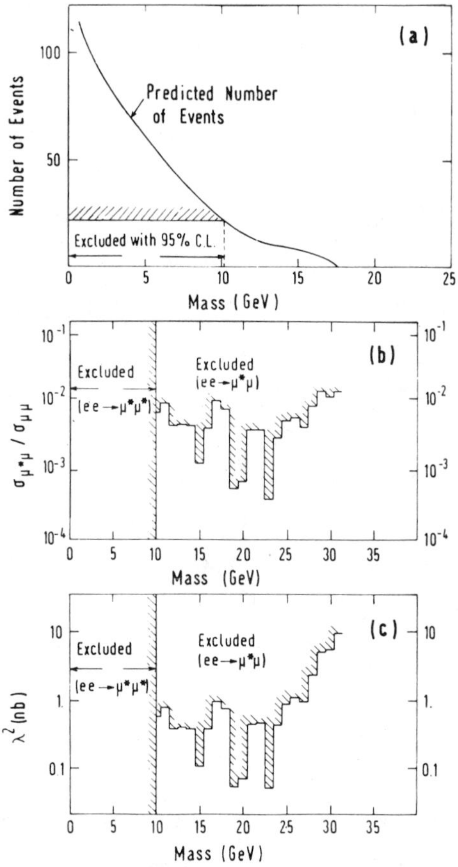

Fig. 22. (a) Predicted number of muon pairs coming from the process $e^+e^- \to \mu^{*+}\mu^{*-} \to \mu^+\mu^-\gamma\gamma$ as a function of M. The horizontal line corresponds to the 95% confidence level upper limit from the measurement of $e^+e^- \to \mu^+\mu^-$. A μ^* with M < 10 GeV is excluded.

(b) 95% Confidence level upper limit on $\sigma_{\mu^*\mu}/\sigma_{\mu\mu}$ as a function of the μ^* mass M.

(c) 95% Confidence level upper limit on λ^2 as a function of M.

$$\frac{d\sigma}{d\Omega} = \lambda^2\alpha^2 \frac{\left(s-M^2\right)^2}{s^3} \{\left(s + M^2\right) - \left(s^2 - M^2\right)\cos\Theta\} , \qquad (10)$$

where M is the mass of the μ^* and λ is the coupling defined by the following phenomenological interaction Lagrangian:

$$L_{eff} = \lambda e\bar{\Psi}_{\mu^*}\sigma_{\beta\delta}\Psi_\mu F^{\beta\delta} + h.c. \qquad (F^{\beta\delta} \text{ is the e.m. field tensor})$$

If the μ^* is massive and decays rapidly via $\mu^* \to \mu\gamma$, $\mu^*\mu$ events would be mixed with the radiative production of muon pairs $e^+e^- \to \mu^+\mu^-\gamma$, and with a small contribution of $e^+e^- \to e^+e^-\mu^+\mu^-$ production where two muons and an electron are observed in the detector. The MARK-J group used the following selection criteria:

- 1 muon with $p_t \gtrsim 1.5$ GeV/c, which penetrates the detector iron and reaches the outer drift chamber; a second muon with $p_t > 0.8$ GeV/c, which reaches the trigger counter midway through the iron;
- $E_{elm} \geq 20\%\ E_{beam}$;
- $\Theta_{elm} \geq 25°$;
- $|p_{\mu_1} + p_{\mu_2} + p_\gamma| < 0.25\sqrt{s}$ (momentum balance);

where E_{elm} and Θ_{elm} are the magnitude and angle (with respect to the beam axis) of the electromagnetic energy measured in the shower counters. With this selection the Monte Carlo calculation predicted 116 events coming from $e^+e^- \to \mu^+\mu^-\gamma$ and $e^+e^- \to e^+e^-\mu^+\mu^-$ [5]. They observed 108 events in their data with $\int Ldt = 43\ pb^{-1}$ for $\sqrt{s} > 30$ GeV.

A Monte Carlo study of the μ^* mass resolution ΔM was performed by computing the two $\mu\gamma$ mass combinations from the muon and photon angles in the final state of $e^+e^- \to \mu^*\mu \to \mu\mu\gamma$. They found $\Delta M \leq 1$ GeV/c^2, independent of M.

The good $\mu\gamma$ mass resolution was exploited by computing n(M), the maximum number of events that can be attributed to a μ^* of mass M, where the measured mass lies in the range from M − 1 to M + 1 GeV/c^2. The MARK-J group computed the corresponding acceptance, A(M), and found upper limits for the production cross section of $\mu^*\mu$:

$$\sigma_{\mu^*\mu} < \frac{n(M)}{\int L dt \cdot A(M)} \qquad (95\%\ \text{C.L.})$$

Their limits on $\sigma_{\mu^*\mu}/\sigma_{\mu\mu}$ are shown in Fig. 22b as a function of M. As seen for M < 30 GeV/c^2, the production cross section for μ^*, $\sigma_{\mu^*\mu}$, is <1% of $\sigma_{\mu\mu}$. From eq. (10) the limit can also be expressed in terms of λ^2, as shown in Fig. 22c. One therefore sees that if the muon is a composite particle, the splitting between ground state and first excited state must be very large: $M/m_\mu \gtrsim 10^2$ (95% C.L.). The interaction binding the constituents must then be very different from the known interactions. Results from the other PETRA groups can be found in Table V.

B. Search for e^*

If there is an excited state of the electron, the process $e^+e^- \rightarrow \gamma\gamma$ would be modified by the exchange of a virtual e^* in the t-channel[36,37,38]. As in the case of the μ^*, the differential cross section[37] depends on the coupling λ' and on the mass M' of the e^*, where λ' is defined by the interaction Lagrangian:

$$L_{eff} = \frac{\lambda' e}{2M'} \bar{\psi}_{e^*} \sigma_{\mu\nu} \psi_e F^{\mu\nu} + h.c.$$

As shown in Fig. 23a, the measurement of the differential cross section agrees quite well with the prediction of pure QED including radiative corrections[5]. By making a two-parameter fit (λ', M') to the data, using the exact form of the cross section derived in Ref. 37, one obtains a 95% C.L. upper limit on λ' as a function of M'. Fig. 23b illustrates the MARK-J result. They find that for λ' = 1, the lower limit on M' is 58 GeV/c^2.

Table V. The results of the particle searches for the PETRA experiments. The mass ranges or lower limits excluded with 95% C.L. in GeV/c^2 are tabulated. The numbers with a * are 90% C.L.

Particle type	CELLO	JADE	MARK-J	PLUTO	TASSO
e^*	43	47	58	46	34
μ^*	30^*	32^*	30	-	-
sequential lepton		18.1	16	14.5	15.5
stable lepton			14		
neutral lepton		3-17 (V-A) 3-20 (V+A)			
scalar leptons					
s_e	2 -16.8	16	-	13	15
s_μ	3.3-15.3	-	3-15	-	14
s_τ	6. -15.3	4-13	14	-	

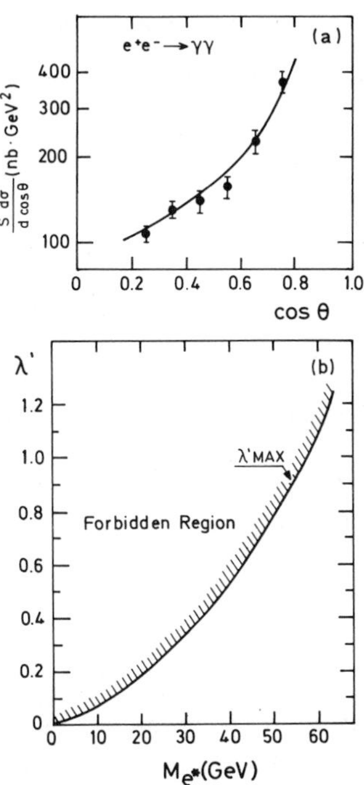

Fig. 23. (a) Differential cross section scaled by s ($\sqrt{s} > 33$ GeV) for the process $e^+e^- \rightarrow \gamma\gamma$, as a function of cosΘ. Θ is the angle of the photons with respect to the beam axis.

(b) 95% Confidence level upper limit on the coupling λ' as a function of the mass of the e^*. A systematic error of 3% on the luminosity measurement and a point-to-point systematic error of 3% were used in the fit.

Therefore their conclusion is that if an e^* exists, its mass must be large or its coupling strength very small: $M'/m_e \geq 10^5$ for $\lambda' = 1$.

C. Heavy Lepton Searches

Following the initial searches[13,14] for the τ-lepton, the discovery[15] and the further study of its properties[16,17,18], there has been great interest in searching for a new heavy lepton, which would extend the series e, μ, τ. Analogous to the τ-lepton one assumes that a new heavy lepton HL couples universally to leptons and quarks according to the standard V-A weak interaction theory and has the following decay modes: (in %), $\tau^- \bar{\nu}_\tau \nu_{HL}$ (9.2), $\mu^- \bar{\nu}_\mu \nu_{HL}$ (10.6), $e^- \bar{\nu}_e \nu_{HL}$ (10.6) and hadrons $+\nu_{HL}$ (69.6). The branching ratios[39] which are mass dependent, are given here for a heavy lepton mass of 14 GeV/c^2. Owing to their large mass and low velocity, the decay products would be expected to have large angles with respect to the HL line of flight. This contrasts to the decay products of the τ at PETRA energies which are tightly collimated.

Heavy lepton production is recognized in the MARK-J detector for events in which one lepton decays into a muon and neutrinos and the other lepton decays into hadrons and neutrinos. Hadrons are detected by their energy deposit, E_{vis}, in the calorimeter[7].

Heavy lepton candidates with masses greater than 6 GeV/c^2 are selected by applying criteria involving cuts on the total deposited energy, the acoplanarity, the charge multiplicity of the events and the total amount of hadronic energy and its direction. No events were observed in the data with $33 < \sqrt{s} < 36.7$ GeV, corresponding to a time integrated luminosity of 6.9 pb^{-1}. The number of events predicted by a Monte Carlo calculation as a function of the heavy lepton mass M_{HL} is shown in Fig. 24 along with the 95% C.L. upper limit from the data. Fig. 24 demonstrates that the existence of a sequential heavy lepton with a mass between 6 and 16 GeV/c^2 is excluded.

Heavy leptons with $M_{HL} < 6$ GeV/c^2 would decay into final states similar in appearance to those from τ-decay, and would tend to be included

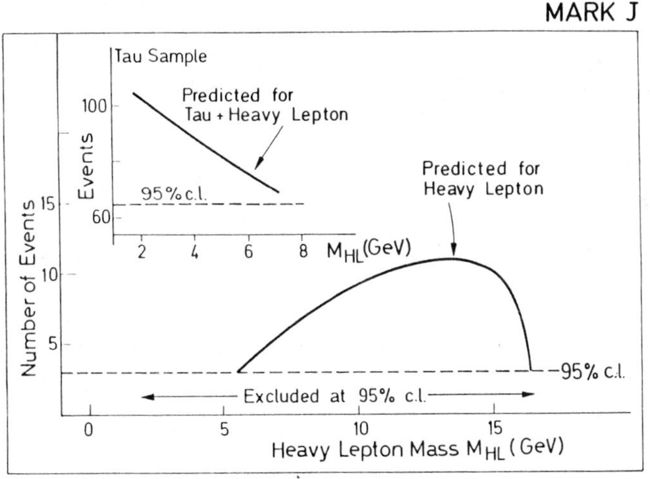

Fig. 24. Number of events expected by the MARK-J group for the production of a new (sequential) heavy lepton as a function of mass. The inset shows the number of events expected in the τ sample from tau and heavy lepton production. A total of 52 τ events is observed. The dashed line corresponds to the 95% confidence upper limits for τ events.

in the sample of $e^+e^- \to \tau^+\tau^-$ events. The inset in Fig. 24 shows the Monte Carlo prediction for the total number of events in the sample from $e^+e^- \to \tau^+\tau^-$[40] and from heavy lepton production, as a function of M_{HL}. The inset demonstrates that the predicted number of heavy lepton events exceeds by more than 2σ the number of observed τ-events. Therefore, with more than 95% confidence they exclude the existence of the heavy lepton with mass $M_{HL} < 6\ \text{GeV}/c^2$. One is thus able to rule out the existence of a new heavy lepton for a $M_{HL} < 16\ \text{GeV}/c^2$.

The criteria employed by the PLUTO group[41] for finding heavy leptons are very similar to the ones mentioned above and they find a limit for M_{HL} of $14.5\ \text{GeV}/c^2$.

The TASSO group[42] searches for a single charged particle recoiling against many hadrons. The angle between the single or "lone" track and the nearest neighbour is plotted in Fig. 25. A new heavy lepton would give an excess of events in this plot at large angles. From the actual observed number, taking into account the background from hadron production, an upper limit on the mass is found to be $M_{HL} < 15.5\ \text{GeV}/c^2$ at 90% C.L.

The JADE group concentrates on the possible hadronic decays of the heavy leptons. They calculate the direction of the two hadron jets and demand that the jets are very acoplanar, excluding by such criteria the events produced by quark-antiquark pairs which give rise to two back-to-back jets. They expect for a $17\ \text{GeV}/c^2$ lepton ~10 events, and they find none[43].

D. Search for a stable heavy lepton[44]

There have been many conjectures concerning the existence of a fourth family of leptons[45]. One interesting possibility is that the neutrino in this fourth family could be more massive than the charged lepton, thus leaving this new heavy lepton relatively stable[46]. Such a new lepton would be pair produced in e^+e^- annihilation via $e^+e^- \to L^+L^-$, and the events would have an appearance similar to muon pair production with a suitably altered momentum spectrum.

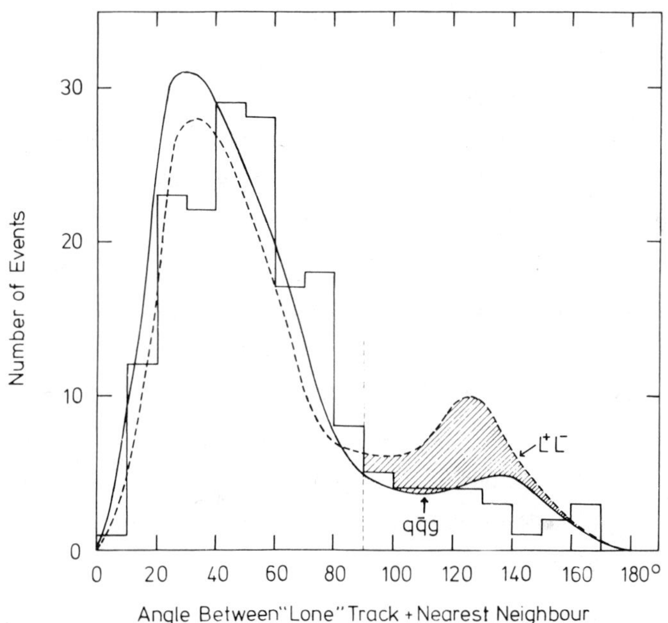

Fig. 25. The distribution of the angle between the "lone" track and the nearest neighbour as measured by the TASSO-Collaboration. The prediction of the production of a heavy lepton is indicated by the dashed line. The QCD expectation is given by the solid curve.

The event selection criteria are similar to those for muon pair selection in the MARK-J detector, namely:

- Two particles must reach the outermost drift chambers within 10 nsec of the beam crossing time.
- Each particle's momentum must be > 50% E_{beam}.
- The acollinearity angle between the particle directions must be <20°.

The number of events expected with these selection criteria is shown in Fig. 26 as a function of the $L^{\pm}$ mass. The prediction for the $L^{\pm}$ momentum spectrum as a function of $L^{\pm}$ mass is shown in the inset. In these predictions, ionization energy loss in the detector[47] and initial state radiative corrections (to order α^3) were taken into account. It should be noted that the genuine muon pairs are included in the expected number of events. The measurement of 57 events in the energy range from 36 GeV to 36.7 GeV corresponding to a time integrated luminosity of 1854 nb^{-1} gives a 95% confidence level lower limit of 14 GeV/c^2 for the mass of $L^{\pm}$.

As a consistency check the MARK-J group also studied the timing of all particles reaching their trigger counters situated 1.35 meters from the interaction region. No event for which the particles were more than 4 ns later than expected for a muon were found. This excludes stable particles with a mass between 9 and 14 GeV/c^2 at the 95% C.L., confirming their previous result. In order for an $L^{\pm}$ to reach the exterior drift chambers, its lifetime in the lab frame should be >> 4 ns.

E. Heavy Neutral Electron

The JADE collaboration has searched for a heavy neutral partner of the electron. These particles appear in non-standard models of the weak interaction[48]. When the neutral current contains non-diagonal terms they can be produced in neutrino and e^+e^- interactions. In models where only diagonal terms are present to describe the neutral current the heavy neutral leptons can only be produced in e^+e^- interactions according to:

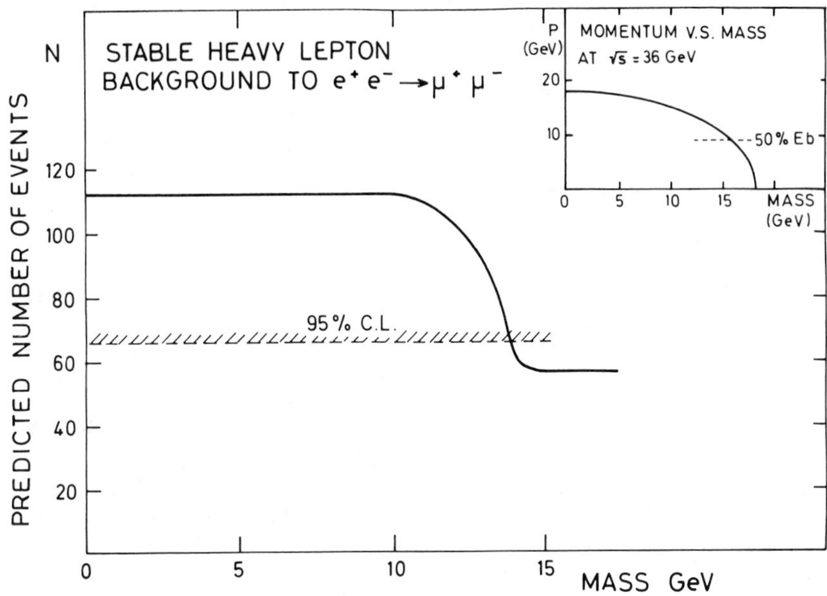

Fig. 26. The expected number of events combined from the processes $e^+e^- \to \mu^+\mu^-$ and $e^+e^- \to L^+L^-$ as a function of the mass of the $L^\pm$. The 95% confidence level limit shown corresponds to the MARK-J measurement of 57 events at $\sqrt{s}$ = 36 GeV. The inset displays the predicted momentum for the $L^\pm$ as a function of its mass at $\sqrt{s}$ = 36 GeV.

$$e^+e^- \rightarrow E^0 + \nu_e \quad \text{and} \quad e^+e^- \rightarrow E^0 + E^0 .$$

Depending on the coupling of the charged current to these particles which can be either V-A or V+A, the differential cross sections for the case of an exchange of a charged weak boson in the t channel can be written as:

$$\frac{d\sigma^{V-A}}{d\Omega_{E^0 (E^0)}} = \frac{G_F^2}{32\pi^2}\left(1 - \frac{M^2}{s}\right)\left(\frac{1}{1-q^2/M_W^2}\right)\left\{2s(1\pm\cos\Theta) - (s-M^2)\sin^2\Theta\right\}$$

$$\frac{d\sigma^{V+A}}{d\Omega_{E^0}} = \frac{G_F^2}{32\pi^2}\left(1 - \frac{M^2}{s}\right)\left(\frac{4s}{1-q^2/M_W^2}\right)$$

where G_F is the Fermi coupling constant, M the mass of E^0 or E^0, M_W the mass of the exchanged charged weak boson, q^2 the momentum transfer in the t-channel and Θ the angle between the beam direction and the produced E^0. The possible decay channels for the E^0 are $E^0 \rightarrow e^+e^-\nu_e$ (10%), $e^-\mu^+\nu_\mu$ (10%), $e^-\tau^+\nu_\tau$ (10%), $e^-u\bar{d}$ (35%) and $e^-c\bar{s}$ (35%) respectively.

The JADE group[49] searched for the events produced via t-channel exchange and for hadronic decays of the lepton. As a ν_e is always produced in such a reaction, the signature of the events is the imbalance of energy and momentum along the line of flight of the neutrino. The experimental cuts were made such that the event had an empty cone (half opening angle 50°) containing no charged particles or neutral energy, opposite to a hadronic jet containing one electron.

The expected number of events as a function of the mass are presented in Fig. 27 for the V+A and V-A case. One event was found which could be due to $\tau^+\tau^-$ production. From the 95% C.L. limit it can be concluded that the neutral heavy leptons do not exist below 18 GeV/c^2 if the coupling is V-A and not below 20 GeV/c^2 if the coupling is V+A[49].

F. Scalar Lepton Searches

In the framework of supersymmetric theories[50], spin zero partners of the electron, the muon and the tau are expected to decay only according to

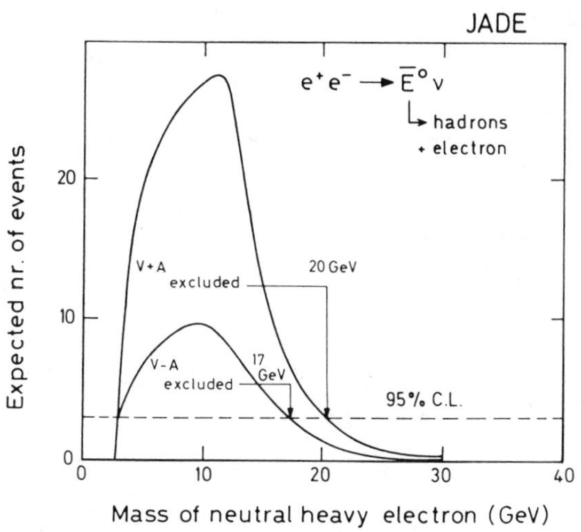

Fig. 27. The number of expected events for heavy neutral lepton production as function of the mass. The 95% C.L. line is also indicated.

the reactions:

$$\bar{s} \rightarrow e^- \; (\mu^-, \; \tau^-) + \text{photino (goldstino)}$$
$$t^- \rightarrow e^- \; (\mu^-, \; \tau^-) + \text{antiphotino (antigoldstino)}$$

where s and t are the spin zero partners of the electron (μ, τ) associated with the left and right handed parts of the electron (μ, τ) field respectively, and the photino and goldstino are the spin ½ partners of the photon and the goldstone boson. Since s and t carry unit electric charge they may be produced in pairs in e^+e^- annihilation according to the cross section:

$$\frac{d(e^+e^- \rightarrow s^-s^+ \text{ or } t^-t^+)}{d(\cos\Theta)} = \frac{\pi\alpha^2\beta^3\sin\Theta}{4s} \; ; \quad \beta = \{1 - (m/E)^2\}^{\frac{1}{2}}$$

which is characteristic of spin zero particle production. Herein, m is the mass of s or t, E is the beam energy, and Θ is the scattering angle.

Because of the uniqueness of the decay reactions and the extremely short lifetimes of s and t and the prediction that the interaction cross section of photino and goldstino are expected to be very small, only electron, muon and tau pairs are observed in the final state. Near threshold production of s and t, the two residual electrons, muons or taus would be produced isotropically in space. Data from SPEAR place a lower limit of 3.5 GeV/c^2 [51] on the mass of s and t. Thus, over the PETRA energy range of 12 and 36.7 GeV an increase in the production of acoplanar electron-positron, muon or tau pairs should be observed if a new threshold is passed.

The event selection criteria used by the PETRA groups to obtain upper limits for the production of these particles are based on the considerations mentioned above. I will describe the procedure followed by the CELLO group in some detail[52]. At an average beam energy $\langle E_b \rangle = 17.3$ GeV the events triggered by at least two charged particles were subjected to the following cuts;

1. The angle between the tracks and the beam direction is constrained

to be $35° < \Theta < 145°$.

2. The acoplanarity is less than 30°.
3. The charged track multiplicity is less than 5.
4. Two tracks have a momentum greater than 0.8 GeV.
5. For e^+e^- pairs the sum of the energies must exceed 5 GeV and for $\mu^+\mu^-$ pairs the momentum of the muon must be greater than 20% of the beam energy.

In the CELLO detector the electrons and photons are identified with a lead-liquid argon calorimeter and the muons with proportional chambers behind 80 cm of iron. No candidates were found. Fig. 28 shows the expected event rate for the e scalar leptons under the abovementioned conditions. At the 95% C.L. the mass range $2 < M_s < 16.6$ GeV/c^2 is excluded. Similar limits were obtained by the other PETRA experiments[53,54].

The supersymmetric particles s_τ and t_τ, associated respectively with the left- and right-handed parts of the tau field, decay rapidly with a branching ratio of 100% into τ's plus photinos or goldstinos. Since photinos and goldstinos are expected to interact very weakly with matter[55], the observed final state would appear as the decay products of a pair of tau leptons.

As in the study of tau pairs from the one photon annihilation channel the MARK-J group searched for the case[56] where one of the taus decays into $\mu\ \bar{\nu}_\mu\ \nu_\tau$ and in which the other tau decays into ν_τ + hadron(s). If the s_τ or t_τ mass M is not too large compared to the tau mass, one would expect a substantial excess of tau pair events from the one photon annihilation channel. In the τ data sample of the process $e^+e^- \to \tau^+\tau^-$ they obtained 269 events with $\sqrt{s} > 30$ GeV, corresponding to a luminosity of 40 pb^{-1}. This yields a 95% upper C.L. of 23 events in excess of the number predicted by QED. The limit takes into account the error on the measured value of the branching ratio for $\tau \to \mu\ \nu_\mu\ \nu_\tau$, the error on the luminosity measurement of 3%, and the statistical error on the Monte Carlo calculation of the acceptance for this reaction. This result, together with the corresponding Monte Carlo prediction for s_τ or t_τ production and decay, is

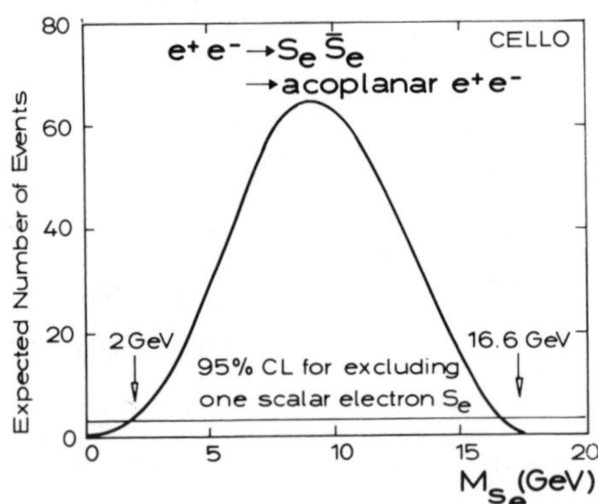

Fig. 28. Number of events expected for the production of a spin zero partner s_e or t_e of the electron as a function of mass. The upper limit of events (95% confidence) and the mass range excluded is also indicated.

shown in Fig. 29a. It can be seen that masses below 7 GeV/c^2 are excluded.

The search for higher mass s_τ and t_τ relied on the expected acoplanarity angle of the tau pairs resulting from the s_τ or t_τ decay. To suppress backgrounds from the reactions $e^+e^- \rightarrow$ hadrons, $e^+e^- \rightarrow e^+e^-\mu^+\mu^-$ and $e^+e^- \rightarrow \tau^+\tau^-$ the following selection criteria were used: (a) a muon; (b) $E_{vis} < 0.75 \sqrt{s}$, where E_{vis} is the total energy measured; (c) $E_H > 0.2\ E_{vis}$, where E_H is the energy deposited in the hadron calorimeter layers which surround the 18 radiation length thick electromagnetic shower counters in the detector; (d) $E_{EM} > 1.5$ GeV, where E_{EM} is the energy deposited in the electromagnetic calorimeter; (e) a hadronic jet with $E_j > 0.1 \sqrt{s}$, $30° < \Theta_j < 150°$; and (f) $\delta > 30°$, where δ is the acoplanarity angle between the muon and the hadron jet axis.

The Monte Carlo prediction for the number of events corresponding to this selection as a function of the mass M is shown in Fig. 29b. No candidate was found in the data. The corresponding 95% confidence level on the number of events is indicated. It can be seen that a mass range from 6 to 14 GeV/c^2 is excluded.

G. Searches for Higgs Particles and Hyper Pions

If new charged scalars exist, they would be produced in e^+e^- annihilation with a rate relative to $e^+e^- \rightarrow \mu^+\mu^-$ production of approximately $\frac{1}{4}\beta^3$ [57]. The dominant decay modes of $H^\pm$ or $\pi'^\pm$ are expected to be $\tau\nu_\tau$ and $c\bar{s}$ [58], where c and s denote the charmed and strange quarks.

To search for $H^\pm$ and π' from the reactions

$$e^+e^- \rightarrow H^+H^- \rightarrow \tau^+\tau^- + \nu_{\tau^+} \nu_{\tau^-} \quad (11)$$

$$e^+e^- \rightarrow \pi'^+\pi'^- \rightarrow (c\bar{s})\tau^- + \nu_{\tau^-} \quad (12)$$

a detailed simulation of the production and decay of such heavy, charged,

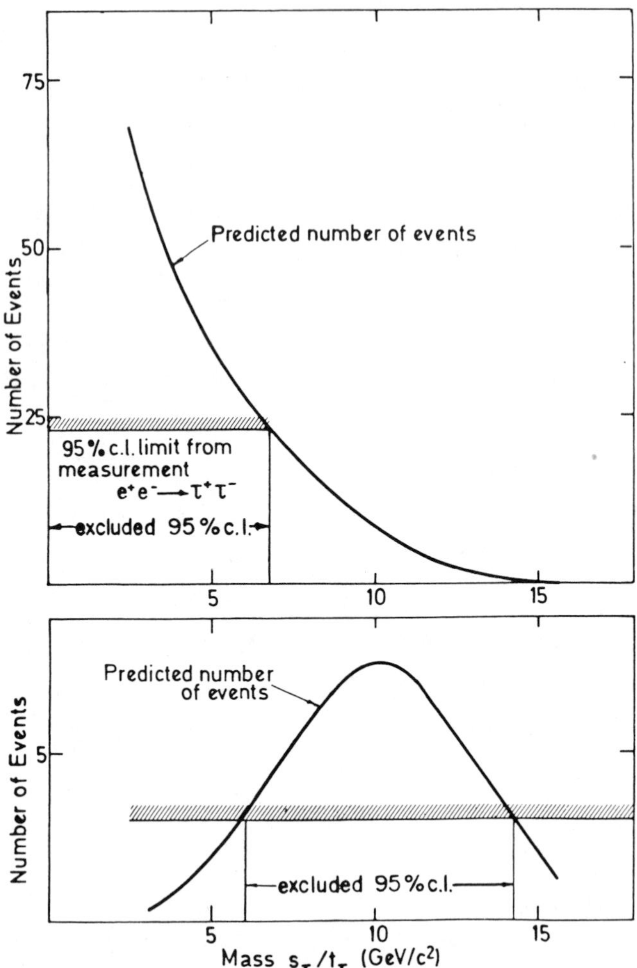

Fig. 29. (a) Predicted excess number of events in the tau pair sample coming from the production of pairs of s_τ or t_τ particles as a function of their mass. The 95% confidence level upper limit from the MARK-J measurement is indicated.

(b) Predicted number of acoplanar muon hadron events as a function of the mass of s_τ or t_τ. The 95% confidence level upper limit corresponding to the measurement is indicated.

spin 0 particles corresponding to various masses and branching ratio to the tau was performed[59]. The kinematics of reaction (11) is identical to the reaction $e^+e^- \rightarrow \tau^+\tau^-$ + photinos or goldstinos and therefore the analysis is identical. The 95% confidence level limit on the branching ratio into τ of Higgs particles from reaction (11) is shown in curve (a) of Fig. 30.

For reaction (12) the MARK-J group selects events with a muon opposite to a hadronic shower, with additional criteria to reject background from the reactions $e^+e^- \rightarrow$ hadrons, $e^+e^- \rightarrow e^+e^-\mu^+\mu^-$ and $e^+e^- \rightarrow \tau^+\tau^-$: (a) $0.3 < E_{vis}/\sqrt{s} < 0.75$; (b) $|\Sigma p_t| > 0.4\ E_{vis}$, where Σp_{out} is the momentum imbalance out of the plane formed by the muon and the beam; (d) $T < 0.93$, where T is the event thrust; (e) more than two tracks found in the inner vertex detector. Criteria a, b, and c are effective for removing muons from hadronic decays, while c, d, and e reduce the background from tau pair events, and criteria a, c, and e suppress the two-photon muon pair contribution.

The Monte Carlo prediction for the number of events corresponding to this selection is calculated as a function of the branching ratio for $H^+ \rightarrow \tau\nu_\tau$. No candidate was found in the data. Using these results, limits on the allowed values of mass and branching ratio were obtained. This, together with results from reaction (11), sets the limits shown as curve (b) in Fig. 30. These limits are extended to lower mass values (below 5 GeV/c^2) by determining the contributions from processes (11) and (12) to the tau pair sample. One can see, for example, that a branching ratio to the tau greater than 25% is excluded for charged Higgs or technipion masses between 5 GeV/c^2 and 13 GeV/c^2.

Contrary to the predictions of many technicolor models, the allowed values for the technipion mass and branching ratio to $\tau\nu_\tau$ are severely restricted.

The results of the other groups at PETRA can be found in Reference 60. The MARK-II group at PEP obtained similar results in their search for charged, spin 0 particles[61].

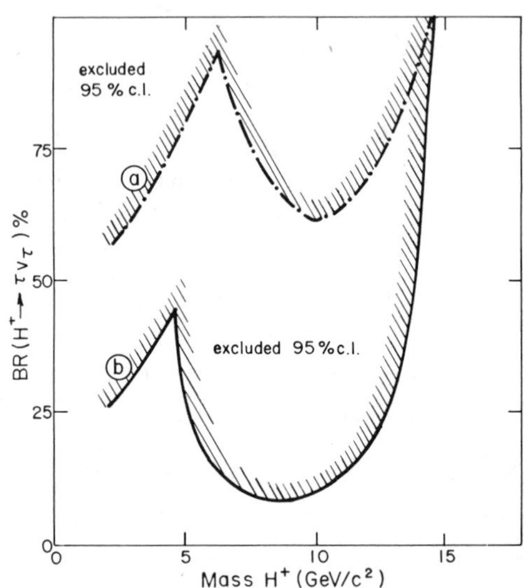

Fig. 30. Limit on the branching ratio of a charged Higgs particle or technipion as a function of the particle mass, under the assumption that the particle decays in either $\tau\nu_\tau$ or $c\bar{s}$. Curve (a) is the limit obtained via the simulation of process (11). Curve (b) is the combined result of processes (11) and (12) with the low mass limit derived from the results in Fig. 29.

VI. CONCLUSIONS

The main results described here can be summarized as follows.

1. Weak effects in the process $e^+e^- \rightarrow \mu^+\mu^-$ are established beyond any doubt. The average muon asymmetry at the highest PETRA energy of $\sqrt{s}$ is measured to be

   $$\langle A_{\mu\mu} \rangle = (-\ 10.4 \pm 1.2)\%,$$

   while - 9.3% is expected from the standard model. A lower limit on the mass of the Z^0 can be derived from this measurement and is found to be $m_Z > 56\ \mathrm{GeV}/c^2$ at 95% C.L.

2. Tau leptons behave like sequential leptons and their measured lifetime τ_τ is in good agreement with the V-A current prediction. The best value measured for τ_τ by the MARK-II group is:

   $\tau_\tau = (3.31 \pm 0.57 \pm 0.60) \times 10^{-13}$ sec. (errors are statistical and systematic respectively).

3. The determination of $\sin^2\Theta_W$ using leptons in the final state is in good agreement with the neutrino experiments in an entirely different kinematic region.

4. The average values for the axial and vector coupling constants for the weak neutral current are determined to be

   $$g_A^2 = 0.28 \pm 0.04 \text{ and}$$
   $$g_V^2 = 0.01 \pm 0.04,$$

   while 0.25 and 0.02 are expected respectively in the standard model with $\sin^2\Theta_W = 0.23$.

5. No evidence has been found for any new particle.

ACKNOWLEDGEMENTS

I wish to thank my colleagues of the MARK-J collaboration and to acknowledge helpful discussions with W. de Boer (CELLO), R. Felst (JADE) and H.U. Martyn (TASSO). I thank Professor A. Zichichi for giving me the opportunity to give this lecture and Professors A.N. Diddens and Samuel C.C. Ting for their continuous support and encouragements; finally I convey thanks to M.A. de Bakker for arranging the lay-out of the manuscript.

REFERENCES

1. S.L. Glashow, Nucl. Phys. 22, 579 (1961); S. Weinberg, Phys. Rev. Lett. 19, 1264 (1967), and Phys. Rev. D5, 1412 (1972); A. Salam and J.C. Ward, Phys. Lett. 13, 168 (1964).
2. There are many excellent review articles on neutrino interactions. For the latest summary, see: Neutrino 81, vols. I and II, Proc. of the 1981 Inter. Conf. on Neutrino Physics and Astrophysics (Hawaii, 1981), edited by R.J. Cence, E. Ma, A. Roberts.
3. R. Budny, Phys. Lett. 45B, 340 (1973); Phys. Lett. 55B, 227 (1975).
4. The calculation is performed via Monte Carlo simulation using $\sin^2\Theta_W = 0.23$ and according to recent work by F.A. Berends, R. Kleiss, S. Jadach, Instituut-Lorentz preprint; J.P. Revol, M.I.T. Ph.D. Thesis, 1981 (unpublished).
5. F.A. Berends and R. Kleiss, Nucl. Phys. B177, 237 (1981).
6. R. Brandelik et al., Phys. Lett. 110B, 173 (1982); H.J. Behrend et al., DESY preprint 82/019; W. Bartel et al., Phys. Lett. 92B, 206 (1980), ibid., 99B, 281 (1981), ibid., 108B, 140 (1982); Ch. Berger et al., Phys. Lett. 99B, 489 (1981), Ch. Berger et al., Z. Phys. C7, 289 (1981); D.P. Barber et al., Phys. Report 63, 337 (1980); D.P. Barber et al., Phys. Rev. Lett. 45, 1904 (1981).

7. D.P. Barber et al., Phys. Rep. 63, 337 (1980).

8. B. Adeva et al., Phys. Rev. Lett. 48, 1701 (1982).

9. B. Adeva et al., Aachen Report PITHA-82-06 (1982).

10. B. Adeva et al., Phys. Rev. Lett. 48, 721 (1982).

11. S.D. Drell, Annals of Physics 4, 75 (1958).

12. M. Davier, Invited talk at International High Energy Physics Conference, Paris (1982).

13. M. Bernardini et al., INFN/AE-67/3 (1967).

14. V. Alles-Borelli et al., Nuovo Cimento Lett. 4, 1156 (1970); M. Bernardini et al., Nuovo Cimento 17A, 383 (1973); see also, S. Orito et al., Phys. Lett. 48B, 165 (1974).

15. M. Perl et al., Phys. Rev. Lett. 35, 1489 (1975); G.J. Feldman et al., Phys. Rev. Lett. 38, 117 (1977).

16. J. Burmester et al., Phys. Lett. 68B, 297 (1977); J. Burmester et al., Phys. Lett. 68B, 301 (1977).

17. W. Bacino et al., Phys. Rev. Lett. 41, 13 (1978).

18. W. Bacino et al., Phys. Rev. Lett. 42, 749 (1979).

19. M. Perl, Proceedings of the International Symposium on Lepton and Photon Interactions at High Energies, Hamburg 1977, edited by F. Gutbrod (Deutsches Elektronen Synchrotron, Hamburg, Germany, 1977; G. Fluegge, Z. Physik C1, 121 (1979); S.C.C. Ting, "Test of Quantum Electrodynamics and the study of Heavy Leptons", International School of Subnuclear Physics 17th Course, Erice-Trapani 1979; G.J. Feldman, Comments Nucl. Part. Phys. Vol. 9, 1, 23 (1979).

20. C.A. Blocker et al., Phys. Lett. 109B, 119 (1982).

21. C.A. Blocker et al., Phys. Rev. Lett. 49, 1369 (1982).

22. R. Brandelik et al., Phys. Lett. 92B, 199 (1980).

23. H.J. Behrend et al., Phys. Lett. 114B, 282 (1982).

24. G.J. Feldman et al., Phys. Rev. Lett. 48, 66 (1982).

25. G.H. Trilling, Contribution to the XXI International Conference on High Energy Physics, Paris, July 1982.

26. D.M. Ritson, Contribution to the XXI International Conference on High

Energy Physics, Paris, July 1982, and; W.T. Ford et al., Phys. Rev. Lett. 49, 106 (1982).

27. D. Lüke, Contribution to the XXI International Conference on High Energy Physics, Paris, July 1982.
28. H.J. Behrend, Contribution to the XXI International Conference on High Energy Physics, Paris, July 1982.
29. A. Böhm, Invited talk at the SLAC Summer Studies Topical Conference, 26-28 August 1982.
30. S.L. Glashow, Nucl. Phys. 22, 579 (1961); S. Weinberg, Phys. Rev. Lett. 19, 1264 (1967); S. Weinberg, Phys. Rev. D5, 1412 (1972); A. Salam, Proc. 8-th Nobel Symposium, Aspenäsgaden, 1968; Almqvist and Wiksell, Stockholm 1968, 367.
31. P. Langacker et al., Proc. Neutrino-79, Bergen Vol. 1, 276 (1979); J.E. Kim et al., Pennsylvania Report UPR-158T (1980); I. Liede and M. Roos, Proc. Neutrino-79, Bergen Vol. 1, 309 (1979); J.J. Sakurai, ibid., 267; L.M. Sehgal, Proc. Symposium on Lepton and Hadron Interactions, Visegard (1979); Ed. F. Csikor et al., (Budapest), Aachen Report PITHA-79/34, 29.
32. G. Ripken, private communication.
33. F.A. Behrends, K.J.F. Gaemers and R. Gastmans, Nucl. Phys. B57, 381 (1973); B63, 381 (1973); F.A. Behrends et al., Phys. Lett. 63B, 432 (1976).
34. H. Faissner, New Phenomena in Lepton-Hadron Physics, eds. D.E. Fries and J. Wess (Plenum, New York, 1979), p. 371; H. Reithler, Phys. Blätter 35, 630 (1979); R.H. Heisterberg et al., Phys. Rev. Lett. 44, 635 (1980); L.W. Mo, Contribution to Neutrino 80 (Erice, 1980); H. Faissner and H. Reithler, private communication.
35. P.Q. Hung and J.J. Sakurai, Phys. Lett. 88B, 91 (1979).
36. P. Duinker, Rev. Mod. Phys. 54, 325 (1982).
37. A. Litke, Harvard University Ph.D. Thesis, 1980 (unpublished).
38. P. Dittman and V. Hepp, Z. Phys. C10, 283 (1981).
39. Y.S. Tsai, SLAC-Preprint, SLAC-PUB-2450, Dec. 1979.
40. D.P. Barber et al., Phys. Rev. Lett. 45, 1904 (1980).

41. Ch. Berger et al., (PLUTO Collaboration), Phys. Lett. 99B, 489 (1981).

42. R. Brandelik et al., (TASSO Collaboration), Phys. Lett. 99B, 163 (1981).

43. D. Cords, 1980, in High Energy Physics - 1980, proceedings of the XX International Conference, Madison, Wisconsin, edited by L. Durand and L.G. Pondrom (AIP, New York, 1981), p. 590.

44. B. Adeva et al., Phys. Rev. Lett., 48, 967 (1982).

45. T. Walsh, Proceedings International Symposium on Lepton and Photon Interactions, Hamburg, 1977.

46. H. Fritzsch, Phys. Lett. 67B, 451 (1977).

47. R.M. Sternheimer et al., Phys. Rev. B3, 3681 (1971).

48. F. Bletzacker and H.T. Nieh, Phys. Rev. D16, 2115 (1977).

49. H. Takeda (JADE Collaboration), private communication.

50. Yu. A. Gol'fand and E.P. Likthman, JETP Letters 13, 323 (1971); J. Wess and B. Zumino, Nucl. Phys. B70, 39 (1974); For a review article, see: P. Fayet and S. Ferrara, Phys. Reports 32C, 249 (1977); P. Fayet, Phys. Lett. 69B, 489 (1977); G.R. Farrar and P. Fayet, Phys. Lett. 76B, 575 (1978); 79B, 442 (1978); P. Fayet, Phys. Lett. 84B, 421 (1979); P. Fayet, Phys. Lett. 86B, 272 (1979); G.R. Farrar and P. Fayet, Phys. Lett. 89B, 191 (1980).

51. F.B. Heiel et al., Nucl. Phys. B138, 189 (1978).

52. H.J. Behrend et al., Phys. Lett. 114B, 287 (1982).

53. D.P. Barber et al., Phys. Rev. Lett. 44, 1722 (1980).

54. Ch. Berger et al., (PLUTO-Collaboration), DESY Report 80/42 (1980).

55. P. Fayet, CERN-Report TH-2864 (1980).

56. B. Adeva et al., Phys. Lett., 115B, 345 (1982).

57. E. Eichten and K. Lane, Phys. Lett. 90B, 125 (1980).

58. S. Weinberg, Phys. Rev. D13, 974 (1976); L. Susskind, ibid. D20, 2619 (1979); E. Farhi and L. Susskind, Phys. Rep. 74, 277 (1981); A. Ali, DESY-Report 81/032 (1981).

59. A. Ali, H.B. Newman, R.Y. Zhu, DESY-Report 80/110 (1980).

60. W. Bartel et al., Phys. Lett. 114B, 211 (1982); H.J. Behrend et al., Phys. Lett. 114B, 287 (1982); R. Brandelik et al., Phys. Lett. 117B, 365 (1982).

61. C.H. Blocker et al., Phys. Rev. Lett. 49, 517 (1982).

D I S C U S S I O N

CHAIRMAN : P. Duinker

Scientific Secretaries : L. Di Ciaccio, L. Petrillo
S. Sherman

DISCUSSION

- *DI CIACCIO* :

I think that it would be interesting to discuss the consequences of the existence of a heavy stable lepton and of a neutral heavy electron. Could you briefly explain the implications of these particles to our present understanding of the e.m. - weak interaction ?

- *DUINKER* :

Neutral heavy leptons were proposed several years ago when the couplings for the weak neutral current were not well determined. There was a paper by Bletzacker and Nieh that tried to incorporate them when other possibilities for the couplings were still possible. Concerning your question about heavy stable lepton, if the neutrino associated with that lepton is very massive the lepton will have a very narrow decay width. I cannot speculate about any other properties of such a lepton.

- *HOFSÄSS* :

You told us that for a sufficiently high c.m. energy one can compute both upper and lower limits on the mass of the Z^o. For what mass of the boson does this become relevant ? Will one be able to do this computation before the Z^o pole energy is acheived ?

- *DUINKER* :

I'll show you figure (10) again. I said that within the framework of the standard model one can make such calculations. It is realistic to think that within the next few years momentum transfers of 3000 GeV^2 will be available at PETRA. The machine can go up to 30 on 30 GeV, which leads to 3600 GeV^2. The standard model with the Z^o at 93 GeV/c^2 predicts an asymmetry of about 30% at that energy. You measure the asymmetry with statistical and systematic errors, and within the framework of the standard model this gives you a range of possible masses for the Z^o.

- *HOFSÄSS* :

In $e^+e^- \rightarrow \mu^+\mu^-$ one can in principle also see parity violation effects. What is the status of these experiments, such as the Iron Ball experiment at SLAC ?

- *DUINKER* :

The Iron Ball experiment at SPEAR tried to measure the asymmetry from parity violation at very low energy. They were in the range of 4 to 8 GeV in the center of mass. This is a range where the effects are very small, so an experiment must have high statistics and very small systematic errors. As far as I know they never saw an effect.

- *HOFSÄSS* :

I thought that one has to measure the helicity of these muons to see the parity violating effects.

- *DUINKEN* :

That is probably not possible for these energetic muons.

- *HOFSÄSS* :

Does one plan on continuing these asymmetry measurements at higher energies, or is it not worth the effort ?

- *DUINKER* :

Of course we intend to continue measurements of muon asymmetry as we go up in energy. We will certainly not stop at PETRA.

- *KAPLUNOVSKY* :

Did your discussion imply that the weak couplings of the tau were supposed to occur via a standard $W_L^{\pm}$, or something different, such as V+A current coupling to an W_R ?

- *DUINKER* :

I gave a short review about what is known about the tau. The tau is coupled with a V-A current to a standard W_L. The data supports the hypothesis that the tau is a normal sequential lepton. I only supported this standard view of the tau with the data presented here.

- *KAPLUNOVSKY* :

What do experiments show about the existence of stable, charged scalar leptons ? In some circumstances the lightest supersymmetry particle should look like a heavy stable muon with spin 0 and not contribute to R.

- *DUINKER* :

The data excludes them for masses ≤ 15 GeV/c^2. The change in cross section should be small, only 25% of σ ($e^+e^- \to \mu^+\mu^-$).

- *PERNICI* :

You have discussed bounds on the masses of scalar leptons associated with ordinary leptons through supersymmetry; you have not found events within a certain energy range, and so deduce that such scalar leptons with mass$\leq$ 16 GeV/c^2 do not exist. Couldn't this be explained by a very weak coupling of the photino or goldstino with ordinary matter, so you wouldn't find any scalar lepton events in your detector.

- *DUINKER* :

We are not trying to find the supersymmetry particles themselves, we are trying to find their decay particles that interact with matter. The scalar leptons are supposed to decay rapidly within the beam pipe and their production is observed as a muon or electron in the detector. We assume the photinos escape the detector. If the particles are normal scalar particles they are produced at a known rate in the interaction region. If the coupling to photons is something different the analysis is no longer valid.

- *LAU* :

You have shown events with a tau decaying into five charged tracks. Have you determined the branching ratio for this decay mode ?

- *DUINKER* :

I displayed a table with results from the CELLO collaboration. They determined the branching ratio to be about 1% (10/868) for tau decay into five prongs.

- *LAU* :

I think this may contradict the PEP results reported by the MAC collaboration. In 30^{-1}pb. of data they do not see any five prong tau decays.

- *DUINKER* :

They have something like 1274 tau events -- I do not know which is a more reliable result. I do not belong to the CELLO group. You would have to discuss the matter with them.

- DUBNIČKOVÁ :

One knows how to calculate coupling constants from a theory. Your results depend on these coupling constants. How are they experimentally determined ?

- DUINKER :

The coupling constants in your postulated Lagrangian carry through into scattering amplitudes, and from these amplitudes into a differential cross section prediction. By fitting the experimentally determined cross section with a function that is derived from the given theoretical model, one obtains values of the coupling constants that are consistent with the data.

- SHERMAN :

Are incorrect charge assignments in $\mu^+\mu^-$ asymmetry measurements truly uncorrelated ? Do they come from multiple scattering in the muon steel or are there extra hits in the chambers ?

- DUINKER :

When a track is very stiff we can sweep it from plus to minus, an effect that is entirely uncorrelated. It is not from extra hits.

- EISELE :

You have shown a table of measurements of $\sin^2 \theta_W$ that looks very suspicious to me, since there is no scatter of the measured points. Which quantities are most sensitive to $\sin^2 \theta_W$? Most quantities are not sensitive to the Weinberg angle.

- DUINKER :

You are referring to Table 1V. The values for $\sin^2 \theta_W$ are close together and the errors are rather large. You would expect them to have more scatter. We are measuring $\sin^2 \theta_W$ which must be a positive quantity, but we measure the square of the couplings in a scattering experiment. If your measurement gives a negative result for g_V^2 the fit pushes $\sin^2\theta_W$ to 0.25 since we do not allow complex values.

The answer to your second question is that the electrons are most sensitive to $\sin^2\theta_W$. I can show you that in figure (18), which shows the angular distribution for Bhabha events. Using the calculation of Behrends et al., the difference of the data from the QED calculation, divided by the QED expectation is plotted, so you expect zero with no weak effects. The Bhabha scattered electrons are most sensitive to g_V^2 which depends on $\sin^2\theta_W$. Here are different curves for different values of g_V^2 and almost the same values of g_A^2, and you can see the sensitivity for determining g_V.

- *FERRARA* :

I believe that there are more stringent bounds on the masses of scalars predicted by supersymmetry theories available from the g-2 measurements conducted with muons at CERN. This is shown in recent papers by Barbieri and Maiani, and also by Ellis and Nanopoulis. This limit is at 20 GeV/c^2 now. In supersymmetry theories g-2 is a measure of the supersymmetry breaking mass scale. In fact, if supersymmetry is not broken the diagrams that contribute to g-2 will exactly cancel. The Dirac theory would be exact. The relevant diagrams are :

QED

Supersymmetry

- *DUINKER* :

Our current limits on the mass of a scalar lepton essentially go to the beam limit. Next year with the increase in beam energy at PETRA we should be able to push on the limit of 20 Ge V/c^2.

THE PHOTON STRUCTURE FUNCTION

Ch. Berger

I. Physikalisches Institut

RWTH Aachen, Germany

I. INTRODUCTION

High-energy electron-positron storage rings offer the opportunity to measure the inclusive reaction

$$e^+e^- \to e^+e^- + X, \tag{1}$$

where X is a leptonic or hadronic final state. This reaction is usually interpreted as lepton or hadron production by two virtual photons which are radiated from the incoming leptons ($e^+e^- \to e^+e^-\gamma^*\gamma^* \to e^+e^-X$). (See Fig.1 for the basic diagram and an explanation of the symbols used in the text.)

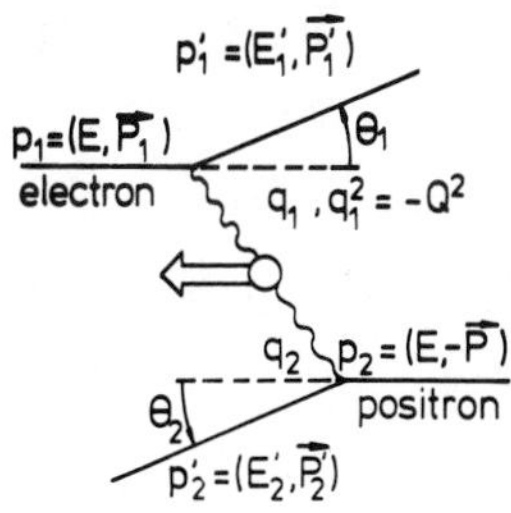

Fig. 1. The basic diagram

Compared to the one-photon reaction, i.e. e^+e^- annihilation ($e^+e^- \to X$), the available invariant mass of the system X is much smaller because of the bremsstrahlung-like energy spectrum of the two virtual photons. On the other hand, the kinematical structure of process(1) is much richer. Even integrating over the kinematical variables of the final state, the cross-section depends on three invariants: the centre-of-mass energy W, and the photon masses squared, q_1^2 and q_2^2. These quantities can be individually tuned by selecting certain intervals in the scattered electron (positron) energies and angles. One can thus study various production mechanisms in one experiment, e.g. the transition from hadron production as expected in vector meson dominance (VMD, $|q_1^2|$, $|q_2^2|$ small) to short-distance phenomena (q_1^2 and/or q_2^2 large) as calculated in the quark-parton model (QPM) and QCD. Because the photons are in a well-defined polarization state, the full experiment is not described by only one cross-section but by a superposition of six cross-sections and interference terms. The situation becomes even more involved if differential cross-sections in the final state are investigated. For a complete discussion of this complicated situation, the reader is referred to other publications (1-3). Fortunately, life is in practice much easier, and all experiments which have been done until now can -- at least with some additional assumptions -- be interpreted in terms of one or two cross-sections and structure functions only.

In principle we have to distinguish between three cases:

i) In 'no-tag experiments' none of the outgoing leptons e^+e^- on the right-hand side of Eq.(1) is detected. The photon flux is completely dominated by transversely polarized photons, which are practically on-mass-shell (θ_1 , θ_2 $\cong m/E$). The use of antitagging, i.e. the exclusion of events with angles $\theta_1, \theta_2 > 20$ mrad (to quote a typical value), restricts $|q_i^2|$ to values $< m_\pi^2$ and makes the interpretation of the results in terms of just one cross-section $\sigma_{\gamma\gamma}(W)$ more reliable. Experimentally, two-photon initiated events can be easily separated from the one-photon channel in cases where the final-state X is completely reconstructed. The main reason is the much smaller total energy in the $\gamma\gamma$ system.

ii) In 'single-tag experiments', either the outgoing e^- or e^+ is detected in a forward spectrometer. Sometimes the tagging information is only used for separating a multihadronic two-photon final state from e^+e^- annihilation states. In this case the physics interest is still in $\sigma_{\gamma\gamma}(W)$ (or its differentials). On the other hand, the information from the forward detectors can be used to investigate the Q^2 behaviour of the cross-section. A combination of tagging on one side with antitagging on the other allows an easy interpretation of the results in terms of electron scattering off a real photon target [4].

iii) Double-tag experiments. In these experiments, both outgoing leptons are measured. In principle, the full kinematical structure of the process can be studied, but we are still a long way from starting to tackle this difficult task. The future study of events with both $|q_1^2|$ and $|q_2^2|$ large will certainly be very interesting from the physics point of view [5].

II. THE TOTAL CROSS-SECTION AND THE PHOTON STRUCTURE FUNCTION

II.1 General remarks

The total hadronic cross-section $\gamma\gamma \to X$ can be studied in single-tag experiments, with antitagging for the other lepton. This set of conditions constrains the kinematics to that of electron scattering off a real-photon target. As in electroproduction experiments, the results are interpreted in terms of two cross-sections σ_T and σ_L, or structure functions F_1 and F_2.

The standard VMD result for the total hadronic scattering cross-section of <u>real</u> photons is [6]

$$\sigma_T(Q^2 = 0,W) = \sigma_{\gamma\gamma} = 240 \text{ nb} + \frac{270 \text{ nb GeV}}{W} \quad (2)$$

For Q^2 values up to 0.5 GeV^2 it is certainly reasonable to assume that the Q^2 dependence of the cross-section is given by the ρ form factor,

$$\sigma_T(Q^2,W) = 240 \text{ nb} + \frac{270 \text{ nb GeV}}{W} F_\rho^2, \tag{3}$$

$$F_\rho = \frac{1}{1 + Q^2/m_\rho^2} \tag{4}$$

Figure 2a contains the data with $1 < W_{vis} < 3.5$ GeV, and Fig. 2b the data with $3.5 < W_{vis} < 10$ GeV obtained by the PLUTO group [7]; W_{vis} is the visible invariant mass. In both cases the Q^2 dependence is well represented by the ρ form factor (solid line in Figs. 2a and 2b), thus justifying the ansatz Eq. (3) . The absolute VMD prediction formulae (2), (3) is well reproduced by the PLUTO data

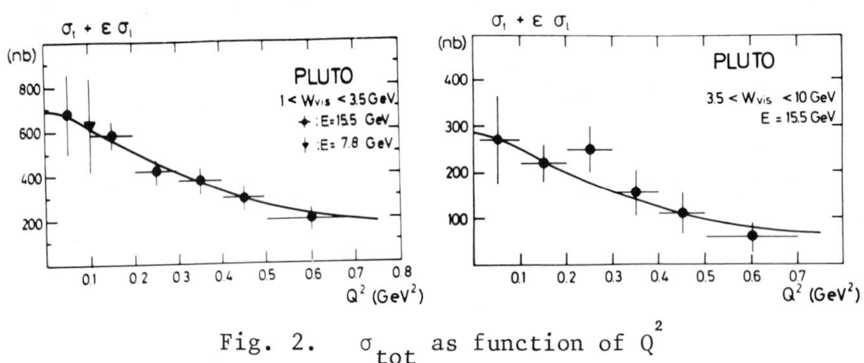

Fig. 2. σ_{tot} as function of Q^2

above 3.5 GeV, whereas for $W_{vis} < 3.5$ GeV the measured cross-section exceeds the prediction. Note also that Brandelik et al.[8] get a different parametrization of σ_T. The data in Figs. 2a and 2b have been obtained by using events with $20 < \theta_1 < 70$ mrad i.e. events with an electron (positron) scattered into the so-called 'small-angle tagger' (SAT) of the PLUTO detector . The Q^2 range has been largely extended by utilizing the 'large-angle tagger' (LAT) ($70 < \theta_1 < 250$ mrad, kinematical Q^2 limit 20 GeV^2). In this angular range it is easily possible to confuse a neutral cluster from a annihilation jet event with an electron tag signal. In order to minimize the jet contributions, the PLUTO group required a tag energy > 8 GeV in the LAT. This cut also helps in reducing the background from radiative annihilation events, where the photon simulates a tag signal in the forward detector. A Monte Carlo study of the annihilation background showed that the expected number of these events (surviving all 2γ cuts) is only 1, which has to be compared to 117 events seen in the experiment.

In Fig. 3, $\sigma_T(Q^2, W_{vis})$[22] of the PLUTO LAT events is plotted versus Q^2 (full squares). The cut in W_{vis} ($3.5 < W_{vis} < 10$ GeV) has been applied because at least at small Q^2 the cross-section is almost constant in this regime [7]. The figure also includes the small Q^2 data of Fig. 2b (full circles) and the ρ form factor curve. The change in the Q^2 behaviour of the data above $Q^2 = 1$ GeV^2 is dramatic. The cross section fall-off is much slower and at $Q^2 = 10$ GeV^2 σ_T is about a factor of 30 higher than the simple VMD expectation.

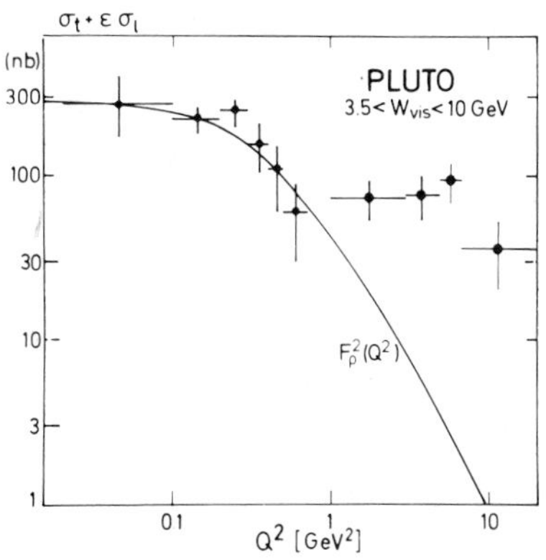

Fig. 3. σ_{tot} up to large Q^2 values

Thus it is very suggestive to explain the cross section in this regime by electron scattering off the quark 'constituents' of the real target photon. Such an explanation is successfull as will be shown below but the reader should be aware of two very interesting papers (ref. 9 and fig.4) which explain the total cross section in the full Q^2 range by a generalized vector meson dominance model.

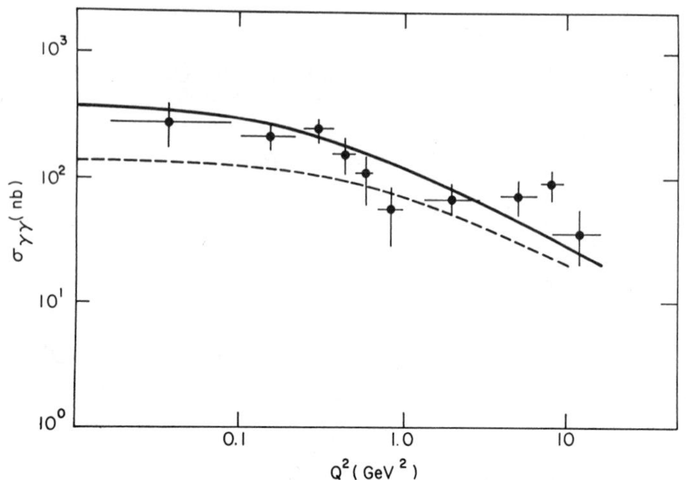

Fig. 4. σ_{tot} in a generalized VMD model (ref. 9)

For the discussion of the deep inelastic scattering phenomena it is more adequate to interpret the data in terms of structure functions instead of cross-sections.

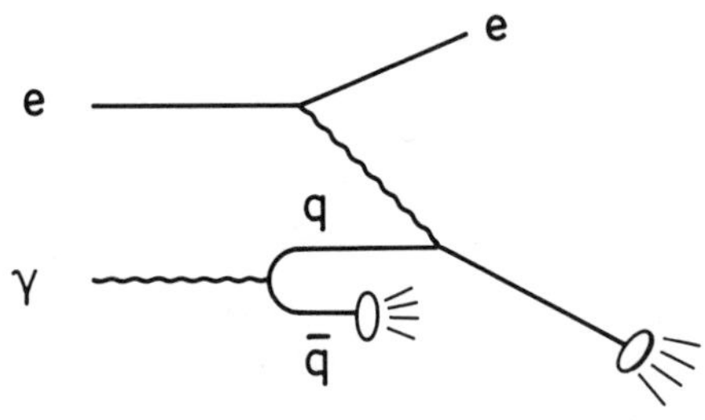

Fig. 5. eγ scattering

The process under consideration can be visualized from Fig.5. The incoming quasi-real photon (radiated from, say, the positron) splits up into a quark-antiquark pair, either bound (ρ meson!) or free. The electron scatters off one of the quarks. After the hard-scattering process, the quarks fragment into final-state hadrons. In a way similar to that for lepton-nucleon scattering, we introduce the scaling variables

$$x = -q_1^2/2q_1q_2 = \frac{Q^2}{Q^2 + W^2} \tag{5}$$

and

$$y = q\ q_2/p_1 \cdot p_2 \cong (E - E_1')/E. \tag{6}$$

The connection between cross-section and structure functions is given by

$$\sigma_T = \frac{4\pi^2\alpha}{Q^2}\ 2xF_T$$
$$\sigma_L = \frac{4\pi^2\alpha}{Q^2}\ F_L \tag{7}$$

Introducing*)

$$F_1 = F_T$$
$$F_2 = 2xF_T + F_L \tag{8}$$

the cross-section for deep inelastic electron-photon scattering reads

$$\frac{d\sigma}{dxdy}_{e\gamma\to eX} = \frac{16\pi\alpha^2 EE_\gamma}{Q^4}\left[(1-y)F_2 + xy^2F_1\right] \tag{9}$$

---

*) This definition is most commonly used. An alternative would be to divide F_1 by 2x, resulting in structure functions which are proportional to absorptive parts of forward helicity amplitudes. I have used this definition in ref. 4a.

Within the experimental acceptance of the present PETRA detectors xy^2 is small and Eq.(9)is well approximated by

$$\frac{d\sigma}{dxdy}\bigg|_{e\gamma\to eX} = \frac{16\pi\alpha^2 EE_\gamma}{Q^4} (1-y)F_2(x,Q^2). \qquad (10)$$

In order to get the cross section for the reaction $e^+e^-\to e^+e^-X$ one has to multiply (10) with the number of photons inside an electron, $f_{\gamma|e}$ (z) dz, where $z = E_\gamma/E$ is the fractional energy of the quasi real target photon.

$$\frac{d\sigma}{dxdy}\bigg|_{e^+e^-\to e^+e^- X} \quad \frac{16\pi\alpha^2 EE_\gamma}{Q^4} (1-y)F_2(x,Q^2)f_{\gamma|e}(z)dz \qquad (11)$$

The standard result for $f_{\gamma/e}$[10] is

$$f_{\gamma|e} = \frac{\alpha}{2\pi} \frac{1+(1-z)^2}{z} \ln \frac{q_2^2 \ \max}{q_2^2 \ \min}$$

$$= \frac{\alpha}{\pi} \frac{1+(1-z)^2}{z} \ln \left(\frac{E}{m} \ \theta_2^{max} \ \frac{1-z}{z} \right) \qquad (12)$$

where the last line has been obtained by calculating the four momentum square of the virtual photon for rather small scattering angles.

II.2 Theoretical Predictions for F_2 (x,Q^2)

Equation (11) describes the elastic electron quark scattering for pointlike quarks and antiquarks, if F_2 is interpreted as the density of quarks inside a photon (weighted by momentum and charge):

$$F_2(x,Q^2) = x \cdot \sum e_q^2 \ (f_{q|\gamma}(x,Q^2) + f_{\bar{q}|\gamma}(x,Q^2)) \qquad (13)$$

As the photon is its own antiparticle this reduces to

$$F_2(x,Q^2) = 2x \cdot \sum e_q^2 \cdot f_{q|\gamma}(x,Q^2) \qquad (14)$$

The different models differ substantially in the predictions for $f_{q|\gamma}(x,Q^2)$. Let us first consider the 'hadronic part' of the photon.

In VMD the photon structure function can be related to the vector meson structure functions. Taking into account the transverse part of the dominant ρ term only we get

$$F_{2,\rho}^{\gamma}(x,Q^2) = \left(\frac{e}{f_\rho}\right)^2 F_2^{\rho^0}(x,Q^2) \tag{15}$$

Isospin invariance allows a connection to measured quantities

$$F_2^{\rho^0}(x,Q^2) = F_2^{\pi^0}(x,Q^2) = F_2^{\pi^-}(x,Q^2)$$

with (16)

$$F_2^{\pi^-}(x,Q^2) = x\left(\frac{4}{9} f_{\bar{u}|\pi^-} + \frac{1}{9} f_{d|\pi^-}\right) = \frac{5}{9} x \cdot f_{\bar{u}|\pi^-}$$

The $\bar{u}$ content of a π^- has been estimated [11] from the Drell Yan reaction $\pi^- p \to \mu^+\mu^- + X$ (see fig. 6). The result can be parametriued as

$$f_{\bar{u}|\pi^-}(x) = 0.52\ \frac{1-x}{x} \tag{17}$$

leading to

$$x^{eff} = \int_0^1 x\, f_{\bar{u}|\pi^-}(x)\, dx = 0.26 \tag{18}$$

This means that $\cong$ 25% of the π^- momentum is carried by the $\bar{u}$ - quark, 25% by the $\bar{d}$ and the rest by gluons, similar to the case of the nucleon structure function.

Inserting (16) and (17) into (15) we get for F_2 (with $f_\rho^2/4\pi$ 2.2, and neglecting longitudinal terms of order $O(1/Q^2)$)

$$F_{2,\rho}^{\gamma}(x,Q^2) = \frac{\alpha}{f_\rho^2/4\pi}\ \frac{5}{9}\ 0.52\ (1-x)$$

(19)

$$F_{2,\rho}^{\gamma}(x,Q^2) = \alpha \cdot 0.13\ (1-x)$$

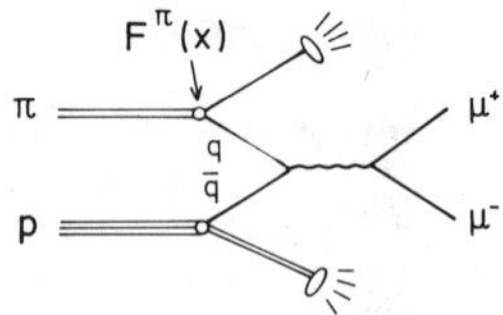

Fig. 6. The Drell Yan process and the structure function of the pion

This calculation has first been presented in ref. 12 based on preliminary data of ref. 13 and came out about 15% lower.

If more vector mesons are taken into account it is probably 'natural' [14] to add the vector mesons coherently i.e. instead of

$$|\gamma\rangle_\rho = \frac{e}{f_\rho}|\rho\rangle = \frac{e}{f\rho} \cdot \frac{1}{\sqrt{2}} (|u\bar{u}\rangle - |d\bar{d}\rangle)$$

we use

$$|\gamma\rangle_{had} = \frac{e}{f_\rho}|\rho\rangle + \frac{e}{f_\omega}|\omega\rangle \tag{20}$$

$$= \frac{e}{f_\rho} \sqrt{2} \left(\frac{2}{3}| u\bar{u}\rangle - \frac{1}{3} |d\bar{d}\rangle - \frac{1}{3}| s\bar{s}\rangle \right)$$

and therefore $F_{2,had}^{\gamma} / F_{2,\rho}^{\gamma} = 1.6$

From these considerations it is clear that the hadronic piece of the photon structure function is rather uncertain and

$$F_{2,had}^{\gamma} = \alpha \cdot (0.18 \pm 0.05)\ (1-x) \tag{21}$$

is probably a good estimate.

Pointlike piece of the photon

In the quark model the photon can split into a quark and an

antiquark with the obvious symmetry condition $f_{q|\gamma}(x) = f_{\bar{q}|\gamma}(1-x)$. Assuming $f_{q|\gamma}(x)$ to be a flat function of x we get a structure function $F_2(x)$ rising ~ x. The function $f_{q|\gamma}(x)$ can be calculated and the most elegant way to do this is to use a reciprocity relation [15]. It relates for each color the probability to find a quark inside a photon $f_{q_c|\gamma}$ to the probability to find a photon inside a quark (Fig. 7).

$$f_{q_c|\gamma}(x) = x\, f_{\gamma|q_c}\left(\frac{1}{x}\right) \tag{22}$$

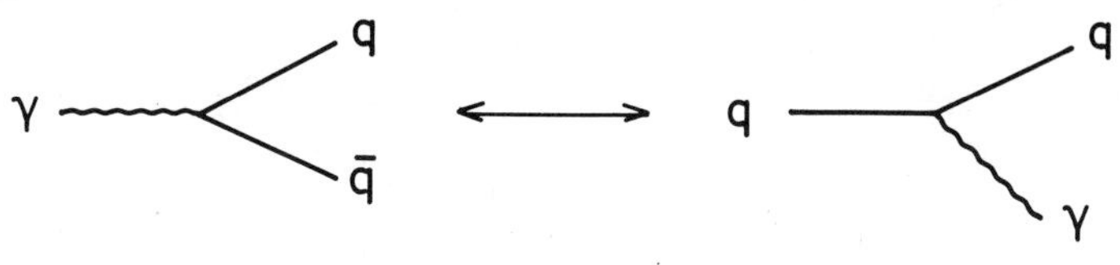

Fig. 7.

We can obtain $f_{\gamma|q_c}(x)$ very easily from eq. 12 by multiplying with the fractional quark charge e_q^2

$$f_{\gamma|q_c}(x) = e_q^2 \frac{\alpha}{2\pi} \frac{1 + (1-x)^2}{x} \ln \frac{t_{max}}{t_{min}} \tag{23}$$

Inserting this into eq. 22 we get after summing over three colors

$$f_{q|\gamma}(x) = 3e_q^2 \frac{\alpha}{2\pi} \left[x^2 + (1-x)^2\right] \ln \frac{t_{max}}{t_{min}}$$

Inserting the minimum and maximum four momentum squared $t_{min,max}$ of the internal quark line one gets

$$\begin{aligned} f_{q|\gamma} &= 3\, e_q^2 \frac{\alpha}{2\pi} \left[x^2 + (1-x)^2\right] \ln \frac{W_{\gamma\gamma}^2}{m_q^2} \\ &= 3\, e_q^2 \frac{\alpha}{2\pi} \left[x^2 + (1-x)^2\right] \ln \frac{Q^2}{m_q^2} \end{aligned} \tag{24}$$

in leading log approximation (LLA) because $W_{\gamma\gamma}^{2} = Q^2 \frac{1-x}{x}$. $W_{\gamma\gamma}$ is the invariant mass of the resulting hadronic system. Our final results now reads:

$$F^{\gamma}_{2,\text{point}}(x,Q^2) = 3\sum_{\text{flavour}} e_q^4 \frac{\alpha}{\pi} x \left[x^2 + (1-x)^2\right] \ln \frac{Q^2}{m_q^{\;2}} \tag{25}$$

This structure function has some remarkable and very unusual properties. First it rises $\sim x$, quite in contrast to all known hadronic structure functions, which decrease with x. Second it shows a strong scale breaking effect. F_2^{γ} is at all x directly proportional to $\ln Q^2$.

For light quark integration down to t_{min} $O(m_q^2)$ corresponds to distances much larger than the confinement radius. Therefore a new cutoff $r = O(\Lambda^{-1})$ has to be introduced and the lowest order structure function in QCD is given by

$$F^{\gamma}_{2,\text{box}} = 3\sum e_q^4 \frac{\alpha}{\pi} x \left[x^2 + (1-x)^2\right] \ln \frac{Q^2}{\Lambda^2} \tag{26}$$

The index box refers to the well known fact that the lowest order diagram (fig.5) is equivalent to the box diagram (fig. 8).

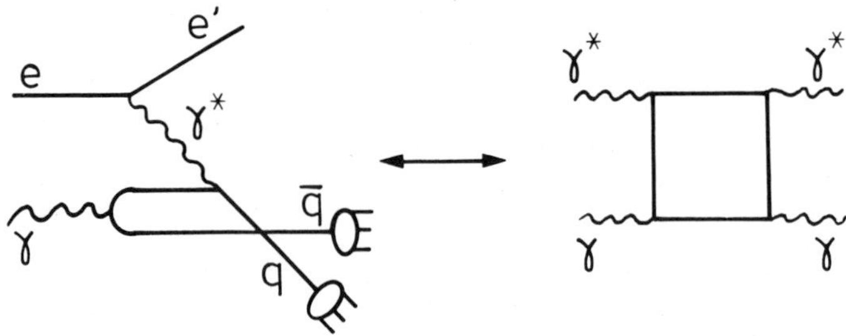

Fig. 8.

Gluon corrections to the box diagram

The gluon corrections can be divided into two classes. If the gluons are radiated by the outgoing quarks the hard scattering process is not affected and the gluon radiation can be included in the fragmentation. But if the gluons are radiated from the internal quark line (fig. 9)

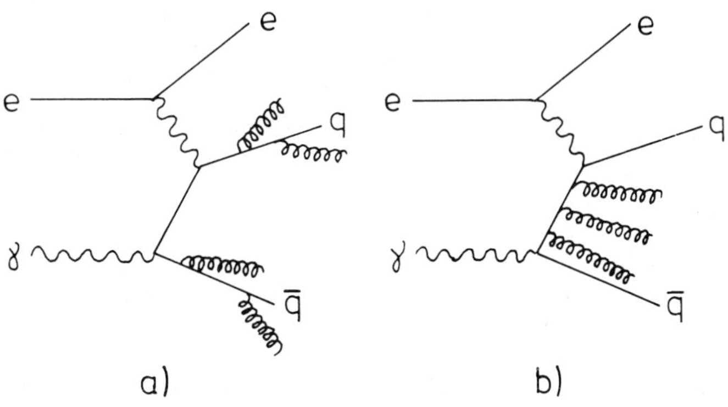

Fig. 9. Gluon Bremsstrahlung

they are involved in the hard scattering process. The effective x as seen by the hard photon is different from the x at the $\gamma \to q\bar{q}$ vertex because the quark looses momentum by gluon radiation. As a result the momentum distribution of the quark inside the photon (and the structure function correspondingly) is shifted to smaller x-values. As the strong coupling constant α_s is large for soft gluon radiation it is not sufficient to calculate the one gluon radiation only. There are three techniques to treat the multi gluon radiation using the renormalisation group equation [15], the Altarelli Parisi equation [16] and diagrammatic methods respectively. The first technique was used in the pioneering work of E. Witten [17] to the leading order and later by Bardeen and Buras [18] and Duke and Owens [19] up to next to leading order QCD. The diagrammtic approach was used by Llewellyn Smith [20] and Frazer and Gunion [21]. The most important result of these calculations is that the structure of eq. 26 does not change, i.e.

$$F_{2}^{\gamma},{}^{LLA} = h(x) \ln \frac{Q^2}{\Lambda^2} \tag{27}$$

The biggest differences between h(x) and $h_{box}(x)$ are at large x values.

If higher order corrections are included there is no longer such a simple Q^2 dependence and the result is expressed in terms of moments of the structure function

$$\int_0^1 dx\, x^{n-2} F_2\,(x,Q^2) = a_n \ln \frac{Q^2}{\Lambda^2} + \tilde{a}_n \ln\ln \frac{Q^2}{\Lambda^2} + b_n + 0\left(\frac{1}{\ln(Q^2/\Lambda^2)}\right) \tag{28}$$

The critical point of the next to leading order calculation is that the constant terms are fixed. This is necessary if one wants to extract a meaningful Λ parameter out of a measurement of photon structure function. That can be seen easily if one changes the Λ parameter from Λ → Λ

$$\ln \frac{Q^2}{\Lambda^2} \to \ln \frac{Q^2}{\Lambda'^2} = \ln \frac{Q^2}{\Lambda^2} + \ln \frac{\Lambda^2}{\Lambda'^2} \tag{29}$$

The two logs differ by a constant term only which by definition is neglected in the LLA in any case. If on the other hand the constant terms are fixed by a higher order calculation then Λ can be determined in principle by the absolute value of F_2^γ at every point (x, Q^2). In Fig.10 the different calculations for $F_2^\gamma\ (x,Q^2)$ are shown for $Q^2 = 20\ \text{GeV}^2$, $\Lambda = 0.2$ GeV, $m_q = 0.3$ GeV. It is interesting to see that the hadronic part is much smaller than the pointlike part, which at least at large x dominates completely.

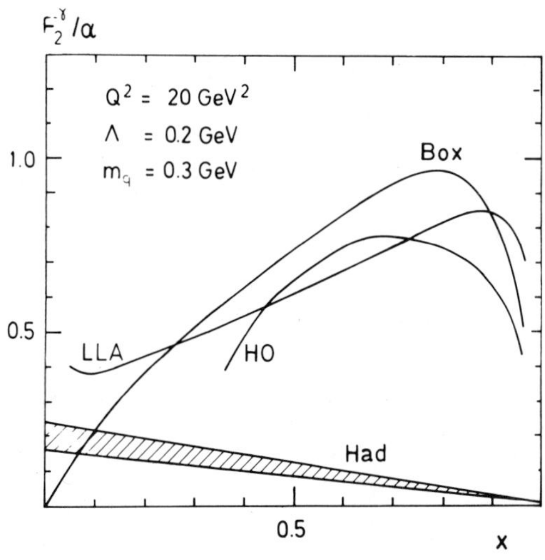

Fig. 10 Theoretical predictions for F_2^γ. Only u,d,s quarks are taken into account for the pointlike piece

II.3 Experimental results on F_2 (x, Q^2)

The PLUTO data shown in Fig.4 have been analysed in terms of the photon structure function [22]. After correcting for the x-x_{vis} difference the result is shown in Fig.11a as a function of x at $\langle Q^2 \rangle \cong 5\ GeV^2$.

In Fig.11b the correlation between Q^2 and x for this experiment is included as calculated from a Monte Carlo model. For most of the x-range the average Q^2 does not vary very much, so that Fig. 11a essentially gives the x dependence of F_2 at a constant $Q^2 \cong 5$ GeV.

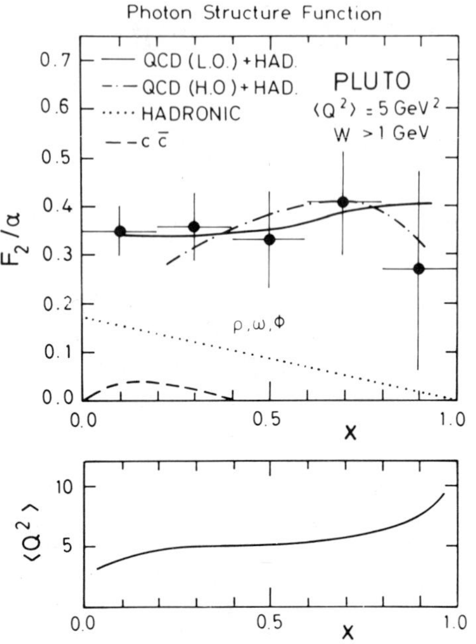

Fig. 11. The PLUTO data for F_2^γ

The data are clearly not described by the hadronic piece alone (dotted line in Fig. 11a, eq. 21), but an excellent agreement can be obtained by adding the leading order QCD prediction [21] for u, d,s quarks with Λ = 0.2 GeV (solid line in Fig. 11a). We thus have a very clear demonstration of the pointlike coupling of (almost) real photons to quarks.

The result of a higher order calculation [25] of $F_2^\gamma(x,Q^2)$ is also included in the figure (dashed dotted line). The good description of the data with $\Lambda_{\overline{MS}}$ = 0.2 GeV is very satisfying. The contribution of charmed quarks to the PLUTO data is small: It has been estimated using formula (25) for c quarks only with m_q = 1.5 GeV (dashed line in fig. 11a).

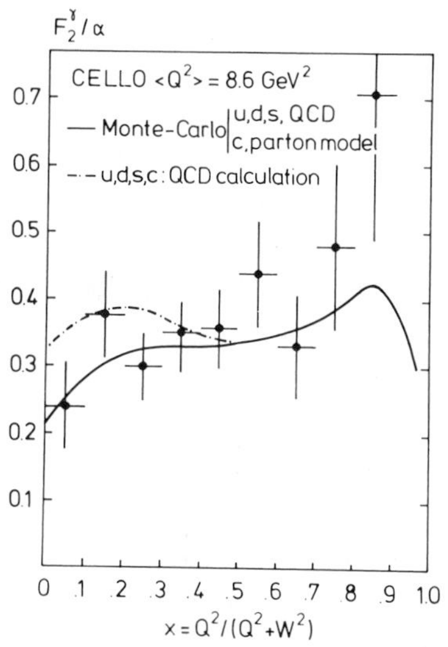

Fig. 12. F_2^γ as measured by the CELLO group [24]

By now there are two further measurements with similar statistics by JADE [23] and CELLO [24]. The CELLO result is based on 267 events with $< Q^2 > = 8.6$ GeV^2, whereas JADE observes 139 events with $< Q^2 > = 23$ GeV^2. Both experiments select events with a single tag in the endcap shower counters of their central detector. The CELLO (fig. 12) data is unfolded and F_2^γ is given as a function of x.

Fig. 12 shows that the CELLO data are in nice agreement with a leading order QCD calculation, using $\Lambda_{LO} = 0.2$ GeV, if the charm contribution as calculated in the quark model ($m_c = 1.8$ GeV) is added (full curve).

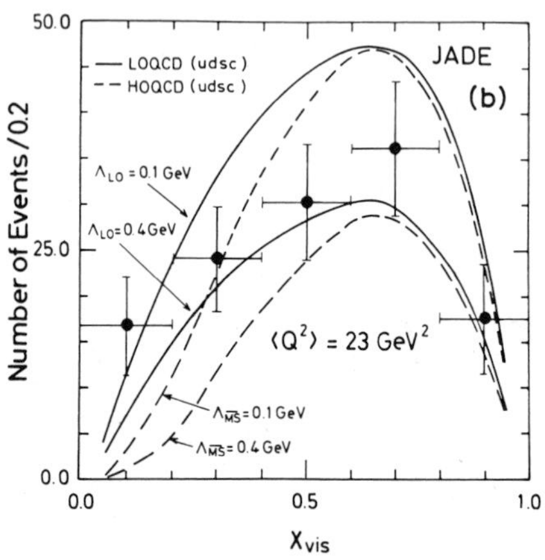

Fig.13. Number of events versus x_{vis} as given by JADE [23]

The JADE data (fig. 13) is not unfolded. In fig. 13 the number of events is plotted versus x_{vis} and compared to different models. The VMD part has not been included in the model calculations. For a more detailed explanation see ref. 23. The curves in fig. 13 using different values of the Λ parameter clearly demonstrate the potential of photon structure function measurements in the future. In fact a fit of the JADE group gives $\Lambda_{\overline{MS}} = 0.22\,^{+.10}_{-.07}$ GeV !

There is no straightforward way to calculate the charm contribution close to the mass threshold in QCD. As a first guess the CELLO and JADE group included only a threshold factor and left Λ unchanged (see e.g. dash dotted line in fig. 12). For a future high precision measurement of the structure function it might be necessary to restrict the data to the W-range $4 < W^2 < 16\ GeV^2$ in order to avoid uncertainties as well in the resonant region as in the charm threshold region. If in addition Q^2 is kept large enough, say $Q^2 > 4$ GeV, the QCD analysis is (hopefully!) on a solid ground so that a meaningful determination of Λ could be possible. With the statistics obtained in the new PLUTO experiment ($\cong$ 4000 events) this should be possible.

Finally the Q^2 dependence of the structure function is investigated in fig.14 using the available PLUTO and JADE data. The PLUTO data are integrated between $0.2 < x_{vis} < 0.8$, the JADE data between $0.3 < x_{vis} < 1$. The data agree qualitatively with QCD pred ctions (ref.14) over two orders of magnitude in Q^2!

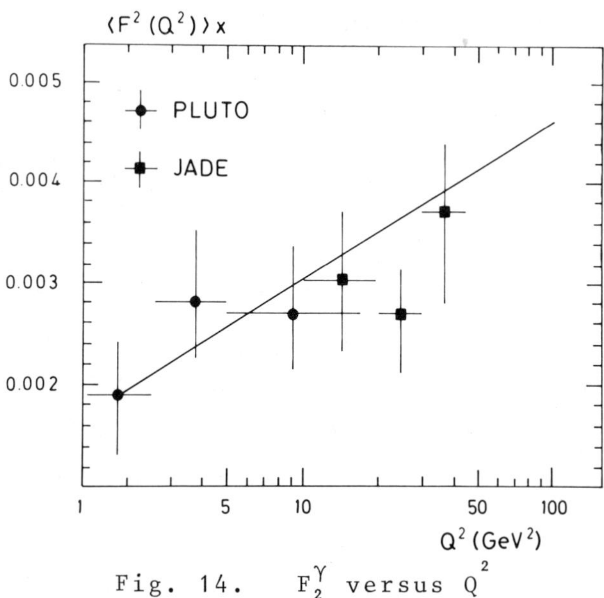

Fig. 14. F_2^γ versus Q^2

I wish to thank Prof. A. Zichichi very much for his warm hospitality during the Erice school. I am indebted to many colleagues from the DESY groups for usefull discussions. Finally I have to thank P. Zerwas from whom I learned a lot about the photon structure functions.

References

(1) V.M. Budnev, I.F. Ginzburg, G.V. Meledin and V.G. Serbo, Phys. Reports 15C, 181 (1975)

(2) G. Bonneau, M. Gourdin and F. Martin, Nucl. Phys. B54,573 (1973)

(3) C. Carimalo, P. Kessler and J. Parisi, Phys. Rev. D20, 1057 (1979) and D 21, 669 (1980)

(4) a) Ch. Berger, γγ Collisions, Proc. International Workshop, Amiens, April 1980, Springer Lecture Notes in Physics, 134, 82 (1980)
b) Ch. Berger and J.H. Field, Nucl. Phys. B187, 585 (1981)

(5) P. Landshoff, Proceedings of the IVth International Colloquium on Photon Photon Interactions World Scientific, Singapore

(6) T. Walsh, J. Phys. (Paris) C-2 Suppl. 3 (1974)

(7) PLUTO Collaboration, Ch. Berger et al., Phys. Lett. 99B, 287 (1981)

(8) TASSO Collaboration, R. Brandelik et al., report DESY 81/052 (1981)

(9) E. Etim and E. Massó, CERN preprint TH 3260 (1982)
E. Etim, E. Massó and L. Schülke, CERN preprint TH 3423

(10) e.g. C. Carimalo, P. Kessler and J. Parisi, Phys. Rev. D21, 669 (1980)

(11) C.B. Newman et al., Phys. Rev. Lett. 42, 951 (1979)

(12) C. Petersson, T.F. Walsh an P.M. Zerwas, Nucl. Phys. B174, 424 (1980)

(13) K.J. Anderson et al., Chicago Princeton Report EFI-78-38

(14) C. Peterson, P. Zerwas and T.F. Walsh, SLAC PUB 2985 (1982)

(15) e.g. E. Reya, Phys. Reports 69, 195 (1981)

(16) G. Altarelli and G. Parisi, Nucl. Phys. B126, 298 (1977)

(17) E. Witten, Nucl. Phys. B120, 189 (1977)

(18) W.A. Bardeen and A.J. Buras, Phys. Rev. D20, 2280 (1980)

(19) D.W. Duke and J.F. Owens, Phys. Rev. D22, 2280 (1980)

(20) Ch. Llewellyn Smith, Phys. Lett. 79B, 83 (1978)

(21) W.R. Frazer and J.F. Gunion, Phys. Rev. D20, 147 (1979)

(22) PLUTO Collaboration, Ch. Berger et al., Phys. Lett. 107B, 168 (1981)

(23) JADE Collaboration, W. Bartel et al., report DESY 82-064

(24) CELLO Collaboration, H.J. Behrend et al., Contributed paper to the XXI International Conference on High Energy Physics

(25) A.J. Buras and D.W. Duke, private communication

(26) J.L. Rosner, University of Minnesota preprint, July 82

DISCUSSION

CHAIRMAN : Ch. Berger

Scientific secretaries : C. De Clercq, M. Pernici

- *LAU* :

Could you briefly describe how to determine the photon structure function from the raw data?

- *BERGER* :

It is essentially done by model fitting. In our analysis we took a constant structure function but put in all the kinematics, flux factors etc. correctly. To describe the decay of hadrons we choose different models; we choose either the fragmentation model or the phase space model. It is important to be sure that the model does not depend heavily on the decay parameter for the hadronic system. Then we have to compare all the experimental distributions we have for example W_{vis}, Q^2, N charged, N neutral, P_T and so on. The P_T distribution is one of the most important distributions because it is very sensitive to what you do and because you loose some of the particles in the forward direction. That done, you can be more confident because there is only one open normalization parameter, directly related to the magnitude of F_2. Consequently if F_2 is a constant, experimentally, you have to stop. If you find something different you reinsert the experimental findings and do the second step and so on.
The PLUTO group has achieved with this model a consistent des-

cription of all the experimental distributions. We only needed to be consistent because of the very limitated statistics. It will be much harder in the experiment we are now analyzing. As shown in one of these transparencies, the data I told you about are based on 125 events; now we will have 4000 events. And things which are consistent with 125 events could be, I think, in clear contradiction with 4000 events. Consequently it will be much harder to do this job.

- *BRANSON* :

You showed a graph of the structure function for different values of Λ and the structure function changes quite a bit. This structure function is proportional to the logarithm of Q^2 over Λ^2. If Λ were zero with α_s zero, what would you get for the structure function and when do you expect this cutoff to take place?

- *BERGER* :

This is a good question. Taken at face value such a function is extremely sensitive to Λ near zero; for example, if Λ were equal to 10 MeV (the value which could not be excluded last year by some of the deep inelastic lepton-nucleon scattering experiments), the structure function would blow up by a factor of 3 to 4. But now you have to remember how this Λ came in. Λ is essentially the p_T^2 cutoff in this area; it does not make sense to take Λ smaller than the quark masses.

- *BRANSON* :

I understand that. But how do you know the quark mass used ? For example, if you take the constituent quark masses, of maybe 300 MeV, then the sensitivity to Λ in the 200 MeV range would be difficult.

- *BERGER* :

Here I have an example from the CELLO experiment (fig. not included in the text). Consider the quark masses of 200 MeV. Here there is a curve, which describes the data available. The skeptic would say that these data do not prove anything; it is just accidental that it is also described by QCD, and there is nothing else in it. On the other hand I think that future experiments will allow us to test the difference between the quark model and QCD. The quark model contains a log W^2 and QCD in LLA contains a log (Q^2), and we will be able to distinguish between these things.

- *BRANSON* :

Couldn't it be possible that the quark parton model has more sensitivity, and that it comes out really $\log(W^2/M^2)$?

- *BERGER* :

Yes. I think that the experiments should be open to that ; it could happen that no real QCD effect is borne out by the data.

- *BRANSON* :

Is a linear combination of $\log(W^2/M^2)$ and $\log(Q^2/M^2)$ a possible solution ? That would be of course much more difficult to determine experimentally.

- *BERGER* :

It is possible, but it is the usual game. In each field of

physics there are different alternative models for QCD. I am personally not worried about this.

- *ETIM* :

I have a comment on the dependence of the structure functions on the value of the quark mass.
I would like to indicate that the dependence of the structure function on the value of the quark mass (i.e. whether current or constituent mass) varies from model to model. In two model calculations we have performed (E. Masso', L. Schülke and myself) we find that in one of these models, which uses vector meson inputs, there is little variation of the structure function with the quark mass. In a second model which uses QCD inputs there is a fairly strong dependence on the quark mass going from current to constituent mass values.

- *KAPLUNOVSKY* :

You probe the photon structure function with another photon. You are interested in the structure function of the real photon and you probe it with a virtual photon. Why do you suppose the virtual photon to be structureless ?

- *BERGER* :

It is only an interpretation of the experimental finding. Consider the total cross section for $e^+e^- \to e^+e^- + X$, where e^+ is scattered at large angle, and e^- is scattered at a very small angle. We interprete this total cross section in terms of the structure function $F_2(x,Q^2)$ of the real photon, where Q^2 is related to the large angle positron. It is also possible to consider the e^- scattered at large angle ; the most general structure function is $F_2(x,Q_1^2,Q_2^2)$, and you could say that in this sense you also get information about the structure of the virtual photon. This may be the most interesting thing to look at in the future.

- *KAPLUNOVSKY* :

My second question concerns the comparison to the Vector Dominance Model. The question is : your photon is at least a bit virtual. So it seems to me that the proper thing to compare it to is not the photoproduction, but rather the electroproduction in the proper kinematical range. Would this change the result ? And could it save the Vector Dominance Model (VDM) ?

- *BERGER* :

Yes, but our Ansatz $\sigma\ (Q,W^2) = \ \sigma\ (0,W^2)F_\rho^2$ has proven to be true experimentally in electroproduction at small Q^2. This answers the first part of your question.
Concerning the second part, it is known from electroproduction that the inclusion of higher vector meson states improves the agreement with data. I think that we can go further in this, and generalize the VDM. I showed you these curves this morning, based on a paper of Etim. With this infinite series of vector mesons, there is a very good description of the data, and Etim even says, which for me is really interesting, that you can get the $\ln Q^2$ factor in the formula. I have always thought that the $\ln Q^2$ does really belong to QCD and Etim says no. It is quite interesting.

- *BELLUCCI* :

Your formulas for the structure functions are affected by radiative corrections of order α^2. Could they modify your estimate of Λ^2, even if they would leave unaffected the logarithmic scaling in Q^2 of the structure functions ?

- *BERGER* :

I did not mention it, but it is true that the data shown up to now do not include radiative corrections. We know from lepton-nucleon scattering that radiative corrections make a big effect. But I

argue that they should not be so important here. I have two arguments. First, there is an explicit calculation of radiative corrections in the regime of W=1.2 GeV, i.e. in the region of the so-called f meson. We have measured the radiative decay of the f meson. It has been shown that radiative corrections are very small. This does not need to hold for the structure functions.
But, secondly, in principle you can get rid of most of the radiative corrections experimentally. What would the radiative corrections be ? The electron in the initial state radiates a hard photon and so you measure Q^2 and W incorrectly. But, because in an ideal detector you can measure the full four-momentum of the final state X you can reconstruct the initial energy. This is the difference with the lepton-nucleon scattering, where one only measures the outgoing leptons and has no measurement of the final state.

- *LOEWE* :

My question is related to Witten's calculation of QCD corrections to the quark box. This morning you said that the result is the same. I think that at least the x distributions have to be completely different. The reason is that you do not only have a quark box with gluons, but you can also have pure gluon contributions. It is a complicated mixture problem, like in DIS. Can you make a comment on this ?

- *BERGER* :

The result of Witten's calculation, which has been confirmed by many other authors, is that h(x) is a little softer than $h_{box}(x)$. This softening is due to the radiation of extra gluons from the incoming quarks. More complicated diagrams are taken into account in higher order calculations.

- *LOEWE* :

The errors on your experimental data are very large. It is very difficult to distinguish between different predictions at this level.

- *BERGER* :

The message is that, at least for $\Lambda \approx .2$ GeV, the QCD higher order corrections are very small. They depend strongly on Λ itself. In the region $.4 \leq x \leq .8$ the cross sections are stable and precise Λ measurements can be made. For $x \leq .4$ the higher order corrections tend to yield negative cross sections.

- *ETIM* :

I have a question about the bump in the $\gamma\gamma \rightarrow 4\pi$ final state. I think that this final state has been analysed after having been regrouped into two ρ's. I think that there must have been a resonance there. This has been interpreted as a glueball. Tell us whether there is a resonance at 2 GeV with J=2,P=+,C=+. What is the experimental status, and how seriously do we have to consider this indications ?

- *BERGER* :

In e^+e^- machines, all sorts of physics can be done at once. One not only has very large time-like Q^2, but also very large space-like Q^2, and in addition, a regime where it is possible to scatter almost real photons on each other.

In DESY there is a lot of activity to look for two real photons going into resonances. The aim is first to measure the decay width of known resonances, and then to search for new resonances. In the last years the TASSO group reported on $\gamma\gamma \rightarrow 4\pi$. They found that this 4π state is almost completely $\rho\rho$. Some people have tried to interpret this as a $\rho\rho$ resonance itself, and even to interpret it as a glueball. TASSO has continued the investigation of this 4π state (this was shown at the Paris conference). What is plotted is the invariant mass of the 4π system, which is almost completely $\rho\rho$.

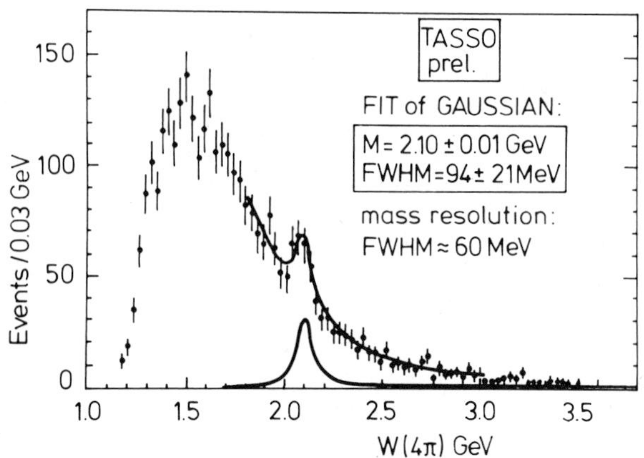

On the falling shoulder of the distribution a new peak is developping. It has a mass of 2.1 $\pm$ .1 GeV, a FWHM of 94 $\pm$ 21 MeV, almost compatible with the experimental resolution of 60 MeV. They did not determine the spin of the system, but they could only give a combination Γ ($\gamma\gamma$ decay)xBRx(2J+1)=1.6$\pm$.4 KeV. From a fit to the falling part of the curve they obtain a 5.2 standard deviation signal. One has to see whether this is confirmed by other experiments. The most surprising thing about this resonance is the rather small width.

- *ETIM* :

Do you have indications from other groups at PETRA ?

- *BERGER* :

No. The new PLUTO data, which contain about the same amount of luminosity, have not yet been analysed.

- *ETIM* :

I have a comment about the use of QCD to fit $\gamma\gamma$ total cross sections, and the possibility to connect the so-called hard part and soft part of the cross section.

The Vector meson Dominance Model (VDM) provides a means to interpolate smoothly between these two regions.

The essential problem of the VDM is an integral representation of the dilatation operator in momentum space,

$$D = Q^2 \frac{d}{dQ^2}$$

For scale invariant structure functions one would have, e.g.

$$DF_2 = 0$$

i.e. F_2 would not depend on Q^2, but only on x. For large Q^2, i.e. $Q^2 \to \infty$, and neglecting masses like $m_p, m_\rho, \ldots$ the D operator has an exact integral representation-called Hilbert transform-, which is defined as

$$Hf(Q^2) := \frac{1}{\pi} \int_0^\infty \frac{ds\ f(s)}{s+Q^2}$$

In the space of homogeneous functions, namely for

$$f(s) = a\ s^{\lambda-1}$$

acting with D on f(s), or with H on f(s), reproduces the function itself, to within a multiplicative constant, i.e.

$$H f(Q^2) = \bar{f}(Q^2) = \Gamma(\lambda)\ \Gamma(1-\lambda)\ (Q^2)^{\lambda-1}$$

This is the basic mathematical relation of the VDM. In addition, D and H commute, therefore

$$DH\ f(Q^2) = \frac{1}{\pi} \int_0^\infty \frac{ds\ s}{(s+Q^2)^2} f(s)$$

and the action of this product reproduces what one would call a vector meson representation, if s is interpreted as a vector meson mass. If one substitutes $f(s) = a\, s^{\lambda-1}$ into the latter equation, i.e.

$$DH\ f(Q^2) = \bar{f}(Q^2) = \Gamma(1+\lambda)\ \Gamma(1-\lambda)(Q^2)^{\lambda-1}$$

the same function as before is reproduced to within a constant. For $\lambda = 0$ one has a scaling theory, the cross section scales with energy. If one knows how a theory behaves asymptotically, one can use this transformation, reproduce the result, and relate it to the low energy behaviour.

- *SHERMAN* :

Could you say what triggers you used to obtain the single tag events ? Did you require charged energy in the central detector ?

- *BERGER* :

That depends. To select high Q^2 physics one requires a tag, and a minimum condition in the central detector, typically one track with $p_T > 300$ MeV. One has to apply harder cuts in the case of no-tag events, e.g. use minimum two tracks, maybe with a harder p_T cut-off, and in addition one has to make a cut on the total energy of the events, e.g. less than 1/3 of the total energy.

- *SAKURAI* :

I have a question on the 2.1 GeV state. Was this glueball not seen in other experiments, such as ψ decay ?

- *BERGER* :

Not as far as I know. Dr. Lindenbaum has a state at 2.1 GeV decaying into $\phi\phi$. I think that there is no relation between the two states. One should rather ask him why his 2.1 GeV state does not decay into 4π.

- *SAKURAI* :

Is his state very narrow ?

- *BERGER*:

I don't know exactly. The analysis in terms of glueballs relies very much on the fact that the production of his state violates very much the OZI rule.

- *BRANSON* :

I heard that if you find e.g. the 2+ state, and you want to think that this is a glueball, that the prime thing is that it should not be produced in two photon interactions. So, if you find a bump in $\gamma\gamma$ interactions it is unlikely that you see a glueball.

- *BERGER* :

This is easy to reconcile, because a glueball has first to go to quarks, which then couple to the photon system. This is the reason why it should be suppressed in $\gamma\gamma$ interactions. I would be surprised if $\gamma\gamma$ physics would be a strong source of glueballs. On the other hand anything that is new is interesting. The two photon decay width can play an important rule in determining the 'glue content' of resonances (see J. Rosner ref. 26).

- *WAGNER* :

Lindenbaum's and your object are quite similar. Both decay into vector mesons, have similar masses, and also the widths are compatible. Lindenbaum's states were also very narrow for a hadronic resonance at that energy.

- *BERGER* :

I would not say that this TASSO state decays into two vector mesons. It is not proven to decay into two rho's. It decays into 4π. They even tried to make a cut against $\rho\rho$.

- *ERREDE* :

The Lindenbaum's wave state at 2160 MeV had a width of 315 MeV.

- *BERGER* :

Lindenbaum did not see it to decay into 4π.

STATUS OF DEEP INELASTIC PHENOMENA

F. Eisele

Institut für Physik der Universität Dortmund
4600 Dortmund 50, W.-Germany

ABSTRACT

The lecture covers the topics of deep inelastic muon and neutrino scattering, the hadronic muon pair production, and aspects of high p_T hadron-hadron reactions.

For deep inelastic lepton scattering, the emphasis is put on a critical comparison of different experiments, the flavour decomposition of the nucleon structure functions and the confrontation of the measurements with QCD. The determination of nucleon and meson structure functions from muon pair production experiments will be reviewed. Finally some basic aspects of high p_T reactions are discussed.

1. INTRODUCTION

Deep inelastic phenomena are supposed to be due to hard scattering processes at the level of pointlike quarks, leptons, gluons and photons. Some examples are given below:

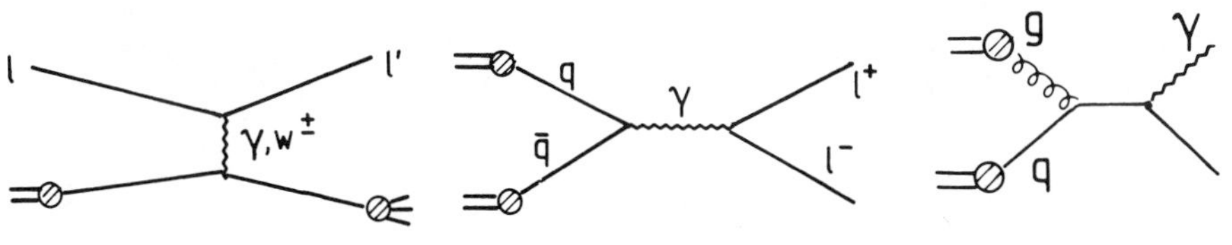

a) deep inelastic scattering b) Drell-Yan process c) single photon production

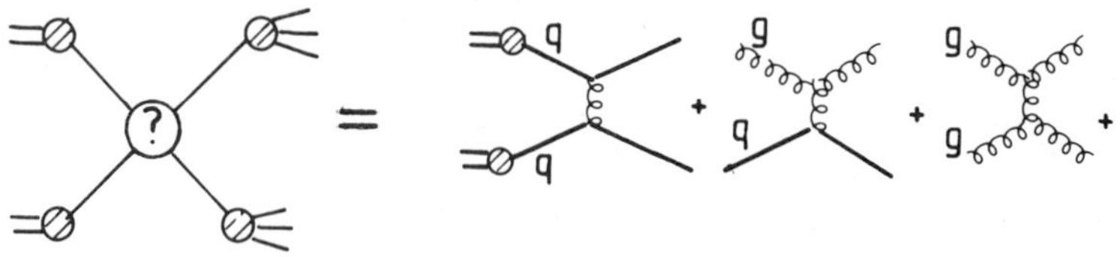

d) leading contributions to high p_T hadron-hadron scattering

The quark parton model (QPM) gives relations between these processes since they are described by the same hadron structure functions and/or the same quark and gluon fragmentation functions. The processes c) and d) involve quark-quark, quark-gluon, and gluon-gluon scattering at short distances such that perturbative quantum chromodynamics (QCD) is applicable. In this case both the absolute magnitude and the differential distributions are predicted in terms of the strong coupling constant $\alpha_s(\mu^2)$ provided that we know the quark, antiquark and gluon distributions in the incoming hadrons. Let me also remind you, that QCD is the only theory which provides an explanation for the success of the QPM.

In this lecture I will concentrate on the determination of structure functions in nucleons and mesons and their application in other hard scattering processes.

2. DEEP INELASTIC LEPTON SCATTERING (DIS)

Deep inelastic scattering experiments with lepton beams are our best method to study the parton distributions in the nucleon. Historically the concept of "partons" evolved from the analysis of inelastic eN-scattering experiments at SLAC.

In 1969, at the electron-photon symposium, R.E.Taylor[1)] presented the result which is shown in fig.1: It shows the ratio of the measured differential cross-section $d\sigma/dQ^2$ to the Mott cross-section for point like scattering for the inelastic scattering of electrons from nucleons. This ratio was found to be approximately constant in striking contrast to the same quantity for elastic scattering which shows a strong decrease with Q^2 as described by the "dipole" formfactor. This result is the basis of the parton picture which turned out to be so successful and is the beginning of structure function measurements in the sense we use it today: structure functions are the momentum distributions of the partons. I have shown this result also because it gives some sort of time scale for experiments: the measurement of structure functions at SLAC involved about a 10 years program, the data are still by far the best in their energy range and we will have to rely on these measurements for quite some time. Also in the present energy range, we need some dedicated experiments and persistent physicists to do a good job on the determination of structure functions.

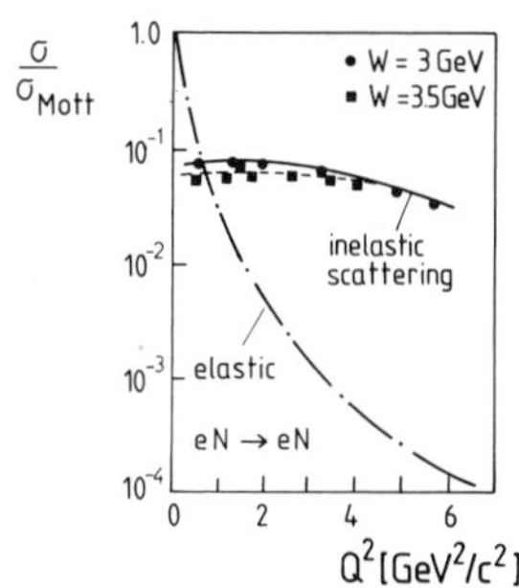

Fig.1:

The differential cross-section $d\sigma/dQ^2$ for the inclusive inelastic scattering of electrons from nucleons divided by the Mott cross-section as measured at SLAC in 1968/69. Also shown is the corresponding distribution for elastic scattering.[1)]

2.1 How do we measure the parton distribution in the nucleon?

Let me remind you shortly how the structure functions are obtained from the measured differential cross-sections and how they are related to the parton distributions:
Experiments on "isoscalar" targets (same number of neutrons and protons) provide the bulk of our present information. By far the best measured structure function is $F2(x,Q^2)$ which is directly proportional to the differential cross-section for electron and muon experiments and to the sum of neutrino and antineutrino cross-section for neutrino experiments.

$$F2^{\mu N} \approx \frac{Q^4}{4\pi\alpha^2 s} \left\{\frac{d^2\sigma^{\mu N}}{dxdy}\right\} \frac{1}{\left[1-y+\frac{y^2}{2(1+R)}\right]} \tag{1}$$

$$F2^{\nu N} \approx \frac{(Q^2+m_W^2)^2}{4\pi g^2 S} \left\{\frac{d^2\sigma^{\nu N}}{dxdy}+\frac{d^2\sigma^{\bar{\nu}N}}{dxdy}\right\} \frac{1}{\left[1-y+\frac{y^2}{2(1+R)}\right]} \tag{2}$$

In the QPM $F2^{\nu N} = x(u+d+s+c+\bar{u}+\bar{d}+\bar{s}+\bar{c})$, where u,d,s,... are the up, down, strange ... quark densities in the proton. The neutrino and muon structure functions are related by $F2^{\nu N}(x,Q^2) \approx 18/5\ F2^{\nu N}$ up to small corrections due to the difference of strange and charmed quark contributions.

Neutrino experiments are able to separate quark and antiquark contributions due to the V-A structure of charged current interactions. The V-A structure is well established also at high energies. This has recently been discussed in contributions to the Paris conference by the CHARM and CDHS collaborations. The structure functions measured are the valence quark distribution $xF3(x,Q^2)$ which is proportional to the difference of neutrino and antineutrino cross-section

$$xF3(x,Q^2) = x(u_v+d_v) \sim \left(\frac{d^2\sigma^\nu}{dxdy} - \frac{d^2\sigma^{\bar\nu}}{dxdy}\right) / [1 - (1-y)^2] \tag{3}$$

and the antiquark distribution $\bar{q}^{\bar\nu}$ which is obtained from antineutrino scattering at high y:

$$\bar{q}^{\bar\nu}(x,Q^2) = x(\bar{u}+\bar{d}+2\bar{s}) \sim \frac{d^2\sigma^{\bar\nu N}}{dxdy} - (1-y)^2\,\frac{d^2\sigma^{\nu N}}{dxdy} \text{ for } y \gtrsim .5. \tag{4}$$

Let me point out, that the determination of F2 and also of $\bar{q}^{\bar\nu}$ requires the knowledge of the longitudinal structure function F_L which in eqs(1) and (2) enters in the quantity $R = \sigma_L/\sigma_T \simeq F_L/F_2$.

For isoscalar targets we cannot separate the contributions of up and down quarks or antiquarks. This is only possible if experiments on hydrogen are also performed. This will be discussed in section 2.5. Measurements of the three structure functions F2, xF3 and $\bar{q}$ for fixed Q^2 are shown in figure 2. The muon and neutrino structure functions F2 are reasonably well related by the QPM factor 18/5.

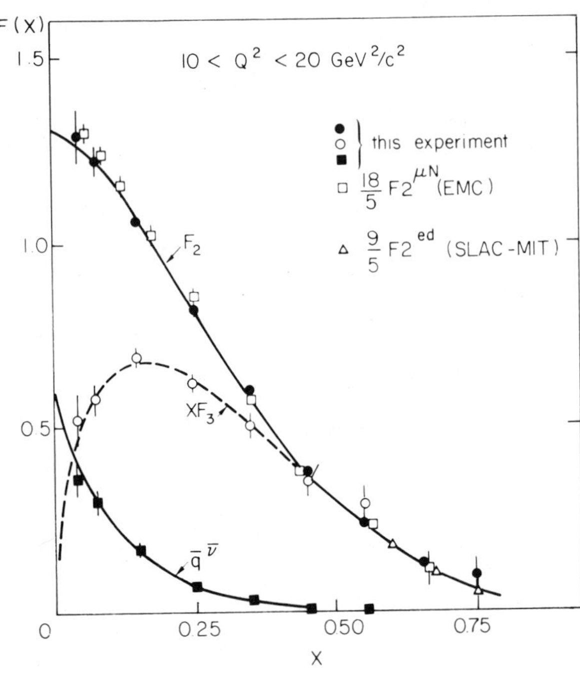

Fig.2:
F2, xF3 and $\bar{q}^{\bar\nu}$ for fixed value of Q^2. The measurements of $F2^{\mu N}$ and $F2^{eD}$ are multiplied by the QPM factors.

2.2 Survey on the high statistics experiments

In this lecture I will concentrate on a few high statistics

experiments. The available kinematic range in (x,Q^2) is shown in figure 3 for the five experiments which will be discussed.

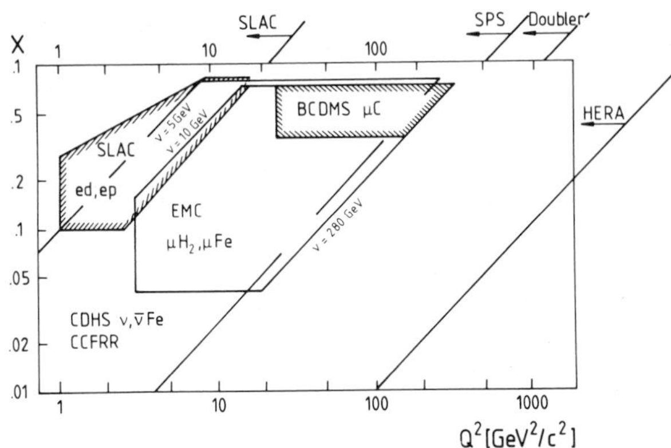

Fig. 3: Kinematic range in $(x-Q^2)$ for five experiments which will be discussed.

At low Q^2 or better low invariant hadron mass $W^2 = Q^2\, x/(1-x)$ the measurements from SLAC[2] provide still by far the most precise data. These data will be used wherever we need the largest possible range in Q^2 and the comparison of these data with the high energy data from CERN will allow a separation of perturbative and non-perturbative contributions. At high energies I will discuss the structure function measurements of the EMC muon experiment[3] and of the CDHS neutrino experiment[4] which cover approximately the same range in x and Q^2 and also the muon experiment of the BCDMS collaboration[5] which is restricted to high Q^2 and $x(x > .3)$ and has by far the highest statistics. Finally I will discuss some preliminary results of the big Fermilab neutrino experiment run by the CCFRR-collaboration[6].

2.3 Comparison of structure function measurements

Let me start with the good news. Figure 4 shows a comparison of the measurements on isoscalar targets at large x for the three high statistics experiments with final results. The measurements of EMC and CDHS agree well with the QPM factor 5/18. The measurements of the BCDMS muon experiment show some deviations at large x at the 5 - 10 % level which however are partly explained by the difference of Fermi motion effects in iron (EMC, CDHS) compared to carbon (BCDMS).

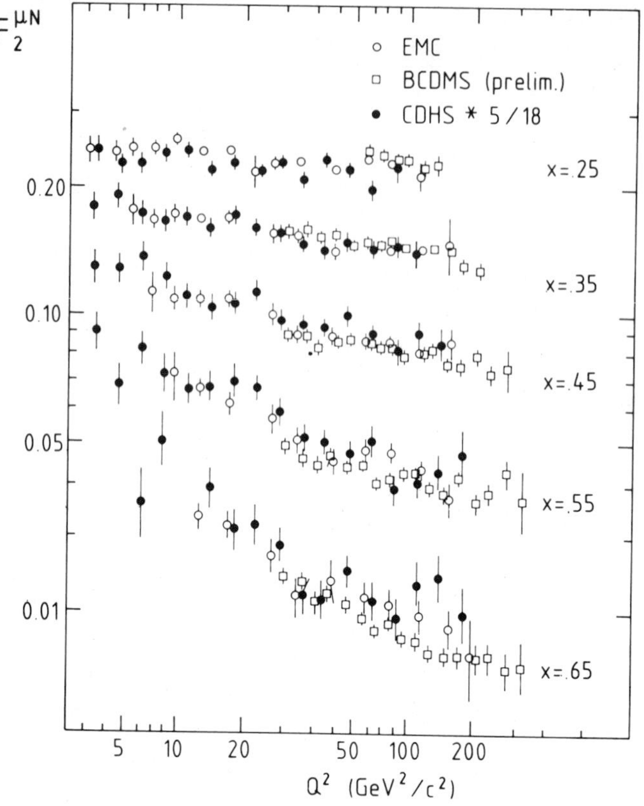

Fig.4: $F2(x,Q^2)$ for the three high statistics experiments.

Figure 5 shows the amount of scaling violations as a function of x for the three experiments. The slopes $d\ell nF2/d\ell nQ^2$ have been obtained by simple power law fits to the data for each bin in x. All three experiments see highly significant scaling violations with a pattern which is well described by QCD and the experiments agree pretty well with each other.

Thus the good new are, that the experiments agree within $\sim$ 10 % both in shape and in the amount of scaling violations. This picture is unchanged if other low statistics experiments are included.

Let me come to the bad news: the experiments do not agree to better than $\sim$ 10 % outside the statistical and sometimes also the given systematic errors. Let me flash a few comparisons to give you an idea about the level of agreement and disagreement. Figure 6 compares the measurements of $F2^{\nu N}(x,Q^2)$ of the CDHS collaboration with the preliminary result of the CCFRR neutrino experiment which has been presented to the Paris conference. Both data sets differ by an overall normalization factor of about 1.18 which is not understood at present. Apart from this problem, the data are in fair agreement at a level of about 10 % with the tendency, that the CCFRR measurements have smaller Q^2-slopes at large x and larger slopes at small x compared to the CDHS measurements.

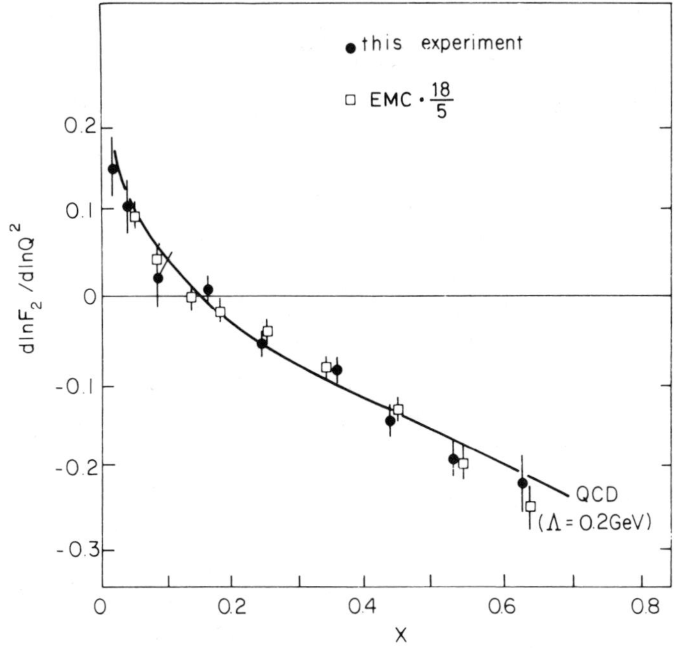

Fig.5:

The slopes $d\ell nF2/d\ell nQ^2$ for two experiments as obtained from power law fits to the whole Q^2-range. The line is the QCD prediction for $\Lambda_{L.O.}$ = 0.2 GeV.

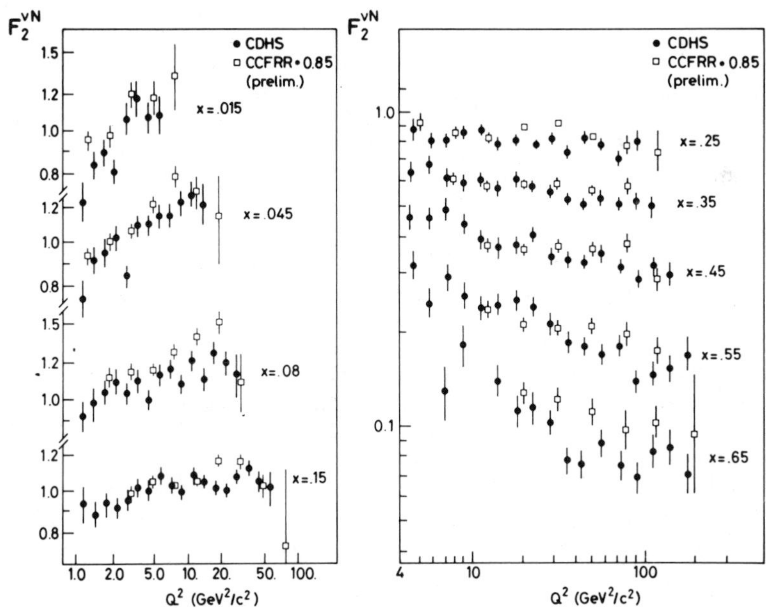

Fig.6: Structure functions $F2^{\nu N}$ for the two high statistics ν experiments using a common overall normalization.

Figure 7 shows the measurements of the structure function F2 for both neutrino and muon experiments. All experiments agree within about 10 % using the QPM factor 5/18 to relate muon and neutrino data. There are however indications of systematic differences which are not covered by statistical errors.

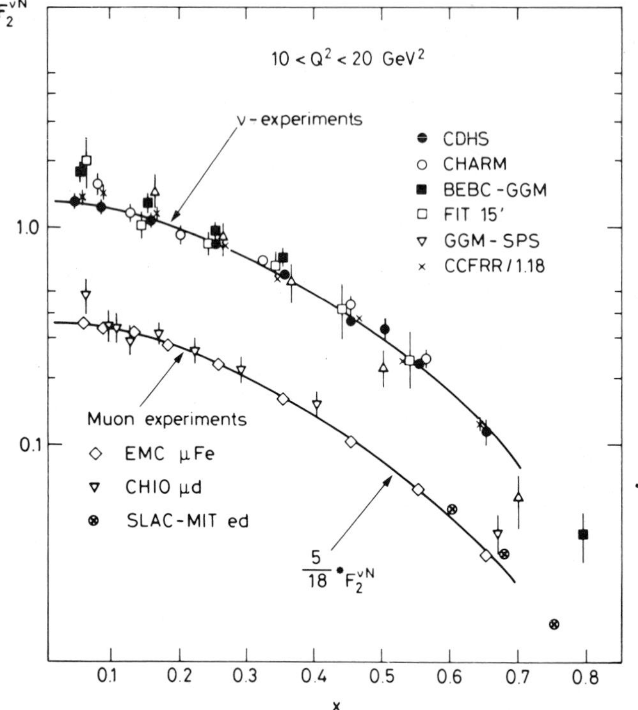

Fig.7:

The structure function F2 for neutrino and muon experiments on isoscalar targets at fixed Q^2. The interpolating curves differ by the QPM factor 5/18.

Finally, figure 8 compares the muon and electron scattering data on hydrogen for fixed Q^2. The most remarkable feature is a 10 % normalization difference between the high energy EMC data and both the SLAC-MIT and CHIO data. Whereas the difference EMC-SLAC may be related to the uncertainty of $R = \sigma_L/\sigma_T$ at large x, the difference EMC-CHIO cannot be explained this way and is barely covered by the assigned systematic errors of the two experiments.

Thus the bad news are: the various experiments disagree at a level of about 10 % outside statistical and sometimes systematic errors. We have an overall normalization problem of about 18 % for the neutrino experiments and some 10 to 20 % normalization differences between the low energy SLAC data and the high energy experiments. The latter problem disappears if $R = \sigma_L/\sigma_T$ is put to zero for the SLAC data at large x. A more reliable measurement of R in this kinematic range would be highly desirable.

The question is then: to what extent can we draw reliable conclusions from this data? A warning is certainly adequate: Systematic errors are by no means negligible and a combined analysis of different experiments is very problematic. Their structure function measurements generally differ in magnitude and shape outside statistical errors.

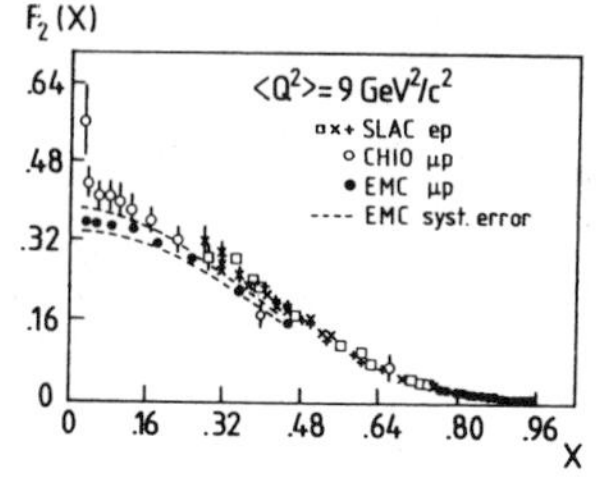

Fig.8:

The structure function $F2^{ep}$ for the three hydrogen experiments at fixed Q^2.

2.4 Discours on experimental problems

In this section I want to give you some ideas about the experimental problems and the present limitations of the structure function measurements.

muon versus neutrino experiments

What are the relative merits of muon and neutrino experiments? Clearly event rates are substantially higher for muon experiments at small and moderate values of Q^2 and thus they are unique to get high statistics data on hydrogen and deuterium targets. Neutrino experiments on the other hand are restricted to heavy targets if large event numbers are required. The main difference between neutrino and muon experiments is however due to the photon propagator effect, which, from an experimental point of view, disfavours muon with respect to neutrino experiments. Loosly speaking, the muon experiments get their large event rate in the wrong kinematic range. At the highest Q^2-values of the neutrino counter experiments the effect of a W-boson propagator with m_W = 80 GeV is only a 10 % correction, whereas the $1/Q^4$ for muon experiments changes the cross-section by more than 3 orders of magnitude in the same kinematic range. This is illustrated in figure 9a) where $d\sigma/dx$ is given for a fixed average energy transfer ν = 125 GeV and in figure 9b) where the cross-section is given for fixed x as a function of Q^2.

For the neutrino experiments, $d\sigma/dx$ is directly proportional to F2 and the acceptance is 100 % in the whole x-range. For muons the $1/Q^4$ propagator gives a decrease of 3 orders magnitude which

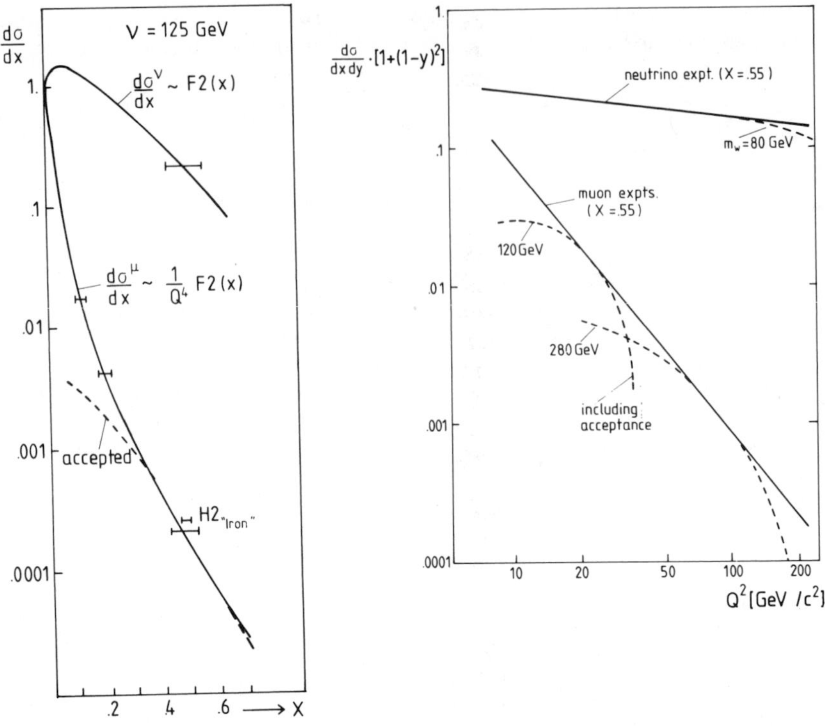

Fig.9a) dσ/dx for neutrino and muon experiments at fixed ν.

Fig.9b) $d\sigma/dxdy(1+(1-y)^2)$ for fixed y versus Q^2.

has to be unfolded. In addition the acceptance of the detectors cuts off at low x both to reduce the event rate and to assure a minimal scattering angle for the muon. As a function of Q^2 (fig.9b) neutrino experiments show a variation which is given by the magnitude of the scaling violations whereas the observed cross-section for muon experiments falls off by 3 orders of magnitude and is seriously distorted by the experimental acceptance. The example given in fig.9b) holds for EMC with measurements at two different energies. As a result, muon experiments need much better understanding of their detector - i.e. calibration of muon energies and angles, acceptance, resolution - compared to neutrino experiments. The measured distributions are subject to very large corrections before the structure function can be obtained. My private conclusion is, that at high energies it is easier to measure structure functions with neutrinos than with muons and as we will see below, the real limitation is not coming from statistics.

neutrino experiments

Neutrino experiments are unique to determine the flavour composition of the nucleon. The distributions of valence quarks, antiquarks and strange quarks are only accessible to neutrino experiments. Also the longitudinal structure function is best measured in neutrino experiments as will be shown later. There are only two high statistics neutrino experiments which are competitive: the CDHS experiment, which up to now has had a unique position and the CCFRR experiment which has presented preliminary results to the Paris conference. Unfortunately the agreement between the structure function measurements of these two groups is very poor, as I have shown before (fig.6). The origin of these differences is however understood and can be related to the difference in the absolute neutrino and antineutrino cross-section and the difference in the energy dependence of the cross-sections.

Averaged over energy the CCFRR collaboration measures ν and $\bar{\nu}$ - cross-sections which are higher by 18 % compared to CDHS. The origin of this difference is not understood at all and can only be resolved by future experiments. It should be noted that a common scale error for both the neutrino and the antineutrino cross-sections does not affect the shape and the Q^2-dependence of the structure functions. Errors in $\sigma^{\nu}/\sigma^{\bar{\nu}}$ and/or the energy dependence of the cross-sections on the other hand affect the shape and slope measurements seriously. The energy dependence of σ^{ν}/E and $\sigma^{\bar{\nu}}/E$ for the two high statistics neutrino experiments is shown in figure 10. The results of the two experiments differ not only in absolute normalization but also in their energy dependence. Whereas CDHS shows significant decrease of $\sigma^{\bar{\nu}}/E_{\nu}$ with energy, the CCFRR measurements show a statistically highly significant rise with energy. As a result, the structure function measurements of the two groups must be different. The most important fact to remember is, that the energy dependence of the cross-sections is obtained in a very different way for the two experiments. For CDHS, the neutrino energy range from 20 to 200 GeV is covered by a single beam setting of 200 GeV/c parents. In this case, the energy dependence is given by the beam optics, decay kinematics and the ratio of pions and kaons in the beam. The data points given in figure 10 include an estimate of systematic point to point errors. The estimated overall flux errors are indicated by the dashed lines. The CCFRR collaboration has the drawback, that their distance to the decay tunnel is about twice that for CDHS. This has the consequence, that a single beam energy setting gives only two rather narrow neutrino energy bands and a

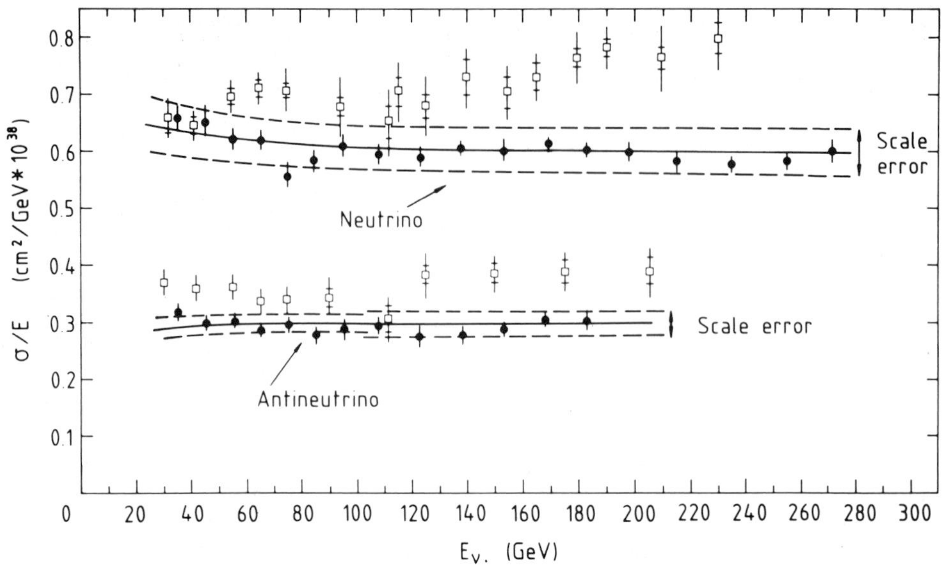

Fig.10: Energy dependence of σ^{ν}/E and $\sigma^{\bar{\nu}}/E$ for the two high statistics neutrino experiments. The data points for CDHS include systematic point to point errors. The overall flux errors are given by the dashed lines. For CCFRR, the inner error bars are statistical, the outer error bars include the flux errors.

total of 5 different beam settings at 120, 140, 168, 200 and 250 GeV is needed to cover the total energy range. The energy dependence of the cross-sections for this experiment is therefore given by 5 absolute flux measurements and 5 pion over kaon ratios.

Given the preliminary nature of the CCFRR data and the fact that the CDHS experiment has much smaller systematic problems to measure the energy dependence, I will not worry further about the present differences except for the overall normalization.

Systematic problems connected to the neutrino detectors are no serious source of worry at present. The most critical quantity to measure is the total hadron energy which is equal to the energy transfer ν. At present the measurement accuracy of E_H limits the accessible range in (x,Q^2). The systematic uncertainties in the hadron energy measurements will however most likely be the limiting factors for future high statistics experiments.

muon experiments

Muon experiments have the nice feature that many systematic effect are selfindicating. It is sufficient to compare data from different muon energies.

This is done for the example of EMC data on iron in figure 11 for the four energies 120, 200, 250 and 280 GeV. The data points clearly differ outside statistical errors, mainly in the overlap region between two energies. Part of this difference may be due to the effect of $R = \sigma_L/\sigma_T$ non zero. Similar effects are observed for the hydrogen data of EMC and the structure function measurement of the BCDMS collaboration.

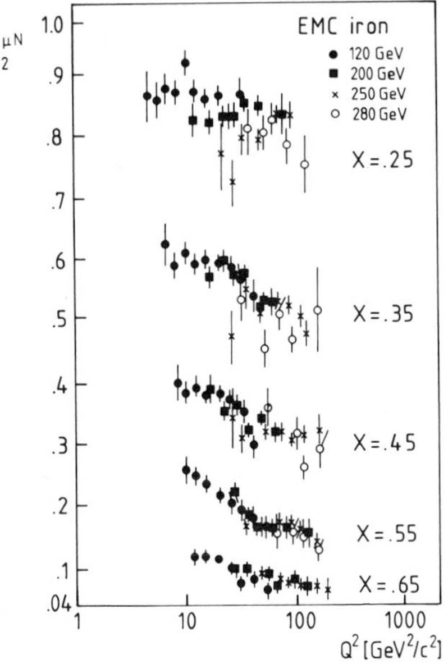

Fig.11:

Measurements of $F2^{\nu N}(x,Q^2)$ for the EMC experiment on iron. Data points with statistical errors are given separately for each energy.

The main origins of systematic problems can be traced most likely to

i) the large variations of the geometric acceptance

ii) time-dependent "electronic" acceptance functions due to chamber inefficiencies etc. and

iii) the strong sensitivity to muon calibration errors mainly to measure $\nu = E_\mu - E_\mu'$.

The experimenters have made a really heroic effort to understand and controll these effects and - from an experimental point of view - they have reached a level of precision and refinement which I can only admire. Calibrations, normalizations etc. are controlled to fractions of a percent - far better than we do for instance in neutrino experiments. However, in spite of these efforts systematic errors are still much larger than present statistical errors and further progress is only possible if they can be substantially reduced.

A-dependence of the structure functions?

In the following I will report on a very interesting new development which has been presented to the Paris conference. The EMC collaboration has measured the structure function $F2^{\nu N}$ both on iron and deuterium targets. The two measurements differ in shape as shown in figure 12 which gives $F2^{iron}(x)/F2^{D2}(x)$ versus x. The structure function for nucleons in iron is more narrow than that for deuterium. This is in contrast to the expectation due to the effects of Fermi motion[7)] and also in contrast to published ideas on collective quark distributions in nuclear matter. Both effects would give a rise of $F2^{iron}/F2^{D2}$ with x and would be restricted to large x.

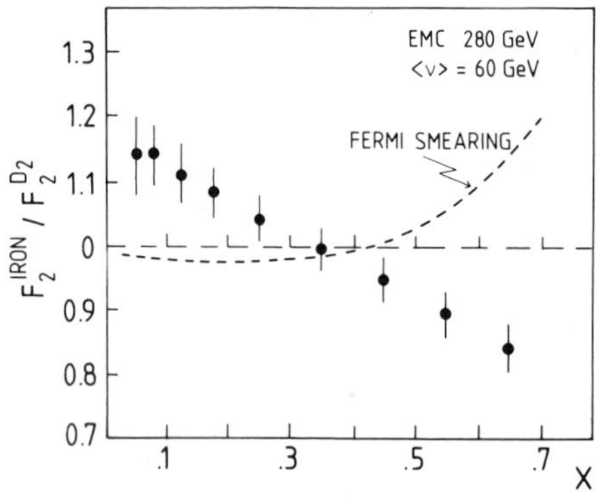

Fig.12:

The ratio of the structure functions $F2^{\nu N}$ as measured on iron and deuterium versus x. (EMC collaboration) The dashed curve gives the expected effect due to Fermi notion

It should be noted, that this data are obtained at 280 GeV muon energy at an average energy transfer of ν = 60 GeV i.e. in the deep inelastic region, where we all believe that the elementary scattering process should be the inelastic scattering of a single pointlike parton. Thus the parton distributions have to be substantially different in nuclear matter compared to isolated nucleons.

As a result, either the systematic errors are larger than the experimenters of EMC believe or they have made a real discovery. In any case the effect needs further study. At present we have to be aware that EMC structure function measurements on H_2, D2 and iron show systematic differences which cannot be consistently described within the standard quark parton model. As an example, if one tries to extract the neutron structure function $F2^{\nu n} = F2^{\nu N} - F2^{\nu p}$ using iron and hydrogen data, $F2^{\nu n}$ becomes negative at small x. It cannot be excluded however, that these differences are due to new physics.

2.5 Shape of structure functions and flavour composition

In the long run this is probably the most important subject for DIS experiments. The knowledge of quark and antiquark distributions in the nucleon is a very important ingredient for the study of parton dynamics.

2.5.1 Longitudinal structure function $F_L(x,Q^2)$

The longitudinal structure function is by far the hardest to measure and a pain in the neck for all experimentalists. It's uncertainty seriously limits our determination of F2 and $\bar{q}^{\nu}$. The measurement of F_L would also allow an important test of QCD: the leading contribution to F_L is calculable in QCD and proportional to $\alpha_s(Q^2)$. In QCD $R = \sigma_L/\sigma_T \approx F_L/F2$ cannot be zero at small x whereas it should go to zero for large x and Q^2.

The CDHS collaboration has obtained a first reliable and precise measurement at large x, which shows that R is near zero for large x and Q^2. This is a major improvement. It is based on the measured quantity $d^2\sigma^{\nu}/dxdy - d^2\sigma^{\bar{\nu}}/dxdy(1-y)^2 \approx \bar{q} + F_L(1-y)$, where the right-hand side is a good approximation for large y. This quantity is found to be very small for large x and can therefore be used to put an upper limit on F_L or R. For $.4 \leq x \leq .7$ and an average Q^2 of 38 GeV^2/c^2 the CDHS collaboration finds $R < .006 \pm .012_{stat.} \pm .025_{syst.}$ including all corrections. A previous measurement of R by CDHS[8] has been based on the analysis of the y-dependence of $d^2(\sigma^{\nu}+\sigma^{\bar{\nu}})/dxdy$ for fixed x and Q^2. This result is shown in figure 13 versus x together with the new upper limits at large x and the QCD expectation for four flavours. Other published results on R from neutrino experiments have even larger errors. Moreover they did not account for the effects of scaling violations.

The EMC collaboration has presented new results on $R = \sigma_L/\sigma_T$ both from their H_2 and iron data[3]. These preliminary results are shown in figure 14 together with published results of the SALC-MIT-experiment.

Let me conclude: The longitudinal structure function is still badly known at small x. At large x and Q^2, R is now well known and near to zero in agreement with the QCD expectation. The SLAC-MIT-experiment at much smaller values of Q^2 finds however non zero values of R at large x, which might be due to non-perturbative "diquark" contributions[9]. The data indicate some Q^2 and x dependence compatible with the QCD expectations, though the errors at present are still very large.

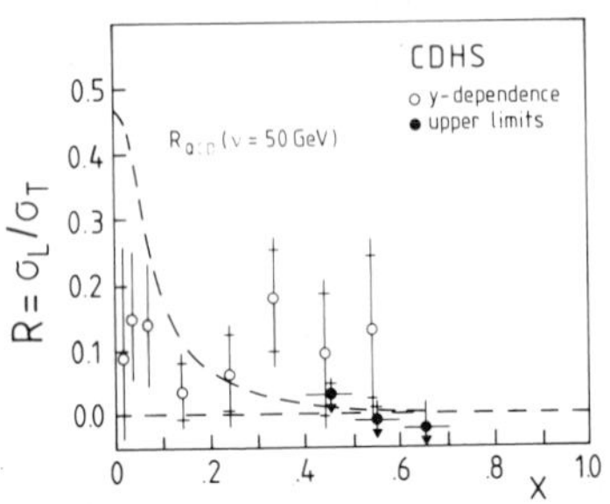

Fig.13:

Measurement of $R = \sigma_L/\sigma_T$ versus x for the CDHS experiment. $\langle\nu\rangle$ = 50 GeV

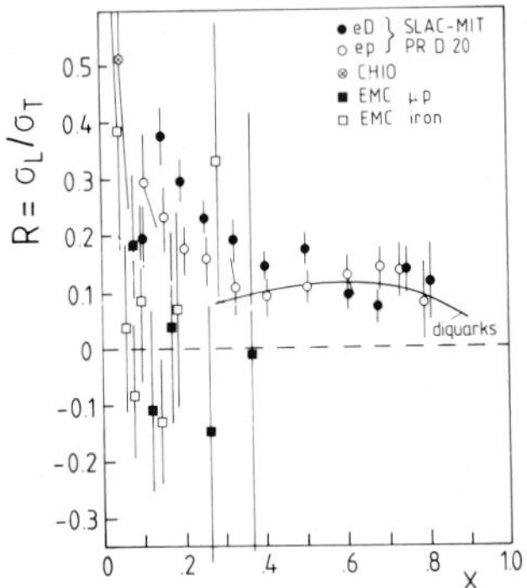

Fig.14:

Measurements of $R = \sigma_L/\sigma_T$ versus x for the EMC experiment (preliminary) and the SLAC-MIT-experiment (statistical errors only).

2.5.2 Valence quark distribution

This measurement is only possible for neutrino experiments. The best measured distribution is $xF3^{\nu N} = x(u_v+d_v)$ which is obtained on isoscalar targets. This distribution is shown in figure 15 for a fixed value of Q^2. The measurement of xF3 can be used for an important test of the quark parton model. The Gross-Llewellyn-Smith sum rule counts the number of valence quarks in the nucleon:

$$\int_0^1 \frac{xF3(x,Q^2)}{x}\,dx = 3\left(1 - \frac{\alpha_s}{\pi} + O(\alpha_s^2)\right)$$

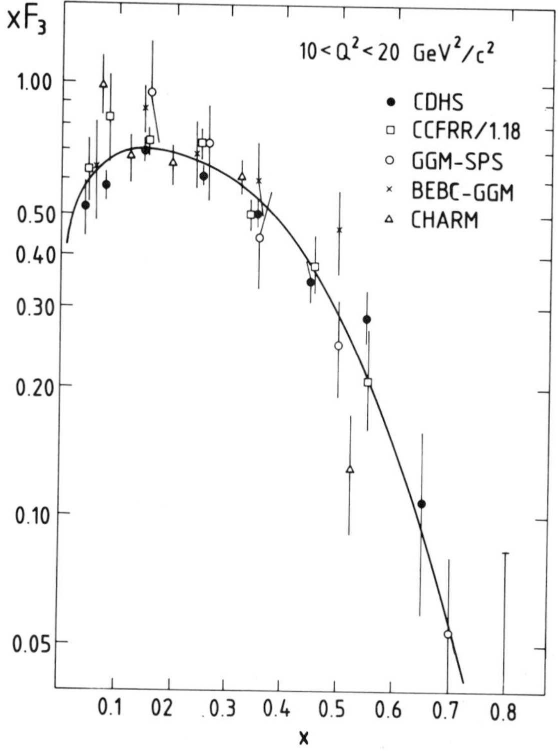

Fig. 15:

Valence quark distribution $xF3 = x(u_v+d_v)$ for fixed Q^2 as measured by neutrino experiments on isoscalar targets.

The evolution of this integral for fixed Q^2 is very problematic however, since a large fraction of the integral comes from the low x region. We expect about 18 % of the integral from the range $0 < x < .01$ and additional 29 % from the range $.01 < x < .06$. To measure in this small x-range for fixed Q^2 we need high energy since the minimal accessible value of x is $x_{min} = Q^2/2mE_\nu$. We need also very good muon angular resolution to measure x precisely. The best experiment to do such a job is the CCFRR counter experiment since it combines high statistics, high energy and excellent muon angular resolution. Preliminary evaluations of $\int_0^1 F3dx$ from the experiment are shown in figure 16 together with results from the ABCDLOS-collaboration[10] versus Q^2.

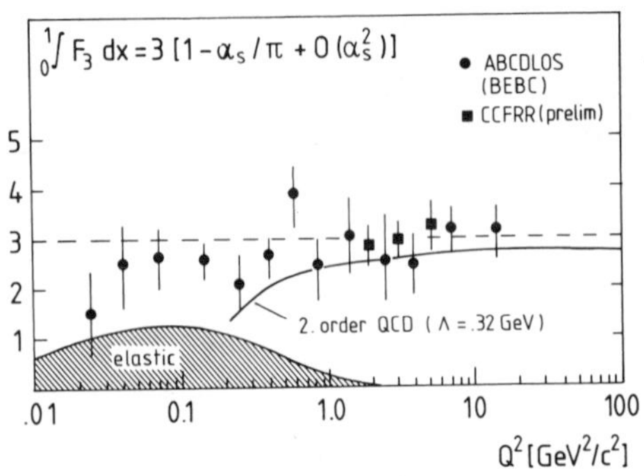

Fig.16: Evaluations of the Gross-Llewellyn-Smith sum rule versus Q^2 for the CCFRR and the ABCDLOS experiments. Also indicated are the 2. order QCD expectation for $\Lambda_{\overline{MS}}$ = .32 GeV and the contribution of elastic scattering.

For the CCFRR points, the integral is mostly measured. They suffer however from uncertainties in the energy dependence of $(\sigma^{\nu} - \sigma^{\bar{\nu}})/E$ at present. The ABCDLOS points at large Q^2 include substantial model dependent corrections at small x. An interesting feature of this data is, that the sum rule seems to be valid down to very low values of Q^2. This is an important experimental fact which is not easily explained in perturbative QCD unless Λ is very small.

The ratio of down and up quark distribution $d_v(x)/u_v(x)$

The separate measurement of up and down quark distributions requires additional data on hydrogen. It is known since a long time from the SLAC experiments on H_2 and D_2, that the x-dependence of the cross-section on neutrons differs from that on protons. The SLAC results are shown in figure 17. In the framework of the quark parton model σ_n/σ_p = [1+4d/u+sea(n)]/[4+d/u+sea(p)] and the SLAC result has to be interpreted, that the down quark distribution in the proton is softer than the up quark distribution. Alternatively it has been proposed[11] that the naive QPM expectation d_v/u_v=0.5 is valid and that the observed cross-section difference is due to substantial diquark contributions. In this case the d_v/u_v = 0.5 would be restored at large Q^2. The EMC collaboration has presented preliminary results from their H_2 and D_2 experiments at much higher values of Q^2 than SLAC. They are also given in figure 17 and are in fair agreement with the low energy data. This observation does not favour diquark models.

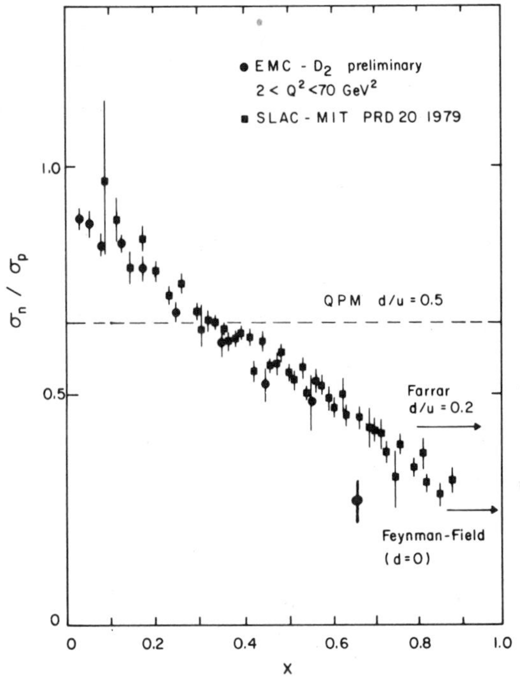

Fig.17:

Ratio of cross-section on neutrons and protons as determined by the SLAC-MIT and the EMC experiments.

Charged lepton scattering experiments cannot separate the effect of the sea quarks. In addition they suffer from substantial Fermi motion corrections at large x since they have to separate neutron and proton contributions in deuterium. Neutrino experiments can get rid of these problems if both neutrino and antineutrino cross-sections on hydrogen are measured.

Figure 18 shown the preliminary results for the CDHS H_2 counter experiment, which for the first time measures d_v/u_v in the whole x-range, compared to new results of the ABCMO collaboration in the valence region. These data are in good agreement and reasonably well described by the simple parametrization $d_v(x)/u_v(x) = .57(1-x)$. This x-dependence differs significantly from the measurements of SLAC-MIT at large x as indicated by the dashed line in figure 18. This difference is however compatible with the expected change due to a QCD evolution of the structure functions with Q^2.

In summary, up-quarks exceed down-quarks at large x by about a factor five. Their ratio is reasonably well described by $d_v(x)/u_v(x) = .57(1-x)$ in the whole x range up to $x \approx .8$. The behaviour for $x \to 1$ is however still not known good enough to distinguish theoretical alternatives. Finally diquark models seem unable to restore the naive QPM relation $d_v(x)/u_v(x) = 0.5$.

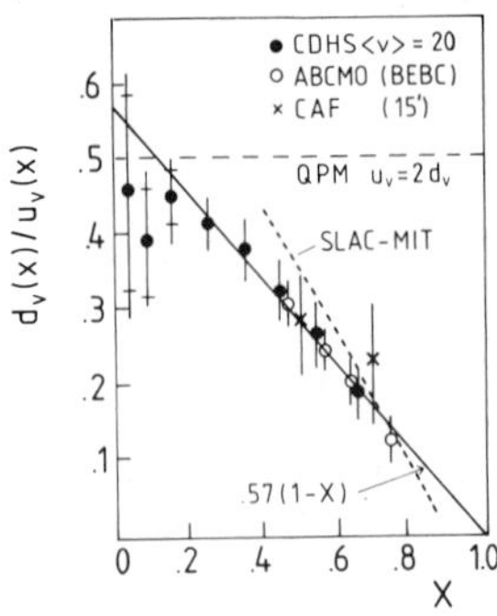

Fig.18:

The ratio of valence down and up quarks versus x measured by neutrino experiments on hydrogen. The dashed line indicates the measurement of SLAC-MIT at lower energy.

2.5.3 Flavour composition of the sea

The x and Q^2-dependence of the antiquark distribution in the nucleon $\bar{q}^{\nu}(x,Q^2) = x(\bar{u}+\bar{d}+2\bar{s})$ has been well measured by the CDHS collaboration based on 175000 antineutrino events from both narrow band and wide band neutrino beams apart from uncertainties due to $R = \sigma_L/\sigma_T$. Their measurement is shown in figure 19 evaluated for a constant value R = .1. Other assumptions about R, i.e. the use of the QCD prediction, change $\bar{q}$ outside the statistical errors. The main question is then, what is the flavour symmetry of the sea?

$\bar{u} = \bar{d}$?

Feynman and Field[12)] have made the educated guess, that $\bar{u} \approx (1-x)^3\bar{d}$ on the basis of the evaluation of the Gottfried sum rule using SLAC data. This notion that the sea is flavour asymmetric has later been supported by the CFS collaboration[13)] based on their evaluation of high mass lepton pair production.

A direct measurement of $\bar{u}/\bar{d}$ is possible using antineutrino scattering on elementary targets. Two experiments have presented results as given below:

Experiment	$(\bar{U}+\bar{S})/(\bar{D}+\bar{S})$
ABPPST(BEBC, $\bar{\nu}D_2$)[14)]	.62 ± .19
CDHS($\bar{\nu}H_2$, $\bar{\nu}Fe$)	1.30 ± .30 (prelim.)

Combining the two results, we have no compelling evidence from neutrino experiments that $\bar{u}$ unequal $\bar{d}$. The errors are however too large to allow definite conclusions.

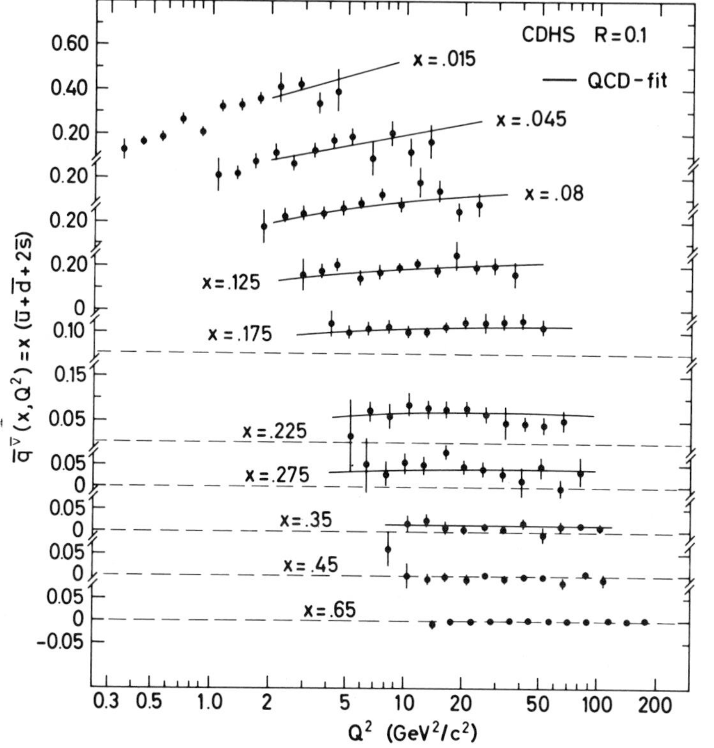

Fig.19: The antiquark distribution in the nucleon $\bar{q}^{\nu} = x(\bar{u}+\bar{d}+2\bar{s})$ as measured by the CDHS collaboration assuming R = 0.1.

The strange sea

The shape and to some extent the magnitude of the strange sea can be measured by a study of single charm production with neutrinos and antineutrinos. The experimental signature which is used are opposite sign dilepton events which are produced i.e. by $\bar{\nu} + \bar{s} \rightarrow \mu^+ + \bar{c} \rightarrow \mu^-$. The CDHS collaboration has presented a detailed study based on 2000 $\bar{\nu}$ and 10000 ν opposite sign dimuon events[15] with the following results:

i) The shape of the strange sea agrees with $x(\bar{u}+\bar{d})$ within the experimental uncertainties. A comparison is shown in figure 20, where the data points are obtained from the observed x-distribution of opposite sign dimuon events from antineutrino scattering and the lines are parametrizations of $\bar{q}^{\nu} = x(\bar{u}+\bar{d}+2\bar{s})$ with and without the effect of charm mass threshold.

ii) The magnitude of the strange sea can be obtained only with additional assumptions about the weak Kobayashi-Maskawa mixing angles[16] and about the effective charm mass. The CDHS collaboration finds $2s/(\bar{u}+\bar{d})= .52 \pm .09$ assuming that θ_2 and θ_3 are small and using $m_c = 1.5$ GeV.

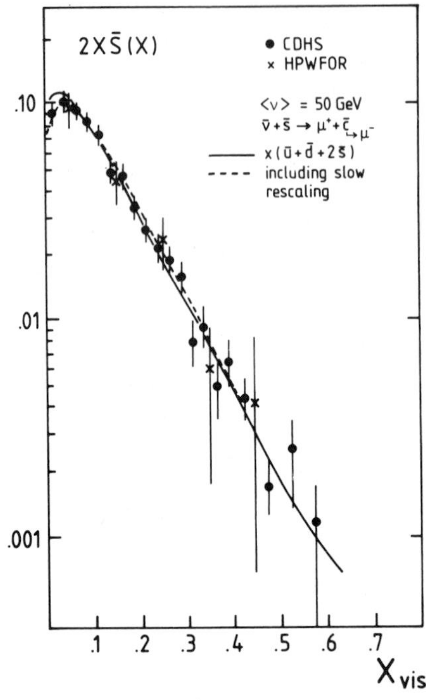

Fig.20:

x-distribution of opposite sign dimuon events from antineutrino scattering (data points) compared to $x(\bar{u}+\bar{d}+\bar{s})$ with and without corrections for the charm mass threshold (lines).

The charmed sea

The contribution of the charmed quarks to muon DIS is well described by the gluon-fusion model which I have no time to explain here. Neutrino experiments can determine the total contribution of an "intrinsic" charmed sea by the study of the reaction $\nu_\mu + c \rightarrow \nu_\mu + c \rightarrow \mu^+$ which leads to wrong sign muon events. The measurement of the CDHS collaboration[17] can be converted to the result:

$$\int_0^1 xc(x)\, dx \leq .5 \int_0^1 xs(x)\, dx \approx .005 \qquad 90 \,\&\, \text{C.L.}$$

Fraction of momentum carried by the constituents

Figure 21 shows the momentum fractions $\int_0^1 F_i(x,\nu)\, dx$ versus ν for different constituents as measured by the CDHS collaboration. Other measurements cannot be easily included, since integrals for fixed ν are normally not given. The curves are derived from parametrizations to the CDHS structure function measurements. Data points are preliminary results from CDHS.

The points to remember are: i) a consistent set of structure functions is measured including their ν-(or Q^2-) dependence; ii) the momentum fractions of sea and valence quarks show a very significant ν-dependence, whereas the momentum sharing between quarks and gluons is more or less independent of ν.

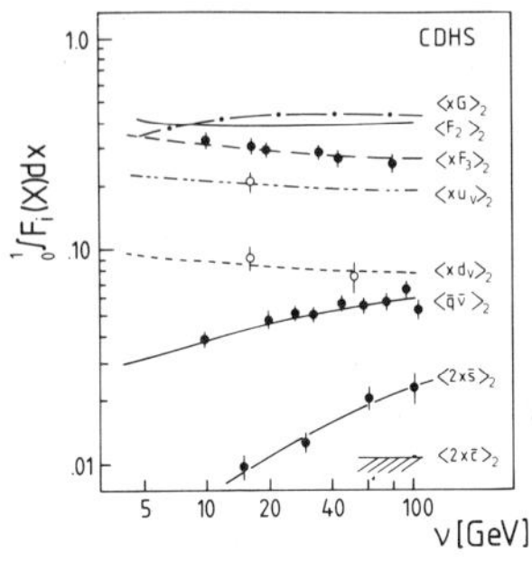

Fig.21:

Total momentum fraction of different constituents versus ν as determined by the CDHS neutrino experiment.

Summary

We have obtained a first precise measurement of $R = \sigma_L/\sigma_T$ at large x. This has to be extended to small x to reduce the uncertainties on F2 and $\bar{q}^{\nu}$. The flavour composition of the nucleon is reasonably well measured for all flavours. The ν-(Q^2-) dependence is well measured and it is very significant for xF3, $\bar{q}^{\nu}$ and xs(x).

2.6 Scaling violations in deep inelastic scattering

The main interest in this subject is the test of perturbative QCD predictions. The first question is, whether the observed scaling violations are mainly due to QCD effects or due to kinematic effects and non-perturbative contributions especially higher twist contributions. In a second step we can then determine the gluon distribution in the nucleon and the QCD scale parameter $\Lambda_{\overline{MS}}$.

Let me shortly remind you, that at present theorists are not yet able to calculate the structure functions from the QCD equations. The perturbative approach allows however to predict the slopes $dF_i/d\ell nQ^2$ of the structure functions, once their shape is experimentally determined. The predicted scaling violations are the combined effect of gluon bremsstrahlung and $q\bar{q}$-pair production from gluons and their magnitude is given by $\alpha_s(Q^2)$, which can be given in terms of the unknown scale parameter Λ. In leading order

$\alpha_s(Q^2) = 12\pi/\ 25[\ell n(Q^2/\Lambda^2_{L.O.})]$. The most direct access for QCD comparisons is via the Altarelli-Parisi equations. The predicted QCD slopes for the structure functions F2 and $\bar{q}^\nu$ are illustrated in figure 22 for fixed Q^2. The contributions of bremsstrahlung and pair production by gluons are given separately. The data points have been obtained from the CDHS data by simple linear fits in $\ell n \ell n Q^2$. They show very well that the observed scaling violations are very significant (scaling would require $dF_i/d\ell n Q^2 \equiv 0$) and in good agreement with the QCD predictions.

Let me start with some general remarks concerning the confrontation of data with QCD:

i) Data of low invariant hadron mass W are suspicious: We expect substantial non-perturbative contributions in this kinematic range and we will see later that we have some experimental evidence that they are present.

ii) Non-singlet structure functions are best. Their QCD interpretation is straightforward since the scale parameter Λ is the only unknown. Best suited is the valence structure function $xF3(x,Q^2)$ which has the additional merit that it is unaffected by uncertainties in R and due to charmed and strange sea.

iii) The singlet structure function F2 and $\bar{q}^\nu$ are best measured. They involve the unknown gluon distribution xG(x) in addition to Λ. We will see, that the combined analysis of F2 and $\bar{q}^\nu$ allows the separation of xG(x) and Λ.

iv) The data do not distinguish between leading and second order QCD. Second order corrections turn out to be small in the $\overline{MS}$ renormalization scheme and $\Lambda_{L.O.} \approx \Lambda_{\overline{MS}}$ for present experiments.

2.6.1 Analysis of the non-singlet structure function $xF3(x,Q^2)$

First I will discuss recent results which are based on the method of Nachtmann moments for fixed Q^2. Historically the first QCD comparisons and the evaluation of Λ have been based on this method which necessarily has to rely heavily on data at large x and low Q^2 i.e. at small W. New results of two bubble chamber experiments are summarized in table I.

Table I: Results of the QCD analysis of xF3 using Nachtmann moments

experiment	Q^2-cut	$\Lambda_{L.O.}$ [GeV]	$\Lambda_{\overline{MS}}$ [GeV]
ABCDLOS[18]	>1.0 GeV^2/c^2	.51 ± .08	.37 ± .03
(BEBC high energy data	>1.5 GeV^2/c^2	.29 ± .08	.22 ± .04
+ GGM low energy data)	>2.0 GeV^2/c^2	.19 ± .08	.14 ± .05
IMF(15')[19]	>2.0 GeV^2/c^2	.42 ± .11	.28 ± .07 ± .10
ABCMO 1977* [20]	>1.0	.74 ± .05	

* This experiment has used the same low energy Gargamelle data as ABCDLOS. For the high energy BEBC data however, 3 times more statistics is availyble now.

$$\frac{dF2}{d\ell nQ^2} = \frac{\alpha_s}{2\pi} \int_0^1 \frac{xdy}{y^2} \quad *$$

$$* \quad [P_{qq}(\frac{x}{y})\ F2\ (y,Q^2)$$

$$+\ 2N_F\ P_{gq}(\frac{x}{y}) yG(y,Q^2)]$$

$$\frac{d\bar{q}}{d\ell nQ^2} = \frac{\alpha s}{2\pi} \int_0^1 \frac{xdy}{y^2} \quad *$$

$$* \quad [P_{qq}(\frac{x}{y})\ \bar{q}\ (y,Q^2)$$

$$+\ N_F\ P_{gq}(\frac{x}{y}) yG(y,Q^2)]$$

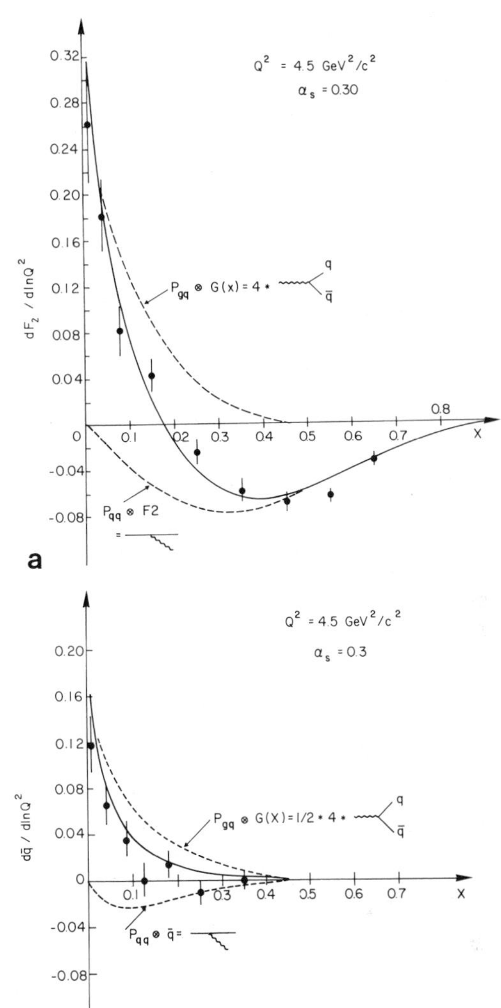

Fig.22:

QCD slopes for F2 and $\bar{q}^{\nu}$ for fixed values of Q^2 for the QCD fits of sect. 2.6. The contributions of bremsstrahlung and pair production are shwon separately. The data points are obtained from linear fits in $\ell n\ell nQ^2$ to the structure function measurements.

What can we conclude from these results? The authors[18)] conclude, that $\Lambda_{\overline{MS}} < .35$ GeV. Their evaluation of the Gross-Llewellyn-Smith sum rule, which has been discussed already before (section 2.3.3) gives some support to this upper limit.

For me the strong Q^2-dependence of the results given in table I are a clear indication that large non-perturbative effects are present at low W. Therefore we should cut out this kinematic region by requiring i.e. $W^2 > 10\ GeV^2$. Such a cut is beautiful also for other reasons: target mass corrections become negligible, remaining data are well above charm threshold and second order corrections are very small. Of course, a moment analysis is no longer possible then. The main drawback of a W-cut is however, that very few experiments survive such a cut - therefore it is not very popular.

Analysis of $xF3(x,Q^2)$ at high W

Such an analysis has been done by the CDHS collaboration, which at present has the only significant measurements at large W. They use $W^2 > 11\ GeV^2$ and $Q^2 > 2\ GeV^2/c^2$ and find $\Lambda_{\overline{MS}} = .20^{+.20}_{-.10}$ GeV. Their result includes a correction for the weak propagator but no Fermi motion correction. The same data have recently been analysed also by Baker et al.[21] For $W^2 > 11\ GeV^2$ these authors find $\Lambda_{\overline{MS}} = .31 \pm .13$ GeV including Fermi motion correction. If they add all other existing data with $W^2 > 11\ GeV^2$, the result is $\Lambda_{\overline{MS}} = .30 \pm .10$ GeV. Let me indicate to you the magnitude of some corrections: Fermi motion corrections increase Λ by + 30 MeV, target mass corrections decrease Λ by 10 MeV and the propagator correction with m_W = 80 MeV decreases Λ by 90 MeV. These results are in line with the previous conclusions based on $\int_0^1 F3\, dx$ and moments which give $\Lambda_{\overline{MS}} < .35$ GeV.

These determinations of Λ are of course only valid if higher twist effects are negligible for $W^2 \gtrsim 11\ GeV^2$. This question will be discussed later.

2.6.2 QCD analysis of singlet structure functions

The aim of this analysis is the determination of the gluon distribution, the investigation of possible higher twist contributions and the best estimate of Λ.

The structure functions F2 and $\bar{q}^{\bar{\nu}}$ are best measured from a statistical point of view. The measurement of the scaling violations suffers however from substantial uncertainties due to $R = \sigma_L/\sigma_T$ and the charm threshold effect. The magnitude of these effects on the slopes of $F2^{\nu N}$ are given in figure 23. The present uncertainty of R at small x affects the slopes of F2 and $\bar{q}^{\nu}$ severely. An improvement would be highly desirable.

The analysis of F2 alone does not allow to separate the effects of gluon bremsstrahlung and quark pair production by gluons. In practice this has the effect, that the value of Λ and the width of the gluon distribution xG(x) are highly correlated. It is easy to show, that the combined analysis of F2 and $\bar{q}^{\nu}$ is able to separate the measurement of Λ and xG(x). If we are interested mainly in Λ, the best method is the following "experimental" approach: We use F2 for large x, say $x \gtrsim .3$ and subtract the sea quark distribution using $\bar{q}^{\nu}$ as measured. This correction is found to be pretty small. We also know that $R = \sigma_L/\sigma_T$ is near zero in the range of x.

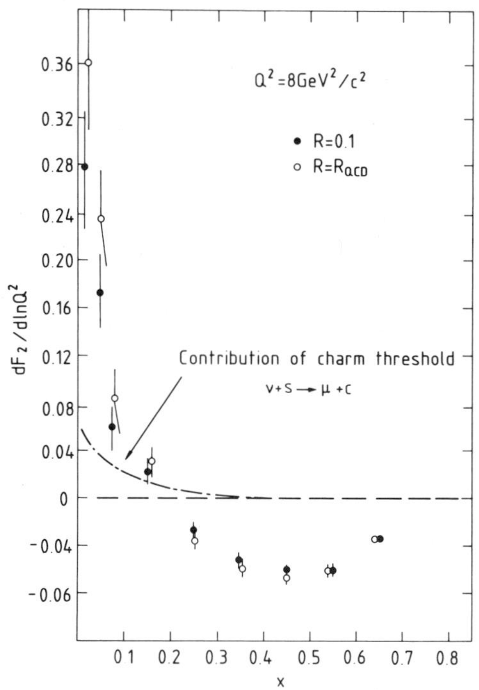

Fig. 23:

The slopes of $F2^{\nu N}$ for fixed Q^2 as obtained from the CDHS measurements versus x. The contribution of the charm threshold in the process $\nu + s \to \mu + c$ and the effect of a change in $R = \sigma_L/\sigma_T$ are given.

Therefore, what remains, is the valence quark distribution which can be analysed as non-singlet structure function. According to my taste, this is the best way to obtain Λ from the F2 structure functions of the muon experiments.

If we are interested mainly in the gluon distribution, the small x region carries very important information, and we have to analyse $F2(x,Q^2)$ and $\bar{q}^{\nu}$ simultaneously. The antiquark distribution is strongly correlated to xG(x). Loosly speaking, we cannot have gluonic contributions to $dF2/d\ell nQ^2$ where the antiquark distribution is zero. The measured slopes of $\bar{q}^{\nu}$ versus x are shown in figure 24 for fixed Q^2, together with the results of the best QCD fits to F2 and $\bar{q}^{\nu}$ described below. The slopes are dominated by the gluon contribution. Experimentally CDHS has found that the antiquark distribution at large x is very small. This is a safe measurement since it is based on the magnitude of the cross-section only and not on its Q^2-dependence.

As a consequence, the width of the gluon distribution is restricted and a rather narrow gluon distribution is favoured. The combination of F2 and $\bar{q}^{\nu}$ gives the best constraint up to now, much better than a combined analysis of F2 and xF3.

Results

Non-singlet fits to F2 at large x have been performed by EMC[22)] and CDHS. I have selected those results where the sea contribution

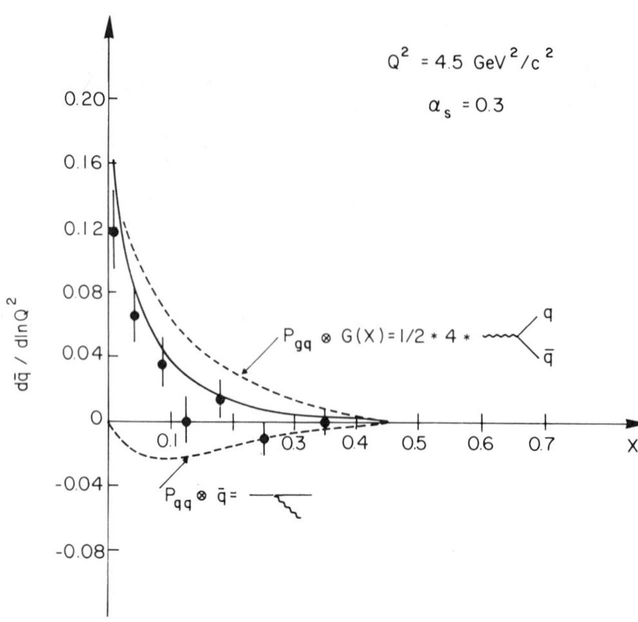

Fig.24:

The slopes of antiquar distribution for fixed Q^2 as determined from the CDHS measurements. The lines are the resu of a QCD fit to $F2^{\nu N}$ and $\bar{q}^{\nu}$.

Table II: Results of the QCD analysis of the singlet structure functions

a) non-singlet analysis after subtraction of a sea contribution			
Data	$\Lambda_{L.O.}$ [GeV]	χ^2/DF	$\Lambda_{\overline{MS}}$ [GeV]
EMC iron: x > .2	$.25 \pm .03 \ ^{+ .16}_{- .10}$ *	1.8	
H_2:	$.15 \ ^{+ .07}_{- .06} \ ^{+ .16}_{- .09}$	1.5	
CDHS (prelim.) x > .3, W^2>11 GeV^2	$.29 \pm .08 \pm .12$	1.0	$.30 \pm .08 \pm .12$
b) singlet fits			
EMC iron: x > .06	$.20 \ ^{+ .03}_{- .03} \ ^{o\ .12}_{- .08}$	1.95	$F2^{CC}$ subtracted $xG(x,Q^2)$ from CDHS fit
H_2:	$.11 \ ^{+ .03}_{- .04} \ ^{+ .16}_{- .04}$	1.90	
CDHS W^2 > 11 GeV^2 $F2^{\nu N}$ and $\bar{q}^{\nu}$	$.18 \pm .03 \pm .1$	1.07	$xG(x,Q_o^2)$ also determined

* statistical and systematic errors are given in turn

has been subtracted using the $\bar{q}^{\bar{\nu}}$ measurement. The results are given in table II. The Λ values of EMC and CDHS are in fair agreement. Including the estimate of systematic errors, both experiments have about the same accuracy. The BCDMS collaboration has published a first QCD analysis of their 120 and 200 GeV data with $\Lambda_{\overline{MS}}$ = 85 MeV.[5] In the meantime they have evaluated their high statistics data at 280 GeV and have recorded new very high statistics data at lower energy which are going to be analysed very soon. They will wait until this analysis is finished before new QCD-comparisons are done. Even so, the addition of the 280 GeV data clearly is in favour of a larger value of Λ.

The determination of the gluon distribution is the main aim of the singlet analysis. This is possible for CDHS by a combined for to F2 and $\bar{q}^{\bar{\nu}}$. The value of Λ is compatible with the results of the non-singlet fits. Also the EMC data can be described by the same gluon distribution and compatible values of Λ.

The gluon distribution

The determination of the gluon distribution from DIS experiments is only possible if the observed scaling violations are predominantly due to QCD effects. Gluons do not contribute as partons to DIS and therefore it is not evident why this is a good place to measure $xG(x,Q^2)$. The reason is simply that the area $\int_0^1 xG(x)dx$ is well defined by the energy momentum sum rule whereas the width is restricted by the absence of antiquarks at large x.

The CDHS collaboration has parametrized the gluon distribution by $xG(x,Q_o^2) = a_g(1-x)^{P_g}(1+C_g+X)$ and obtained the shape as given in figure 25.[23] This is for R = .1 and various assumptions about charm threshold. It is now known that R is zero at large x. A reanalysis

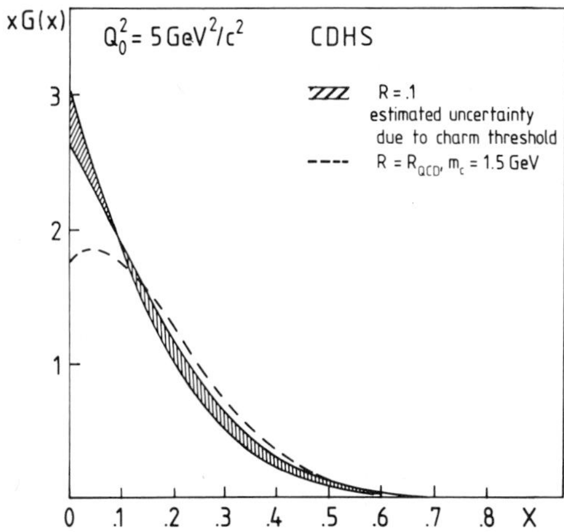

Fig.25:

Gluon distribution for $Q_o^2 = 5\ GeV^2/c^2$ as determined by the CDHS collaboration[23] from a QCD analysis to $F2^{\nu N}$ and $\bar{q}^{\nu}$.

of the same data using R = R_{QCD} gives a somewhat broader gluon distribution as given in figure 25 and an increase of $\Lambda_{L.O.}$ to .29 GeV. This has to be regarded as the best estimate from the present data. The effect of second order corrections has been evaluated by Baulieu and Kounnas[24] and also by Duke and Roberts[25] and found to be comparable to the experimental uncertainties. Thus my conclusion is, that for the first time we have a good constraint to get a separate measurement of Λ and the gluon distribution and the gluon distribution is reasonable well determined.

Higher twist contributions?

I have not yet discussed the question, if the scaling violations at large W are really due to QCD effects. It has been suggested that Λ could be very small and all observed scaling violations are due to higher twist effects i.e. contributions which have a Q^2-evolution $\sim (1/Q^2)^n$ in contrast to $1/\ell nQ^2$ for QCD. The fact is, that we can never be sure. A power series in $(1/Q^2)^n$ can fit an elephant! Actually in the present Q^2-range a functional form $F2(x,Q^2) = a(x)/Q^2 + b(x)/Q^4$ (i.e. twist 4 + twist 6 term) does all the job with $\Lambda \equiv 0$[26])! The higher twist terms have only to mimic the x-dependence of the QCD slopes.

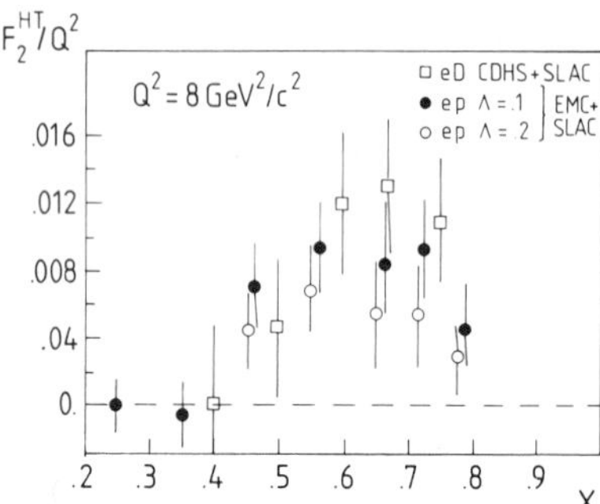

Fig.26:

Shape of a twist 4 contribution as determined from fits to SLAC-MIT and high energy muon and neutrino data.

A more interesting statement is, that QCD does not seem to be able to describe the observed scaling violations in the whole (x,Q^2) range if the precise data from SLAC at low energy are included even after correction for target mass effects. We seem to need substantial non-perturbative contributions at large x and low Q^2 i.e. at small invariant hadron mass. In this kinematic range a possible and natural assumption is that they are due to higher twist contributions. Finally this is the natural place where they have to show up.

Assuming a single higher twist term proportional $1/Q^2$ CDHS and EMC have determined it's shape and magnituge by a combined anaylsis of their high energy data with SLAC-MIT data. These determinations use only the <u>slopes</u> of the SLAC data for each bin in x, not the measured relative normalization. The results are shown in figure 26. Both experiments obtain similar conclusions: Higher twist contributions are needed only for a large x ($x \gtrsim .5$). This shape would be well compatible with the expectation for diquark contributions[27].

Let me repeat: As long as high values of Λ are disfavoured by the high energy data we need non-perturbative contributions at large x and low W. The minimal and "natural" solution is the addition of a single higher twist term as given in figure 26. In this case about 30 % of the observed scaling violations in the SLAC-range would be due to the higher twist contribution whereas the slopes for $W^2 \gtrsim 10$ GeV would not be affected significantly.

What is the best value of Λ?

This is not the main issue for deep inelastic scattering experiments. To my taste we get a lot of important and interesting results and I am annoyed by the fact that most people just ask: What is Λ? Deep inelastic scattering experiments are not terribly good to measure Λ but others are even worse. Also a word of warning: there is an inflation of papers which make "QCD-comparisons" to DIS data. In many cases the uncertainties due to the analysers are larger or equal to the experimental uncertainties. Much of the confusion in the past and probably in the future is due to this.

Let me be positive: We do see QCD effects unless higher twist contributions mimic QCD. For "reasonable" assumptions about higher twist contributions these are negligible for $W^2 \gtrsim 10$ GeV^2. In this case my best estimate for Λ is $.1 < \Lambda_{\overline{MS}} < .35$ GeV including the estimate of systematic effects.

Finally: do not ask subtle questions to the data. They are not good enough for that.

2.6.3 Summary for deep inelastic scattering

1. The experimental situation as a whole is quite satisfactory. Exceptions are the poor knowledge of $R = \sigma_L/\sigma_T$ at small x and absolute normalization problems.
2. The flavour composition of the nucleon and the gluon distribution are reasonably well measured. This is very important input for the analysis of other hard scattering processes.
3. QCD is the natural and successful theory to describe the observed scaling violations at large Q^2 and in the sea region. The best estimate of Λ is then $.1 < \Lambda_{\overline{MS}} < .35$ GeV under "reasonable" assumptions about higher twist contributions.
4. The progress of the last years is enouraging. There is however room and need for improvement.

3. MASSIVE LEPTON PAIR PRODUCTION BY HADRONS

A typical mass spectrum of the lepton pair is shown in figure 27 for $p + N \rightarrow \mu^+\mu^- + X$. It has two components: the continuum which is due to the electromagnetic Drell-Yan process and the bumps which

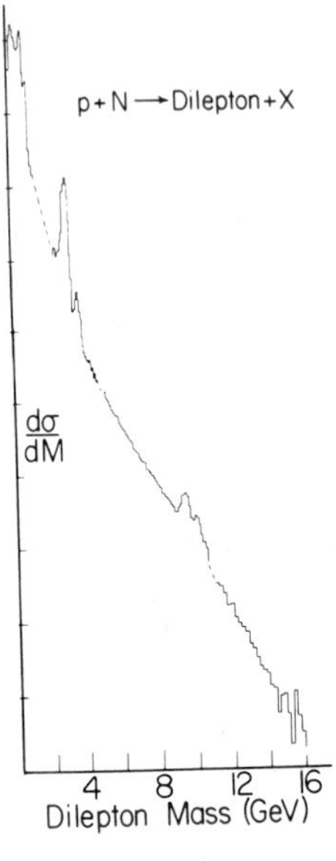

Fig.27:

Mass-spectrum of the muon pairs in $p + N \rightarrow \mu^+\mu^- + X$.

are of hadronic origin. The origin is well checked for instance by the study of particle ratios like π^+/π^-, $\bar{p}/p$ etc. The differential cross-section in the parton picture is proportional to the product of structure functions $f_i(x)$ of the two initial partons and the parton-parton differential cross-section:

$$\frac{d^2\sigma}{dx_1 dx_2} \sim \sum_{ij} c_{ij}\, f_i^{H1}(x_1) \cdot f_j^{H2}(x_2) \cdot \sigma_{ij}$$

If nucleons are involved, we can use the structure functions as measured in DIS. For incident mesons we can use these processes to measure their structure function.

The Bjorken x values of the incident partons are related to the muon mass and the Feynman x_F of the muon pair. We have $M_{\mu\mu}^2 = s \cdot x_1 \cdot x_2$, $x_F = x_1 - x_2$ ignoring transverse momenta of the partons. We can also use the scaling variable $\sqrt{\tau} = M_{\mu\mu}/\sqrt{s}$ which is equal $x_{BJORKEN}$ for symmetric production i.e. $x_F \approx 0$.

3.1 Continuum lepton pair production

Drell and Yan have proposed a long time ago, that the basic process which leads to lepton pairs of high invariant mass is due to the annihilation of a quark and antiquark to a virtual photon and successive creation of the lepton pair. The basic process is therefore just the inverse of $q\bar{q}$-production in e^+e^--annihilation. It is assumed that the other partons in the hadrons are only spectators which do not influence the interacting partons.

Let me start by saying that this basic Drell-Yan mechanism is well established except for overall normalization. The measured A-dependence, the angular distribution of the muon pair, the beam particle ratios and the approximate scaling are in agreement with the Drell-Yan prediction. This has been summarized by A.Michelini last year at the Lisbon conference[28].

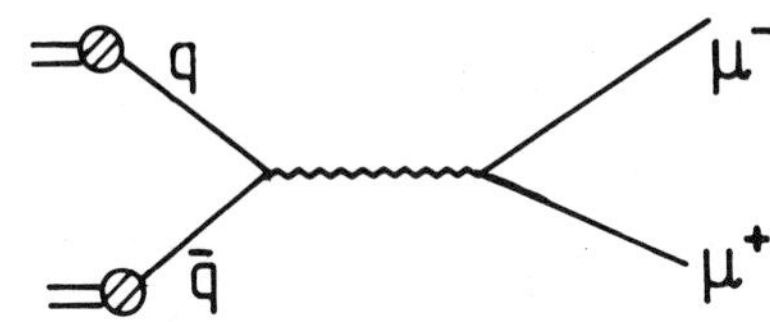

Drell-Yan diagram for continuum lepton-pair production

The structure function measurements have been summarized for the last time by F. Lefrançois at the Madison conference[29]. Not much new has happened since.

Before going to a summary of present knowledge about nucleon and meson structure functions let me shortly remind you of the K-factor:

K-factor: all Drell-Yan cross-sections are higher by about a factor 2.2 compared to the QPM prediction based on the nucleon structure functions as measured in DIS. For this reason only the shape of structure functions can be determined in the Drell-Yan process not their absolute magnitude. Moreover, the shape determination depends on the working hypothesis that the K-factor does not depend on $\sqrt{\tau}$ i.e. on Bjorken x.

The real world should consist of the QPM plus the QCD corrections:

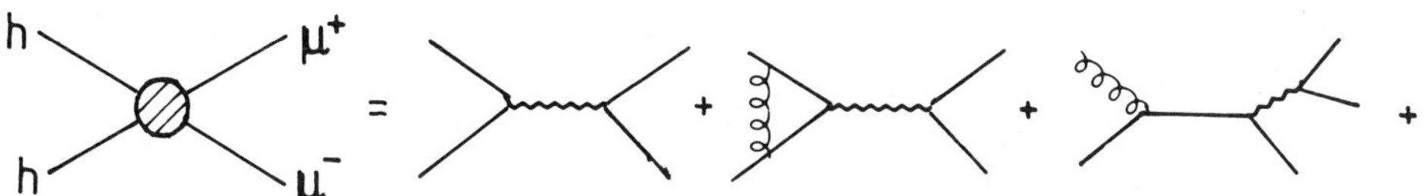

The evaluation of the first order QCD corrections has suggested that the origin of the K-factor may be understood. A very large (∿ 100 %) correction is calculated for the gluon vertex correction because the Drell-Yan process used structure functions at time-like values of Q^2 in contrast to DIS, where they are measured in the space-like region. There is the conjecture, that these large corrections can be summed to all orders thus that the K-factor would be well under control.

However this conjecture has not yet been proven and we have also an open debate about the validity of factorisation in the process. This has been discussed in detail by Landshoff at this school.

In the following I will simply rely on the working hypothesis, that K is independent of $\sqrt{\tau}$ and show than this is supported by the analysis of processes where only nucleons are involved.

3.2 Results on nucleon structure functions

The antiquark distribution in the nucleon

The Drell Yan process gives a good handle to understand the antiquark distribution in the nucleon since precise measurements can be obtained up to very large x-values and we have a chance to test the hypothesis that $\bar{u} = \bar{d}$. The flavour symmetry of the sea can be tested because we can compare data on pp-scattering which is mainly due to $\bar{u} * u_v \cdot 4/9$ to data on pn-scattering which is due to $\bar{d} * u_v \cdot 1/9 + \bar{u} * d_v \cdot 4/9$.

The major improvement since 1980 has been the measurement of $\bar{q}^{\bar{\nu}}(x,Q^2)$ in deep inelastic antineutrino scattering. Figure 28 shows a comparison of the antiquark distribution $x(\bar{u}+\bar{d})$ as published by the CFS experiment[13)] (mainly from 400 GeV proton-nucleon scattering) to the measurements of the CDHS neutrino experiment evaluated for

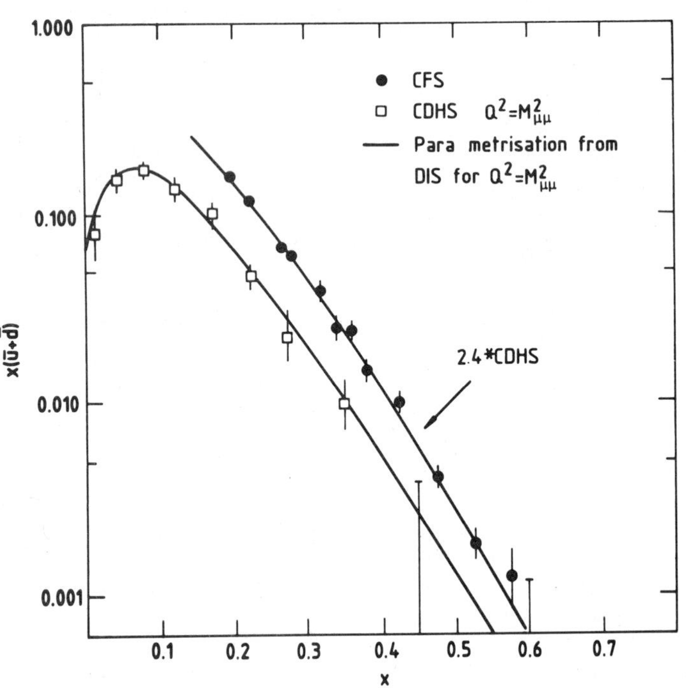

Fig.28: The antiquark structure function $x(\bar{u}+\bar{d})$ as published by the CFS collaboration compared to the DIS measurement of the CDHS collaboration for $Q^2 = M^2_{\mu\mu}$ and $R = R_{QCD}$.

$Q^2 = M^2_{\mu\mu}$ and $R = R_{QCD}$. The shape of the two distributions is in excellent agreement even at large x values. We need however a K-factor of 2.4 to account for the absolute magnitude of the Drell-Yan result. The effect of scaling violations cannot be seen in the x-range covered by the CFS experiment, since they are large only at small x and Q^2. The (x,Q^2)-range can be extended by adding also p-p scattering data from the ISR. This is done in figure 29. The invariant cross-section of the CFS experiment for p-n scattering at $\sqrt{s}$ = 27 GeV and the ISR results of the CCAHNP collaboration[30] are well described by the predictions based on the new CDHS structure functions $xu_v(x,Q^2)$, $xd_v(x,Q^2)$ and $\bar{q}^{\nu}(x,Q^2)$ for a K-factor K = 2.0 and $Q^2 = M^2_{\mu\mu}$ provided we use a symmetric sea $\bar{u} = \bar{d}$. If I assume however an asymmetric sea of the form $\bar{u} = \bar{d}(1-x)^{3.5}$ as favoured by the CFS collaboration[13] and suggested by Feynman and Field[12] the CDHS structure functions are no longer able to describe the p-p and p-N results simultaneously. The CFS result alone is compatible with an asymmetric sea however a factor K = 3.5 would be needed. This is an interesting result which should be studied further.

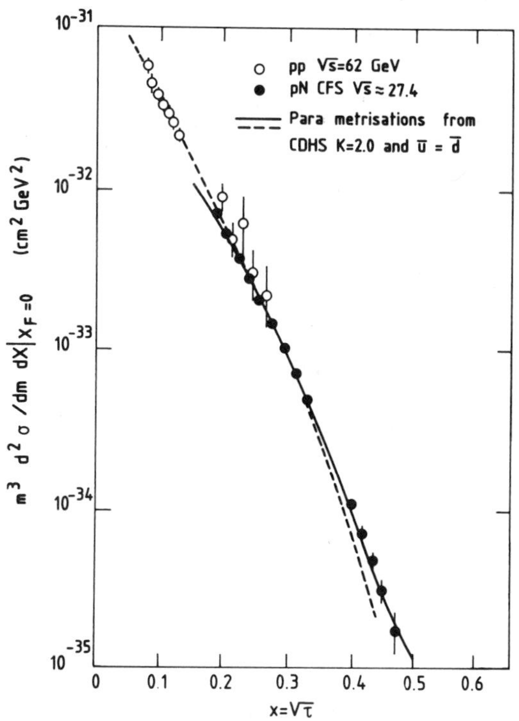

Fig.29:

Invariant differential cross-section for muon pair production at $x_F = 0$ versus $\sqrt{\tau} = x$ for p-N[13] and p-p experiments[30].
The curves are calculated using the new CDHS structure functions with $\bar{u} = \bar{d}$ for $Q^2 = M^2_{\mu\mu}$.

The main differences between the new predictions and the published CFS analysis are i) the inclusion of the Q^2-dependence of the structure function, ii) new up and down valence distributions which differ substantially from those which have been assumed in the CFS analysis. This results also in slightly different values for the K-factor.

The valence quark distributions in the nucleon

The NA 3 collaboration has published an analysis which is based on the cross-section difference $\sigma(\bar{p}N) - \sigma(pN)$ using 300 $\mu^+\mu^-$ events in the continuum observed in an antiproton beam[31]. In the Drell-Yan model, this quantity depends only on the valence quark distribution in the nucleon.

Their result is in good agreement with the valence distribution from DIS (CDHS) with a factor K = 2.5 ± .4.

Summary

The observed differential distributions for muon pair production involving nucleons only are well described by the Drell-Yan formula using DIS structure functions and a constant value of $K \simeq 2 \div 2.5$. This gives us some confidence that we can use the Drell-Yan measurements to study also the shape of meson structure functions.

I have also given first indications that a combined analysis of the Drell-Yan data and the new precise structure function measurements from DIS may give interesting new results about the sea quark distributions which cannot be obtained easily by other means. A more refined analysis would be worth while.

3.3 Meson structure functions

The Drell-Yan process using meson beams gives the best chance to study the quark distribution in the mesons. The following strategy is applied:

i) measure muon pair production in $\pi^{\pm}N$, $K^{\pm}N$ reactions which, in the Drell-Yan model, are mainly due to $\bar{q}_V^{Meson}(x_1) * q_V^N(x_2)$.

ii) use nucleon structure functions as measured in DIS

iii) use simple parametrizations for the meson structure functions: $x\bar{u}_V^{\pi}(x) \sim x^{\alpha}\ (1-x)^{\beta}$, $\bar{q}^{\pi}(x) \sim (1-x)^{\gamma}$

iv) get the normalization for the meson structure functions by the requirement $\int_0^1 \bar{u}_V(x)dx = 1$ i.e. require one valence antiquark in the meson

Results

The first results have been obtained by the CIP experiment[32] (published 79). They found that the valence quark distribution for mesons are substantially harder than for nucleons. Using $xu_V^{\pi}(x) \sim \sqrt{x}(1-x)^{\beta}$ their result was $\beta = 1.27 \pm .06$ compared to $\beta \simeq 3$ to 4 for nucleons.

A summary of present knowledge is given in figure 30 which compares the shape of the pion structure function as determined by three experiments and in table III. All experiments confirm the first result of the CIP experiment.

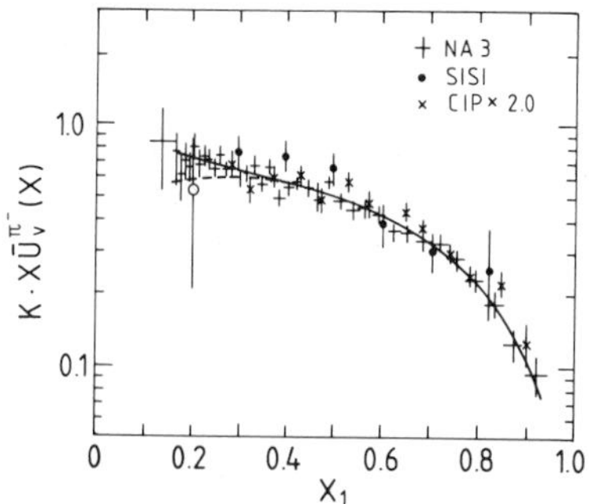

Fig.30:

Shape of the meson structure function as obtained by three experiments using pion beams. The result of the CIP experiment has been multiplied by a factor 2 to correct approximately for their assumed A-dependence $\sigma \sim A^{1.12}$.

The most precise results have been obtained by the NA3 experiment, which has been able to determine the valence and the sea distribution in the pion and the K-factor. Their structure functions are measured at $\langle Q^2 \rangle \approx 25\ \mathrm{GeV}^2/c^2$.

The NA3 experiment has also studied the difference between pion and kaon structure functions[36)] by comparing K^- and π^- data. They find $\bar{u}_v^{K^-}(x)/\bar{u}_v^{\pi^-}(x) < 1$ for $x > .7$, or $\bar{u}_v^K(x)/\bar{u}_v^\pi(x) \approx (1-x)^{0.15}$.

This observed difference is in line with the expectations based on the momentum sharing between light and strange quarks in the kaon[37)].

3.4 Gluon distribution in the mesons

The observation of massive lepton pairs in meson-hadron collisions allows also to get information about the gluon distribution in the mesons. The production of ψ-particles in hadronic interactions

Table III: Pion structure functions as determined by Drell-Yan experiments using the parametrizations $x\bar{u}_v(x) \sim x^\alpha(1-x)^\beta$ and $\bar{q}^\pi(x) \sim (1-x)^\gamma$

Experiment	α	β	γ	K
NA3[31)]	.38 ± .04	.94 ± .06	8.7 ± 2.7	2.48 ± .30
CIP[32)]	.5 fixed	1.27 ± .06		
SISI[33)]	.5 fixed	1.57 ± .18		
BCE[34)]	.44 ± .12	.98 ± .15	5.0 fixed	
E 326[35)] (preliminary)	.5 fixed	1.01 ± .1	5.4 fixed	

is well described by the gluon-gluon fusion model in which the basic diagrams are supposed to be:

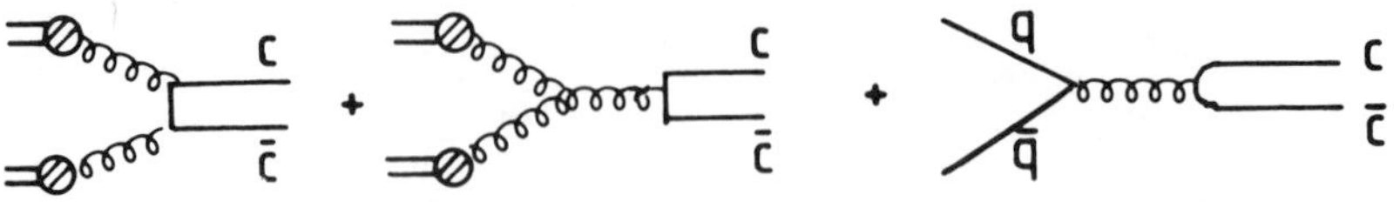

c c̄ + c c̄ + q q̄ c c̄

I do not have the time to discuss the subject in any detail. Let me just refer you to a recent review article[38] on the subject.

The net conclusion is, that the gluon distribution in the mesons is most likely broader than in the nucleons. If the simple parametrization $xG(x) \approx (1-x)^{P_g}$ is used, we find $P_g \approx 3$ for pions and kaons whereas $P_g \approx 5$ for nucleons in the same Q^2-range. This observation is not unexpected since we have found that also the valence quark distributions are broader in the mesons.

3.5 Future experiments

New experiments to study muon pair production are under way both at Fermilab and at CERN. They all provide substantial enlargements of the accessible kinematic range in x_F, $M^2_{\mu\mu}$ and in p_T. The main aims are:

i) a more accurate measurement of meson structure functions,
ii) an extension to large $M^2_{\mu\mu}$ due to higher luminosity and energy,
iii) detailed studies of high p_T effects for which rather reliable QCD predictions can be tested.

Muon pair production experiments will remain an important tool in the future since they are one of the best ways to study parton dynamics.

4. HIGH p_T HADRON-HADRON SCATTERING

The production of high p_T hadrons in hadron-hadron collisions has been consistently described as the hard scattering of two partons leading to four jet events in the final state.

A schematic diagram for these processes is shown in figure 31. The main interest in these processes is the determination of the parton-parton differential cross-section $d\sigma/dt$. To get this information, we have to know the structure functions of quarks, antiquarks and gluons to describe the initial state and for most experiments also the fragmentation functions to describe the production of the trigger particles. QCD tells us that at high p_T the scale μ^2 should be large enough such that $\alpha_s(\mu^2)/2\pi << 1$ and perturbative QCD is applicable. In this case the differential cross-section is given by the diagrams:

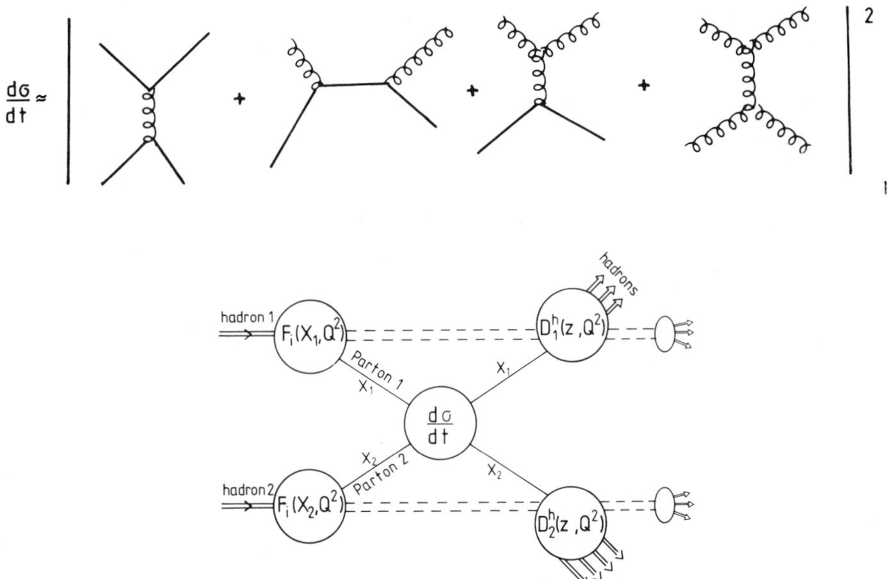

Fig.31: Schematic diagram for high p_T hadron-hadron scattering

The aim of the experiments is to study the quark-quark, quark-gluon anf gluon-gluon scattering contributions separately and to determine their differential scattering cross-section.

How do the structure functions enter the analysis? A typical kinematic configuration is shown below:

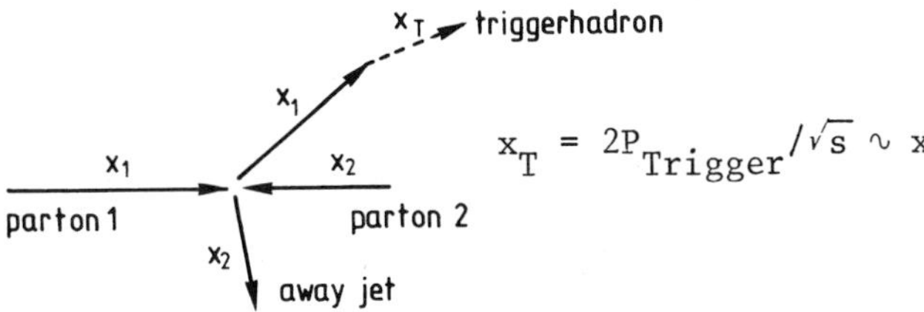

$$x_T = 2P_{Trigger}/\sqrt{s} \sim x$$

In the ISR energy range where most of the studies have been done up to now, the trigger is generally based on a single high p_T hadron. This trigger selects hard scattering processes very effectively in contrast to transverse energy triggers. The variable, which is then most directly related to $x_{Bjorken}$, is the scaling variable $x_T = 2\, p_{Trigger}/\sqrt{s}$. If p_T is increased, the trigger selects incident partons which carry larger and larger $x_{Bjorken}$ (for triggering angles which are not 90°). This is illustrated in figure 32.

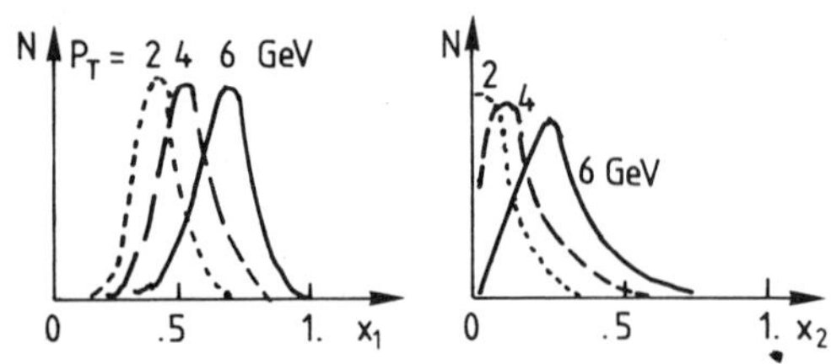

Fig.32: x-distribution of the two partons for trigger particles at different transverse momentum. Valid for ISR energies and a triggering angle of 20°.

We can draw an immediate conclusion: We know that the gluon distribution is much larger than the quark distribution at small x. Therefore gluon processes will dominate at small x_T (i.e. small p_T) whereas quark-quark scattering will dominate at large p_T. A good example are the new results from the $\bar{p}p$ collider which have been discussed for instance by Odorico during this school. In the present p_T-range x_T is very small and the hard parton collisions should be dominated entirely by gluon-gluon collisions.

Some selected results

Most of the systematic experimental work on high p_T scattering has been done in the last years at the intersecting p-p storage ring (ISR) at CERN. Let me just discuss a few examples to demonstrate to you, that we start to learn some things about the differential parton-parton cross-sections and the relative contributions of the subprocesses. An interesting result has recently been obtained by the CERN-Columbia-Oxford-Rockefeller(CCOR) collaboration[39] at the ISR at CM-energies $\sqrt{s}$ = 44 and 63 GeV. They trigger on two high p_T π^o's in opposite hemispheres in the range $2.5 < p_T(\pi^o) < 8$ GeV^2/c^2 around 90°. This trigger selects hard scattering processes and the two π^o's remember the CM-angle θ^* of the scattered partons. They evaluate the differential cross-section $d^2\sigma/dm_{\pi^o\pi^o} d\cos\theta^*$ and compare it to the QCD expectation. The agreement is quite satisfactory.

A systematic study of high p_T jet physics has been performed by the CERN-Dortmund-Heidelberg-Warsaw collaboration using the split field magnet facility at the ISR[40]. This experiment has the unique feature, that hadrons can be measured and to some extent also identified over 4π. This allows also the study of particle correlations which can be used to help identify the partons and the underlying subprocesses.

Let me indicate just some of their results. In the experiment a charged high p_T trigger particle is required under an angle of 50°. Pions and kaons (+ protons) can be identified by threshold Č-counters. Fig.33 shows their measured invariant single particle cross-section for π^- and K^- triggers compared to the leading order QCD expectations for the various subprocesses. The QCD predictions have been obtained with a Monte Carlo simulation which includes the nucleon structure functions as determined by CDHS and our best knowledge about the fragmentation functions of quarks and gluons.

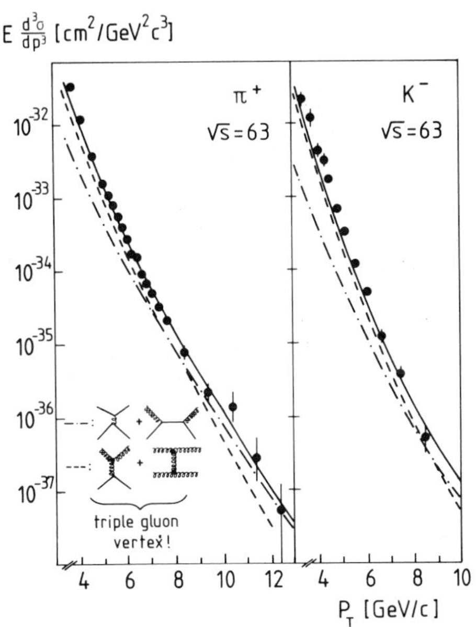

Fig.33: Invariant inclusive single particle spectra for π^- and K^- triggers as observed by the CDHW-collaboration[40]. The expected contributions of the QCD-subprocesses are indicated.

The overall agreement of about a factor 2 is quite satisfactory, given the uncertainties of the experimental input on structure functions and fragmentation functions. Moreover we should not be surprised if the normalization is off by a factor 2 - we have seen a similar factor in the Drell-Yan process.

Figure 33 underlines what I said before about the relative contributions of gluons and quarks. At small p_T, g-g and g-q scattering dominates because gluons are abundant at small x whereas q-q scattering takes over at large p_T. Another interesting feature is that K^- triggers have a much larger probability to select gluon jets on the trigger side than pion triggers. This is because gluons are flavour neutral whereas up and down quarks fragment preferentially into pions. Figure 34 a)shows the ratio of K^- triggers to π^--triggers versus p_T compared to Monte Carlo predictions for pure q-q scattering and including also the gluon contributions. If only quark scattering takes place, then $N(K^-)$ over $N(\pi^-)$ is expected to be nearly independent of p_T whereas the data rise towards low p_T in reasonably good agreement with the QCD expectation. (Part of the remaining difference is probably due to the contribution of antiprotons to $N(K^-)$.) If this interpretation is correct, then we expect a substantial amount of gluon jets for K^- triggers at moderate p_T. This is confirmed by the study of particle correlations. Therefore this result gives a quite convincing evidence that gluon-quark and gluon-gluon scattering contribute to high p_T scattering. This is very interesting since both diagrams involve the triple gluon vertex. The authors have also made an attempt to learn something about the differential cross-section. Figure 34 b)shows the rapidity distribution for particles H_i in the away jet for all triggers with $4 < p_T^{trigger} < 6$ GeV/c and $p_T^{H_i}/p_T^{trigger} > .5$. This distribution is related to the CM angle of the away side parton and therefore is a measure of $d\sigma/dt$. The measured distribution is compared to a QCD model calculation and a model which uses scalar gluons (and therefore has no gluon-gluon coupling). This at least shows that the data are sensitive to changes in the differential cross-section and that the QCD description is clearly favoured.

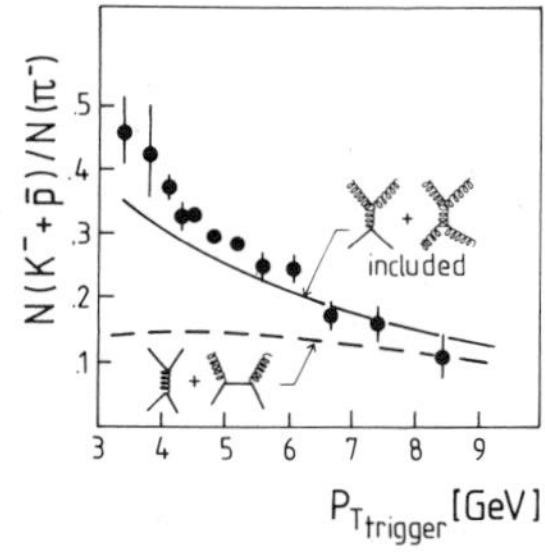

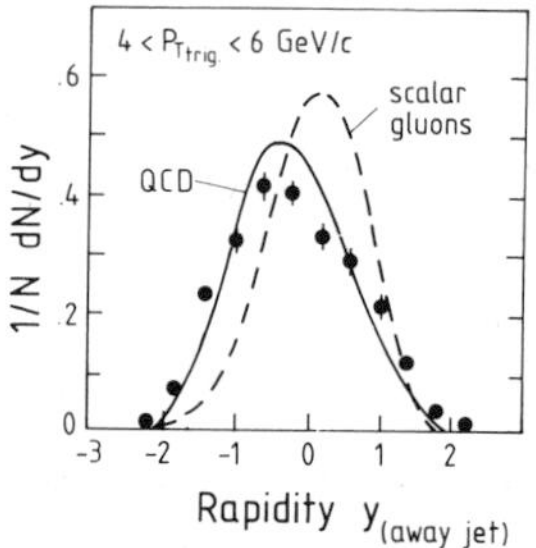

Fig.34: a) Ratio of high p_T events for K^- and π^- triggers as measured by the CDHW experiment. Monte Carlo prediction for q-q scattering and QCD are also given.

b) Rapidity distribution for particles in the away jet with p_T/p_T trigger > .5. Predictions for QCD and a model with scalar gluons are also given.

I find these results very encouraging and further improvements are certainly possible. The $\bar{p}p$-collider should give a unique chance to study parton-parton scattering under much better experimental conditions in the near future.

5. CONCLUSIONS

i) We have accumulated a rather complete set of parton distributions of nucleon, mesons and the photon.

ii) The experimental situation as a whole is quite satisfactory - further process is possible. The experiments are difficult and need encouragement. Present experimental problems are of minor importance compared to the big achievements of the last years.

iii) The parton picture (including gluons) works surprisingly well over the whole range of the hard scattering processes. QCD is unique in providing a basis for a parton model. It has been subject to quantitative, semiquantitative and qualitative tests. They are all successful.

iv) As long as the theorists are unable to apply QCD in the non-perturbative regime, our confidence in QCD has to rely on the study of many different processes and many different kinds of tests. Deep inelastic phenomena have played a very important role and will undoubtedly do so in the future.

Acknowledgements

The content of these lectures overlaps very much with the rapporteur talk which I gave at the Paris conference immediately before the school. The preparation profited a lot from discussions with many of my experimental colleagues from the various experiments. I have also profited from many discussions with my colleagues M. Glück and E. Reya about aspects of QCD and related subjects.

References

1. R.E.Taylor, proceedings of the 4th symposium on electron and photon interactions, Liverpool 1969, Daresbury Nucl.Phys.Lab. p.251.
2. A. Bodek et al., Phys.Rev. D 20 (1979) 1471.
3. J.J.Aubert et al., Phys.Lett. 105 B(1981) 315; 105 B (1981) 322, and contributions to the XXI int.conf. on HEP, Paris 1982, presented by A.Edwards.
4. H.Abramowicz et al., Phys.Lett. 107 B (1981) 141; Z.Physik C 12 (1982) 225; C 12 (1982) 289; C 13 (1982) 199 and contributions to the XXI int.conf. on HEP, Paris 1982, presented by J.Merlo.
5. D.Bollini et al., Phys.Lett. 104 B (1981) 403.
6. Caltech-Columbia-Fermilab-Rockefeller-Rochester collaboration, contribution to the XXI int.conf. on HEP, Paris 1982, presented by M. Shavitz.
7. See for example: A.Bodek and J.L.Ritchie, Phys.Rev. D 27 (1981) 1070.

8. H.Abramowicz et al., Phys.Lett. 107B (1981) 141.
9. L.F.Abbot et al., Phys.Lett. 88 B (1979) 157.
10. T.Bolognese et al., preprint CERN/EP 82-78.
11. A.Donnachie and P.V.Landshoff, Phys.Lett. 95 B (1980) 437.
12. R.D.Field and R.P.Feynman, Phys.Rev. D 15 (1977) 2590.
13. A.S.Ito et al., Phys.Rev. D 23 (1981) 604.
14. J.Hanlon et al., Phys.Rev.Lett. 45 (1980) 1817.
15. H.Abramowicz et al., Z.Phys. C 15 (1982) 19.
16. See for example lecture of J.J.Sakurai during this school.
17. M.Holder et al., Phys.Lett. 74 B (1978) 277.
18. P.C.Bosetti et al., preprint PITHA 82/04 RWTH Aachen.
19. V.V.Ammosov et al., contribution Nr. 482 to the XXI int.conf. on HEP, Paris 1982.
20. P.C.Bosetti et al., Nucl.Phys. B 142 (1978) 1.
21. J.S.Baker et al., contribution Nr.710 for the XXI int.conf. on HEP, Paris 1982.
22. J.J.Aubert et al., contribution Nr. 749 to the XXI int.conf. on HEP, Paris 1982.
23. H.Abramowicz et al., Z.Physik C 12 (1982) 289.
24. L.Baulieu and C.Kounnas, preprint Ref. TH.3266-CERN
25. D.Duke, private communication
26. F.Eisele et al., Phys.Rev. D 26 (1982) 41.
27. I.A.Schmidt and R.Blankenbecher, Phys.Rev. D 16 (1977) 1318.
28. A.Michelini, rapporteur talk at the EPS int.conf.HEP, Lisbon 1981 preprint CERN EP/81-128.
29. J.Lefrançois, rapporteur talk, proceedings XX int.conf. on HEP, Madison 1980, p.1318.
30. D.Antreasyan et al., Phys.Rev.Lett. 45 (1980) 863.
31. J.Badier et al., preprint CERN-EP/80-148 and J.Boucrot, private communication.
32. E.P.Newman et al., Phys.Rev.Lett. 42 (1979) 951.
33. R.Barate et al., Phys. Rev.Lett. 43 (1979) 1541.
34. M.Corden et al., Phys. Lett. 96 B (1980) 417.
35. H.J.Frisch et al., contribution Nr. 169 to the XXI int.conf. on HEP, Paris 1982.
36. J.Badier et al., Phys.Lett. 93 B (1980) 354.
37. See for example: F.Martin, proc. Moriond workshop on lepton pair production, Les Arcs 1981, ed. Frontières, p.331.
38. R.J.N. Phillips, rapporteur talk in proc. XX int.conf. on HEP, Madison 1980, p.1470.
39. A.L.S.Angelis et al., CCOR-collaboration, contributed paper Nr. 703 to the XXI int.conf. on HEP, Paris 1982.
40. D.Drijard et al., contributions Nr. 836, 837, 838 to the XXI int.conf. on HEP, Paris 1982.

DISCUSSIONS

CHAIRMAN : F. Eisele

Scientific Secretaries : M. Jändel, A. Ogawa, P. L'Heureux

DISCUSSION 1

- *KAPLUNOVSKY* :

Because there is more u than d in the proton, one would expect, according to the Fermi statistics, to see less $\bar{u}$ than $\bar{d}$ produced from the sea. How big is this effect, and why would one <u>ever</u> expect isospin symmetry between $\bar{u}$ and $\bar{d}$?

- *EISELE* :

Referring to my discussion of the flavor composition of the sea, recall that, based on the Gottfried sum rule, Feynman had suggested that there should be fewer $\bar{u}$ than $\bar{d}$. This is not unnatural since, by the Pauli exclusion principle, valence u's would suppress u's from the sea, and therefore $\bar{u}$'s. Thus, since there are more u's than d's, there should be fewer $\bar{u}$'s than $\bar{d}$'s. However at x = 0, $\bar{u}$ and $\bar{d}$ should be equal. What remains is that the $\bar{u}$ should be steeper than the $\bar{d}$ distribution. To my knowledge there is no quantitative estimate to that, but it should be true at small Q^2, or energy.
Also, if you think that the u quark has a different mass than the d quark, you would expect to see a difference, and this had been

suggested by some QCD papers. But at high enough Q^2, such effects would diminish to nothing. And now in the context of QCD, if the sea evolves with Q^2 according to the Altarelli-Parisi equations, even if you start with an asymmetrical sea, as you go to higher Q^2 the sea rapidly becomes symmetrical.
In the discussion of Drell-Yan processes, I will show some evidence that the sea is symmetric in $\bar{u}$ and $\bar{d}$, in a model dependent way.

- *KAPLUNOVSKY* :

This is in reference to the possible A-dependence of the structure function. If there really are nuclear effects as the data suggest, does this not cast doubt on the validity of extracting the neutron structure function from hydrogen and deuterium data ? Considering that we do not know how to connect H, D, and Fe data, how is one to proceed to calculate structure functions ?

- *EISELE* :

First of all we need at least one other experiment which supports this effect. These experiments are subject to quite large systematic uncertainties. At the present time I am not quite willing to believe that this effect is real. Note that in the data on F^{Fe}/F_2^D the errors are purely systematic. You know that if you are flooded with systematic effects, you do not know how to estimate them reliably.
The thing to emphasize is that this is really deep, inelastic scattering and that everything you know of, like Fermi motion, points in the opposite sense : F_2^{Fe}/F_2^D should increase with x. There are speculations that in heavy nuclei, the quark distribution functions of the nucleons should extend to higher x, like $x \cong 2$. But if you think about it this should also cause the ratio to increase with x. Thus to my knowledge, there is no theoretical explanation for the effect, which makes it all the more interesting.
I think that what has to be done is to perform different experi-

ments with different systematic errors, different target materials and so on, to find out if the effect is real or not.

- *TRIPICCIONE :*

A few years ago DIS was used to make tests of QCD, but now we use e^+e^- since it was found that DIS has certain systematic problems. Do you think that this is going to last? Or maybe in the future when an ep machine is, if ever, built, it will be a chance to do DIS better, so that you can really make quantitative comparisons with QCD?

- *EISELE :*

I don't agree with your statement. DIS is not a particularly beautiful way to test QCD, since, as I will show in the second lecture, you do not even need QCD to explain DIS data. On the other hand, I know of no better way. For instance, in e^+e^- you are probably thinking of three-jet events. There I don't agree.
As for the other half of your question, I believe as soon as the theorists are able to calculate structure functions, yes, we will be in better shape.
As for going to higher energies, you really need a big step. These are log (Q^2) effects. Going to the energy doubler at Fermilab you gain practically nothing in log (Q^2). Really you have to go to HERA where you double your log (Q^2) range. Still basically, the situation will not change. One of the problems goes under the name of higher twist. You will always be able to explain scaling violations by an infinite series in $1/Q^2$, with arbitrary coefficients. Of course if the data extend to higher Q^2, you start to get more confidence. My understanding is that QCD works on a semi-quantitative basis over an enormous range of phenomena, Drell-Yan, high P_T scattering, gluon fusion models, e^+e^-, you name it. Every single one of these can be

debated as to whether or not it proves QCD, probably not. But I know of no other simple and elegant way to describe nature as coherently as QCD does.

- *MORSE* :

The EMC experiment finds that F_2^{Fe}/F_2^D varies as a function of x. You also showed tha F_2^n/F_2^p varies in x by a factor of 5. Could the F_2^{Fe}/F_2^D data be explained by just a couple percent excess of neutrons over protons in their iron target?

- *EISELE* :

Remember that iron is nearly isoscalar. The data have already been corrected for the small surplus of neutrons over protons in the iron target, the so-called isoscalar correction. This amounts to at most a few percent correction to the data. Note that the effect is also seen for muon scattering off carbon, which truly has the same number of neutrons and protons.

- *FLURI* :

May the uncertainty of the energy reconstruction be responsible for the discrepancy in normalisation between the CDHS (CERN) and CCFRR (Fermilab) data?

- *EISELE* :

Electronic detectors measure $\nu = E_H$ and $E_\nu \mp E_\mu + E_H$, so the hadron energy enters twice. In a calorimeter, the energy resolution gets better and better as the energy increases. For this type of detectors, typically, $\Delta E_H/E_H = 0.7/\sqrt{E_H}$ (E_H in GeV). At present we understand the energy calibration good enough so that this is not a major factor.

- *FLURI* :

But, specifically could there be a systematic bias in the reconstructed energy of one of the experiments, leading to effective biases in the cross section?

- *EISELE* :

I do not think so. To my understanding the origin of the discrepancies in the two experiments is in the flux measurement. Looking again at the illustration of ν cross section measurement, we see that the Fermilab experiment needed five different beam energy settings (and therefore five flux readings and five K/π ratios) to cover the entire kinematic range. By comparison, the CERN 300 GeV beam, where the distance from the decay volume to the detector is shorter by a factor of 2, covers the entire neutrino energy spectrum. There is even an overlap between K and π neutrinos. The energy dependence of the cross section is therefore measured using only one beam energy setting. For such a beam the energy dependence is just obtained by calculating the K and π decays and geometry of your beam. So this is a completely different experimental situation. I think it is clear that if you are in a situation such as CDHS you are in much better shape to measure the energy dependence. If one includes the normalisation errors of CCFRR it is not true to say that these experiments have a systematic difference in the energy dependence. Including the overall normalisation errors, they are compatible. But on a statistical level, and this is the thing which enters in structure functions, the difference is significant.

- *LOEWE* :

How are the flavor distributions derived from the actual experimental data?

- *EISELE* :

Linear combinations of various cross sections are taken as follows. We first take neutrino scattering from isoscalar targets:

$$\frac{d\sigma^{\nu N}}{dx\,dy} + \frac{d\sigma^{\bar{\nu} N}}{dx\,dy} \sim F_2 = x\cdot(u(x) + d(x) + s(x) + c(x) + \bar{u}(x) + \bar{d}(x) + \bar{s}(x) + \bar{c}(x)),$$

where u(x), d(x), s(x),... are the densities of u, d, s,... quarks in the nucleon.

$$\frac{d\sigma^{\nu N}}{dx\,dy} - \frac{d\sigma^{\bar{\nu} N}}{dx\,dy} \sim x\cdot F_3 = x\cdot(u_v(x) + d_v(x)),$$ for the valence quark

distributions. Then you can separate the antiquark content, $\bar{q}_{\bar{\nu}} = x\cdot(\bar{u}(x) + \bar{d}(x) + 2\bar{s}(x))$. This is obtained by measuring antineutrino scattering at high y:

$$\frac{d\sigma^{\bar{\nu} N}}{dx\,dy} \sim \bar{q}(x) + q(x)\ (1 - y^2),$$

so, if you go to high y, then you suppress the scattering of the quarks, the left-handed constituents, and you remain with the scattering of the antiquarks. Thus by going to high y, antineutrino-nucleon scattering, you can measure directly the antiquark content of the nucleon. But, this is possible only in this specific flavor combination.

Another thing to do is to look for the strange quarks. This is best obtained by investigating the basic scattering process $\bar{\nu} + \bar{s} \to \mu^+ + \bar{c}$ ($\bar{c} \to \mu^-$). You probably know that there is plenty of single-charm production seen in neutrino experiments by the flavor changing charged currents. In the case of antineutrino, this has to go by scattering off the strange sea. We have studied this with a few thousand events and we have really established that this GIM prescription is valid. We can describe this opposite sign dimuon production, both neutrino and antineutrino, consistently in this model. Consequently we can use this to measure the strange sea. Charm sea is much more difficult, and there I do not want to elaborate.

This is all obtained so far from complex targets. If you want to separate u(x) and d(x), then you have at least to add one more piece of information. This is best done by muon DIS off hydrogen.

There you measure mostly the up quarks. This enables one to separate u_v from d_v. This is only possible by using elementary (proton) targets.

- *TZENG* :

For the measurement of the strange quark sea, the CDHS looks for the process $\bar{\nu}_\mu + \bar{s} \to \mu^+ + \bar{c}_{\hookrightarrow \mu^-}$ where the event signature is opposite sign dimuons in the final state. But, is single charm production the only source of opposite sign dimuons?

- *EISELE* :

First of all, they don't all come from single charm production. It depends on the density of the detector: if you have a heavy, high density detector, about ten percent of them come from ordinary charged current interactions with K or π decay. This, however, can be simulated by Monte Carlo calculation and subtracted.
Your question was a very hot subject at the beginning of neutrino DIS some years ago. Events with many muons are always a sign of something new happening. At that time people thought that this could be bottom production, heavy neutral leptons, or whatever. A lot of study has gone into that. I think we have no evidence, up to now, that we have any source other than charm production and π/K decay for opposite sign dimuon production.
These early claims were based partly on statistical fluctuations and partly on wrong interpretation of data. CDHS has about 15,000 opposite sign dimuon events and we can explain all distributions in detail by just assuming it's from π/K decay and single charm production. We have even used this data to place limits on beauty production, although these limits are not terribly good ones.

- *DAWSON* :

Recently there has been a claim that the physics you extract from DIS experiments depend upon the W cuts you make. Do the data support this claim.

- *EISELE* :

You are referring to M. Barnett? I will discuss this point in the second lecture under scaling violations.

- *HOFSÄSS* :

Is it possible that multiple scattering is the cause of the observed A-dependence of the structure functions?

- *EISELE* :

I don't know. There is a rescattering effect which is A-dependent and which is called shadowing, but this is restricted to low x and low Q^2, that is, the diffractive scattering region. (Note added: As long as the scattering reaction is of short range, i.e. $^1/Q$ less than one fermi, multiple scattering does not affect the inclusive cross section, which, in the case of muon experiments, is measured by looking at the muons only.)

- *MARTIN* :

Can you place limits on the mass of the W?

- *EISELE* :

Yes but they are not terribly interesting, something like 50 GeV. We hoped once to do better than that but we could not.

DISCUSSION 2

- *DOBREV* :

Why do you consider higher twist effects only for singlet distributions?

- *EISELE* :

For the muon experiments it is clear - they have nothing else. So EMC has to use F_2 because it is the only quantity they measure.

In addition, you must include the SLAC/MIT data, which supplies the necessary low Q^2 data. They of course measure only singlet, F_2 structure functions.

Now for neutrino experiments, we can also do it for $x \cdot F_3$. However, statistically the data are not so very well determined. It has been tried by many people but most of the attempts to look for higher twist were to assume a specific functional form like:

$$F_2(x,Q^2) = F_2^{QCD}(x,Q^2) \cdot (1 + \mu^2/((1 - x)Q^2)),$$

that is, a specific ansatz for the x dependence. The singlet analysis which I have described gives the possibility to see, without any prejudice about the shape, what magnitude and shape the data require for non-perturbative contributions. This could be higher twist.

It turns out that all the standard Ansätze do not work. What is required, if you insist on this form, is something like

$$1 + 1.5 \cdot x^{3.6}/((1 - x)Q^2)$$

This is a completely different shape from what people have used before. In conclusion, F_2 is by far the most distinguishable, and therefore it is best suited to produce some reasonably reliable statement that non-perturbative contributions are needed.

- *DOBREV* :

You parametrize the valence u quark distribution as $x \cdot u_v(x) = x^{\alpha}(1 - x)^{\beta}$, with the power of x to be around 1/2. But in this case $u_v(x)$ diverges like $x^{-1/2}$ near 0. Either the parametrization is incorrect or $x \cdot u(x)$ is the only relevant quantity.

- *EISELE* :

The only requirement on u(x) is that the integral be 2, reflecting the presence of two u quarks in the proton. The probability distribution need not be finite everywhere.

- *ERREDE* :

During the first lecture, you briefly mentioned limits on charm changing neutral currents and the observation of wrong sign muons. Would you care to discuss this further?

- *EISELE* :

The process in question is $\nu_\mu + c \rightarrow \nu_\mu + c_{\hookrightarrow \mu^+}$, where the c in the initial state is a charm quark from the sea, or "intrinsic" charm as proposed by Brodsky, et al. The signature of this process is the observation of a wrong sign lepton in the final state, produced in the semileptonic decay of a charmed meson. The main motivation to look for wrong sign muons has been to look for flavor changing neutral currents and wrong sign heavy leptons. It can, however, also be used to put limits on the total fractional momentum carried by charm quarks.

- *ERREDE* :

Have there been any recent improvements to these limits, or are they the same as before?

- *EISELE* :

We had hoped to improve the limits substantially as we now have much higher statistics. But, unfortunately, we are limited by unavoidable backgrounds.

- *KREMER* :

There have been theoretical attempts to calculate higher twist effects in DIS (Shuryev and Vainshteyn in the framework of SVT sum rules and Jaffe and Soldate in the bag model). How do these (semi-theoretical) results compare with your results?

- *EISELE* :

According to Jaffe and Soldate, higher twist effects are

negligible in our Q^2-range. It is not clear whether what we see are higher twist effects or the result of an ensemble of small, dirty effects. For example the way target mass corrections are done, or, near x = 1, phase space effects, or the choice of kinematic variable. So, higher twist is just one possibility natural in this region, and our result is more or less in line with that estimate. Clearly, to get any further however, more theoretical input is needed: e.g. what is the reasonable shape for higher twist effects?

- *KREMER* :

The contribution of the three gluon vertex to a physical process depends on the gauge chosen for the calculation: in some gauge choice it could be unimportant. What then, do you mean when you say that one could possibly see the effects of the three gluon vertex in the evolution of the gluon distribution and in jets?

- *EISELE* :

What I said was that there is no experimental evidence yet for the evolution of the gluon distribution function. However, if you believe in QCD, then you measure the gluon distribution function at one value of Q^2, and use the Altarelli-Parisi equations to predict it at all other values of Q^2. In doing this Q^2 evolution, one sees that the gluon distribution becomes softer much faster than any quark distribution function. This is due to the gluon self-coupling, or the triple gluon vertex, since the gluon can split into two further gluons. Unfortunately we have no experimental evidence for supporting this idea. There have not been many proposals how to get hold of the triple gluon vertex, so as to see some experimental manifestation of it. One example would be in high P_T hadron-hadron scattering. From the mere fact that there is a lot of gluon jets produced (and I think there is some evidence for that), the parton scattering picture can only hold if diagrams including the triple gluon vertex are included.

- *HOFSÄSS* :

You mentioned that the inclusion of higher twist effects may bring the Λ parameter down to zero. Does this also work in the opposite direction, that is, to increase Λ ? Can one obtain an upper limit on Λ this way?

- *EISELE* :

It works both ways. In one paper we generalized this approach by including at least two higher twist operators. We obtained two solutions: the first one is the one favored by the data, and the second one leads to a negative twist four contribution and a positive twist six contribution, the effect being that Λ becomes larger. Therefore, it is possible to obtain a large Λ from higher twist effects. By including a twist four effect, say, which gives positive slope at large x, and overcompensating it with a much larger value of Λ, one can fit the measured distributions.

- *HOFSÄSS* :

Prof. Landshoff talked about this K factor in the Drell-Yan process and said it was a big problem. From your talk I get the impression that the one-photon exchange process works well and that there is no problem. Could you comment?

- *EISELE* :

There is indeed an experimentally clearly established K factor. The only thing we can do at this point is to take the working hypothesis that the K factor does not depend on x. We can check this by using the Drell-Yan processes which only involve nucleons, such as $pp \to \mu^+\mu^- x$. We should then obtain structure functions consistent with the shape of those obtained from deep inelastic lepton scattering. It turns out that the shapes are indeed consistent. The K factor does not seem to depend on x. Consequently, this is a good working hypothesis.

I have mentioned that the QCD corrections to the Drell-Yan diagram are large and may be the origin of the K factor. It has been conjectured that the large correction can be summed to all orders and are thus under control, but this is not yet proven.

- *ITOYAMA* :

Is the data which you showed sufficient to exclude a theory with non-trivial fixed point at small Q^2, or not?

- *EISELE* :

I have avoided discussing this topic, but we did write a paper claiming that fixed point theories are not compatible with our data. I am no longer sure that this is a relevant question. For instance, at the recent Paris conference, Politzer made the point that these fixed point theories are just paper tigers, because they are not based an a solid theoretical basis. Thus you can't be sure what you are comparing the data with.

Nevertheless if you wish you can make the comparison. If you look at xF_3, the non-singlet distribution, you can use this to distinguish between fixed point theories vector gluons, and scalar gluons.

It turns out that with xF_3 alone, you can't distinguish much. The vector theory is favored, but not at a high degree of credibility. This is of course, with the W cut I mentioned in the talk.

Singlet distributions present a completely different picture. We compared the data with QCD, a scalar fixed point theory, and a vector fixed point theory. The fixed point theories describe the observed scaling violations at small x reasonably well. At high x, however, they give approximate scaling, in disagreement with the data.

So as long as you believe that scaling violations are due to perturbative effects, they are only compatible with QCD, and are incompatible with fixed point theories.

- *MORSE* :

Your final graph showed that in single particle inclusive production, the ratio of K to π falls rapidly with P_T. Does this imply that gluon jets are more important at low P_T?

- *EISELE* :

Yes. I discuss this point in greater detail in the lecture under high P_T inclusive reactions.

- *ODORICO* :

How much of the effects attributed to higher twists in the combined analysis of SLAC-MIT and CDHS data for structure functions can be reproduced by suitably modifying the argument of α_s. I refer to the variable suggested by Barnett et al. some time ago and which is reputed to better fit the data near x = 1.

- *EISELE* :

The prescription is: near x = 1, you replace Q^2 by something like $Q^2(1 - x)$. We have tried this and our result is that it goes in the right direction, but the effect is too small. It is not big enough to resolve the discrepancy between slopes of SLAC data and the data at high energy.

- *ODORICO* :

I have a comment concerning the curves that you showed in the second talk, namely probing structure functions by varying the hadron P_T in hard hadronic collisions. I point out that initial gluon bremsstrahlung, i.e. the mechanism responsible for the scaling violations of structure functions, also generates transverse momentum for the colliding partons. From a recent calculation I show that on the average about 30 percent of the total transverse energy is due to initial gluon bremsstrahlung at CERN $\bar{p}p$ collider energies. Large P_T hadrons can therefore originate not only from initial partons at large x, but also from partons with moderate, or small x and large k_T due to this mechanism. As a consequence, the connection between the hadron P_T and the x of initial partons is considerably diluted.

- *DUBNIČKOVA'*:

Is the observation of jets an indication that QCD is the underlying dynamics and that quarks and gluons do exist? In other words, is QCD the only explanation for jets?

- *EISELE* :

If you believe that quarks and gluons do exist and do act as elementary scatterers in DIS, then you expect to see jets. However, the converse is not true: seeing jets does not prove that QCD is the correct theory.

- *KAPLUNOVSKY* :

What are typical properties of a recoil jet in a high P_T process? Are they mainly quark jets or gluon jets or both?

- *EISELE* :

What I know is what we expect in this case. It should be mostly gluon jets. Since there is no particle identification at the SFM on the away side, it is hard to distinguish between quark jets and gluon jets. This situation will hopefully be different at the $\bar{p}p$ collider.

- *RAUH* :

How does the value of Λ extracted from neutrino data compare with the value obtained from that of other types of experiments?

- *EISELE* :

Within the experimental uncertainties, all are compatible with a value of about .25 GeV.

EVIDENCE FOR EXPLICIT GLUEBALLS FROM THE REACTION $\pi^- p \to \phi\phi n$

S.J. Lindenbaum

Brookhaven National Laboratory, Upton, New York 11973

and City College of New York, New York, New York 10031

INTRODUCTION

In a pure Yang-Mills theory[1] where $SU(3)_c$ has local gauge symmetry, all hadrons would be glueballs[2] (i.e., multi-gluon resonances). This is due to the self-coupling of the gluons which becomes stronger with decreasing energy and color confinement.

But what do we find experimentally? The hadronic sector is dominated by $q\bar{q}$ and qqq and there is yet no prior "experimental" demonstration of the explicit existence of glueballs, although there were a number of glueball candidates and extensive discussion of them.[3-17] Thus the quarks which are a source* of particles for gluons in QCD appear to have completely taken over the hadronic sector.

Therefore finding glueballs is crucial to QCD, Grand Unification Schemes and Partial Unification Schemes which utilize $SU(3)_c$. In fact it has been the author's opinion for some time[5,6] that if we don't establish glueballs, QCD is in serious trouble. On the other hand, the explicit establishment of glueball's would indeed be a great triumph for QCD.

* Gluons are also a source of gluons due to the self-couplings in a non-Abelian Gauge Theory.

HOW DO YOU FIND GLUEBALLS?

It is obvious from prior experimental observation that if glueballs exist they are essentially masked in the vast collection of meson nonets, existing in the mass range where one would expect to find them ($\sim$ 1-3 GeV).

Pattern Recognition of a Decuplet

The direct approach is a complicated pattern recognition problem. One must find a nonet with a glueball with the same quantum numbers near enough to the singlets in the nonet to mix with them. Thus one would have

$$\text{nonet} + \text{glueball} \rightarrow \text{decuplet}$$

with characteristic mixing and splitting (and other special characteristics of glueballs). Calculations have shown that the ideal mixing observed in a great deal of nonets would be affected in these decuplets, and pattern recognition would have to be used.[16,17] A glueball candidate of this type is the SLAC $J^{PC} = 0^{-+}$ $\iota(1440)$,[8-10,16] which could be the tenth member of a ground state 0^{-+} decuplet.* Another glueball candidate of this type is the BNL/CCNY $J^{PC} = 0^{++}$ $g_s(1240)$.[15] This would make a new 0^{++} multiplet with apparently the right characteristics: Of course one must realize that there are many other possible explanations for these states.**

Look in an OZI Suppressed Channel with a Variable Mass

In an OZI suppressed channel with variable mass glueballs with the right quantum numbers should break down the OZI suppression in the mass regions where they exist, and dominate the reaction channel. Thus the OZI suppression can act as a filter for letting

* The SLAC $\iota(1440)$ is thought to be in a channel where glueballs are enhanced since it is found in J/ψ radiative decay.

** One could for example inadvertently mix states from the basic nonet with those of a radial excitation.

glueballs pass while suppressing other states. Furthermore, the breakdown of the OZI suppression can serve as a clear signal that one or more glueballs are present in the mass region. The basic idea is that according to present concepts in QCD, the OZI suppression is due to the fact that two or more hard gluons are needed to bridge the gap in an OZI disconnected or hairpin diagram,* and the early onset of asymptotic freedom leads to a relatively weak coupling constant for these gluons, which then causes the OZI suppression. On the other hand, if the glue in the intermediate state resonates to form a glueball, the effective coupling constant (as in all resonance phenomena) must become strong, and the OZI suppression should disappear in the mass range of the glueball. This should allow hadronic states with the glueball quantum numbers to form with essentially no OZI suppression. This argument has been made previously by the author at ERICE[5] and elsewhere.[6,13-14] Thus the OZI suppression essentially acts as a filter which lets glueballs pass and suppresses quark states.

THE $\pi^-p \to \phi\phi n$ (OZI FORBIDDEN CHANNEL)

The BNL/CCNY collaboration had shown several years ago that in the OZI forbidden[18] (or suppressed) reaction $\pi^-p \to \phi\phi n$ at an incident pion energy of 22.6 GeV, that the OZI suppression was essentially absent[4] in this OZI forbidden process. This was quantitatively demonstrated, and interpreted by the author as evidence for glueballs in the $\phi\phi$ system.[5,6] However, initially 100, and later 170, events were obtained and this small number did not allow a viable convincing partial wave analysis to explicitly identify the glueball candidates quantum numbers, mass, width, etc. The observed $\phi\phi$ mass spectrum in other later low statistics measurements were consistent with our results.[19]

* Three gluons are needed if the hairpin is a vector, two are needed if the hairpin is a scalar.

However, BNL/CCNY planned a new experiment to obtain > than an order of magnitude more data which would hopefully allow a significant partial wave analysis. In order to accomplish this the BNL MPS (Multiparticle Spectrometer) was being redesigned so that a new novel high speed drift chamber system replaced the spark chambers and thus allowed gathering data at an order of magnitude faster rate.[20]

This spring we successfully commissioned the new MPS II and in a 2-3 week run obtained 1203 $\pi^- p \to \phi\phi n$ events even though the visible cross section is only $\sim$ 6 nanobarns.

A partial wave analysis of this data which was just presented at the Paris Conference,[21] yields two explicit strong glueball candidates in the $\phi\phi$ system with all quantum numbers, mass and width determined. The experiment was done at an incident π^- energy of 22 GeV.

Let us now look at the diagrams in Fig. 1 which represent the various channels studied in the experiment.

The basic reaction observed is given by the OZI allowed reaction (Fig. 1a) $\pi^- p \to K^+K^-K^+K^- n$.

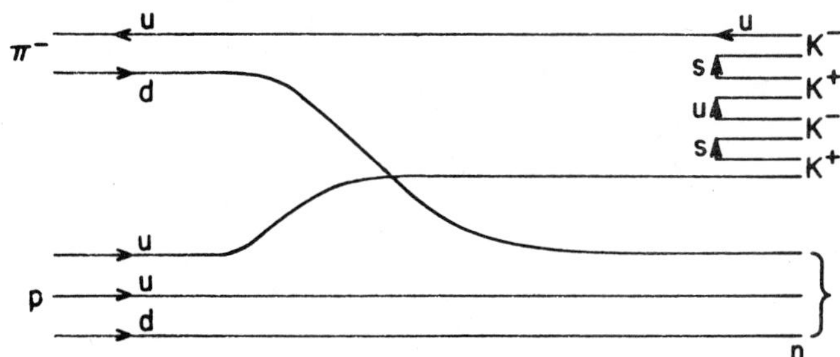

Figure 1a: The quark line diagram for the reaction $\pi^- p \to K^+K^-K^+K^- n$, which is connected and OZI allowed.

In QCD[22] one considers these OZI allowed reactions to proceed by a continuous series of exchanges of single and perhaps some low energy multiple gluons which have relatively strong effective coupling constants and thus proceed as strong interactions. The poorly understood hadronization process can to a large extent occur near the outer regions of the confinement region and have unsuppressed cross sections.

If one K^+K^- pair forms a ϕ we have the reaction $\pi^-p \rightarrow \phi K^+K^-n$ (see Fig. 1b) which is still a connected diagram and OZI allowed.

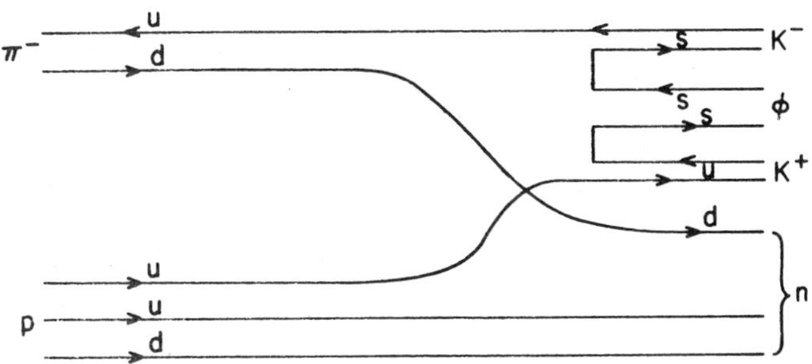

Figure 1b: The quark line diagram for the reaction $\pi^-p \rightarrow \phi K^+K^-n$, which is connected and OZI allowed.

However if both K^+K^- pairs form ϕ's we have a disjoint (hair-pin) diagram which is OZI forbidden. Thus $\pi^-p \rightarrow \phi\phi n$ as shown in Fig. 1c is an OZI forbidden diagram and should exhibit the OZI suppression.

This has been clearly shown for $\pi^-p \rightarrow \phi n$ where the OZI suppression factor has been found to be ~ 100.[23] Thus typically one finds

$$\frac{\sigma(\pi^-p) \rightarrow \omega n}{\sigma(\pi^-p) \rightarrow \phi n} \sim 100$$

reflecting the OZI suppression factor.

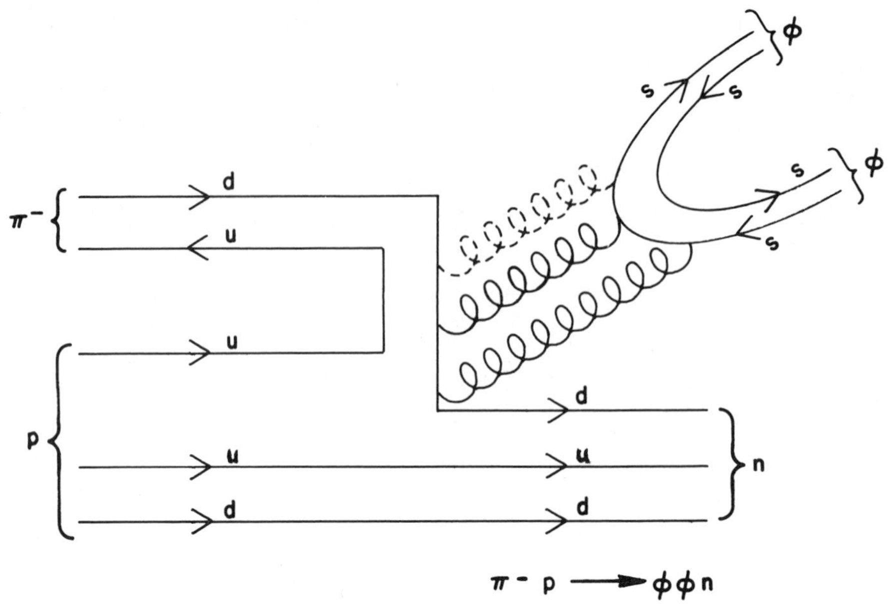

Figure 1c: The quark line diagram for the reaction $\pi^- p \to \phi\phi n$ which is disjoint (i.e. a hairpin diagram) and is OZI forbidden. Two or three gluons are shown connecting the disconnected parts of the diagram depending upon the quantum numbers of the $\phi\phi$ system.

However, another OZI allowed process corresponding to a connected diagram is $K^- p \to \phi\Lambda$. The ratio

$$\frac{\sigma(K^- p) \to \phi\Lambda}{\sigma(\pi^- p) \to \phi n} \approx 60$$

again showing the typical OZI suppression.[24]

The decay matrix element squared of $\phi \to K^+K^-$ (shown in Fig. 2a), an OZI allowed process, is $\sim$ 100 times that for $\phi \to \rho^+\pi^-$ which is OZI suppressed.[18a] Hence in both the production and decay, a single ϕ hairpin (disjoint diagram, see Fig. 2b), corresponds to an OZI suppression factor $\sim$ 100.

Now you may ask is it as legitimate to consider $\pi^- p \to \phi\phi n$ also as a disjoint diagram subject to the OZI suppression. The answer is clearly yes as I shall now demonstrate. Each of the two ϕ's

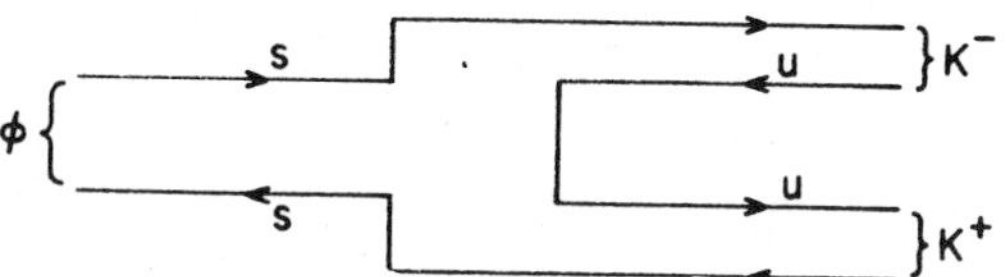

Figure 2a: The quark line diagram for the reaction $\phi \rightarrow K^+K^-$ which is connected and thus OZI allowed.

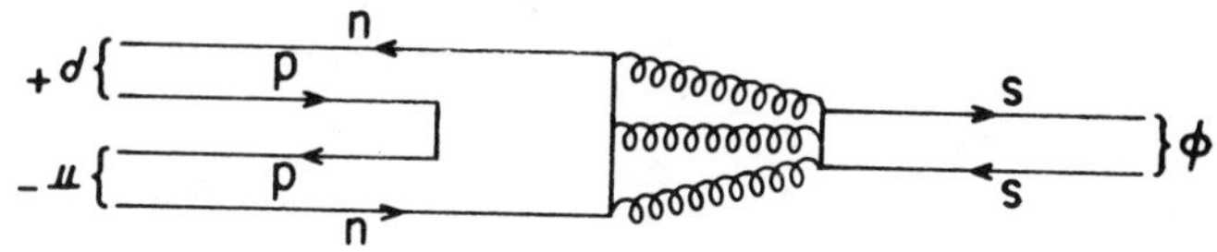

Figure 2b: The quark line diagram for the reaction $\phi \rightarrow \rho^+\pi^-$ which is disjoint (i.e. hairpin) and OZI forbidden.

is an almost pure $s\bar{s}$ meson system. If you look at Fig. 1c from right to left you have two $s\bar{s}$ states disjoint from the π^-, p and n part of the diagram which is connected, and contains only u and d quarks. This is basically no different a disjoint diagram than the single ϕ production case (Fig. 2c) where the OZI suppression is

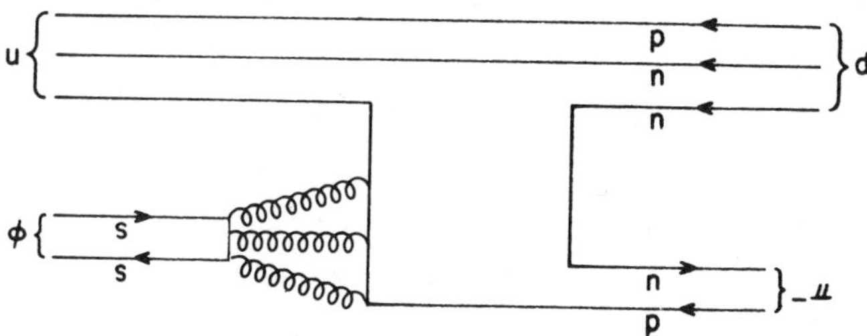

Figure 2c: The quark line diagram for the reaction $\pi^-p \rightarrow \phi n$ which is a disjoint (i.e. hairpin) and OZI forbidden.

calculated based on experimental results to be just what is expected from the ϕ decay, a factor ~ 100.

The following is an experimental example of another case where a disjoint diagram formed by two particles in the final state composed of new types of quarks and their antiquarks leads to OZI suppression.[25]

$$\psi(3685) \to J/\psi\ \pi^+\pi^-\ (33 \pm 2)\%, \text{ or } J/\psi\ \pi^o\pi^o\ (17 \pm 2)\%$$

The full width of the $\psi(3685)$ is 0.215 MeV clearly showing the strong OZI suppression corresponding to the fact that the initial state contains $c\bar{c}$ quarks only, whereas the final state still contains the $c\bar{c}$ quarks but the diagram becomes disconnected when u and d quarks and their antiquarks (to form the two pions) are included in the final state.

But you might say what if I introduce two-step processes or other complicated intermediate states or processes, other than hard multigluons to jump the disjointed part of the diagram. The author has discussed this[6] and shown that the OZI rule is peculiar in that you can defeat it by two-step processes (Fig. 3) or in QCD language changing the nature of the multigluon exchange needed in the one-step diagrams to a series of the ordinary OZI allowed gluon exchanges.

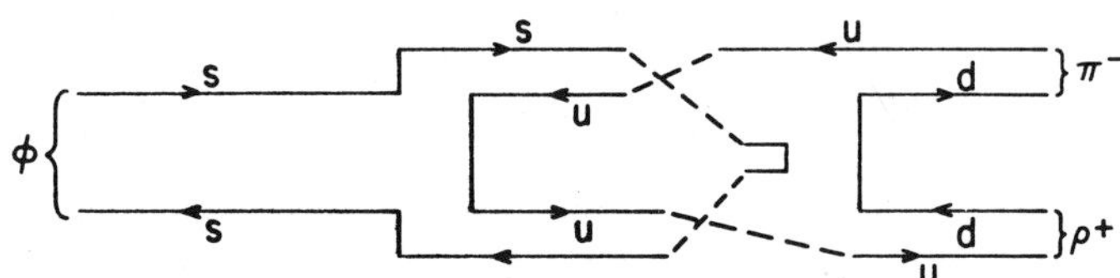

Figure 3a: The reactions $\phi \to K^+K^- \to \rho^+\pi^-$, a two-step connected diagram which appears to be OZI allowed but is not (according to experiment). Proposed cancellations[18b] of artificial nature were proposed to eliminate this problem, but the author believes the simple Ansatz that they are highly suppressed is much more likely to be consonant with observations and QCD.

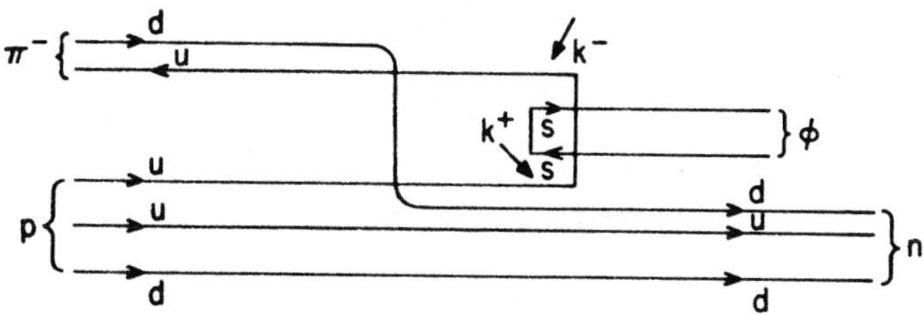

Figure 3b: The reaction $\pi^- p \to K^+ K^- n \to \phi n$, a two-step connected diagram which appears to be OZI allowed but is not (according to experiment).

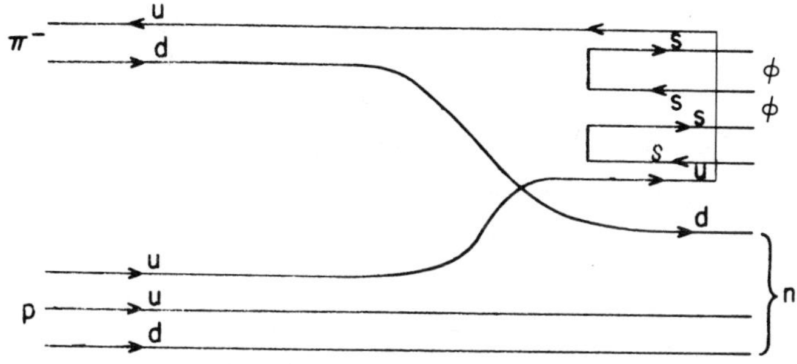

Figure 3c: The reaction $\pi^- p \to \phi K^+ K^- n \to \phi\phi n$ via a two-step connected diagram which appears to be OZI allowed but is not.

In other words, Zweig's diagrams are, based on all experimental observations, to be taken literally as one-step processes and the multigluon exchanges needed to connect disconnected parts of the diagrams are not to be tampered with. Why are these peculiarities observed? I cannot answer that. Neither can I answer why color exists, why confinement? Why quarks? etc. etc. etc. These are all concepts based on observation.

It is certainly consistent with all experimental observations in the ϕ, J/ψ and Υ systems that the OZI rule[18] works very well,* and as I have pointed out if one wants to invent appropriate multi-step processes or tamper with the nature of a gluon exhcange in the one-step Zweig diagrams one can defeat the rule.** Therefore I assume the OZI rule (with the caveats that I have mentioned) as an input assumption, and of course QCD as an input assumption in drawing my conclusions in this paper. If you grant me QCD and OZI as valid assumptions, I will later conclude that we have discovered one or more glueballs. If you quarrel with assuming QCD, there is absolutely no point in discussing glueballs. If you quarrel with assuming OZI (as caveated) we will have to demote our conclusion of glueball discovery to discovery of strong glueball candidates, and suggest you explain why the assumption of the OZI rule which has been consistently observed to work be replaced by complicated alternatives. Remember the name of the physics game is simplicity when it works.

THE NEW BNL/CCNY $\pi^- p \to \phi\phi n$ EXPERIMENT

We utilized the new BNL MPS II, the experimental arrangement of which is shown in Fig. 4. The major changes from the MPS I experiments to MPS II experiments was to replace the spark chambers with drift chambers with ten times more data-gathering rate capability and to improve the (charged particle and γ) veto box around the liquid hydrogen target to obtain an even cleaner neutron signal.

* The fact that the OZI rule works well for the single ϕ, J/ψ and Υ is understandable if there are no glueballs with the right quantum numbers at their masses.

** These restrictions apparently violate crossing and unitarity. I consider these peculiarities as another fact of life of quarks, gluons, and color confinement.

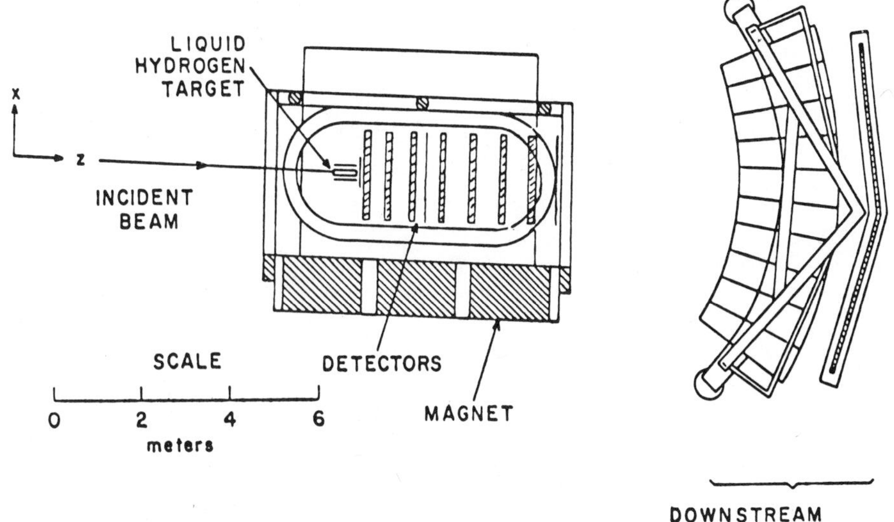

Figure 4: The MPS II and the experimental arrangement (see Ref. 3 for further details).

Figure 5 is a scatter plot of the mass of one K^+K^- pair versus the mass of the second K^+K^- pair. Each event has two points on the plot since there are four possible combinations. One clearly notices the two ϕ bands standing out over the 4-kaon background. However, where the two ϕ bands cross we find a black spot whose peak intensity (corrected for resolution and double counting) is greater than 1,000 times that of the adjoining 4-kaon event intensity. The ϕ band intensity (corrected for resolution) is about a factor of 20 higher than the adjoining 4-kaon event intensity. Where the two ϕ bands cross, the $\phi\phi$ intensity (corrected for resolution) is $\sim$ 50 times greater than the $\phi(K^+K^-)$ intensity. If the OZI suppression were working very little enhancement would be seen here. Thus we have a patent violation of the OZI suppression. This effect was already clearly noted by us in 1978.[4] The speaker has previously shown that if one uses the isobar model,[26] which is well known to work well and has no provision for OZI suppression

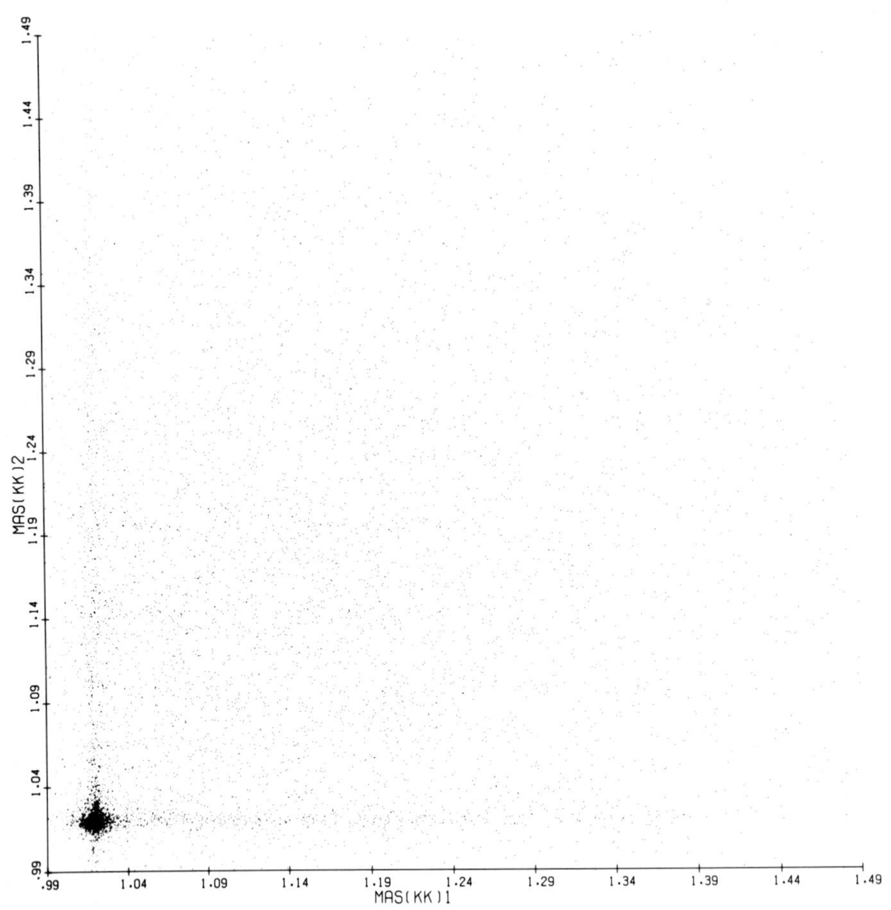

Figure 5: Scatter plot of K^+K^- effective mass, two randomly chosen mass combinations are plotted for each event. Clear bands of $\phi(1020)$ are seen with an enormous enhancement (black spot) where they overlap (i.e. $\phi\phi$).

in it, one can quantitatively explain[5,6] the behavior of this scatter plot within a factor of 2. The new greater statistics experimental data are consistent with the earlier experiment in this regard.

Independent evidence of the breakdown of the OZI suppression is given[27] by a study of the reaction $K^-p \rightarrow \phi\phi\Lambda$ or $\phi\phi\Sigma^o$. This is an OZI allowed reaction and yet the cross section obtained is only a factor ∿ 4 larger than the cross section for $\pi^- p \rightarrow \phi\phi n$ which is an OZI forbidden reaction. We also have kaons in our beam and have studied this reaction and from preliminary results obtain the same factor ∿ 4. One should divide the 4 by a factor of 2 since in the π^- case only n is allowed to accompany the $\phi\phi$, whereas in the K^- case, either a Λ or Σ^0 is accepted.

Thus within a factor of 2 the two cross sections are equal, showing little difference between the OZI allowed and forbidden reactions.

In contrast to this, the ratio[24]

$$\frac{\sigma(K^- p) \rightarrow \phi\Lambda}{\sigma(\pi^- p) \rightarrow \phi n} \approx 60$$

showing the typical OZI suppression of the forbidden to the allowed reaction, and as is also well known;

$$\frac{\sigma(\pi^- p) \rightarrow \omega n}{\sigma(\pi^- p) \rightarrow \phi n} \sim 100$$

shows the typical OZI suppression ∿ 100.

Hence we have shown in a number of ways that the large OZI suppression ∿ 100 expected in single ϕ production is present, whereas that expected in $\phi\phi$ production is broken down to within a factor of 2 of OZI allowed processes which is within the uncertainties of the comparisons. Thus we can clearly conclude on a number of grounds that the expected OZI suppression is essentially entirely absent in the $\pi^- p \rightarrow \phi\phi n$ OZI forbidden process. Figure 6 shows the mass spectrum of the other K^+K^- pair in one event whenever one K^+K^- pair falls in the ϕ mass band (1014.6 ± 14 MeV) and clearly indicates the huge $\phi\phi$ signal. Figure 7 shows a very clear neutron recoil from the $\phi\phi$ with an estimated contamination of non-neutron events in our data sample ∿ 3% which should have a negligible effect on our analysis. In the mass region where we did our partial wave analysis, the ϕK^+K^- background was small (approximately 10%) and it was included in our analysis.

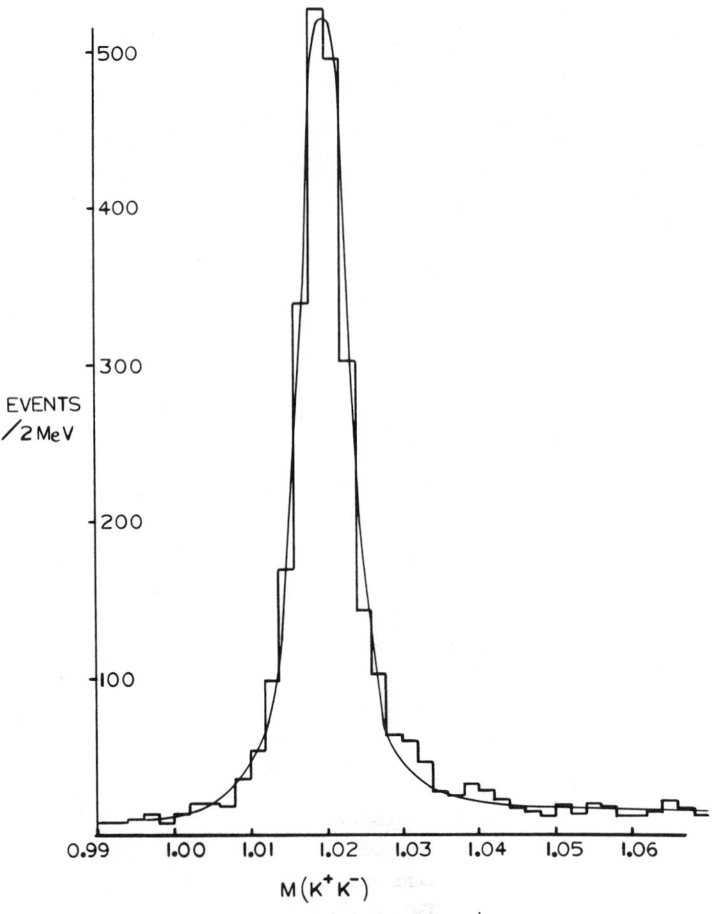

Figure 6: The effective mass of each K^+K^- pair for which the other pair was in the ϕ mass band.

The histogram in Fig. 8 shows the detected $\phi\phi$ effective mass spectrum for 1203 $\pi^-p \to \phi\phi n$ with an estimated background of 130 events from ϕK^+K^- ($\approx$ 10%) and $\approx$ 40 events of non-neutron recoil. The dashed line is the Monte Carlo determined acceptance of the apparatus of our partial wave analysis solution to be discussed later. However one should note that the result obtained for the acceptance is close to that one would obtain from phase space. Furthermore the results of the partial wave analysis are insensitive to considerable changes in the acceptance. The observed

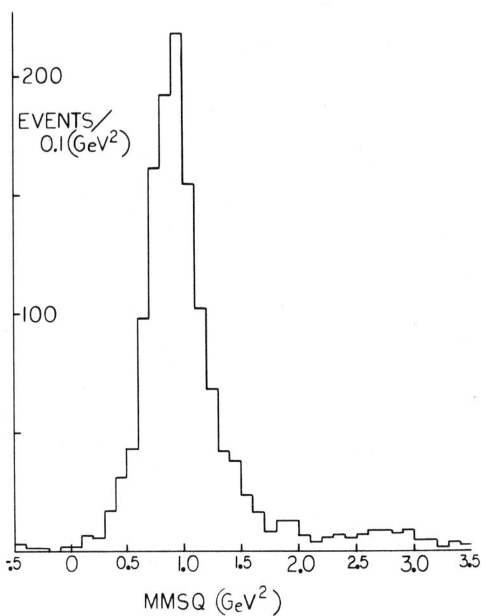

Figure 7: The missing mass squared for the neutral recoiling system from the $\phi\phi$.

spectrum is consistent with that of Ref. 3 and other subsequent low statistics $\phi\phi$ experiments.[19] One should note that the $|t'| \lesssim 0.3$ GeV^2 the t' distribution is consistent with $e^{(9.4 \pm 0.7)t'}$. It

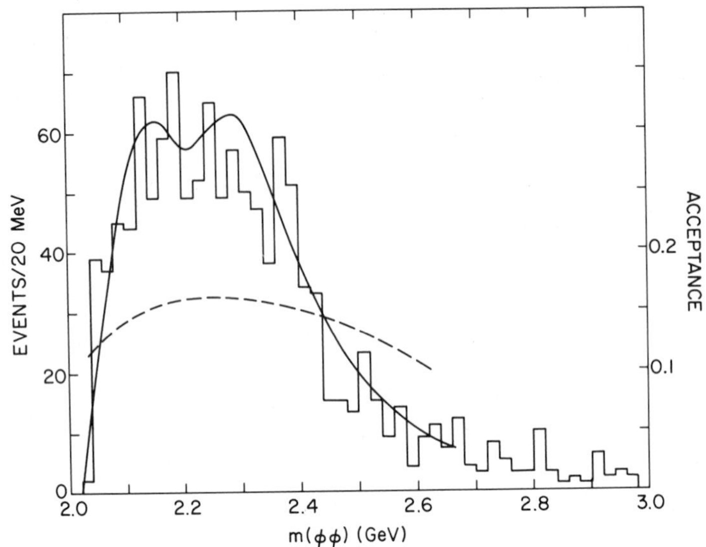

Figure 8: The observed $\phi\phi$ effective mass spectrum. The dashed line is the Monte Carlo calculated acceptance. The solid line is the P.W.A. fit to be discussed later.

should be noted that the $\phi\phi$ mass spectrum from $K^-p \rightarrow \phi\phi\Lambda/\Sigma$ [27] is much broader and extends to much higher masses (see Fig. 9).

THE PARTIAL WAVE ANALYSIS

In order to perform the partial wave analysis (PWA) we used six angles to specify all kinematic and other characteristics of the $\phi\phi$ system with each ϕ decaying into a K^+K^- pair.

Figure 10 shows the Gottfried-Jackson frame (which is the rest frame for the $\phi\phi$ system). The usual GJ angles, β(polar) and γ azimuthal, were employed. We then considered the rest frame of each ϕ.

Figure 11 shows the rest frame of ϕ_1. In it we label the polar angle of the decay of the K_1^+ relative to the ϕ direction as θ_1, and the azimuthal angle of the decaying K_1^+ as α_1. There is

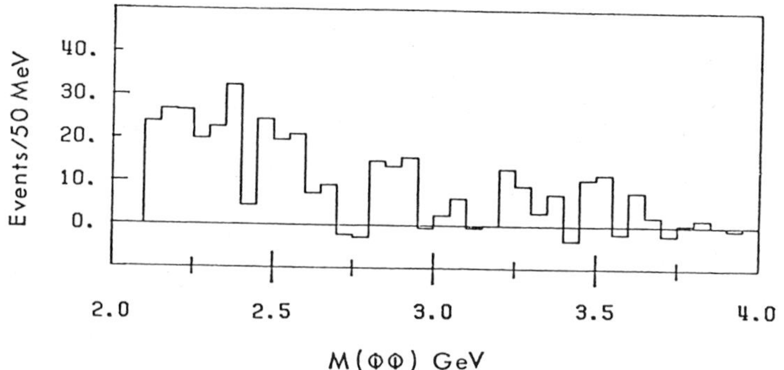

Figure 9: The $\phi\phi$ effective mass spectrum for $K^-p \rightarrow \phi\phi\ \Lambda/\Sigma$.

a similar rest frame (not shown) for the second ϕ with corresponding polar angle for K_2^+ of θ_2 and azimuthal angle α_2. Since the azimuthal angles α are the same in either ϕ rest frame, α_2 is shown in the rest frame of α_1. However, of course, θ is different in the two rest frames.

Since the decay kaons are spinless, these six angles specify everything for the $\phi\phi$ system decaying into kaons. These angles and relevant combinations of them were used in the partial wave analysis. For the $\phi\phi$ system, $I = 0$ and $C = +$.

The partial waves considered were all waves with $J = 0,1,2,3,4$; $L = 0,1,2,3$; $S = 0,1,2$; $-J \leq M \leq J$, $P = \pm$, $\eta = \pm$ where J is the total angular momentum of the $\phi\phi$ system. L is the orbital angular momentum, M is J_z, P is the parity and η the exchange naturality of the wave.

Due to the identity of the two ϕ mesons bose statistics requires that L + S = even number, and this was an imposed requirement. The above criteria led to a group of 52 independent waves. The maximum likelihood method was used for the PWA. In order to determine the partial waves playing a major role in the $\phi\phi$ system, the events in the mass region 2.1 to 2.3 GeV were fitted with an

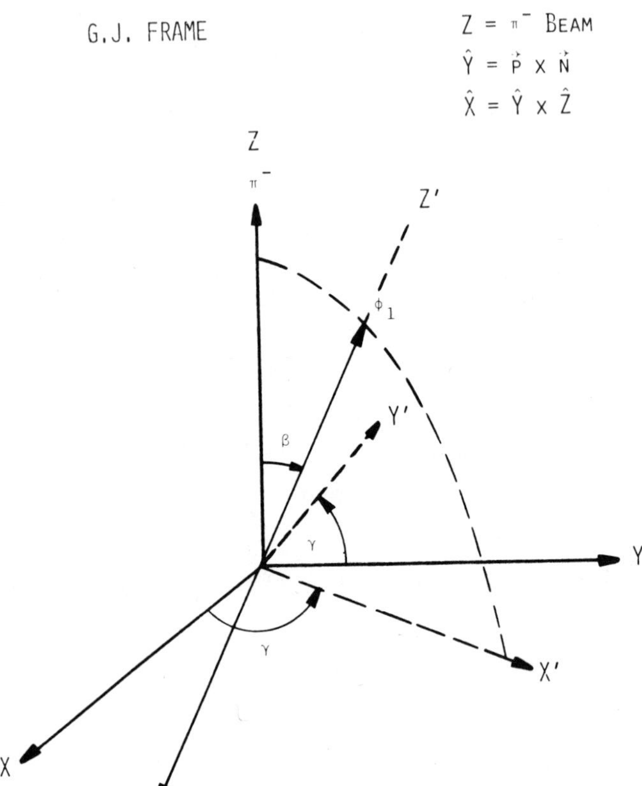

ϕ_1 AND ϕ_2 LIE IN (Z,X') PLANE

Figure 10: The Gottfried-Jackson frame with polar angle β and azimuthal angle γ.

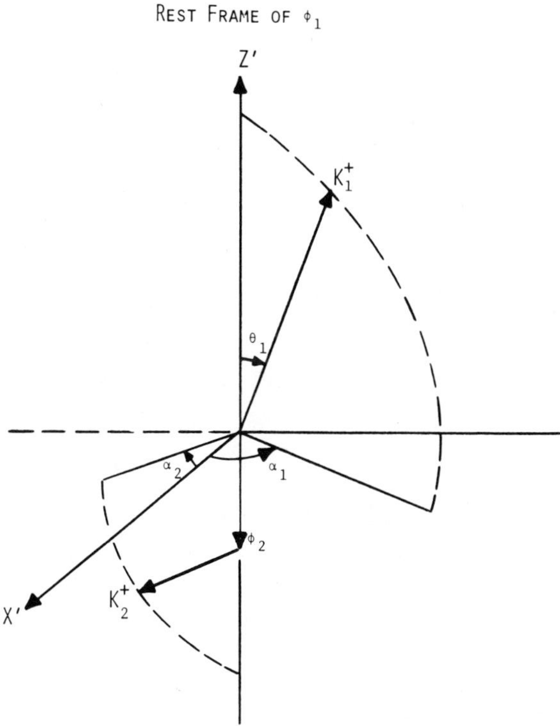

Figure 11: The ϕ rest frame with the polar angle θ_1 of the decay K_1^+ (relative to ϕ direction) and the azimuthal angle α_1 of the decay K_1^+.

incoherent background plus one additional partial wave of specific J^P, S, L, M and η, cycling through each of the 52 waves described above. The largest, and only significant contribution came from $J^{PC}SLM^{\eta} = 2^{++}200^-$. This wave was retained and in order to search for other waves each of the other 51 were added one at a time in turn. The only significant additional contribution came from $J^P SLM^{\eta} = 2^+220^-$. These two waves were then retained and each of the remaining fifty were added one at a time in turn. No significant contribution from any other wave was found. The $\phi\phi$ data were then divided up into five adjoining 100 MeV wide bins starting from threshold, so that we could explore the mass dependence of the partial wave structure. The bin size was chosen because about 200 events per bin are needed to obtain reliable solutions.

The background from ϕK^+K^- events ($\approx$ 11%) was estimated from an examination of the regions adjacent to the $\phi\phi$ peak. There was no evidence of any angular structure, so this background was represented by a flat distribution in all angles. A maximum likelihood fit to the five bins using the two $J^P = 2^+$ waves described gave a very good fit with χ^2/D.F. $\approx$ 1 when the statistics and systematic errors were considered. To ensure that no other combination of two waves would give an equivalent fit, each possible combination of two waves, i.e. 52 x 51/2 = 1326, were tried in the central bin where the S and D waves found had a significant overlap. The closest one came to a fit was 5σ away from the original fit. These $\sim 5\sigma$ fits always involved the S-wave originally found as one of the two waves. Hence the original two-wave fit is clearly selected.*

Therefore the mass independent solutions (i.e. no parameterization chosen) for the $J^P SLM^{\eta} = 2^+200^-$ S-wave and the 2^+220^- D-wave are shown in Fig. 12. The lower half of the figure shows the $|S|^2$

* One might perhaps expect a background of the L = 0, $J^P = 0^+$ wave at threshold, but this wave contributes only 10 ± 5% of the events in the lowest mass bin. Furthermore one should recall backgrounds do not break down OZI suppression.

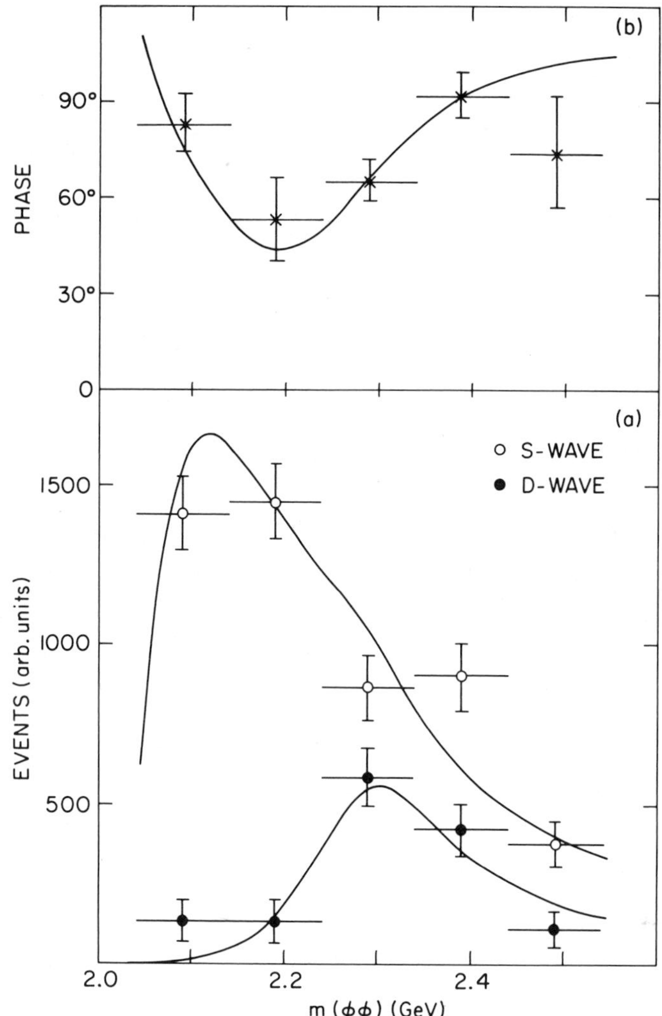

Figure 12: (a) The points show the intensity ($|S|^2$ and $|D|^2$) and for the best mass-independent two-wave fit described in the text.

(b) The D-S phase difference (mass-independent) for the best two-wave fit described in the text. The curves show the resultant best maximum likelihood fits for the parameterization of two interfering Breit-Wigner resonances.

normalized to events as the open circle points and the corresponding $|D|^2$ amplitudes squared are shown as the closed points. The x points on the top part of the figure show the D-S phase difference. A natural parameterization for these data is one or two Breit-Wigners. A one Breit-Wigner fit was rejected by $> 10\sigma$, primarily because of the phase difference. A two Breit-Wigner fit on the other hand was very good with $\chi^2/DF \approx 1$. The solid lines show this fit and the quantum numbers and parameters of the two resonances to be discussed later are shown in Table I.

It at first appears remarkable that we can demonstrate such selectivity (i.e. 2 waves selected out of 52). However this results from the fact that the $\phi\phi$ system is a very powerful analysis system for picking particular waves. In order to see how this comes about let us look at Figs. 13a and 13b where angular variables for numerous allowed (i.e. L + S = even) pure waves up to $J^{PC} = 4^{++}$ are shown. Our data shows a flat distribution in γ, all M = 0 waves which are shown have this feature. Therefore there is no need to plot γ.

We can notice from these figures which show the behavior of α, $\alpha_1' - \alpha_2'$, $\alpha_1' + \alpha_2'$, $\cos\beta$, $\cos\theta$, $\cos\theta_1' + \cos\theta_2'$, $\cos\theta_1' - \cos\theta_2'$, that each wave has its own characteristic signature in the various variables shown. The primes (i.e. $\alpha_1' - \alpha_2'$) were modifications to

TABLE I

Quantum numbers and parameters of the Breit-Wigner resonance fit to the S- and D-wave amplitudes (and phase difference) of Fig. 12.

	$g_T(2160)$	$g_T(2320)$
$I^G J^{PC}$	0^+2^{++}	0^+2^{++}
Mass (GeV)	2.16 ± 0.05	2.32 ± 0.04
Γ_{tot}	0.31 ± 0.07	0.22 ± 0.07
Ratio of Partial Widths	$\Gamma_D/\Gamma_S \approx 0.02$	$\Gamma_S/\Gamma_D \approx 0.04$

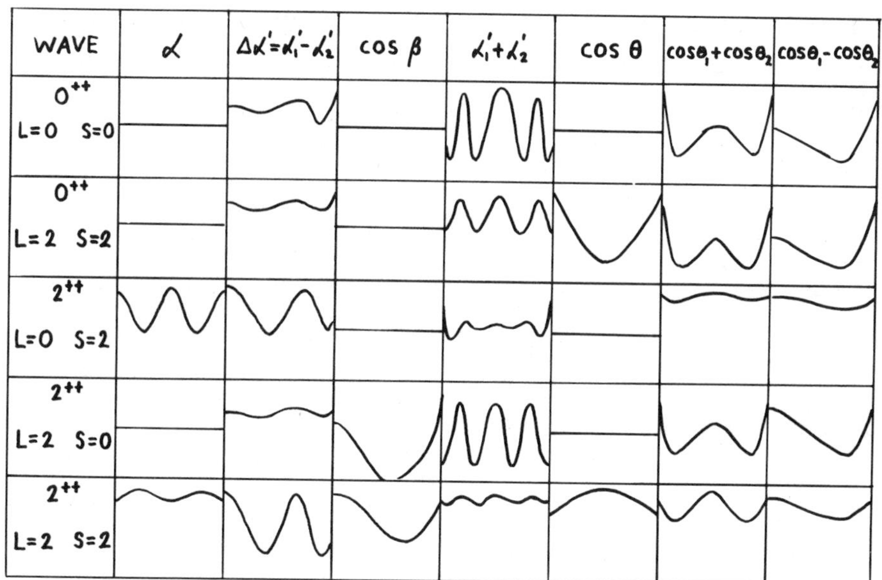

Figure 13a: Various pure waves from $J^{PC}=0^{++}$ to $J^{PC}=2^{++}$ with M = 0.

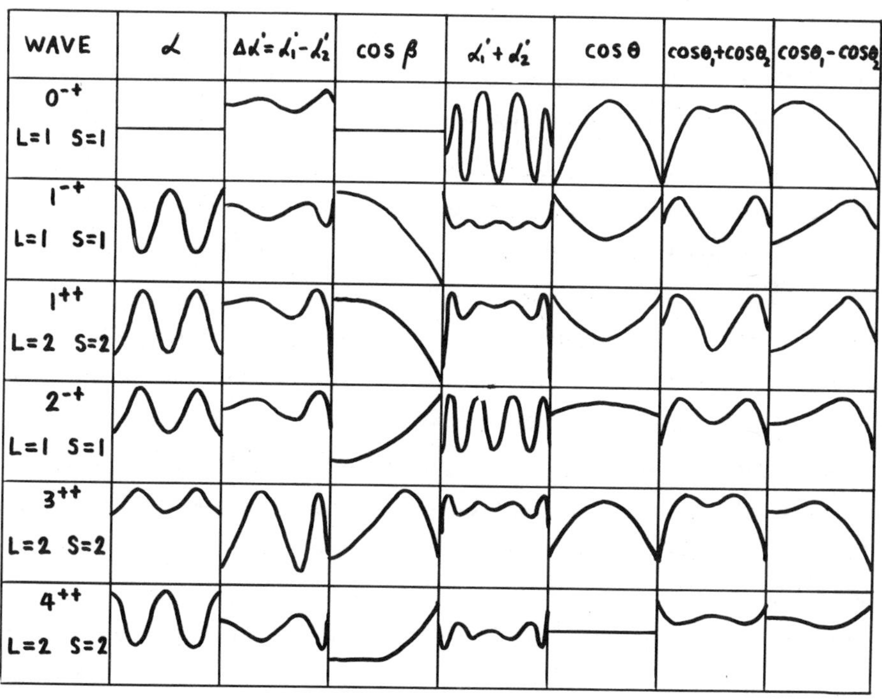

Figure 13b: Various pure waves from $J^{PC}=0^{-+}$ to $J^{PC}=4^{++}$ with M = 0.

the variables for display and comparison purposes so as to equalize the phase space in each histogram bin (to be shown later). For example

$$\Delta\alpha' = \frac{(\alpha_1 - \alpha_2)}{\pi} \; \frac{(1 - |\alpha_1 - \alpha_2|)}{4\pi} .$$

Similar equalizations in phase space were made per bin wherever primes are shown. Due to the inherent symmetry, $\cos\beta$, $\Delta\alpha' = \alpha'_1 - \alpha'_2$, and $\cos\theta'_1 - \cos\theta'_2$ have been folded, the data for α_1 and α_2 added, and the data for $\cos\theta_1$ and $\cos\theta_2$ added. The very characteristic signature for particular pure waves in these angular variables give us the great selectivity we have found. For example, notice that the two $J^P = 2^+$ (S and D) waves that we have found in the partial wave analysis (the third and the fifth from the top in Fig. 13a) have similar very characteristic large structure in $\alpha'_1 - \alpha'_2$ and the S-wave has a characteristic structure in α whereas the D-wave does not. Thus $\alpha'_1 - \alpha'_2$ and α are the most important variables in selecting the $J^P = 2^+$ waves we found in our partial wave analysis.

With this introduction I now turn to a detailed comparison (in 3 mass bins) of the data and the Monte Carlo generated prediction for our fit from the partial wave analysis. The Monte Carlo results are acceptance-corrected and are based on over 14,000 events, more than an order of magnitude more statistics than the data (for which the actual number of events are shown in the plots). Thus the statistical fluctuations in the Monte Carlo results will be small compared to those in the data. Furthermore we determined that the angular variables and correlations were not sensitive to the acceptance except in the case of the G.J. angle β.

Figure 14a shows a comparison of the data and Monte Carlo for γ (the G.J. azimuthal angle) and the polar angle θ. The agreement is excellent.

Figure 14b shows a comparison of the data and Monte Carlo for $\cos\beta$, where β is the G.J. polar angle. Even though $\cos\beta$ is sensitive to the acceptance, we obtain a quite reasonable agreement.

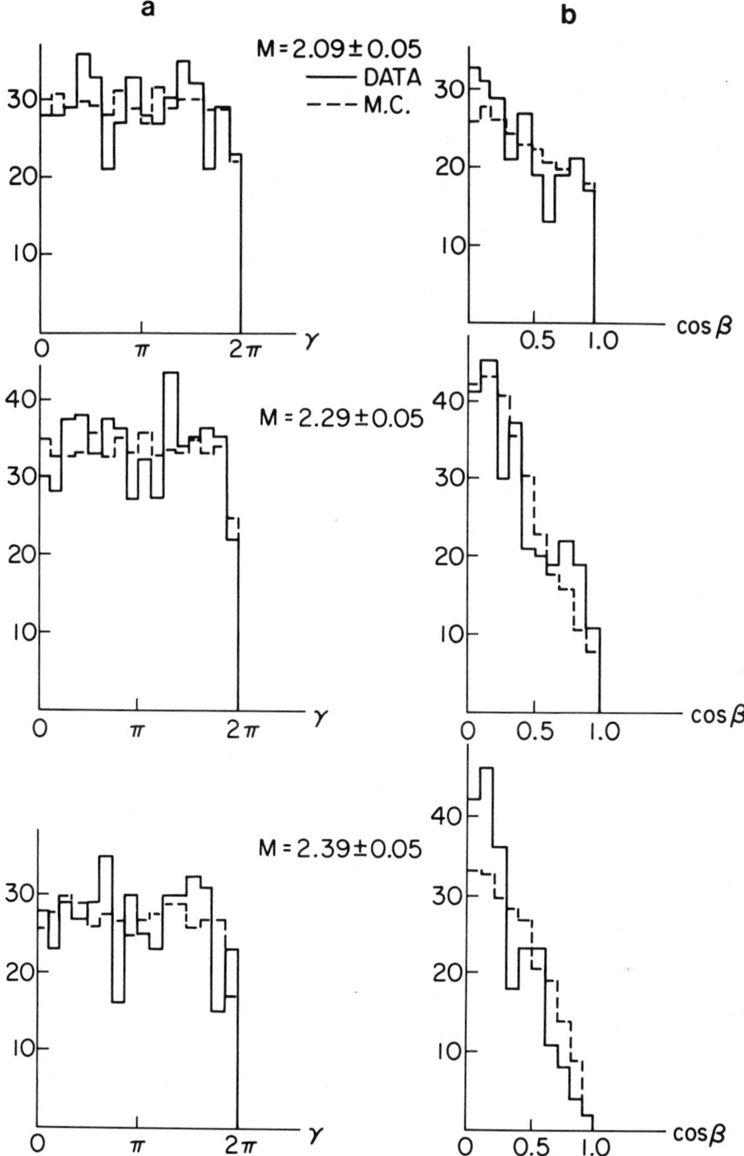

Figure 14: (a) Comparison of the data and the acceptance-corrected Monte Carlo for the fit in G.J. azimuthal angle γ.
(b) Comparison of the data and the acceptance-corrected Monte Carlo for the fit in G.J. polar angle function cosβ.

Figure 15 shows the data and Monte Carlo predictions of the fit for α and $\Delta\alpha' = \alpha_1' - \alpha_2'$. The agreement between the data and the Monte Carlo prediction based on the fit is most impressive for $\Delta\alpha'$ since there are large factors $\gtrsim$ 3 between peaks and valleys.

In the case of α the agreement is also quite good. The first bin shows large structure, characteristic of the S-wave, as we have remarked previously. The next bin shown (third bin) is where the D-wave is very important and shows very little structure in α_1, which, as we pointed out previously, is a feature of the D-wave. The agreement is good. The next bin shows the structure returning as the D-wave drops down and again indicating good agreement.

Figure 16 shows the comparison of the data with the Monte Carlo for $\alpha_1' + \alpha_2'$ and $\cos\theta$. Here again the agreement is generally quite good and there is no sizeable structure in these variables.

Figure 17 shows the comparison with the Monte Carlo for $\cos\theta_1' - \cos\theta_2'$ and $\cos\theta_1' + \cos\theta_2'$. Here again the agreement is quite good and there is no sizeable structure in these variables.

Thus we have made ten* characteristic angular correlations for six independent variables and found good agreement - striking at times in $\Delta\alpha'$ and α for example. The data and Monte Carlo agree in all mass bins in all variables.

The next question is how does our fit compare with the observed $\phi\phi$ mass spectrum? This is shown in Fig. 18 where the solid line is the fit prediction. The agreement here is also quite good. However, I must remark that in dealing with the $\phi\phi$ system, its myriad and characteristic angular distributions and angular correlations are much more important tests of the significance of the fit, than the mass spectrum. Thus we can feel quite confident that our two Breit-Wigner fits are in excellent agreement with the data.

---

* α represents the data for α_1 and the data for α_2 added due to symmetry. Cosθ represents the data for $\cos\theta_1$ and the data for $\cos\theta_2$, added due to symmetry.

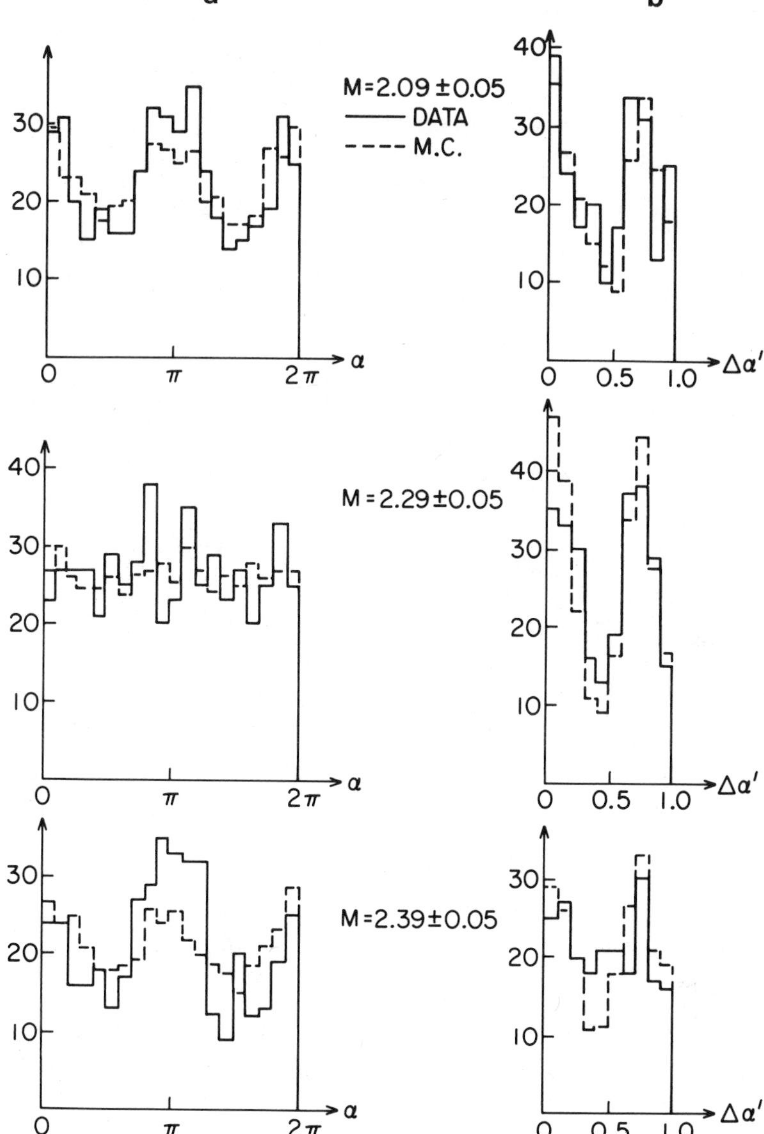

Figure 15: (a) Comparison of the data and the acceptance-corrected Monte Carlo for the azimuthal angle α of the decay K^+ in the ϕ rest frame.

(b) Comparison of $\Delta\alpha' = \alpha_1' - \alpha_2'$ with the acceptance-corrected Monte Carlo for the fit.

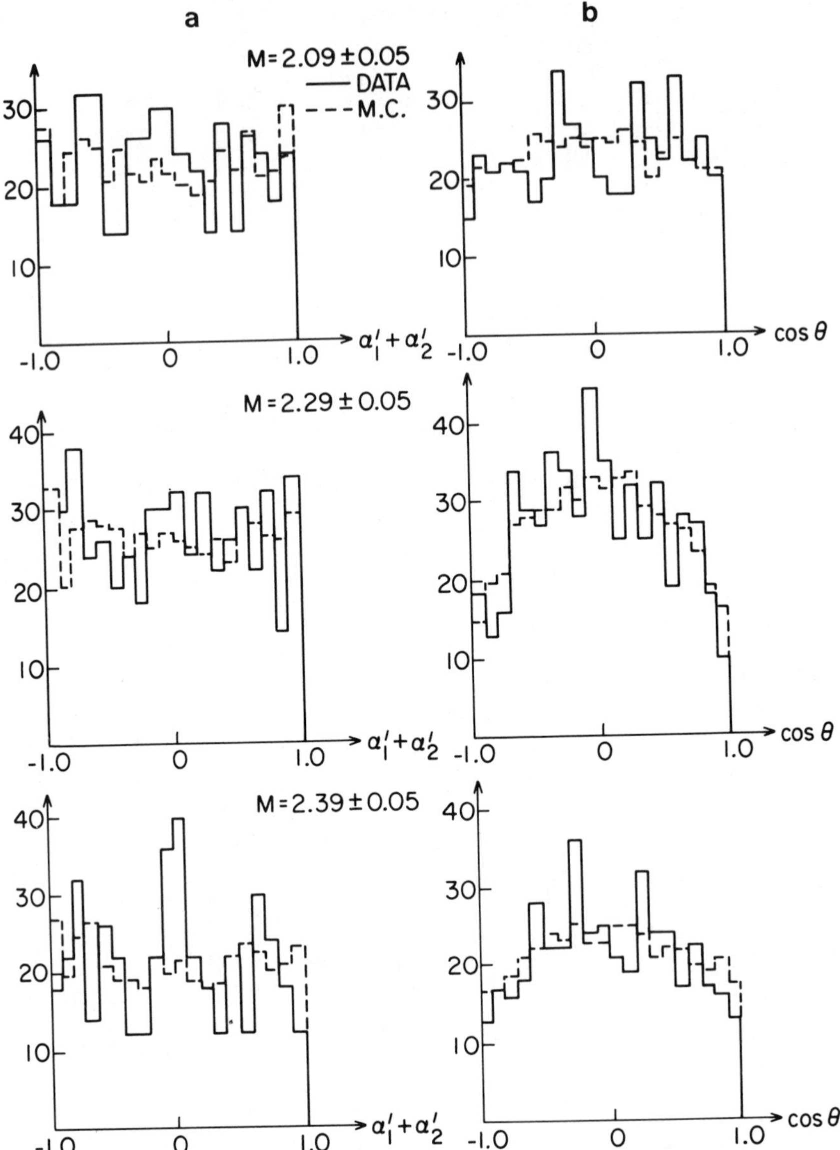

Figure 16: (a) Comparison of the data and the acceptance-corrected Monte Carlo for $\alpha_1' + \alpha_2'$.

(b) Comparison of the data and the acceptance-corrected Monte Carlo for cosθ.

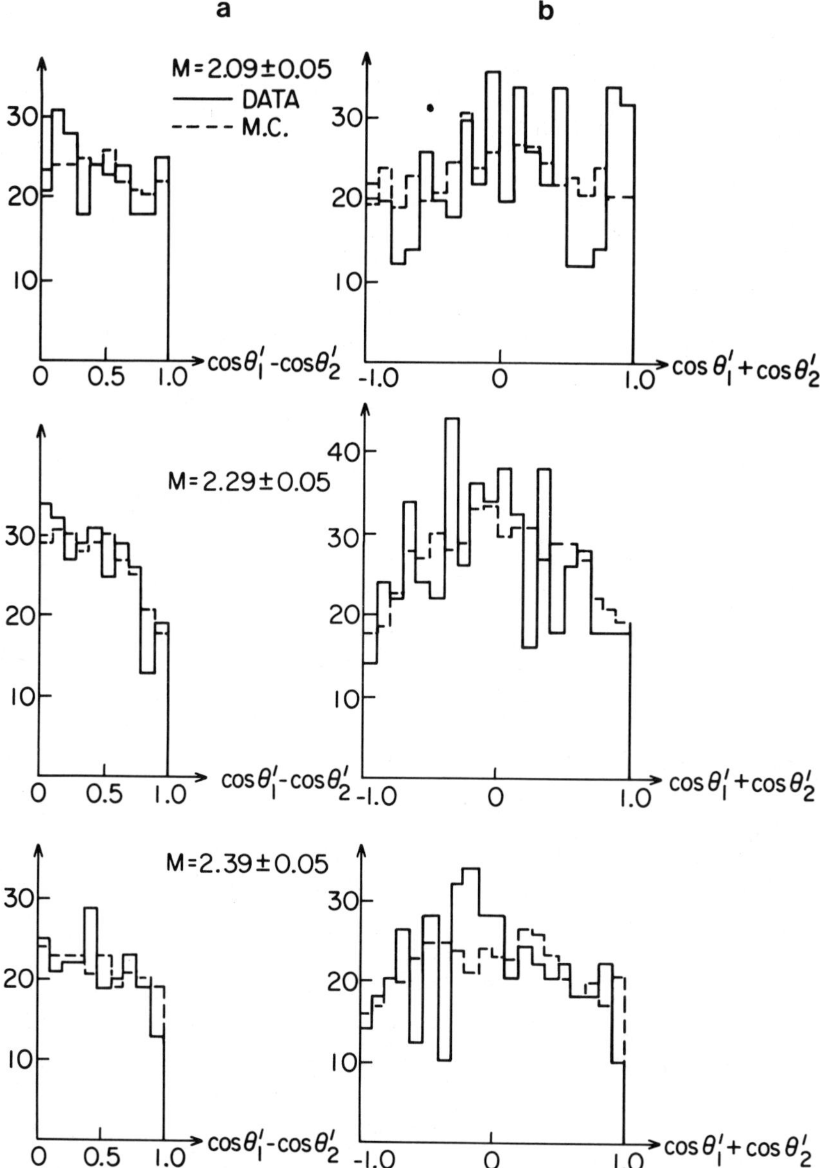

Figure 17: **(a) Comparison** of the data and the acceptance-corrected **Monte Carlo** for $\cos\theta_1' - \cos\theta_2'$.

(b) Comparison of the data and the acceptance-corrected **Monte** Carlo for $\cos\theta_1' + \cos\theta_2'$.

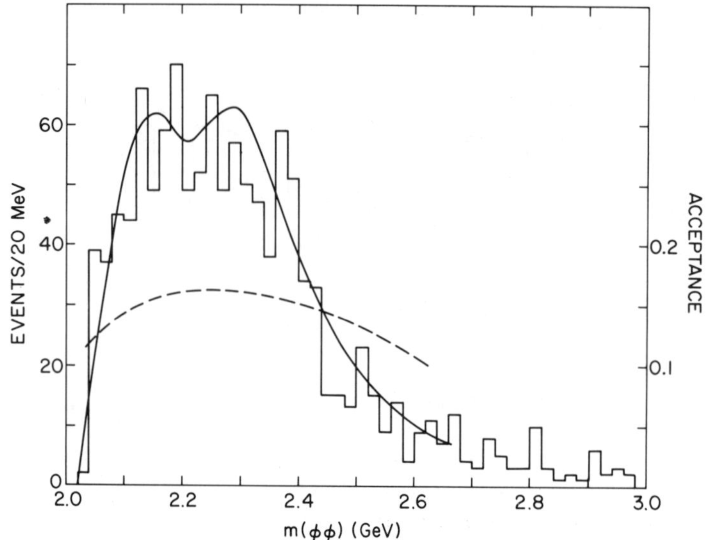

Figure 18: The observed ϕϕ mass spectrum compared to the predicted (solid line curve) mass spectrum from the acceptance-corrected fit. The dashed line is the acceptance.

The mass and full width, partial width ratios, and all the quantum numbers for these two Breit-Wigner resonances are given in Table I. They are at the very least strong glueball candidates due to the breakdown of the OZI suppression, and the striking selectivity of 2 out of 52 possible waves selected.

In fact if one assumes:

1. The correctness of QCD;

2. The universality of the OZI rule (with the necessary caveats) as described previously. Namely, the necessary one-step requirement which is equivalent to the requirement that there is no change in the nature of a gluon exchange. The above is equivalent to the statement that a disjoint Zweig diagram, due to introduction of a new type of $q\bar{q}$ pairs, must involve a multigluon exchange. This leads to OZI suppression due to a weakly coupled multigluon exchange. As I have pointed out previously, a glueball in which the gluons resonate would lead to effectively strongly-coupled glue, and break down the OZI suppression.

This leaves me as the only explanation of the OZI suppression breakdown and the observed selectivity the presence of one or two primary glueballs in the mass region with these quantum numbers. Impure $q\bar{q}$ intermediate states, 4 quark states, etc. are ruled out by the above assumptions (assumption 2).

Why do I say one or two primary glueballs? Because one primary glueball could break down the OZI suppression and possibly mix with a nearby quark state with the same quantum numbers yielding two states very rich in resonating glue. Of course both states could come from different primary glueballs* since we expect that there is a glueball spectrum of states - not just a single glueball.

It should be noted that in a number of papers it was concluded that the width of a glueball should be narrower than hadronic resonances typically by a factor $\sim \sqrt{\text{OZI suppression factor}}$. These considerations were based on treating the quark-glue, glue-glue coupling as weak, and clearly do not apply if the glue-glue coupling becomes strong enough to form a resonance, in which case we are generally dealing with a very strongly interacting multigluon resonance.

In fact the glue-glue coupling is effectively stronger than the quark-glue coupling, and therefore, glueballs should be as wide, or wider than, typical hadronic resonances in the mass region.

In Table II we list some typical resonance widths from the particle data group tables and widths for other glueball candidates. We see that $\Gamma \sim (200\text{-}300) \pm 100$ MeV are reasonable values for glueballs.

MASS AND J^{PC} OF THE GLUEBALLS FROM VARIOUS PHENOMENOLOGICAL APPROACHES

In constituent glueball models[12] due to confinement, the gluon is considered to have an effective mass $m_g \gtrsim 0.75$ GeV.[28]

* They might also eventually dress themselves to some degree with $q\bar{q}$ pairs.

TABLE II

Past Resonance Widths for Some Hadronic Resonances from the Particle Data Group Table[25]

State	$I^G(J^P)C_n$	Full Width Γ in MeV
g(1690)	$1^+(3^-)^-$	200 ± 20
ρ'(1600)	$1^+(1^-)^-$	300 ± 100
f(1270)	$0^+(2^+)^+$	179 ± 20
Resonance Widths for Other Glueball Candidates[9,10,15]		
SLAC ι(1440)	$(0^-)+$	55^{+20}_{-30}
θ(1640)	$(2^+)+$	220^{+100}_{-70}
BNL/CCNY $g_s(1240)$	$0^+(0^{++})$	140 ± 10

Thus we might expect to be in the three-gluon sector. One should note that due to the self-coupling between the gluons and their splittings that a gauge invariant description with a definite number of gluons is not possible. Nevertheless it is physically appealing and reasonable to expect in constituent gluon models that the lowest lying ground state would be mostly composed of 2 gluons and have a mass $\approx 2 \times 0.75$ GeV ≈ 1.5 GeV. One would expect another ground state in the 3g sector mostly composed of 3 gluons with a mass $\approx 3 \times 0.75$ GeV ≈ 2.25 GeV.

The MIT bag calculations of glueballs[29] assume massless gluons and obtain predictions for quantum numbers and masses of various states.[29] The masses do not fit some present glueball candidates. Hyperfine energy shifts that depend on α_s have been put into the bag calculations to allow such fits.[30] Adapting these methods, we have derived m_g for two-gluon states as a function of α_s. The SLAC ι(1440) and θ(1640) glueball candidates, and the BNL/CCNY $g_s(1240)$ glueball candidate, were used as inputs to derive the results shown in Fig. 19. As you can see, we can obtain a $J^{PC} = 2^{++}$ $g_T(2160)$ at about the right mass as an excited state in the 2g

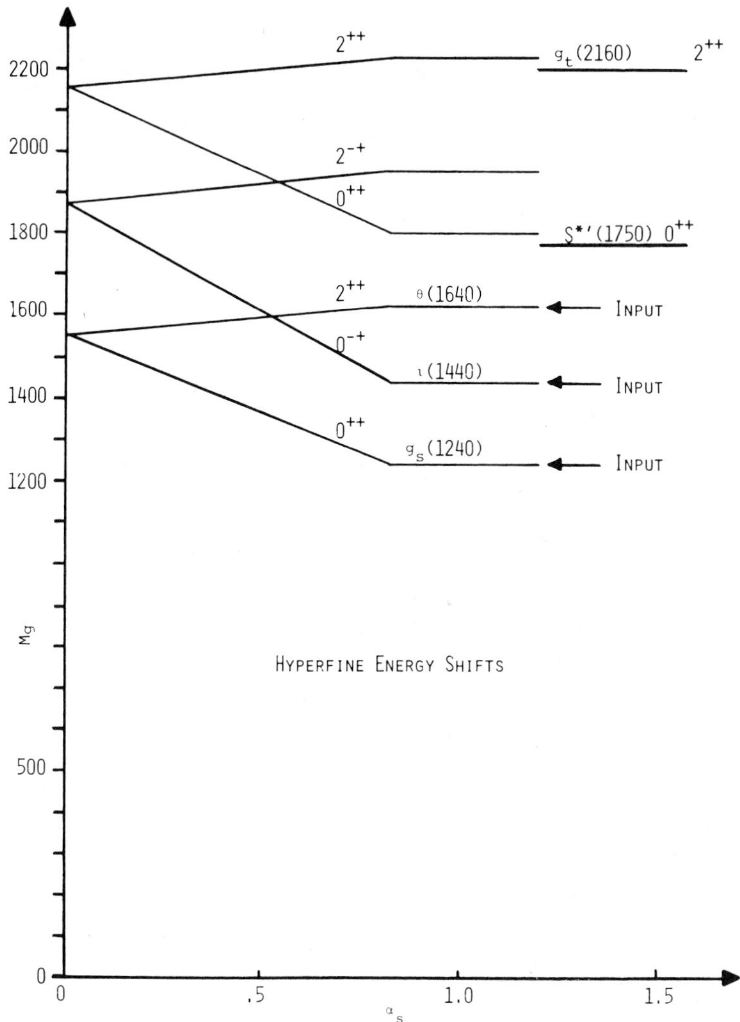

Figure 19: The predicted mass and quantum numbers of the 2g glueballs from adopting the methods of Ref. 30. The $\iota(1440)$, $\theta(1640)$ and $g_s(1240)$ were used as input to determine the overall mass level, the spacing between the levels, and α_s (till break in lines).

sector.* However, the massless assumption for gluons in the bag does not allow $J^{PC} = 2^{++}$ for low-lying 3-gluon states, in contrast to the constituent gluon model which allows all J^{PC} for 3g states and all J^{P+} (i.e., C = +) for 2g states.

So far, lattice calculations[31] have concentrated mainly on the glueball ground state getting $J^{PC} = 0^{++}$, $M \sim 0.8 - 1.0$ GeV. Recently they have begun to attack higher spin states.[31] The work is still preliminary, but indications are that higher spin states could well show up in our mass region. Thus, in summary, one finds that the phenomenological models are generally compatible with our results, except for the possibility of the MIT bag calculations if we are in the 3g sector.

CONCLUSIONS

Given QCD as an Ansatz and OZI (with appropriate restrictions) as a second Ansatz, I conclude we have discovered either two glueballs with characteristics described below, or two states very rich in resonating glue formed from one primary glueball mixing with a nearby quark state of the same quantum numbers, thus forming two states.

The quantum numbers and characteristics of these states are:

	I^G	J^{PC}	Mass (MeV)	Γ (MeV)
$g_T(2160)$	0^+	2^{++}	2160 ± 50	310 ± 70
$g_T(2320)$	0^+	2^{++}	2320 ± 40	220 ± 70

* However, one should be aware that perturbative treatments are not justifiable at high values of α_s.

REFERENCES

1. C.N. Yang and R.L. Mills, Phys. Rev. 96, 191 (1954).
2. a) Fritzch and Minkowski, Nuovo Cimento 30A, 393 (1975).
   b) R.P. Freund and Y. Nambu, Phys. Rev. Lett. 34, 1645 (1975).
   c) R. Jaffe and K. Johnson, Phys. Lett. 60B, 201 (1976).
   d) Kogut, Sinclair and Susskind, Nucl. Phys. B114, 199 (1975).
   e) D. Robson, Nucl. Phys. B130, 328 (1977).
   f) J. Bjorken, SLAC Pub. 2372.
3. A. Etkin, J.J. Foley, J.H. Goldman, W.A. Love, T.W. Morris, S. Ozaki, E.D. Platner, A.C. Saulys, C.D. Wheeler, E.H. Willen, S.J. Lindenbaum, M.A. Kramer, U. Mallik, Phys. Rev. Lett. 40, 422 (1978).
4. A. Etkin, K.J. Foley, J.H. Goldman, W.A. Love, T.W. Morris, S. Ozaki, E.D. Platner, A.C. Saulys, C.D. Wheeler, E.H. Willen, S.J. Lindenbaum, M.A. Kramer and U. Mallik, Phys. Rev. Lett. 41, 784 (1978).
5. S.J. Lindenbaum, Hadronic Physics of $q\bar{q}$ Light Quark Mesons, Quark Molecules and Glueballs. Lecture presented at XVIII Course: The High Energy Limit, 31 July - 11 August 1980, The International School of Subnuclear Physics, Erice (to be published in proceedings); Also BNL 28498, October 1980.
6. S.J. Lindenbaum, Il Nuovo Cimento, 65A, 222 (1981).
7. P. Fishbane, Glueballs, A Little Review. Talk presented at the 1981 ORBIS Scientiae, Ft. Lauderdale, Florida (to be published).
8. D.L. Scharre, Glueballs, A Status Report, ORBIS Scientiae 1982, Miami, Florida (to be published), and SLAC Report 2880 (1982).
9. D. Antreasyan, Y.F. Gu, J. Irion, W. Kollman, M. Richardson, K. Strauch, K. Wacker, A. Weinstein, D.A. Schman, T. Burnett, M. Cavalli-Sforza, D. Coyne, C. Newman, H.F. Sadrozinski, D. Gelphman, R. Hofstadter, R. Horisberger, L. Kirkbride, H. Kolanoski, K. Konigsman, R. Lee, A. Liberman, J. O'Reilly, A. Osterheld, B. Pollock, J. Tompkins, E.D. Bloom, F. Bulos,

R. Chestnut, J. Gaiser, G. Godfrey, C. Kiesling, W. Lockman, M. Oreglia and D.L. Scharre, Phys. Rev. Lett. 49, 259 (1982).

10. a) M. Chanowitz, Phys. Rev. Lett. 46, 981 (1981).
b) C.E. Carlson, J. Coyne, P.M. Fishbane, F. Gross and S. Meshkov, Phys. Lett. 98B, 110 (1981).

11. J.F. Donoghue, K. Johnson and B. Li, Phys. Lett. 99B, 416 (1981).

12. a) C. Carlson, J. Coyne, P. Fishbane, F. Gross, S. Meshkov, Phys. Rev. D23, 2765 (1981).
b) J. Coyne, P. Fishbane and S. Meshkov, Phys. Lett. 91B, 259 (1980); C. Carlson, J. Coyne, P. Fishbane, F. Gross and S. Meshkov, Phys. Lett. 99B, 353 (1981).

13. S.J. Lindenbaum, Proc. Sixteenth Rencontre De Moriond, "New Flavours and Hadron Spectroscopy," Vol. II, pg. 187, Ed., J. Tran Thanh Van (Editions Frontieres, France, 1981); H.J. Schnitzer, ibid. pg. 648; K.R. Schubert, ibid. pg. 635.

14. S.J. Lindenbaum, "Evidence for Glueballs", Proc. 1981 EPS Int. Conf. on High Energy Physics, July 9-15, 1981 (Calouste Gubenkian Foundation, Av. Berne, Lisbon, Portugal) (to be published); F.E. Close, "Glueballs, Hemaphrodites and QCD Problems for Baryon Spectroscopy", ibid.

15. A. Etkin, K.J. Foley, R.S. Longacre, W.A. Love, T.W. Morris, S. Ozaki, E.D. Platner, V.A. Polychronakos, A.C. Saulys, Y. Teramoto, C.D. Wheeler, E.H. Willen, K.W. Lai, S.J. Lindenbaum, M.A. Kramer, U. Mallik, W.A. Mann, R. Merenyi, J. Marraffino, C.E. Roos, M.S. Webster, Phys. Rev. D25, 2446 (1982).

16. E. Bloom, talk presented at the XXI Int. Conf. on High Energy Physics, Paris, France, July 26-31, 1982 (to be published); J. Donoghue, ibid.

17. a) J.F. Donoghue, Experimental Meson Spectroscopy - 1980, Sixth Int. Conf., Brookhaven National Laboratory, April 25-26, 1980, Eds. S.U. Chung and S.J. Lindenbaum, AIP Conf. Proc. #67, Particles and Fields Subseries #21, pg. 1040.

b) G. Bhanot, Phys. Lett. 101 B, 95 (1981).

c) G. Bhanot and C. Rebbi, Nuc. Phys. B180, 469 (1981).

d) H. Hamber and G. Parisi, Phys. Rev. Lett. 47, 1792 (1981).

18. a) S. Okubo, Phys. Lett. 5, 165 (1963); G. Zweig, CERN Report TH 401 and 412 (1964); J. Iizuba, Prog. Theor. Phys. Suppl. 37-38, 21 (1966); J. Iizuba, K. Okuda and O. Shito, Prog. Theor. Phys. 35, 1061 (1966); S. Okubo, "A Survey of the Quark Line Rule," Univ. Rochester Report UR 641 (1977).

b) S. Okubo, Phys. Rev. D 16, 2336 (1977).

c) T. Applequist, K. Kane and M. Barnett, Ann. Rev. Nucl. Sci. 28, 387 (1978).

d) I.J. Muzinich and F.E. Paige, Phys. Rev. D 21, 1151 (1980).

e) S.J. Lindenbaum, "Quark Line Diagrams, Rules, and Some Recent Data", BNL 50812 (December 1977).

19. T.A. Armstrong et al., Nucl. Phys. B 196, 176 (1982); C. Daum et al., Phys. Letts. 104B, 246 (1981). The narrow enhancement in ϕ pairs produced from Be as reported in the latter paper is not confirmed by higher statistics results reported by C. Daum et al., and by B.R. French et al. at the XXI Int. Conf. on High Energy Physics, Paris, France, July 26-31, 1982 (to be published).

20. S. Eiseman, A. Etkin, K.J. Foley, R.S. Longacre, W.A. Love, T.W. Morris, S. Ozaki, E.D. Platner, V.A. Polychronakos, A.C. Saulys, C.D. Wheeler, S.J. Lindenbaum, M.A. Kramer, Y. Teramoto. "The MPS II Drift Chamber System", paper submitted to the IEEE 1982 Nuclear Science Symposium, October 20-22, 1982, Washington D.C. (to be published in IEEE Trans. on Nucl. Sci.).

21. S.J. Lindenbaum, C. Chan, A. Etkin, K.J. Foley, M.A. Kramer, R.S. Longacre, W.A. Love, T.W. Morris, E.D. Platner, V.A. Polychronakos, A.C. Saulys, Y. Teramoto, C.D. Wheeler, "A New Higher Statistics Study of $\pi^- p \to \phi\phi n$ and Evidence for Glueballs". Paper presented at the XXI Int. Conf. on High Energy Physics, Paris, France, July 26-31, 1982 (to be published).

22. H. Fritzch and M. Gell-Mann, XVI Int. Conf. on High Energy Physics, Chicago-Batavia, 1972, Vol. 2, p. 135; H. Fritzch, M. Gell-Mann and H. Leutwyler, Phys. Lett. 47B, 365 (1973); S. Weinberg, Phys. Rev. Lett. 31, 494 (1973); S. Weinberg, Phys. Rev. D8, 4482 (1973); D.J. Gross and F. Wilczek, ibid, 3633 (1973).

23. D. Cohen et al., Phys. Rev. Lett. 38, 269 (1977).

24. D.S. Ayres et al., Phys. Rev. Lett. 32, 1463 (1974).

25. Particle Data Group Tables, Phys. Letts. 111B, 12 (1982).

26. S.J. Lindenbaum and R.M. Sternheimer, Phys. Rev. 105, 1874 (1957); 106, 1107 (1957); 109, 1723 (1958); 123, 333 (1961).

27. T. Armstrong et al., CERN EP/82-103 (1982).

28. G. Parisi and R. Petronzio, Phys. Lett. 94B, 51 (1980).

29. J.F. Donoghue, K. Johnson and Bing An Li, CTP #891 UN HEP-139 October 1980.

30. T. Barnes, F.E. Close and S. Monoghan, Phys. Lett. 110B, 159 (1982).

31. C. Rebbi, talk presented at the XXI Int. Conf. on High Energy Physics, Paris, France, July 26-31, 1982 (to be published).

DISCUSSION

CHAIRMAN: S.J. Lindenbaum

Scientific Secretaries: S. Errede and D.M. Errede

DISCUSSION

- *WAGNER:*

What does the t-distribution look like? Is it consistent with pion exchange?

- *LINDENBAUM:*

The t-distribution agrees well with what one expects from pion exchange.

- *HOFSÄSS:*

What do the model calculations that you mentioned give for the glueball widths?

- *LINDENBAUM:*

In the Bag Model, the calculations for even the glueball masses are uncertain, and the decay widths are not determined. As far as the calculations of Meshkov, Fishbane, et al. are concerned, they do not take the masses in their calculations too seriously. They just take the masses of the confined gluons and add them to get the glueball mass. Their predictions for the decay widths are based on perturbation theory which may not be applicable for a glueball resonance, especially if J^{PC} is not exotic.

- *PERNICI:*

Couldn't the glueball candidate be really a color singlet $q^2\bar{q}^2$ exotic state?

- *LINDENBAUM:*

Our states do not have exotic J^{PC}. Since simplicity is the name of the physics game when it works and there is not serious evidence for cryptoexotic $q^2\bar{q}^2$ states, why assume this. If they existed, the OZI rule could not be universal -- which is generally assumed. There is no reason why the resonance width should be so for $q^2\bar{q}^2$. The principal reason why the glueball candidate cannot be a $q^2\bar{q}^2$ is that the OZI rule will not be violated for an ordinary $q^2\bar{q}^2$ exotic state.

- *WIGNER:*

Can your theory reproduce the positions of some of the resonances on the energy scale?

- *LINDENBAUM:*

No, the positions of the resonances for glueballs would have only second order impact on the understanding of $q\bar{q}$ resonances that I can foresee at this time. The MIT Bag people have done a lot of work on the positions of resonances; our results may encourage them to re-evaluate how they do their work at some point in the future.

- *ZICHICHI:*

I believe Prof. Wigner's questions was: Can you predict the position of the glueball resonances? The answer, I think is no. Probably the key point is: you observe an effect which is OZI-forbidden by a large factor. If it was not for this, you would not observe the effect. Is this right?

- *LINDENBAUM:*

Yes.

- *KAPLUNOVSKY:*

What are the main decay channels for the glueballs you have found, besides $\phi\phi$?

- *LINDENBAUM:*

The glueball is flavor neutral and in principle can decay into many channels. The $\phi\phi$ channel is one which is experimentally very clean due to OZI suppression whereas for other channels, the chances of finding a glueball signal in them is very small.

- *DEVCHAND:*

I have a question which perhaps Prof. Yang could answer. Could glueballs actually be solutions of the loop space equations for the pure gauge theory?

- *YANG:*

That's too difficult a question, I don't know how to answer that.

Since I have the floor, I would like to ask Prof. Lindenbaum a question: Can you vary the energy of the π^- beam and make a similar analysis which might increase the credibility of your results?

- *LINDENBAUM:*

Yes, in principle we could, although it is a matter of how much running time we could get to do the experiment. However, I'm not convinced that it would add to the credibility of our results, because the partial wave analysis depends only peripherally on the beam energy. Our present plans are to gather more statistics at the same beam energy. If one goes to higher beam energies, the flux drops down; if one goes to lower energies, the kinematic range drops down for this exclusive channel.

- *ZICHICHI:*

A question for Prof. Yang: are there any theoretical predictions for the lowest glueball mass?

- *YANG:*

No. The calculations are very difficult and there are many unknowns.

- *KLEINERT:*

In the Bag picture, the QCD vacuum is a condensate of colored magnetic monopoles with constant pressure which squeezes the color electric flux between two opposite charges into a small bag whose volume is proportional to the square of the flux. Correspondingly, the energy is roughly proportional to the square of the charges. Now the quarks in the lowest ordinary meson (the pion) have fractional charge, whereas the gluons in a glueball have integer charge. This leads to the estimate for the lowest glueball mass of $M_G \simeq 3^2 M_\pi$.

- *HEUSCH:*

A principal feature of gluonium states is flavor independence.

So, a simple expectation tied to the observed candidate gluonium state decaying into $\phi\phi$ would be that this state should be seen in other channels. One such channel, which does not even necessarily imply different flavors is $\eta\eta$. The Crystal Ball experiment has precise data on $\psi \to \gamma\eta\eta$ and does not observe a gluonium state at 2.1 GeV or 2.3 GeV. Shouldn't it?

- *LINDENBAUM:*

For one thing, if we assume that this state we postulate comes from the three-gluon sector, it cannot be seen in $\psi \to \gamma G$. If it belongs to the two-gluon sector it may be harder to observe than the $\theta(1640)$ due to its higher mass and I believe low energy kaons are hard to detect in the SLAC MK II apparatus. But I have not looked into this question in detail.

- *S. ERREDE:*

A short comment. The decay of an S-wave glueball state ($J^{PC}LSM = 2^{++}000$) cannot go to $\eta\eta$ because the η is a pseudoscalar particle ($J^{PC} = 0^{-+}$). The decay of a D-wave glueball state ($J^{PC}LSM = 2^{++}200$) to $\eta\eta$ will be suppressed by the angular momentum barrier. Thus, it would be unlikely for the Crystal Ball to have observed these two states in $\psi \to \gamma G$ through the $\eta\eta$ decay channel.

- *WAGNER:*

Do you really need both resonances? The D-wave looks like a nice Breit-Wigner, whereas the S-wave does not.

- *LINDENBAUM:*

Yes, both are required. The two resonance fit is much better than a single resonance fit by more than 10 standard deviations, primarily due to the phase difference.

- *RICHARDSON:*

Have you tried repeating your analysis with uncorrelated $\phi\phi$ pairs, i.e. by combining ϕ's from different events together and forming for example, the $\phi\phi$ mass spectrum?

- *LINDENBAUM:*

No, we have not tried this because it is not relevant to our results. Furthermore those so-called uncorrelated $\phi\phi$ pairs are, in my opinion, quite correlated.

- *MARX:*

The spin 1/2 - spin 0 system partial wave analysis is known to have many ambiguities in J^P assignments. I would expect a spin 1 - spin 1 system to have similar ambiguities. How do you know that your assignments are unique?

- *LINDENBAUM:*

We stopped with J = 4, L^{max} = 3 and let all other quantum numbers vary. In that sense it is clearly unique because there is nothing in the remaining 50 waves which gives anything like the signatures we obtain for α_1 and $\alpha_1 - \alpha_2$ and acceptable results for other variables. To go to higher spins is another story. We are relatively near threshold for this reaction and therefore it is unlikely to produce a high angular momentum state at this energy.

- *S. ERREDE:*

How sensitive are your results to changes in cuts, for example, varying the cuts made in obtaining the $\phi\phi$ mass distributions? Have you investigated this?

- *LINDENBAUM:*

Our experiment is as cut-less as possible. The only cuts made are those in obtaining the $\phi\phi$ signal and the cut on the neutron. The result is entirely insensitive to those cuts. The ϕK^+K^- background is the most sensitive to these cuts, but is small. With the statistics we have, as best we can tell, this background is completely incoherent and small. Therefore, it would not affect our analysis at all.

HADROPRODUCTION OF HEAVY FLAVOURS

F. Muller

CERN
Geneva, Switzerland

1. INTRODUCTION

Hidden charm in the form of the J/ψ was discovered in 1974[1], naked charm in the form of the D meson in 1976[2] (though the first example[3] of Λ_c^+ was probably seen in 1975), hidden beauty in the form of the Υ in 1977[4], and naked beauty through its decay kaons and electrons in 1980[5]. Even though the partner of the beauty, the top, is still to be discovered, and heavier flavours may exist, there is now a tendency, in these days when enterprising people are hunting for intermediate vector bosons, Higgs bosons, SUSY particles, proton decay, neutrino oscillations, etc., to take the world of heavy flavours for granted, or at best as a subject for e^+e^- colliders. However, there are several compelling reasons to pursue a vigorous effort in this field with proton machines:

i) A heuristic reason is that, contrary to prevailing prejudice, most of the first evidence for new flavours was found, directly or indirectly, in hadronic interactions. This was the case for pions and strange particles (cosmic rays) and all their resonances (bubble chambers) -- for lack of competition, one may say -- but, more recently, this was true also for the J, the Υ, the Λ_c, the Σ_c, the F (as opposed to the Ψ, D, D^*, and B for the e^+e^- colliders) and there are still many charmed particles and all of the B particles to find! As for the naked top the competition is open between the SPS $p\bar{p}$ collider and the boosted-up PETRA. If the mass of the top is too high and/or if heavier flavours exist, only the $p\bar{p}$ colliders (at CERN or FNAL) remain in the game -- until the advent of TRISTAN, SLC, and LEP.

ii) Another heuristic reason is that fixed-target experiments with high-energy beams (γ or hadrons) are the only way, thanks to the Lorentz dilation of time, to measure very small lifetimes, as are expected for beauty particles ($\tau \lesssim 10^{-14}$ cm $\rightarrow \lambda \lesssim 100$ μm at p $\simeq$ 150 GeV/c). In the same line of thought, hyperon and kaon beams should be a good way to produce charmed-strange particles (A, F^+, ...).

iii) Finally, a more profound motivation to study hadroproduction of heavy flavours lies in the fact that it falls in a domain of relatively large Q^2's where QCD should apply -- and hence, as for the Drell-Yan process or the high-p_T phenomena, QCD should have a predictive power which can be tested by experiment. As we shall see, this is not yet quite the case for charm production, though qualitatively our understanding has progressed a lot over the last two years; experimental data on beauty production, which involves higher Q^2's, should be extremely valuable, as well as more complete results on charm production versus energy.

These lectures are essentially devoted to the third point and will present the latest results on charm hadroproduction and the evolution in theoretical ideas which they have brought about. In the light of this confrontation, the possibility of finding b and t particles in hadronic interactions [point (i)] will be discussed. But first, new results on properties of heavy-flavoured particles, originating either from proton [point (ii)] or e^+e^- machines, will be presented, since they have an important bearing on the interpretation of production data or on design of new experiments -- and also for their own interest.

2. PROPERTIES OF HEAVY-FLAVOURED PARTICLES

2.1 Charmed Particles

The experimental results available at the end of 1980 have been excellently summarized by Trilling[6]. The data were rather extensive for the D^+, D^0, and D^* mesons, meagre for the Λ_c and Σ_c, scanty and controversial for the F meson, and non-existing for other charmed particles. This 1980 information on mass, decay modes, and branching ratios is still valid; on these points nothing has been added to our knowledge of the D mesons, and rather little to that of the charmed baryons, but a few decay channels of the F meson seem to be identified now with more certainty. The main progress lies in the better determination of lifetimes, especially of the D^0 and D^+ mesons, owing to the increased use of new (or revived) techniques.

Since results of several of those experiments will be mentioned at various places -- and also because some of these techniques should be more and more important in the near future -- let us describe them briefly. All use a vertex detector able to separate the interaction point from the decay apex of the charmed particle and to measure the distance of flight $\ell = c\tau p/m \simeq 300\ \mu m$ for $p = 20$ GeV/c and $\tau = 10^{-13}$ s (however, up to now, this characteristic property of charm has not been used as a trigger). The various techniques used are:

i) Visual identification of decays, either with emulsions, high-resolution bubble chambers, or streamer chambers, usually followed by a spectrometer measuring and (ideally) identifying the secondary particles.

The emulsion experiments are E531 [7], in a neutrino-beam at FNAL, and WA58 [8] in a 40-70 GeV/c photon beam at CERN (the spectrometer is the Ω). An example of $\Lambda_c\bar{D}^0$ production in WA58 is shown in Fig. 1; the Λ_0 from the Λ_c is seen in the Ω.

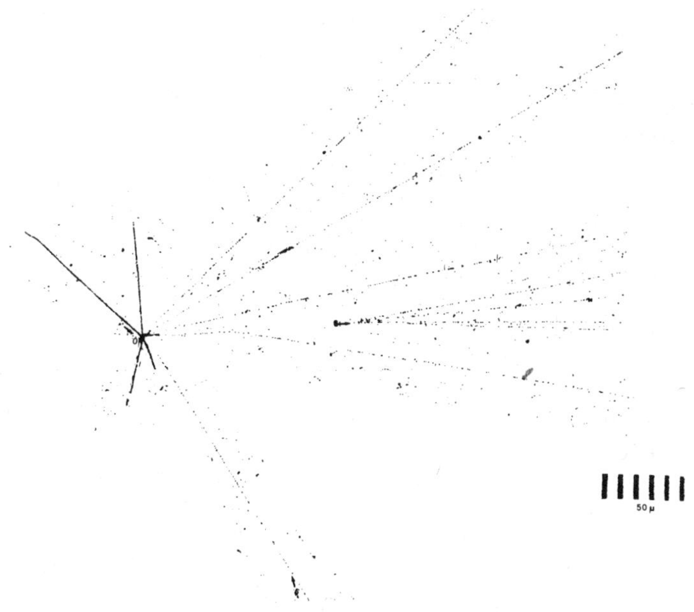

Fig. 1. A $\Lambda_c\bar{D}^0$ event in emulsion (Ref. 8).

The bubble-chamber experiments are BC73 [9], using the SLAC 40 ft hydrogen chamber in a 19.5 GeV/c photon beam, NA16 [10], where LEBC, a small rapid-cycling hydrogen chamber in front of the (incomplete) EHS spectrometer, was exposed to 340 GeV/c π's and 360 GeV/c protons, and NA18 [11], which exposed BIBC, a very small heavy liquid bubble chamber followed by a streamer chamber, to a 340 GeV/c π^- beam. Figure 2 shows the photoproduction of a charmed pair, as seen in experiment BC73.

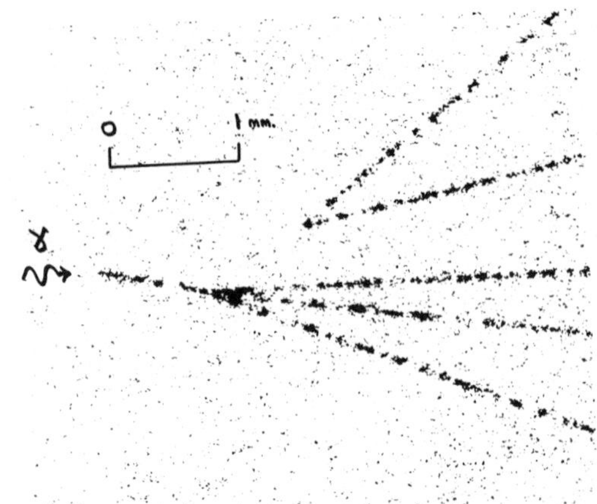

Fig. 2 A charmed pair in a bubble chamber (Ref. 9)

ii) Detection of the increase of charged multiplicity due to D decay(s), downstream from the interaction. This was achieved by the photoproduction experiment NA1 [12], using an active silicon target in front of the FRAMM spectrometer. Figure 3 clearly shows two steps $\Delta n_{ch} = 2$, as expected from two successive charm decays (however background from secondary interactions and photon conversion is not negligible). Quite a few groups plan to use a similar technique in the years to come -- and possibly to incorporate it in a "charm trigger".

iii) Reconstruction of vertices (interaction and decay) by extrapolation of tracks measured in precise vertex detectors. For the moment, this approach has been used only by the MARK II group[13] with the help of a small high-precision ($\sim$ 50 μm) cylindrical drift chamber surrounding the vacuum tube of PEP. Extrapolation gives a longitudinal error of about 700 μm on the decay vertex, good enough to measure the lifetime of the τ lepton[14]: $\tau(\tau) = 3.31 \pm 0.57\ (\pm 0.6) \times 10^{-13}$ s, in agreement with the theoretical value ($\sim 2.8 \times 10^{-13}$ s). Several groups are working on vertex detectors built of silicon microstrips, with which a resolution of several microns can in principle be achieved; a more ambitious goal would be, with the help of fast processors, to use them in the trigger.

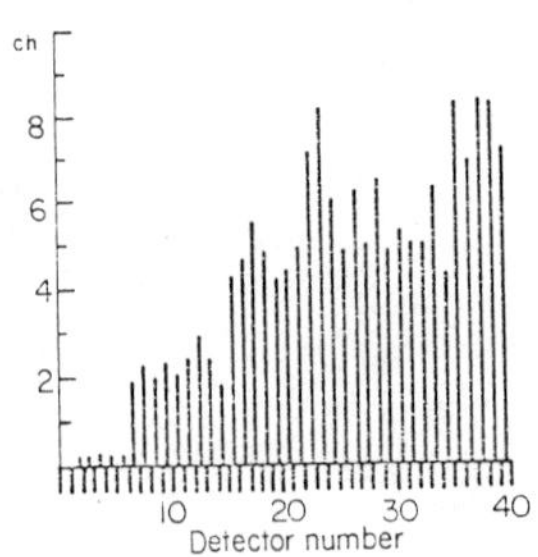

Fig. 3 Multiplicity steps in a silicon active target (Ref. 12).

2.1.1 The D mesons. Table 1 gives the properties of the D mesons most useful to the experimentalist.

Except for the lifetimes, all values are from Ref. 6, in which branching ratios for a few other exclusive decay channels can be found (all of the order of a few per cent, with rather large errors). The mean charged multiplicity is 2.16 ± 0.16 for the D^+ and 2.46 ± 0.14 for the D^0. Finally, the measured inclusive branching ratios (BR) of the D mesons into $\bar{K}$ are, within errors, compatible with the expected value $\cos^2 \theta_c$ ($\sin \theta_c \simeq 0.22$) -- i.e. Cabibbo-favoured decays do dominate.

Table 1. Masses, Lifetimes, and Branching Ratios (BR) of D Mesons

	M (MeV)	τ (10^{-13} s)	BR(D → e^+)	Main Hadronic Channels	BR (%)
D^0	1863.7 ± 0.4	$4.0^{+1.2}_{-0.9}$	< 5	$K^-\pi^+$	2.6 ± 0.4
			5.5 ± 3.7	$K^0\pi^+\pi^-$	3.9 ± 0.9
D^+	1868.4 ± 0.4	$9.3^{+2.7}_{-1.8}$	24 ± 3.7	$\bar{K}^0\pi^+$	1.8 ± 0.5
			16.8 ± 6.4	$K^-\pi^+\pi^+$	4.7 ± 0.8
D^{*0}	2007.0 ± 1.2			$D^{*0} \to \gamma D^0$	∿ 40
				$D^{*0} \to \pi^0 D^0$	∿ 60
D^{*+}	2010.1 ± 0.8			$D^{*+} \to \pi^+ D^0$	∿ 63
				$D^{*+} \to \pi^0 D^+$	∿ 30
				$D^{*+} \to \gamma D^+$	∿ 7

Quoted lifetimes are the world averages of the recent experiments described above as reported at the 1982 Paris Conference[15]. The results of individual experiments are given in Fig. 4, and are seen to agree fairly well with each other (as a comparison, at the 1980 Madison Conference, the result of the statistically most significant experiment, E531, was $\tau_0 = 1.0^{+0.43}_{-0.27}$ s and $\tau^+ = 10.3^{+10.5}_{-4.0}$ s). This leads to a ratio

$$R = \tau(D^+)/\tau(D^0) = 2.2^{+0.9}_{-0.6} \tag{1}$$

(instead of 10 or 5, two years ago). As observed by Treiman and Pais[16], for Cabibbo-favoured semi-leptonic decays, which are isospin 0 transitions between two isospin ½ states, the relation $\Gamma(D^+ \to \ell^+ \ldots) = \Gamma(D^0 \to \ell^+ \ldots)$ should hold, and hence

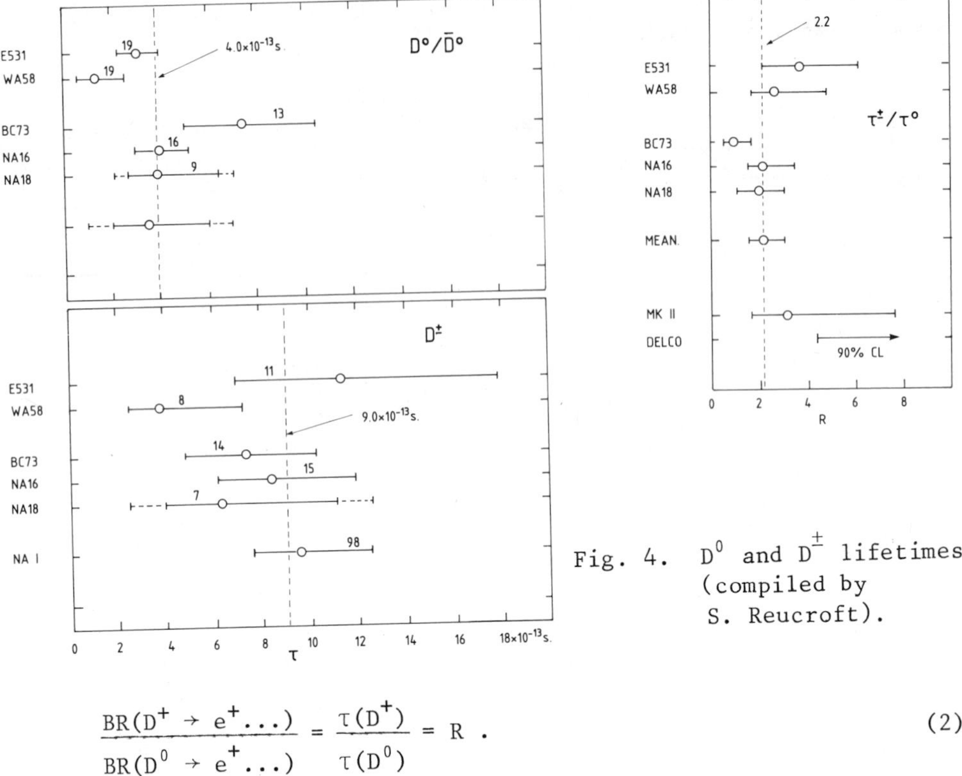

Fig. 4. D^0 and $D^{\pm}$ lifetimes (compiled by S. Reucroft).

$$\frac{BR(D^+ \to e^+ \ldots)}{BR(D^0 \to e^+ \ldots)} = \frac{\tau(D^+)}{\tau(D^0)} = R \ . \tag{2}$$

Results on BR from two different experiments at SPEAR are given in Table 1; obviously more precise data are needed to check relation (2). On the other hand, the data on the inclusive electron energy spectrum from D decay, from both SPEAR and DORIS, are precise enough and allow it to be interpreted in terms of an $\sim$ 50-50 mixture of $Ke\nu$ and $K^*e\nu$ final states[6].

At this point, it is necessary to stress that the commonly used value $BR(D \to e^+) \simeq 8\%$ applies to the mixture of various charmed particles, as produced in e^+e^- collisions with $3.77 < \sqrt{s} < 8$ GeV. If, as is the case at the Ψ'' mass (M = 3.77 GeV), only D^0 and D^+ are produced, in approximately equal numbers, one can then derive -- with all due caution! -- $BR(D^0 \to e^+) \simeq 5\%$ and $BR(D^+ \to e^+) \simeq 11\%$, as values which could be used in the analysis of experimental data on D^0 and D^+ production, when triggered by an electron (or a muon). These values of $BR^{\pm}$, together with the measured values of $\tau^{\pm}$ lead to $\Gamma(D \to e^+) \simeq 1.2 \times 10^{11}\ s^{-1}$, in good agreement with the theoretical[17] value $\Gamma \simeq (G_F^2/192\pi^3)M_c^5$, with $M_c \simeq 1.5\ GeV/c^2$. This value is calculated using the so-called spectator model, illustrated in Fig. 5a for Cabibbo-favoured decays. That model predicts identical lifetimes for D^+, D^0 (and also F and Λ_c). The observed value R = $= \tau_{D^+}/\tau_{D^0} \simeq 2$ is qualitatively explained by the presence for the D^0

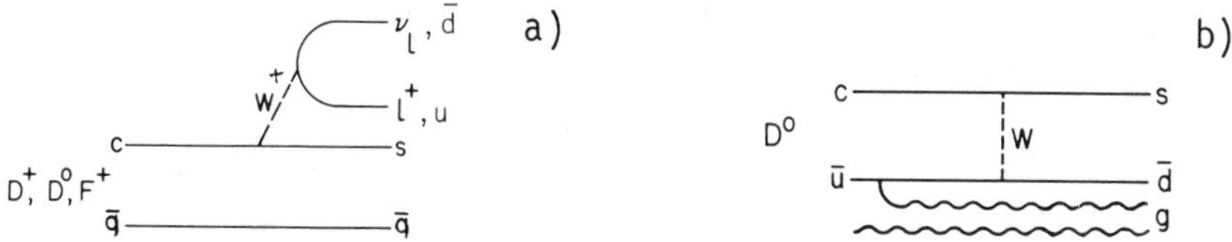

Fig. 5. a) Spectator model diagram; b) Decay of D^0 by W exchange (Ref. 6).

of a W-exchange diagram (Fig. 5b), which does not exist (for Cabibbo-favoured transitions) for the D^+, and which leads to extra hadronic final states. (Gluon radiation is necessary to avoid suppression by helicity arguments.)

2.1.2 <u>The F meson.</u> Evidence for the F^+ meson (c, s, u) lies mainly (apart from a few well-reconstructed events from the emulsion-neutrino experiment E531[7] and some, not unambiguous, events in the bubble-chamber experiments at CERN[10,11]) in mass peaks observed in photoproduction experiments. Figure 6 shows the $\eta(5\pi)$, $\eta(3\pi)$, and $\eta\pi$ peaks seen in experiment WA14[18] with the 40-70 GeV γ-beam in the CERN Ω spectrometer, and Fig. 7 the $\eta(4\pi)$ and KK(2π) peaks in experiment NA1 [12], already described. Mass ($M \simeq 2.05$ GeV/c) and branching ratios are still poorly known.

Lifetime measurements from four of the five experiments just quoted are given, together with those of the Λ_c, in Fig. 8. The average is $\tau(F) = (2.9 {}^{+1.8}_{-0.9}) \times 10^{-13}$ s subject to an uncertainty due to a possible misidentification of F decays, especially in the bubble-chamber experiments; the same remark applies to Λ_c decays, for which $\tau(\Lambda_c^+) = (2.2 {}^{+0.9}_{-0.5}) \times 10^{-13}$ s.

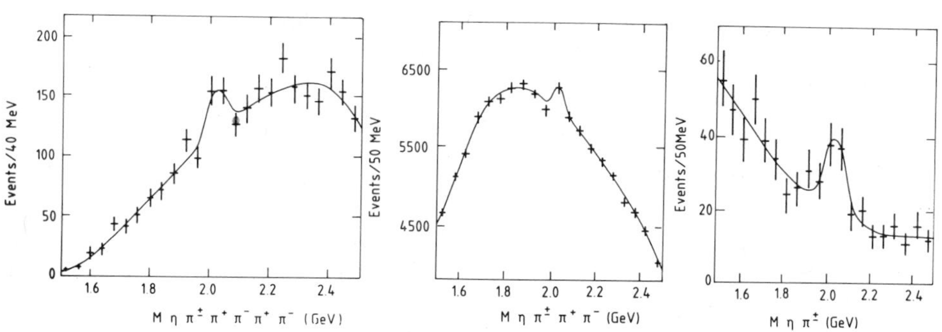

Fig. 6. Some evidence for F photoproduction (Ref. 18).

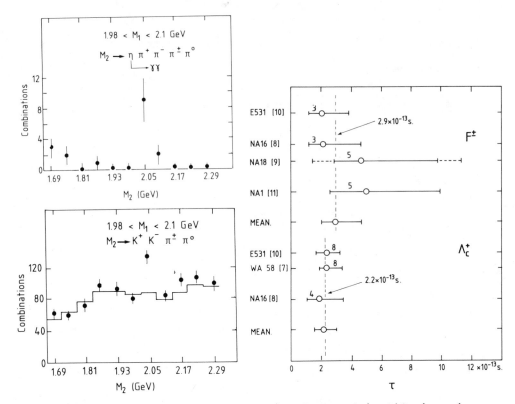

Fig. 7 Other evidence for F photoproduction (Ref. 12)

Fig. 8 F and Λ_c lifetimes (compiled by S. Reucroft)

With this caveat in mind, one notes that the F and Λ_c lifetimes seem to be smaller than that of the D^+ ($\sim 9 \times 10^{-13}$ s). As in the case of the D^0, this may be explained by annihilation or exchange diagrams, as shown in Figs. 9a and b for Cabibbo-favoured transitions. Hence one expects $\tau(F) \approx \tau(\Lambda_c) \simeq \tau(D^0) < \tau(D^+)$, in agreement with present data [more detailed theoretical considerations lead to the expectation of $\tau(F) > \tau(D) > \tau(\Lambda_c)$]. It should be remarked that

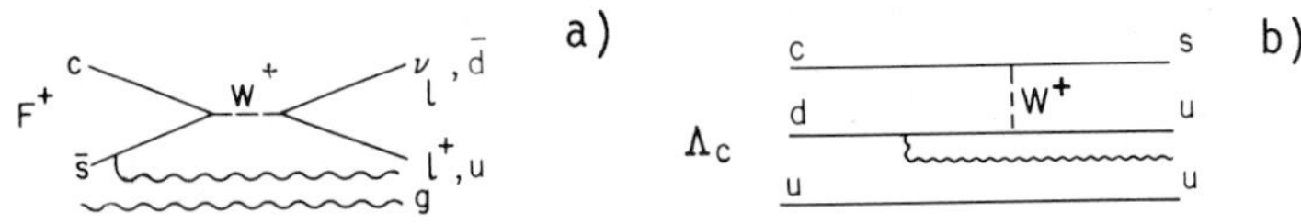

Fig. 9. a) F^+ annihilation diagram, b) Λ_c exchange diagram (Ref. 6).

the dominant final states resulting from the annihilation diagram (Fig. 9a) will not be $s\bar{s}$ like those expected from the spectator model in Cabibbo-favoured transitions (Fig. 5a). Experimentally the proportion of $s\bar{s}$ final states ($K\bar{K}$, $\eta\pi\pi$, ...) in F decays is still an open question.

2.1.3 Charmed baryons. The best known is the singlet Λ_c (c, u, d), with the following characteristics[6]:

$$M(\Lambda_c) = 2285 \pm 5 \text{ MeV}$$

$$BR(\Lambda_c \to K^- p\pi^+) = (2.2 \pm 1.0)\% \ .$$

Other observed decay modes ($K^0 p$, $\Lambda\pi^+$, $\Sigma^0\pi^+$) seem to have branching ratios of roughly one-half the above value. Upper limits of suspected decay modes ($\Lambda^0\pi^+\pi^+\pi^-$, $K^0 p\pi^+\pi^-$) are of the same order. The bulk of the Λ_c decay modes are consequently unknown.

A new result[19] has come from MARK II, using a procedure similar to that which gave the branching ratio into $K^- p\pi^+$:

$$BR(\Lambda_c \to e^+ \ldots) = (4.5 \pm 1.7)\%$$

including

$$BR(\Lambda_c \to pe^+ \ldots) = (1.8 \pm 0.9)\%$$

$$BR(\Lambda_c \to \Lambda^0 e^+ \ldots) = (1.1 \pm 0.9)\% \ .$$

The Σ_c (I = 1) has been seen via its (strong) decay modes: $\Sigma_c^{++} \to \Lambda_c^+\pi^+$ and $\Sigma_c^+ \to \Lambda_c^+\pi^0$. Its mass is such that $M(\Sigma_c) - M(\Lambda_c) = 168 \pm 3$ MeV, as compared with an expected theoretical value of ~ 160 MeV.

At the time of that write-up, preliminary evidence was presented for the charmed strange hyperon A^+ (c, s, u), in the form of an $\sim 5\ \sigma$ peak in the $\Lambda K^-\pi^+\pi^-$ mass spectrum (Fig. 10) obtained[20] with a spectrometer in a 135 GeV/c Σ^- beam [however, the corresponding production cross-section seems rather large and the A^0 (c, s, d) member of the doublet has not been seen].

2.2 Beauty

Since the first hints[5] of naked beauty in e^+e^- collisions, via formation of a broad resonance Υ''' with abundant K and electron yields (as expected from $\Upsilon''' \to B\bar{B}$, $B \to C \ldots$, $C \to K \ldots$, and

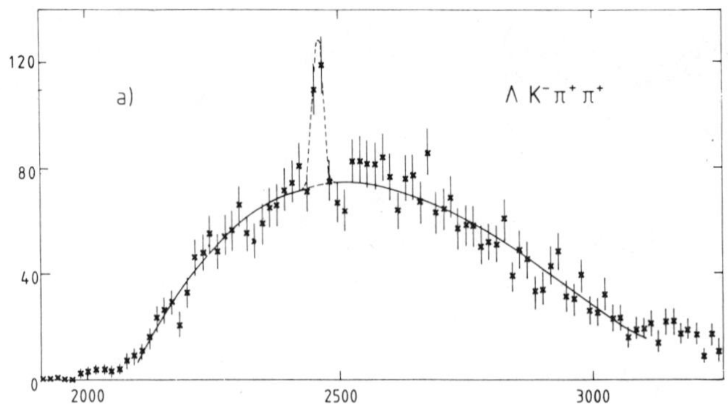

Fig. 10. First evidence for the A^+ baryon (Ref. 20).

$B \to C + e + \ldots$), some properties of the B meson have been more precisely measured (see Ref. 15 for more details).

From the non-observation of the ~ 50 MeV γ's expected to arise from $B^* \to B\gamma$, leading to the inequality $2M_B > M(\Upsilon''') - 50$ MeV, together with the obvious inequality $2M_B < M(\Upsilon''')$, one estimates $M(B) = (5256 \pm 7)$ MeV.

The energy spectrum of the decay electrons has a shape in agreement with that predicted for a decay $B \to e^-\bar{\nu}X$, with $M(X) \approx 2$ GeV, as expected if the favoured decay mode is $b \to ce^-\bar{\nu}$ rather than $b \to ue^-\bar{\nu}$. More quantitatively, from the decay spectrum, $|b \to u|^2/|b \to c|^2 < 0.042$. The branching ratio for electron decay (average of four experiments) is found to be $(13 \pm 2)\%$, and $BR(B \to \mu^- \ldots)$ has a comparable value.

The dominance of $b \to c$ decays, followed by Cabibbo-favoured $c \to s$ decays implies the presence in the final state of almost 2K's (1K, $1\bar{K}$) per $B\bar{B}$ event. The present experimental results on K^0 and $K^\pm$ yield $(b \to c)/\text{all} = 0.74 \pm 0.18$, a less precise number than that derived from the electron spectrum. The mean charged multiplicities per B decay are 4.1 ± 0.35 for semi-leptonic decays and 6.3 ± 0.3 for hadronic final states.

Finally, only an upper limit ($\tau_B < 1.4 \times 10^{-12}$ s) has been measured for the lifetime. Theoretical expectations[21] based on the spectator model (annihilation diagrams should be less important than for charm, because of the higher mass) are around $\tau \simeq 10^{-14}$–10^{-15} s (assuming $b \to c$ dominance), leading to a decay length visible in emulsions, but very difficult, if not impossible, for other techniques.

2.3 Top

The top quark should build with the beauty quark the 3^d quark doublet, corresponding to the 3^d lepton doublet (τ, $\bar{\nu}_\tau$). The only thing we know for sure about it, if it exists, is that its mass is higher than ≈ 20 GeV (from the non-observation of toponium at PETRA). Theoretical upper limits for the mass are reviewed elsewhere[21]; a firm one[22] is ≈ 200 GeV, which leaves ample room (however, the higher the mass, the more difficult it will be to find it).

The lifetime ($\sim m^{-5}$) will be unmeasurable. Semi-leptonic decays, with BR ≃ 12% each and $\langle p_T \rangle \gtrsim 10$ GeV, should be relatively easy to detect, if not to identify. Favoured hadronic decays should be into $bc\bar{s}$ and $bu\bar{d}$, the first one leading to three kaons in the final state. If the mass is high enough, decays such as $t \to bH^+$, $t \to bW^+$ should occur.

3. HADROPRODUCTION OF CHARMED PARTICLES

3.1 Experimental Methods

As will be seen below in more detail, the production of a pair of charmed particles, C and $\bar{C}$, via the reaction:

$$p(\text{or } \pi^-) + p \to C\bar{C}'X \tag{3}$$

has a total cross-section, summed up over the various possible $C\bar{C}'$ pairs, equal to about 10^{-3} the total inelastic pp cross-section at the maximum SPS energy ($\sqrt{s} \simeq 27$ GeV), rising to somewhat more than 10^{-2} at the ISR ($\sqrt{s} = 52$ or 63 GeV).

To separate the charmed signal from the dominant background, various characteristic features of charmed decays are used at the analysis stage and in some cases the trigger level.

3.1.1 Lepton emission. This proceeds according to the $\Delta Q = \Delta C$ rule:

$$C \to \ell^+\nu \ldots, \quad \bar{C}' \to \ell^-\bar{\nu} \ldots \quad (\ell = e \text{ or } \mu) \; . \tag{4}$$

In the charmed particle c.m. the spectrum of E^*, the lepton-energy, ends at about 1 GeV for the D meson. Hence, in the laboratory system, if $p_T(D) < M(D)$, most leptons will have $p_T < 1$ GeV/c and their average energy will be $\langle E \rangle \simeq x_F \sqrt{s}/2M \langle E^* \rangle$ at the ISR and $\langle E \rangle \approx x_F\, p/M \langle E^* \rangle$ in a beam of momentum p, x_F and M being the Feynman x and the mass of the charmed particle, respectively. For $M \simeq 2$ GeV, and with $\langle E^* \rangle \simeq M/4$, $\langle E \rangle \approx 4 x_F$ at the ISR and $\langle E \rangle \approx 100\, x_F$ at the SPS (in GeV). The high values of $\langle E \rangle$ at the SPS, together with the small value of p_T, makes the detection of muons and neutrinos relatively easy and efficient in a small-angle forward cone but not that of electrons, which are accompanied in that cone by other particles which cannot be filtered out, as they are in beam-dump experiments designed to search for prompt μ's or ν's. On the other hand, μ or

ν detection is impractical or impossible at the ISR, where only inclusive yields of electrons at large angles have been measured.

Besides the experimental problem of subtraction of leptons of non-charm origin (electrons from γ conversion, μ's and ν's from π and K decays), the main drawback of this approach is that it gives only, so to say, second-hand information on charm production: the observed lepton spectrum results from the (unknown) charm-production spectrum through a convolution with the charm-decay spectrum.

Moreover the nature of the parent charmed particle (D^+, D^0, Λ_c, ...) is not directly known -- and, as seen above, the semi-leptonic branching ratio, and hence the inferred production cross-section, will depend on the assumption made about it.

The method has been extended to dilepton ($e^{\pm}\mu^{\mp}$, $\mu^{\pm}\mu^{\mp}$) detection, reducing the background problems, but "squaring" the uncertainties in interpretation.

3.1.2 <u>Narrow mass peaks.</u> These are looked for in exclusive hadronic decay channels, such as the Cabibbo-favoured $\Delta C = \Delta S$ decays (see Table 1):

$$D^0 \to K^-\pi^+ \quad \text{or} \quad \Lambda_c^+ \to K^-p\pi^+ \ . \tag{5}$$

Since the decay is a weak process, the width of the peak should be equal to the experimental resolution. Hence resolutions of a few megaelectronvolts allow charmed decays to be distinguished from strange resonances, which, in that mass range, have a width of the order of 100 MeV. Supplementary confirmation of the charm nature may be provided

i) by direct quantum number assignment, as in

$$D^+ \to K^-\pi^+\pi^+ \quad \text{or} \quad \Sigma_c^{++} \to \Lambda_0\pi^-\pi^+\pi^+\pi^+ \tag{6}$$

which clearly cannot be $I = \frac{1}{2}$ or $I = 1$ resonances of strangeness -1;

ii) by the presence in the interaction (through the trigger, for instance) of a lepton of the right sign, i.e. an e^- with a C particle, an e^+ with a $\bar{C}$;

iii) by the observation of a distribution of small, but non-zero, decay lengths associated with the mass peaks, as in Ref. 12 or 13.

Usually, the sought-after charmed mass peak, if any, will sit on a much bigger non-charmed background, owing to the many non-charmed combinations in the same mass range. This background is reduced (but not completely suppressed) by particle identification -- a comprehensive charm production study thus calls for an ideal 4π spectrometer (which up to now does not exist). The signal can also be enhanced by selecting *a priori* a region of phase space (for instance, "diffractive" production or high p_T) where intuition or theory points to a better signal/background ratio -- of course this procedure introduces a strong bias.

In both cases (lepton or mass-peak detection) the experimental result is a $dN/dp_T\,dx$ distribution (background subtracted) over a limited region of phase space. with acceptance varying over that region. To extract from it a production cross-section, one has to assume a charm production law, calculate from it the expected $d\sigma/dp_T\,dx$ (of the lepton or the mass peak), and compare it with the observed one (for most experiments, statistics do not allow one to distinguish between various possible production distributions and one can then only quote the values of the total production cross-section under the various hypotheses). The most commonly used production law is of the form:

$$E\,\frac{d^3\sigma}{dp^3} \sim (1-x)^n\, e^{-bp_T} \quad (\text{or } e^{-b' p_T^2}) \,, \tag{7a}$$

which describes a central production process if $n \approx 5$ and a forward process if $n \approx 1$ (see later). Other phenomenological distributions have been used, especially for the early ISR data:

$$\text{a flat-y distribution:} \quad \frac{d\sigma}{p_T dp_T dy} \sim e^{-bp_T} \tag{7b}$$

$$\text{a flat-x distribution:} \quad \frac{d\sigma}{p_T dp_T dx} \sim e^{-bp_T} \,. \tag{7c}$$

When two leptons or a mass peak associated with leptons are observed, the same procedure is applied, starting from an assumed double distribution law at production, which may be a product of two distributions of type (7) (uncorrelated production), or the distribution resulting from the decay of a $C\bar{C}$ system produced according to some law of type (7) (correlated production).

3.1.3 Small decay length. Up to now, as seen in Section 2.1, this method was used mainly for lifetime determinations. In the case of emulsions or bubble chambers, the primary flux is limited, yielding a rather small sample of charmed events, but these events are essentially background free and usually well identified, and moreover, the acceptance is almost constant* (for $x_F > 0$). Hence,

* The charm decay is detected during scanning by the presence of a secondary track missing the interaction vertex by a distance d, whose most probable value is $d \approx c\tau$, for $x_F > 0$.

they are perfectly suited for production studies and, indeed, as will be seen below, recent results from bubble-chamber experiments at CERN have brought novel information on D-production. One may hope that, in the near future, charm triggered microstrip detectors (or high-precision drift or streamer chambers) will provide decent samples of rather pure charmed events. Of course, as outlined above, a good spectrometer behind is a must.

Finally, one word of caution should be said about results extracted from experiments with a heavy target; it is now customary, for the naked charm production cross-section σ_c, to assume that $\sigma_c(pA) = A^{\alpha}\sigma_c(pp)$, with $\alpha \approx 1$ as found experimentally for μ-pair or J/ψ production[23]. An $A^{2/3}$ law, as used for total cross-section or diffraction, would lead to charm cross-sections about 4 times higher with a Cu target (this was partly responsible for some high cross-sections reported from the first beam-dump experiments[24]). Recent experimental data[25] on inclusive particle production as a function of Feynman x, shown in Fig. 11, indicate that the exponent α decreases when x increases, roughly in the same way for all non-charmed particles. Knowledge of the corresponding curve for charm would be most useful and should be a goal of future experiments; for lack of it, one can only make an educated guess: $\alpha \approx 1$ for small x, dropping perhaps to about ⅔ for high x. In what follows, cross-sections will be calculated with $\alpha = 1$, except otherwise mentioned, but this word of caution should be kept in mind. Note also that a dependence of α on x, such as in Fig. 11, will make the inclusive x-distribution obtained from a heavy target softer than the one seen in pp collisions, since small values of x will be more favoured (a similar remark[26] applies to the p_T distribution, which becomes wider when A increases).

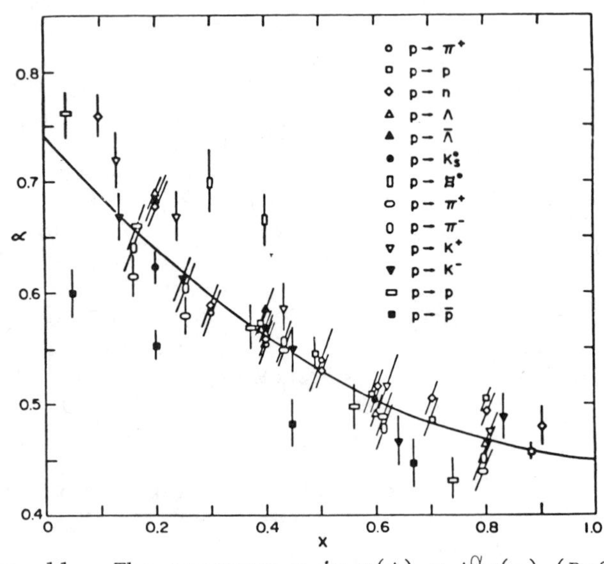

Fig. 11. The exponent α in $\sigma(A) = A^{\alpha}\sigma(p)$ (Ref. 27).

3.2 First Results and Theoretical Background

Let us briefly summarize the evolution of data and theoretical ideas up to 1981 (for more complete experimental surveys, see Ref. 27, and for a lucid theoretical appraisal see Ref. 28).

First QCD calculations for charm production (see for instance, Ref. 29) were based on the diagrams of Fig. 12a ($q\bar{q}$ annihilation) and 12b (gluon-gluon fusion), the last one being dominant in pp collisions. They predicted (Fig. 13) total charm production cross-sections of a few microbarns at the SPS. The first (1978) convincing evidence for charm hadroproduction came from beam-dump experiments[24] and pointed to a much higher cross-section ($\sim$ 30 μb in the CDSH experiment). Figure 14 shows an attempt[30], by varying parameters to accommodate that value within the framework of the gluon-gluon fusion model; in particular the mass of the c quark had to be chosen equal to 1.15 GeV, whereas m_c = 1.5 GeV is required[28] for a good fit of the data on charm photoproduction to the gluon-photon fusion model[31], which is very similar to the gluon-gluon fusion model. Except for the proviso about absolute cross-sections, most data, old and recent, from proton beam-dump experiments seem to fit well a $D\bar{D}$ central production model [Eq. (7a)] with $n \approx 4$-6, $b \approx 2$ GeV, and $\sigma(D\bar{D}) \approx 20$ μb and leave little room for production of charmed particles at large x values (see later for more details).

Soon after (1979) three different ISR experiments[32] detected large mass peaks in the (until then unknown) $\Lambda_c \to K^- p\pi^+$ decay

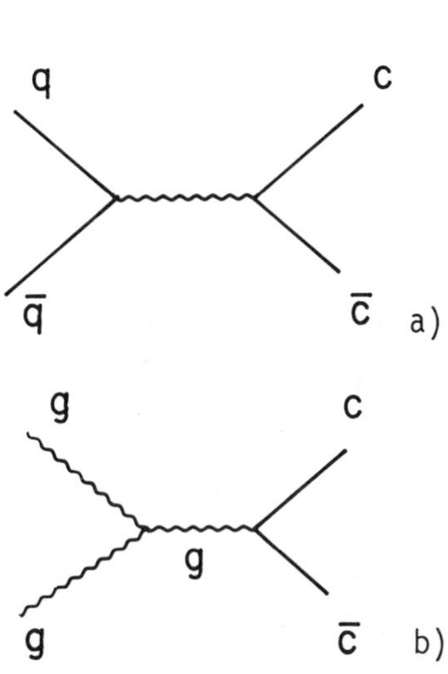

Fig. 12 $c\bar{c}$ creation diagrams (Ref. 29)

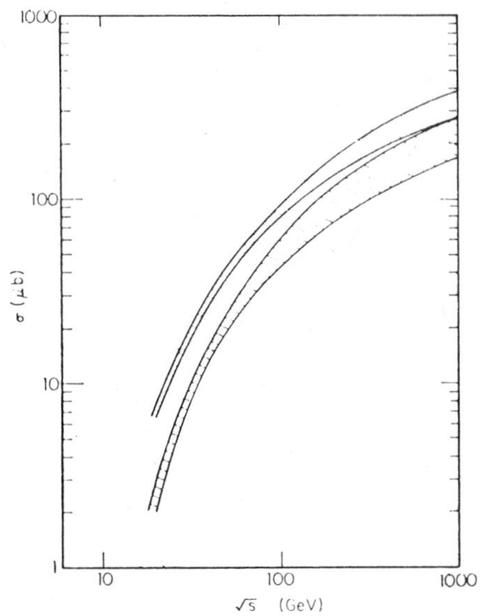

Fig. 13 Charm production (upper curves: $p\bar{p}$, lower: pp) via $c\bar{c}$ creation (Ref. 29)

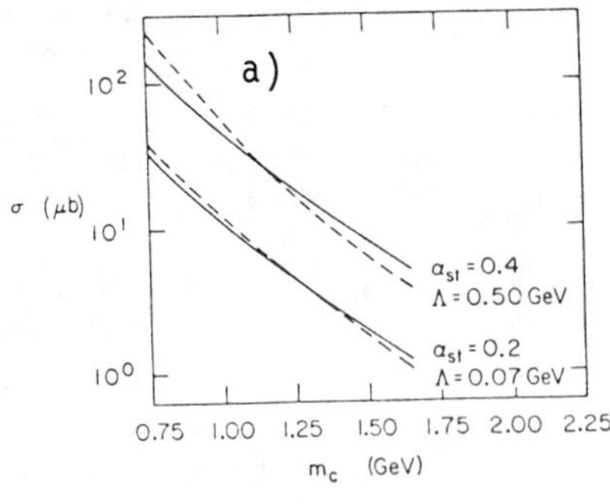

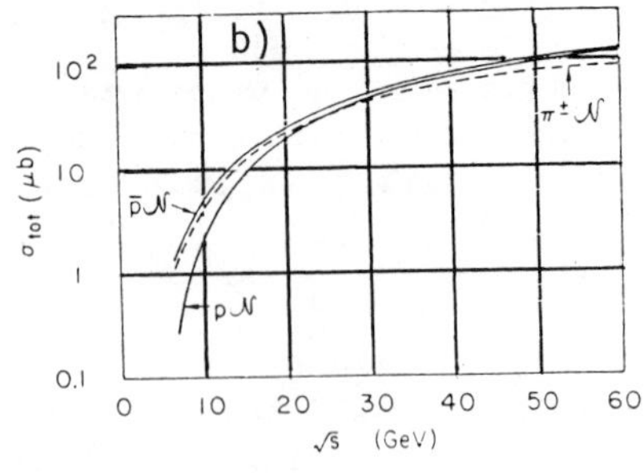

Fig. 14 Gluon fusion: a) at $\sqrt{s}$ = 27 GeV, b) with m_c = 1.1 GeV, Λ = 0.5 GeV (Ref. 30).

channel and one of them[33] also reported D^+ production. These results had two important common features: the total inclusive cross-section was large (200-400 μb for either the Λ_c^+ or the D^+), and an abundant production of the Λ_c^+ and the D^+ was observed at large Feynman x values. These two characteristics have since then been confirmed by other ISR experiments; Fig. 15, from Ref. 34, shows that the x behaviour of Λ_c^+ production is in close analogy with that of the Λ^0, which is well known. Clearly some other process must exist, besides gluon-gluon fusion which predicts much smaller cross-sections (Fig. 14) and central production (reflecting the soft distribution of the gluons).

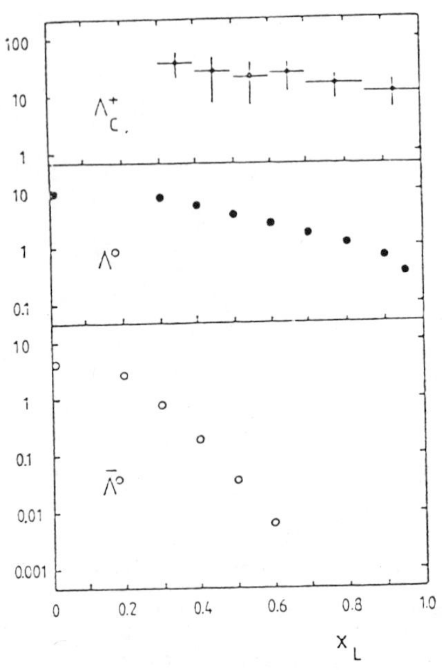

Fig. 15 Λ_c^+ production at the ISR, compared to Λ^0 and $\bar{\Lambda}^0$ (Ref. 34)

Actually, as first pointed out by Combridge[35], another possibility for charm production is charm excitation, as illustrated in Figs. 16a and b, whereby a parton from one nucleon excites a charmed quark from the sea of the other one. The charm cross-section versus energy, as calculated in Ref. 35 using the then standard parton distributions and evolution laws[36] and m_c = 1.87 GeV, is shown in Fig. 17, together with the experimental results available in 1981 (black dots) and some new results (open dots), which will be discussed below. The cross-section is smaller than that of the former gluon-fusion model[30] (with m_c = 1.15 GeV), but it has the interesting feature of growing much faster (using the same *ad hoc* trick as in Ref. 30, i.e. putting m_c = 1.3 GeV, one even obtains a rather good fit of the data). This indicates that the diagram of Fig. 16b (which dominates over that of 16a) should have not been forgotten, even though in its original formulation[35] it did not explain the large x_F values (the knocked-out c quark should give a charmed particle in the central region, the other charm will be in the spectator jet, with an x_F resulting from the usually low x of the $\bar{c}$ quark via recombination with appropriate normal quarks).

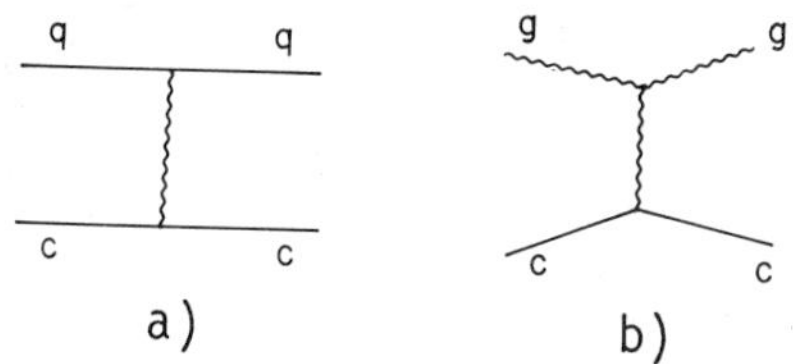

Fig. 16. Charm excitation diagrams (Ref. 35).

A first attempt at explaining charm production at large x (besides earlier models, such as those for diffractive production[37] or gluoproduction[38]) was the intrinsic charm model[39]. The hypothesis -- which is based on debated theoretical grounds -- is that the ground state of the proton is a superposition of several states: $|\psi\rangle = (1-\varepsilon)|uud\rangle + \varepsilon_1|uudc\bar{c}\rangle + \ldots$, i.e. that a $c\bar{c}$ pair has a probability $|\varepsilon_1|^2$ of being present in the proton -- hence the name "intrinsic". This is in contrast with $c\bar{c}$ pairs from the QCD sea, which start to be generated only when $Q^2 > Q_0^2$ (they are sometimes called "extrinsic"). An immediate consequence of the hypothesis is that the c-quarks, in order to remain bound with the other ones, must have the same velocity: hence, since they are heavier, the c and $\bar{c}$ quarks share between them most of the momentum of the proton. Figure 18a illustrates this fact, and Figs. 18b and c give the x_F distributions for Λ_c and D production which result from the c-quark x distribution. Qualitatively at least, a 1-2% intrinsic charm component could thus explain the main features of charm production at the ISR, as well as a D^+ (D^-) signal observed[40] in 217 GeV/c π^-p

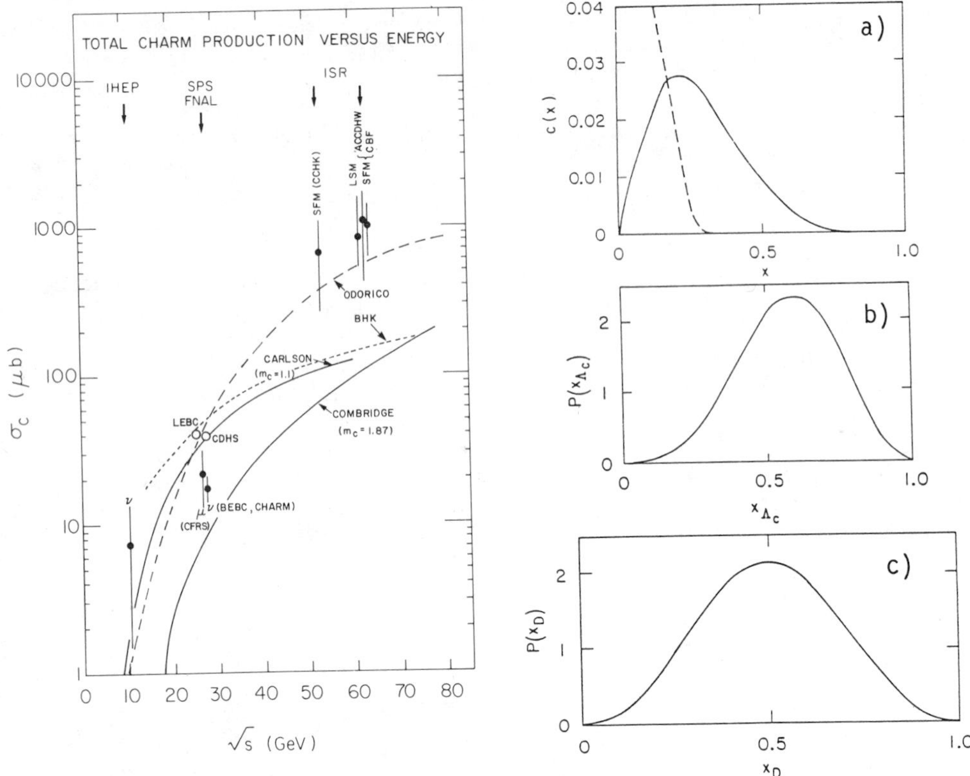

Fig. 17 Charm pp production cross-sections compared to various models (see text)

Fig. 18 Intrinsic charm model predictions (Ref. 39)

diffractive interactions* at FNAL, and reproduced in Fig. 19. However, in this model, the cross-section is proportional to log s (as a diffractive cross-section), not enough to account for the big ratio (> 10) between SPS and ISR. A more stringent argument against intrinsic charm may be the fact that it does not seem to manifest itself in the charm structure function derived from charm muoproduction experiments. Figure 20 shows this structure function as obtained by the EMC experiment[41] from dimuon events produced in the (virtual) photoproduction process $\mu Fe \to \mu C \ldots$, $C \to \mu \ldots$. It fits rather well the gluon fusion model (GFM) used by the authors (plain curves) in the acceptance region, giving an upper limit of 0.28% at 90% c.l. for intrinsic charm. However, the deviation from GFM at high Q^2 may indicate the onset of an intrinsic charm component (IC) with a strong threshold suppression[42].

---

* Detection of a recoil proton signalled the diffractive interaction; one of the D decays was tagged by the presence of a muon, the other one was seen as a mass peak in the $D^{\pm} \to K^{\mp}\pi^{\pm}\pi^{\pm}$ mass distributions. In the natural hypothesis of $X^- = D^+D^-\pi^-$ diffractive production, one finds $\sigma(p\pi^- \to pX^-) = (7\text{-}10) \pm 4$ μb, using $BR(D^+ \to \mu^+) = 23\%$ (Table 1).

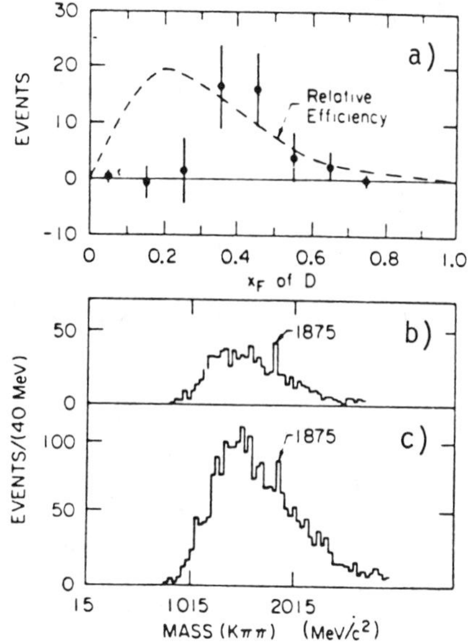

Fig. 19 Diffractive production of $D^{\pm}$ (Ref. 40)

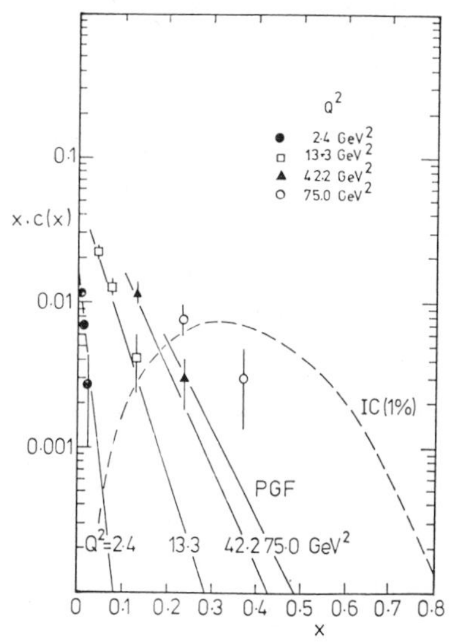

Fig. 20 Measured charm structure function (Ref. 41) versus two models (see text)

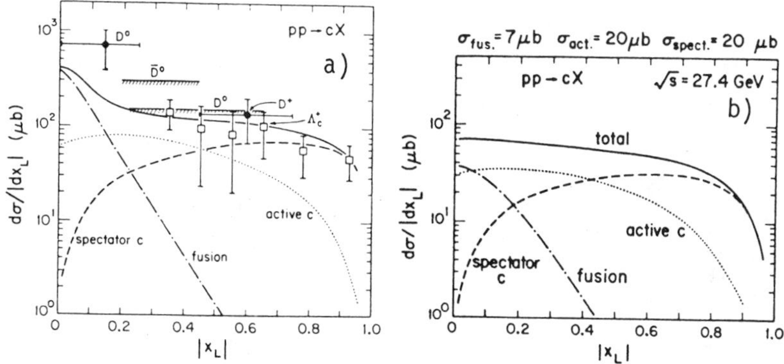

Fig. 21. Calculated x-distributions of charm a) at ISR, b) at SPS (Ref. 43).

A more recent model[43] tries to avoid the theoretical and experimental objections raised by the intrinsic charm model. While retaining the idea of charm excitation from an evolutive sea as in Ref. 35, it gives the charm quarks a hard distribution $xC(x) \sim x(1-x)$ similar to that resulting from the intrinsic charm model[39]. In the diagram of Fig. 16b, then, the knocked-out or "active" quark carries away its fractional momentum, while the spectator quark may recombine with its original proton. The resulting x_L distributions at ISR and SPS energies are shown in Figs. 21a and b. The agreement with ISR results looks good, but the total experimental ISR cross-section (Λ_c + D^+ + D^0) is several times higher than the calculated one (see Fig. 17) (this is because the excitation cross-section is roughly proportional to $\log^2$ s only).

3.3 Recent Results and Developments

They can be summarized in one sentence: several SPS or FNAL experiments, using very different techniques, have now observed the large x component of charm production, which was one of the surprises brought about by the ISR, with the following caveat: this effect is seen mainly in π^- interactions, only two experiments have indirect evidence for it in proton interactions (contrary to the ISR, Λ_c production remains elusive at SPS energies). Results are now reported, starting with those most easy to interpret.

3.3.1 Results from visual detectors. In 360 GeV π^-p interactions, the LEBC experiment NA16 has reported results[44] of a new kind on D and $\bar{D}$ production: for the first time, the production of the various charged states could be observed in the same experiment, and, despite the small statistics (18 D decays) new information has emerged. Figures 22a and b show the x distribution of D^- and D^0 events on the one hand, and D^+ and $\bar{D}^0$ on the other. It is clear that the first one extends to high x, whereas the second does not; qualitatively, this may be explained by the fact that a spectator c or $\bar{c}$ can recombine with the $\bar{u}$ or d quark of the π^- to give a D^0 or

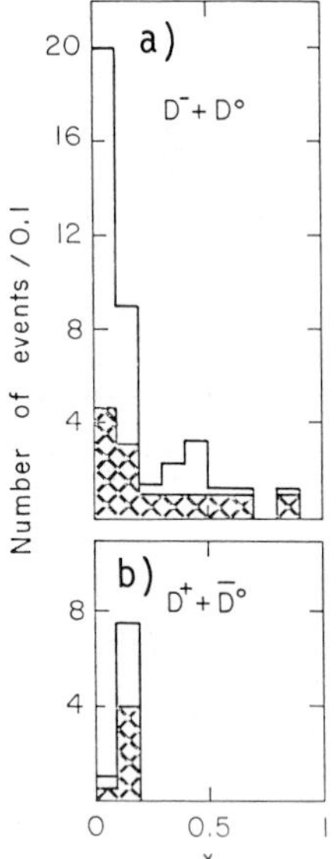

Fig. 22 x-distributions of various D mesons in LEBC (Ref. 44)

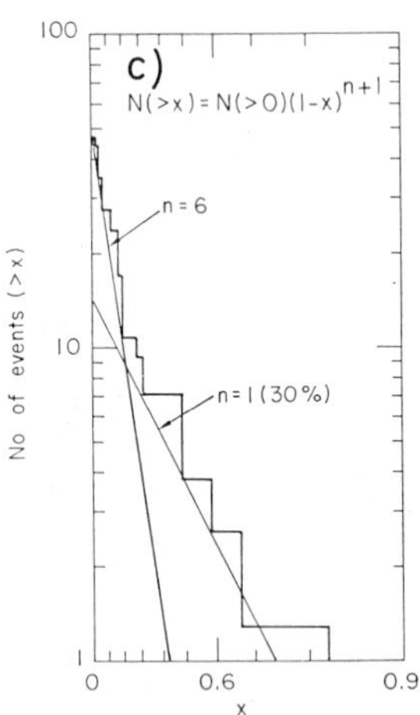

Fig. 23 Integral x-distribution of all D mesons in LEBC (Ref. 44)

D^-, but cannot give in the same way a D^+ or $\bar{D}^0$ -- lending support to the diagram in Fig. 16b, as treated in Refs. 39 or 43 (see Ref. 45 for a more complete discussion). Figure 23 shows the integral x distribution of all observed D and $\bar{D}$ particles, where two components may be seen, one (central) with slope $n \approx 6$, the other one with $n \simeq 1$ containing about 30% of the D's. The p_T distribution (not shown) is $\sim \exp(b' p_T^2)$ with $b' = (1.1 \pm 0.3)$ GeV^{-2}, and the total cross-section for inclusive $(D + \bar{D})$ production is $40 {}^{+15}_{-8}$ μb for $x > 0$ (≈ 31 μb for $D + D^0$, ≈ 9 for $D^+ + \bar{D}^0$). Another interesting result is that $D\bar{D}$ observed in pairs seem to be strongly correlated: $\langle \Delta y \rangle \approx 0.5$ (which should be considered as a lower limit, since the pair is not always seen).

The same experimental set-up was used with a 360 GeV/c proton beam[46]. Here no difference in the x distribution was seen between the various D particles; the inclusive D distribution fitted a $(1-x)^{1.8\pm0.8}\ e^{-(1.1\pm0.3)p_T^2}$ law*, with $\sigma(D+\bar{D}) = 56 {}^{+25}_{-12}$ μb (for all x). If the excess of $\bar{D}$ is interpreted as due to $\bar{\Lambda}_c\bar{D}$ production, then $\sigma(D\bar{D}) = 19 {}^{+13}_{-5}$ μb, and $\sigma(\Lambda_c\bar{D}) = 18 {}^{+5}_{-10}$ μb (only 3 ambiguous Λ_c decays have been detected, one of which is $\Lambda_c \to K^- p\pi^+$, corresponding to $\sigma \sim 40$ μb). Finally no strong correlation between D and $\bar{D}$ is observed ($\langle \Delta y \rangle > 1.2$), unlike what happens in $\pi^- p$ interactions.

In another experiment[11], NA18 (340 GeV/c π^- in BIBC, a small C_3F_8 bubble chamber), 14 well-identified D candidates were found and reconstructed (in a streamer chamber downstream). Their x distribution is given in Fig. 24, together with the acceptance; the average value of p_T is 0.95 GeV/c. Using a central production

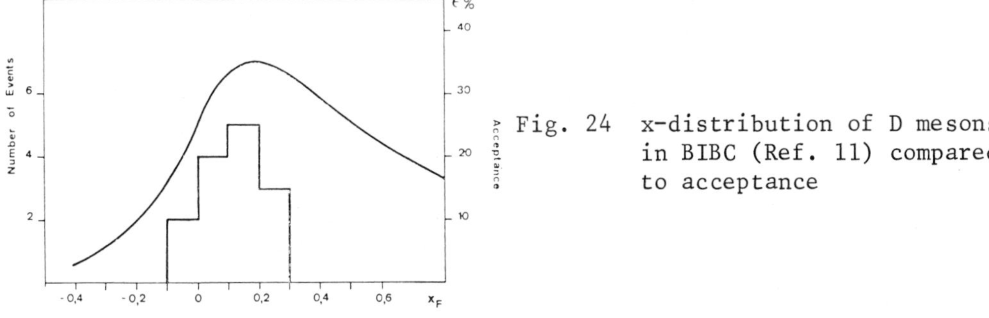

Fig. 24 x-distribution of D mesons in BIBC (Ref. 11) compared to acceptance

---

* One would not expect an $n = 1$ value similar to that observed for D^- in $\pi^- p$ interactions (or expected for Λ_c in pp, as at the ISR): $\bar{D}$'s can be formed by using one quark of the proton, and, according to quark counting rules[47], an $n \approx 3$ value would be expected in that case, versus $n \approx 5$ (truly central production) for D's.

model, which from Fig. 24 seems to be favoured*, the authors obtain, from ∿ 35 (background subtracted, but non-identified) double decays, $\sigma(\pi^- p \rightarrow D\bar{D}) = 28 \pm 11$ μb (73 ± 27 if $\sigma \sim A^{2/3}$).

Finally an experiment in which a stack of 600 μm thick emulsions was exposed to 400 GeV/c protons reported[48] a background-subtracted signal of eight 3-prong but otherwise unidentified charm decays. Assuming these are $\Lambda_c \rightarrow K^- p \pi^+$ decays ($D^\pm$ decays would give a much higher cross-section, because of the large escape probability due to the bigger $D^\pm$ lifetime), a cross-section $\sigma(pp \rightarrow \Lambda_c \ldots) \sim 100$ μb is obtained.

3.3.2 Results from mass peaks. They come essentially from experiment NA11 at CERN. This experiment[49] uses an electron trigger ($p_T > 0.3$ GeV) to enhance the charm signal in π^- (or p) Be interactions detected in a spectrometer with Čerenkov identification and photon calorimetry (Fig. 25). With incoming 180 GeV/c π^-'s, a clear $D^0 \rightarrow K^-\pi^+$ (and c.c.) peak is observed (Fig. 26a), when applying the requirement $143 < m(K^-\pi^+\pi^+) - m(K^-\pi^+) < 148$ MeV, i.e. requiring that the D^0 stems from a D^*. The acceptance-corrected x_F distribution of the D^* events, shown in Fig. 26b, agrees well with a $(1-x)^n$ law, with $n = 1 \pm 0.6$; an $e^{-p_T^2}$ dependence accounts

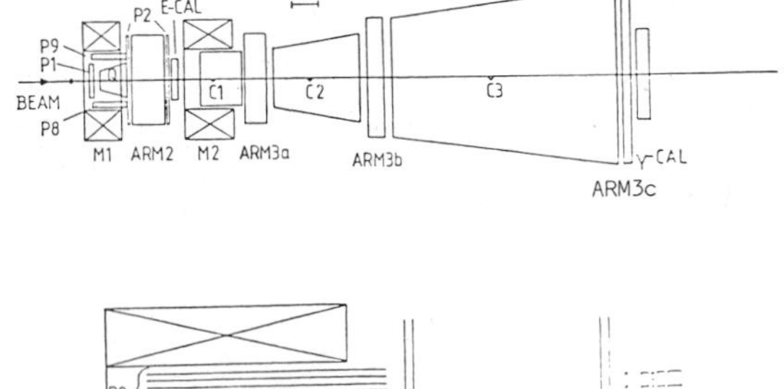

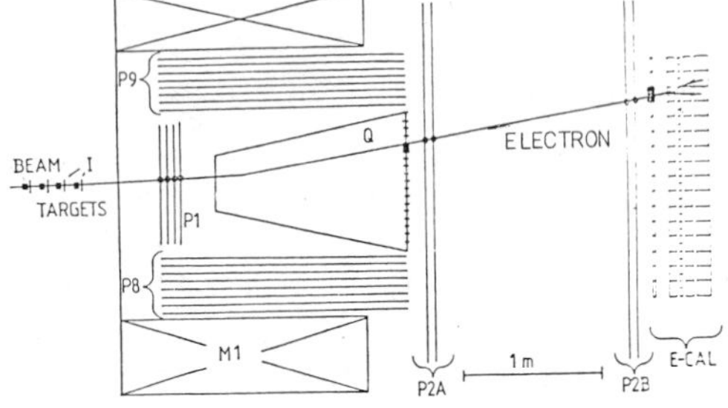

Fig. 25. The ACCMOR spectrometer (Ref. 49).

---

* This seems in contradiction with the result of NA16, reported above. However a possible high-x tail may well be suppressed by statistical fluctuations or the reverse may have taken place (fitting the distribution of Fig. 23 to a single $(1-x)^n$ law yields $n = 2.8 \pm 1.2$).

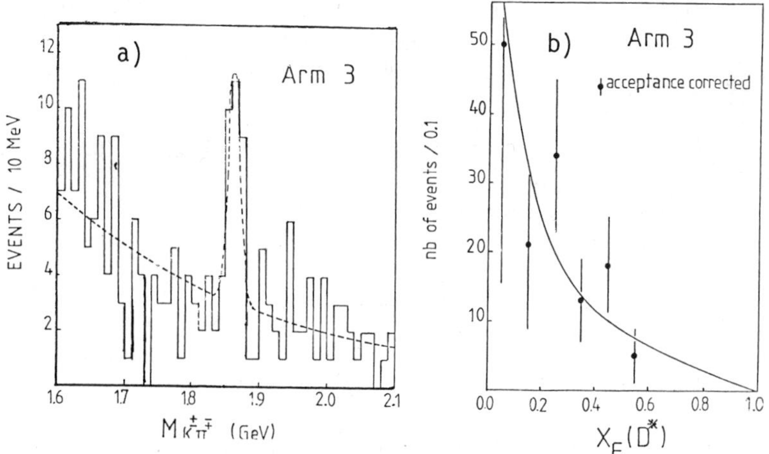

Fig. 26. a) D^0 peak from $D^{*\pm}$ production, b) $X(D^*)$ distribution versus $(1-x)$ law (Ref. 49).

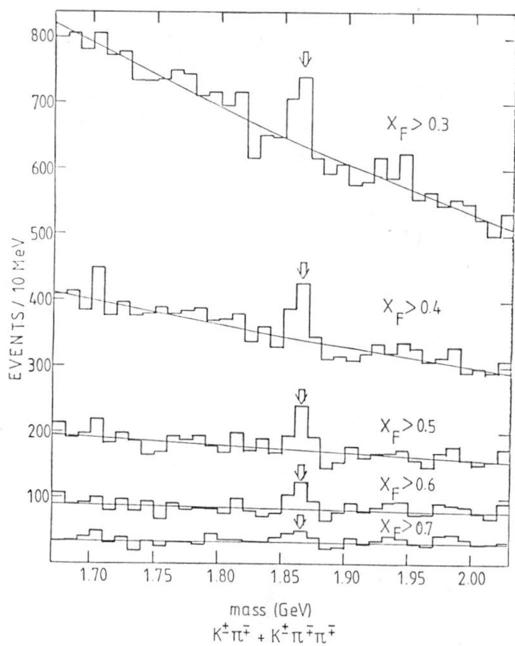

Fig. 27. Inclusive D production in experiment NA11 (Ref. 49).

for the $d\sigma/dp_T^2$ distribution. Using these laws for D^{*+} and D^{*-} production, values approximately as in Table 1 for the various branching ratios, and $\sigma(\pi^- A) = A\sigma(\pi^- p)$, the authors obtain

$$\sigma(D^{*+} \ldots) + \sigma(D^{*-} \ldots) = (17 \pm 3)\ \mu b \pm 10\ \mu b$$

Inclusive D production is also directly observed, as shown in Fig. 27, where it is seen to extend also to rather high x-values. Assuming a D/D^* ratio of 1 at production, one gets $\sigma(D\bar{D})$ = = (34 ± 8) μb ± 24 μb.

D^{*+} and D^{*-} production occurs in a similar way with 120 GeV/c π^-'s, from which one gets σ(120 GeV)/σ(180 GeV) = 0.41 ± 0.15 (the first measurement of energy dependence in the same experiment, hence without systematic errors).

Finally, with 150 GeV/c protons, no charm signal was observed, yielding an upper limit at 90% c.l. of (24 ± 17) μb for $\sigma(pp \rightarrow \Lambda_c \ldots)$, when one uses production laws of type (7a) for $\Lambda_c D$ production, with n = 2.5 (4.5) and b′ = 2.5 (2.5) for the $\Lambda_c(\bar{D})$.

The results of NA11 on $D^{*\pm}$ production recall to mind those of an older 200 GeV π^- Be experiment at FNAL[50], which first used the trick of recognizing charm production by the small Q value of the decay $D^{*+} \rightarrow \pi^+ D^0$. The x_F (D^*) distribution, shown in Fig. 28, fits best a $(1-x)^3$ law, without excluding (1-x). With $(1-x)^3$, a cross-section $\sigma(\pi^- p \rightarrow D^*) \sim 4.4$ μb was calculated.

No charge asymmetry between D^{*+} and D^{*-} production was mentioned in Refs. 49 and 50, as was found between D^+ and D^- in Ref. 44 (but not in the diffractive production of Ref. 40). The common feature of these experiments is the indication of a rather broad x(D) distribution [n = 1 - 3 in Eq. (7a)], in qualitative agreement with the models of Refs. 39 and 43, and with the latest ISR results[34], rather than with a purely central production (n ≈ 5).

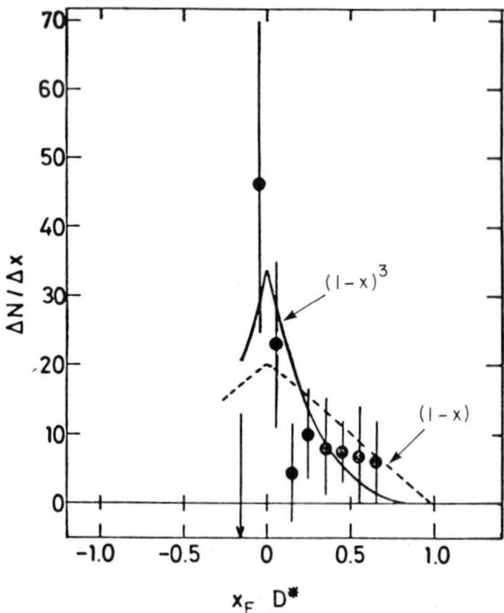

Fig. 28. $x(D^*)$ distribution (Ref. 50).

Finally a spectrometer experiment in a 58 GeV/c neutron beam at Serpukhov[51] has observed mass peaks in the $K^0 p\pi^+\pi^-$ and $\Lambda^0\pi^+\pi^+\pi^-$ final states, corresponding to $\sigma(np \to \Lambda_c \ldots) \approx 44\ \mu b$, a value which needs confirmation.

3.3.3 Results from single lepton experiments

i) Neutrino beam-dump experiments

A recent experiment, E613[52] at FNAL, used a tungsten target (instead of copper as employed at CERN) and a detector much nearer to the target (allowing the recording of ν interactions up to 40 mrad, instead of $\sim$ 2 mrad at CERN). The energy and angle distributions derived from ν_e ($\bar{\nu}_e$) charged-current events, shown in Fig. 29, agree with a $D\bar{D}$ production model $\sim (1-x)^n\, e^{-bp_T}$. With $n = 3$, $b = 2\ \text{GeV}^{-1}$, a cross-section $\sigma(pp \to D\bar{D}) = 18 \pm 4\ \mu b$ is obtained*, using $\sigma(pA) \sim A\sigma(pp)$ and $BR(D \to \nu) = 16.4\%$ (twice the average value for $D \to e$). This can be compared with the result from the CHARM experiment at CERN[53], illustrated in Fig. 30, which can be translated into $\sigma(pp \to D\bar{D}) \simeq 19 \pm 6\ \mu b$ (with $n = 4$). For "central" $D\bar{D}$ production, one expects the yields of $\bar{\nu}$ and ν to be equal (in Ref. 52, $\bar{\nu}_\mu/\nu_\mu = 0.65 \pm 0.3$ for $E_\nu > 25$ GeV).

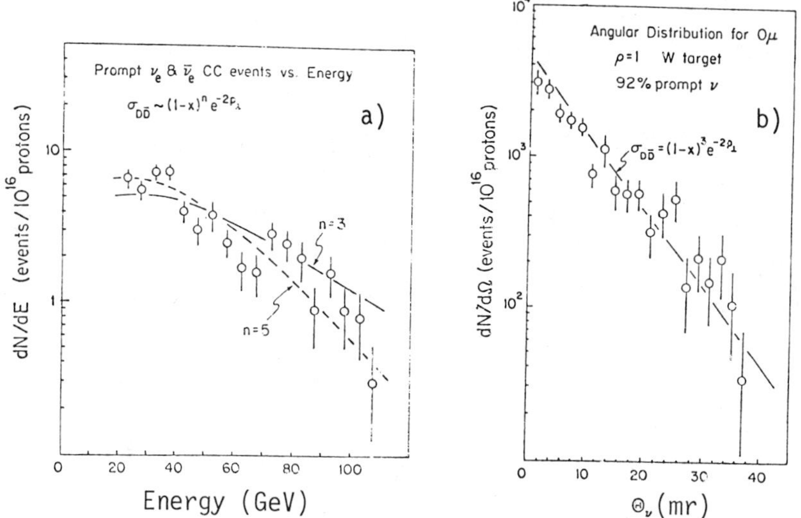

Fig. 29. Distributions of ν events from FNAL experiment E513 (Ref. 52).

---

* Also, by considering the number of events in Fig. 29 with E > > 120 GeV, an upper limit at 90% c.l. for diffractive production $pp \to \Lambda_c D$ is obtained: $\sigma < 7.3\ \mu b$ [using $\sigma(pA) \sim A^{2/3}$ and $BR(\Lambda_c \to e) = 4.5\%$].

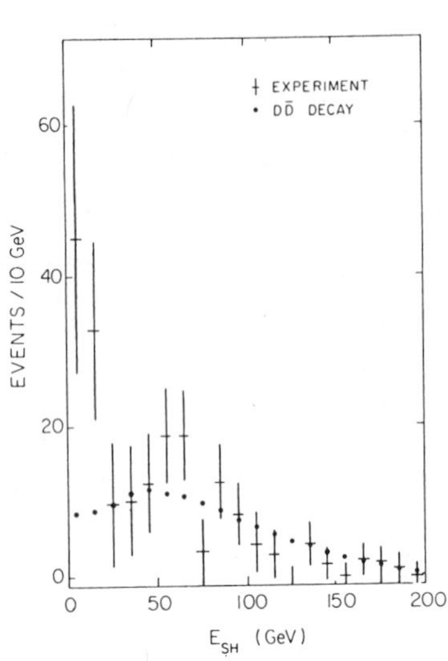

Fig. 30 Distribution of shower energy from the CHARM experiment (Ref. 53)

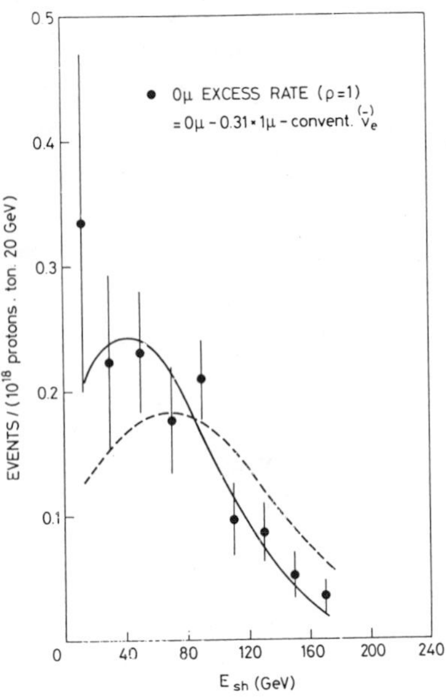

Fig. 31 Distribution of shower energy from the CDHS experiment (Ref. 54)

On the other hand, the CDHS experiment at CERN[54] published recently a new interpretation of their data on prompt neutrino production. They explain their experimental result $\bar{\nu}_\mu/\nu_\mu = 0.46 {}^{+0.21}_{-0.16}$ by $\Lambda_c\bar{D}$ production: the Λ_c being produced more forward than the $\bar{D}$, the higher acceptance at large x values allows the detection of more ν_μ's (from Λ_c) than $\bar{\nu}_\mu$'s (from $\bar{D}$), whereas the reverse would be expected for $D\bar{D}$ production, if, according to quark-counting rules[47], D^-'s are produced more forward than D^+'s. Figure 31 shows the shower energy distribution for ν_e's, compared with predictions from $\Lambda_c\bar{D}$ production (unbroken line n = 3 for Λ_c^+, n = 5 for $\bar{D}$; broken line n = 1 and 3), assuming the same semi-leptonic branching ratio for Λ_c and D [actually, as seen in Section 2.1.3, $BR(\Lambda_c \to e^+) \approx$ $\approx 0.6\ BR(\bar{D} \to e^-)$, which should depress the high-energy tail of the calculated distribution, improving the agreement of the broken line]. Using for the Λ_c^+ a $(1-x)^3\ e^{-2p_T}$ distribution, and assuming a linear A dependence, the authors obtain from the rate of $\nu_\mu \to \mu^-$ events a value $\sigma(p + N \to \Lambda_c^+ \ldots) \cdot BR(\Lambda_c \to \nu_\mu \ldots) = (3.9 \pm 1.0)\ \mu b$/nucleon, from which, with $BR(\Lambda_c^+ \to \nu_\mu \ldots) \cong BR(\Lambda_c^+ \to e^+ \ldots) \simeq 4.7\%$, one can derive

$$\sigma(pp \to \Lambda_c^+ \ldots) \approx 40\ \mu b\ .$$

As outlined in Section 3.1.1, the interpretation of beam-dump experiments in terms of charm production, is not unambiguous! A potentially bigger question mark arises from the fact that the measured ratio of ν_e to ν_μ yields -- which should be equal to 1 if lepton universality holds, as we believe it does in charm decay-- seem to be less than 1. The experimental numbers are: $0.56 \pm ^{0.35}_{0.21}$ (BEBC), $0.64 \pm ^{0.22}_{0.15}$ (CDHS), $0.48 \pm ^{0.12}_{0.10}$ (CHARM), 0.78 ± 0.19 (E613). Except for one of them they are all individually consistent with 1 (and E613 outlines that $\nu_e/\nu_\mu = 1 \pm 0.3$ for $E_\nu > 30$ GeV), but the general trend clearly favours $\nu_e/\nu_\mu < 1$. The best way to check this surprising result would be to measure prompt electron and muon production in the same experiment.

ii) Muon beam dump

The CFRS experiment detecting prompt muons produced in an iron target has recently published results[55] from 350 GeV/c protons and 278 GeV/c π's. The proton data are in general agreement with neutrino beam-dump results (except Ref. 54), in that the ratio μ^-/μ^+ is found to be compatible with 1 (1.1 ± 0.2) and the μ^+ or μ^- energy distributions (Fig. 32a) fit well a central D or $\bar{D}$ production model ($n = 6 \pm 0.8$ for D or $\bar{D}$ inclusive production at $x > 0.3$), yielding $\sigma(pp \to D) = 24.6 \pm 2.1\ (\pm 3.3)\ \mu b$, with the usual assumptions $[BR(D \to \mu) = 8\%,\ \sigma \sim A]$.

On the other hand, the recent π^- exposure has produced significantly different results for μ^- and μ^+: $\mu^-/\mu^+ = 2.23 \pm 0.29$ and different energy spectra (Fig. 32b). More precisely, the exponent n of the $(1-x)^n$ D distribution is found to be $n = 3.4 \pm 1.0$ for $D \to \mu^+$ and $n = 1.3 \pm 0.8$ for $\bar{D} \to \mu^-$ (for $x > 0.3$), yielding similar values for the inclusive D and $\bar{D}$ production cross-sections, $\sim 10 \pm 1\ (\pm 3)\ \mu b$/nucleon. As for the NA16 results[44], the difference in the n values may be attributed to preferred D^- (and D^0) forward production, because of quark-counting rules[47].

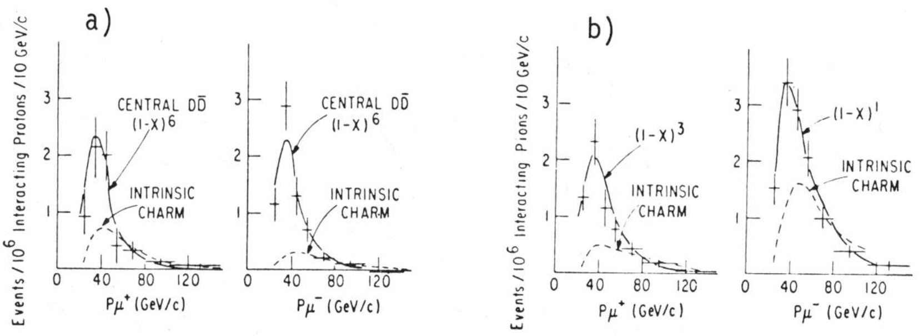

Fig. 32. Distribution of μ momenta: a) with protons, b) with pions (Ref. 55).

Finally, the maximum intrinsic charm contribution is shown (dotted line) in Fig. 32, as it results from fitting the high-energy tail of the spectrum; the authors conclude that the intrinsic charm component in the proton and the pion is $\lesssim 2 \times 10^{-4}$.

iii) Single electrons

As mentioned in Section 3.1.1, prompt electrons have been the object of searches at the ISR -- ever since an e/π ratio of $\sim 10^{-4}$ for $p_T > 1$ GeV/c was found at FNAL (1974). The first ISR experiments confirmed this result, but also indicated[56] a rise of the e/π ratio (at $\theta = 30°$ and $\sqrt{s} = 53$ GeV) with decreasing p_T (up to $\sim 5 \times 10^{-4}$ for $p_T \sim 0.4$ GeV/c). This controversial finding, which implied[57] a large (several hundred microbarns) charm cross-section has now been confirmed[58] in another ISR experiment (at $\theta = 90°$ and $\sqrt{s} = 62$ GeV). A summary of ISR measured values of the e/π ratio in the central region versus p_T is given in Fig. 33; the curves indicate the contribution of various processes, in particular charm production, calculated for a 200 μb D production according to a $(1-x)^3$ law, and beauty production (1 μb). It is clear that the e/π ratio in the region $0.4 < p_T < 1$ GeV/c, where charm electrons dominate, can be accounted for by an ∿ 500 μb charm-production cross-section, as directly measured from D^+ and D^0 mass peaks[34] -- there is no electron crisis. On the other hand, a further increase below $p_T \approx 0.4$ GeV/c, as observed[59] in 20 GeV/c π beams, would need another explanation[60].

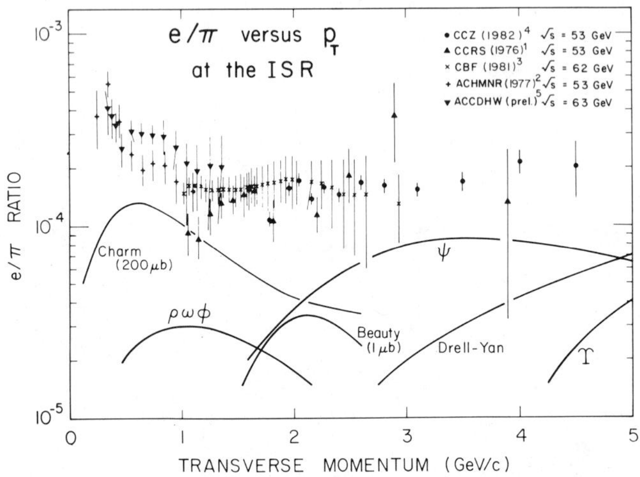

Fig. 33. e/π ratios measured at the ISR (compiled by T. Ekelöf).

3.3.4 Summary. The variation of prompt lepton emission versus incoming proton energy is displayed in Fig. 34, which gives the measured values of ℓ/π ($\ell = \mu, \nu, e$) in the central region ($x \approx 0$) and for $\langle p_T \rangle \approx 0.5$ GeV/c. The factor of 10 increase between top SPS and ISR energies reflects the increase in charm production cross-section* displayed in Fig. 17 (linear A-dependence was used in both figures to extract pp results from beam-dump data).

The large increase of cross-section from SPS to ISR seems thus to be definitively established. Also, despite many discrepancies in details, the leading-particle effect suggested by quark-counting rules, has now been observed not only in Λ_c^+ production at the ISR, but also in D production with π^- beams (and possibly Λ_c^+ production with p beams). Theoretical models emphasizing the importance of flavour excitation (Fig. 16b) manage to predict x distributions (see for instance Fig. 21a) which seem in reasonable agreement with the available data (actually, no complete set of data from a single experiment allows a really meaningful comparison). The big rise in cross-section from SPS to ISR energies seems harder to account for; Fig. 17 displays a recent, apparently more successful, attempt[61]

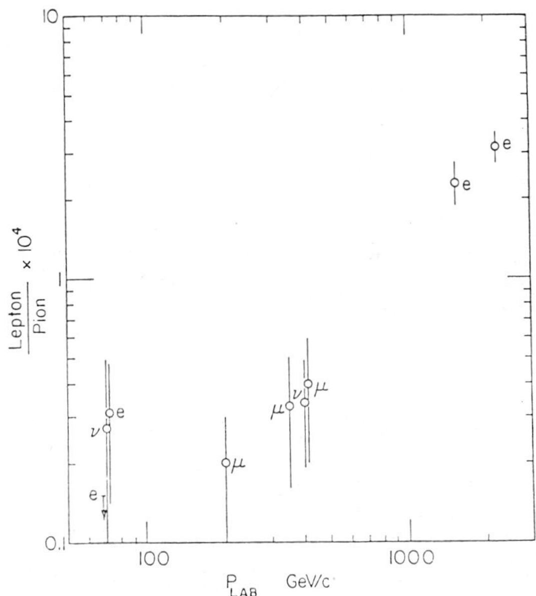

Fig. 34. Measured ℓ/π ratios versus energy (Ref. 58).

---

* Note that, within the SPS and ISR energy ranges, the ℓ/π ratio -- measured inside each region with similar techniques -- exhibits a clear trend to increase with energy (this is not a surprise, but a check of the relative quality of the measurements).

(for a more optimistic appraisal, see Ref. 45). In the exoneration of theory it should be stressed again that experimental results on cross-sections are marred by systematic uncertainties, resulting from model dependence and/or A dependence (beam-dump experiments and most mass peak searches with spectrometers), and/or statistical errors (bubble chambers).

In conclusion, we now have a qualitative understanding of charm production, at least for D and Λ_c production. More reliable and precise results are needed for a truly quantitative description with predictive power, which would be most helpful for heavier flavour searches (and for searches for other charmed particles, such as F^+, A^+, etc.). Reliability implies good acceptance over most of phase space and small background (as in bubble chambers), precision requires good statistics. 4π spectrometers combined with an accurate vertex detector and provided with a high data-taking rate capability (and/or with an efficient and unbiased trigger) are being developed for that purpose both at CERN and FNAL (where they will operate at $\sqrt{s} \approx 40$ GeV, bridging the gap between SPS and ISR). In a few years, good samples of 10^3-10^4 charmed events, instead of 10^1-10^2 now, should bring some answers to the pending problems.

4. HADROPRODUCTION OF HEAVIER FLAVOURS

Data are scarce, and mostly negative, on beauty production and non-existing for top; theoretical predictions are rather uncertain, as seen in the case of charm production. So this section should be considered not as a review, but rather as an overview of present knowledge, theoretical prejudice and prospects for the future.

4.1 Beauty

The most stringent limits on beauty production come from multi-muon final states, often as a by-product of spectrometer studies of the production of J/ψ, Υ, and Drell-Yan pairs. If a $B\bar{B}$ pair is produced, it may lead via the decays $B \to C + \mu^- + \ldots$ and $\bar{B} \to \bar{C} + \ldots$, $\bar{C} \to \mu^-$ to a pair of same-sign muons, or even to 3 (or 4) muons if the C or (and) the $\bar{B}$ have a μ^+ decay. To go back from the observed numbers of 2, 3, or 4 muon events to the $B\bar{B}$ production cross-section per nucleon, one has to conjecture a production model (usually a central one is used), use the B and C semileptonic branching ratios and decay spectra, and assume an A-dependence: $\sigma(A) \sim A^\alpha \sigma(\text{nucleon})$. The uncertainties introduced by such a procedure have been amply outlined before, in the case of single-lepton emission as a signature of charm production; here they are even greater (pair production and cascade decay, instead of inclusive production and single decay).

The most significant result obtained in this way is due to experiment NA3 at CERN; using a 280 GeV/c π^- beam on a platinum target and assuming $\alpha = 1$, NA3 obtains[62] $\sigma(\pi^- p \to B\bar{B} \ldots) < 2$ nb for central production, 10 nb for diffractive production, at 90% c.l. With 400 GeV protons, the lowest published[63] upper limit is 33 nb for central production; a limit of 40 nb has been reported[64] for diffractive production by 350 GeV/c protons (on iron, using $\alpha = 2/3$). The only possibly positive evidence for like-sign muon pairs originating from beauty comes from the EMC experiment[65] using a 250 GeV/c muon beam on iron, i.e. production by virtual photons; they observe three events, from which a photoproduction cross-section of about 1 nb can be derived, i.e. $\sim 10^{-3}$ of the charm photoproduction cross-section. On the other hand, it has been proved (see, for instance, Ref. 45) that the relatively abundant same-sign muon pairs observed in neutrino interactions cannot originate from beauty. One should also be aware of the fact that same-sign muon pairs may originate from D^0-$\bar{D}^0$ mixing; a limit of 10^{12} $\hbar$ s^{-1} has been found[64] in this way, for the possible mass difference, Δm. If Δm is indeed not zero, then only a part of the same-sign muon-pairs come from beauty decays and the upper limits quoted above should be even lower.

Another means used to search for beauty was to look for decays of the type $B \to J/\psi + C$, $C \to K$ or μ, as suggested by Fritzsch[66]. Here also results have been negative; using $BR(B \to J/\psi) = 1\%$ (the measured[67] upper limit is 1.4%), upper values for $\sigma(pp \to B\bar{B} \ldots)$ ranging from ~ 10 nb to ~ 30 nb have been found[62,68,69] with 200-300 GeV/c π-beams on complex targets. Events with two J/ψ's have been observed by NA3, but their rate is too high to assign them to $B\bar{B}$ production[62].

Experiment NA19 at CERN tried to detect beauty decays in emulsion, the production of beauty being tagged by the presence of three muons in the final state. No event was found, leading to an upper limit[70] at 90% c.l. of 90 nb for $\sigma(\pi^- p \to B\bar{B})$ at 350 GeV/c.

At the ISR, an estimate of central $B\bar{B}$ production can be obtained from the measured e/π ratio at 90° (Fig. 34) in the region $1 < p_T < 2$ GeV/c, which is compatible with a cross-section of a few (< 5) microbarns. Two SFM experiments have looked for forward Λ_B production, via the decay mode $\Lambda_B \to D^0 p\pi^-$: a signal at $m(\Lambda_B) \simeq$ $\simeq 5.42$ GeV was reported by one of them[34], corresponding with a flat-$x(\Lambda_B)$ distribution, to $\sigma(pp \to \Lambda_B \ldots)BR(\Lambda_B \to D^0 p\pi^-) =$ $= 27 {}^{+17}_{-11}$ μb; no signal was found by the other one[58] ($\sigma \cdot BR \lesssim 5$ μb). Experiments to obtain more conclusive results are in progress at the ISR.

In summary, experimental upper limits on beauty production cross-sections are of the order of a few nanobarns at SPS energies and a few microbarns at the ISR. The original calculations of Combridge[35] gave ~ 0.1 nb and ~ 0.1 μb, respectively at the SPS and the ISR, whereas the more recent model of Barger et al.[43] yields $\sim$ 100 nb at the SPS, 1 μb at the ISR, and 20 μb at the CERN $p\bar{p}$ collider ($\sqrt{s}$ = 540 GeV). The energy dependence of the various processes involved in charm, bottom, and top production in that model is displayed in Fig. 35, for $p\bar{p}$ collisions (essentially not different from pp collisions, since the $q\bar{q}$ contribution is very small). As outlined for charm production, the energy dependence might well not be strong enough -- at least for the dominating flavour excitation diagram -- so that a value of 1-10 nb, compatible with experimental data, seems more realistic at SPS energy.

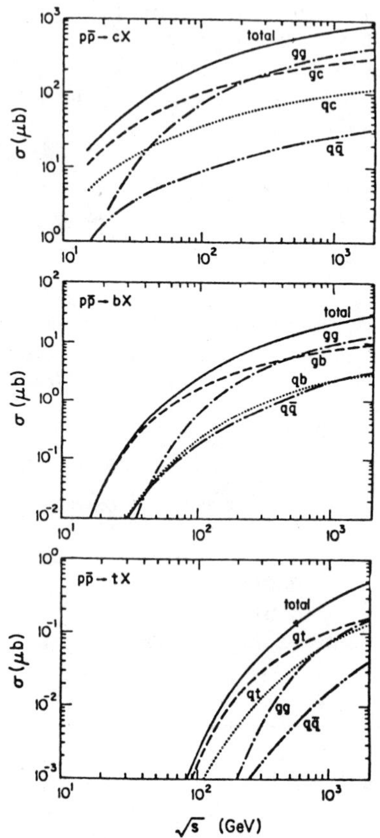

Fig. 35 Predictions for a) charm, b) beauty, c) top production (Ref. 43).

Several experiments are being prepared, both at CERN and FNAL, to detect beauty production by observing decays in emulsions. All associate with the emulsion a vertex detector accurate enough to select charm-decay candidates, a spectrometer for event reconstruction, and a trigger (multiplicity step, multimuon, or multikaon). Figure 36 shows sketches of two of those experiments, WA71 and WA75, both in 350 GeV/c π^- beams, which will be running at CERN in 1983 and aim at sensitivities of the order of a few events per nanobarn.

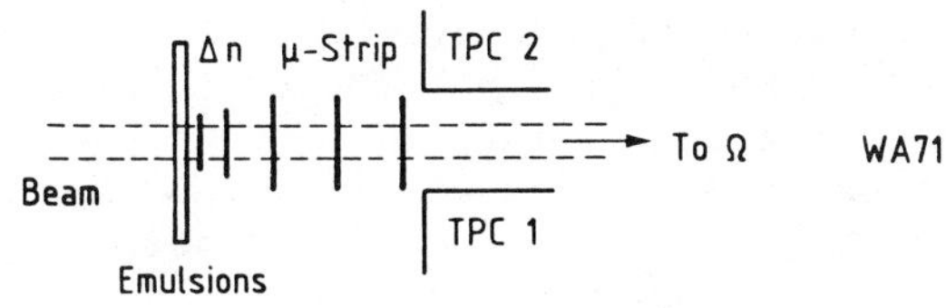

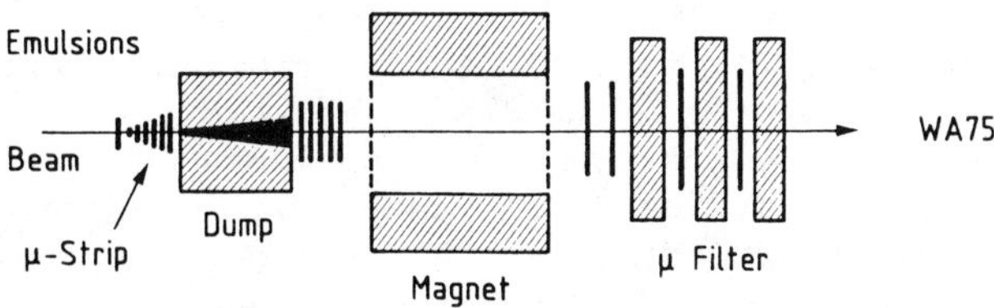

Fig. 36. Two 1983 CERN experiments looking for beauty (see text).

4.2 Top

As mentioned before, the only reasonable hope for hadroproduction of top lies in the $p\bar{p}$ colliders ($\sqrt{s}$ = 540 GeV at CERN, working; $\sqrt{s}$ ∿ 1000 GeV at FNAL, in 1986) or possibly later in the pp collider planned at BNL ($\sqrt{s}$ = 800 GeV, much higher luminosity) or eventually in a very high energy fixed-target machine (20 TeV → $\sqrt{s}$ ≃ 200 GeV).

According to Fig. 35, the production cross-section at the CERN collider would be ∿ 0.1 μb. This value, for m(t) = 20 GeV, is comparable to that formerly calculated at $\sqrt{s}$ = 800 GeV in Ref. 35; it drops when the mass increases, roughly by a factor of 10 when m = 40 GeV*.

Even with such small cross-sections and the limited luminosity of the SPS collider ($L \to 10^{29}\ cm^{-2}\ s^{-1}$ in 1983), top production is rather abundant (∿ 1000 $t\bar{t}$/day). The challenge is to fish it out of the sea of other flavours. As was the case for charm and beauty, the main problem is the trigger: i.e. reduce the counting rate by a factor $\gtrsim 10^{-3}$ while keeping most of the top particles. The fact that this problem has not yet been solved satisfactorily for charm and beauty shows its difficulty. Among the various methods envisaged for beauty, i.e. multiplicity steps, multikaon, and multilepton triggers, only the last one is technically feasible with the present detectors at the collider. A single lepton (e or μ) trig-

* Above a value of m of the order of 50 GeV, one can then predict that production of top via the weak process $p\bar{p} \to W^{\pm}$ ..., $W^{+} \to t\bar{b}$ will be more important than hadroproduction (such events will look like jet-jet events[71] and the presence of the heavy quarks may be shown by an abundance of strange particles.)

ger, with $E_T \gtrsim m(t)/4$, to eliminate most of the background*, would keep about 5% of the centrally produced $t\bar{t}$ events. The leptonic decay of one of the top particles, $t \to b\ell\nu$, would give rise to a jet from the beauty decay (in which a charmed combination should be found), and to a measurable missing energy corresponding to the neutrino, in which case the top decay could be reconstructed (the other top particle would probably be difficult to find).

Another possibility is to look for forward production, as was done with success for the Λ_c (and possibly the Λ_B) at the ISR. In that case one could trigger on a high-p_T lepton at angles around 20-30° (as suggested by Horgan and Jacob[74]), accompanied by a forward jet from beauty decay; the other top particle would most of the time give rise to a hadronic jet: $t \to bu\bar{d}$ or $t \to bc\bar{s}$ which may, especially in the last case, be identifiable.

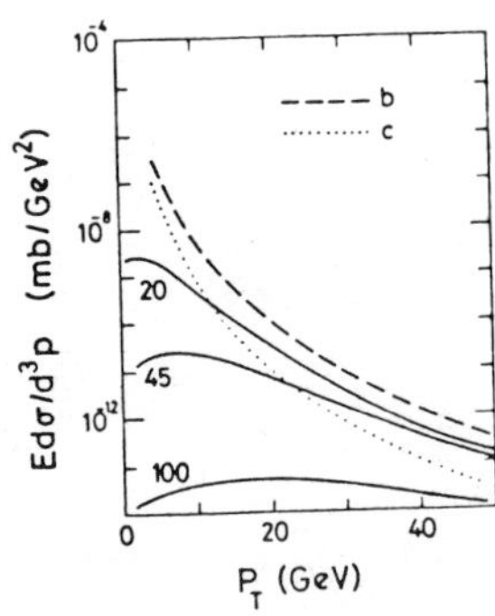

Fig. 37 Single lepton yields at 90° at $\sqrt{s}$ = 540 GeV, from b, c, and t (m_t = 20, 45, 100 GeV) production (Ref. 72).

---

* It is interesting to note[72] that, besides the trivial backgrounds which mar lepton triggers, at collider energy the yield of single leptons from beauty decay is higher than that due to top decay, whatever the p_T value (Fig. 37). Various methods have been proposed[73], using some characteristics of lepton-pair production as a signature of $t\bar{t}$ production, but they need higher luminosities than available at the $p\bar{p}$ collider.

To conclude on the subject of heavier flavours, there is hope to have rather soon some evidence for beauty and top hadronic production (possibly in 1983, with some optimism), but further progress will be rather slow -- until the advent of a new generation of detectors (for beauty) and machines (for top and still heavier flavours). This was the case for charm production, a domain where the first rather contradictory results of 1978 (beam dump) and 1979 (ISR) are just beginning to be qualitatively understood -- and where improved detectors should lead to more confidence in results and trust in their interpretation.

ACKNOWLEDGEMENTS

The author wishes to thank B. Combridge, W. Geist, M. Jacob and S. Reucroft for interesting discussions, and these same people, as well as G. d'Ali, D. Cline, L. Foà, D. Green, C. Fisher, A. Martin, P. Petroff, P.G. Rancoita, J. Sandweiss, T. Sloan, K.C. Stanfield, M.-C. Toutboul, G. Vanderhaeghe and P. Weilhammer, for communication of recent information. The hospitality of the Centro Ettore Majorana was highly appreciated.

REFERENCES

1. J. J. Aubert et al., Phys. Rev. Lett. 33:1404 (1974).
   J. E. Augustin et al., Phys. Rev. Lett. 33:1406 (1974).
2. G. Goldhaber et al., Phys. Rev. Lett. 37:255 (1976).
3. E. G. Cazzoli et al., Phys. Rev. Lett. 34:1125 (1975).
4. S. W. Herb et al., Phys. Rev. Lett. 39:252 (1977).
5. CLEO Collaboration (presented by E. H. Thorndike), "First results on bare b-physics", Proc. XX ICHEP*, Madison, Wisc., 1980 (L. Durand and L. G. Pondrom, eds.) (AIP, New York, 1981), p. 705.
6. G. H. Trilling, Phys. Reports 75:57 (1981).
7. N. Ushida et al., Papers 547 and 835 submitted to the XXI ICHEP, Paris (1982).
   N. Ushida et al., Phys. Rev. Lett. 48:844 (1982).
8. M. Adamovitch et al., Paper 843 submitted to the XXI ICHEP, Paris (1982).
9. K. Abe et al., Paper 735 submitted to the XXI ICHEP, Paris (1982).
10. M. Aguilar-Benitez et al., preprint CERN-EP/82-188 (1982), to be published in Phys. Letters.
11. A. Bardetscher et al., Paper 789 submitted to the XXI ICHEP, Paris (1982), and to be published in Phys. Letters.
12. E. Albini et al., Phys. Lett. 110B:339 (1982).
    S. R. Amendiola et al., Paper 742 submitted to the XXI ICHEP, Paris (1982).
13. Mark II Collaboration, "Measurement of the D^0 lifetime", contributed paper to the XXI ICHEP, Paris (1982).
14. G. J. Feldman et al., Phys. Rev. Lett. 48:66 (1982).
15. G. Kalmus, Rapporteur's talk at the XXI ICHEP, Paris, 1982 (P. Petiau and M. Porneuf, eds.) (J. Phys. 43, Coll. C3, Paris, 1982), p. 431.
16. S. B. Treiman and A. Pais, Phys. Rev. D 15:2529 (1977).
17. L. Maiani and N. Cabibbo, Phys. Lett. 73B:418 (1978).
18. D. Aston et al., Phys. Lett. 94B:113 (1980) and 100B:91 (1981).
19. E. Vella et al., Phys. Rev. Lett. 48:1515 (1982).
20. S. F. Biagi et al., preprint CERN-EP/83-09 (1983).
21. L. Maiani, Rapporteur's talk at the ICHEP, Paris, 1982 (P. Petiau and M. Porneuf, eds.) (J. Phys. 43, Coll. C3, Paris, 1982), p. 631.
22. M. Veltman, Nucl. Phys. B123:89 (1977).
    N. Cabibbo et al., Nucl. Phys. B158:295 (1979).
23. J. G. Branson et al., Phys. Rev. Lett. 38:334 (1977).
    J. Badier et al., Proc. XX ICHEP, Madison, Wisc. 1980 (L. Durand and L. G. Pondrom, eds.), (AIP, New York, 1981), p. 201.

---

* ICHEP = International Conference on High Energy Physics.

24. P. C. Bosetti et al., Phys. Lett. 74B:143 (1978);
P. Alibran et al., Phys. Lett. 74B:134 (1978);
T. Hansl et al., Phys. Lett. 74B:139 (1978).
25. D. S. Barton et al., preprint FNAL-Pub 82/64-EXP, submitted to Phys. Rev. D.
26. W. Geist, private communication.
27. S. Wojcicki, Proc. XX ICHEP, Madison, Wisc., 1980 (L. Durand and L. G. Pondrom, eds.) (AIP, New York, 1981), p. 1430.
F. Muller, "Status of charmed particles", in The high-energy limit (A. Zichichi, ed.) (Plenum Press, New York, 1982), p. 627.
F. Muller, Proc. IV Warsaw Smyposium on Elementary Particle Physics, Kazimierz, 1981 (Z. Ajduk and K. Doroba, eds.) (Inst. of Theor. Phys., Warsaw, 1981), p. 141.
28. R. J. N. Phillips, Proc. XX ICHEP, Madison, Wisc., 1980 (L. Durand and L. G. Pondrom, eds.) (AIP, New York, 1981), p. 1470.
29. A. M. Georgi et al., Ann. of Phys. 114:273 (1978).
30. C. E. Carlson and R. Suaya, Phys. Lett. 81B:329 (1979).
31. M. Gluck and E. Reya, Phys. Lett. 79B:453 (1978) and 83B:98 (1979).
32. K. L. Giboni et al., Phys. Lett. 85B:437 (1979);
W. Lockman et al., Phys. Lett. 85B:443 (1979);
D. Drijard et al., Phys. Lett. 85B:452 (1979).
33. D. Drijard et al., Phys. Lett. 81B:250 (1979).
34. A. Zichichi, "Heavy flavours at the ISR", Proc. ICHEP, Lisbon, 1981 (J. Dias de Deus and J. Soffer, eds.) (EPS, Lisbon, 1982), p. 1167.
35. B. L. Combridge, Nucl. Phys. B151:426 (1979).
B. L. Combridge, "Theoretical approaches to charm hadroproduction", Proc. Moriond Workshop on New Flavours. Les Arcs, 1982 (J. Tran Thanh Van and L. Montanet, eds.) (Ed. Frontières, Gif-sur-Yvette, 1982), p. 451.
36. A. J. Buras and K. J. F. Gaemers, Nucl. Phys. B132:148 (1978).
37. G. Gustafson and C. Peterson, Phys. Lett. 67B:81 (1977).
38. H. Fritzsch and K. H. Streng, Phys. Lett. 78B:447 (1978).
39. S. J. Brodsky et al., Phys. Lett. 93B:451 (1980) and Phys. Rev. D 23:2745 (1981).
40. L. J. Koester et al., "Diffractive hadronic production of D-mesons", Proc. XX ICHEP, Madison, Wisc., 1980 (L. Durand and L. G. Pondrom, eds.) (AIP, New York, 1981), p. 190.
41. J. J. Aubert et al., "Production of charmed particles in 250 GeV μ^+-iron interactions", preprint CERN-EP/82-153 (1982);
J. J. Aubert et al., Phys. Lett. 110B:73 (1982).
42. D. P. Roy, "On the indication of hard charm in the latest EMC dimuon data", Bombay preprint TIFR/TH/83-1 (1983).
43. V. Barger et al., Phys. Rev. D 25:112 (1982) and D 24:1428 (1981).
44. M. Aguilar-Benitez et al., CERN-EP/82-203 (1982), to be published in Phys. Lett.

45. F. Halzen, rapporteur's talk at the XXI ICHEP, Paris, 1982 (P. Petiau and M. Porneuf, eds.) (J. Phys. 43, Coll. C3 Paris, 1982), p. 381.
46. M. Aguilar-Benitez et al., preprint CERN-EP/82-204 (1982), to be published in Phys. Lett.
47. J. Gunion, Phys. Lett. 88B:150 (1979).
48. T. Aziz et al., Nucl. Phys. B199:424 (1982).
49. C. Daum et al., "Inclusive charm production in high energy $\pi^-/p+Be$ interactions", contributed paper to the XXI ICHEP, Paris, 1982.
50. V. L. Fitch et al., Phys. Rev. Lett. 46:761 (1981).
51. A. N. Aleev et al., preprint JINR-E1-82-759 (1982).
52. R. C. Ball et al., "Beam dump neutrino experiment", contributed paper to the XXI ICHEP, Paris, 1982.
53. M. Jonker et al., Proc. XX ICHEP, Madison, Wisc. 1980 (L. Durand and L. G. Pondrom, eds.) (AIP, New York, 1981), p. 242.
54. H. Abramowicz et al., CERN-EP/82-17 (1982) and Z. Phys. C13:179 (1982).
55. A. Bodek et al., "Pion and proton beam dump", contributed paper to the XXI ICHEP, Paris, 1982.
56. M. Barone et al., Nucl. Phys. B132:29 (1978), and references therein.
57. M. M. Block et al., Nucl. Phys. B140:525 (1978).
58. W. Geist, Proc. Moriond Workshop on New Flavours, Les Arcs, 1982 (J. Tran Thanh Van and L. Montanet, eds.) (Ed. Frontières, Gif-sur-Yvette, 1982), p. 407.
M. Heiden, Thesis, University of Heidelberg (1981).
59. D. Blockus et al., Nucl. Phys. B201:197 (1982).
60. R. Rückl, Phys. Lett. 64B:39 (76).
61. R. Odorico, Phys. Lett. 107B:231 (1981) and Bologna preprint IFUB 82/3 (1982).
See also P. Mazzanti and S. Wada, Bologna preprint IFUB 82/9 (1982), and K. Kurek and L. Lukaszuk, "The hadroproduction of heavy flavours", Warsaw preprint (1982).
62. J. Badier et al., preprint CERN-EP/83-12 (1983), submitted to Phys. Lett.
63. A. Diamant-Berger et al., Phys. Rev. Lett. 44:82 (1982).
64. A. Bodek et al., Phys. Lett. 113B:82 (1982).
65. J. Coignet, Proc. ICHEP, Lisbon, 1981 (J. Dias de Deus and J. Soffer, eds.) (EPS, Lisbon, 1982), p. 789.
66. H. Fritzsch, Phys. Lett. 94B:84 (1980).
67. D. Andrews et al., Cornell report CLNS 82/547 (1982).
68. R. N. Coleman et al., Phys. Rev. Lett. 44:1313 (1980).
69. R. Barate et al., preprint CERN-EP/82-171 (1982), to be published in Phys. Lett. B.
70. J. P. Albanese et al., preprint CERN EP/82-183 (1982), to be published in Phys. Lett. B.
71. G. Arnison et al., CERN-EP/83-02 (1983), to be published in Phys. Lett. B.

72. D. M. Scott, "Getting to the top", Cambridge preprint DAMTP 82/1 (1982).
73. W.-Y. Keung, L.-L. Chau and S. C. C. Ting, "Multilepton study for search of new flavors and the Higgs boson", Brookhaven preprint BNL 29598 (1981).
S. Pakvasa et al., Ph. Rev. D 20:2862 (1979).
74. R. Horgan and M. Jacob, Phys. Lett. 107B:395 (1981).

D I S C U S S I O N S

CHAIRMAN: F. Muller

Scientific Secretaries: S. Bellucci, S. Errede, L. Fluri.

DISCUSSION

- *RAUH:*

The lifetime of the F meson seems to be similar to the D^o lifetime suggesting that the annihilation diagram may give an important contribution. Then one should not especially expect $K\bar{K}$ pairs in the final state. The emulsion events however do contain $K\bar{K}$ pairs in the final state. The annihilation mechanism may also lead to glueballs in semi-leptonic decays. Are there any indications of this?

- *MULLER:*

The annihilation mechanism (see fig. 9a) occurs when the c and $\bar{s}$ annihilate to a W^+ which can then "decay" to a u $\bar{d}$ pair (a Cabibbo-favored transition) or leptonically to a $\ell^+\nu$ pair. The bare annihilation diagram is helicity suppressed, however with the emission of a gluon, this process is no longer suppressed. The radiated gluon can combine with (initial state) gluons to give pions. It is true that in the case of semi-leptonic decays that occur via the annihilation mechanism that there may well be glueballs in the final state for the F meson. But you must remember that there are only 9 observed decays of the F meson in the world. Thus, it may be some time before this question can be answered experimentally.

- *RAUH:*

The present experimental data suggest that

$$\frac{Br(D^+ \rightarrow e^+ + X)}{Br(D^o \rightarrow e^+ + X)} \neq \frac{\tau(D^+)}{\tau(D^o)}$$

How do you explain this?

- *MULLER:*

The experimental errors on the measurement of the semi-electronic branching ratios for D^+, D^o decays and the D^+, D^o lifetimes are such that the ratio of branching ratios and the ratio of D^+, D^o lifetimes are not incompatible with each other. Perhaps Professor Sakurai would like to comment on this point?

- *SAKURAI:*

The Cabibbo favored charm-changing current ($\bar{s}c$) is isoscalar. In the semi-leptonic decay of the D^+, D^o mesons which have $I = \frac{1}{2}$, the final hadronic system must also have $I = \frac{1}{2}$, like the $\bar{K}^o$, $\bar{K}^-$. From this follows immediately the Pais-Treiman relation $\Gamma(D^+ \to e^+ + X) = \Gamma(D^o \to e^+ + X)$. This relation is based purely on symmetry arguments. It is independent of whether the spectator mechanism (fig.5a) dominates or the annihilation mechanism (fig.5b) dominates in D meson decays.

So we must have $\dfrac{Br(D^+ \to e^+ + X)}{Br(D^o \to e^+ + X)} = \dfrac{\tau(D^+)}{\tau(D^o)}$.

A violation of this relation would mean that there is something fundamentally wrong in the standard model.

DISCUSSION II (Scientific Secretaries: S. Bellucci, S. Errede, L. Fluri).

- *SHERMAN:*

Professor Zichichi has shown the compatibility of the e^+e^- collisions and pp collisions. At Paris three groups have shown that $e^+e^- \to D^* + X$ at large x is a large fraction of charm production at high energy machines. Several groups at ISR see a D peak in the $K\pi\pi$ systems. Have any D^* signals at ISR been observed?

- *MULLER:*

D^* signals have been observed - some at large x - at the SPS and FNAL with π^- beams, but not at the ISR. This does not necessarily mean they do not exist.

HIGH-ENERGY SOFT (pp) INTERACTIONS COMPARED WITH (e^+e^-) AND DEEP-INELASTIC SCATTERING

M. Basile, G. Bonvicini, G. Cara Romeo, L. Cifarelli, A. Contin, M. Curatolo, G. D'Ali, C. Del Papa, B. Esposito, P. Giusti, T. Massam, R. Nania, F. Palmonari, G. Sartorelli, M. Spinetti, G. Susinno, L. Votano and A. Zichichi

CERN, Geneva, Switzerland
Dipartimento di Fisica dell'Università, Bologna, Italy
Istituto Nazionale di Fisica Nucleare, Laboratori Nazionali di Frascati, Italy
Istituto Nazionale di Fisica Nucleare, Sezione di Bologna, Italy

(presented by A. Zichichi)

1. INTRODUCTION

Last year I presented the first set of data[1] obtained with a new way of studying high-energy, but soft (i.e. low-p_T), interactions between two protons. The study was performed at the CERN ISR using protons with different nominal total centre-of-mass energies, $(\sqrt{s})_{pp}$, i.e.

$$(\sqrt{s})_{pp} = 30,\ 44,\ \text{and}\ 62\ \text{GeV}\ . \tag{1}$$

These nominal energies produce the same effective hadronic energies, where we define

$$\text{Effective hadronic energy} = \sqrt{(q_{tot}^{had})^2}\ . \tag{2}$$

It is this quantity which represents the total energy effectively available for the production of new particles.

The reason is as follows. In a (pp) collision, on the average, the two incident protons carry away about 50% of the incident energy. We have investigated[2,3] this phenomenon as thoroughly as possible — using all published data — and we have reached the following conclusion: no matter if the interaction is strong, electromagnetic, or weak, when a hadron is in the initial state, the energy distribution in the final state will be such as to give a privileged sharing to the initial hadron. This is the "leading" effect. When, in the final state, the initial hadron transforms into another hadron, losing one or more quarks, the "leading" effect will be less. The more quarks are lost from the initial to the final state, the less will be the "leading" effect.

These results are summarized in Figs. 1 and 2.

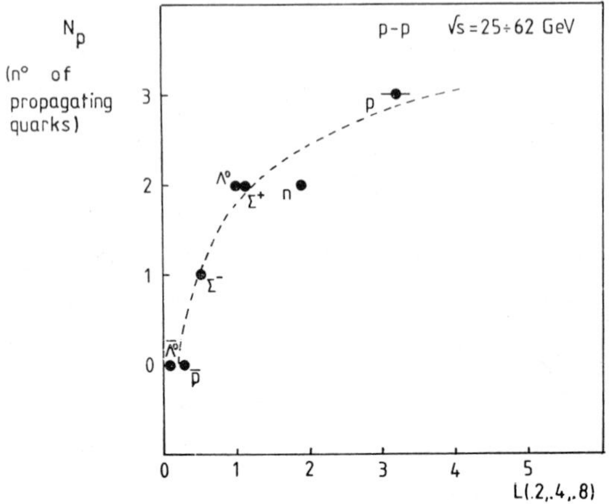

Fig. 1 The leading quantity L(0.2, 0.4, 0.8), for various final-state hadrons in (pp) collisions at ISR energies (25 to 62 GeV), is plotted against the number of propagating quarks from the incoming into the final-state hadrons. $L(x_0, x_1, x_2)$ is defined as $L(x_0, x_1, x_2) = \int_{x_1}^{x_2} F(x)\, dx / \int_{x_0}^{x_1} F(x)\, dx$, where $F(x) = (1/\pi) \int [(2E/\sqrt{s})\,(d^2\sigma/dxdp_T^2)]\, dp_T^2$. The dashed line is obtained by using a parametrization of the single-particle inclusive cross-section, as described in Refs. 28 and 29.

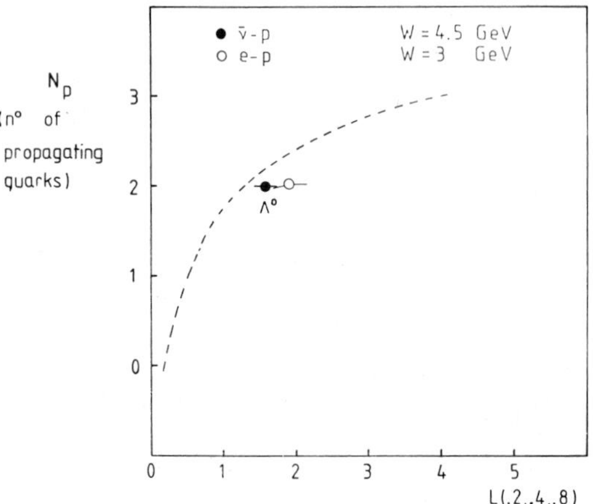

Fig. 2 L(0.2, 0.4, 0.8) for Λ^0 production in ($\bar{\nu}$p) and (ep) reactions. The dashed line is the same as for Fig. 1.

The existence of the "leading" effect has important consequences in the study of hadronic interactions, as we will see in the next section.

2. THE CHOICE OF THE CORRECT VARIABLES

Let us consider a (pp) interaction in terms of the four-momenta involved in the process. This is shown in Fig. 3. There are two incident four-momenta

$$q_1^{inc} \text{ and } q_2^{inc} \tag{3}$$

associated with the two incident protons. The centre-of-mass (c.m.) system is defined as that where the vector sum of the momenta of the two protons is zero

$$(\vec{q}_1^{\,inc} + \vec{q}_2^{\,inc}) = \vec{0} \ . \tag{4}$$

This, at the ISR, coincides with the laboratory system [apart from a small correction due to the fact that the (pp) collision is not a head-on collision]. Here, the nominal c.m. energy is

$$\sqrt{(q_1^{inc} + q_2^{inc})^2} = (\sqrt{s})_{pp} \ . \tag{5}$$

As shown in Fig. 3, the four-momenta, associated with the two "leading" protons in the final state, play a basic role in the choice of the correct variables. In fact, it is not q_1^{inc} which interacts with q_2^{inc}. It is the four-momentum

$$q_1^{had} = q_1^{inc} - q_1^{lead} \tag{6}$$

which interacts with the four-momentum

$$q_2^{had} = q_2^{inc} - q_2^{lead} \ . \tag{7}$$

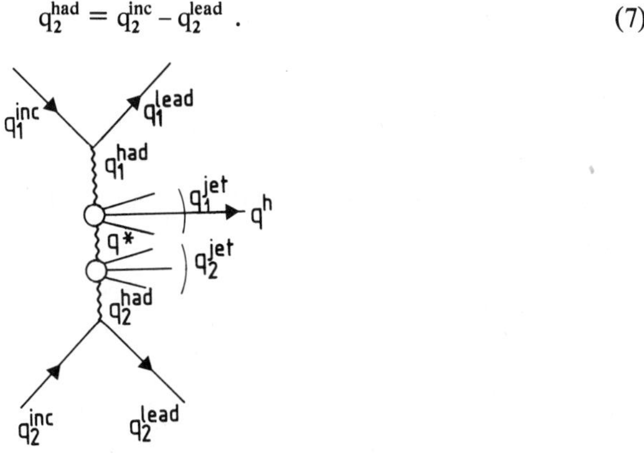

Fig. 3 Schematic diagram for the (pp) interactions.

The effective total hadronic energy available for particle production is given by

$$\sqrt{(q_1^{had} + q_2^{had})^2} = \sqrt{(q_{tot}^{had})^2} . \tag{8}$$

Moreover, the correct reference frame where the hadronic particle production and its properties can be studied is that where the vector parts of the two space-like momenta are equal and opposite, i.e.

$$(\vec{q_1}^{\,had} + \vec{q_2}^{\,had}) = \vec{0} . \tag{9}$$

The quantity (8) is the correct quantity to compare with the total energy in (e^+e^-) annihilation, i.e.

$$\sqrt{(q_{tot}^{had})^2} = (\sqrt{s})_{e^+e^-} . \tag{10}$$

In fact, in (e^+e^-) annihilation q_1^{inc} and q_2^{inc} fully contribute to $(s)_{e^+e^-}$, i.e.:

$$[(q_1^{inc})_{e^+} + (q_2^{inc})_{e^-}]^2 = (s)_{e^+e^-} . \tag{11}$$

In all (e^+e^-) colliding-beam machines, where the two beam energies are equal,

$$(s)_{e^+e^-} = (E_{e^+}^{beam} + E_{e^-}^{beam})^2 = (2E^{beam})^2 . \tag{12}$$

So far, when comparing (pp) interactions with (e^+e^-) annihilation, the basic quantities used were

$$(\sqrt{s})_{pp} \text{ and } (\sqrt{s})_{e^+e^-} . \tag{13}$$

While for the (e^+e^-) case the total c.m. energy measured in the laboratory system from the two incident four-momenta [see expression (11)] is the effective energy available for particle production, for the case of the (pp) interactions this is not true because of the "leading" effect.

Let us now consider the "fractional energy", $(x)_{pp}^{had}$, of the particles produced in a (pp) interaction. From all we have said so far (see Fig. 3), this quantity is

$$(x)_{pp}^{had} = 2\,(q^h \cdot q_{tot}^{had})/(q_{tot}^{had} \cdot q_{tot}^{had}) , \tag{14}$$

where q^h is the four-momentum of the hadron "h" produced and the other quantities have already been defined.

It is now straightforward to compare the fractional energy (14) with that measured in (e^+e^-) annihilation:

$$(x)^{had}_{e^+e^-} = 2\left[q^h \cdot (q^{had}_{tot})_{e^+e^-}\right] / \left[(q^{had}_{tot})_{e^+e^-} \cdot (q^{had}_{tot})_{e^+e^-}\right] \ . \tag{15}$$

The subscripts e^+e^- in formula (15) are there to make it clear that these quantities are measured in (e^+e^-) annihilation and are the quantities equivalent to those measured in (pp) interactions.

For clarity, let us illustrate in Fig. 4 the (e^+e^-) annihilation in terms of the four-momenta involved. From Fig. 4 it is clear that in (e^+e^-) annihilation the total hadronic four-momentum is the sum of two time-like four-momenta; comparing Fig. 4 with Fig. 3 we can see that in (pp) interactions, when studied in the correct way, the total hadronic four-momentum is the sum of two space-like four-momenta.

The same diagram can be used to work out the key quantities needed if we want to compare (pp) physics with deep-inelastic scattering (DIS) processes. In fact (see Fig. 5) in DIS there is an incident four-momentum, q_1^{inc}, associated with a lepton (e, μ, ν) and a four-momentum q_1^{lead}, associated with the same lepton, called "leading" for reasons of analogy with the case of (pp) interactions.

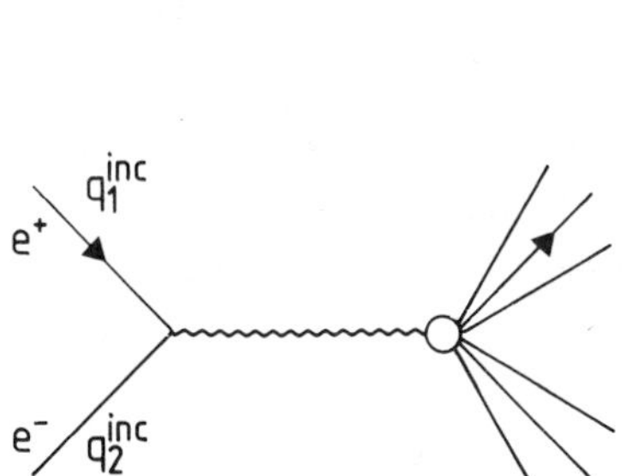

Fig. 4 Schematic diagram for the (e^+e^-) annihilations.

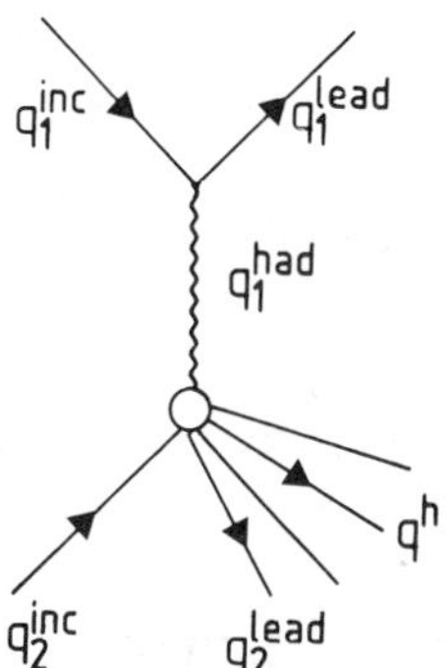

Fig. 5 Schematic diagram for the DIS processes.

These two four-momenta produce q_1^{had}, i.e.

$$(q_1^{inc} - q_1^{lead}) = q_1^{had} , \tag{16}$$

which is space-like. The interaction of this four-momentum q_1^{had}, with the other time-like four-momentum q_2^{inc} produces what is called the total hadronic energy

$$(W^2)_{DIS} = (q_1^{had} + q_2^{inc})^2_{DIS} \ . \tag{17}$$

A comparison of Fig. 5 with Fig. 3 shows that a (pp) collision is a source of four-momenta which can be combined in such a way as to reproduce $(W^2)_{DIS}$ [see expression (17)]. In fact

$$(q_1^{had} + q_2^{inc})^2_{pp} = (W^2)^{had}_{pp} \tag{18}$$

Deep-inelastic scattering processes at given $(W^2)_{DIS}$ can be compared with (pp) interactions using the correspondence

$$\sqrt{(W^2)_{DIS}} = \sqrt{(W^2)^{had}_{pp}}\ , \tag{19}$$

instead of $(\sqrt{s})_{pp} = \sqrt{(W^2)_{DIS}}$.

Let us now consider the fractional energy, $(z)_{DIS}$, of the particles produced in DIS processes (see Fig. 5):

$$(z)_{DIS} = (q^h \cdot q_2^{inc})/(q_1^{had} \cdot q_2^{inc})\ . \tag{20}$$

The corresponding quantity for the (pp) interactions is not $(x)^{had}_{pp}$ of formula (14). The correct fractional energy of a particle produced in a (pp) interaction, analysed "à la" DIS is

$$(z)^{had}_{pp} = (q^h \cdot q_2^{inc})/(q_1^{had} \cdot q_2^{inc})\ . \tag{21}$$

Note that in $(W^2)_{DIS}$ the four-momentum q_2^{lead} is not subtracted. This is the reason for the difference found in the comparison between DIS data and (e^+e^-) data. In fact $(W^2)_{DIS}$ is not the effective total energy available for particle production because it contains the leading proton. Moreover, the W-reference frame is not the correct one if we want to compare DIS processes with (e^+e^-) annihilations.

3. EXPERIMENTAL RESULTS

Last year I reported[1] the results obtained for:

i) the inclusive single-particle fractional momentum distribution of the produced particles[4,5];
ii) the inclusive single-particle transverse-momentum distribution of the produced particles[6–9];
iii) the average charged-particle multiplicity[7,10,14];
iv) the ratio of "charged" to "total" energy[12] of the multiparticle hadronic system produced;
v) the planarity of the multiparticle hadronic system produced.

These quantities were compared with those measured in (e^+e^-) annihilation. Agreement was found in all these comparisons.

This year I will give more results on the comparison of (pp) interactions with (e^+e^-) data. Furthermore, I will present the results obtained in comparing (pp) interactions with DIS processes.

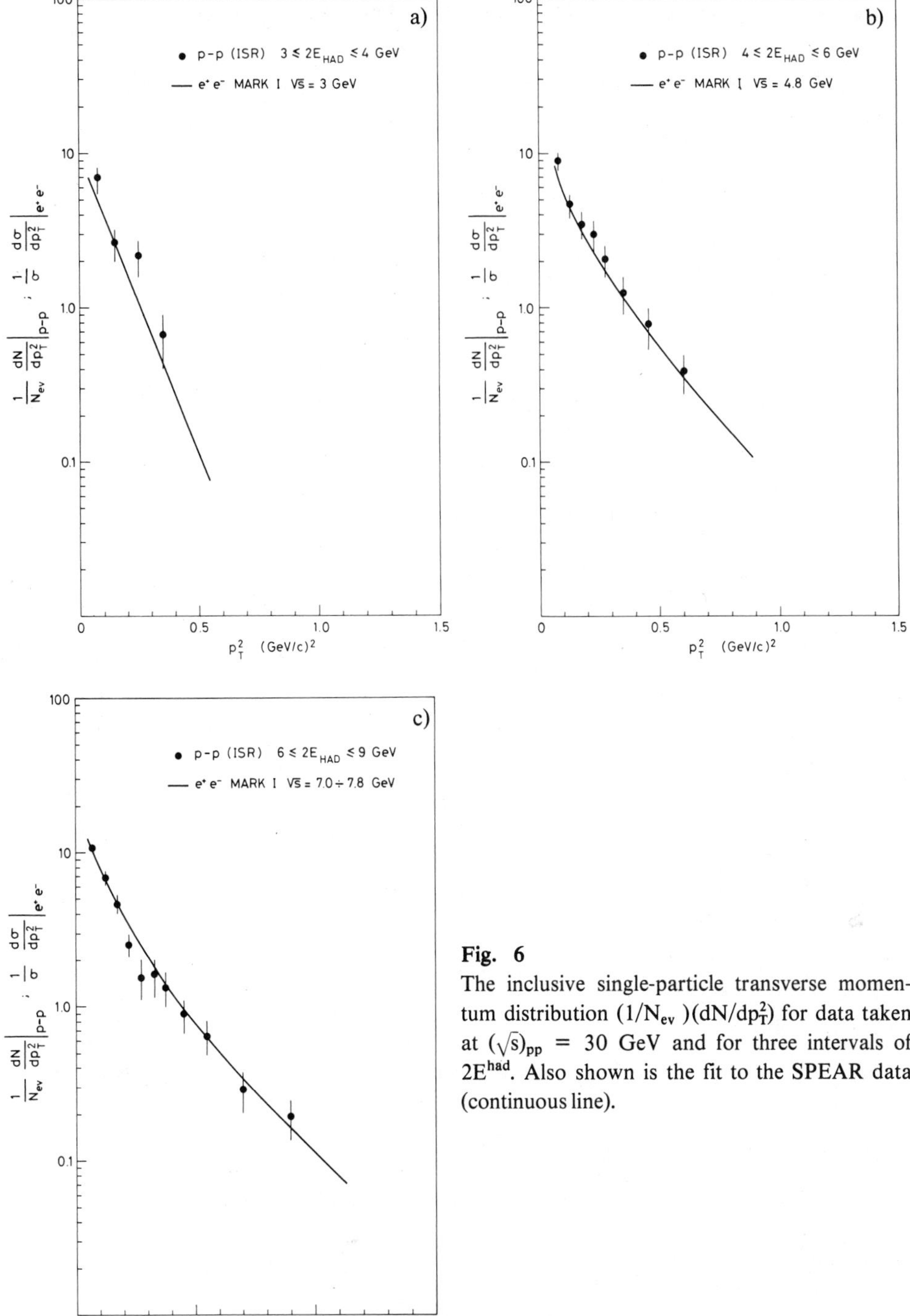

Fig. 6
The inclusive single-particle transverse momentum distribution $(1/N_{ev})(dN/dp_T^2)$ for data taken at $(\sqrt{s})_{pp} = 30$ GeV and for three intervals of $2E^{had}$. Also shown is the fit to the SPEAR data (continuous line).

3.1 More data on the comparison with (e^+e^-) annihilations

These data concern:

i) the inclusive single-particle transverse-momentum distributions[6–8)] at different values of

$$\sqrt{(q_{tot}^{had})^2} \simeq 2E^{had} ,$$

from the lowest (3 GeV) to the highest (32 GeV). (See Figs. 6a,b,c and Fig. 7).

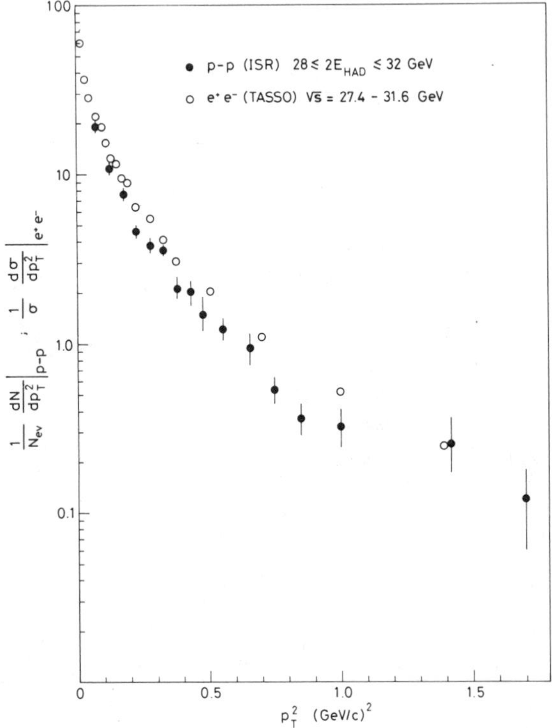

Fig. 7 The inclusive single-particle transverse momentum distributions $(1/N_{ev})(dN/dp_T^2)$ for the range $28 \leq 2\,E_{had} \leq 32$ GeV. Also shown are data from TASSO at PETRA.

ii) The short-range correlations[15,16]. So far, it has been stated that "short-range correlations" in the final states produced in (pp) interactions do not agree with the short-range correlations measured in (e^+e^-) annihilations. This is true if (pp) interactions [and also ($\bar{p}p$) interactions at the ISR] are analysed without using the correct variables (see Ref. 15 and references quoted therein), but, if our new method of studying (pp) interactions is adopted, the short-range correlations measured in (pp) and in (e^+e^-) do agree

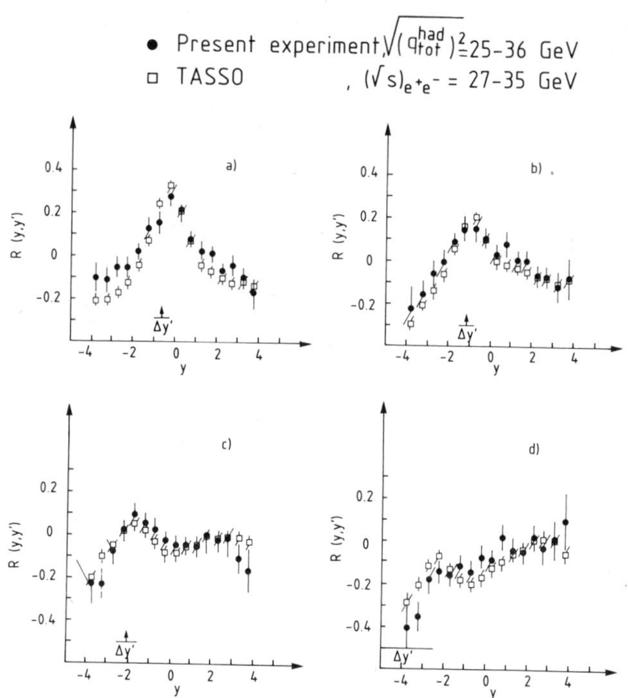

Fig. 8 Two-particle correlation in rapidity space R(y,y′) for different y′ intervals, as measured in the present experiment after the leading proton subtraction in the $\sqrt{(q_{tot}^{had})^2}$ range 25 to 36 GeV (black points), compared with the results obtained by the TASSO Collaboration at $(\sqrt{s})_{e^+e^-}$ between 27 and 35 GeV (open squares).

quite well (see Fig. 8)[15]. Moreover, we can measure the correlation function, R(y,y′)[16] which is defined as

$$R(y,y') = [\rho_2(y,y')/f\rho_1(y)\rho_1(y')] - 1 \ ,$$

where $\rho_1(y)$ is the rapidity single-particle density

$$\rho_1(y) = (1/\sigma_{in}) \ [d\sigma(y)/dy] \ ,$$

$\rho_2(y,y')$ is the rapidity two-particle density

$$\rho_2(y,y') = (1/\sigma_{in})\ [d\sigma(y,y')/dy\, dy'] ,$$

and, finally, f is a normalization parameter

$$f = \langle n_{ch}(n_{ch} - 1)\rangle / \langle n_{ch}\rangle^2 .$$

In Fig. 9, R(0,0) is shown as a function of $\sqrt{(q_{tot}^{had})^2}$. There are no data to compare with (e^+e^-) annihilation because the variation of the short-range correlations as a function of $(\sqrt{s})_{e^+e^-}$ has not been measured so far.

Notice that in Fig. 9 the short-range correlations for the (pp) data as a function of $(\sqrt{s})_{pp}$ are also reported in order to emphasize the difference from the data analysed in the correct way.

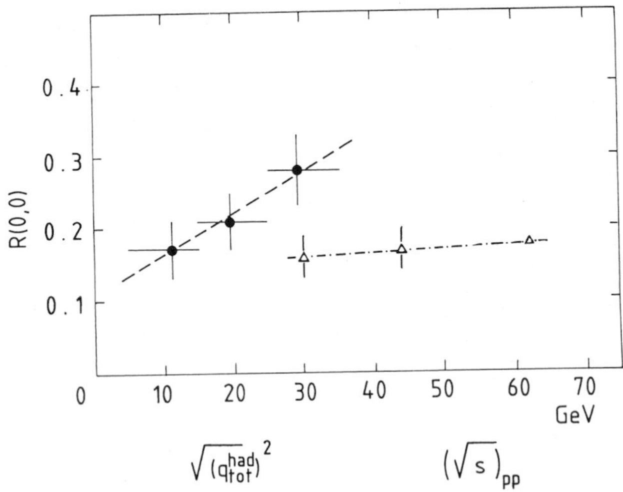

Fig. 9 Correlation function R(y,y′), measured at y = y′ = 0, as a function of $\sqrt{(q_{tot}^{had})^2}$. The data are indicated as black points. The broken line is the best fit. For comparison, the open triangles show R(0,0) as a function of $(\sqrt{s})_{pp}$. In this case the analysis of the final state is made without subtracting the leading proton effects. The dash-dotted line is the best fit to these data. Notice that the abscissa for the black points is $\sqrt{(q_{tot}^{had})^2}$; for the open triangles it is $(\sqrt{s})_{pp}$.

3.2 Comparison with DIS

The analysis "à la" DIS of the final states produced in (pp) interactions was performed using the following quantities:

3.2.1 The inclusive fractional energy distributions of the particles produced[17]

The data are reported in Figs. 10a and b. In Fig. 10a the values of W^2,

$$81 \leq W^2 \leq 225\ \mathrm{GeV}^2 \quad \text{for (pp) interactions}$$

are compared with those having an average value

$$\langle W^2 \rangle = 140\ \mathrm{GeV}^2 \quad \text{for the } (\mu p) \text{ DIS processes.}$$

In Fig. 10b the data with the values of W^2

$$225 \leq W^2 \leq 529\ \mathrm{GeV}^2 \quad \text{for (pp) interactions}$$

are compared with those having an average value

$$\langle W^2 \rangle = 350\ \mathrm{GeV}^2 \quad \text{for the } (\mu p) \text{ DIS processes.}$$

The agreement is remarkable in spite of the fact that different $(q_1^{had})^2$ values are involved in the (pp) and (μp) cases, i.e. those in (pp) reactions are much lower than those in (μp) processes.

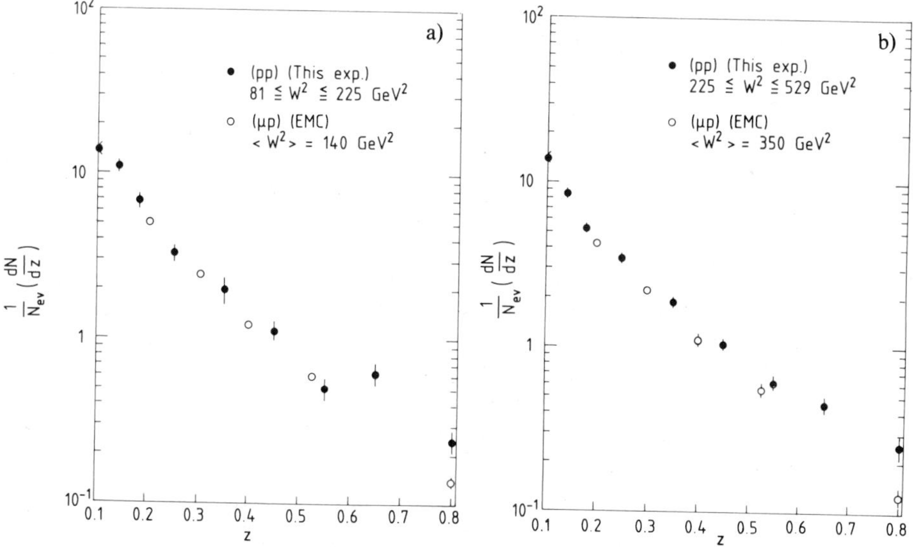

Fig. 10 The inclusive distribution of the fractional energy z for (pp) reactions: a) in the energy interval $(81 \leq W^2 \leq 225)\ \mathrm{GeV}^2$ compared with the data from (μp) reactions at $\langle W^2 \rangle = 140\ \mathrm{GeV}^2$; b) in the energy interval $(225 \leq W^2 \leq 529)\ \mathrm{GeV}^2$, compared with the data from (μp) reactions at $\langle W^2 \rangle = 350\ \mathrm{GeV}^2$.

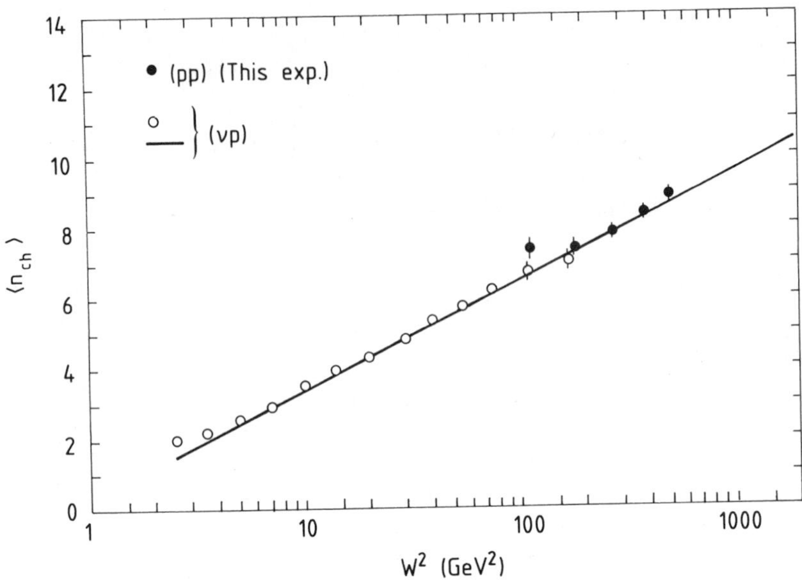

Fig. 11 The average charged-particle multiplicities $\langle n_{ch} \rangle$ measured in (pp) at $(\sqrt{s})_{pp} = 30$ GeV, using a DIS-like analysis, are plotted against W^2 (black points). The open points are the (νp) data and the continuous line is their best fit.

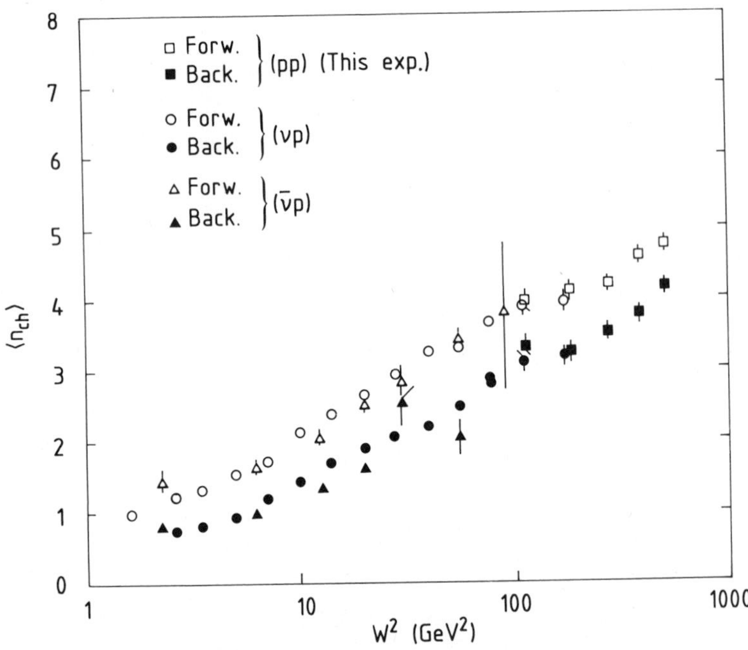

Fig. 12 The mean charged-particle multiplicities $\langle n_{ch} \rangle_{F,B}$ in the forward and backward hemispheres as a function of W^2, in (νp), ($\bar{\nu}$p), and (pp) interactions.

3.2.2 The average charged multiplicity[18]

The data are shown in Fig. 11. Here the comparison is done using (νp) DIS data, i.e. the interaction which produces the final-state hadronic system is weak. Our values of W^2 are higher than those in (νp) data. Nevertheless, it is significant that our data on $\langle n_{ch} \rangle$ lie on the extrapolation of the (νp) DIS data and they are in agreement, within errors, in the overlapping range.

3.2.3 The forward-backward average charged multiplicities[19], $\langle n_{ch} \rangle_{F,B}$

In (νp) and ($\bar{\nu}$p) interactions it has been reported that the "forward" and the "backward" hemispheres do not show the same average charged-particle multiplicities. Using our method we can simulate (νp) DIS processes and measure the average charged-particle multiplicities in both hemispheres. The "forward" hemisphere in (νp) and ($\bar{\nu}$p) is where the lepton is, i.e. where the "leading" four-momentum is. In exact analogy we repeat the analysis made in DIS processes by studying the average charged-particle multiplicities associated with the two hemispheres. The results are shown in Fig. 12.

4. CONCLUSIONS

The use of the correct variables and of the correct reference frame in (pp) interactions at the ISR allows us to establish a common basis of comparison between processes such as (e^+e^-) annihilation, deep-inelastic scattering, and (pp) reactions. The multi-hadronic final states produced in these processes show a set of impressive similarities. Notice that:

- DIS means either weak (νp, $\bar{\nu}$p) or electromagnetic (ep, μp) signals of space-like nature exchanged with a hadron;
- (e^+e^-) annihilation means a time-like electromagnetic signal producing hadrons;
- (pp) interaction means two space-like hadronic signals producing hadrons.

In the last case a comparison can be established with (e^+e^-) annihilation . The (pp) interactions can be studied also in terms of a space-like hadronic signal exchanged with a hadron. In this case a comparison can be established with DIS processes, no matter if they are of electromagnetic or weak origin. In all cases, the basic principle is to use relativistic invariant quantities, no model of any sort.

The hadronic final states, produced in all these apparently different ways, show remarkable analogies as if the source of all these production mechanisms were the same.

REFERENCES

1. M. Basile, G. Bonvicini, G. Cara Romeo, L. Cifarelli, A. Contin, M. Curatolo, G. D'Ali, B. Esposito, P. Giusti, T. Massam, R. Nania, F. Palmonari, A. Petrosino, V. Rossi, G. Sartorelli, M. Spinetti, G. Susinno, G. Valenti, L. Votano and A. Zichichi, "Proc. 19th Int. School of Subnuclear Physics 'Ettore Majorana', Erice, 1981: The Unity of the Fundamental Interactions", Plenum Press, New York–London, p. 695.
2. M. Basile, G. Cara Romeo, L. Cifarelli, A. Contin, G. D'Ali, P. Di Cesare, B. Esposito, P. Giusti, T. Massam, R. Nania, F. Palmonari, V. Rossi, G. Sartorelli, M. Spinetti, G. Susinno, G. Valenti, L. Votano and A. Zichichi, *Nuovo Cimento* **66A**: 129 (1981).
3. M. Basile, G. Cara Romeo, L. Cifarelli, A. Contin, G. D'Ali, P. Di Cesare, B. Esposito, P. Giusti, T. Massam, R. Nania, F. Palmonari, V. Rossi, G. Sartorelli, M. Spinetti, G. Susinno, G. Valenti, L. Votano and A. Zichichi, *Nuovo Cimento Lett.* **32**: 321 (1981).
4. M. Basile, G. Cara Romeo, L. Cifarelli, A. Contin, G. D'Ali, P. Di Cesare, B. Esposito, P. Giusti, T. Massam, F. Palmonari, G. Sartorelli, G. Valenti and A. Zichichi, Phys. Lett. **92B**: 367 (1980).
5. M. Basile, G. Cara Romeo, L. Cifarelli, A. Contin, G. D'Ali, P. Di Cesare, B. Esposito, P. Giusti, T. Massam, R. Nania, F. Palmonari, G. Sartorelli, G. Valenti and A. Zichichi, *Nuovo Cimento* **58A**: 193 (1980).
6. M. Basile, G. Cara Romeo, L. Cifarelli, A. Contin, G. D'Ali, P. Di Cesare, B. Esposito, P. Giusti, T. Massam, R. Nania, F. Palmonari, G. Sartorelli, M. Spinetti, G. Susinno, G. Valenti, L. Votano and A. Zichichi, *Nuovo Cimento Lett.* **31**: 273 (1981).
7. M. Basile, G. Cara Romeo, L. Cifarelli, A. Contin, G. D'Ali, P. Di Cesare, B. Esposito, P. Giusti, T. Massam, R. Nania, F. Palmonari, V. Rossi, G. Sartorelli, M. Spinetti, G. Susinno, G. Valenti, L. Votano and A. Zichichi, *Nuovo Cimento* **65A**: 400 (1981).
8. M. Basile, G. Cara Romeo, L. Cifarelli, A. Contin, G. D'Ali, P. Di Cesare, B. Esposito, P. Giusti, T. Massam, R. Nania, F. Palmonari, V. Rossi, G. Sartorelli, M. Spinetti, G. Susinno, G. Valenti, L. Votano and A. Zichichi, *Nuovo Cimento* **65A**: 414 (1981).
9. M. Basile, G. Cara Romeo, L. Cifarelli, A. Contin, G. D'Ali, P. Di Cesare, B. Esposito, P. Giusti, T. Massam, R. Nania, F. Palmonari, V. Rossi, G. Sartorelli, M. Spinetti, G. Susinno, G. Valenti, L. Votano and A. Zichichi, *Nuovo Cimento Lett.* **32**: 210 (1981).
10. M. Basile, G. Cara Romeo, L. Cifarelli, A. Contin, G. D'Ali, P. Di Cesare, B. Esposito, P. Giusti, T. Massam, R. Nania, F. Palmonari, G. Sartorelli, G. Valenti and A. Zichichi, *Phys. Lett.* **95B**: 311 (1980).

11. M. Basile, G. Cara Romeo, L. Cifarelli, A. Contin, G. D'Ali, P. Di Cesare, B. Esposito, P. Giusti, T. Massam, R. Nania, F. Palmonari, G. Sartorelli, G. Valenti and A. Zichichi, *Nuovo Cimento Lett.* **29**:491 (1980).
12. M. Basile, G. Cara Romeo, L. Cifarelli, A. Contin, G. D'Ali, P. Di Cesare, B. Esposito, P. Giusti, T. Massam, R. Nania, F. Palmonari, G. Sartorelli, M. Spinetti, G. Susinno, G. Valenti and A. Zichichi, *Phys. Lett.* **99B**:247 (1981).
13. M. Basile, G. Cara Romeo, L. Cifarelli, A. Contin, G. D'Ali, P. Di Cesare, B. Esposito, P. Giusti, T. Massam, R. Nania, F. Palmonari, G. Sartorelli, M. Spinetti, G. Susinno, G. Valenti and A. Zichichi, *Nuovo Cimento Lett.* **30**:389 (1981).
14. M. Basile, G. Bonvicini, G. Cara Romeo, L. Cifarelli, A. Contin, M. Curatolo, G. D'Ali, P. Di Cesare, B. Esposito, P. Giusti, T. Massam, R. Nania, F. Palmonari, A. Petrosino, V. Rossi, G. Sartorelli, M. Spinetti, G. Susinno, G. Valenti, L. Votano and A. Zichichi, *Nuovo Cimento* **67A**:244 (1982).
15. J. Berbiers, G. Bonvicini, G. Cara Romeo, L. Cifarelli, A. Contin, M. Curatolo, G. D'Ali, C. Del Papa, B. Esposito, P. Giusti, T. Massam, R. Nania, F. Palmonari, G. Sartorelli, G. Susinno, L. Votano and A. Zichichi, *Nuovo Cimento Lett.* **36**:563 (1983).
16. J. Berbiers, G. Bonvicini, G. Cara Romeo, L. Cifarelli, A. Contin, M. Curatolo, G. D'Ali, C. Del Papa, B. Esposito, P. Giusti, T. Massam, R. Nania, G. Natale, F. Palmonari, G. Sartorelli, G. Susinno, L. Votano and A. Zichichi, *Nuovo Cimento Lett.* **37**:246 (1983).
17. G. Bonvicini, G. Cara Romeo, L. Cifarelli, A. Contin, M. Curatolo, G. D'Ali, C. Del Papa, B. Esposito, P. Giusti, T. Massam, R. Nania, G. Sartorelli, G. Susinno, L. Votano and A. Zichichi, *Nuovo Cimento Lett.* **37**:289 (1983).
18. M. Basile, G. Bonvicini, G. Cara Romeo, L. Cifarelli, A. Contin, M. Curatolo, G. D'Ali, C. Del Papa, B. Esposito, P. Giusti, T. Massam, R. Nania, F. Palmonari, G. Sartorelli, M. Spinetti, G. Susinno, L. Votano and A. Zichichi, *Nuovo Cimento Lett.* **36**:303 (1983).
19. G. Bonvicini, G. Cara Romeo, L. Cifarelli, A. Contin, M. Curatolo, G. D'Ali, C. Del Papa, B. Esposito, P. Giusti, T. Massam, R. Nania, F. Palmonari, G. Sartorelli, G. Susinno, L. Votano and A. Zichichi, *Nuovo Cimento Lett.* **36**:555 (1983).

D I S C U S S I O N S

CHAIRMAN: Prof. A. Zichichi

Scientific Secretaries: Z.A. Dubničkova, M. Sheaff, and G. D'Ali

\- Morse:

Is there a simple way to understand why there is a crossing in the $d\sigma/dx$ distribution when going from low to high energy?

\- Zichichi:

No. There are qualitative QCD arguments which predict a rise of the cross-section in the low-x region and a depletion in the high-x region, with increasing energy, as is indeed seen. The predictions disagree, however, on the cross-over point, and they are, all model-dependent.

\- Kremer:

Your $d\sigma/dx$ curves for pp agree very well with e^+e^- data for small x, but for larger x they are significantly above the e^+e^- data. What is the reason for this?

\- Zichichi:

We believe that this is due to some experimental bias, which we are not able to correct. We are investigating this and hope to solve the problem in the future.

\- Kremer:

Should the analysis you presented to us also be applied to $p\bar{p}$ experiments later?

- Zichichi:

Definitely yes. The results should be the same. We plan to do this on the ISR $p\bar{p}$ data we have already collected. It would be interesting if we could also apply it to the $p\bar{p}$ collisions at collider energy.

- Etim:

(Comment on Kremer's question that there appears to be a systematic difference between the $d\sigma/dx$ distribution obtained from pp scattering and e^+e^- annihilation for large x.) I would like to mention that if one uses a modified x-variable in the pp scattering distribution of $d\sigma/dx$, then the observed difference at large x tends to disappear.

- Zichichi:

Your indication appeared to be valid when we observed only one of the two produced jets, and we were not able to define, in a fully correct way, a Lorentz-invariant quantity for the available hadronic energy. Now that our data take into account both leading protons this is no longer true, and the difference in the $d\sigma/dx$ distribution for large x persists. I repeat that we are confident that we understand the origin of this "difference". This difference is not physics.

- Wigner:

Is there a correlation between the energies of the two "leading protons" resulting from the same collision?

- Zichichi:

No. They are uncorrelated.

- Mukhi:

In comparing pp cross-sections (with the leading particle subtracted) and e^+e^- cross-sections you find a $d\sigma/dx$ about the same for the two processes. At higher energies will gluon-gluon interactions change the pp cross-section compared with e^+e^-? Or do we already have significant gluon-gluon interactions included in the $d\sigma/dx$ you presented?

- Zichichi:

Gluons are produced in e^+e^- reactions only by radiative QCD effects. Protons, on the other hand, contain three quarks and gluons, with the gluons sharing about one-half the total energy. Thus, if effects of gluon-gluon interactions are to be seen, you would expect them to show up in proton-proton machines. We tried to disentangle some gluon-induced effects from our data. For example, QCD predicts that gluon jets will have much higher multiplicities, but we saw no evidence for these. The e^+e^- experiments are also unable to separate gluon jets from quark jets using multiplicities.

- Pernici:

Is it possible to have two leading particles in high-energy pp interactions?

- Zichichi:

Yes. But they are uncorrelated.

- Devchand:

You had some formula for the object L. Could you explain why this measures the "leading effect"?

- Zichichi:

The quantity L is defined by

$$L = \int_{0.4}^{0.8} F(x_L)dx_L \Big/ \int_{0.2}^{0.4} F(x_L)dx \ ,$$

where

$$F(x_L) = 1/\pi \int \left[(2E_c/\sqrt{s})(d^2\sigma/dxdp_T^2)\right]dp_T^2, \quad x_L = 2p_L/\sqrt{s} \ ,$$

and E_c is the energy of the produced particle. It represents a suitable parameter to quantify the difference in the x-distribution between leading and non-leading particles. Moreover, as I have shown this morning, it is connected to the quark flow going from the initial state to the final states. When applied to inclusive proton production it yields 3, compared to the value 0.16 obtained in e^+e^- reactions where there is no quark flow, and to a value of less than 0.5 in pp reactions when the inclusively produced particle contains no quarks in common with the incident proton.

- Dobrev:

How do you determine the ranges of integration in the definition of the quantity L?

- Zichichi:

We make a cut at 0.2 to exclude the central region $0 \leq x \leq 0.2$ and at 0.8 to exclude diffractive production. The choice of 0.4 to divide the regions of integration is made arbitrarily. It has no effect on the actual data analysis.

- Van Baal:

Why do you not use moments of the function F(x), that is, $\int_0^1 x^n F(x)dx/\int_0^1 F(x)dx$, since your definition of L is so arbitrary and might depend on energy (concerning the interpretation).

- Zichichi:

As I stressed before, we used the parameter L only as a tool for our main game; that is, to renormalize the energy available for multiparticle production in soft pp interactions. We can discuss it in more detail privately.

As I have shown this morning, "L" is a parameter which allows us to measure how much "leading" is the leading hadron, which is present in any interaction involving a hadron in the initial state. After the leading hadron has been isolated, we forget about it and continue the analysis on the associated event, having renormalized the effective energy available for particle production.

- Turok:

Could you include some arbitrary parameter in your definition of the central cut in the definition of L? This could then give an idea of how sensitive the graph plots were to such variation.

- Zichichi

Since we never use the definition of L in the analysis, this is obviously unnecessary.

- Duinker:

Do you reject the events where one of the protons has x = 0.9 and the opposite-side proton has x = 0.2?

- Zichichi:

Yes. These events are rejected.

- Yang:

What happens when on the same side you have a pion of x = 0.4 and a proton of x = 0.3?

- Zichichi:

Since we define the "leading" particle to be the particle with the highest x, we would take the pion at x = 0.4 as the leading particle. This is a possible cause of the disagreement between our data and the e^+e^- results at high x in the $d\sigma/dx$ distribution.

- Eisele:

It seems to me that you generally underestimate the leading particle effect because many of the leading baryons should be in an excited state, i.e. N^* etc. Actually, all your comparisons of e^+e^- and pp data show some systematic differences at large x_R^*, which could be due to this effect.

- Zichichi:

We have seen N^* production in our data, and from this signal we were able to establish that N^*'s do not affect our results significantly. Quantitatively, they are produced in $\sim$ 7% of the events. As you pointed out, we slightly underestimate the energy of the "leading" proton in this fraction of the events.

- Lykken:

Does the good agreement between your data and the e^+e^- data imply that final-state interactions between particles in the jet and the leading particle are not important?

- Zichichi:

Yes, otherwise the leading particle analysis would not work.

- Lykken:

Would you expect such effects to show up in the next generation of experiments?

- Zichichi:

It depends on the accuracy. So far we have established that the "leading" particle correlation is below the 2% level. Within this degree of uncorrelation, the leading particle just flies away without seeing the particles in the jet at all.

- Kaplunovsky:

It has been known for a long time that all jets -- e^+e^- jets, DIS jets (both e^- and μ-induced and ν-induced), as well as high-p_T jets -- are similar and have the same fragmentation properties (at the same energy). In fact, all jets, except forward jets in hadronic collisions, show these similarities. Now, after you subtract leading particles, would the forward jets become similar to the other ones?

- Zichichi:

I am reluctant to use the term "jet". I would prefer to call what remains, after the leading particle subtraction, the "multiparticle hadronic system". As we have shown, we can compare the multiparticle hadronic systems from our pp data with the multiparticle hadronic system in e^+e^- and in DIS data via a great number of variables (see the summary given at the beginning of the lecture). As far as these variables are concerned, we conclude that they look similar. To follow your terminology: forward jets look similar to the other ones.

- Kaplunovsky:

At high enough energies in e^+e^- machines, one sees three-jet events as much. Did you see them at the ISR at an effective energy of about 30 GeV? It could be sufficient.

- Zichichi:

Had we seen them clearly, I would have shown them in the lecture. There are indirect indications for the existence of a three-jet contribution. Unfortunately, it is harder to analyse three-jet events on pp machines than on e^+e^- ones. The main reason is the absence of a 4π detector in our case.

- Loewe:

People at PLUTO measured the so-called acolinearity in two-jet events which is related to the so-called DDT form factor in QCD. Is it possible to make such measurements in the two jets you produce in the pp reaction?

- Zichichi:

We have not measured such a quantity. It is very difficult to measure because of the leading proton effect which is not present in the e^+e^- case.

- Heusch:

Just a point of information: I was astonished at the small error bars for your leading effect. From where do you take your data on electroproduction? I an not aware of any large body of experimental evidence.

- Zichichi:

I will give you the references from our paper on the subject.

- Neumann:

How large is the remaining cross-section after restriction to two leading protons so that you get a hadronic system within 50% to 70% of the available energy?

- Zichichi:

The event rate is not limited by the cross-section, but by the maximum possible data acquisition rate for the experimental set-up.

- Neumann:

How well can the hadronic system be reconstructed in its centre-of-mass system (compared to jet analysis, as it is done in e^+e^-)? Is this affected by the Lorentz boost of the event?

- Zichichi:

It cannot be reconstructed as well as in e^+e^- because of the leading particle effect and our acceptance (we do not have a 4π detector). The Lorentz boost does not play an important role in the energy range where we select the leading particles ($x = 0.35-0.86$).

- Van de Ven:

You tell us that your measurements relate the three processes: pp scattering, e^+e^- annihilation, and DIS, and that there is no theoretical understanding of this. Gribov and Lipatov have shown in their fundamental work that there are very precise relations between e^+e^- annihilation and DIS (structure functions). I do not see any way to extend their work to pp scattering.

- Zichichi:

I am sorry, but I hope you agree that priority has to be given to experimental results. Theoretical models come later. The only thing we emphasize is that we observe close similarities in the data, when comparing pp soft processes with e^+e^- and DIS, once the leading particle effect is correctly taken into account.

SPECIAL SESSION ON SYMMETRIES AND GAUGE INVARIANCE

CHAIRMAN: A. Zichichi

Speakers: P.A.M. Dirac, S. Ferrara, H. Kleinert, A. Martin, E.P. Wigner, C.N. Yang, and A. Zichichi

Scientific Secretaries: C. Devchand, V. Dobrev, S. Mukhi, and J.E. Nelson

A. ZICHICHI

The motivation for this special session is based on two facts:

i) we have here in Erice the highest available concentration of experts in this field;

ii) it is true that the same terminology has been used to mean different things.

In 1933, *Pauli* defined a gauge transformation of the *first kind* as that acting on the field ψ, and a gauge transformation of the *second kind* as that acting on the potential A_μ, whereas in 1963 *Coleman* (here in Erice) defined gauge transformations of the *first kind* as global, i.e. the same at each space-time point; and those of the *second kind* as local, i.e. different at each space-time point.

We live in an era of local gauge theories; we believe that the fundamental forces of nature originate from local gauge invariances. Furthermore, the only quantum numbers that are taken seriously, i.e. absolutely conserved, are those associated with a local gauge group. It is not a question of fashion -- it is a basic principle.

I was also asked what is the role of lepton number conservation in the era of local gauge theory. The absolutely conserved charges are at present believed to be the electric charge and the colour charge. What about the baryonic charge? Nobody believes at present in the infinite lifetime of the proton $\tau_p = \infty$, first, because the baryonic charge is not associated with any local gauge-invariant transformation. Second, because if we think that all forces of nature originate from the same source, then quarks and leptons are bound to stay in the same multiplets, and transitions like those in Fig. 1 are bound to take place. So we are ready to believe that $\tau_p \neq \infty$. On the other hand, I want to point out a very important limit. If $\tau_p = 10^{33}$ years, a detector with 10^4 tons of sensitive matter can observe just five events per year. Nobody has yet been able to reach this limit.

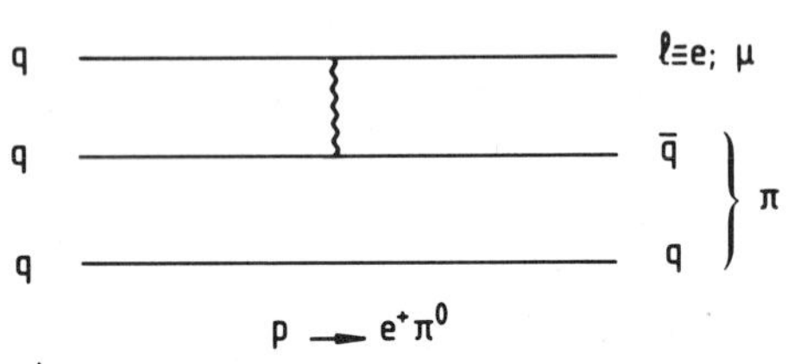

Fig. 1 Diagram illustrating the transitions quark → lepton, and quark → antiquark, which induce the proton to decay into $e^+\pi^0$.

Let me now turn to the main question. At present, what is believed to be true is that local gauge transformations, in one real dimension U(1) produce the electromagnetic forces; in two complex dimensions, SU(2), they produce the weak forces; and in three complex dimensions, SU(3), the superstrong colour forces. The basic property is that all these transformations are local: it is the "locality" of the transformation that explains why all gauge particles are vectors. In fact, the photon, the $W^{\pm}$ and Z^0, and the gluons, are all vector particles. All basic processes of nature can be described by one diagram (see Fig. 2).

What produces the forces of nature is the fact that, by applying a rotation -- in these above-mentioned single or multiple, real and complex, intrinsic dimensions -- you produce no observable effects.

Let me give a schematic view (see Fig. 3) of the present possibilities to let us believe that we can understand such a tremendous range of energy, from 10^0 TeV up to 10^{16} TeV. Conclusion: local gauge invariances produce the fundamental forces of nature.

What about "flavour" symmetry, such as the $SU(3)_f$ of Gell-Mann and Ne'emann? The reason why this global symmetry shows up is, again, the flavour invariance of the gauge forces amongst quarks and gluons. So the conclusion of my present understanding is that

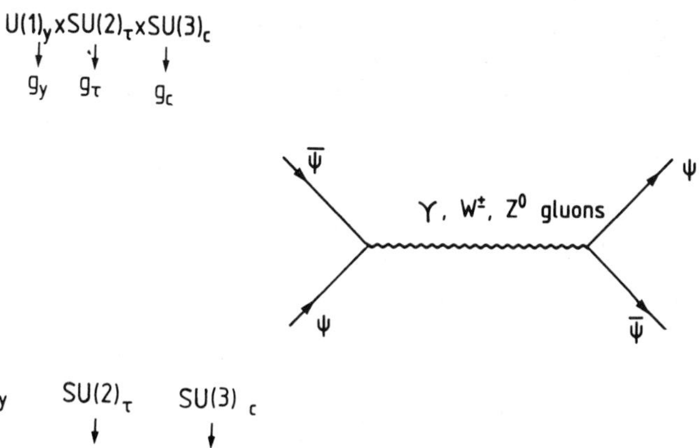

Fig. 2 All "gauge" particles are vector; the diagram illustrates all possible interactions mediated by "gauge" particles.

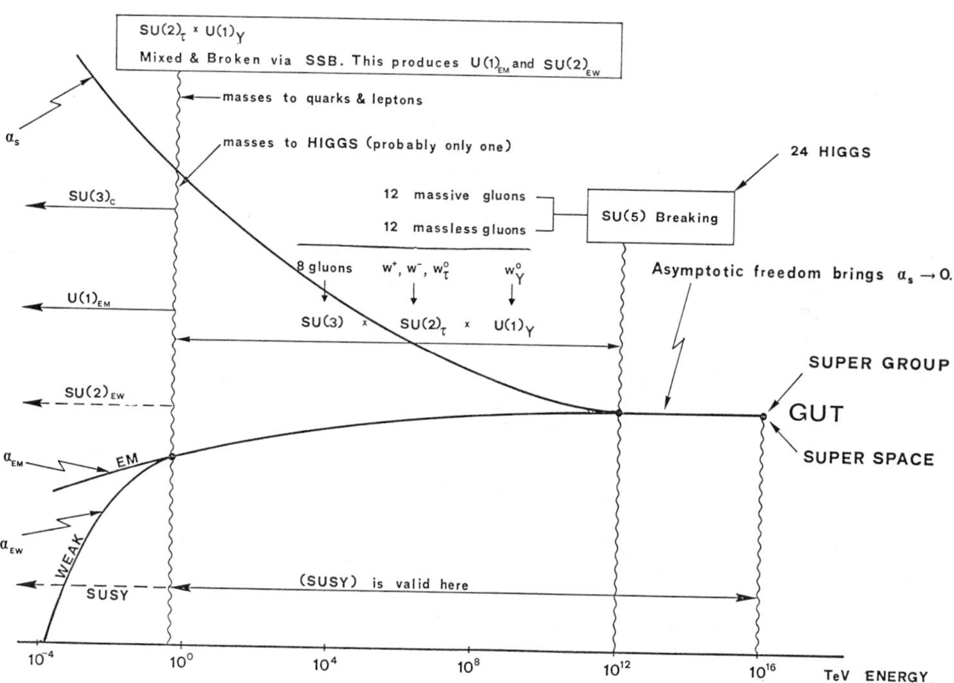

Fig. 3 Our present understanding of the ways in which global symmetries and local gauge invariances span the energy range from 10^0 TeV to the Planck limit of 10^{16} TeV.

the colour local gauge invariance $SU(3)_c$ produces, in addition to the colour forces, the global flavour symmetry $SU(3)_f$. Examples of "global" symmetries, involving 4 flavours are shown in Figs. 4 and 5.

I would like to make a proposal: to use the word *"symmetry"* for the *global invariance* such as $SU(3)_f$ and the word *"gauge invariance"* for the *local invariance* such as $SU(3)_c$.

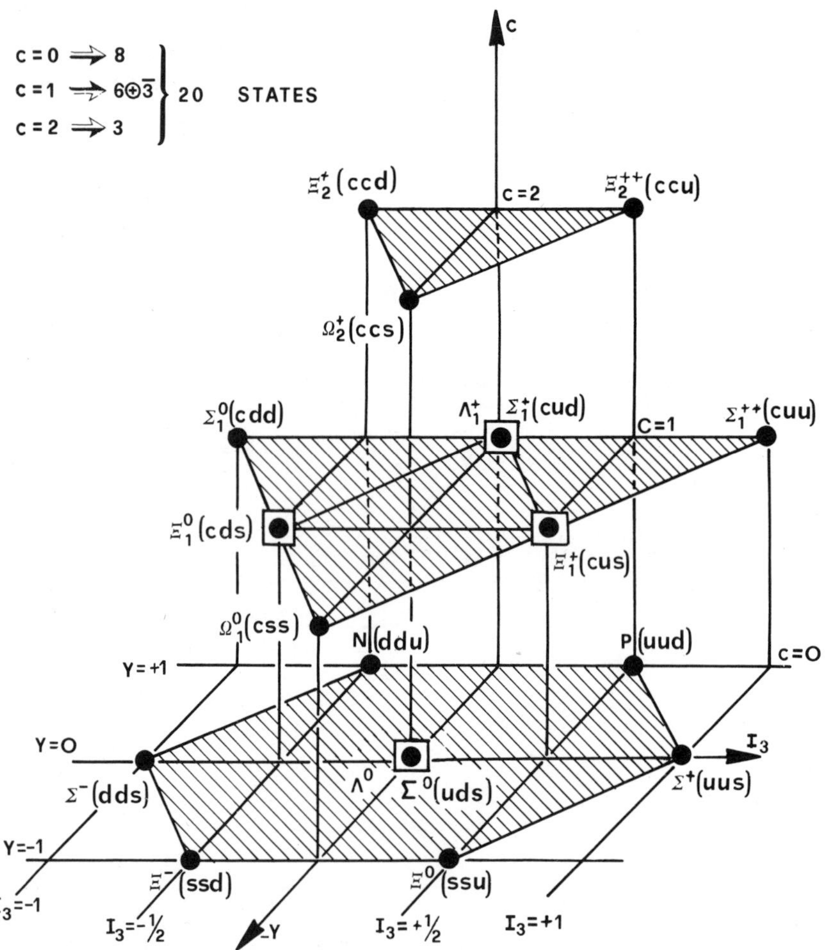

Fig. 4 An example of global flavour symmetry based on the four flavours: udcs. The SU(3) baryon octet is the first plane.

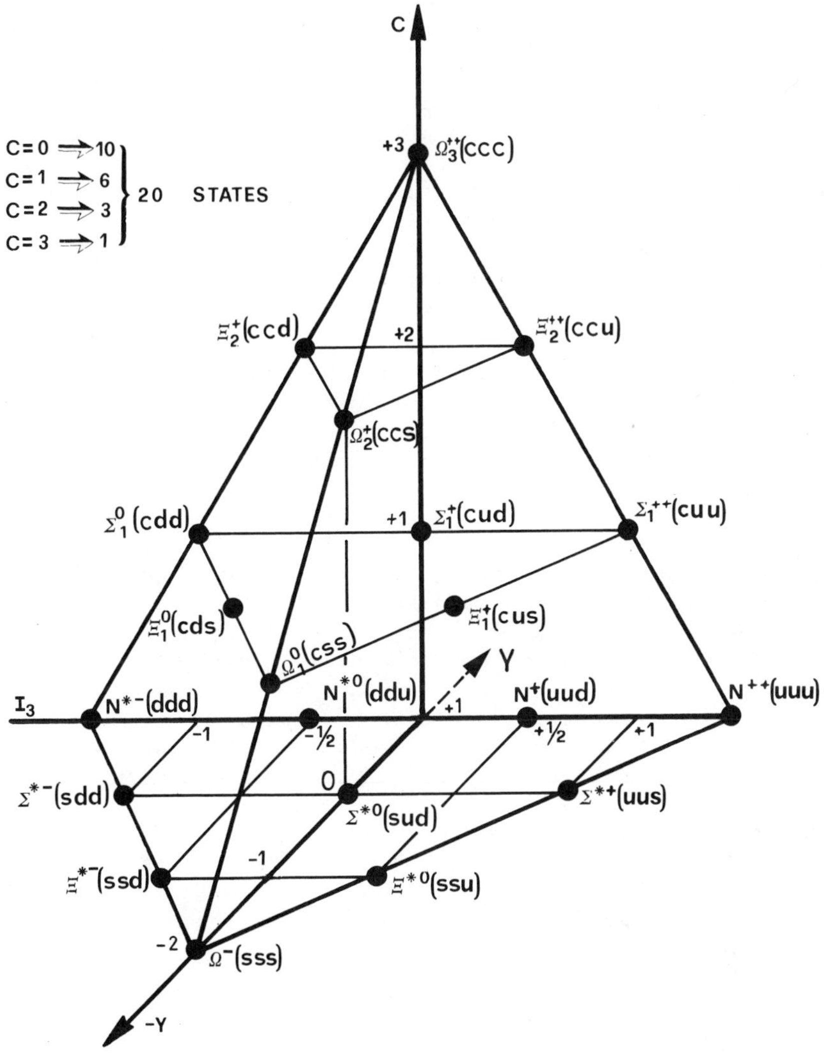

Fig. 5 Another example of global flavour symmetry, based on four flavours: udcs. The first floor is the baryon decuplet.

E. WIGNER

We are fundamentally discussing a question of language: what we now call symmetry. But questions of language are not unimportant. After all, we want to communicate with each other. In physics, the word symmetry has been used, in my opinion, in three different senses. I will try to describe them and tell you why I do not like the idea of gauge invariance being a symmetry principle. But perhaps I am wrong and I will be contradicted. Well, in the ordinary sense, symmetry means symmetry of objects, mainly of crystals. You know that there are 32 crystal classes, 230 space groups; and you know that crystals have remarkable symmetries even though the symmetry is not usually quite as good as it is postulated. But the internal structure obeys the principle of symmetry really amazingly well, and if we were not discussing a very general question, I would have gone into more details of this because there are many interesting things that can be said about it. This is the first concept of symmetry, just a symmetry of objects.

But a more important symmetry is the symmetry of the laws of nature. I used to like to call them invariances, but if we call them symmetries of the laws of nature, I have no objection. It is connected with observations in quantum mechanics; with paths of particles which are real in classical physics. If somebody is displaced in space or time, or rotated, or put into uniform straight motion, the laws of nature will appear the same to him as they appear to an observer here, now, and at rest. In other words, the connection between subsequent states of a system, or between the possibility of producing such systems, will be the same, no matter whether I am here or there, excepting of course that my environment is not symmetric. This is naturally an invariance of the laws of nature, which I call Poincaré transformations (perhaps incorrectly; somebody told me recently that Poincaré did write a paper in which he denied his agreement with Poincaré invariance as I have defined it -- but that does not matter much). These play a role with respect to the laws of nature similar to that which the laws of nature play with respect to events -- they establish properties of these, assert or deny the probability of their correctness. Similarly, the laws of nature, to a certain degree, control the invariance.

I had a discussion with Dr. Kleinert and he suggested that we give a less general but more concrete formulation to these symmetries or invariances. He suggested, for example, that we rely on Heisenberg's conception that what quantum mechanics furnishes are the collision matrices. He therefore suggested that we define as invariances or symmetries of the laws of nature the unitary operators which commute with all collision matrices. I quite agree that this is all right and that it is a much more mathematical definition of the invariance operators in quantum mechanics than the one given by me. But, if I am honest, I prefer a definition which is more abstract and applies to all physical theories. However, I will not forget, I hope, Dr. Kleinert's idea.

These are the first two meanings of symmetry, and even though I prefer the word "invariance" for the second kind; I do not object to the use of the word "symmetry" because I can distinguish the two by calling the first "symmetry of objects" and the second "symmetry of the laws of nature".

The situation is different with respect to the third use of the word "symmetry", for gauge transformations. These express the fact that one description of the physical state of a system is not uniquely determined by the state thereof. The oldest and most commonly known example for this is the description of the electromagnetic field by a "vector potential" usually denoted by A. The electromagnetic fields are given by

$$F_{\mu\nu} = \frac{\partial A_\mu}{\partial x_\nu} - \frac{\partial A_\nu}{\partial x_\mu},$$

and the physical quantities, the $F_{\mu\nu}$, are the same for any two vector potentials A and A′ which differ by a gradient

$$A'_\mu = A_\mu + \frac{\partial \phi}{\partial x_\mu},$$

as one sees at once by inserting A′ for A in the above equation. This does mean that the equations must be gauge invariant -- that is, invariant in their physical content with respect to the gauge transformations just considered -- and this is proven in all textbooks. Thus, for instance, the wave function solving the equations of motion also undergoes a "gauge transformation" in order to satisfy the equations of motion, but this does not change its meaning. This is, in my opinion, an entirely different thing from the invariance of the equations of motion with respect to the Poincaré transformations, which do not change the physical content of the equations of motion. The invariance with respect to gauge transformations only tells us that we should have essentially the same equations no matter whether we use one description of the same physical situation or another.

Perhaps I should also observe that there is another, even simpler, transformation which does not change the system which the function subjected to the transformation describes: this is the multiplication of the wave function by a constant factor of modulus 1. But the existence of neither of these transformations gives any physical information on the properties of the laws of nature -- as the Poincaré invariance gives, for instance. They are transformations which give different descriptions of the same systems, and have no real physical meaning for the properties of the systems, but only for their descriptions. I think they should not have the same name as those which describe true invariances of the laws of nature, since they give only invariances of our description of states.

Perhaps I may admit at this point that I found it unattractive to have fundamental equations referring to quantities which are not observable, such as the electromagnetic vector potential which remains uncertain up to a gradient, which can be added but does not change its physical meaning. I have therefore tried to transform the Dirac equation, for instance, so that the electromagnetic field appearing therein is given by the field strengths, not by the vector potential. But I must admit I did not succeed. This is a rather serious matter.

Well, these are the three types of symmetries that people talk about, but I would prefer not to call the gauge invariance a symmetry, because it does not express anything physical. It only tells us something about our mode of description of the physical situation. This is what I wanted to communicate to you, but let me repeat that what I am arguing about is a question of language. It may be a not entirely unimportant question of language, since we do not want to use words in two very different senses.

C.N. YANG

A few days ago when Professor Wigner discussed the point which he has now made more explicit, I said I was not absolutely sure that you cannot construct an experimental apparatus which will illustrate the symmetry which is implied by gauge invariance. But I have since thought about it and have come to the conclusion that he is entirely right. So what I would say is merely a footnote to what he has just said and I think it is a very interesting point which I certainly did not realize before. So, with Professor Wigner's permission, I will explain to you my understanding of what he meant.

A traditional symmetry, for example spherical symmetry, implies that if you find a trajectory of motion you can generate other trajectories of motion from that. For example, since the Coulomb field is spherically symmetric, if you find an orbit which is elliptical, it follows from spherical symmetry that there are other orbits which are obtained by rotation of that orbit. So when the symmetry argument is applied it generates, from one orbital motion, a string of other orbital motions, except for degenerate cases like circular motion. Similarly, if we observe the decay $\pi^+ \rightarrow \mu^+\nu$ of π^+ at rest, we can predict, just from the Lorentz symmetry of forces involved in the decay, properties such as the lifetime of a π^+ and the angular distribution of its decay products when π^+ is in motion. So the usual symmetries, such as the Lorentz symmetry, have the specific property that, given some experimental apparatus which detects something, we can construct another "transformed" experimental apparatus for which we can predict the results. This is precisely the principle with which parity was tested. The parity test in which

Dr. Hayward, who is here, participated, was essentially two pieces of equipment which were mirror images of each other and the test was whether they gave the same result.

So now the question is: If gauge invariance is satisfied, can we construct two experiments which are gauge transforms of one another? And if so, will it be similar to the parity test? If we cannot do that, then clearly gauge invariance is a different kind of symmetry. I thought about this and I came to the conclusion that Professor Wigner is entirely right. If you have any experimental set-up, and you make a gauge transformation of it, it is still the same experimental set-up and you will get the same result. So in that sense, gauge invariance in terms of its realization on experimental equipment is different from the parity test experiment. I hope that Professor Wigner agrees with what I have said.

- WIGNER: Absolutely.

- YANG: So what, then, is gauge invariance? It is in some sense a symmetry which really tells us how interactions are formed. I have called this "symmetry dictates interactions". There is some similarity, in the mathematical structure of gauge invariance, with the "usual" kinds of symmetries, but physically it is really quite different. Professor Zichichi has proposed that we call the usual type "symmetry", and that we do not say "gauge symmetry" but "gauge invariance". I think it will be a good terminology to adopt.

A. MARTIN

I am afraid that what I want to say is rather orthogonal to what was said before. I will comment on the connection between gauge invariance of the strong interactions, QCD, and the approximate flavour symmetries that we observe in nature. Before QCD we had flavour symmetries such as isospin invariance, namely the group SU(2); then we had another flavour invariance, the group SU(3) introduced by Gell-Mann and Ne'eman; and later on, the symmetry which is not really a symmetry, SU(6), inspired by the supermultiplet theory of Professor Wigner. Then suddenly came QCD, and since then we no longer speak about flavour symmetries as basic symmetries. Why? We have this SU(3) group of colour invariance, which is non-Abelian and therefore imposes some constraints on the nature of the interaction of quarks and gluons. Crudely speaking, we have the interactions of the quarks with the gluons, but the gluons also have a self-interaction which is controlled by the same coupling constant α_s. This same α_s controls the interactions of all quarks of different flavours and thus reproduces the flavour symmetry. So, now we have a much better understanding of flavour symmetry than we had previously. As we discover new generations of quarks it is not necessary to change our picture, but only to incorporate the new quarks into the basic SU(3) colour interactions of quarks and gluons.

I will now give a brief summary of the successes of this approach. I remind you that by using this QCD approach we rediscover all the basic results of SU(3) flavour, such as the equal spacing of the baryon decuplet, and the Gell-Mann - Okubo formula for the octet. All we have to do is decide that the symmetry is broken by the masses of the quarks taken phenomenologically.

The miracle of QCD is that the interactions of the quarks and gluons are constructed in such a way that when you have a meson represented by a $q\bar{q}$ colour singlet there is an attractive force between the q and the $\bar{q}$ -- and similarly for a baryon made of three quarks, which is globally a colour singlet. We understand how the symmetry is broken, and in particular the successes of the spin-dependent forces. In the one-gluon exchange approximation these forces are essentially of the Fermi type: and in this way, for the first time, we understand the mass difference $M_\Sigma - M_\Lambda$:

$$M_\Sigma - M_\Lambda = \frac{2}{3}\left(1 - \frac{m_u}{m_s}\right)\left(M_\Delta - M_N\right).$$

This naïve formula works extremely well. Now, for the Σ_c and Λ_c observed last year, we obtain from this formula:

$$M_{\Sigma_c} - M_{\Lambda_c} = 170 \text{ MeV},$$

whereas experimentally it is 166 MeV. Similarly we predict that

$$M_{\Sigma_b} - M_{\Lambda_b} = 190 \text{ MeV}.$$

In the case of mesons we understand very well in a quantitative way the mass differences between vector and pseudo-scalar particles, and we can predict the mass difference

$$m_B^* - m_B = 50 \text{ MeV}.$$

But in my opinion the most spectacular success of flavour symmetry occurs when it is applied to systems composed entirely of "heavy" quarks, namely $b\bar{b}$, $c\bar{c}$, and even $s\bar{s}$ and $c\bar{s}$. Then, since the interaction is the same between all the constituents (because of flavour independence *induced* by colour invariance) we predict the existence of a universal potential for all these systems. Various people have proposed various potentials, but all look alike at distances between 0.1 and 1 fm. I have my own naive version of colour potential (Fig. 6):

$$V = -8.064 + 6.870\ r^{0.1} ,$$

and with this potential we can reproduce the whole spectrum of $c\bar{c}$, $b\bar{b}$, and $s\bar{s}$ systems (Table 1). For $c\bar{c}$, we fit the mass of the charmed quark to the ground state of the system and predict the masses of ψ', ψ'', etc., accurately to within 15 MeV. With the same potential fitting the Υ mass, we predict the $b\bar{b}$ spectrum accurately to within 20 MeV (Table 1). At last week's Paris conference, the CUSB (Columbia-Stony Brook) group announced that they have observed the χ_b' with a mass of 10.247 GeV, whereas the theory predicted 10.24 GeV. We can do the same for the $s\bar{s}$ system and predict $m_{\phi'} - m_{\phi} = 0.615$ GeV, whereas experiment gives 0.630 GeV. We also get a good prediction for m_F (Table 1). Finally, we can try to make predictions for baryons. When the force is restricted to one-gluon exchange, we know that V_{QQ} inside the baryon equals $\frac{1}{2}\ V_{Q\bar{Q}}$ inside the meson. If we take this as a strict rule, we can calculate (from the previously determined potential in the meson sector) the binding energies of the baryons. For instance, Richard has calculated $M_{\Omega^-} = 1.665$ GeV, whereas the experiment gives 1.672 GeV.

It seems that we now understand that colour invariance of the theory produces flavour invariance of the forces because different flavours possess the same colour charges. In the case of heavy

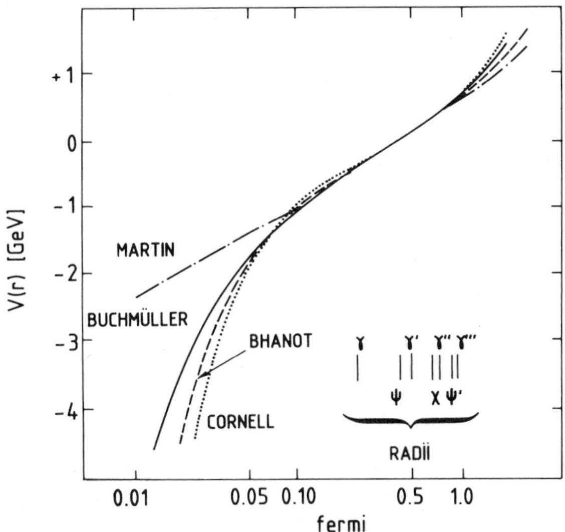

Fig. 6 Several potentials reproducing the levels of the $b\bar{b}$ and $c\bar{c}$ system. One notices that they approximately coincide from 0.1 to 1 fermi.

Table 1.

Excitation energies and relative leptonic widths for the $s\bar{s}$, $c\bar{c}$ and $b\bar{b}$ systems compared with the theoretical predictions of Buchmüller et al. and Martin. Also given are A. Martin's predictions for m_F and m_F^*.

	$c\bar{c}$			$b\bar{b}$			$s\bar{s}$	
1^{--} ground state	3095 MeV			9460 MeV			1020 MeV	
Quark mass	Exp.	Buchmüller 1480 MeV	Martin 1800 MeV	Exp.	Buchmüller 4870 MeV	Martin 5174 MeV	Exp.	Martin 518 MeV
2S - 1S	589	600	592	561	560	560	630	615
$\Gamma_{ee}(2S)/\Gamma(1S)$	0.46 ±0.06	0.45	0.40	0.43	0.44	0.41		
3S - 1S	935	1020	937	890	890	900	$c\bar{s}$ No free parameter	
$\Gamma_{ee}(3S)/\Gamma(1S)$	0.16?	0.32	0.25	0.32	0.32	0.35		
4S - 1S	1319?	1380	1185	1116	1160	1140		
$\Gamma_{ee}4S/\Gamma(1S)$				0.21	0.26	0.27	Exp.	Th.
1P - 1S	425	420	407	440	430	400	$m_F = \begin{cases} 2020 \\ 1970 \end{cases}$	1990
2P - 1S				796	790	780	$m_F^* = 2140?$	2110
1D	677	710	692					

quarks, it is most simply described by a universal potential between them. So we have made tremendous progress because we no longer need to speak of flavour symmetry groups. I do not like the idea of using SU(4) to incorporate charmed quarks, because the symmetry breaking by the mass is very large, whereas the forces are the same. And if someone discovers a new quark we only need to iterate.

As Professor Wigner and Professor Yang said, changing the gauge leaves the physical structure unchanged. Nevertheless, this structure is so constrained in the case of a non-Abelian gauge group that we see beautiful physical consequences.

S. FERRARA

Concerning the definition of symmetry: what Professor Wigner and Professor Yang were talking about is related to the fact that we have an S-matrix in which only physical states appear, or, in terms of quantum field theory, where only asymptotic states appear. Then we can discuss the symmetry of the S-matrix -- for instance, its Poincaré invariance. In this case the symmetry is related to something which can be observed. However, the word symmetry is sometimes used nowadays with a different meaning, namely in connection with an action principle or with a certain Lagrangian. A symmetry is something which leaves the action invariant. This concept of symmetry is more general than the invariance of the S-matrix, because we could have an S-matrix without any Lagrangian field theory underlying it. So, if I define a symmetry as some operation which leaves the action invariant, then all symmetries are unified in the same mathematical sense. Poincaré symmetry, gauge invariance, and even general coordinate transformations have the same meaning; the action can be written as a space-time integral of a Lagrangian density. What is meant by invariance is that a certain operation on the Lagrangian transforms it into a total divergence. We want the most general transformation of the Lagrangian which leaves the action unchanged. Defined in this way, symmetry could be physical like the Poincaré symmetry applied to the S-matrix, or non-physical like local gauge or general coordinate transformations. For instance in Einstein's theory, when I write the Einstein Lagrangian as a scalar density the action is invariant under general coordinate transformations. However, in quantum gravity where there are gravitons (analogous to photons in QED), I could never interpret, from an S-matrix point of view, the general coordinate transformations. So, from my point of view, the general coordinate transformations are exactly like the Yang-Mills gauge symmetry but in a more general sense. Poincaré symmetry is also of this type but with a different interpretation; Poincaré and flavour symmetries are symmetries of asymptotic states. They give information about the S-matrix. It is clear that this concept of invariance applied to a Lagrangian or an action plays a crucial role in Professor Yang's discussion. I think that symmetries do indeed dictate interactions. If I have particles of spin 1, spin 2, spin 3/2, etc., from the S-matrix viewpoint there are no problems in describing Poincaré states of any spin. On the other hand, in a Lagrangian field theory there are problems in describing interactions of higher spin particles. And in fact the interactions of spin 1 and spin 2 particles are uniquely fixed by gauge invariance: for spin 1 by Yang-Mills gauge invariance, and for spin 2 by the gauge invariance of Einstein, namely, general coordinate transformations.

H. KLEINERT

I think we all agree with Professor Wigner when he said: "Symmetry is the invariance of the laws of nature". The question is, What is meant by invariance of the laws of nature? To that question Professor Wigner said that according to Heisenberg the symmetry of the laws of nature is the symmetry of the scattering matrix. In history this concept has been generalized in two ways. One is really a generalization of the concept of symmetry to include unphysical transformations of the S-matrix, for example isospin symmetry. If you perform an isospin transformation on a proton p you get $p' = p \cos \alpha + n \sin \alpha$, but there is no way of making such a mixed state. So, we talk about an isospin-symmetric world even though the mixed states cannot be experimentally observed. The second case is: invariance of the laws of nature means invariance of the field equations. But field equations involve Green functions, which are off-shell quantities, whereas scattering amplitudes are on-shell. I will therefore conclude by saying that there are structural invariances which are not directly observable in nature, but whose consequences we can test. For example, colour gauge invariance directly implies quark confinement.

A. ZICHICHI

I want to make a remark on the last point. The fact that quarks do not escape is certainly not the most direct check of the existence of colour and of its conservation. For the existence of "colour", there are precise measurements that one can make, such as the π^0 lifetime, the ratio R of hadronic to muon pair production in (e^+e^-) annihilation,and the Drell-Yan effect. There is a factor of 27 between the two extreme measurements. It might be that confinement is a low-energy phenomenon; low, when compared with the GUT energy levels. In this case it could be that we will never be able to be sure about the real meaning of confinement.

C.N. YANG

I want to make some rather general, perhaps philosophical, remarks. It is very obvious that symmetry is playing an essential role in our concept of the structure of the physical universe. It is extremely important to realize, as Professor Wigner first pointed out, that the meaning of gauge symmetry is different from previous uses of the word "symmetry". It seems to me that precisely what symmetry is and how it is to be applied to fundamental physics is bound to be an evolving process. In fact, the very word "symmetry" has, in some sense, almost emotional connotations, and I am not talking about the emotions of physicists; I am talking about the emotions of mankind. After all, when early man first found that babies have two hands which are marvelously symmetrical, that clearly generated a feeling of respect or awe or whatever it is. It is also extremely interesting that the early philosophers did seize upon the concept of symmetry as something which might play a role in the human understanding of the universe. Plato seized upon the platonic solids which are, in fact, beautiful symmetries. He wanted to identify four of the five symmetrical solids (regular polyhedrons) with what were thought to be the four fundamental elements: earth, water, air, and fire. In China, in the philosophy of I Ching, a different track was followed. The Chinese philosophers had a tendency to think more in algebraic terms. So the I Ching is built on an algebraic binary system of 0's and 1's at once symmetrical and complementary.

These ancient Greek and Chinese symmetries are very different from what we are now talking about. But I submit that there is clearly a deep relationship between the ancient concept that symmetry is related to the structure of the universe and our present concept that *symmetry dictates interactions*. And if you go to more recent history, e.g. the discovery of the symmetry of space groups to which Professor Wigner referred, a beautiful and profound discovery which had a tremendous impact in those days, was that the same symmetry we are talking about today? No. Is it related? Of course -- it is deeply related.

Let us take next the case of CPT symmetry. Is it a symmetry? Yes. But CPT involves much more complicated things -- it involves causality and analyticity and it is, therefore, very different from simple Lorentz symmetry.

I would submit that it is very likely that the basic concept of symmetry will continue to evolve in its applications to physics. In which direction would this evolution proceed? What new mathematics would it require? No one knows the precise answers to these questions, but that is why I asked Professor Dirac yesterday whether he thought that the new mathematics would come from the realm of analysis, geometry, algebra, or topology. And he said he thought it would be from analysis.

E. WIGNER

I have something very remarkable to point out, namely that there are approximate symmetries. The existence of approximate symmetries is really in some ways a miracle, and is closely connected with the fact that there are four different kinds of interactions. As you know, one of Professor Yang's most important accomplishments was to point out that reflection symmetry is only approximate. But the other symmetries which Dr. Martin discussed are even more approximate and in some ways very crude. But they exist and are useful and we should learn about them. The question naturally arises, Is any symmetry better than approximate? The great philosopher Mach said No, everything that we know is only approximate. I hope you will work on these problems and try to clarify them and find out whether Mach was right or whether anybody else was right, and whether there is a reason for so many approximate, very useful symmetries, and whether any other symmetries are really exact.

P.A.M. DIRAC

I would like to say that I have been very much impressed by the recent developments in fundamental physics based on symmetries and gauge theories. The various successes of these theories have come to me as a succession of surprises. I think that this is the best way to put it. Each success is a surprise. At the same time I am unhappy about the situation. The reason is that there is a basic problem which is unsolved -- the problem of explaining the number hc/e^2, 137. Why does it have this value instead of some neighbouring value? Well, we have no idea how this question is to be answered and I feel that, until it is answered, all the rest of physics is uncertain. This number, 137, is *the* main problem of physics, the most important problem of physics, and we are working in the dark with respect to it. When this problem is solved we shall have a different base in physics -- we shall have different fundamental fields -- and then our theory will be able to take on a more precise form.

There are so many uncertainties still surviving in modern developments of symmetries and gauge theories, and these questions and uncertainties will then have a definite answer. We should be able to give a definite answer as to whether the proton is stable or not. I feel that the present kind of development is the best that one can do at the present time, and I do not want to discourage it in any way, but people should keep their eyes open for the possible arrival of a new theory in which the number 137 is explained.

A. ZICHICHI

The conclusion of this interesting and illuminating round table is as follows:

1933 - Pauli defined the gauge transformations to be of

- 1st kind - acting on the field ψ,
- 2nd kind - acting on the potential A_μ.

1963 - Coleman (Erice) defined the gauge transformations to be of

- 1st kind $\equiv$ global,
- 2nd kind $\equiv$ local.

1982 - Erice proposal

- Symmetry $\equiv$ global symmetry,
- Gauge invariance $\equiv$ local gauge invariance.

Never use "gauge" when the symmetry is global. Never use the word "symmetry" when the invariance law is local.

THE GLORIOUS DAYS OF PHYSICS:

PROFESSOR DIRAC'S BIRTHDAY

- A. Zichichi

We now come to the formal part of the celebration of Professor Dirac's 80th birthday. This celebration will have many components.

You probably do not know that Professor Dirac, among his very many activities, collaborates with the weekly page on science which a newspaper of Italy, called 'Il Tempo', publishes every Saturday. The new feature of this page is that only scientists are allowed to write -- not journalists -- and Professor Dirac is one of the most distinguished contributors to our Science Page. So in order to mark this anniversary, the newspaper has decided to attribute the gold medal of "Modern Culture" to Professor Dirac. This is not all. There is also a notebook on the back of which is printed "Eightieth birthday of Professor Paul Adrien Maurice Dirac, Erice, 8 August 1982". As you know, Professor Dirac does not like to be the centre of attraction or of any attention. So this is why the Director of the newspaper has decided to offer the same gift to all of you; it will be given to you at the end of the festival to keep as a souvenir of this great evening.

I would like now to give the floor to Professor Wigner. As you know, this year we had in the programme "The Glorious Days of Physics". It is very important for the new generations to know what was going on in the past days: in fact our admiration for Professor Dirac is due to his extremely imaginative and outstanding contributions to Physics. When you read his papers you realize the tremendous difficulties he had to face. He had to understand what negative energy means. He showed an incredible fantasy in postulating, for example, the uniform sea of negative energy electrons in which we are supposed to be. We know it is not right, but it led to the idea of antiparticles, and if you read his Nobel prize lecture he even conjectured the existence of antigalaxies. On this occasion we therefore want to express to Professor Dirac our great esteem for his work. As the present School shows, his interventions during discussions demonstrate that what he has done is still alive; and it is not easy for a scientific work to be alive after so many years. On the occasion of this great celebration I would like, on behalf of all the students here -- and I am sure of all physicists in the world -- to express to Professor Dirac our deepest admiration for his tremendous fantasy and deep insight into the search for the fundamental understanding of the Logic of Nature.

- E.P. Wigner:

It is pleasure and honour to greet Paul, "my famous brother-in-law", on his 80th birthday.

I think I know Paul longer than any of you, and will therefore concentrate on reviewing the early years of our friendship.

Of course, I knew about Paul and his work a good deal before we met. Perhaps I will tell you what brought me back to physics. I was a chemical engineer in a leather tannery but subscribed to the Zeitschrift für Physik. When I read the article of Born and Jordan I was overwhelmed, and accepted an invitation to be an assistant at the Technische Hochschule in Berlin.

There I soon found, as did many others, the articles of Schrödinger and of Paul Dirac. (This was in 1926.) We read them with enthusiasm. I must admit that for me it was easier to read Schrödinger's papers, partly because they were in German. But we realized the importance of Paul's work, and the great significance of his articles.

The first time I met Paul was in Göttingen, but this was a short encounter. Later in the yeat I worked a good deal with Jordan, and we wrote an article together. After that, we started to work on developing a relativistic equation for the spin ½ particle. We worked hard on this modification of the Klein-Gordon equation, but we were not satisfied with our results. Then, suddenly, Max Born recieved a two-page letter from Paul -- the first was referring to his visit, the second told about what is now called the Dirac equation for the electron. It was not much more than half a page, but it gave the idea and the equation clearly. Born showed the letter to Jordan and also to me, and Jordan commented: "I wish we had come that idea ourselves -- I am sorry we did not. But is is wonderful that someone did have it." I believe I have a copy of Paul's letter to Born in Princeton.

The next article of Paul's which came to our attention about a year later -- of course in the meantime we did read the article on "the Dirac equation" in the Royal Society Proceedings -- contained the foundation of the quantum field theories. By that time we could read the English article a bit more easily, and it was again a wonderful surprise. I was back in Berlin when this happened. And there was another surprise: one day I had a visitor from Göttingen, Victor Weisskopf. He suggested that we use Paul's work and ideas and try to describe more completely the emission of photons by atoms in excited states. And we did that; Weisskopt worked mainly in Göttingen, I in Berlin, and we wrote two articles on the subject. I consider both of interest although of course not nearly as interesting or important as Paul's basic work.

My later contacts with Paul were more personal -- I spent half my time in Princeton, half in Berlin, and he visited Princeton several times for half a year. We saw each other a great deal -- we had almost all our meals together in restaurants.

And at one half-yearly of his visits my sister Margrit happened to accompany me to Princeton and Paul entered the restaurant where we were having our luncheon. He looked at us in surprise and appeared a bit confused. My sister noticed this. When I told her about Paul and our earlier sharing of meals, she asked me to invite him to have his lunch with us and they became acquainted. You known how this ended -- very, very happily.

So far I have only told you of the effect that two of Paul's papers had on me. But he wrote a great many others -- as you may remember, the subject during about two days' of our meeting was based on his article on magnetic monopoles. And he made a number of other important contributions to physics, many of which are still of interest to us, I believe. His books on quantum mechanics excel in clarity and conciseness, and are useful in several ways.

In the course of time I came to admire Paul more and more, not only for his ideas, but also for his modesty, his interest in several subjects, his ability to communicate his ideas in spite of his constant desire -- and the resulting inclination -- to keep silent. He has great imagination and also a wealth of knowledge.

I admire all this and wish him luck, happiness, and many good days here and elsewhere. And the same wishes go to his wife.

- C.N. Yang

Twenty years ago I was learning to ski in Stowe, Vermont, with my family, and one day, in the cafeteria, I saw a man coming in with a big muffler around his head. I thought to myself that he looked very much like Professor Dirac and, after a few minutes, he turned around -- and sure enough it was he. So I asked him what he was doing there and he told me he had taken his daughter to ski. He then asked me what I was doing there. I explained that my family was skiing and I was also trying to learn. He asked if I found it difficult. I said I thought it was difficult. He then said "Do you think it is a good thing?" and I said "Yes". The next day I went to the slopes and found Dirac with skis. I asked him what had happened and he answered: "If you can learn skiing at forty, I can learn skiing at 60!". But I suspect that neither of us did learn.

I am deeply honoured today to be asked to speak on this occasion. Dirac is one of the physicists that I admire most, ever since I became a student of physics. I still remember how much I learned from his book, that great classic on quantum mechanics. There is a purity of style in his work which goes directly to the heart of the subject. I also remember reading, as a student in China, his great paper -- one of the greatest in this century -- on the relativistic equation of the electron. To pull out of pure numbers the magnetic moment of the electron -- it was absolutely thrilling when I learned it. The "infinite sea" paper was also extremely striking. I had in the fifties, in a lecture, likened that development to the introduction of the negative numbers. In fact in some philosophical sense, that concept, that most daring, most crazy concept of Dirac's, has changed the human perception of what is the vacuum. We are still struggling with the complicated vacuum today.

Not all of Dirac's papers are as famous as the ones that I just mentioned but each one is in some sense a gem. I still remember, as a graduate student in China, studying a paper which is little noticed, on classical electrodynamics, in which by some very elegant methods he tried to put tubes around the world lines of the electrons and then shrunk the tubes to really get at the self-energy of the electrons.

Dirac's papers have a distinctive style which is unique. I would characterize this unique style by the following words: purity, elegance, cogency, and restraint.

In a book entitled "Physics and Beyond", published about ten years ago by Heisenberg, there was a description of conversations between Heisenberg and Dirac when they met in front of a hotel in the Yellowstone Park one day in the late 1920's and then embarked on a trip together through Japan back to Europe. Heisenberg wrote that Dirac and he talked about their respective experiences in America and their ideas about the future of atomic physics. Heisenberg implied that Dirac's style was "pragmatic". I, of course, do not know the actual content of their conversations during that trip, but I find pragmatism to be very, very far from Dirac's style: I would almost say that pragmatism is the opposite of what characterized Dirac's papers. One finds in his papers an economy of words. It is as if in writing them down he was carefully sculpturing a condensation of thought and of logic. Those of us fortunate enough to have been personally acquainted with Professor Dirac know that that style in which he writes his papers also characterizes his daily statements. There was, for example, a story about a student who at the end of a lecture by Professor Dirac stood up and said, "I did not understand your explanation of that equation on the top of the blackboard". So Dirac took a look at the equation and gave an explanation. At the end of it the student said, "But, Professor Dirac, that is exactly the same explanation you gave before". Dirac said, "Yes, I gave it again because that is the best explanation".

Professor Dirac, as I have said, I have been an admirer of yours ever since my student days. It is a true privilege for me to participate in the celebration of your eightieth birthday. And I am happy to say that Nino has assured me that he will invite me to Erice in ten years' time to celebrate your ninetieth birthday!

- P.A.M. Dirac

The title of this set if taks is "The Glorious Days of Physics". What that really means is the glorious days for physicists. They occurred from 1925 onwards, for a few years, and the reason why they were glorious days was that a big new field had been opened up by Heisenberg's discovery of the use of non-commutative algebra for describing dynamical systems. A vast new field was opened up. This meant that there were many problems to be solved, and at that time it was easy for quite a second-rate physicists to pick one of these problems, do a little work on it, publish a paper about his results and get a substantial amount of credit for that work. In that way they were glorious days for physicists. Glorious days which had not occurred before, and I doubt if they will occur again, but one can look back on the physicists of those days with envy because they had such wonderful opportunities open to them which physicsts nowadays do not have.

GREETINGS TO FRANK YANG IN MID-COURSE

(AUGUST 12, 1982)

Edward Teller

Senior Research Fellow
Hoover Institution
Stanford, California 94305

It is a really exceptional pleasure to talk on the sixtieth birthday of my best student. Now the first point is that it isn't his sixtieth birthday. His sixtieth birthday is supposed to be at some uncertain time. I heard clear proof that it is on the twenty-second of September, and I have heard further proof that it is on the first of October. But I wouldn't amount even to any approximation of a theoretical physicist if I could not find one more difficulty. You see, in the birth of a theoretical physicist the important thing is not the calendar. The important thing is when his thoughts started, and that is very hard to find out. And in the case of Frank, we used to call him that, in the case of Frank Yang I wonder when he started first to think about the difference between right and left, even if not between matter and antimatter. So we can say by definition that this is his sixtieth birthday.

Now I found out that he has already told you about all his life so there is nothing more of substance that I can say. But one day Frank did arrive in Chicago, and he seemed to have the intention of becoming a theoretical physicist and was sort of interested in my advice. So I proposed to him a great number of excellent topics. I am not sure what they were. One was I believe the K capture in beryllium and how it changes if you compress beryllium which was at the time a little unconventional. Frank worked it out and reported to a group including Fermi and others, but he didn't use that as a thesis. And then we considered the evaporation of a nucleus heated by a strong impact. And then we considered other crazy things like the magnetic moment of a deuteron and lithium 6 and beryllium 10, differing always by an alpha particle.

Then Frank came to me one day and said, "I will have to go back to China, and when I go back to China I don't know that theoretical physics will be any good. Experimental physics that is what the Chinese will need and that is what I want to work on." So Frank deserted me and went to work with Allison. Now very soon a legend grew around Frank. In theoretical physics, he already had started to do wonderful things and he started to do wonderful things of a different kind in experimental physics which was summarized by a short poem which I still can recite. The poem was: If there is a bang, it's Yang. And very soon he became for this reason known as the Yellow Peril. I heard these rumors. While I didn't know it at that time, I had two other theoretical students who also were very good and who got linked up with experiment. In their cases, it was for life.

One was Frank's friend, Marshall Rosenbluth, who wrote a paper about something ancient and antediluvian like mesons. We no longer talk about them except if they are charming. Marshall finished a little after Frank, and Frank helped him a little. He not only helped him in his thesis but by the time Marshall got his thesis written, Frank was also already a member of the faculty and on his examination board. But Marshall did not stay in pure physics. He got mixed-up deeply in plasmas and in controlled fusion, and today he is the best theorist in that remarkable field. If anybody today can understand the experiments on controlled fusion, it is Marshall Rosenbluth.

I had still another student who actually became an experimental physicist: Arthur Kantrowitz. He is a little older. He came to me at Columbia, a few years before Frank and I met. Kantrowitz told me something very interesting about shock waves in carbon dioxide. That was at a time when no theoretical physicist knew anything about shock waves. I knew a little about it because another Hungarian, Theodore von Karman, who had a lot to do with flying, told me about them. So I listened to Kantrowitz, and I said, "This is very interesting." And Arthur said, "Do you think that would be all right for a thesis?" I said, "That certainly would be all right for a thesis." And he asked, "Would you want to be my advisor?" And I said, "You know a hundred times more about shock waves than I do. How can I be your advisor?" He said, "But you see, I went to everybody else, and nobody would even listen to me. So would you please be my advisor."

I became his advisor. He got his Ph.D. He worked on hydrodynamics and later he worked on lasers, and he became the best man in practical lasers in the United States. And now he has stopped doing lasers, because he had an idea, a very good idea. We are fighting about all kinds of things--like genetic engineering, flouridation of water, low level radioactivity and a number of other things. We can't agree and everybody gets confused. "Let's have," Kantrowitz says, "a science court where we discuss these things in an interesting way so people will listen. It will be a great benefit." He is working on that, and he is no longer working on lasers.

With Frank it turned out differently. But after a time he came back to me and said, "You know, you and Konopinski wrote a paper some time ago in which you said that you suspected there is a relation between the angular distribution of particles coming out of a nuclear reaction and the angular momentum change. And I can prove your statement." And I said, "All right. Prove it." He went to the blackboard and proved it in a few minutes. In fact it was very clear, and I said, "Look I heard some stories that your experiments don't go so well. Would you mind writing that up for a thesis." Frank hesitated a little bit, and then he said, "I'll try."

He came back in two days with a thesis that was three pages long. One, two, three pages. Quite long. But you know we had a strange custom in Chicago at that time which suggested that a thesis should be even longer than that. So I told Frank, "Look this is a nice thesis, but could you make it a little longer? For instance, could you extend it to half integral angular momentum changes?" So Frank went away and came back with a paper which was seven pages long, and it included half integral angular momenta. I was very rude and argued that his proof could be explained more clearly and in more detail. That was not true. It was already clear enough. But after a lot of fighting, Frank went away and took an enormously long time, maybe ten days, and came back with a thesis eleven pages long. At that point I gave up. He got a well deserved Ph.D.

Now I must say, in his subsequent work Yang maintained the density of ideas per page. In spite of this he wrote longer papers for the simple reason that he had much more to say. The little question of right/left symmetry or asymmetry I have already mentioned.

I cannot talk about the work of Frank without reminding you unnecessarily of the famous paper of Yang and Mills about the generalization of gauge invariance on isotopic spin and on noncommuting variables. This after all has become the basis of almost every advanced discussion. And I would like to propose that he deserves for that a second Nobel Prize. There are other topics--for instance equations of state--very interesting discussions.

But there is a birthday present for Frank in the making. You know he has been interested, and so have many of us, in the magnetic monopole invented originally by Dirac. You may have heard, and I want to remind you in detail, that one physicist at Stanford, Blas Cabrera, seems to have found the magnetic monopole. And this is the experiment. He has a loop of a superconductor. I will save the details and just say, if it were a single loop then the current in the superconductor should jump causing a change of two flux quanta. But Cabrera did not have one loop. He had four loops, and therefore he expected a jump by eight quanta. And he didn't observe it. He waited, and he got little jumps that one can understand because the apparatus is not perfect. But these were jumps of one-third of a quantum or one-half of a quantum as it seemed. But then he found one jump after about half a year that was just a little short of eight quanta. Among the smallest ones, the biggest he saw was maybe .4, but then he saw one that was 7.6. It's very hard to understand such a jump except if you believe that a magnetic monopole actually passed through the loop. Several people including Cabrera are constructing bigger loops to catch more magnetic monopoles. If we could find a magnetic monopole or many of them there would be really wonderful things that could be done.

I want to talk about the magnetic monopole now for a while,

not for very long. You all heard Frank suggest that if we had only a magnetic monopole, then we could catch an electron and positron and make a perpetual motion machine by expending $2mc^2$ and getting back a little more. Well that is by far the best proposal that I have yet heard for solving the energy crisis. Now if I were a little inaccurate in calling it a perpetual motion machine, you will forgive me; you have already discussed this topic. But I want to talk about other experiments on magnetic monopoles which are also interesting.

One is a proposal by my friend, Luis Alvarez who said it would be so nice to have a magnetic monopole. I could take hold of it somehow and accelerate it in a magnetic field and in a cheap apparatus get enormous energies. Then I could catch it in a little swimming pool full of water, and bring it back and repeat the experiment and so we can go up easily to 10^{12}eV and really make a little progress in physics with slightly fewer dollars than are usually spent.

There is one difficulty. We now believe, and I am not sure that we are right, that this magnetic monopole is a little heavy. I think the usual figure is 10^{17} gigavolts. Now a magnetic monopole is not easily lost once you have it, except if it finds an antipole. Then it disintegrates, and the disintegration energy is equivalent to approximately, I am not sure that I am accurate, 9 kilograms of TNT. That would be a little disagreeable, and we would have lost the monopole and the antipole. So since I have become (I am very sorry to say) something of a politician in the meantime, I have thought of a political solution.

Let's have a treaty with the Soviets, and the West will have all the monopoles, and the Soviets will have all the antipoles. Then we cannot get into trouble. Well it's always possible to get into trouble with the Soviets, but if there is an iron curtain in between then perhaps we will avoid the trouble. And there is a further advantage. If we make that arrangement precisely as I said it, then for the first time there will be full justification on the part of the Soviets to call us monopolists.

Well, I would like to go a little deeper into physics. Not as deep as Frank did over in that big auditorium. With that I shall not try to compete. He has found the bound state of the electron in the Dirac equation. Now it turns out that it is quite easy to discuss this same problem in the Schroedinger equation. And, at least, the more highly excited states might have some meaning in the Schroedinger equation. And the following result turns out. In the Schroedinger equation, you can handle easily problems where the attraction goes as $1/r^2$. In fact when I was very young I once wrote a paper proving that if a potential is less deep everywhere than $-\hbar^2/8mr^2$ where m is the reduced mass, then you get no bound solutions. But if it is everywhere deeper than that, then you get infinitely many bound states both at increasing distances and decreasing distances. Of course, the bound states at decreasing distances cannot be taken seriously because sooner or later they go relativistic. Then the Schroedinger equation is no longer valid. But up to that point, it is as I told you. And $1/r^2$ is just the interaction between the monopole and the dipole of the electron. Incidentally, it is also the interaction between the electric field of the electron or the proton and the electric dipole that the monopole should have if it is described as by the Dirac equation. And the expression remains then in the Schroedinger approximation the same 1/8 of $\hbar^2$ over mr^2 with m as the reduced mass.

It turns out that for an electron this interaction is not strong enough. And in the Schroedinger approximation, you don't get any bound states. Furthermore, you don't get any bound states either if you just take the magneton $e\hbar/2mc$ for the magnetic moment of the electron. You need at least two magnetons or rather slightly more than two magnetons to get a bound state. But then you get infinitely many for the lowest possible angular momentum which as Frank has proved is $1/2\hbar$. Thus the electron in the Schroedinger equation cannot be bound.* The proton can be bound because it has 2.7 magnetons and not only one of them but in the lowest angular momentum state as many radially excited states as you like. Only their binding energy decreases quite rapidly. It so happens that the neutron with 1.8 magnetons cannot be bound. But the proton can, and the neutron while it is not bound may eventually be attracted by the protons.

Now this may just conceivably explain another experiment of Luis Alvarez. He had the really wonderful idea that if in this ashtray a monopole is hiding, then by whirling a loop around this ashtray, you should get an electric current that you can observe. And he collected all kinds of objects, and never found a monopole. He even got several samples of moon rock, and he could not find the monopole. Then why is it that Cabrera finds a monopole while Alvarez could not? I do not know, because most people say that probably Cabrera somehow made a mistake. Incidentally, I wish for Frank's official birthday a second monopole to be found because then everybody will believe it.

But still, if there are monopoles, why didn't Alvarez find them. I have a crazy suggestion which is probably wrong. Protons can be bound, and they can be bound in low energy states and in

---

*This seems to agree with a result from the Dirac equation that pure magnetic fields represented by vector potentials do not give bound states for electrons.

high energy states which are unrelativistic in a region where there is no question that the proton can be handled as a proton rather than three quarks, where you can think of the proton as a point with a dipole. Actually, a few protons could be bound this way before the energy falls low enough that thermal interactions or other effects will get rid of them. But once you have a few protons clustered around a monopole, the neutrons, which cannot be bound by the magnetic field alone will be attracted, and then they will be bound by the protons. In this way, you could begin to grow a nucleus, an actual atomic nucleus with a not so slightly irregular central region around a monopole. As this thing grows, you capture more neutrons. If you capture too many neutrons, they will make a beta decay and become protons. Then you can get to uranium or beyond. And in the end you are bound to get a fission.

A fission in a region of a micron may melt the crystal, and the monopole which is contained in one of the fragments will be set free. If it is set free it will move a small distance along the earth's magnetic field and eventually will get into the interior of the earth. Perhaps there will be places like the molten core of the earth where there are lots of monopoles around, and we just occasionally can snatch one of them. If we can do so--can get our hands on it--it will be a very useful and interesting object.

Now, I see that I have talked just a little bit more than I wanted to talk, and therefore, I will not tell you more nonsense but only something very real and very serious. It is an old Jewish custom to say at a birthday, I look forward to congratulate you on your hundred-twentieth birthday.[**] So I am particularly happy to greet Frank just in his mid-career.

[**]Developments in molecular biology may justify this old custom.

THE REQUIREMENTS OF A BASIC PHYSICAL THEORY

P.A.M. Dirac

The Florida State University
Department of Physics
Tallahassee, Florida 32306

I would like to talk to you about just one idea. It is a very important idea. But there are many physicists who do not understand its importance, or at any rate do not appreciate its importance. The idea is that the Heisenberg dynamics is a very good dynamics. It is extremely good and we have to keep to it, at all costs.

Now, first I should explain just what I mean by the Heisenberg dynamics. It is a dynamics in which one works with dynamical variables which do not satisfy commutative multiplication. They can be more general mathematical quantities subject only to the requirement that their multiplication shall obey the associative law.

We have these dynamical variables as the basis of our theory and it is necessary that they should have a reality condition. If we have any dynamical variable u then there is a meaning to its complex conjugate $\bar{u}$, and if $\bar{u}$ is equal to u then we say that u is real. We have these dynamical variables which we work with, and we need also a Hamiltonian. A Hamiltonian is a real dynamical variable and it provides equations of motion. It provides the equations of motion for all the dynamical variables and I write down, here below, the equation of motion for any dynamical variable u,

$$i\hbar \frac{du}{dt} = uH - Hu.$$

This equation of motion is an essential part of the Heisenberg dynamics.

Now, I am saying that this dynamics is a very good theory. Why is it a good theory? Well it was introduced by Heisenberg in 1925 at a time when we had only Bohr orbits to work with, and we did not understand how to bring in the interaction between the electrons. It was a wonderful advance to have the Bohr orbits replaced by this Heisenberg dynamics and it provided great excitement. When the idea was first introduced it was not properly understood, but it needed only one or two years before one had worked out the mathematics and it needed slightly longer to get a physical interpretation for this mathematics. It led to a wonderful opening up of physical theory. It led to what is now referred to as the Golden Age of physics. We suddenly had a really satisfactory theory to replace the Bohr orbits.

Now how did the development of this theory go? One soon found that there was a close connection between this Heisenberg dynamics and the old classical dynamics. It was such a close connection that if we had equations in the old dynamics one could transfer them and get the corresponding equations in the new dynamics. If one were dealing with a certain dynamical system according to the old dynamics one could give a meaning to the corresponding dynamical system with the new mechanics. That was a very satisfactory connection. But in one way it was a bit unfortunate because people attached rather too much importance to this connection between the old dynamics and the new Heisenberg dynamics. What people did was the following : they took the old dynamical theory of particles interacting with fields and then they worked out the corresponding theory in the new mechanics of Heisenberg and obtained the Hamiltonian H from this procedure. Then they proceeded to apply the Heisenberg dynamics with this H which was supplied by analogy with the classical theory. The result of doing this was to give one equations of motion, according to the equation written previously, which were not sensible. The equations of motion involved infinities. The natural conclusion to draw from that, I would say, is that one has the wrong Hamiltonian.

However, most physicists did not accept that point of view. They thought that this Hamiltonian, provided by classical mechanics, was correct and one had to modify the Heisenberg scheme in some way. What they did was to set up a theory, called renormalization, in which the infinite terms which one gets from this straightforward application of the Heisenberg theory were cancelled out. This meant that one was really departing from the Heisenberg scheme. One was replacing the Heisenberg equations by a set of working rules. This renormalization procedure gave a very good agreement with experiment under certain conditions, especially for electrodynamics. The agreement with observation was amazingly good and many physicists were satisfied with that situation. They said that the object of physics is to provide a set of rules which will enable one to calculate the results of observation.

Now I want to say emphatically that that is a wrong point of view. We do not merely want to have a set of rules giving results which agree with observation. After all, the Bohr orbit theory gave results in agreement with observation in certain simple cases. Even so, the Bohr orbit theory was a very primitive theory, and was not a correct theory. We want to have more than mere agreement between our working rules and the results of observation. We want to understand how Nature works. We want to understand the basic ideas which Nature is using, and for that purpose it is not sufficient to have rules giving results in agreement with observation. Only when we have understood the basic ideas which Nature is using can we have a reliable basis on which to proceed with the further development of physics.

I might give an example of this. Let us go back to the early stages of Quantum Mechanics and consider the problem of getting a satisfactory one-electron theory. There you all know what the answer to that is. We have to introduce dynamical variables corresponding to 4 x 4 matrices. Such dynamical variables have to appear in the theory. These dynamical variables were not suggested by any analogy with the classical theory. They have to come out from some mathematical development. When we have introduced these dynamical variables, corresponding to 4 x 4 matrices, we can set up a good theory of the one-electron problem. By a good theory I mean a theory in agreement with the Heisenberg formalism. A theory in which the Heisenberg equations of motion are fulfilled. Now that, I think, provides a good example to show how we must rely on the Heisenberg formalism and look for the correct mathematics to provide us with dynamical variables which are suitable.

Now I would like you to appreciate the enormous power of the Heisenberg formalism. We can use any dynamical variables satisfying non-commutative algebra, but they must satisfy associative algebra, and these dynamical variables may lie at the basis of our theory. That provides us with an extremely great power, much more power than we have in the classical mechanics, much more power than we can get by working from analogy with classical mechanics. We can have a wide variety of new dynamical variables appearing in the theory; for example, they could be the elements of some group. Any group would do, in principle. It may very well be a group which physicists have not yet thought of. Or it may be that some new kind of dynamical variable is needed. In any case, we do need a new Hamiltonian, and we have this wide power available to us in the search for a new Hamiltonian.

Of course, this problem that I am talking about will not be solved until the new Hamiltonian is discovered. That might take a long time - one does not know; to discover it is a mathematical problem. I do not think experimental physics can provide any help in obtaining a solution, just like experimental physics did not provide any help in thinking of the use of 4 x 4 matrices in the one-electron problem. One just had to proceed by looking for beautiful mathematics, mathematics of such beauty that one feels very happy with it and one thinks that Nature may have chosen it. I have been working for many years trying to find a new Hamiltonian scheme, and without success up to the present. But still one must go on in this direction.

I would like to refer to one idea that I put forward in 1970 in which I proposed a new kind of wave equation for the electron, a wave equation in which only positive energy states can occur. The usual wave equation allows both positive and negative energy states, and this was a new wave equation in which only positive energy states can occur.

I would like to mention at this point that there is an earlier theory which was discovered by Majorana in 1932. I am very happy to have this opportunity of speaking about the genius of Majorana in this School. Majorana proposed a wave equation which also allowed only positive energy solutions. Now Majorana's equation is related to the equation which I proposed in 1970, but there are essential differences, and the main difference is that Majorana's equation allows various possible values for the mass of the electron. The mass can have a certain maximum value or it may have a lesser value, and there is a whole series of values which the mass may have going down to zero. Now each mass corresponds to a different value for the spin. The maximum mass gives the spin zero, then the lesser masses have values for the spin, equal to 1, 2, 3, and so on, up to infinity. That is what the theory gave, and that was a bit unfortunate from the point of view of the physicist because the physicist likes to think of a particle of zero spin as being a simple kind of particle and the particles of higher spins one would expect to be more complicated particles and to have higher masses. Majorana's theory gave quite definitely the answer that the masses get less and less as the spin goes up to infinity. That was a situation which physicists did not like and for that reason they did not pay very much attention to Majorana's theory, although it was a really very interesting theory mathematically and it needed a great deal of genius to think of it.

Well, the theory that I proposed in 1970 differs from the Majorana theory in that it allows only one value for the mass and zero spin. It also requires that the wave function must satisfy certain linear conditions. You cannot just have any wave function and say that that corresponds to a physical state, only wave functions that satisfy certain linear conditions correspond to physical states. Now this bringing in of conditions into the Heisenberg theory is quite permissible. It does not change any of the essential features of the theory. Of course, it is necessary that the linear conditions on the wave function, if they are initially satisfied, must remain satisfied under the action of the equations of motion.

Well that was the situation with this 1970 theory and I worked with it for quite a while. But it did not seem to lead to any solution of the fundamental problems. There was a difficulty that the equations are very restrictive, you cannot do very much with them without getting inconsistencies and, in particular, one could not introduce the electromagnetic field to interact with the particle in the usual way. It was not possible to develop the theory in a way analogous to the way one was used to in the old wave equation which used the 4 x 4 matrices. I should say that this new wave equation does not have the 4 x 4 matrices in it at all, instead it has just one component for the wave function and there are some internal degrees of freedom for the particle, internal degrees of freedom which correspond to the particle having an internal motion corresponding to two harmonic oscillators.

Well, I was not able to get very far with the development of this theory, but recently it has been taken up by Sudarshan and his co-workers in Austin, and they have found a way of developing the theory, bringing in more dynamical variables, such that one can introduce the possibility of fields, like one has in the ordinary electron theory.

Well there is this old theory which I felt very excited about at the time, but so far I have not found how to get a solution to any of the fundamental problems with its help, and I am not sure whether it is following the right line or not. In any case one must continue to search for a scheme of equations following the Heisenberg formalism.

Now a very great deal of modern physics is concerned with the modified scheme in which one departs from the Heisenberg equation of motion and goes over essentially to a set of working rules. These working rules involve the problem of handling the infinities which would arise from the Heisenberg equations. One handles them by a process known as renormalization which provides one with definite rules for subtracting out the infinities. Now I am rather surprised at the successes of this theory. I should also say that I am gratified by it but I am also a bit disturbed, because I feel that it will not lead to any solution of the fundamental problems. Our main problem is to explain the reason for the reciprocal of the fine structure constant $\alpha^{-1} = \hbar c/e^2$ having a value close to 137.

This I would say is the most important problem of physics, and until it is solved there will not be any substantial progress. One is working with imperfect theories, one is working in the dark. To continue with the renormalization theory can be justified on the grounds that it is the best that we can do with the existing ideas. That is really a very good reason for continuing with this kind of work, and it is this kind of work which has dominated this School. It is a very good reason but at the same time one should not be too complacent about it. If one has successes one can look on them with a feeling of happy surprise. It may very well be that there are cases in which it fails. One should then not be too much disturbed. If it is a failure it comes about because we do not know the basic ideas. That is essentially what I wanted to say, and I feel that many people will disagree with what I said and if anybody likes to make comments or ask questions I will do my best to answer them.

D I S C U S S I O N

CHAIRMAN: P.A.M. DIRAC

DISCUSSION

- *C.N. YANG:*

I would like to ask Prof. Dirac a historical question about the Heisenberg equation. If my reading of what happened in the 1920's is correct, I thought when Heisenberg wrote down his algebra, later amplified and completed by Jordan and Born and then by Born, Jordan and Heisenberg, there did not occur any complete dynamical system. It was a few months later when you found the relationship between the Poisson bracket and the commutator that this equation was born. Was that correct?

- *P.A.M. DIRAC:*

I didn't hear the beginning of your remarks because this microphone wasn't working. Could you repeat your question?

- *C.N. YANG:*

Yes, I was saying that, if I read history correctly, I thought that when Heisenberg developed his algebra, a profound and important development, he didn't develop it in full, and that algebra of the commutator between p and q was later completed in the so-called two-man and three-man papers. But those three papers, the one-man two-man and three-man papers, did not really have the whole dynamics. The whole dynamics appeared really for the first time a few months later when you developed the relationship between the commutator and the classical Poisson bracket. I want to know whether that was the right history.

- *P.A.M. DIRAC:*

I think that is correct, yes.

- *E.P. WIGNER:*

I am often wondering whether it is reasonable to use the variables x, y, z, t, because in the ordinary quantum mechanics we define the wave function, for a definite t, as a function of x, y, z. But it is not possible to observe such a wave function because the signal from a distant point doesn't come in on time and one would have to wait really infinitely long to have a full information. Has it been considered to have the wave function defined on the negative light-cone or something like that? And what do you think? Would that be reasonable?

- *P.A.M. DIRAC:*

It is possible but not necessary that that should be the case. All that is really necessary is that the Heisenberg equation shall hold, and I am not sure whether the variables x, y, z, t, are really fundamental. Of course, in some way we have to bring in Lorentz transformations in connection with the equation of Heisenberg and there is no obvious way in which that can be done. That is really the main problem. We must find some new scheme in which the Lorentz transformations will appear. Maybe in some new way, an unexpected way, just as in the one-electron theory, when the 4 x 4 matrices were introduced, the Lorentz transformations then again appeared in a different way, an unexpected way, and a way in which one would not have thought of if one had insisted on keeping to the old requirements of Lorentz invariance. It could very well be that Lorentz invariance will re-appear in the new theory in some quite unexpected way.

- *E.P. WIGNER:*

Thank you very much. I would not think that entirely unexpected if we defined it on the negative light-cone but it is not the customary thing now.

- *P.A.M. DIRAC:*

You are restricting yourself unnecessarily if you talk about the negative light-cone, and I would prefer to work without this restriction.

- *H. KLEINERT:*

A long time ago Dyson suggested that the vacuum of quantum electrodynamics would become unstable if the fine structure constant is negative. How are the infinities related to this ....?

- *P.A.M. DIRAC:*

Negative?

- *H. KLEINERT:*

According to Dyson if α were negative, then by pair-production at far distances the universe would just blow up, and so the point $\alpha = 0$ is a singular point of the theory. How are the infinities, which bother you so much, related to the fact that when you do a perturbative expansion, you expand around a singular point of the theory?

- *P.A.M. DIRAC:*

I would like to have a theory which does not allow one to work with arbitrary values for α. I would like to have a theory where α is forced to have just one particular value. That is what we are striving for.

- *S. FERRARA:*

I would like to ask two questions. The first is, why you attach so great importance to understanding the number 137 and not other fundamental numbers which appear in physics, like the Fermi coupling constant, which fixes the strength of weak interactions, or the Planck mass, which is related to the gravitational interactions?

- *P.A.M. DIRAC:*

Can I answer the questions one at a time please?

- *S. FERRARA:*

Yes, this is my first question.

- *P.A.M. DIRAC:*

Yes. Now, it is just that this number has been with us for a longer time. We had many decades to think about it. These new coupling constants are probably important, but I have the feeling that they are less important. Maybe I am wrong there. I wouldn't want to attach much importance to that. But it is just that the 137 problem has been with us for such a long time, and no one has any good ideas for dealing with it.

I wonder if I might mention the theory of Euler and Cockle, which was brought out in the 1930's. Maybe people have forgotten it now. Euler and Cockle pointed out that we have to modify the Maxwell electrodynamics for very strong fields. A possible way of this modification was suggested by a theory of Max Born, an interesting possible way.

Also one gets a modification of the Maxwell equations if one takes into account pair-production and recombination of electrons. The result is that two photons passing by have a chance of scattering each other. Now one can equate the modification in Maxwell's electrodynamics provided by pair-production with that provided by Born's electrodynamics, and the result of that equation is to provide you with a value for the fine structure constant. Euler and Cockle worked it out, by following straightforward methods, and they got a result somewhere around 104, I think, not a very good result. But then someone else (was it Infeld? I am not quite sure who it was) worked out a modified theory in which he took into account other possibilities and was able to get a number somewhere around 130, which was very much better. That provided some hope that there might be some justification for this kind of procedure. But in any case it was very artificial. If you are interested in this work, I would suggest that you look up a little paper about it that was published in Nature, and the reference is Infeld, Nature, volume 137.

- *S. MUKHI:*

Yes, I would like to know what you think of the idea that space-time has more than four dimensions, either physically or in a purely mathematical sense.

- *P.A.M. DIRAC:*

It is quite possible. I don't mind how many dimensions you have, provided you keep to the Heisenberg equation of motion.

- *S. MUKHI:*

Do you think that this might lead to some new physical development?

- *P.A.M. DIRAC:*

I think it is rather unlikely, it seems too crude an idea. I think it will need some more subtle mathematics, and I am inclined to think that the mathematics will be connected with complex variable theory, simply because complex variable theory is the most beautiful and the most powerful branch of mathematics.

- *C. DEVCHAND:*

I would like to ask, is there any fundamental reason for the dynamical variables being associative?

- *P.A.M. DIRAC:*

I wouldn't say it is very fundamental but it is necessary for the Heisenberg scheme. Some people have considered the possibility of non-associative multiplication. Already in the 1920's, as soon as the value of non-commutative multiplication was discovered, people wondered if one could go on to non-associative multiplication, and several people worked on that idea, but nobody got very far with it. I doubt if any work has been published about it. It is not a new idea, I would say, quite an old idea, which people have not dealt with very successfully.

- *V. DOBREV:*

I would like to ask one question. Now we have quantum mechanics but we still have problems with the correspondence between the classical variables and the operators. Do you think that your new Hamiltonian, that you have been talking about, will have a classical correspondence, and whether this new mathematics will solve also the problem of correspondence between the classical and the quantum states of the theory?

- *P.A.M. DIRAC:*

Well, the new mathematics must have mass parameters occurring in it and there are mass parameters also occurring in Nature. They both have to be finite quantities; the difference between them, therefore, will be finite. We cannot allow an infinity there.

- *E. TELLER:*

I also would like to ask a historical question of a somewhat different kind, because I remember that it made a great impression on me, a little more than thirty years ago. We had a physics conference, a theoretical physics conference and the comment of Niels Bohr was; we are not going to make any progress until we have discovered a contradiction. To my mind this means that words like you use are hard to understand, because what is beautiful mathematics may be subjective. But if you have a contradiction, like the contradiction that was clear in 1905 between the wave nature and the particle nature of light, this has forced us or forced some people into new ideas, which later were recognized to be beautiful. And it is not the beauty that came first but the contradiction. I wonder whether you would like to comment on this.

- *P.A.M. DIRAC:*

I would not work on those lines at all. The problem is to find the right equations. If you just discussed physical ideas, I don't think that will help you to find the right equations. In looking for the right equations, one pays a great deal of attention to beauty and power, and that, I would say, is a subjective thing. As a result of thinking over these problems for many decades, I have come to the conclusion that complex variable theory is the most beautiful and most powerful branch of mathematics, and I think Nature will have made use of it.

- *A. ZICHICHI:*

Of course, you were able to find the right equation, but it is not easy.

- *E. TELLER:*

Just to continue along the same lines, Nature may have made use of complex variables. Nature made use of space-time curvature. Nature has not yet made clear use of the magnetic monopole, and you have postulated in 1932 that it would be astonishing if Nature would not make use of it.

- *P.A.M. DIRAC:*

That is so; I thought so in 1932. I think that the final answer will not be obtained until the problem of 137 is solved. One should keep an open mind about it; that is, about the monopole.

- *(UNIDENTIFIED):*

What is your view of the idea raised by Stückelberg in 1941 that in using the Heisenberg equation of motion in the relativistic domain, one should not take H to be the energy of the system because the energy is the 4-th component of a 4- vector and is, in fact, an observable, and one should not take t as the evolution parameter, but rather other quantities, in order to get the right generalization of this equation to the relativistic case.

- *P.A.M. DIRAC:*

These are possibilities, but they are not fundamental. A fundamental thing is to have equations which conform to the Heisenberg scheme, and this H does not need to be the total energy. It may be the total energy as modified by some new kind of field, or some new kind of phenomenon which is not completely understood at the present time.

- *T. HOFSÄSS:*

I would like to know your opinion on the following point of view, concerning the infinities occurring in the renormalization procedure. It might be that Nature has provided us with some substructure at very small space-time distances, which we do not yet know anything about, and which, if we would know it, we would not have the trouble with these infinities.

- *P.A.M. DIRAC:*

I agree with that possibility. It may very well be that there is a substructure which has not yet been discovered, which it would be necessary to discover before one finds the correct Hamiltonian.

- *(UNIDENTIFIED):*

I just wanted to ask you the following question. Do you think that it is necessary to include the treatment of gravity in the full theory in order to explain the value 137?

- *P.A.M. DIRAC:*

Probably not; I am not sure, but I think probably not. Gravity seems to be rather outside atomic physics, because it plays so little a role. I wouldn't exclude it, but I think, probably not.

- *H. KLEINERT:*

Do you believe that there ever will be a final field theory answering our questions about Nature to arbitrarily short distances?

- *P.A.M. DIRAC:*

I wouldn't like to use the word "final". I think, maybe, physicists will always have problems. But I think that there will be a theory which is a vast improvement on the present one.

- *H. KLEINERT:*

But if it is not final in that sense then there always will be ignorance about the very short distances and, therefore, there will always be an infinity corresponding to the unknown structures at very short distances.

- *P.A.M. DIRAC:*

Well, I will say that there may always be uncertainties about very short distances but there will not necessarily be an infinity connected with these short distances. One can hope that there wouldn't be.

- *A. OGAWA:*

I am not sure that this is on the same level as the α, but it seems to be a problem of a similar kind of nature. The self-energy of the electron, even in classical terms, is infinite, and I know that that problem has been considered by some people without success and is still considered by them to be a very important question.

- *P.A.M. DIRAC:*

The self-energy is not infinite if you take the electron to be a little sphere, Abraham's sphere or a Lorentz sphere or something like that, so that this kind of infinity is not fundamental.

CLOSING CEREMONY

The Closing Ceremony took place on Friday, 13 August 1982. The Director of the School presented the prizes and scholarships as specified below.

PRIZES AND SCHOLARSHIPS

Prize for *BEST STUDENT*

awarded to Leonardo CASTELLANI - State University of New York USA.

Fourteen Scholarships were open for competition among the participants. They were awarded as follows :

Patrick M.S. BLACKETT Scholarship to

Leonardo CASTELLANI - State University of New York, USA.

James CHADWICK Scholarship to

Vadim KAPLUNOVSKY - Tel-Aviv University, Israel.

Amos-de-SHALIT Scholarship to

Neil TUROK - Imperial College, London, UK.

Gunnar KÄLLEN Scholarship to

Arthur OGAWA - SLAC, USA.

André LAGARRIGUE Scholarship to

Sunil MUKHI - International Centre for Theor. Phys., Trieste, Italy.

Giulio RACAH Scholarship to

Vladimir DOBREV - Institute for Nuclear Research, Sofia, Bulgaria.

Giorgio GHIGO Scholarship to

Pierre VAN BAAL - Institute for Theoretical Physics, Utrecht, The Netherlands.

Enrico PERSICO Scholarship to

Manfred KREMER - University of Heidelberg, FRG.

Peter PREISWERK Scholarship to

Thomas HOFSÄSS - Freie Universität, Berlin, FRG.

Gianni QUARENI Scholarship to

Joseph LYKKEN - MIT, Cambridge, MA, USA.

Antonio STANGHELLINI Scholarship to

Miss Marie Luise SCHÄFER - University of Bremen, FRG.

Alberto TOMASINI Scholarship to

Andreas ALBRECHT - University of Pennsylvania, Philadelphia, USA.

Ettore MAJORANA Scholarship to

Steven ERREDE - University of Michigan, Ann Arbor, USA.

Benjamin W. LEE Scholarship to

Stefano BELLUCCI - University of Rome, Italy.

Prize for *BEST SCIENTIFIC SECRETARY* awarded to

Miss Maria Luise SCHÄFER - University of Bremen, FRG.

HONORARY MENTIONS awarded to

Steven S. SHERMAN - CALTEC, USA.

Klaus-Georg RAUH - University of Heidelberg, FRG.

Marcelo LOEWE - University of Hamburg, FRG.

Jochum J. VAN DER BIJ - University of Michigan, Ann Arbor, USA.

Zuzana DUBNIČKOVÁ - Comenius University, Bratislava, Czechoslovakia.

Francesco ANTONELLI - Scuola Normale Superiore, Pisa, Italy.

Chandrashekar DEVCHAND - University of Durham, UK.

Yeshayahu LAVIE - Tel-Aviv University, Israel.

Mario PERNICI - State University of New York, USA.

David RICHARDS - University of Cambridge, UK.

Anton VAN DE VEN - State University of New York, USA.

The following participants gave their collaboration in the scientific secretarial work :

Andreas ALBRECHT
Stefano BELLUCCI
Alexander CALOGERACOS
Leonardo CASTELLANI
John COLE
Giacomo D'ALI'
Catherine DE CLERCQ
Chandrashekar DEVCHAND
Lucia DI CIACCIO
Vladimir DOBREV
Zuzana DUBNIČKOVA
Steven ERREDE
Louis FLURI
Andreas GOCKSCH
José Luis GOITY
Thomas HOFSÄSS
Joey HUSTON
Magnus JÄNDEL
Wolfhard JANKE
Vadim KAPLUNOVSKY
Manfred KREMER
Yves LEBLANC
Pierre L'HEUREUX
Joseph LYKKEN
Deborah MILLS ERREDE
Sunil MUKHI
Jeanette E. NELSON
Arthur OGAWA
Mario PERNICI
Laura PETRILLO
Klaus-Georg RAUH
David RICHARDS
Marie Luise SCHÄFER
Marleigh SHEAFF
Steven S. SHERMAN
Neil TUROK
Pierre VAN BAAL
Anton VAN DE VEN

PARTICIPANTS

José ABDALLA HELAYËL-NETO	Scuola Internazionale Superiore di Studi Avanzati Strada Costiera 11 34014 TRIESTE, Italy
Andreas ALBRECHT	University of Pennsylvania Department of Physics PHILADELPHIA, PA 19104, USA
Francisco ALONSO	Instituto de Estructura de la Materia Consejo Superior de Investigaciones Cientificas Serrano, 119 MADRID 6, Spain
Miriana ANTIC	Max-Planck Institut für Physik und Astrophysik Föhringer Ring 6 8000 MÜNCHEN 40, FRG
Francesco ANTONELLI	Scuola Normale Superiore Piazza dei Cavalieri, 7 56100 PISA, Italy
Stefano BELLUCCI	Istituto di Fisica dell'Università Piazzale Aldo Moro, 2 00185 ROMA, Italy
Christoph BERGER	I Physikalisches Institut der RWTH Physikzentrum 5100 AACHEN, FRG
Diego BETTONI	Istituto Nazionale di Fisica Nucleare Via Livornese 56010 S. PIERO A GRADO (Pisa), Italy

James G. BRANSON — Massachusetts Institute of Technology
Department of Physics
CAMBRIDGE, MA 02139, USA

Alexander CALOGERACOS — University College London
Department of Physics
Gower Street
LONDON WC1E 6BT, UK

Roberto CAROSI — Istituto Nazionale di Fisica Nucleare
Piazza Torricelli, 2
56100 PISA, Italy

Leonardo CASTELLANI — State University of New York
Institute for Theoretical Physics
STONY BROOK, NY 11794, USA

Paolo CEA — Istituto di Fisica dell'Università
Via Amendola, 123
70126 BARI, Italy

John COLE — University of Sussex
School of Mathematical and
Physical Sciences
BRIGHTON, BN1 9QH, UK

Michael COOKE — University of Manchester
The Schuster Laboratory
Department of Theoretical Physics
MANCHESTER, M13 9PL, UK

Luigi DADDA — Politecnico di Milano
Piazza Leonardo da Vinci, 32,
20123 MILANO, Italy

Giacomo D'ALI — Istituto di Fisica
Università di Bologna
Via Irnerio, 46
40126 BOLOGNA, Italy

Sally DAWSON — Fermi National Accelerator Laboratory
Theory Department, MS-106
P.O. Box 500
BATAVIA, IL 60510, USA

Wim DE BOER — Max-Planck Institut
für Physik und Astrophysik
Föhringer Ring 6
8000 MÜNCHEN 40, FRG

Catherine DE CLERCQ	Inter-University Institute for High Energies V.U.B. - U.L.B. Pleinlaan, 2 1050 BRUSSEL, Belgium
Jean Pierre DERENDINGER	Université de Genéve Département de Physique Théorique 32, Boulevard d'Yvoy 1211 GENEVE 4, Switzerland
Chandrashekar DEVCHAND	University of Durham Department of Mathematics Science Labs. South Road DURHAM DH1 3LE, UK
Lucia DI CIACCIO	Istituto di Fisica dell'Università Piazzale Aldo Moro, 2 00185 ROMA, Italy
Paul A.M. DIRAC	The Florida State University Department of Physics TALLAHASSEE, Florida 32306, USA
Vladimir DOBREV	Bulgarian Academy of Sciences Institute of Nuclear Research and Nuclear Energy Boulevard Lenin 72 SOFIA 1184, Bulgaria
Zuzana DUBNIČKOVÁ	Comenius University Katedra Teoretickej Fysiky Mlynska Dolina 81631 BRATISLAVA 16, Czechoslovakia
Pieter DUINKER	NIKHEF-H Kruislaan 409 P.O. Box 41882 1009 DB AMSTERDAM, The Netherlands
Franz EISELE	University of Dortmund Postfach 500500 4600 DORTMUND 50, FRG

Steven ERREDE	University of Michigan The Harrison M. Randall Laboratory of Physics ANN ARBOR, MI 48109, USA
Etim ETIM	Laboratori Nazionali di Frascati C.P. 13 00044 FRASCATI, Italy
Sergio FERRARA	CERN TH Division 1211 GENEVE 23, Switzerland
Louis FLURI	Université de Neuchâtel Institut de Physique Rue A.-L. Brequet 1 2000 NEUCHATEL, Switzerland
Daniel FROIDEVAUX	CERN EP Division 1211 GENEVE 23, Switzerland
Francesco GIANI	Istituto di Fisica dell'Università Corso M. D'Azeglio, 46 10125 TORINO, Italy
Andreas GOCKSCH	New York University Department of Physics 4 Washington Place NEW YORK, NY 10003, USA
José Luis GOITY	Max-Planck Institut für Physik und Astrophysik Föhringer Ring 6 8000 MÜNCHEN 40, FRG
Giuseppe GRELLA	Istituto di Fisica dell'Università Via Vernieri, 42 84100 SALERNO, Italy
Bohdan GRZADKOWSKI	University of Warsaw Institute of Theoretical Physics Ul. Hoza 74 00682 WARSZAWA, Poland

Raymond W. HAYWARD	National Bureau of Standards Center for Absolute Physical Quantities WASHINGTON, DC 20234, USA
Clemens A. HEUSCH	University of California High Energy Physics Natural Science II SANTA CRUZ, CA 95060, USA
Thomas HOFSÄSS	Freie Universität Berlin Institut für Theoretische Physik Fachbereich Physik Arnimallee 14 1000 BERLIN 33, FRG
Joey HUSTON	Fermi National Accelerator Laboratory MS 221 - P.O. Box 500 BATAVIA, IL 60510, USA
Hiroshi ITOYAMA	Columbia University Department of Physics Box 87 NEW YORK, NY 10027, USA
Magnus JÄNDEL	Royal Institute of Technology Department of Theoretical Physics 10044 STOCKHOLM, Sweden
Wolfhard JANKE	Freie Universität Berlin Institut für Theoretische Physik Arnimallee, 3 1000 BERLIN, FRG
G.C. JOSHI	University of Melbourne School of Physics PARKVILLE, Victoria 3052, Australia
Kiyoshi KAMIMURA	Istituto Nazionale di Fisica Nucleare Largo E. Fermi, 2 Arcetri 50125 FIRENZE, Italy
Vadim KAPLUNOVSKY	Tel Aviv University Department of Physics and Astronomy Ramat-Aviv TEL-AVIV, Israel

Hagen KLEINERT	Freie Universität Berlin Institut für Theoretische Physik FB 20 WE 4 Arnimallee, 3 1000 BERLIN, FRG
Manfred KREMER	Universität Heidelberg Institut für Theoretische Physik Philosophenweg 16 6900 HEIDELBERG, FRG
Peter V. LANDSHOFF	University of Cambridge Department of Applied Mathematics and Theoretical Physics Silver Street CAMBRIDGE, CB3 9EW, UK
Kwong LAU	Stanford University Stanford Linear Accelerator Center P.O. Box 4349 STANFORD, CA 94305, USA
Yeshayahu LAVIE	Tel Aviv University Department of Physics and Astronomy Ramat-Aviv TEL-AVIV, Israel
Yvan LEBLANC	University of Alberta Institute of Theoretical Physics Department of Physics EDMONTON, Alberta T6G 2JI, Canada
Pierre L'HEUREUX	McGill University Rutherford Physics Building 3600 University St. MONTREAL, Quebec H3A 2T8, Canada
Cesare LIGUORI	Istituto Nazionale di Fisica Nucleare Via Celoria, 16 20133 MILANO, Italy

Samuel J. LINDENBAUM	Brookhaven National Laboratory Associated Universities, Inc. Department of Physics UPTON, NY 11973, USA and City College of New York Physics Department Convent Ave at 138th St. NEW YORK, NY 10031, USA
Marcelo LOEWE	II Institut für Theoretische Physik Universität Hamburg Notkestrasse 85 2000 HAMBURG 52, FRG
Joseph LYKKEN	Massachusetts Institute of Technology 6-415 Center for Theoretical Physics 77 Massachusetts Avenue CAMBRIDGE, MA 02139, USA
Michael MACDERMOTT	Rutherford Appleton Laboratory Chilton DIDCOT, Oxon, OX11 OQX, UK
Ezio MAINA	Rutgers University Department of Physics PISCATAWAY, NJ 08854, USA
André MARTIN	CERN TH Division 1211 GENEVE 23, Switzerland
Michael MARX	State University of New York Department of Physics STONY BROOK, NY 11794, USA
Mustafa MEKHFI	Scuola Internazionale Superiore di Studi Avanzati Strada Costiera 11 34014 TRIESTE, Italy
Deborah MILLS ERREDE	University of Michigan The Harrison M. Randall Laboratory of Physics ANN ARBOR, MI 48109, USA

Madjid MOBAYYEN	Imperial College of Science and Technology Department of Physics The Blackett Laboratory Prince Consort Road LONDON SW7 2BZ, UK
William MORSE	Brookhaven National Laboratory Physics Department UPTON, NY 11973, USA
Urs MOSER	Universität Bern Abt. Hochenergiephysik Sidlerstrasse 5 3012 BERN, Switzerland
Sunil MUKHI	International Centre for Theoretical Physics Strada Costiera 11 P.O. Box 586 34100 TRIESTE, Italy
Francis MULLER	CERN EP Division 1211 GENEVE 23, Switzerland
Demetrios NANOPOULOS	CERN TH Division 1211 GENEVE 23, Switzerland
Jeanette E. NELSON	University of London Queen Mary College Department of Physics Mile End Road LONDON E1 4NS, UK
Bernhard NEUMANN	Universität - Gesamthochschule Siegen Fachbereich 7 - Physik Naturwissenschaften I Postfach 210209 5900 SIEGEN 21, FRG
Roberto ODORICO	University of Bologna Via Irnerio, 46 40126 BOLOGNA, Italy

Arthur OGAWA
Stanford Linear Accelerator Center
SLAC
P.O. Box 4349
STANFORD, CA 94305, USA

Gülsen ÖNENGUT
Middle East Technical University
Physics Department
ANKARA, Turkey

Mario PERNICI
State University of New York
Institute for Theoretical Physics
STONY BROOK, NY 11794, USA

Laura PETRILLO
Istituto di Fisica dell'Università
Piazzale Aldo Moro, 2
00185 ROMA, Italy

Mario RACITI
Istituto di Fisica dell'Università
Via Celoria, 16
20133 MILANO, Italy

Klaus-Georg RAUH
Universität Heidelberg
Institut für Theoretische Physik
Philosophenweg 16
6900 HEIDELBERG, FRG

Roberto REMMERT
Università di Torino
Istituto di Fisica Teorica
Corso M. d'Azeglio 46
10125 TORINO, Italy

David RICHARDS
University of Cambridge
Department of Applied Mathematics
and Theoretical Physics
Silver Street
CAMBRIDGE, CB3 9EW, UK

Jean RICHARDSON
CERN
EP Division
1211 GENEVE 23, Switzerland

Jun John SAKURAI
Max-Planck Institut für
Physics und Astrophysics
Föhringer Ring 6
Postfach 401212
8000 MÜNCHEN 40, FRG

Marie Luise SCHÄFER	Universität Bremen Kufsteinerstrasse 1 - NW I 2800 BREMEN 1, FRG
Marleigh SHEAFF	University of Wisconsin Department of Physics 1150 University Avenue MADISON, WI 53705, USA
Steven S. SHERMAN	California Institute of Technology CALTEC High Energy Physics, 256-48 PASADENA, CA 91125, USA
Alexander William SMITH	Universität Karlsruhe Institut für Theoretische Physik Kaiserstrasse, 12 Physikhochhaus Postfach 6380 7500 KARLSRUHE, FRG
Bjarne STUGU	CERN EP Division 1211 GENEVE 23, Switzerland
Giancarlo SUSINNO	CERN EP Division 1211 GENEVE 23, Switzerland
Edward TELLER	University of California Lawrence Livermore Laboratory P.O. Box 808 LIVERMORE, CA 94550, USA
Dieter TEUCHERT	DESY/Group F13 Notkestrasse 85 2000 HAMBURG 52, FRG
Raffaele TRIPICCIONE	Scuola Normale Superiore Piazza dei Cavalieri, 7 56100 PISA, Italy
Neil TUROK	Imperial College of Science and Technology Department of Theoretical Physics The Blackett Laboratory Prince Consort Road LONDON, SW7 2BZ, UK

Liang David TZENG	Yale University J.W. Gibbs Laboratory P.O. Box 6666 NEW HAVEN, CT 06511, USA
Pierre VAN BAAL	Institute for Theoretical Physics Princetonplein 5 P.O. Box 80006 3508 TA UTRECHT, The Netherlands
Jochum J. VAN DER BIJ	University of Michigan The Harrison M. Randall Laboratory of Physics ANN ARBOR, MI 48109, USA
E. VAN DER SPUY	Atomic Energy Board Private Bag X256 PRETORIA 0001, South Africa
Anton VAN DE VEN	State University of New York Institute for Theoretical Physics STONY BROOK, NY 11794, USA
Luciano VANZO	Libera Università degli Studi di Trento Dipartimento di Fisica 38050 POVO (Trento), Italy
F. WAGNER	Max-Planck Institut für Physik und Astrophysik Föhringer Ring 6 Postfach 401212 8000 MÜNCHEN 40, FRG
Roland WALDI	Universität Heidelberg Institut für Hochenergiephysik Schröderstrasse, 90 6900 HEIDELBERG, FRG
Johann WEIGL	Institut für Theoretische Physik II Technische Universität Wien Karlsplatz 13 1040 WIEN, Austria

Eugene P. WIGNER | Princeton University
Department of Physics
P.O. Box 708
PRINCETON, NJ 08540, USA

Marc WINTER | CERN
EP Division
1211 GENEVE 23, Switzerland

Günter WOLF | DESY
Notkestrasse, 85
2000 HAMBURG, 52, FRG

Chen Ning YANG | State University of New York
Institute for Theoretical Physics
STONY BROOK, NY 11749, USA

Fabio ZWIRNER | Istituto di Fisica dell'Università
Via Marzolo, 8
35100 PADOVA, Italy

INDEX

Abelian
  factor, 117
  gauge theory, 304, 317, 322
ACCMOR spectrometer, 680
Acollinearity
  angle, 463
  distribution, 373, 374, 407, 461
  measurement, 722
Acoplanar muon hadron events, 506
Acoplanarity angle, 505
Additive constant, 102
ALEPH detector, 340-343, 364
Altarelli-Parisi equations, 270, 535, 578, 600, 609
Angular distribution, 409, 413
Angular momentum, 35, 40, 55
Annihilation mechanism, 698
Antigoldstino, 502
Antineutrino scattering, 574, 588, 604
Antiphotino, 502
Antiquark, 533
  distribution, 574, 575, 582, 588
  scattering, 604
  structure function, 588
Antisymmetrical matrices, 44
Approximate symmetries, 739
Automatic color neutralization, 261
Auxiliary fields, 95, 96, 103

B meson, 453, 668
b quark, 453
Bag model, 653, 655
Baryon
  decay, 144, 145
  production, 452
  recombinations, 221, 224, 227
Baryonic charge, 725
Baryons
  fast charmed, 223, 225
  parton fragmentation into, 401
  qqq, 151
Beam-dump experiments, 673, 683, 688
Beauty, 667-668
  production, 688-690
  quark, 668
BGO shower counter material, 362
Bhabha scattering, 372-374, 406, 408-409, 455, 484
Binding energy, 271, 297
Bohm-Aharonov
  effect, 32
  experiment, 15
Born term, 269, 271
Boson
  diagrams, 89
  fields, derivatives of, 71
  masses, 120
Bosons
  bosons to, 112
  couplings related with fermions, 103
  distinguishing from fermions, 91, 94
  Goldstone, 80
  Higgs, 99, 145, 156
  interactions with fermions, 106
  Nambu-Goldstone, 302

Bosons (continued)
  neutral intermediate vector, 484
  reality of, 95
  relations with fermions, 111
  superheavy vector, 105
  symmetry, 92, 119
  ultraviolet divergencies, 66
  weak vector masses, 67, 97
Bound-state
  diagram, 46
  energy, 46
Branching
  fractions, 437
  ratio, 507, 518, 667, 699
Breit-Wigner resonance, 635, 636
Bremsstrahlung, 298, 385, 458, 535, 578, 580, 613
Bubble-chamber experiments, 661, 679, 688

C parameter, 406, 424-425, 452
c quark, 221, 224
c_{active}, 224
$c_{spectator}$, 220, 224, 231
c_{struck}, 224
Cabibbo
  angle, 120
  -favored transitions, 666-667, 698
Cabrera experiment, 19-27, 29, 33, 35, 57
Cartesian product, 105
CELLO detector, 503
Centre-of-mass system, 703
Čerenkov
  identification, 680
  photons, 356
Charge asymmetry, 413-422, 434, 462
Charged
  multiplicity, 662, 713
  particles, 712
Charm
  -changing current, 699
  cross-section, 206
  decay, 431, 433, 671
  excitation diagrams, 675
  model, 675-677
Charm (continued)
  production, 231-234, 242, 673, 675, 676, 678, 684, 688, 699
  structure function, 677
  trigger, 662
  x-distributions, 677
Charmed
  baryons, 223, 225, 660, 667
  hadrons, 198, 211, 212-228
  meson, 608
  particles, 450, 660-667, 669-688
  quarks, 201, 208, 211, 212, 222, 229, 230, 539, 608, 736
  sea, 202-206, 234, 576, 604
Chiral
  multiplets, 115
  superfields, 128-129
  symmetry, 99
Christoffel symbols, 6
Collider energies, 231-234
Color, 738
  charge, 725
  interactions, 734
  invariance, 734, 735
  potential, 735
Complex
  manifolds, 11, 15
  variable theory, 38
Composites, 9
Confinement, 738
Connection concept, 14
Constituent
  Interchange Model (CIM), 160-163, 166, 187, 189
  models, 405
Correlation function, 709, 710
Cosmic rays, 465
Cosmological constant, 101, 103
Cosmology, 16
Coulomb
  energy, 47
  nuclear interference, 247
  peak, 247
  potential, 310
Coupling constants, 519
CPT
  symmetry, 739
  theorem, 14, 58

CPTM theorem, 58
Critical Pomeron, 249, 250, 258
Cross section
  asymmetry, 391, 392
  measurement, 603
Crystal Ball experiment, 656

D mesons, 230, 232, 236, 434, 660, 663-666, 678, 679
Dalitz plot, 385, 386
DDT form factor, 722
Decay
  identification, 661
  lengths, 478
  modes, 437, 475-477, 505-507, 518
  particles, 518
Decuplet pattern recognition, 616
Deep inelastic scattering (DIS)
  compared with high-energy soft (pp) internations, 701-715
    discussion, 721, 723
  scaling violations, 577-585
  status of, 555-598
    discussion, 599-614
  summary of, 585
DELPHI detector, 337, 353-358, 364, 366
Detector
  acceptance, 464
  asymmetry, 464
Dimensional
  counting rule, 160
  reduction, 9
Dipole
  Electric Moment of the Neutron (DEMON), 111, 144
  form-factor, 556
Diquarks, 188, 189
Dirac
  equation, 39, 40, 59, 732
  magnetic charge, 19
  monopole, 31, 39-64, 319
  quantization rule, 26, 48
Disorder fields, 303
Double scattering, 161
Drell-Yan
  diagram, 611
  formula, 180
Drell-Yan (continued)
  pairs, 202
  process, 179, 181, 184, 194, 531, 532, 587, 600, 610, 738
Drift chamber, 346, 349, 351, 361
D-S phase difference, 635, 636
D-type breaking, 117
Dual field strength, 305, 308
Dual gauge theory, 317

e*, search for, 491-494
$e^+e^-\to\mu^+\mu^-$ process, 329-334, 456-470, 516
$e^+e^-\to\tau^+\tau^-$, 478-480
Effective mass spectrum, 628, 630, 631
e-g-γ system, 47, 48, 55, 59, 61
Elastic scattering, 275, 276, 279, 298
  diffraction picture of, 186
  large t, 158-168
ELECTRA detector, 349-353, 362, 364
Electric
  charge, 725
  dipole moment, 35, 156
Electrodynamics, 5, 19
Electromagnetic
  current, 450
  field, 52, 731
  shower counter, 343
  vector potential, 732
Electromagnetism, 2
Electron
  -monopole bound states, 43
  -positron annihilation, 202, 213
  -positron interaction, 370-374
  -positron storage rings, 523
  scattering, 424, 527, 562
Electronic detectors, 602
Electrons
  axial vector coupling, 415
  energy resolution, 367
  identification in CELLO detector, 503
  prompt, 669-670
  single, 686
  spin zero partners, 500

Electroproduction, 288, 290, 291
Electroweak
  interaction, 6, 404-408, 449, 480-486
  interference, 450
  reactions, 426-428
  scale, 137-146
  solutions, 471
  theory, 456, 469
Elementary
  emission probability, 202
  particles, 3, 9, 16, 129, 376
Ellis-Karliner angle, 387
Emulsion experiments, 661
Energy
  conservation, 385
  -energy correlation, 391, 443, 445, 448
  flow pattern, 446
Equation of motion, 757
Excitation energies, 736

F mesons, 285, 299, 549, 665-667, 698
Fayet-Iliopoulos
  constant, 70
  term, 70, 75
Fermi
  motion, 600
  statistics, 599
Fermion
  -antifermion pairs, 58, 437
  covariant derivatives, 73
  doublets, 405
  mass differences, 76
  mass hierarchy problem, 145
  mass matrix, 121
  masses, 99, 146
  pair production, 370
Fermionic
  spin, 75
  terms, 72
  wave function, 29
Fermions
  and boson interactions, 106
  couplings, 103
  distinguishing from bosons, 91, 94
  fermions to, 112
  Goldstone, 74
Fermions (continued)
  Higgs field as partner, 89
  Majorana, 80, 81
  relations with bosons, 111
  search for structure, 434-435
  symmetry, 92, 119
  ultraviolet divergencies, 66
Feynman
  diagram, 49, 61, 370, 377, 457, 459
  -Field fragmentation model, 389, 395, 396, 443, 444, 448, 544
  path integral, 62
  rules, 52, 55, 56
Field theories, 12-13
Final-state interactions, 179-184
5-dimensional space, 30
Fixed point theories, 611, 612
Flavor
  changing neutral currents (FCNC), 120
  composition, 569-577
    of the sea, 574-577, 599
  determination, 446
  distributions, 603
  electric particles, 318
  excitation calculation, 206-213
  excitation diagrams, 198, 199
  excitation mechanism, 241
  identification, 433
  independence, 655
  invariance, 733, 735
  production, 453
  symmetry, 726, 733, 734, 736
Flux measurement, 603
Forward-backward
  average charged multiplicities, 713
  charge asymmetry, 333, 335, 410, 412
Forward jets, 721
Fractional energy, 704, 711
Fragmentation model, 389, 395, 396, 443, 444, 448, 544
Froissart bound, 248-250, 253, 256
F-type breaking, 117
Fusion
  excitation diagrams, 199
  mechanism, 241

$g_A{}^2$, determination of, 484-486
Gauge fields
  differential formalism, 5
  integral formalism of, 5
  non-Abelian, 3-5, 9, 12, 15, 54
  origin of concept, 1
  self-dual, 10
Gauge group, 114, 450
  choice of, 8, 9
  non-Abelian, 736
Gauge hierarchy problem, 110, 117, 119, 140
Gauge invariance, 1, 13, 16, 725-740
Gauge theory, 8, 449, 655, 725, 740
  Abelian, 304, 317, 322
  definition of, 303
  free gravity as, 10
  non-Abelian, 34, 54
Gauge transformation, 3, 13, 15, 726, 731, 733
  first kind, 725, 740
  second kind, 725, 740
Gauginos, 81, 132
Gell-Mann
  -Goldberger-Thirring dispersion relation, 281
  -Okubo formula, 734
General relaticity, 6
Georgi-Weinberg-Salam theory, 323
G.J. azimuthal angle, 638, 639
G.J. polar angle, 638, 639
Glasgow-Weinberg-Salam (GWS) theory, 6, 404, 416, 480
Global
  flavor symmetry, 728, 729
  invariance, 728
  supersymmetry, 109-150, 726-728, 740
Glueball
  models, 645
  resonance, 653
Glueballs, 553, 554, 615-652
  calculations of, 646
  decay channels, 654
  discussion, 653-657
  how to find, 616-617
  lowest mass, 655
Glueballs (continued)
  mass and J^{PC} from various phenomenological approaches, 645-648
  predicted mass and quantum numbers, 647
  primary, 645
  resonances, 654
  S-wave, 656
  variable mass, 616
Glue-glue coupling, 645
Gluinos, 81, 139, 151, 152
Gluon
  bremsstrahlung, 578, 580, 613
  contributions, 595, 596, 609
  corrections, 187, 535
  distribution, 270, 581, 583-584, 591-592
  exchange, 261, 280, 283, 644
  fragmentation, 393-398
  fusion model (GFM), 207, 676
  -gluon fusion, 674
  -gluon interactions, 176, 178, 717, 718
  initiated cascades, 204, 205, 224
  jets, 613, 718
  models, 646
  /quark coupling, 182
  radiation, 535
Gluonium states, 655
Gluons, 257, 265, 726
  attaching to spectator, 183
  color interactions, 734
  colored, 318
  as elementary scatterers, 613
  fixed point theories vector, 611
  generation, 220, 221
  hard, 384
  scalar, 389, 446, 611
  self-coupling, 615, 646
  soft, 388
Goldberger sum rule, 252
Goldstino, 97, 100, 117, 118, 152, 502, 507
Goldstone fermion, 74
Gottfried-Jackson frame, 630, 632
Gottfried sum rule, 574, 599
Graded group, 87

Graded Lie Algebra (GLA), 12, 92, 112
Grand Unification Schemes, 615
Grand Unified Theories (GUTs), 57, 82, 109, 115, 122-128, 134-137, 154, 405
Grassman variables, 106
Gravitino, 74, 76, 77, 79, 81, 83, 96, 97, 100, 106-107, 131, 140
Graviton, 131
Gravity, 10, 11
Green function, 96, 738
Gross-Llewellyn-Smith sum rule, 572, 579
g_V^2, 520
  determination of, 484-486

Hadron-hadron
  collision, 259
  scattering, 257, 592-597
  transformation, 702
Hadron production, 334, 375-403, 418
Hadronic
  charm production, 198, 242
  energy, 701, 704
  events, 428, 434
  interactions, 659
  jets, 380-384
  particle production, 704
  photon, 265, 269-273
  system, 723
Hadronization model, 213
Hadrons, 151, 183, 195, 230, 266, 267, 287, 334-339, 494, 507, 612, 615
  decay, 544
  evolution, 271
  final state, 398-403
  massive lepton pair production by, 586-592
Hamiltonian, 39, 40, 42, 61, 62, 102, 757, 758, 759
Heavy flavor production, 197-240
Heavy-flavored particles, properties of, 660-669
Heavy flavors, hadroproduction of, 659-697
Heavy lepton searches, 494, 496
Heavy neutral electron, 498-500
Heavy quarks, 734
  production asymmetry, 433
Heavy stable lepton, 515
Heisenberg
  dynamics, 757-762
    discussion, 763-770
  equation, 761, 763, 764
Hidden charm, 217, 659
Hierarchy problem, 66, 154
Higgs
  coupling, 436
  disorders fields, 331-317
  doublets, 124
  effect, 74, 76, 80, 302
  fields, 12, 35, 36, 57, 60, 66, 110, 115, 134, 302, 306, 311-317, 321, 325, 339
  mass, 323
  mechanism, 435
  particles, 66, 301-302, 505-507
    discussion, 321-326
  potential, 323
High energy e^+e^--physics, 369-442
High-energy soft interactions, 701-715
High statistics experiments, 558-559
Higher twist effects, 158, 584-585, 606, 608, 610
Hilbert space, 26, 27
Hoyer model, 449
Hubble constant, 155
Hyper pions, 505-507

Inelastic
  Compton graph, 287
  scattering, 200, 213, 528, 529
Integral formalism, 5-6
Integration ranges, 719
Isoscalar
  correction, 602
  targets, 557
Isospin
  invariance, 531, 733
  transformation, 737

K factor, 180, 587, 610, 611
K mesons, 402

Kähler
  gauge symmetry, 129
  gauge transformations, 129, 130
  manifold, 70, 71, 129, 130
  metric, 70
  potential, 70, 71, 74, 129
  transformation, 70, 71
Kaons, 401, 402
Kinematic reflection, 451

L3 detector, 346-349
Lagrangian, 5, 47, 61, 62, 66, 67, 68, 70, 71, 73, 83, 95, 97, 99, 100, 101, 103, 107, 116, 129, 130, 318, 491, 519, 737
Large-angle tagger, 526
Large-multiplicity mechanism, 175
Large t elastic scattering, 158-168
Lasers, 365
Lattice
  derivative, 305
  gauge theory, 16, 303-311, 315
Laws of nature, 730, 731, 737, 738
Leading
  effect, 702-704, 718, 722
  hadron, 719
  log approximation (LLA), 444, 534, 536, 546
  particle, 236, 720, 721
  protons, 717, 720, 722
  twist, 158
LEP (large electron positron storage ring), 327-361
  central role, 339
  design parameters, 329
  details of, 327-329
  detectors proposed for, 340
  discussion, 362-367
  physics motivation of, 366
  physics topics, 329-340
Lepton
  emission, 669, 687
  experiments, 683-688
  -nucleon scattering, 529, 548
  number conservation, 725
  pair production, 574, 586-592
  production, 428-434
Lepton (continued)
  scattering, 556-585, 610
  triggers, 692
  yields, 692
Leptonic
  data interpretation, 422-425
  reactions, 422
  widths, 736
Leptons, 671
  charged, 414
  discussion, 97, 99, 100, 105, 515-521
  incoming, 523
  lightness of, 145
  outgoing, 524
  point-like nature of, 486
  scalar, 500-505, 517, 521
  searches for new, 486-507
  status of, 455-514
  supersymmetric partners of, 438
Lie
  algebras, 12, 92, 112
  groups, 8, 88
Lipkin-Weisberger-Peshkin problem, 31, 40-59
Local
  invariance, 15, 728, 740
  supersymmetry, 73, 109-150
LOGIC detector, 356, 359
Loop space equations, 655
Lorentz
  boost, 723
  force equation, 47, 48
  invariance, 63, 394, 764
  symmetry, 732
  transformations, 764
Low energy data, 451
Low-energy effects, 66
Low energy potential, 138
Low-Energy Supersymmetry Breaking, 111
Low-energy theory, 76, 78, 96, 120-122, 124, 127, 137
Lund model, 394, 395, 396, 448

Magnetic
  fields, 22, 28, 30, 731
  lines of force, 23, 24, 28-38, 54, 58, 63, 319, 768
Mandelstam representation, 248

Mass
  determination, 472-473
  difference, 734
  effects, 241, 242
Maxwell's
  equation, 6, 14, 47, 48, 766
  theory, 5, 14
Meissner
  effect, 302
  -Higgs effect, 302
Meson
  distributions, 224
  lifetime limit, 428
  nonets, 616
  structure functions, 590
Mesons
  charged, 176
  flavored vestor, 318
  gluon distribution in, 591-592
  qq, 151
  quasi-elastic vector, 276
  soft charmed, 228-231
M_{LESB}, 117-118
Momentum
  distribution, 401
  fractions, 576-577
  measurement, 464
Monopole
  disorder, 315
  -electron interaction, 39
  -electron system, 50
  formation, 321
  harmonics, 25, 26, 30, 39
  problem, 154
  velocity, 30
Motivations, 243
μ*, search for, 486-491
Multiparticle
  hadronic system, 721
  production, 719
  Spectrometer, 618, 624
Multiple
  gluon exchange, 261
  scattering, 189, 606
Multiplicity distribution, 398, 399
Muon
  asymmetry, 471, 517
  beam dump, 685-686
  experiments, 562, 563, 566-568, 606
Muon (continued)
  pair events, 462, 465-467
  pair production, 372, 406, 407, 409-413, 592
Muons
  angular distribution, 334
  asymmetry values, 468
  axial vector coupling, 415
  branching ratio of production of charmed particles to, 450
  charge measurement, 465
  charged momentum for forward vs. backward, 417
  decay, 500
  energy loss of, 274
  helicity, 516
  lepton decay into, 494
  outgoing, 462
  partial separation from botton decay, 430
  pointlike production cross section, 473
  transverse momentum distribution, 431, 432
  under magnetic field, 418, 464
  wrong sign, 608
Muoproduction, 291, 292

Nachtmann moments, 578
Naked
  beauty, 667
  charm, 659, 672
Nambu-Goldstone bosons, 302
Neutral
  currents, 608
  heavy electron, 515
  recoiling system, 629
Neutrino
  beam-dump experiments, 683
  experiments, 424, 563, 565, 571, 607, 613
  production, 684
  scattering, 603
Neutrinos
  lepton decay into, 494
  mass, 156
  structure function, 562
Neutron, 4, 156
  structure function, 600

Non-Albelian gauge
fields, 3-5, 9, 12, 15, 54
group, 736
invariance, 13
theory, 34, 54, 322
Non-integrable phase factor, 5, 16
Non-linear systems, 325
Nonperturbatively, 101
Non-renormalization, 135, 143
theorems, 66, 119
Normalization parameter, 710
Nucleon-nucleon collisions, 221
Nucleons, 221, 231, 274
parton distribution, 557-558
structure functions, 588-590
valence quark distribution in, 590

Odderon 1, 247
Odderon 2, 247
Odderon 3, 248
Omega Spectrometer, 285
One-electron
problem, 759, 760
theory, 764
OPAL detector, 343-346 364
Opposite helicities, 115
Order field, 301
Order parameter, 301
Ordinary field theory, 52
OZI
allowed reactions, 619, 626, 627
forbidden channel, 617-624
forbidden diagram, 619
forbidden reaction, 627
rule, 624, 644, 654
suppressed channel, 616-617
suppression, 619, 620, 622, 625-627, 634, 644

Pair-production, 766
Pair-Treiman relation, 699
Parity nonconservation, 61
Partial
Unification Schemes, 615
wave analysis, 630-645, 657
Particle
identification, 367
Particle (continued)
interactions, 65-86
spectrum, 142
Partition function, 306, 307, 309, 311, 319
Parton
charges, 287
distribution, 557-558
jet calculation, 202
subprocess, 208
Partons
concept of, 556
fragmentation, 401, 443
likelihood of finding in nucleon, 288
target, 266
x-distribution, 594
Path
-dependent phase factor, 5, 11
integral quantization, 55
integrals, 63
Pauli
exclusion principle, 599
matrices, 44
Perturbation theory, 101
Phase
angle, 46
change factor, 3
factor, 5, 63
space, 458
space diagram, 460
space model, 544
transformation, 3
ϕ-isotropy, 191
Photinos, 81, 152, 502, 507
Photoelectrons, 356
Photon
beam, 298
calorimetry, 680
evolution, 273-274, 288
-gluon fusion process, 207, 284
pair production, 372
scattering, 265-295, 298
structure function, 272, 523-543
discussion, 544-554
Photons
energy resolution, 367
final state, 448
from decaying Ti^o,s, 477

Photons (continued)
  identification in CELLO
    detector, 503
  initial state, 447
  intermediate, 373
  nature of, 266-269
  virtual, 176, 208, 484
  and Z^o exchange, 329
Photoproduction, 267, 275, 277,
    278, 281, 282, 285, 293,
    298
Photosensors, 365
π^2-problem, 181
Pion
  energy spectra, 474
  structure functions, 591
Pions, 152, 179, 298, 401, 473,
    595, 685, 720
Planck
  limit, 727
  mass, 100
  scale, 65
Plasma of e^+e^-pairs around
    monopole, 47, 54
Poincaré
  algebra, 122
  invariance, 736
  symmetry, 737
  transformations, 730, 731
Poisson bracket, 763
Poisson's formula, 307-309
Polonyi potential, 131
Pomeranchuk
  singularity, 258
  theorem, 245-247, 252, 255
Pomeron, 194
  exchange, 269, 280, 296, 298
pp and $p\bar{p}$ diffraction scattering
    at high energies, 245-251
  discussion, 252-263
Primakoff-effect, 297
Proton
  decay, 57, 127, 155, 156, 726
  production, 719
  structure function, 272
Protons, 4, 177, 263, 599, 607,
    685, 703, 720
Psuedorapidity distribution, 235
ψ field, 322
Q^2 dependence, 290, 292
QCD, 268, 379
  charm production calculations,
    673
  corrections to Drell-Yan
    diagram, 611
  discussion, 186-195, 541, 546,
    547, 549-551, 733, 734
  first-order corrections, 587
  in hadron production, 384-403
  in hadronic calculation, 540,
    595
  higher-order effects, 445, 446
  importance of glueballs to, 615
  leading-order calculations,
    540, 595
  lowest-order, 168
  Monte Carlo calculations,
    202-206, 444
  OZI allowed reactions, 619
  perturbative, 158, 159, 172,
    173, 176, 198, 201, 236,
    242, 262, 389, 390, 453
  pp and $p\bar{p}$ results, 257
  predictions for heavy flavor
    production, 197-240
  radiative effects, 718
  singlet structure function
    analysis, 580-585
  soft (noperturbative), 298
  structure functions from, 577
  test method, 601, 602
  theory of strong interactions,
    261
QCD
  cascades, 205
  comparisons, 585
  coupling, 177
  coupling constant, 297
  degradation, 201
  differential cross section,
    385-389
  fusion diagrams, 198
  inclusive processes, 169-184
  model difficulties, 165-168
  vacuum, 655
Quantization condition, 52, 53,
    62, 319, 325
Quantum electrodynamics, (QED),
    47, 265, 467, 469, 480,
    484, 520

Quantum electrodynamics, (QED) (continued)
  $\alpha 4$, 373
  asymmetry, 450
  contributions to $A_{\mu\mu}^{4}$, 457-458
  couplings, 268
  higher-order effects, 416
  low q^2, 372
  lowest order, 371
  photon bremsstrahlung calculation, 385
  vacuum of, 765
Quantum
  field theory, 52, 60
  gravity, 66, 737
  mechanics, 2, 14, 16, 53, 56, 62, 146, 759, 764, 767
  numbers, 102, 104, 105, 266, 620, 636, 645, 648, 657
  physics, 19
  theory, 90
Quark
  -antiquark pair, 528
  asymmetries, 452
  beams, 283
  charges, 275, 333, 430
  confinement, 738
  constitutents, 527
  decay, 451
  distribution, 572-573
  evolution, 270
  flavors, 377, 383
  fragmentation, 213, 393-398
  helicities, 269
  interactions, 733, 734
  jets, 446, 447, 613
  line diagram, 618-621
  masses, 97, 99, 545-547
  model, 532
  pair production, 371, 372, 380, 580
  parton model (QPM), 382, 444, 524, 546, 556
  plasma around monopoles, 57
  production, 433
  -quark scattering, 192
Quarks, 141, 145, 175-179, 319, 335, 384, 595, 615, 702, 726
  active, 182, 183, 212
Quarks (continued)
  additive model, 276
  charm, 200, 205, 211, 212
  discussion, 100, 102, 105, 152, 156, 242, 257, 263
  electroweak reactions, 426-428
  free, 365
  initial, 206
  quasi-valence, 273
  separation, 286
  spectator c, 224
  uncharmed, 220
Quaternions, 63

R ratio, 375-379, 426-427, 443
Radiative correction, 410, 458, 459, 461, 491, 548, 549
Recombination
  model, 211, 218, 228
  probability, 224
  process, 219
Reflection symmetry, 739
Regge
  behavior, 257
  poles, 188
Renormalizability, 6-7, 12, 91, 324
Renormalization, 48, 54, 60
  group equation, 535
  theory, 761
Resonance, 376, 550, 554
R-hadrons, 151, 152
Riemann curvature tensor, 6
Reimannian covariant derivative, 6
Ring imaging Cerenkov counters (RICH), 353, 355, 356, 358, 361, 362, 366, 367
R-parity, 151
Rutherford scattering cross-section, 43

S-matrix, 736, 737
s quark, 120, 139, 153, 402
Scaler
  fixed point theory, 611
  leptons, 500-505, 517, 521
  particle, 435
  potential, 71, 74, 78, 79
Scale change, 1, 2

Scaling violations, 577-585, 611
Scattering
  amplitude, 254
  angle, 461
Schrodinger equation, 25, 40
Sea quark, 228
Seagull diagrams, 316
Semi-transparent centre, 260
'Set it and forget it', 119, 133
Sfermion, 121, 122
Shadowing, 606
Short-range correlations, 708
Shower counter, 353, 356, 361
Silicon active target, 662
$\sin^2\theta_W$, 427, 480-484, 519, 520
Sine-Gordon-Thirring model, 60
Single particle spectra, 595
Slepton, 120, 139
Smuon, 153
Soft breaking, 119, 122
Soft charmed mesons, 228-231
Solid state theory, 16
Space groups, 739
Space-like secondary, 204
Space-time-dependent
  fields, 8
  scale change, 1, 2
Spectator
  charm quark, 220, 227
  model diagram, 665
  recombinations, 224
  sea quarks, 230
Spectrometers, 688
Spherical
  harmonics, 25, 26
  symmetry, 732
Spin, 90-92, 104, 106, 144-115, 281, 296, 319, 326, 371, 472-473
Spinor, 104
Spontaneous Breaking (SB), 110
SPS Collider, 172, 175, 178, 184
Stable heavy lepton search, 496-498
Sterman-Weinberg cross section, 443
Strange quarks, 604
Strange sea, 575-576, 604, 605
Streamer chamber, 661, 679
Strong interaction coupling constant, 389-393
Struck charm quark, 230
Structure function, 612
  A-dependence of, 568-569, 606
  longitudinal, 569
  non-singlet, 578-581
  shape of, 569-577
  singlet, 580-585
Structure function measurements, 559-563
Sturm-Liouville
  analysis, 45
  problem, 44
  theory, 45, 46
Sudakov
  effect, 187
  factor, 167
  factors, 182
  form factors, 187
SUDEC, SUSY DeCoupling mechanism, 117-118
Superconductivity, 301, 353
Supergravity, 65, 66-83, 114, 146
  discussion, 91, 93, 101-103
  physical structure, 128-133
  simple (N=1), 134-146
Supergravity
  hierarchy problem, 133, 137
  (SUGAR) models, 130, 131, 133, 137, 139, 140, 143, 144
SuperHiggs effect, 74, 76, 77, 100, 131
Superpotential, 70, 71, 115
Supersymmetric GUTs, 405
Supersymmetry, 65-86, 112-114
  discussion, 87-108, 151-156
  global, 109-150, 726-728, 740
  local, 73, 109-150
  motivation(s) for, 109
  physical properties of, 114-119
Supersymmetry breaking, 76, 77, 79, 96, 98, 100-102, 105, 106, 116, 119, 144, 155, 156
Supersymmetry particles, 518
Symmetry, 6-7
  and charge conservation, 15
  choice of, 12
  discussion, 9, 14, 29, 53, 725-740

Symmetry (continued)
gauge, 15
global, 15
local gauge, 14
Symmetry breaking, 10, 12, 435-438, 451, 736
Symmetry dictates interactions, 6, 15

Tau, 500, 517, 518
Tau asymmetry values, 481
τ-lepton, 470-480
τ-lifetime measurements, 477, 480
Tau pair
charge asymmetry, 421
events, 480
production, 406, 409-413, 474
Tau pairs, 465, 505, 506
t-distribution, 653
Technicolor models, 405
Technipion, 507
θ-angle, 66
'tHooft-Polyakov
monopole, 31, 34, 36, 54
solutions, 29, 54
Threshold effects, 242
Thrust, 381-383
distribution, 383, 384
Timelike secondary, 204
Top production, 668-669, 691-693
Total cross-section, 525, 542, 547
TPC (Time Projection Chamber), 353, 362, 366
Transverse
-energy triggers, 173-175
momentum distribution, 431, 432, 707, 708
Transition radiation detector, 352, 362
Transverse V^o propagator term, 290
Trigger inefficiency, 465
Triple scattering, 161, 162-163, 166, 167
Triplicity versus energy, 381

u quarks, 607
Ultraviolet
divergences, 155

Ultraviolet (continued)
-finite model, 88
-finite theory, 89
Unification, 82

Valence
quark distribution, 220, 571-573, 581, 590, 607
quarks, 227
Vector
dominance model (VDM), 271, 278, 290, 292, 548
Meson Dominance Model (VDM), 295, 528, 531, 540, 552
potential, 731
Vertex
detectors, 662
uncertainties, 479
Villain approximation, 312
Visual detectors, 678
Vulcanization processes, 107

Wave
equation, 760
function, 39-41, 43
Wess-Zumino model, 106

x-variable, 241

Yang-Mills
gauge symmetry, 737
group, 95
system, 67-69, 71, 74, 77
theory, 10, 11, 615
-Higgs systems, 31
Yukawa
coupling, 138, 139, 141
interactions, 116

Z boson, 405, 406
Z^o exchange, 329
Zweig diagram, 623, 624, 644